HANDBOOK

of

BASIC TABLES

for

CHEMICAL ANALYSIS

Third Edition

HANDBOOK
of
BASIC TABLES
for
CHEMICAL
ANALYSIS

Third Edition

Thomas J. Bruno
Paris D. N. Svoronos

CRC Press
Taylor & Francis Group
Boca Raton   London   New York

CRC Press is an imprint of the
Taylor & Francis Group, an **informa** business

CRC Press
Taylor & Francis Group
6000 Broken Sound Parkway NW, Suite 300
Boca Raton, FL 33487-2742

© 2011 by Taylor and Francis Group, LLC
CRC Press is an imprint of Taylor & Francis Group, an Informa business

No claim to original U.S. Government works

Printed in the United States of America on acid-free paper
10 9 8 7 6 5 4 3 2 1

International Standard Book Number: 978-1-4200-8042-1 (Hardback)

Library of Congress Cataloging-in-Publication Data

Bruno, Thomas J.
CRC handbook of basic tables for chemical analysis / Thomas J. Bruno, Paris D.N. Svoronos. -- 3rd ed.
p. cm.
Includes bibliographical references and index.
ISBN 978-1-4200-8042-1 (hardcover : alk. paper)
1. Chemistry, Analytic--Tables. I. Svoronos, Paris D. N. II. Title.

QD78.B78 2011
543.02'1--dc22

2010021865

Visit the Taylor & Francis Web site at
http://www.taylorandfrancis.com

and the CRC Press Web site at
http://www.crcpress.com

Dedication

We dedicate this work to our children, Kelly-Anne, Alexandra, and Theodore.

Notice

Certain commercial equipment, instruments, or materials are identified in this handbook in order to provide an adequate description. Such identification does not imply recommendation or endorsement by the National Institute of Standards and Technology, the City University of New York, or Georgetown University, nor does it imply that the materials or equipment identified are necessarily the best available for the purpose. The authors, publishers, and their respective institutions are not responsible for the use in which this handbook is made. Occasional use is made of non-SI units, in order to conform to the standard and accepted practice in modern analytical chemistry.

Preface to the First Edition

This work began as a slim booklet prepared by one of the authors (TJB) to accompany a course on Chemical Instrumentation presented at the National Institute of Standards and Technology, Boulder Laboratories. The booklet contained tables on chromatography, spectroscopy, and chemical (wet) methods, and was intended to provide the students with enough basic data to design their own analytical methods and procedures. Shortly thereafter, with the co-authorship of Professor Paris D.N. Svoronos, it was expanded into a more extensive compilation entitled Basic Tables for Chemical Analysis, published as a National Institute of Standards and Technology Technical Note (number 1096). That work has now been expanded and updated into the present body of tables.

Although there have been considerable changes since the first version of these tables, the aim has remained essentially the same. We have tried to provide a single source of information for those practicing scientists and research students who must use various aspects of chemical analysis in their work. In this respect, it is geared less toward the researcher in analytical chemistry than to those practitioners in other chemical disciplines who must make routine use of chemical analysis. We have given special emphasis to those "instrumental techniques" which are most useful in solving common analytical problems. In many cases, the tables contain information gleaned from the most current research papers, and provide data not easily obtainable elsewhere. In some cases, data are presented which are not available at all in other sources. An example is the section covering supercritical fluid chromatography, in which a tabular P-Δ-T surface for carbon dioxide has been calculated (specifically for this work) using an accurate equation of state.

While the authors have endeavored to include data, which they perceive to be most useful, there will undoubtedly be areas which have been slighted. We therefore ask you, the user, to assist us in this regard by informing the corresponding author (TJB) of any topics or tables which should be included in future editions.

The authors would like to acknowledge some individuals who have been of great help during the preparation of this work. Stephanie Outcalt and Juli Schroeder, chemical engineers at the National Institute of Standards and Technology, provided invaluable assistance in searching the literature and compiling a good deal of the data included in this book. Teresa Yenser, manager of the NIST word processing facility, provided excellent copy despite occasional disorganization on the part of the authors. We owe a great debt to our board of reviewers, who provided insightful comments on the manuscript: Professors D.W. Armstrong, S. Chandrasegaran, G.D. Christian, D. Crist, C.F. Hammer, K. Nakanishi, C.F. Poole, E. Sarlo, Drs. R. Barkley, W. Egan, D.G. Friend, S. Ghayourmanesh, J.W. King, M.L. Loftus, J.E. Mayrath, G.W.A. Milne, R. Reinhardt, R. Tatken, and D. Wingeleth. The authors would like to acknowledge the financial support of the Gas Research Institute and the United States Department of Energy, Office of Basic Energy Sciences (TJB) and the National Science Foundation, and the City University of New York (PDNS). Finally, we must thank our wives, Clare and Soraya, for their patience throughout the period of hard work and late nights.

Preface to the Third Edition

Although only seven years have elapsed since the publication of the second edition of this book, we have undertaken this revision for several reasons. In the intervening years, we have published the *CRC Handbook of Fundamental Spectroscopic Correlation Charts*, a laboratory tool that was intended to "go to work" each day with the user. Many of our readers have expressed the desire for a single volume with the information presented in both of these volumes. The third edition fulfills this desire. In addition, the content has been significantly expanded with new information in the chapters on gas, liquid, and thin layer chromatography; nuclear magnetic resonance spectrometry (especially for high field applications); infrared spectrophotometry; and mass spectrometry. Responding to user input from the second edition, we have added information on the detection of outliers in experimental data, basic information on thermocouples, chemical indicators, chromatographic column regeneration, and some of the latest stationary phases for chromatographic methods and extractions.

Reflecting the growing emphasis on laboratory safety, this topic is now treated far more in depth. We provide information on many kinds of chemical hazards, electrical hazards in the analytical laboratory, and information to aid the user in selecting laboratory gloves, apparel, and respirators. In this edition, we have treated radiation and laser hazards, and the carcinogen information has been updated to reflect the current opinions. This aspect of the book is unique, since no other handbook of analytical chemistry provides a self-contained source of information that covers not simply carrying out a lab procedure, but carrying it out safely.

Our philosophy in preparing this book has been to include information that will help the user make decisions. In this respect, we envision each table as being something the user will consult when reaching a decision point in designing an analysis or interpreting results. We have deliberately chosen to exclude information that is merely interesting, but of little value at a decision point.

Similarly, it has occasionally been difficult to strike an appropriate balance between presenting information that is of general utility and information that is highly specific and perhaps simply a repetition of what is contained in vendor catalogs, promotional brochures, and web sites. In this respect, we have tried to keep the content as generic and unbiased as possible. Thus, some specific chromatographic phases and columns, available only under trade names, have been excluded. The same is true regarding many of the materials available for electrophoresis. This must not be regarded as a value judgment, but simply a reflection of our philosophy.

Acknowledgments

The authors would like to acknowledge some individuals who have been of great help during the preparation of this work. We owe a great debt to our board of reviewers: Mr. T. Grove, Professors M. Jensen, A.F. Lagalante, D.C. Locke, K.E. Miller, D.F. Steadman, Drs. W.C. Andersen, A.M. Harvey, D. Joshi, T.M. Lovestead, M.M. Schantz, S. Rudge, S. Ringen, D.G. Friend, M.L. Huber, D. Smith, S. Ghayourmanesh, J.A. Widegren, and B. Windom. Finally, we must again thank our wives, Clare and Soraya, and our children, Kelly-Anne, Alexandra, and Theodore, for their patience and support throughout the period of hard work and late nights.

Authors

Thomas J. Bruno, PhD, is a group leader in the Thermophysical Properties Division at the National Institute of Standards and Technology, Boulder, CO. Dr. Bruno received his BS in chemistry from the Polytechnic Institute of Brooklyn, and his MS and PhD in physical chemistry from Georgetown University. He served as a National Academy of Sciences-National Research Council postdoctoral associate at NIST, and was later appointed to the staff. Dr. Bruno has done research on properties of fuel mixtures, chemically reacting fluids, and environmental pollutants. He is also involved in research on supercritical fluid extraction and chromatography of bioproducts, the development of novel analytical methods for environmental contaminants and alternative refrigerants, novel detection devices for chromatography, and he manages the division analytical chemistry laboratory. In his research areas, he has published approximately 200 papers, seven books, and holds seven patents. He was awarded the Department of Commerce Bronze Medal in 1986 for his work on the thermophysics of reacting fluids. He has served as a forensic consultant and an expert witness for the U.S. Department of Justice, and received a letter of commendation from the DOJ for these efforts in 2002. He was awarded the Department of Commerce Silver Medal in 2010 for the development of a new method for analyzing complex fluid mixtures that facilitates the introduction of new fuels into the U.S. energy infrastructure.

Paris D.N. Svoronos, PhD, is professor of chemistry and department chair at Queensborough Community College (QCC) of the City University of New York. In addition, he holds a continuing appointment as visiting professor in the Department of Chemistry at Georgetown University. Dr. Svoronos obtained a BS in chemistry and a BS in physics at the American University of Cairo, and his MS and PhD in organic chemistry at Georgetown University. Among his research interests are synthetic sulfur and natural product chemistry, organic electrochemistry, and organic structure determination and trace analysis. He also maintains a keen interest in chemical education, and has authored several widely sought laboratory manuals used at the undergraduate levels. In his fields of interest, he has approximately 70 publications. He has been in the *Who's Who of America's Teachers* three times in the last five years. He was selected as the 2003 Outstanding Professor of the Year by the Council for the Advancement and Support of Education (CASE) and the Carnegie Foundation. He was the general chairman of the American Chemical Society 40th Middle Atlantic Regional Meeting in Bayside, New York in 2008, the first one ever in New York City, and has been on the governing board of the American Chemical Society-Long Island section since 2002. He is particularly proud of his students' successes in research presentations, paper publications, and professional accomplishments.

Contents

CHAPTER 1

Gas Chromatography

CONTENTS

CARRIER GAS PROPERTIES

The following table gives the properties of common (and less common) gas chromatographic carrier and make-up gases. These properties are those used most often in designing separation and optimizing detector performance. With few exceptions (indicated by an asterisk), all the properties provided in this table have been derived from the Helmholtz free energy equation of state and appropriate transport property models that are based upon the most reliable experimental measurements [1]. Unless indicated, the pressure in each case was 101.325 kPa or standard atmospheric pressure. For gases in which the heat capacity showed a very slight temperature dependence, only selected values are provided.

In addition to the pure gases, two mixtures are included. The hydrogen + helium mixture is sometimes used to obtain positive peaks when using the thermal conductivity detector in the analysis of mixtures containing hydrogen. The argon + methane mixture is sometimes used with electron capture detection.

REFERENCES

1. Lemmon, E. W., M. L. Huber, and M. O. McLinden. REFPROP, Reference Fluid Thermodynamic and Transport Properties, NIST Standard Reference Database 23, Version 8. Gaithersburg, MD: National Institute of Standards and Technology, 2007.

Carrier Gas Properties

Carrier Gas	Relative Molecular Mass (RMM)	Density (kg/m³)	Thermal Conductivity (mW/m-K)	Viscosity (µPa-s)	Sound Speed (m/s)	Heat Capacity, C_p (kJ/kg-K)	Heat Capacity Ratio, C_p/C_v
Hydrogen	2.016						
25 °C		0.08235	184.88	8.9154	1315.4	14.306	1.4054
100 °C		0.06580	221.42	10.389	1468.2		
200 °C		0.05190	267.83	12.206	1652.0		
Helium	4.003						
25 °C		0.16353	155.31	19.846	1016.4	5.1930	1.6665
100 °C		0.13068	181.41	23.154	1137.0		
200 °C		0.10307	213.93	27.294	1280.2		
Nitrogen	28.016						
25 °C		1.1452	25.835	17.805	352.07	1.0413	1.4013
100 °C		0.91469	31.038	21.101	393.73		
200 °C		0.72124	37.418	25.066	442.56		
Argon	39.94						
25 °C		1.6339	17.746	22.624	321.67	0.52156	1.6696
100 °C		1.3048	21.311	27.167	359.92		
200 °C		1.0288	25.635	32.684	405.31		
Methane	16.04						
25 °C		0.65688	33.931	11.067	448.47	2.2317	1.3062
100 °C		0.52429	45.282	13.395	495.46	2.4433	1.2711
200 °C		0.41324	63.134	16.214	548.70	2.7999	1.2280
Ethane	30.07						
25 °C		1.2385	20.984	9.3541	311.31	1.7572	1.1939
100 °C		0.9857	31.643	11.485	344.31	2.0668	1.1576
200 °C		0.7757	48.771	14.111	383.29	2.4866	1.1264
Ethene	28.05						
25 °C		1.1533	20.326	10.318	329.92	1.5373	1.2461
100 °C		0.9187	30.730	12.769	363.37	1.7981	1.2006
200 °C		0.7234	45.734	15.758	403.21	2.1414	1.1620

Substance	Molar mass						
Propane	44.09						
25 °C		1.8320	18.310	8.1463	248.63	1.6847	1.1364
100 °C		1.4516	27.362	10.129	276.92	2.0121	1.1074
200 °C		1.1398	41.954	12.643	310.06	2.4404	100854
n-Butane	58.12						
25 °C		2.4493	16.564	7.4054	210.49	1.7317	1.1053
100 °C		1.9252	24.869	9.2534	236.8	2.0436	1.0807
200 °C		1.5064	38.722	11.625	266.67	2.4596	1.0638
Oxygen	32.00						
25 °C		1.3088	26.340	20.550	328.72	0.91963	1.3967
100 °C		1.0452	32.061	24.507	366.63	0.96365	1.3701
200 °C		0.8241	39.206	29.283	410.47		
Carbon dioxide	44.01						
25 °C		1.8080	16.643	14.932	268.62	0.85085	1.2941
100 °C		1.4407	22.875	18.475	297.63		
200 °C		1.1346	31.274	22.891	332.02	0.99708	1.2356
Carbon monoxide	28.01						
25 °C		1.1453	26.478	17.649	352.03	1.0421	1.4013
100 °C		0.9147	31.411	20.813	393.58	1.0590	1.3904
200 °C		0.7212	37.631	24.592	442.07		
Sulfur hexafluoride	146.05						
25 °C		6.0383	12.06 (0 °C)*	1.42 (0 °C)*	134.98	0.6690	1.0984
100 °C		4.7954	14.06 (25 °C)*	1.61	150.81	0.7702	1.0718
200 °C		3.7698		(25 °C)*	169.55	1.8617	
Hydrogen in helium (8.5 % mol/mol[a])							
25 °C		0.1566	162.65	19.159	1027.7	5.6002	1.6319
100 °C		0.1252	194.87	22.180	1149.2		
200 °C		0.0987	237.32	25.880	1293.8		
Methane in argon (5 % mass/mass)							
25 °C		1.5215	20.053	22.049	325.12	0.6055	1.5883
100 °C		1.2149	24.777	26.490	361.97		
200 °C		0.9582	31.121	31.882	404.40		

a Note for this mixture, the composition specification is provided on a molar basis, which is how the mixture is typically sold. The reader should be aware that the composition specification is also commonly expressed on a volume basis, making the ideal gas assumption. If there is ambiguity about a particular mixture, the user should discuss the specific composition specification with the supplier.

CARRIER GAS VISCOSITY

The following table provides the viscosity of common carrier gases, in $\mu Pa \cdot s$, used in gas chromatography [1,2]. The values were obtained with a corresponding states approach with high accuracy equations of state for each fluid. Carrier gas viscosity is an important consideration in efficiency and in the interpretation of flow rate data as a function of temperature. In these tables, the temperature, T, is presented in ^BC, and the pressure, P, is given in kilopascals and in pounds per square inch (absolute). To obtain the gauge pressure (that is, the pressure displayed on the instrument panel of a gas chromatograph), one must subtract the atmospheric pressure. Following the table, the data are presented graphically.

1. Lemmon, E. W., A. P. Peskin, M. O. McLinden, and D. G. Friend. Thermodynamic and Transport Properties of Pure Fluids, NIST Standard Reference Database 12, Version 5.0. Gaithersburg, MD: National Institute of Standards and Technology, 2000.
2. Lemmon, E. W., M. L. Huber, and M. O. McLinden. REFPROP, Reference Fluid Thermodynamic and Transport Properties, NIST Standard Reference Database 23, Version 8. Gaithersburg, MD: National Institute of Standards and Technology, 2007.

P = 204.8 kPa, 29.7 psia

T[C]	He	H_2	Ar	N_2	Air	Ar/CH$_4$ (90/10)	Ar/CH$_4$ (95/5)
0	18.699	8.3996	20.979	16.655	17.277	20.013	20.505
10	19.163	8.6088	21.625	17.129	17.775	20.625	21.134
20	19.621	8.8154	22.264	17.597	18.266	21.229	21.755
30	20.076	9.0197	22.894	18.058	18.75	21.826	22.369
40	20.527	9.2218	23.517	18.513	19.228	22.415	22.975
50	20.974	9.4216	24.133	18.962	19.699	22.998	23.574
60	21.418	9.6194	24.742	19.404	20.165	23.573	24.166
70	21.858	9.8152	25.344	19.842	20.624	24.142	24.751
80	22.294	10.009	25.939	20.273	21.078	24.705	25.329
90	22.727	10.201	26.527	20.7	21.526	25.261	25.901
100	23.157	10.391	27.109	21.121	21.969	25.811	26.467
110	23.583	10.58	27.685	21.538	22.407	26.355	27.027
120	24.007	10.767	28.255	21.949	22.84	26.893	27.581
130	24.427	10.952	28.819	22.357	23.268	27.426	28.129
140	24.845	11.136	29.378	22.759	23.691	27.953	28.671
150	25.26	11.318	29.931	23.157	24.11	28.474	29.209
160	25.672	11.498	30.479	23.552	24.524	28.991	29.74
170	26.082	11.678	31.021	23.942	24.934	29.502	30.267
180	26.489	11.856	31.558	24.328	25.34	30.008	30.788
190	26.894	12.033	32.09	24.71	25.742	30.51	31.305
200	27.296	12.208	32.618	25.089	26.14	31.006	31.817
210	27.696	12.382	33.14	25.464	26.534	31.499	32.324
220	28.094	12.555	33.658	25.835	26.924	31.986	32.826
230	28.49	12.727	34.172	26.203	27.311	32.47	33.325
240	28.883	12.898	34.681	26.568	27.695	32.949	33.818
250	29.274	13.068	35.186	26.93	28.075	33.424	34.308
260	29.664	13.236	35.687	27.288	28.451	33.894	34.793
270	30.051	13.404	36.183	27.644	28.825	34.361	35.275
280	30.436	13.571	36.676	27.996	29.195	34.824	35.752
290	30.82	13.736	37.164	28.346	29.562	35.284	36.226
300	31.201	13.901	37.649	28.692	29.927	35.739	36.696

P = 308.2 kPa, 44.7 psia

T[C]	He	H$_2$	Ar	N$_2$	Air	Ar/CH$_4$ (90/10)	Ar/CH$_4$ (95/5)
0	18.704	8.4024	21.001	16.672	17.296	20.033	20.527
10	19.167	8.6114	21.647	17.146	17.794	20.644	21.155
20	19.625	8.8179	22.285	17.613	18.284	21.248	21.775
30	20.08	9.0222	22.915	18.074	18.767	21.844	22.388
40	20.531	9.2241	23.537	18.528	19.244	22.433	22.993
50	20.978	9.4239	24.152	18.977	19.715	23.015	23.592
60	21.421	9.6217	24.76	19.419	20.18	23.59	24.183
70	21.861	9.8174	25.361	19.856	20.639	24.158	24.768
80	22.297	10.011	25.956	20.287	21.092	24.72	25.346
90	22.73	10.203	26.544	20.713	21.54	25.276	25.917
100	23.159	10.393	27.126	21.134	21.982	25.825	26.483
110	23.586	10.582	27.701	21.55	22.42	26.369	27.042
120	24.009	10.769	28.271	21.962	22.852	26.907	27.596
130	24.43	10.954	28.835	22.369	23.28	27.439	28.143
140	24.847	11.137	29.393	22.771	23.703	27.966	28.685
150	25.262	11.319	29.945	23.169	24.121	28.487	29.222
160	25.675	11.5	30.493	23.563	24.535	29.003	29.754
170	26.084	11.68	31.035	23.953	24.945	29.514	30.28
180	26.491	11.857	31.572	24.338	25.351	30.02	30.801
190	26.896	12.034	32.103	24.72	25.752	30.521	31.317
200	27.298	12.21	32.631	25.099	26.15	31.018	31.829
210	27.698	12.384	33.153	25.474	26.544	31.51	32.336
220	28.096	12.557	33.671	25.845	26.934	31.997	32.838
230	28.492	12.729	34.184	26.213	27.321	32.48	33.336
240	28.885	12.899	34.693	26.577	27.704	32.959	33.829
250	29.276	13.069	35.198	26.939	28.084	33.434	34.319
260	29.666	13.238	35.698	27.297	28.46	33.904	34.804
270	30.053	13.405	36.194	27.652	28.834	34.371	35.285
280	30.438	13.572	36.687	28.005	29.204	34.834	35.763
290	30.822	13.738	37.175	28.354	29.571	35.293	36.236
300	31.203	13.903	37.66	28.701	29.935	35.749	36.706

P = 446.1 kPa, 64.7 psia

T[C]	He	H$_2$	Ar	N$_2$	Air	Ar/CH$_4$ (90/10)	Ar/CH$_4$ (95/5)
0	18.71	8.406	21.032	16.696	17.322	20.061	20.556
10	19.172	8.6149	21.676	17.169	17.818	20.671	21.183
20	19.63	8.8213	22.313	17.636	18.307	21.274	21.802
30	20.085	9.0254	22.942	18.096	18.79	21.869	22.414
40	20.535	9.2273	23.563	18.549	19.266	22.457	23.019
50	20.982	9.427	24.178	18.997	19.736	23.038	23.616
60	21.425	9.6246	24.785	19.439	20.2	23.612	24.207
70	21.865	9.8203	25.385	19.875	20.658	24.18	24.79
80	22.301	10.014	25.979	20.306	21.111	24.741	25.368
90	22.734	10.206	26.567	20.731	21.558	25.296	25.939
100	23.163	10.396	27.148	21.152	22	25.845	26.504

(Continued)

P = 446.1 kPa, 64.7 psia (Continued)

T[C]	He	H_2	Ar	N_2	Air	Ar/CH$_4$ (90/10)	Ar/CH$_4$ (95/5)
110	23.59	10.584	27.723	21.567	22.437	26.388	27.062
120	24.013	10.771	28.292	21.978	22.869	26.925	27.615
130	24.433	10.956	28.855	22.385	23.296	27.457	28.163
140	24.851	11.14	29.413	22.786	23.719	27.983	28.704
150	25.266	11.322	29.965	23.184	24.137	28.504	29.24
160	25.678	11.502	30.512	23.578	24.55	29.02	29.771
170	26.088	11.682	31.053	23.967	24.96	29.53	30.297
180	26.495	11.86	31.59	24.353	25.365	30.036	30.818
190	26.899	12.036	32.121	24.734	25.766	30.537	31.334
200	27.302	12.212	32.648	25.113	26.164	31.033	31.845
210	27.701	12.386	33.17	25.487	26.557	31.524	32.352
220	28.099	12.559	33.687	25.858	26.947	32.012	32.854
230	28.495	12.731	34.2	26.226	27.334	32.494	33.351
240	28.888	12.901	34.709	26.59	27.717	32.973	33.844
250	29.279	13.071	35.213	26.951	28.096	33.447	34.333
260	29.668	13.24	35.713	27.309	28.472	33.918	34.818
270	30.056	13.407	36.209	27.664	28.845	34.384	35.299
280	30.441	13.574	36.702	28.016	29.215	34.847	35.776
290	30.824	13.74	37.19	28.366	29.582	35.306	36.25
300	31.206	13.904	37.674	28.712	29.946	35.761	36.719

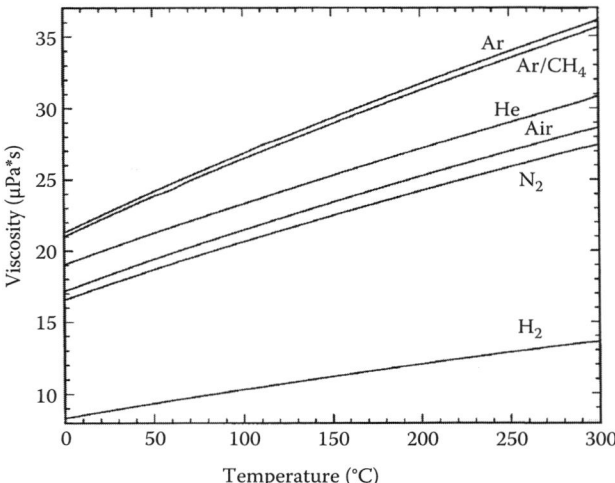

Figure 1.1 Plot of viscosity as a function of temperature (°C) at 29.7 psia.

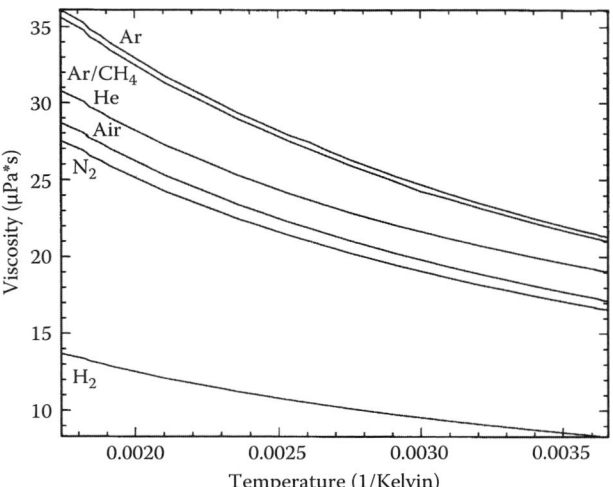

Figure 1.2 Viscosity vs. temperature at 29.7 psia.

GAS CHROMATOGRAPHIC SUPPORT MATERIALS FOR PACKED COLUMNS

The following table lists the more common solid supports used in packed column gas chromatography and preparative scale gas chromatography, along with relevant properties [1–5]. The materials are also used in packed capillary columns. The performance of several of these materials can be improved significantly by acid washing and treatment with DMCS (dimethyldichlorosilane) to further deactivate the surface. The nonacid washed materials can be treated with hexamethyldisilane to deactivate the surface, however the deactivation is not as great as that obtained by an acid wash followed by DMCS treatment. Most of the materials are available in several particle size ranges. The use of standard sieves will help insure reproducible sized packings from one column to the next. Data are provided for the Chromosorb family of supports since they are among the most well characterized. It should be noted that other supports are available to the chromatographer, with a similar range of properties provided by the Chromosorb series.

REFERENCES

1. Poole, C. F., and S. A. Schuette. *Contemporary Practice of Chromatography*. Amsterdam: Elsevier, 1984.
2. Gordon, A. J., and R. A. Ford. *The Chemist's Companion.* New York: John Wiley and Sons, 1972.
3. Heftmann, E., ed. *Chromatography: A Laboratory Handbook of Chromatographic and Electrophoretic Methods.* 3rd ed. New York: Van Nostrand Reinhold, 1975.
4. Poole, C. F. *The Essence of Chromatography*. Amsterdam: Elsevier, 2003.
5. Grant, D. W. *Gas-Liquid Chromatography.* London: Van Nostrand Reinhold, 1971.

Gas Chromatographic Support Materials For Packed Columns

Support Name	Support Type	Density (Free Fall) g/mL	Density (Packed) g/mL	pH	Surface Area m²/g	Maximum Liquid Loading	Color	Notes
Chromosorb A	diatomite	0.40	0.48	7.1	2.7	25 %	pink	Most useful for preparative gas chromatography; high strength; high liquid phase capacity; low surface activity.
Chromosorb G	diatomite	0.47	0.58	8.5	0.5	5 %	oyster white	High mechanical strength; low surface activity; high density.
Chromosorb P	diatomite firebrick	0.38	0.47	6.5	4.0	30 %	pink	High mechanical strength; high liquid capacity; moderate surface activity; for separations of moderately polar compounds.
Chromosorb W	diatomite	0.18	0.24	8.5	1.0	15 %	white	Lower mechanical strength than pink supports; very low surface activity; for polar compound separation.
Chromosorb 750	diatomite	0.33	0.49		0.75	7 %	white	Highly inert surface; useful for biomedical and pesticide analysis; mechanical strength similar to Chromosorb G.
Chromosorb R-4670-1	diatomite				5-6	low	white	Ultra-fine particle size used to coat inside walls of capillary columns; typical particle size is 1-4 μm.
Chromosorb T[a]	polytetrafluoroethylene	0.42	0.49		7.5	5 %	white	Maximum temperature of 240 °C; handling is difficult due to static charge; tends to deform when compressed; useful for analysis of high polarity compounds.

(*Continued*)

Gas Chromatographic Support Materials For Packed Columns (Continued)

Support Name	Support Type	Density (Free Fall) g/mL	Density (Packed) g/mL	pH	Surface Area m²/g	Maximum Liquid Loading	Color	Notes
Kel-F[a]	chlorofluorocarbon				2.2	20 %	white	Hard, granular chlorofluorocarbon; mechanically similar to chromosorbs; generally gives poor efficiency; use below 160 °C, very rarely used.
Fluoropak-80[a]	fluorocarbon resin				1.3	5 %	white	Granular fluorocarbon with sponge-like structure; low liquid phase capacity; use below 275 °C.
Teflon-6[a]	polytetrafluoroethylene				10.5	20 %	white	Usually 40–60 (U.S.) mesh size; for relatively nonpolar liquid phases; low mechanical strength; high inert surface; difficult to handle due to static charge; difficult to obtain good coating of polar phases due to highly inert surface.
T-Port-F[a]	polytetrafluoroethylene	0.5					white	Use below 150 °C.
Porasil (Types A through F)	silica				2–500, type dependent	40 %	white	Rigid, porous silica bead; controlled pore size varies from 10–150 mm; highly inert; also used as a solid adsorbent.

[a] The fluorocarbon supports can be difficult to handle since they develop electrostatic charge easily. This is especially problematic in dry climates. It is generally advisable to work with them below 19 °C (solid transition point), using polyethylene laboratory ware.

MESH SIZES AND PARTICLE DIAMETERS

The following tables give the relationship between particle size diameter (in μm) and several standard sieve sizes. The standards are as follows:

United States Standard Sieve Series, ASTM E-11-61
Canadian Standard Sieve Series, 8-GP-16
British Standards Institution, London, BS-410-62
Japanese Standard Specification, JI S-Z-8801
French Standard, AFNOR X-11-501
German Standard, DIN-4188

Particle Size, μm	U.S. Sieve Size	Tyler Mesh Size	British Sieve Size	Japanese Sieve Size	Canadian Sieve Size
4000	5	—	—	—	—
2000	10	9	8	9.2	8
1680	12	10	—	—	—
1420	14	12	—	—	—
1190	16	14	—	—	—
1000	18	16	—	—	—
841	20	20	18	20	18
707	25	24	—	—	—
595	30	28	25	28	25
500	35	32	—	—	—
420	40	35	36	36	36
354	45	42	—	—	—
297	50	48	52	52	52
250	60	60	60	55	60
210	70	65	72	65	72
177	80	80	85	80	85
149	100	100	100	100	100
125	120	115	120	120	120
105	140	150	150	145	150
88	170	170	170	170	170
74	200	200	200	200	200
63	230	250	240	250	240
53	270	—	300	280	300
44	325	—	350	325	350
37	400	—	—	—	—

French and German Sieve Sizes

Particle Size, μm	Sieve Size
2000	34
800	30
500	28
400	27
315	26
250	25
200	24
160	23
125	22
100	21
80	20
63	19
50	18
40	17

Mesh-Size Relationships

Mesh Range	Top Screen Opening, μm	Bottom Screen Opening, μm	Micron Screen, μm	Range Ratio
10/20	2000	841	1159	2.38
10/30	2000	595	1405	3.36
20/30	841	595	246	1.41
30/40	595	420	175	1.41
35/80	500	177	323	2.82
45/60	354	250	104	1.41
60/70	250	210	40	1.19
60/80	250	177	73	1.41
60/100	250	149	101	1.68
70/80	210	177	33	1.19
80/100	177	149	28	1.19
100/120	149	125	24	1.19
100/140	149	105	44	1.42
120/140	125	105	20	1.19
140/170	105	88	17	1.19
170/200	88	74	14	1.19
200/230	74	63	11	1.17
230/270	63	53	10	1.19
270/325	53	44	9	1.20
325/400	44	37	7	1.19

PACKED COLUMN SUPPORT MODIFIERS

During the analysis of strongly acidic or basic compounds, peak tailing is almost always a problem, especially when using packed columns. Pretreatment of support materials, such as acid-washing and treatment with DMCS will usually result in only modest improvement in performance. A number of modifiers can be added to the stationary phase (in small amounts, 1–3 %) in certain situations to achieve a reduction in peak tailing. The following table provides several such reagents [1,2]. It must be remembered that the principal liquid phase must be compatible with any modifier being considered. Thus, the use of potassium hydroxide with polyester or polysiloxane phases would be inadvisable since this reagent can catalyze the depolymerization of the stationary phase. It should also be noted that the use of a tail-reducing modifier may lower the maximum working temperature of a particular stationary phase.

REFERENCES

1. Poole, C. F., and S. A. Schuette. *Contemporary Practice of Chromatography.* Amsterdam: Elsevier, 1984.
2. Poole, C. F. *The Essence of Chromatography.* Amsterdam: Elsevier, 2003.

Compound Class	Modifier Reagents	Notes
Acids	phosphoric acid, FFAP (carbowax-20 m-terephthalic acid ester), trimer acid	These modifiers will act as subtractive agents for basic components in the sample; FFAP will selectively abstract aldehydes; phosphoric acid may convert amides to the nitrile (of the same carbon number), desulfonate sulfur compounds, and may esterify or dehydrate alcohols.
Bases	potassium hydroxide, polyethyleneimine, polypropyleneimine, N,N′-bis-l-methylheptyl-p-phenylenediamine, sodium metanilate, THEED (tetrahydroxyethylenediamine)	These modifiers will act as subtractive agents for acidic components in the sample; polypropyleneimine will selectively abstract aldehydes, polyethyleneimine will abstract ketones.

PROPERTIES OF CHROMATOGRAPHIC COLUMN MATERIALS

The following table provides physical, mechanical, electrical, and (where appropriate) optical properties of materials commonly used as chromatographic column tubing [1–6]. The data will aid the user in choosing the appropriate tubing material for a given application. The mechanical properties are measured at ambient temperature unless otherwise specified. The chemical incompatibilities cited are usually only important when dealing with high concentrations, which are normally not encountered in gas chromatography. Caution is urged nevertheless.

REFERENCES

1. Materials Engineering—Materials Selector. Cleveland: Penton/IPC, 1986.
2. Khol, R., ed. Machine Design. *Materials Reference Issue* 58, no. 8, 1986.
3. Polar, J. P. *A Guide to Corrosion Resistance.* Greenwich: Climax Molybdenum Co., 1981.
4. Fontana, M. G., and N. D. Green. *Corrosion Engineering.* New York: McGraw-Hill Book Co., 1967.
5. Shand, E. B. *Glass Engineering Handbook.* New York: McGraw-Hill Book Co., 1958.
6. Fuller, A. Corning Glass Works, Science Products Division, Corning, NY, Private Communication, 1988.

Aluminum (Alloy 3003)

Density	2.74 g/mL
Hardness (Brinell)	28–55
Melting range	643.3–654.4 °C
Coefficient of expansion (20–100 °C)	2.32×10^{-5} °C^{-1}
Thermal conductivity (20 °C, annealed)	193.14 W/(m·K)
Specific heat (100 °C)	921.1 J/(kg·K)
Tensile strength (hard)	152 MPa
Tensile strength (annealed)	110 MPa

Notes: Soft and easily formed into coils; high thermal conduction; incompatible with strong bases, nitrates, nitrites, carbon disulfide, and diborane.
Actual alloy composition: Mn = 1.5 %; Cu = 0.05–0.20 %; balance is Al.

Copper (Alloy C12200)[a]

Density	8.94 g/mL
Hardness (Rockwell-f)	40–45
Melting point	1082.8 °C
Coefficient of expansion (20–300 °C)	1.76×10^{-5} °C^{-1}
Thermal conductivity (20 °C)	339.22 W/(m·K)
Specific heat (20 °C)	385.11 J/(kg·K)
Tensile strength (hard)	379 MPa
Tensile strength (annealed)	228 MPa
Elongation (in 0.0508 m annealed) %	45

Notes: Copper columns often cause adsorption problems; incompatible with amines, anilines, acetylenes, terpenes, steroids, and strong bases.
[a] High purity phosphorus deoxidized copper.

Borosilicate Glass

Density	2.24 g/rnL
Hardness (Knh 100)	418
Young's modulus (25 °C)	62 GPA
Poisson's ratio (25 °C)	0.20
Softening point	806.9 °C
Annealing point	565 °C
Melting point	1600 °C
Strain point	520 °C
Coefficient of expansion (av.)	3×10^{-6} °C^{-1}
Thermal conductivity	1.26 W/(m·K)
Specific heat	710 J/(kg·K)
Refractive index[a]	1.473
Normal service temperature (annealed)	215 °C
Extreme service temperature (annealed)	476 °C
Critical surface tension	750 mN/m

Notes: Has been used for both packed columns and capillary columns; incompatible with fluorine, oxygen difluoride, and chlorine trifluoride.
[a] Clear grade, at 588 mm.

Fused Silica (SiO$_2$)

Density	2.15 g/mL
Hardness (Moh)	6
Young's modulus (25 °C)	72 GPa
Poisson's ratio (25 °C)	0.14
Softening point	1590 °C
Annealing point	1105 °C
Melting point	1704 °C
Strain point	1000 °C
Coefficient of expansion (av.)	5×10^{-7} °C^{-1}
Thermal conductivity	1.5 W/(m·K)
Specific heat	1000 J/(kg·K)
Refractive index (588 nm)	1.458
Normal service temperature (annealed)	886 °C
Extreme service temperature (annealed)	1086 °C
Critical surface tension	760 mN/m

Notes: Used for capillary columns; typical inside diameters range from 5 to 530 μm; coated on outside surface by polyimide or aluminum to prevent surface damage; incompatible with fluorine, oxygen difluoride, chlorine trifluoride, and hydrogen fluoride.

Nickel (Monel R-405)

Density	8.83 g/mL
Hardness (Brinell, 21 °C)	110–245
Melting range	1298.89–1348.89 °C
Coefficient of expansion (21–537 °C)	1.64×10^{-5} °C^{-1}
Thermal conductivity (21 °C)	21.81 W/(m·K)
Specific heat (21 °C)	427.05 J/(kg·K)
Tensile strength (hard)	483 MPa
Tensile strength (annealed)	793 MPa
Elongation (in 2 in, 21.1 °C)	15–50 %

Notes: Provides excellent corrosion resistance; no major chemical incompatibilities. Actual alloy composition: Ni = 66 %; Cu = 31.5 %; Fe = 1.35 %; C = 0.12 %; Mn = 0.9 %; S = 0.005 %; Si = 0.15 %.

Polytetrafluorethylene (Teflon)

Specific gravity	2.13–2.24
Hardness (Rockwell-d)	52–65
Melting range	1298.89–1348.89 °C
Coefficient of expansion	1.43×10^{-4} °C^{-1}
Thermal conductivity (21 °C)	2.91 W/m·K
Specific heat (21 °C)	1046.7 J/kg·K
Tensile strength	17–45 MPa
Refractive index[a]	1.35

Notes: Flexible and easy to use; cannot be used above 230 °C; thermal decomposition products are toxic; tends to adsorb many compounds that may increase tailing. No major chemical incompatibilities; Static charge can be problematic.

[a] Using sodium-D line, as per ASTM standard test D542-50.

Stainless Steel (304)

Density	7.71 g/mL
Hardness (Rockwell B)	149
Melting range	1398.9–1421.1 °C
Coefficient of expansion (0–100 °C)	1.73×10^{-5} °C^{-1}
Thermal conductivity (0 °C)	16.27 W/(m·K)
Specific heat (0–100 °C)	502.42 J/(kg·K)
Tensile strength (hard)	758 MPa
Tensile strength (annealed)	586 MPa
Elongation (in 2 in)	60 %

Notes: Good corrosion resistance; easily brazed using silver bearing alloys; high nickel content may catalyze some reactions at elevated temperatures. No major chemical incompatibilities.

Actual alloy composition: C = 0.08 %; Mn = 2 % (max); Si = 1 % (max); P = 0.045 % (max); S = 0.030 (max); Cr = 18–20 %; Ni = 8–12 %; balance is Fe. The low carbon alloy, 304L, is similar except for C = 0.03 % max, and is more suitable for applications involving welding operations and where high concentrations of hydrogen are used.

Stainless Steel (316)

Density	7.71 g/mL
Hardness (Rockwell B)	149
Melting range	1371.1–1398.9 °C
Coefficient of expansion (0–100 °C)	7.17×10^{-5} °C^{-1}
Thermal conductivity (0 °C)	16.27 W/(m·K)
Specific heat (0–100 °C)	502.42 J/(kg·K)
Tensile strength (annealed)	552 MPa
Elongation (in 2 in)	60 %

Notes: Best corrosion resistance of any standard stainless steel, including the 304 varieties, especially in reducing and high temperature environments.

Actual alloy composition: C = 0.08 % (max); Mn = 2 % (max); Si = 1 % (max); P = 0.045 % (max); S = 0.030 (max); Cr = 16–18 %; Ni = 10–14 %; Mo = 2–3 %; balance is Fe. The low carbon alloy, 316L, is similar except for C = 0.03 % max, and is more suitable for applications involving welding operations and where high concentrations of hydrogen are used.

PROPERTIES OF SOME LIQUID PHASES FOR PACKED COLUMNS

The following table lists some of the more common gas-chromatographic liquid phases that have been used historically, along with some relevant data and notes [1–3]. Most of these have been superseded by silicone phases used in capillary columns, but these liquid phases still find application in some instances. This is especially true with work involving established protocols, such as ASTM or AOAC methods. Moreover, the data are still useful in interpreting analytical results in the literature. The minimum temperatures, where reported, indicate the point at which some of the phases approach solidification, or when the viscosity increases to the extent that performance is adversely affected. The maximum working temperatures are determined by vapor pressure (liquid phase bleeding) and chemical stability considerations. The liquid phases are listed by their most commonly used names. Where appropriate, chemical names or common generic names are provided in the notes.

The McReynolds constants (a modification of the Rohrschneider constant) tabulated here are based on the retention characteristics of the following test probe samples:

Constant	Test Probe
X	benzene
Y	1-butanol
Z	3-pentanone
U	1-nitropropane
S	pyridine

Compounds that are chemically similar to these probe solutes will show similar retention characteristics. Thus, benzene can be thought of as representing lower aromatic or olefinic compounds. Higher values of the McReynolds constant usually indicate a longer retention time (higher retention volume) for a compound represented by that constant, for a given liquid (stationary) phase.

Solvents: Ace: acetone   MeCl: methylene chloride
     Chlor: chloroform   Tol: toluene
     Pent: n-pentane   MeOH: methanol
     DMP: dimethylpentane H_2O: water
     EAC: ethyl acetate

Polarity: N: nonpolar
    P: polar
    I: intermediate polarity
    HB: hydrogen bonding
    S: specific interaction

REFERENCES

1. McReynolds, W. O. "Characterization of Some Liquid-Phases." *Journal of Chromatographic Science* 8 (1970): 685.
2. McNair, H. M., and E. J. Bonelli. *Basic Gas Chromatography.* Palo Alto: Varian Aerograph, 1968.
3. Heftmann, E. *Chromatography: A Laboratory Handbook of Chromatographic and Electrophoretic Methods.* 3rd ed. New York: Van Nostrand Reinhold Co., 1975.

Properties of Some Liquid Phases for Packed Columns

Liquid Phase	T_{min}, °C	T_{max}, °C	Polarity	Solvents	McReynolds Constant					Notes
					X	Y	Z	U	S	
Acetonyl acetone (2,5-hexanedione)	-4	25	I	Ace						
Acetyl tributyl citrate	25	180	I	Ace	135	268	202	314	233	
Adiponitrile	5	50	I	Chlor MeCl						1,4-dicyanobutane
Alka terge-T, amine surfactant	59	75	I	Chlor MeCl MeOH						60 % oxazoline, weakly cationic
Amine 220	0	180	P	Chlor MeCl	117	380	181	293	133	2-(8-heptadecenyl)-2-imidazoline-ethanol
Ansul ether		80	P	MeOH						tetraethylene glycol dimethyl ether, used for hydrocarbons
Apiezon H	50	275	N	Chlor	59	56	81	151	129	low vapor pressure hydrocarbon oil
Apiezon J	50	300	N	Chlor MeCl	38	36	27	49	57	low vapor pressure hydrocarbon oil
Apiezon L	50	300	N	Chlor MeCl	32	22	15	32	42	low vapor pressure hydrocarbon oil
Apiezon M	50	275	N	Chlor MeCl	31	22	15	30	40	low vapor pressure hydrocarbon oil
Apiezon N	50	300	N	Chlor MeCl	38	40	28	52	58	low vapor pressure hydrocarbon oil
Apiezon K	50	300+	N	Chlor						low vapor pressure hydrocarbon oil
Apiezon W	50	275	N	Chlor	82	135	99	155	154	low vapor pressure hydrocarbon oil
Apolane-87	30	280	N	Tol	21	10	3	12	35	24,24-diethyl-19,29-dioctadecyl heptatetracontane, C-87 hydrocarbon
Armeen SD		100	P,S,HB	Chlor MeCl						primary aliphatic amine
Armeen 12D		100	P,HB	Chlor MeCl						
Armeen		125	P,HB	Tol						secondary aliphatic amine
Armeen 2HT		100	P,HB	Chlor						
Arneel DD		100	P	MeOH						aliphatic nitrile
Arochlor 1242		125		Ace						chlorinated polyphenyl, used for gases, may be carcinogenic
Asphalt		300	N	Chlor MeCl						complex mixture of aliphatic, aromatic and heterocyclic compounds
Atpet 80			I	Chlor						sorbitan partial fatty acid esters
p,p-Azoxydiphenetol	130	140	I	Chlor						
Baymal		300		Tol						colloidal alumina
Beeswax		200		Chlor	43	110	61	88	122	for essential oils

Name	Min	Max	Type	Solvent	1	2	3	4	5	Notes
Bentone-34	20	200	S	Tol						dimethyl dioctadecylammonium bentonite
7,8-Benzoquinoline		150	I	Chlor						for hydrocarbons, aromatics, heterocycles, and sulfur compounds
Benzylamine adipate		125	I	Chlor						
Benzyl cellosolve		50	I	Ace						2-(benzyloxy ethanol), for hydrocarbons
Benzyl cyanide		50	I	MeOH						phenyl acetonitrile
Benzyl cyanide-AgNO$_3$		25	S	MeCl						
Benzyl diphenyl		100	I	Ace						
Benzyl ether		50	I	Chlor MeCl						dibenzyl ether
Bis (2-butoxyethyl) phthalate		175	I	MeOH	151	282	227	338	267	
Bis (2-ethoxyethyl) phthalate					214	375	305	446	364	
Bis (2-ethoxyethyl) sebacate					151	306	211	320	274	
N,N-bis (2-cyanoethyl formamide)	0	125	I	MeOH	690	991	853	110	000	
Bis (2-ethoxyethyl) adipate	0	150	I	Ace						
Bis (2-methoxyethyl) adipate	20	150	I	Ace Chlor						
Bis (2-ethylhexyl) tetrachlorophthalate	0	150	I	Chlor MeCl	112	150	123	108	181	
Butanediol adipate	60	225	I,P	Chlor MeCl						
Butanediol 1,4-succinate		225	I,P	Chlor	370	571	488	651	611	(BDS) craig polyester, for alcohols, aromatics, heterocycles, fatty acids and esters, and hydrocarbons
Bis[2-(2-methoxy-ethoxy) ethyl] ether		50	I	Chlor						tetraethylene glycol dimethylether
Carbitol		100	P	Ace						glycol ether (mol. mass 134) for aldehydes, ketones
Carbowax 300	10	100	P	MeCl						polyethylene glycol, av. mol. mass < 380
Carbowax 400	10	125	P	MeCl	333	653	405			polyethylene glycol, av. mol. mass 380–420
Carbowax 400 mono-oleate	10	125	P	MeCl						

(Continued)

Properties of Some Liquid Phases for Packed Columns (Continued)

Liquid Phase	T_{min}, °C	T_{max}, °C	Polarity	Solvents	McReynolds Constant X	Y	Z	U	S	Notes
Carbowax 550	20	125	P	MeCl						
Carbowax 600	30	125	P	MeCl	323	583	382			polyethylene glycol, av. mol. mass 570–630
Carbowax 600 monostearate		125	P	MeCl						
Carbowax 750	25	150	P	MeCl						methoxy polyethylene glycol, av. mol. mass 715–785
Carbowax 1000	40	175	P	MeCl	347	607	418	626	589	polyethylene glycol, av. mol. mass 950–1050
Carbowax 1500 (or Carbowax 540)	40	200	P	MeCl						polyethylene glycol, av. mol. mass 500–600
Carbowax 1540	40	200	P	MeCl	371	639	453	666	641	polyethylene glycol, av. mol. mass 1300–1600
Carbowax 4000 (or 3350)	60	200	P	MeCl	317	545	378	578	521	polyethylene glycol, av. mol. mass 3000–3700
Carbowax 4000 TPA		175	P	MeCl MeOH						terminated with terephthalic acid
Carbowax 4000 monostearate	60	220	P	MeCl	282	496	331	517	467	
Carbowax 6000	60	200	P	MeCl	322	540	369	577	512	polyethylene glycol, av. mol. mass 6000–7500
Carbowax 8000	60	120	P	Chlor	322	540	369	577	512	polyethylene glycol, av. mol. mass 7000–8500
Carbowax 20M	60	250	P	MeCl	322	536	368	572	510	polyethylene glycol, av. mol. mass 15,000–20,000
Carbowax 20M-TPA	60	250	P	MeCl	321	537	367	573	520	terminated with terephthalic acid
Castorwax	90	200	P	MeCl	108	265	175	229	246	triglyceride of 12-hydroxysteric acid (hydrogenated castor oil)
Chlorowax 70		130	P	MeCl						chlorinated paraffin, 70 % (wt/wt) Cl; for hydrocarbons
1-Chloronaphthalene		75	—	Tol						
Citroflex A-4		150	—	MeOH	135	286	213	324	262	tributyl citrate
Cyanoethyl sucrose	20	175	P	Ace	647	919	043	976		vitrifies at –10 °C

Name										Applications
Cyclodextrin acetate	100	250	I	Ace	269	446	328	498	481	for fatty acids and esters
Cyclohexane dimethanol succinate		210	I	Chlor						
n-Decane		30	N	MeCl						for inorganic and organometallic compounds
Di (ethoxyethoxyethyl) phthalate					233	408	317	470	389	
Di(butoxyethyl) adipate	-10	150	P	Ace	137	278	198	300	235	
Di(butoxyethyl) phthalate	-30	200	P	Tol	157	292	233	348	272	
Di-n-butyl cyanamide		50		MeOH						for gases
Di-n-butyl maleate	0	50	P,I	Tol						for halogenated compounds
Di-n-butyl phthalate	-20	100	I	Tol						for aldehydes, ketones, halogenated compounds, hydro-carbons, phosphorus compounds
Dibutyltetrachloro-phthalate	0	150	I	Tol						
Didecyl phthalate	20	150	I	Tol	136	255	213	320	235	
Dicyclohexyl phthalate		200			146	257	206	316	245	
Diethylene glycol adipate	0		I	MeCl	378	603	460	665	658	DEGA, for aldehydes, ketones, esters, fatty acids, and pesticides
Diethylene glycol glutarate		225	I	MeCl						
Diethylene glycol sebacate	80	190	I	MeCl						DEGSB
Diethylene glycol succinate	20	190	P	MeCl	496	746	590	837	835	DEGS; for alcohols, aldehydes, ketones, amino acids, essential oils, steroids, esters, phosphorus compounds, and sulfur compounds
Diethylene glycol stearate					64	193	106	143	191	
Di(2-ethylhexyl) phthalate	20	150	P	Tol	135	254	213	320	235	
Di(2-ethylhexyl) adipate	-30	250	P	Ace	76	181	121	197	134	dioctyl adipate
Di(2-ethylhexyl) sebacate	-20	125	I	Tol	72	168	108	180	125	for alcohols, drugs, alkaloids, esters, fatty acids, halogenated compounds, blood gases
Diethyl-D-tartarate		125	P,S	MeCl						for alcohols
Diglycerol	20	120	HB	MeClMeOH	371	826	560	676	854	for alcohols, aldehydes, ketones, aromatics, heterocycles, hydrocarbons

(Continued)

Properties of Some Liquid Phases for Packed Columns (Continued)

Liquid Phase	T_{min}, °C	T_{max}, °C	Polarity	Solvents	McReynolds Constant					Notes
					X	Y	Z	U	S	
Dilauryl phthalate		150	I	Tol	79	158	120	192	158	
Disodecyl adipate	−10	175	P	Ace	71	171	113	185	128	
Disooctyl adipate	90	150	P	Ace	78	187	126	204	140	
Disodecyl phthalate	0	150	I	TolAce	84	173	137	218	155	for alcohols, aromatics, heterocycles, essential oils, esters, halogen and sulfur compounds, hydrocarbons
Disoctyl sebacate		175	I	Ace						for aldehydes, ketones, hydrocarbons
2,4-Dimethyl sulfolane	0	50	P	Chlor						for hydrocarbons, inorganic and organometallic compounds
Dimer acid		100	I	MeCl						C_{36} dicarboxylic acid
Disooctyl phthalate	0	175	I	Tol	94	193	154	243	202	
Dimethyl formamide	−20	20	P	Ace						DMF
Dimethyl sulfoxide	20	30	P	MeCl						DMSO, for gases
Dinonyl phthalate	20	150	I	Tol	83	183	147	231	159	for aromatics, heterocycles, halogen compounds
Dioctyl phthalate	−20	150	I	Tol	92	186	150	230	167	for aromatics, heterocycles, halogen compounds
Dioctyl sebacate		100	I	MeCl	72	168	108	180	123	
Diphenyl formamide	75	100	I	Tol						
Di-n-propyl tetrachloro phthalate	10	75	I	Tol						
Ditridecyl phthalate	−10	225	P	Tol	75	156	122	195	140	
Emulphor ON-870	0	200	I	Chlor	202	395	251	395	344	aryloxy polyethylene oxyethanol; for aromatics, heterocycles, essential oils, halogen compounds
EPON 1001	60	225	P	MeCl hot	284	489	406	539	601	epichlorohydrin-bisphenol A resin, av. mol. mass 900; for steroids and pesticides
Ethofat 60/25	50	125	I	MeCl hot	191	382	244	380	333	polyethylene oxyglycol stearate; for aldehydes and ketones
Ethomeen S/25		75	P	MeCl	186	395	242	370	339	polyethoxylated aliphatic amine for hydrocarbons
Ethyl benzoate		150	I	MeOH						

Ethylene glycol adipate	100	225	I,P	MeCl	372	576	453	655	617	for alcohols, aromatics, heterocycles, bile/urinary compounds, drugs, alkaloids, essential oils, nitrogen and sulfur compounds
Ethylene glycol phthalate	100	200	I,P	Tol	453	697	602	816	872	for nitrogen compounds, steroids
Ethylene glycol succinate	100	200	I,P	Ace	537	787	643	903	889	
Ethylene glycol glutarate		225	I,P	MeCl						
Ethylene glycol sebacate		200	I,P	MeCl hot						
Ethylene glycol tetrachlorophthalate	120	200	P	Tol	307	345	318	428	466	
Ethylene glycol		30	HB	MeOH						
Ethylene glycol silver nitrate		30	S	Ace						
Eutectic (LiNO$_3$-NaNO$_3$-KNO$_3$/27.3–18.2–54.5)		400	—	H$_2$O						for aromatic hydrocarbons and heterocycles
Eutectic (KCl-CdCl$_2$/33–67)		400	—	H$_2$O						for aromatic hydrocarbons and heterocycles
Eutectic (NaCl-AgCl/41–59)		400	—	H$_2$O						for aromatic hydrocarbons and heterocycles
Eutectic (BiCl$_3$-PbCl$_3$/89-11)		400	—	H$_2$O						for aromatic hydrocarbons and heterocycles
FFAP	50	250	P,S	Chlor	340	580	397	602	627	Carbowax 20M nitro terephthalic acid ester; for aldehydes, ketones
Flexol 8N8		180	P	Ace	96	254	164	260	179	2,2'-(2-ethyl hexynamido)-diethyl-di-2-ethylhexanoate, for alcohols and nitrogen compounds
Fluorolube HG-1200		100	—	Ace	51	68	114	144	116	polymers of trifluoro vinyl chloride; for halogenated compounds
Formamide	20	50	—	MeOH						for alcohols
Glycerol	20	100	HB	MeOH						
Fluorad FC-431	40	200	—	EAC	281	423	297	509	360	fluorocarbon surfactant
Hallcomid M-18	40	150	—	MeCl	79	580	397	602	627	dimethylsteramide; for alcohols, ketones, aldehydes, esters
Hallcomid M-18-OL	8	150	—	MeCl	89	280	143	239	165	dimethyloleamide; for alcohols, ketones, aldehydes, fatty acids
Halocarbon 10-25	20	100	—	Chlor	47	70	108	113	111	

(Continued)

Properties of Some Liquid Phases for Packed Columns (Continued)

Liquid Phase	T_{min}, °C	T_{max}, °C	Polarity	Solvents	McReynolds Constant					Notes
					X	Y	Z	U	S	
Halocarbon K 352	0	250	I		47	70	73	238	146	
Halocarbon W9X (600)	50	150		Ace	55	71	116	143	123	
Halocarbon-1321	0	100		Ace						
Halocarbon-11-14	0	100		Ace						
HMPA	20	35	P	Chlor						hexamethylphosphoramide
Hi-Eff-1 AP	20	210	I,P	Chlor	378	603	460	665	658	diethyleneglycol adipate
Hi-Eff-2 AP	100	210	I,P	Chlor	372	576	453	655	617	cyclohexane dimethanol succinate
Hi-Eff-8 BB	100	250	I,P	Chlor	271	444	333	498	463	diethylene glycol succinate
Hi-Eff-1 BP	20	200	I,P	Chlor	499	751	593	840		ethylene glycol succinate
Hi-Eff-2 BP	100	200	I,P	Chlor	537	787	643	903	889	neopentyl glycol adipate
Hi-Eff-3 AP	50	230	I,P	Chlor						cyclohexane dimethanol adipate
Hi-Eff-8 AP	100	250	I,P	Chlor						tetramethyl cyclobutanediol adipate
Hi-Eff-9 AP	100	250	I,P	Chlor						neopentyl glycol succinate
Hi-Eff-3 BP			I,P							butane-1,4-diol succinate
Hi-Eff-4 BP	50	230	I,P	Chlor						phenyl diethanolamine succinate
Hi-Eff-10 BP	20	230	I,P	Chlor						ethylene glycol sebacate
Hi-Eff-2 CP	100	200	I,P	Chlor						neopentyl glycol sebacate
Hi-Eff-3 CP	50	230	I,P	Chlor						ethylene glycol isophthalate
Hi-Eff-2 EP	100	210	I,P	Chlor						ethylene glycol phthalate
Hi-Eff-26 P	100	210	I,P	Chlor						octakis (2-hydroxy propyl) sucrose
Hyprose-SP-80		225	P	MeOH	336	742	492	639	727	
1,2,3,4,5,6 Hexakis-(2-cyanoethoxy-cyclohexane)	125	150	I,P	Tol	567	825	713	978	901	
Hercoflex 600		150	P	MeCl	112	234	168	261	194	high boiling ester of pentaerythritol and a saturated aliphatic acid
n-Hexadecane	20	50	N	Pent						
Hexadecane	20	50	N	Pent						isomeric mixture
1-Hexadeconal		35	I	MeOH						cetyl alcohol; for halogenated compounds and hydrocarbons
Hexatricontane	80	150	N	MeCl	12	2	-3	1	11	$C_{36}H_{74}$

Stationary phase	Min	Max	Class	Solvent						Application
IGEPAL CO-880	100	200	I	MeCl hot	259	461	311	482	426	nonyl phenoxypolyethyleneoxy ethanol n = 30; for alcohols
IGEPAL CO-990	100	200	I	MeCl hot	298	508	345	540	475	nonylphenoxypoly(ethyleneoxy-ethanol) n = 100; for alcohols
IGEPAL CO-630	100	200	I	MeCl hot	192	381	253	382	344	nonylphenoxy poly(ethylene oxyethanol) n = 9; for alcohols
IGEPAL CO-730	100	200	I	—	224	418	279	428	379	
IGEPAL CO-710	100		I	—	205	397	266	401	361	
β,β-Iminodipropio-nitrile		110	I,P	MeOH						for halogenated compounds
Isoquinoline		50	P	MeCl						for hydrocarbons
Lexan	220	270	P	DMP hot						polycarbonate resin
Mannitol	170	200	HB	H₂O						for sugars
Montan wax		175	I	Chlor	19	58	14	21	47	for halogenated compounds
Naphthylamine		150	I	Chlor						for aromatics and heterocycles
Neopentylglycol adipate	50	240	I	MeCl	234	425	312	402	438	NPGA; for amino acids, drugs, alkaloids, pesticides, steroids
Neopentylglycol isophthalate	50	240	I	MeCl						
Neopentylglycol sebacate	50	225	I	MeCl	172	327	225	344	326	NPGSB; for amino acids, steroids
Neopentylglycol succinate	50	225	I	MeCl	272	469	366	539	474	NPGS; for amino acids, bile and urinary compounds, esters, inorganics
Nitrobenzene		150	I	MeOH						for hydrocarbons, inorganic and organometallic compounds
Nujol		100	N	Pent	9	5	2	6	11	paraffin oil, mineral oil; for hydrocarbons
n-Octadecane	30	55	N	Pent						for inorganic and organometallic compounds
Octyl decyl adipate		175	P	Ace	79	179	119	193	134	
Oronite NIW		170	P	Ace	180	370	242	370	327	complex mixture of petroleum liquids
β,β′-Oxydipropio-nitrile		100	P	Ace						for halogenated compounds
Phenyl diethanolamine succinate		225	P	Ace	386	555	472	674	654	for drugs, alkaloids, and hydrocarbons
Polyethylene imine	0	250	P	MeOH	322	800	—	573	524	
Poly-m-phenylxylene	125	375	I	Tol	257	355	348	433	—	PPE-20
Poly-m-phenyl ether		250	I	Tol	176	227	224	306	283	5 rings; for aromatics and heterocycles

(Continued)

Properties of Some Liquid Phases for Packed Columns (Continued)

Liquid Phase	T_{min}, °C	T_{max}, °C	Polarity	Solvents	McReynolds Constant					Notes
					X	Y	Z	U	S	
Poly-m-phenyl ether	0	300	I	Ace, Tol	182	233	228	313	293	6 rings; for alcohols, essential oils, and esters
Poly-m-phenyl ether	50	400	I	Tol						high polymer
Poly-m-phenyl ether with squalane	50	100	I	MeCl						6 rings
Polypropylene glycol	0	150	HB	MeOH	128	294	173	264	226	av. mol. mass = 2000; for drugs, alcohols, alkaloids
Polypropylene glycol sebacate	20	225	I	Chlor	196	345	251	381	328	
Polypropylene glycol silver nitrate	20	75	S	MeCl						PEG/AgNO$_3$-3/1; for unsaturated hydrocarbons
Polypropylene imine	0	200	I,P	Chlor	122	425	168	263	224	
Propylene carbonate	0	60	P	MeCl						1,2 propanediol cyclic carbonate; for gases and hydrocarbons
Polysulfone	0	315	I	Ace						
Polyvinyl pyrrolidone	80	225	HB	MeOH						
Quadrol	0	150	HB	Chlor	214	571	357	472	489	Consists of N,N,N',N'-tetrakis (2-hydroxy-propyl) ethylenediamine; for alcohols, aldehydes, ketones, amino acids, essential oils
Reoplex 400	0	200	I	MeCl	364	619	449	647	671	poly(propylene glycol adipate); for aromatics, heterocycles, vitamins, sulfur and phosphorus compounds
Reoplex 100	0	200	I	MeCl						poly(propylene glycol sebacate)
Renex-678	0	200	I	MeOH	223	417	278	427	381	ethylene oxide-nonylphenol surfactant; for alcohols
Sebaconitrile		150	P							
Squalane	20	100	N	Pent	0	0	0	0	0	for hydrocarbons, organic vapors, nitrogen, sulfur and phosphorus compounds
Squalene	0	100	N,I	Pent	152	341	238	329	344	for hydrocarbons, gases, nitrogen, sulfur and phosphorus compounds
Sorbitol	15	150		Chlor	232	582	313			hexahydric alcohol, C$_6$H$_6$(OH)$_6$

Phase	Min T	Max T	Type	Solvent						Application / Notes
STAP	100	255	P	Chlor	345	586	400	610	627	steroid analysis phase
Siponate-DS-10	20	210	I,P	MeOH						sodium dodecylbenzene sulfonate
Sorbitan monooleate	20	150	P	Chlor	97	266	170	216	268	SPAN-80
Sorbitol hexaacetate			P		335	553	449	652	543	
Sucrose acetate isobutyrate	0	200	I,P	MeCl	172	330	251	378	295	SAIB; for alcohols, essential oils
Sucrose octaacetate	90	250	I,P	Ace	344	570	461	671	569	
Tergitol Nonionic NP-35	10	175	P	Chlor	197	380	258	389	351	Surfactant mixture
TCEPE	30	175	P,S	MeCl	526	782	677	920	837	tetracyanoethylatedpentaery-thritol; for fatty acids and esters
Terephthalic acid	100	250	P,I	Tol						
Tetraethylene glycol		70	P	MeCl						for hydrocarbons
Tetraethylene-pentamine		150	HB	MeOH						for nitrogen compounds
1,2,3,4-Tetrakis-(2-cyanoethyl)butane	110	200	I,P	Chlor	617	860	773	048	941	
THEED	0	125	HB	Chlor	463	942	626	801	893	tetrahydroxyethylenediamine; for alcohols, hydrocarbons, nitrogen compounds
β,β'-Thiodipropio-nitrile		100	P	MeOH						for hydrocarbons
Triacetin		60	P	MeOH						for gases
Tributyl phosphate	20	125	I	Ace						for gases
Tricresyl phosphate	20	215	I	MeCl	176	321	250	374	299	tritolyl phosphate
Triethanolamine	20	100	HB	MeOH						for alcohols, gases
Trimer acid	20	200	HB	MeOH	94	271	163	182	378	C_{54} tricarboxylic acid; for alcohols
1,2,3-Tris(2-cyano-ethoxy)propane	30	150	P	MeOH	594	857	759	031	917	for alcohols, aldehydes, ketones, halogen compounds, inorganic and organometallic compounds
Tris(tetrahydrofur-furyl) phosphate	20	125	I	Ace						
Tris(2-cyanoethyl) nitromethane	20	140	I,P	Chlor						
Triton X-100	20	190	P	MeCl	203	399	268	402	362	octylphenoxypolyethyl ethanol, for aromatics and heterocycles
Triton X-305	20	250	P	Ace	262	467	314	488	430	octylphenoxypolyethyl ethanol, for alcohols

(Continued)

Properties of Some Liquid Phases for Packed Columns (Continued)

Liquid Phase	T_{min}, °C	T_{max}, °C	Polarity	Solvents	McReynolds Constant					Notes
					X	Y	Z	U	S	
Trixylol phosphate	20	250	I,P	Ace						
TWEEN 20	20	150	P	MeOH						polyethoxysorbitan mono-laurate; for essential oils
TWEEN 80	20	160	P	MeOH	227	430	283	438	396	polyethoxysorbitan mono-oleate; for fatty acids, esters, pesticides
UCON LB-550-X	0	200	P	Chlor	118	271	158	243	206	10 % polyethylene glycol, 90 % propylene glycol
UCON 50-HB-280-X	0	200	P	Chlor	177	362	227	351	302	30 % polyethylene glycol, 70 % propylene glycol; for alcohols, fatty acids, esters
UCON 50-HB-2000	0	200	P	Chlor	202	394	253	392	341	40 % polyethylene glycol, 60 % propylene glycol; for alcohols, aldehydes, and ketones
UCON 50-HB-5100	20	200	P	MeCl	214	418	278	421	375	50 % polyethylene glycol, 50 % propylene glycol
UCON LB-1715	20	200	I	MeCl	132	297	180	275	235	for alcohols, ketones, nitrogen compounds
UCON 75-H-90,000	20	200	P	MeCl	255	452	299	470	406	80 % polyethylene glycol, 10 % propylene glycol
Versamide 900	190	250	P	MeCl	109	314	145	212	209	polyamide resin; for alcohols
Versamide 940	115	200	P	MeCl	109	313	144	211	209	polyamide resin; for alcohols
Versamide 930	115	150	P	MeCl	109	313	144	211	209	polyamide resin
Versamide 940		200	P	See Notes	109	314	145	212	209	soluble in hot chloroform butanol, 50/50 v/v, for aromatics, heterocycles, pesticides, nitrogen compounds
Xylenyl phosphate		175	I							
Zonyl E7		200	I	MeCl	223	359	468	549	465	fluoroalkyl ester
Zonyl E91		200	I	MeCl	130	250	320	377	293	fluoroalcohol camphorate
Zinc stearate	135	175	I	Ace (warm)	61	231	59	98	544	

SELECTION OF STATIONARY PHASES FOR PACKED
COLUMN GAS CHROMATOGRAPHY

The following stationary phases have been shown to be of value in the separation of these major classes of compounds, using packed columns of typical dimensions (4–10 m in length, 0.32 cm diameter) [1–10]. They have also been useful in packed capillary columns. The resolution will undoubtedly be lower than that obtainable with capillary columns, which have superseded packed columns in many applications. The two main exceptions are in the analysis of permanent gases and preparative scale gas chromatography. Data on the packed column stationary phases are included since they still find use in many laboratories. This table is meant to provide only a rough guide. The additional data, which can be found in the preceding stationary phase data table, will aid in determining the final choice.

REFERENCES

1. Heftmann, E., ed. *Chromatography: A Laboratory Handbook of Chromatographic and Electrophoretic Methods.* 3rd ed. New York: Van Nostrand Reinhold Co., 1975.
2. Grant, D. W. *Gas-Liquid Chromatography.* London: Van Nostrand Reinhold Co., 1971.
3. McNair, H. M., and E. J. Bonelli. *Basic Gas Chromatography.* Palo Alto: Varian Aerography, 1969.
4. Grob, R. L., ed. *Modern Practice of Gas Chromatography.* 2nd ed. New York: Wiley Interscience, 1985.
5. Poole, C. F., and S. A. Schuette. *Contemporary Practice of Chromatography.* Amsterdam: Elsevier, 1984.
6. Mann, J. R., and S. T. Preston. "Selection of Preferred Liquid Phases." *Journal of Chromatographic Science* 11 (1973): 216.
7. Coleman, A. E. "Chemistry of Liquid Phases, Other Silicones." *Journal of Chromatographic Science* 11 (1973): 198.
8. Yancey, J. A. "Liquid Phases Used in Packed Gas Chromatographic Columns. Part 1: Polysiloxane Phases." *Journal of Chromatographic Science* 23 (1985): 161.
9. Yancey, J. A. "Liquid Phases Used in Packed Gas Chromatographic Columns. Part 2: Use of Liquid Phases Which Are Not Polysiloxanes." *Journal of Chromatographic Science* 23 (1985): 370.
10. McReynolds, W. O. "Characterization of Some Liquid Phases." *Journal of Chromatographic Science* 8 (1970): 685.

Selection of Stationary Phases for Packed Column Gas Chromatography

Compound	Suggested Stationary Phases
Alcohols C_1–C_5	Apiezon L; Apiezon M; benzyldiphenyl; butane diol succinate (Craig polyester); Carbowax 400, 600, 750, 1000, 1000 (monostearate), diethylene glycolsuccinate; di(2-ethylhexyl) sebacate; diethyl-d-tartrate; di-n-decyl-phthalate; diglycerol; diisodecyl phthalate; dinonyl phthalate; ethylene glycol succinate; Flexol 8N8; Hallcomid M-18-OL; quadrol; Renex 678; sorbitol; tricresyl phosphate; triethanolamine
C_5–C_{18}	Butane diol succinate (Craig polyester); carbowax 1500, 1540, 4000, 4000 (dioleate), 4000 (monostearate), 6000, 20M, 20M-TPA; ethylene glycol adipate; Igepal series; Ucon series; Versamid series
Aldehydes (and ketones)	Apiezon L, M; carbowax 400, 750, 1000, 1500, 1540; di-n-butyl phthalate; diethylene glycol succinate; ethylene glycol succinate; Hallcomid M18; squalene; tricresyl phosphate; 1,2,3-tris (2-cyanoethoxy) propane; Ucon series
Alkaloids (includes drugs and vitamins)	Apiezon L; carbowax 20M; di(2-ethylhexyl) sebacate; ethylene glycol adipate; ethylene glycol succinate, neopentyl glycol adipate; phenyldiethanolamine succinate; SE-30 (methyl silicone phases)
Amides	Carbowax 600 (on chromosorb T); diethylene glycol succinate; ethylene glycol succinate; neopentyl glycol sebacate; versamid 900; SE-30 (methyl silicone phases)
Amines	Penwalt 213; Chromosorb 103 (see support modifiers)
Amino acids (and derivatives)	Carbowax 600; diethylene glycol succinate (stabilized); Ethofat (on chromosorb T); ethylene glycol succinate; neopentyl glycol adipate; SE-30; XE-60 (methyl silicone phases)
Boranes	Apiezon L; beeswax; carbowax 400, 1540, 4000, 20M; castorwax; diethylene glycol succinate; di-n-decyl phthalate; diisodecyl phthalate; Emulphor-ON-870; ethylene glycol adipate; FFAP, polyphenyl ether (5 or 5 ring); quadrol; reoplex 400; SE-30; XE-60; sucrose acetate isobutyrate; tricresyl phosphate; Ucon series
Esters	Apiezon L; benzyldiphenyl; carbowax 20M; cyclodextrin acetate; diethylene glycol adipate; di(2-ethylhexyl) sebacate; diisodecyl phthalate; dimer acid/OV-1 (50/50, V/V); Hallcomid M18; neopentyl glycol succinate; propylene glycol; SE-30; SE-52; XE-60; Friton X-100; Tween-80
Ethers	Apiezon L; carbowax 1500, 1540, 4000, 20M; diethylene glycol sebacate, ethylene glycol adipate
Glycols	Porapak-Q, Porapak-1S; QF-1
Halogenated compounds	Bentone 34; benzyldiphenyl; butanediol succinate (Craig polyester); carbowax 400, 1000, 4000, 20M; dibutyl phthalate; diethylene glycol succinate; di (2-ethylhexyl) sebacate; di-n-decyl phthalate; dinonyl phthalate; dioctyl phthalate; β,β'-iminodipropionitrile; β,β'-oxydipropionitrile; SE-30; squalane; Tween-80
Hydrocarbons C_1–C_5 (aliphatic)	Carbowax 400–1500; most branched and substituted phthalate, sebacate, succinate and adipate phases; octadecane; squalane (boiling point separations); methyl silicones
Inorganic Compounds (includes organometallic compounds)	n-decane; di-n-decyl phthalate; dimethyl sulfolane; neopentyl glycol succinate; 1,2,3-Trix (2-cyanoethoxy) propane; SE-30 (methyl silicone phases)
above C_5 (aliphatic)	Apiezon phases; carbowax 1500, 1540, 4000, 6000, 20M; most of the high temperature substituted adipates, phthalates, succinates and sebacates (boiling point separations); methyl silicones
(aromatic)	Apiezon phases; bentone-34; carbowax phases; substituted adipates, phthalates, succinates and sebacates; tetracyanoethylated pentaerythritol; liquid crystalline phases; phenyl methyl silicone phases
Nitrogen compounds	Apiezon L; Armeen SD; butanediol succinate (Craig polyester); carbowax 400, 1500, 20M; ethylene glycol adipate; propylene glycol; tetraethylene glycol dimethylether; THEED; Ucon phases
Pesticides	Carbowax-20M; diethylene glycol adipate; Epon 1001; neopentyl glycol adipate; methyl silicone phases, including gum viscosities

Selection of Stationary Phases for Packed Column Gas Chromatography (Continued)

Compound	Suggested Stationary Phases
Phosphorous Compounds	Apiezon L; carbowax 20M; di-n-butyl phthalate; diethylene glycol succinate; Emulphor-ON-870; ethylene glycol succinate; Reoplex-400; methyl silicone phases, including gum viscosities; squalane; STAP
Silanes	Methyl silicone phases; STAP
Sugars	Apiezon L; butanediol succinate; carbowax 4000, Hyprose SP80; mannitol; methyl silicone phases
Sulfur Compounds	Apiezon L; 7,8-benzoquinoline; carbowax 1500, 20M; diethylene glycol succinate; diisodecyl phthalate; methyl silicone phases; Reoplex-400; tricresyl phosphate
Urinary and Bile Compounds	Ethylene glycol adipate; methyl silicone-nitrile phases

ADSORBENTS FOR GAS-SOLID CHROMATOGRAPHY

The following table lists the more common adsorbents used in gas-solid chromatography, along with relevant information on separation and technique [1–3]. The adsorbents are used chiefly for the analysis of gaseous mixtures. The maximum temperatures listed represent the point of severe resolution loss. The materials are often chemically stable to much higher temperatures. The 60–100 mesh sizes (U.S.) are most useful for chromatographic applications. All of these materials must be activated before being used, and the degree of activation will influence the retention behavior. The user should also be aware that the adsorption of water during use will often change retention characteristics dramatically, sometimes resulting in a reversal of positions of adjacent peaks. Due to surface adsorption of solutes, some experimentation with temperature may be necessary to prevent tailing or to avoid statistical correlation (or a propagating error) among replicate analyses [4–5].

REFERENCES

1. Jeffery, P. G., and P. J. Kipping. *Gas Analysis by Gas Chromatography.* Oxford: Pergamon Press, 1972.
2. Cowper, C. J., and A. J. DeRose. *The Analysis of Gases by Chromatography.* Oxford: Pergamon Press, 1983.
3. Breck, D. W. *Zeolite Molecular Sieves.* New York: John Wiley and Sons, 1973.
4. Bruno, T. J. "An Apparatus for Direct Fugacity Measurements on Mixtures Containing Hydrogen." *Journal of Research of the National Bureau of Standards* (U.S.) 90, no. 2 (1985): 1127.
5. Bruno, T. J., K. H. Wertz, and M. Caciari, "Kovats Retention Indices of Halocarbons on a Hexafluoropropylene Epoxide Modified Graphitized Carbon Black," *Analytical Chemistry* 68, no. 8 (1996): 1347–1359.

Adsorbents for Gas-Solid Chromatography

Packing Name	Max.Temp.°C	Separation Affected	Notes
Silica gel	300	H_2, Air, CO, C_1 to C_4, normal hydrocarbons, alkenes and alkynes	Used often as a second column (with a molecular sieve); very hydrophilic; requires activation; can be unpredictable; largely replaced by porous polymers
Porous silica	300	same as silica gel	Higher surface area than silica gel; often used with a humidified carrier gas; can be coated with a conventional liquid phase; Spherosil and Porasil are examples
Alumina	300	light hydrocarbons at ambient temperature (C_1–C_5), H_2 and light hydrocarbons at subambient temperature	Often useful with controlled water preadsorption after activation; can be coated with a conventional liquid phase, a variety of polarities are available depending on the washing technique used.
Activated carbon	300	H_2, CO, CO_2, C_1–C_3 alkanes, alkenes, and alkynes	Requires oxygen-free carrier gas; largely replaced by porous polymers
Cyclodextrin	260	light hydrocarbons (C_1–C_{10}) and halocarbons	α and β cyclodextrins have been used; care should be taken with halocarbon analysis, due to the potential of HF contamination of the sample
Graphite	300	light hydrocarbons, H_2S, SO_2, CH_3SH, sour gas	Often modified with small quantities (1.5–5 %) of conventional liquid phases; requires oxygen-free carrier
Graphitized carbon black	300	light hydrocarbons, halocarbons, alternative refrigerants	Often modified with hexafluoropropylene epoxide fluid for the analysis of halocarbons, since this modifier can withstand trace HF contamination.
Carbon molecular sieve	300	H_2 (O_2, N_2 co-elute), CO, CH_4, H_2O, CO_2, C_1–C_3 alkanes, alkenes, alkynes	High affinity for hydrocarbons; requires oxygen-free carrier
Molecular sieve, 5A	225	air and light gas analysis; H_2, O_2, N_2, (CH_4, CO, NO, SF_6 co-elute)	Synthetic calcium alumino-silicate (zeolite) having an effective pore diameter of 0.5 nm CO_2 is adsorbed strongly; 5A usually gives the best results of all synthetic zeolites; should be activated before use, and used above critical adsorption temperature; 21.6 % (mass/mass) water capacity.
Molecular sieve, 13X	200	same as 5A, but with C_1–C_4 alkanes, alkenes, and alkynes being separated as well	Sodium alumino-silicate (zeolite), having a larger pore size than 0.5 nm, thus producing lower retention times and less resolution; 28.6 % (mass/mass) water capacity
Molecular sieve, 3A	200	light permanent gases	Potassium alumino-silicate (zeolite) 20 % (mass/mass) water capacity, smaller pore size than 0.5 nm, thus different retention characteristics
Molecular sieve, 4A	200	light permanent gases	Sodium alumino-silicate (zeolite); 22 % (mass/mass) water capacity; retention characteristics differ from 5A due to smaller pore size

POROUS POLYMER PHASES

Porous polymer phases, first reported by Hollis [1], are of great value for a wide variety of separations. They are usually white in color, but may darken during use especially at higher temperatures. This darkening does not affect their performance. High temperature conditioning is required to drive off solvent and residual monomer. The polymers may either swell or shrink with heating; thus, flow rate changes must be anticipated. The retention indices reported here are from the work of Dave [2]. The use of these indices is the same as for the packed column liquid phases, provided in an earlier table.

Index	Test Probe
W	benzene
X	t-butanol
Y	2-butanone
Z	acetonitrile

The physical property data were taken from the work of Poole and Schuette [3].

REFERENCES

1. Hollis, O. L. "Separation of Gaseous Mixtures Using Porous Polyaromatic Polymer Beads." *Analytical Chemistry* 38 (1966): 309.
2. Dave, S. "A Comparison of the Chromatographic Properties of Porous Polymers." *Journal of Chromatographic Science* 7 (1969): 389.
3. Poole, C. F., and S. A. Schuette. *Contemporary Practice of Chromatography.* Amsterdam: Elsevier, 1984.

Phase Name	Maximum Temperature °C	Material Type	Free Fall Density g/cm^3	Surface Area m^2/g	Pore Diameter Av., μm
Chromosorb 101	275	styrene-divinylbenzene copolymer	0.30	< 50	0.3–0.4

Retention Indices

W	X	Y	Z	Separation Affected	Notes
745	565	645	580	free fatty acids, glycols, alcohols, alkanes, esters, aldehydes, ketones, ethers	Hydrophobic; condition at 250 °C; not recommended for amines or anilines, lower retention times than obtained with Chromosorb 102

Phase Name	Maximum Temperature °C	Material Type	Free Fall Density g/cm^3	Surface Area m^2/g	Pore Diameter Av., μm
Chromosorb 102	250	styrene-divinylbenzene copolymer	0.29	300–500	0.0085

Retention Indices

W	X	Y	Z	Separation Affected	Notes
650	525	570	460	subambient temperature: H_2, O_2, N_2, Ar, NO, CO; ambient temperature: H_2, (Air + Ar + NO + CO), CH_4, CO_2, H_2O, N_2O, C_2H_6; above ambient temperature: C_1–C_4 hydrocarbons, H_2S, COS, SO_2, esters, ethers, alcohols, ketones, aldehydes, glycols	May entrain some species; hydrophobic; condition at 225 °C; not recommended for amines or nitriles; little tailing of water or oxygenated hydrocarbons

Phase Name	Maximum Temperature °C	Material Type	Free Fall Density g/cm³	Surface Area m²/g	Pore Diameter Av., µm
Chromosorb 103	275	polystyrene cross-linked	0.32	15–25	0.3–0.4

Retention Indices

W	X	Y	Z	Separation Affected	Notes
720	575	640	565	ammonia, light amines, light amides, alcohols, aldehydes, hydrazines	Hydrophobic; high affinity for basic species; not recommended for acidic species, glycols, nitriles, nitroalkanes

Phase Name	Maximum Temperature °C	Material Type	Free Fall Density g/cm³	Surface Area m²/g	Pore Diameter Av., µm
Chromosorb 104	250	acrylonitrile divinyl-benzene copolymer	0.32	100–200	0.06–0.08

Retention Indices

W	X	Y	Z	Separation Affected	Notes
845	735	860	885	sulfur gases, ammonia, nitrogen oxides, nitriles, nitroalkanes, xylenols, water in benzene	Hydrophobic; condition at 225 °C; not recommended for glycols and amines; moderately polar

Phase Name	Maximum Temperature °C	Material Type	Free Fall Density g/cm³	Surface Area m²/g	Pore Diameter Av., µm
Chromosorb 105	250	acrylic ester (polyaro-matic)	0.34	600–700	0.04–0.06

Retention Indices

W	X	Y	Z	Separation Affected	Notes
635	545	580	480	permanent and light hydrocarbon gases; aqueous solutions of light organics such as formalin	Hydrophobic; less polar than Chromosorb 104; condition at 225 °C; not recommended for acidic species, glycols, amines and amides

Phase Name	Maximum Temperature °C	Material Type	Free Fall Density g/cm³	Surface Area m²/g	Pore Diameter Av., µm
Chromosorb 106	250	polystyrene cross-linked	0.28	700–800	0.05

Retention Indices

W	X	Y	Z	Separation Affected	Notes
605	505	540	405	fatty acids from fatty alcohols, up to C5; benzene from nonpolar organic compounds	Hydrophobic; not recommended for glycols and amines

Phase Name	Maximum Temperature °C	Material Type	Free Fall Density g/cm³	Surface Area m²/g	Pore Diameter Av., µm
Chromosorb 107	250	acrylic ester cross-linked	0.30	400–500	0.8

Retention Indices

W	X	Y	Z	Separation Affected	Notes
660	620	650	550	aqueous solutions of formaldehyde; alkynes from alkanes	Hydrophobic; moderately polar; not recommended for glycols and amines

Phase Name	Maximum Temperature °C	Material Type	Free Fall Density g/cm³	Surface Area m²/g	Pore Diameter Av., µm
Chromosorb 108	250	acrylic ester cross-linked	0.30	100–200	0.25

Retention Indices

W	X	Y	Z	Separation Affected	Notes
710	645	675	605	polar materials such as water, alcohols, aldehydes, glycols	Hydrophobic; condition at 250 °C

Phase Name	Maximum Temperature °C	Material Type	Free Fall Density g/cm³	Surface Area m²/g	Pore Diameter Av., µm
Haysep A	165	divinylbenzene/ ethyleneglycol- dimethacrylate (high purity)	0.356	526	—

Retention Indices[a]

W	X	Y	Z	Separation Affected	Notes
				separates permanent gases at ambient temperatures, and is useful for hydrocarbons to C_2, H_2S, H_2O at elevated temperatures	Relatively high polarity

Phase Name	Maximum Temperature °C	Material Type	Free Fall Density g/cm³	Surface Area m²/g	Pore Diameter Av., µm
Haysep B	190	divinylbenzene/ polyethyleneimine	0.330	608	—

Retention Indices[a]

W	X	Y	Z	Separation Affected	Notes
				separates C_1 and C_2 amines and trace levels of NH_3 and H_2O	High polarity

[a] Retention indices are not available for these porous polymers, but a table of relative retentions on some representative solutes is included at the end of this section.

Phase Name	Maximum Temperature °C	Material Type	Free Fall Density g/cm^3	Surface Area m^2/g	Pore Diameter Av., μm
Haysep C	250	divinylbenzene/ acrylonitrile	0.322	442	—

Retention Indices[a]

W	X	Y	Z	Separation Affected	Notes
				separates polar hydrocarbons, also HCN, NH$_3$, H$_2$S, H$_2$	Moderate polarity, with separation characteristics similar to Chromosorb 104

Phase Name	Maximum Temperature °C	Material Type	Free Fall Density g/cm^3	Surface Area m^2/g	Pore Diameter Av., μm
Haysep D	290	divinylbenzene (high purity)	0.3311 (av.)	795 (av.)	0.0308–0351

Retention Indices[a]

W	X	Y	Z	Separation Affected	Notes
				separates light gases; CO, CO$_2$, C$_2$H$_2$, C$_2$ hydrocarbons, H$_2$S, H$_2$O	Low polarity polymer available in four formulations of different surface area (771–802 m^2/g), density (0.3283–0.3834 g/mL), and porosity (64.2–70.4 %)

[a] Retention indices are not available for these porous polymers, but a table of relative retentions on some representative solutes is included at the end of this section.

Phase Name	Maximum Temperature °C	Material Type	Free Fall Density g/cm^3	Surface Area m^2/g	Pore Diameter Av., μm
Haysep N	165	divinylbenzene/ ethyleneglycol-dimethacrylate (high purity)	0.355	405	—

Retention Indices[a]

W	X	Y	Z	Separation Affected	Notes
				separation similar to Porapak materials; moderately high H$_2$O retention; see retention table	Low polarity polymer

Phase Name	Maximum Temperature °C	Material Type	Free Fall Density g/cm^3	Surface Area m^2/g	Pore Diameter Av., μm
Haysep P	250	divinylbenzene/styrene	0.420	165	—

Retention Indices[a]

W	X	Y	Z	Separation Affected	Notes
				separation of low molecular mass materials containing halogens, sulfur, water, aldehydes, ketones, alcohols, esters and fatty acids	Moderate to low polarity

[a] Retention indices are not available for these porous polymers, but a table of relative retentions on some representative solutes is included at the end of this section.

Phase Name	Maximum Temperature °C	Material Type	Free Fall Density g/cm³	Surface Area m²/g	Pore Diameter Av., µm
Haysep Q	275	divinylbenzene	0.351	582	—

Retention Indices[a]

W	X	Y	Z	Separation Affected	Notes
				separation similar to Haysep P; see retention table	Low polarity

Phase Name	Maximum Temperature °C	Material Type	Free Fall Density g/cm³	Surface Area m²/g	Pore Diameter Av., µm
Haysep R	250	divinylbenzene/ N-vinyl-2-pyrollidinone	0.324	344	—

Retention Indices[a]

W	X	Y	Z	Separation Affected	Notes
				separation similar to Haysep P; see retention table	Moderate polarity

[a] Retention indices are not available for these porous polymers, but a table of relative retentions on some representative solutes is included at the end of this section.

Phase Name	Maximum Temperature °C	Material Type	Free Fall Density g/cm³	Surface Area m²/g	Pore Diameter Av., µm
Haysep S	250	divinylbenzene/4-vinyl-pyridine	0.334	583	—

Retention Indices[a]

W	X	Y	Z	Separation Affected	Notes
				separation similar to Haysep P; see retention table	Moderate polarity

Phase Name	Maximum Temperature °C	Material Type	Free Fall Density g/cm³	Surface Area m²/g	Pore Diameter Av., µm
Haysep T	165	ethyleneglycol-dimethacrylate (high purity)	0.381	250	—

Retention Indices[a]

W	X	Y	Z	Separation Affected	Notes
				see retention table	High polarity

[a] Retention indices are not available for these porous polymers, but a table of relative retentions on some representative solutes is included at the end of this section.

Phase Name	Maximum Temperature °C	Material Type	Free Fall Density g/cm³	Surface Area m²/g	Pore Diameter Av., µm
Porapak-Q	250	ethylvinylbenzene-divinyl benzene copolymer	0.35	500–700	0.0075

Retention Indices

W	X	Y	Z	Separation Affected	Notes
630	538	580	450	similar to Chromosorb 102	Similar to Chromosorb 102; condition at 250 °C; most popular of all porous polymer phases

Phase Name	Maximum Temperature °C	Material Type	Free Fall Density g/cm³	Surface Area m²/g	Pore Diameter Av., µm
Porapak-P	250	styrene-divinyl benzene copolymer	0.28	100–200	—

Retention Indices

W	X	Y	Z	Separation Affected	Notes
765	560	650	590	similar to Porapak-Q	Hydrophobic; low polarity; larger pore size than Porapak-Q, thus lower retention times are observed; not recommended for amines or anilines; condition at 250 °C

Phase Name	Maximum Temperature °C	Material Type	Free Fall Density g/cm³	Surface Area m²/g	Pore Diameter Av., µm
Porapak-N	200	vinylpyrolidone	0.39	225–350	—

Retention Indices

W	X	Y	Z	Separation Affected	Notes
735	605	705	595	similar to Chromosorb 105; high water retention; CO_2, NH_3, H_2O, C_2H_2, from light hydrocarbons	Condition at 175 °C; not recommended for glycols, amines, or acidic species

Phase Name	Maximum Temperature °C	Material Type	Free Fall Density g/cm³	Surface Area m²/g	Pore Diameter Av., µm
Porapak-R	250	vinylpyrolidone	0.33	300–450	0.0076

Retention Indices

W	X	Y	Z	Separation Affected	Notes
645	545	580	455	ethers, esters, H_2O from chlorine gases (HCl, Cl_2) nitriles and nitroalkanes	Moderately polar; condition at 250 °C; not recommended for glycols and amines

Phase Name	Maximum Temperature °C	Material Type	Free Fall Density g/cm³	Surface Area m²/g	Pore Diameter Av., µm
Porapak-S	250	vinyl pyridine	0.35	300–450	0.0076

Retention Indices

W	X	Y	Z	Separation Affected	Notes
645	550	575	465	normal and branched alcohols, aldehydes, ketones, halocarbons	High polarity; not recommended for acidic species and amines; condition at 250 °C

Phase Name	Maximum Temperature °C	Material Type	Free Fall Density g/cm³	Surface Area m²/g	Pore Diameter Av., µm
Porapak-T	200	ethylene glycol-dimethacrylate	0.44	250–300	0.009

Retention Indices

W	X	Y	Z	Separation Affected	Notes
–	675	700	635	water in formalin (and other aqueous organic mixtures), retention characteristics similar to Chromosorb 107	Condition at 180 °C; highest polarity of Porapak series; not recommended for glycols and amines

Phase Name	Maximum Temperature °C	Material Type	Free Fall Density g/cm³	Surface Area m²/g	Pore Diameter Av., µm
Porapak-QS	250	ethylvinylbenzene-divinyl-benzene copolymer	—	—	—

Retention Indices

W	X	Y	Z	Separation Affected	Notes
625	525	565	445	similar to Porapak-Q at lower operating temperatures, but useful for higher molecular weight solutes	Silanized Porapak-Q, reduces tailing of high polarity compounds; condition at 250 °C

Phase Name	Maximum Temperature °C	Material Type	Free Fall Density g/cm³	Surface Area m²/g	Pore Diameter Av., µm
Porapak-PS	250	styrene-divinyl-benzene copolymer	—	—	—

Retention Indices

W	X	Y	Z	Separation Affected	Notes
—	—	—	—	similar to Porapak-P	Silanized Porapak-P; condition at 250 °C

Phase Name	Maximum Temperature °C	Material Type	Free Fall Density g/cm^3	Surface Area m^2/g	Pore Diameter Av., μm
Tenax-GC	375	p-2,6 diphenyl-phenylene oxide polymer	0.37	18.6	—

Retention Indices

W	X	Y	Z	Separation Affected	Notes
—	—	—	—	similar to Porapak-Q	Highest thermal stability of all porous polymers

RELATIVE RETENTION ON SOME HAYSEP POROUS POLYMERS

The following table provides relative retention values for Haysep polymers N, Q, R, S, and T. These data were obtained using a 2 m long, 0.32 cm O.D. stainless steel column, using helium as the carrier gas.

Haysep Polymer

Compound	N	Q	R	S	T
Hydrogen	0.19	0.143	0.17	0.19	0.21
Air	0.23	0.186	0.2	0.21	0.25
Nitric oxide	0.25	0.217	0.21	0.23	0.33
Methane	0.30	0.256	0.28	0.3	0.35
Carbon dioxide	0.71	1.15	0.50	0.52	0.85
Nitrous oxide	0.80	1.43	0.59	0.59	—
Ethylene	0.83	0.74	0.78	0.78	0.9
Acetylene	1.41	0.74	1.0	0.87	2.11
Ethane	1.0	1.0	1.0	1.0	1.0
Water	10.1	1.45	0.68	4.12	19.1
Hydrogen sulfide	2.1	1.40	1.73	1.87	2.88
Hydrogen cyanide	1.93	2.31	15.6	8.26	28.8
Carbonyl sulfide	2.82	2.33	2.46	2.63	3.4
Sulfur dioxide	12.0	3.05	9.78	17.8	19.0
Propylene	4.66	3.20	3.45	3.65	4.91
Propane	4.66	3.67	3.88	4.1	4.63
Propadiene	6.50	4.12	4.39	4.7	7.55
Methylacetylene	9.5	4.12	4.84	5.14	11.3
Methyl chloride	7.43	3.93	4.67	4.92	9.2
Vinyl chloride	14.9	6.04	9.04	9.7	17.3
Ethylene oxide	17.7	6.06	8.78	9.7	23.3
Ethyl chloride	35.0	12.25	19.3	20.7	43.2
Carbon disulfide	—	32.4	—	—	40.7

STATIONARY PHASES FOR POROUS LAYER OPEN TUBULAR COLUMNS

The practical application of solid adsorbents is commonly in porous layer open tubular (PLOT) columns. In this table, several of the more common PLOT column stationary phases are listed, along with the separations that may be affected and some additional information [1–4]. The maximum temperatures listed represent the point of severe resolution loss. The materials are often chemically stable to much higher temperatures. The user should also be aware that the adsorption of water during use will often change retention characteristics dramatically, sometimes resulting in a reversal of positions of adjacent peaks. Due to surface adsorption of solutes, some experimentation with temperature may be necessary to prevent tailing or to avoid statistical correlation (or a propagating error) among replicate analyses [5].

REFERENCES

1. Jeffery, P. G., and P. J. Kipping. *Gas Analysis by Gas Chromatography*. Oxford: Pergamon Press, 1972.
2. Cowper, C. J., and A. J. DeRose. *The Analysis of Gases by Chromatography*. Oxford: Pergamon Press, 1983.
3. Breck, D. W. *Zeolite Molecular Sieves*. New York: John Wiley and Sons, 1973.
4. Poole, C. F. *The Essence of Chromatography*. Amsterdam: Elsevier, 2003.
5. Bruno, T. J. "An Apparatus for Direct Fugacity Measurements on Mixtures Containing Hydrogen." *Journal of Research of the National Bureau of Standards* (U.S.) 90, no. 2 (1985): 1127.

Stationary Phases for Porous Layer Open Tubular Columns

Phase	Max. Temp.°C	Separation Affected	Notes
Silica gel	250	H_2, Air, CO, C_1–C_4, normal hydrocarbons, alkenes and alkynes, inorganic gases, volatile ethers	Very hydrophilic; requires activation; can be unpredictable; largely replaced by porous polymers; bonded versions are suitable for use with GC-MS, because of the absence of particles
Alumina, deactivated with KCl	300	C_1–C_8 hydrocarbons, especially useful for resolution of propadiene and butadiene from ethylene and propylene	Least polar of the alumina phases; lowest retention of olefins relative to the corresponding paraffin; specified in many standard methods
Alumina, deactivated with Na_2SO_4	300	C_1–C_8 hydrocarbons, resolves acetylene from n-butane and propylene from isobutane	Medium and high polarity phases are available among the alumina phases; specified in many standard methods
Cyclodextrin	260	light hydrocarbons (C_1–C_{10}) and halocarbons	α and β cyclodextrins have been used; care should be taken with halocarbon analysis, due to the potential of HF contamination of the sample
Styrene – divinyl benzene	250	C_1–C_3 hydrocarbons; paraffins up to C_{12}; CO from air, ethers, sulfur gases, water	See the information on porous polymers
Divinyl benzene ethylene glycol dimethacrylate		C_1–C_7 hydrocarbon isomers; CO_2, CH_4, amines, common solvents, alcohols, aldehydes, ketones	More polar than styrene—divinyl benzene phases
Molecular sieve, 5A	350	air and light gas analysis; H_2, O_2, N_2 (CH_4, CO, NO, SF_6 co-elute); thick film phase can resolve Ar from O_2 at 35 °C	Synthetic calcium alumino-silicate (zeolite) having an effective pore diameter of 0.5 nm CO_2 is adsorbed strongly; 5A usually gives the best results of all synthetic zeolites; thick and thin film columns are available
Molecular sieve, 13X		same separations as those performed on 5A, but with C_1–C_4, alkanes, alkenes, and alkynes being separated as well	Sodium alumino-silicate (zeolite), having a larger pore size than 0.5 nm, thus producing lower retention times and less resolution; 28.6 % (mass/mass) water capacity
Monolithic carbon	350	C_1–C_5 hydrocarbon isomers; acetylene in ethylene; methane	Phase consists of a bonded carbon monolith; suitable for use with GC-MS, because of the absence of particles

SILICONE LIQUID PHASES

The following table lists the chromatographic properties of some of the more popular polysiloxane-based liquid phases [1–8]. The polysiloxanes are the most widely used stationary phases in gas chromatography and are especially applicable to capillary columns. The listing provided here is far from exhaustive. Since it is impractical to present the structures of all polysiloxane-based phases, the OV phases have been chosen as representative since their properties are among the most well characterized. The phases that are listed in the notes as "similar phases" have thermal and chromatographic properties that are similar to the phase described. In modern applications of capillary column gas chromatography, silicone phases are cross-linked to provide stability. Cross-linking can change the properties of a phase to some extent, but often this is relatively minor.

The reader should note that there are many commercial variations of silicone liquid phases available. In compiling properties such as those listed in this table, one must strike a balance between general usefulness and simply providing information that is contained in vendor catalogs, promotional brochures and Web sites. In that context, this table is meant to serve as a starting point for the design of an analysis.

The McReynolds constants are indices with respect to the following test probe compounds:

McReynolds Constant	Test Probe
1	benzene
2	1-butanol
3	2-pentanone
4	1-nitropropane
5	pyridine
6	2-methyl-2-pentanol
7	1-iodobutane
8	2-octyne
9	1,4-dioxane
10	cis-hydrindane

The use of these constants is described in the table entitled "Properties of Some Liquid Phases for Packed Columns." The viscosity data, where available, are presented in cSt, which is 10^{-6} m^2/s. Cross-linked silicone phases based on the silicones are especially valuable for capillary gas chromatography. They are not specifically treated in this table since the differences in many properties are quite often subtle. The cross-linked phases have much longer lifetimes due to the effective immobilization.

Abbreviations:

Solvents:

    Ace: acetone   Chlor: chloroform
    Tol: toluene
    (when a silicone fluid is cross-linked, it will be insoluble.)

Note: N denotes a phenyl group in a structure.

    N = nonpolar
     I = intermediate polarity
    P = polar

REFERENCES

1. Yancey, J. A. *Journal of Chromatographic Science* 23 (1985): 161.
2. McReynolds, W. O. *Journal of Chromatographic Science* 8 (1970): 685.
3. Mann, J. R., and S. T. Preston. *Journal of Chromatographic Science* 11 (1973): 216.
4. Trash, C. R. *Journal of Chromatographic Science* 11 (1973): 196.
5. McNair, H. M., and E. J. Bonelli. *Basic Gas Chromatography.* Palo Alto: Varian Aerograph, 1969.
6. Heftmann, E. *Chromatography: A Laboratory Handbook of Chromatographic and Electrophoretic Methods.* 3rd ed. New York: Van Nostrand Reinhold, 1975.
7. Grant, D. W. *Gas Liquid Chromatography.* London: Van Nostrand Reinhold, 1971.
8. Coleman, A. E. *Journal of Chromatographic Science* 11 (1973): 198.

Silicone Liquid Phases

Liquid Phase	Solvent	Av. Mol. Mass	Viscosity	T_{min}, °C	T_{max}, °C	Polarity	McReynolds Constants										Notes
							1	2	3	4	5	6	7	8	9	10	
OV-1, dimethyl silicone (gum) 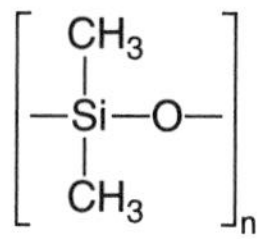	Tol	$>10^6$	gum	100	350	N	16	55	44	65	42	32	4	23	45	−1	100 % methyl, low selectivity, boiling point separations; similar phases: UCC-L45, UCC-W-98, SE-30, DB-1, HP-1
OV-101, dimethylsilicone fluid 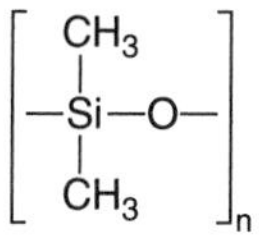	Tol	3×10^4	1500	20	350	N	17	57	45	67	43	33	4	23	46	−2	100 % methyl, low selectivity, boiling point separations; similar phases: DC-11, DC-200, DC-550, SF-96, SP-2100, STAP

(Continued)

Silicone Liquid Phases (Continued)

Liquid Phase	Solvent	Av. Mol. Mass	Viscosity	T_{min}, °C	T_{max}, °C	Polarity	McReynolds Constants										Notes
							1	2	3	4	5	6	7	8	9	10	
Phenylmethyl-dimethyl silicone 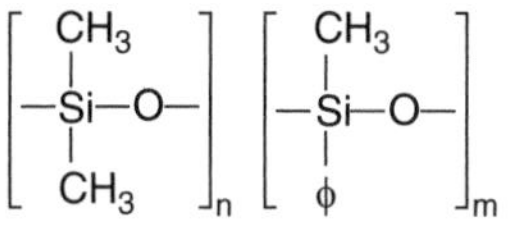	Tol	2×10^4		20	350	I	33	72	66	99	67	46	24	36	68	10	5 % phenyl methyl, boiling point separations; similar phases: DB-5, HP-5. This does not have a corresponding OV identification number because it was formulated later than the other fluids. This phase is probably the most common starting phase for most analyses, and one that is specified in many standard protocols; similar to SE-54
OV-3, phenylmethyl-dimethyl silicone 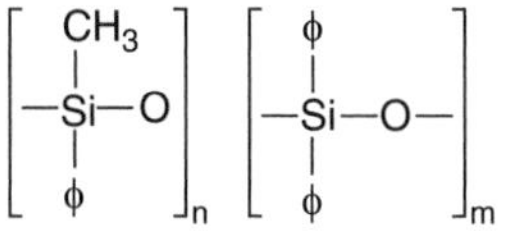	Ace	2×10^4	500	20	350	I	44	86	81	124	88	55	39	46	84	17	10 % phenylmethyl; similar to SE-52
OV-7, phenylmethyl-dimethyl silicone 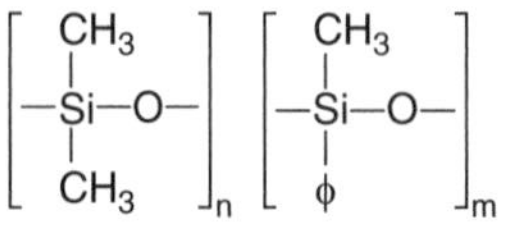	Ace	1×10^4	500	20	350	I	69	113	111	171	128	77	68	66	120	35	20 % phenylmethyl

OV-11, phenylmethyl-dimethyl silicone 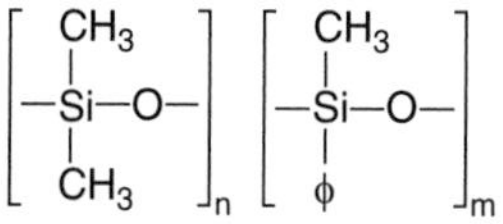	Ace	7×10^3	500	0	350	—	102	142	145	219	178	100	103	92	164	59	35 % phenylmethyl; similar phases: DC-710
OV-17, phenylmethyl silicone 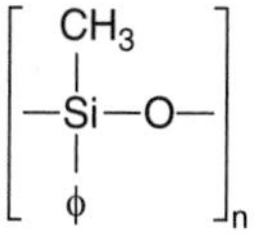	Ace	4×10^3	1300	20	350	—	119	158	162	243	202	112	119	105	184	69	50 % methyl, similar phases: SP-2250
OV-22, phenylmethyl diphenyl silicone 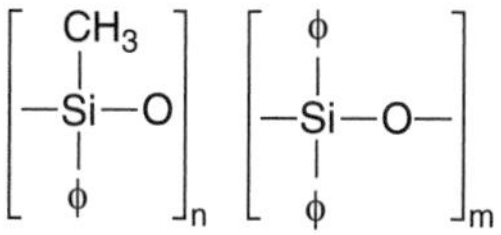	Ace	8×10^3	>50,000	20	350	—	160	188	191	283	253	133	152	132	228	99	65 % phenyl

OV-11 structure:

$$\left[\begin{array}{c} CH_3 \\ -Si-O- \\ CH_3 \end{array} \right]_n \left[\begin{array}{c} CH_3 \\ -Si-O- \\ \phi \end{array} \right]_m$$

OV-17 structure:

$$\left[\begin{array}{c} CH_3 \\ -Si-O- \\ \phi \end{array} \right]_n$$

OV-22 structure:

$$\left[\begin{array}{c} CH_3 \\ -Si-O- \\ \phi \end{array} \right]_n \left[\begin{array}{c} \phi \\ -Si-O- \\ \phi \end{array} \right]_m$$

(Continued)

Silicone Liquid Phases (Continued)

Liquid Phase	Solvent	Av. Mol. Mass	Viscosity	T_{min}, °C	T_{max}, °C	Polarity	McReynolds Constants										Notes
							1	2	3	4	5	6	7	8	9	10	
OV-25, phenylmethyl diphenyl silicone 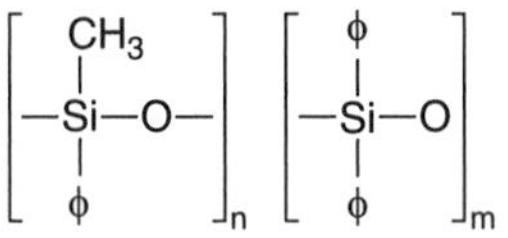	Ace	1×10^4	100,000	20	350	—	178	204	208	305	280	144	169	147	215	113	75 % phenyl
OV-61, diphenylldimethyl silicone 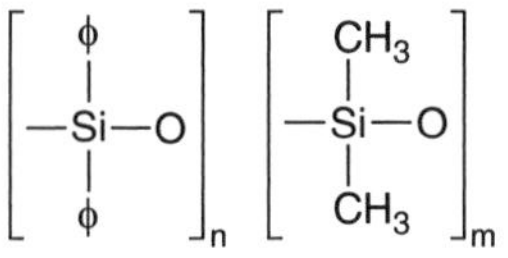	Tol	4×10^4	>50,000	20	350	—	101	143	142	213	174	99	—	86	—	—	33 % phenyl
OV-73, diphenylldimethyl silicone gum 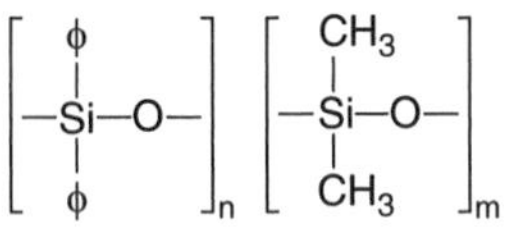	Tol	8×10^5	gum	20	350	—	40	86	76	114	85	57	—	39	—	—	5.5 % phenyl, similar phases: SE-52, SE-54

Phase	Solvent															Notes
OV-105, cyano propylmethyl-diemthyl silicone 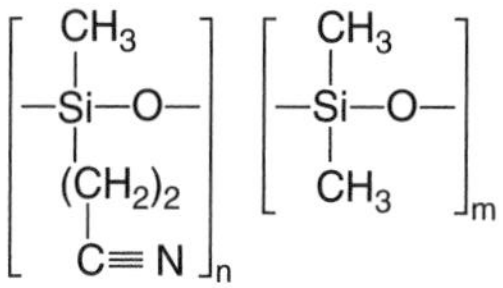	Ace	1500	20	250	N,I	36	108	93	139	86	74	—	29	—	—	
OV-202, trifluoropropyl-methyl silicone 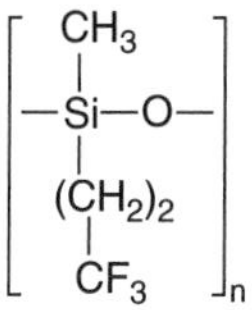	Chlor	1×10^4	500	0	275	I,P	146	238	358	468	310	202	139	56	283	60	50 % trifluoropropyl fluid, similar phases: SP-2401; phases can be prone to oxidation at the Si-C bond
OV-210, trifluoropropyl-methyl silicone 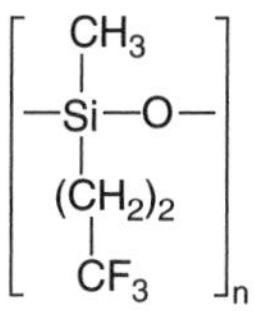	Chlor	2×10^5	10,000	20	275	I,P	146	238	358	468	310	206	139	56	283	60	50 % trifluoropropyl, similar phases: QF-1, FS-1265, SD-2401; phases can be prone to oxidation at the Si-C bond

OV-105, cyano propylmethyl-diemthyl silicone:

$$\left[\begin{array}{c} CH_3 \\ | \\ -Si-O- \\ | \\ (CH_2)_2 \\ | \\ C\equiv N \end{array} \right]_n \left[\begin{array}{c} CH_3 \\ | \\ -Si-O- \\ | \\ CH_3 \end{array} \right]_m$$

OV-202, trifluoropropyl-methyl silicone:

$$\left[\begin{array}{c} CH_3 \\ | \\ -Si-O- \\ | \\ (CH_2)_2 \\ | \\ CF_3 \end{array} \right]_n$$

OV-210, trifluoropropyl-methyl silicone:

$$\left[\begin{array}{c} CH_3 \\ | \\ -Si-O- \\ | \\ (CH_2)_2 \\ | \\ CF_3 \end{array} \right]_n$$

(Continued)

Silicone Liquid Phases (Continued)

Liquid Phase	Solvent	Av. Mol. Mass	Viscosity	T_{min}, °C	T_{max}, °C	Polarity	McReynolds Constants										Notes
							1	2	3	4	5	6	7	8	9	10	
OV-215, trifluoropropylmethyl silicone gum			gum			I,P	149	240	363	478	315	208	—	56	—	—	50 % trifluoropropyl; phases can be prone to oxidation at the Si-C bond
OV-225, cyanopropyl-methylphenyl methylsilicone	Ace	8×10^3	9000	20	275	I,P	228	369	338	492	386	282	226	150	342	117	25 % phenyl, 25 % cyanopropylmethyl; similar phases: EX-60, AN-600
OV-275, dicyanoallyl silicone	Ace	5×10^3	20,000	20	275	P	781	1006	885	1177	1089	—	—	—	—	—	
Dexsil 300 copolymer	Chlor	16,000–20,000	waxy solid	50	450	I	47	80	103	148	96	—	—	—	—	—	Carborane-methyl silicone; siloxane to carborane ratio, 4:1; used for methyl esters, aromatic amines, halogenated alcohols, pesticides, polyphenyl ethers, silicone oils

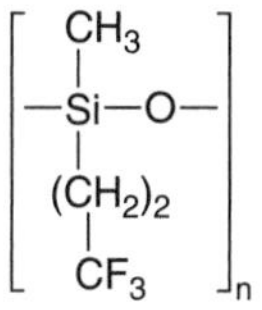

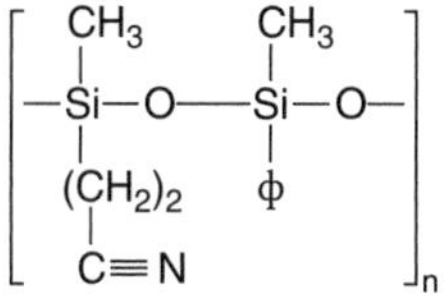

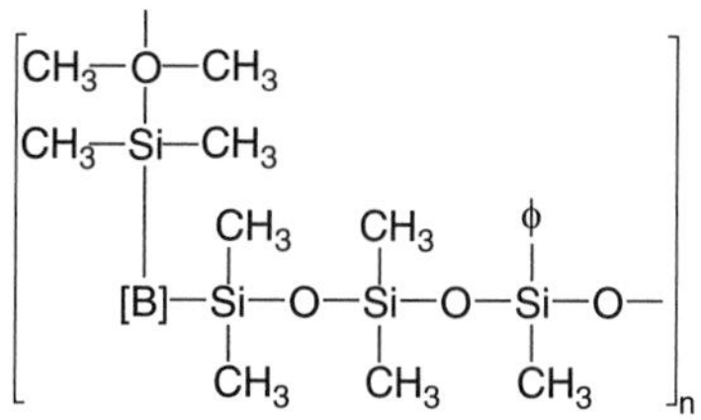

Dexsil 400

[B] = CB$_{10}$H$_{10}$C, meta-carborane

Chlor 12,000–16,000 — 20 375 — 60 115 140 188 174 — — — — Carborane-methyl phenyl silicone copolymer; siloxane to carborane ratio, 5:1

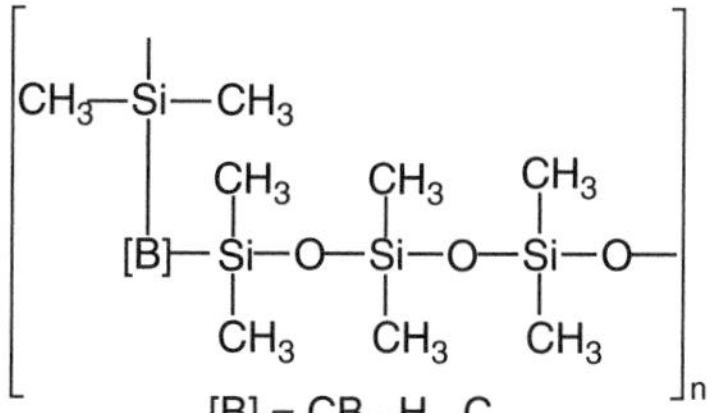

Dexsil 410

[B] = CB$_{10}$H$_{10}$C meta-carborane

Chlor 9000–12,000 20 375 — 85 165 170 240 180 — — — — Carborane-methyl-β-silicone cyanoethyl copolymer; siloxane to carborane ratio, 5:1

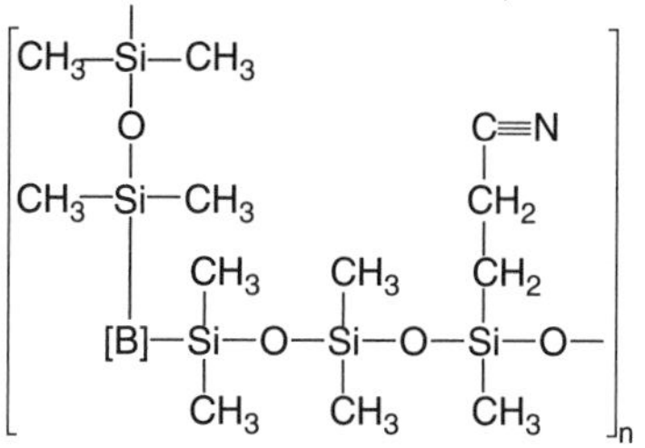

[B] = CB$_{10}$H$_{10}$C

PROPERTIES OF COMMON CROSS-LINKED SILICONE STATIONARY PHASES

The preceding table on the silicone stationary phases provides a useful means of comparing the various silicone stationary phases that are the most widely used in gas chromatography. As was noted, the fluids that were used for those measurements were not cross-linked, but rather were coated on a packing. Cross-linking or application to an open tubular (or capillary) column will not change the chromatographic behavior (that is reflected in retention indices such as those of Kovats and McReynolds) to any significant extent; however, in the following table we provide chromatographic data on the two most common cross-linked phases [1]. We note that the efficiency (in terms of the number of theoretical plates) of a typical commercial open tubular column is much higher than that of a column prepared with a coated packing. The retention indices presented in these tables were measured at 120 °C isothermally. Retention indices are temperature dependent; the temperature dependence of the Kovats indices have been studied for many compounds [2].

REFERENCES

1. Vickers, A. Life Sciences and Chemical Analysis, Agilent Technologies, Folsom, CA, personal communication, 2009.
2. Bruno, T. J., K. H. Wertz, and M. Caciari. "Kovats Retention Indices of Halocarbons on a Hexafluoropropylene Epoxide Modified Graphitized Carbon Black." *Analytical Chemistry* 68, no. 8 (1996): 1347–59.

Phase: 5 % phenyl dimethylpolysiloxane

Temperature Ranges:

–60–325°C isothermally, –60–350°C programmed for < 0.32 mm I.D. columns
–60–300°C isothermally, –60–320°C programmed for 0.53 mm I.D. columns
–60–260/280°C for > 2.0 µm films

Similar Phases

DB-5, Ultra-2, SPB-5, CP-Sil 8 CB, Rtx-5, BP-5, OV-5, 007-2 (MPS-5), SE-52, SE-54, XTI-5, PTE-5, HP-5MS, ZB-5, AT-5, MDN-5

Notes: This phase is probably the most commonly used stationary phase in gas chromatography, since it combines boiling point separation with a minor contribution of a specific interaction; typically used as the first phase in any method development; versatile for hydrocarbons and more polar compounds; other varieties of this phase.

Probe Compound	McReynolds Constant	McReynolds Code	Kovats's Retention Index
n-Hexane			600
1-Butanol	66	y′	656
Benzene	31	x′	684
2-Pentanone	61	z′	688
n-Heptane			700
1,4-Dioxane	64	L	718
2-Methyl-2-pentanol	41	H	731
1-Nitropropane	93	u′	745
Pyridine	62	s′	761
n-Octane			800
Iodobutane	22	J	840
2-Octyne	35	K	876
n-Nonane			900
n-Decane			1000

Phase: dimethylpolysiloxane

Temperature Range:

−60–325°C for normal operations, periodic operation to 350°C can be used to facilitate column clean-up

−60 to 260–280°C for > 2.0 μm films

Similar Phases

DB-1, OV-1, HP-1, DB-1ms, HP-1ms, Rtx-1, Rtx-1ms, CP-Sil 5 CB Low Bleed/MS, MDN-1, AT-1

Notes: Useful for the separation of hydrocarbons, pesticides, PCBs, phenols, sulfur compounds, flavors and fragrances, and some amines; columns are typically stable and low bleed; a good all-purpose column used to begin method development protocols.

Probe Compound	McReynolds Constant	McReynolds Code	Kovats's Retention Index
n-Hexane			600
1-Butanol	54	y′	644
Benzene	16	x′	669
2-Pentanone	44	z′	671
n-Heptane			700
1,4-Dioxane	46	L	700
2-Methyl-2-pentanol	31	H	721
1-Nitropropane	62	u′	714
Pyridine	44	s′	743
n-Octane			800
Iodobutane	3	J	821
2-Octyne	23	K	864
n-Nonane			900
n-Decane			1000

MESOGENIC STATIONARY PHASES

The following table lists the liquid crystalline materials that have found usefulness as gas chromatographic stationary phases in both packed and open tubular column applications. In each case, the name, structure, and transition temperatures are provided (where available), along with a description of the separations that have been done using these materials. The table has been divided into two sections. The first section contains information on phases that have either smectic or nematic phases or both, while the second section contains mesogens that have a cholesteric phase. It should be noted that each material may be used for separations other than those listed, but the listing contains the applications reported in the literature.

It should be noted that some of the mesogens listed in this table are not commercially available, and must be prepared synthetically for laboratory use. The reader is referred to the appropriate citation for details.

REFERENCES

1. Panse, D. G., K. P. Naikwadi, B. V. Bapat, and B. B. Ghatge. "Applications of Laterally Mono and Disubstituted Liquid Crystals as Stationary Phases in Gas Liquid Chromatography." *Indian Journal of Technology* 19 (December 1981): 518–21.
2. Grushka, E., and J. F. Solsky. "p-Azoxyanisole Liquid Crystal as a Stationary Phase for Capillary Column Gas Chromatography." *Analytical Chemistry* 45, no. 11 (1973): 1836.
3. Witkiewicz, Z., and P. Stanislaw. "Separation of Close-Boiling Compounds on Liquid-Crystalline Stationary Phases." *Journal of Chromatography* 154 (1978): 60.
4. Naikwadi, K. P., D. G. Panse, B. V. Bapat, and B. B. Ghatge. "I. Synthesis and Application of Stationary Phases in Gas-Liquid Chromatography." *Journal of Chromatography* 195 (1980): 309.
5. Dewar, M., and J. P. Schroeder. "Liquid Crystals as Solvents. I. The Use of Nematic and Smectic Phases in Gas-Liquid Chromatography." *Journal of the American Chemical Society* 86 (1964): 5235.
6. Dewar, M., J. P. Schroeder, and D. Schroeder. "Molecular Order in the Nematic Mesophase of 4,4′-di-n-Hexyloxyazoxybenzene and Its Mixture with 4,4′-Dimethoxyazoxybenzene." *Journal of Organic Chemistry* 32 (1967): 1692.
7. Naikwadi, K. P., S. Rokushika, and H. Hatano. "New Liquid Crystalline Stationary Phases for Gas Chromatography of Positional and Geometrical Isomers Having Similar Volatilities." *Journal of Chromatography* 331 (1985): 69.
8. Richmond, A. B. "Use of Liquid Crystals for the Separation of Position Isomers of Disubstituted Benzenes." *Journal of Chromatographic Science* 9 (1971): 571.
9. Witkiewicz, Z., M. Pietrzyk, and R. Dabrowski. "Structure of Liquid Crystal Molecules and Properties of Liquid-Crystalline Stationary Phases in Gas Chromatography." *Journal of Chromatography* 177 (1979): 189.
10. Ciosek, M., Z. Witkiewicz, and R. Dabrowski. "Direct Gas: Chromatographic Determination of 2-Napthylamine in 1-Napthylamine on Liquid-Crystalline Stationary Phases." *Chemia Analityczna* 25 (1980): 567.
11. Jones, B. A., J. S. Bradshaw, M. Nishioka, and M. L. Lee. "Synthesis of Smectic Liquid-Crystalline Polysiloxanes from Biphenylcarboxylate Esters and Their Use as Stationary Phases for High-Resolution Gas Chromatography." *Journal of Organic Chemistry* 49 (1984): 4947.
12. Porcaro, P. J., and P. Shubiak. "Liquid Crystals as Substrates in the GLC of Aroma Chemicals." *Journal of Chromatography* 9 (1971): 689.
13. Witkiewicz, Z., Z. Suprynowicz, J. Wojcik, and R. Dabrowski. "Separation of the Isomers of Some Disubstituted Benzenes on Liquid Crystalline Stationary Phases in Small-Bore Packed Micro-Columns." *Journal of Chromatography* 152 (1978): 323.
14. Dewar, M., and J. P. Schroeder. "Liquid-Crystals as Solvents. II. Further Studies of Liquid Crystals as Stationary Phases in Gas-Liquid Chromatography." *Journal of Organic Chemistry* 30 (1965): 3485.

15. Witkiewicz, Z., J. Szule, R. Dabrowski, and J. Sadowski. "Properties of Liquid Crystalline Cyanoazoxybenzene Alkyl Carbonates as Stationary Phases in Gas Chromatography." *Journal of Chromatography* 200 (1980): 65.

16. Witkiewicz, Z., Z. Suprynowicz, and R. Dabrowski. "Liquid Crystalline Cyanoazoxybenzene Alkyl Carbonates as Stationary Phases in Small-Bore Packed Micro-Columns." *Journal of Chromatography* 175 (1979): 37.

17. Lochmüller, C. H., and R. W. Souter. "Direct Gas Chromatographic Resolution of Enantiomers on Optically Active Mesophases." *Journal of Chromatography* 88 (1974): 41.

18. Markides, K. E., M. Nishioka, B. J. Tarbet, J. S. Bradshaw, and M. L. Lee. "Smectic Biphenylcarboxylate Ester Liquid Crystalline Polysiloxane Stationary Phase for Capillary Gas Chromatography." *Analytical Chemistry* 57 (1985): 1296.

19. Vetrova, Z. P., N. T. Karabanov, T. N. Shuvalova, L. A. Ivanova, and Ya. I. Yashin. "The Use of p-n-Butyl Oxybenzoic Acid as Liquid Crystalline Sorbent in Gas Chromatography." *Chromatographia*, 20 (1985): 41.

20. Cook, L. E., and R. C. Spangelo. "Separation of Monosubstituted Phenol Isomers Using Liquid Crystals." *Analytical Chemistry* 46, no. 1 (1974): 122.

21. Kong, R. C., and L. L. Milton. "Mesogenic Polysiloxane Stationary Phase for High Resolution Gas Chromatography of Isomeric Polycyclic Aromatic Compounds." *Analytical Chemistry* 54 (1982): 1802.

22. Bartle, K. D., A. I. El-Nasri, and B. Frere. *Identification and Analysis of Organic Pollutants in Air.* Ann Arbor, MI: Ann Arbor Science Publishers, 1984.

23. Finklemann, H., R. J. Laub, W. E. Roberts, and C. A. Smith. Use of Mixed Phases for Enhanced Gas Chromatographic Separation of Polycyclic Aromatic Hydrocarbons. II. Phase Transition Behavior, Mass-Transfer Non-Equilibrium, and Analytical Properties of a Mesogen Polymer Solvent with Silicone Diluents, Polynuclear Aromatic Hydrocarbons: Physics, Biology, and Chemistry, 6th International Symposium M. Cooke, ed. 275–85, (1982).

24. Janini, G. M. "Recent Usage of Liquid Crystal Stationary Phases in Gas Chromatography." *Advances in Chromatography* 17 (1979): 231.

25. Witkiewicz, Z., and A. Waclawczyk. "Some Properties of High-Temperature Liquid Crystalline Stationary Phases." *Journal of Chromatography* 173 (1979): 43.

26. Zielinski, W. L., R. Johnston, and G. M. Muschik. "Nematic Liquid Crystal for Gas-Liquid Chromatographic Separation of Steroid Epimers." *Analytical Chemistry* 48 (1976): 907.

27. Smith, and Wozny, M. E. "Gas chromatographic separation of underivatized steroids using BPhBT liquid crystal stationary phase, HRC CC." *Journal of High Resolution Chromatography & Chromatography Communications* 3 (1980): 333.

28. Barrall, E. M., R. S. Porter, and J. F. Johnson. "Gas Chromatography Using Cholesteryl Ester Liquid Phase." *Journal of Chromatography* 21 (1966): 392.

29. Heath, R. R., and R. E. Dolittle. "Derivatives of Cholesterol Cinnamate. A Comparison of the Separations of Geometrical Isomers when Used as Gas Chromatographic Stationary Phases." *Journal of High Resolution Chromatography & Chromatography Communications* 6 (1983): 16.

30. Sonnet, P. E., and R. R. Heath. "Aryl Substituted Diastereomeric Alkenes: Gas Chromatographic Behavior on a Non-Polar Versus a Liquid Crystal Phase." *Journal of Chromatography* 321 (1985): 127.

Mesogenic Stationary Phases

Name: 2-chloro-4'-n-butyl-4-(4-n-butoxybenzoyloxy) azobenzene
Structure:

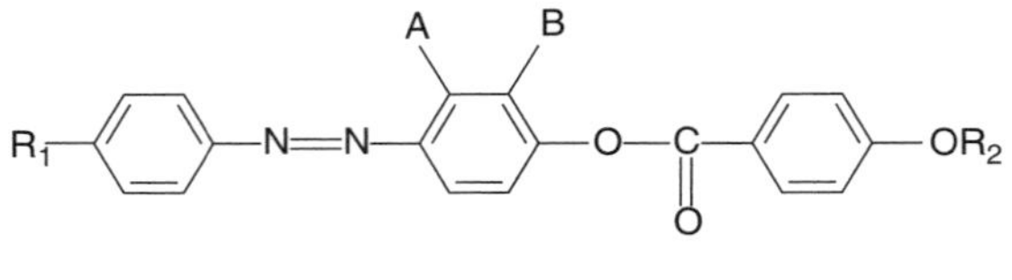

$R_1 = n–C_4H_9$
$A = Cl$
$B = H$
$R_2 = n–C_4H_9$

Thermophysical Properties:

solid → nematic 87.2 °C
nematic → isotropic 168 °C

Analytical Properties: Separation of close-boiling disubstituted benzenes
Reference 1

Name: p-azoxyanisole (4,4'-dimethoxyazoxybenzene)
Structure:

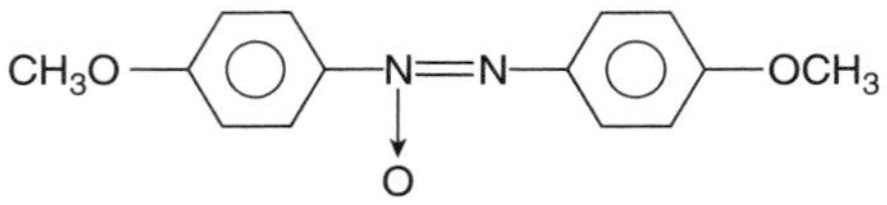

Thermophysical Properties:

solid → nematic 118 °C
nematic → isotropic 135 °C

Note: Supercooling has been noted at 110 °C by observing nematic-like properties.
Liquid crystalline behavior can sometimes persist to 102 °C

Analytical Properties: Separation of xylenes, separation of lower molecular weight aromatic hydrocarbon isomers, especially at the lower area of the nematic region.
Reference 2

Name: 2-chloro-4'-n-butyl-4-(4-methylbenzoyloxy) azobenzene
Structure:

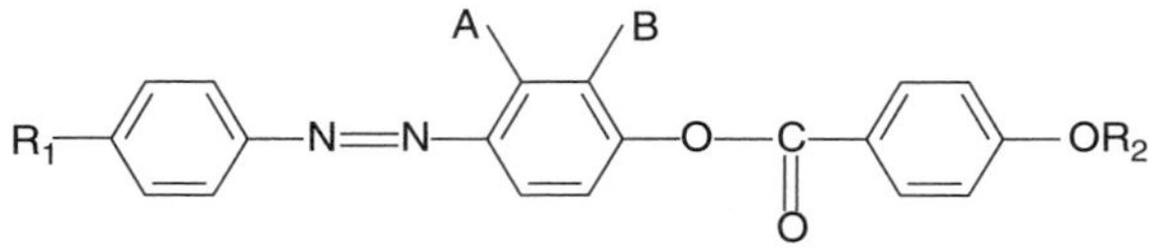

$R_1 = n–C_4H_9$
$A = Cl$
$B = H$
$R_2 = CH_3$

Thermophysical Properties:

solid → nematic 92.5 °C
nematic → isotropic 176 °C

Note: Supercooling has been noted at 110 °C by observing nematic-like properties.
Liquid crystalline behavior can sometimes persist to 102 °C

Analytical Properties: Separation of close boiling disubstituted benzenes.
Reference 1

Name: 2-chloro-4'-ethyl-4-(4-n-butoxybenzoyloxy) azobenzene
Structure:

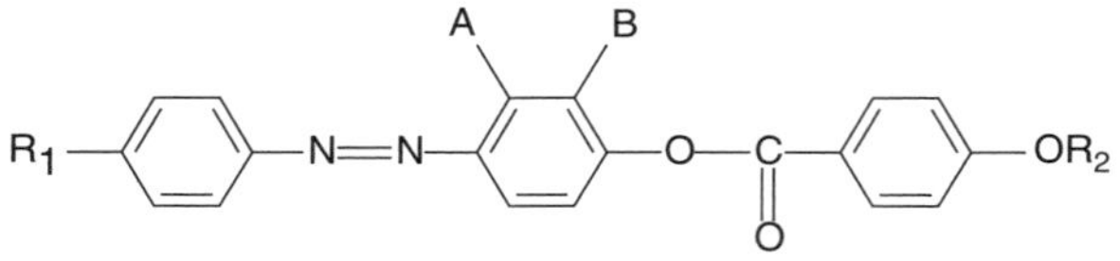

$R_1 = C_2H_5$
$A = Cl$
$B = H$
$R_2 = n–C_4H_9$

Mesogenic Stationary Phases (Continued)

Thermophysical Properties:

solid   → nematic 117 °C

nematic   → isotropic 172 °C

Note: Supercooling has been noted at 110 °C by observing nematic-like properties.
Liquid crystalline behavior can sometimes persist to 102 °C

Analytical Properties: Separation of close boiling disubstituted benzenes.

Reference 1

Name: 2-chloro-4′-n-butyl-4-(4-ethoxybenzoyloxy) azobenzene

Structure:

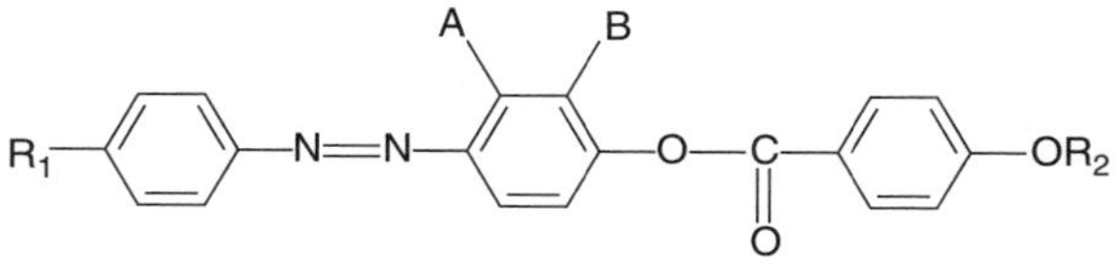

$R_2 = n-C_4H_9$
$A = Cl$
$B = H$
$R_2 = C_2H_5$

Thermophysical Properties:

solid   → nematic 89.7 °C

nematic   → isotropic 170 °C

Analytical Properties: Separation of close-boiling disubstituted benzenes

Reference 1

Name: 2-chloro-4′-methyl-4(4-n-butoxybenzoyloxy) azobenzene

Structure:

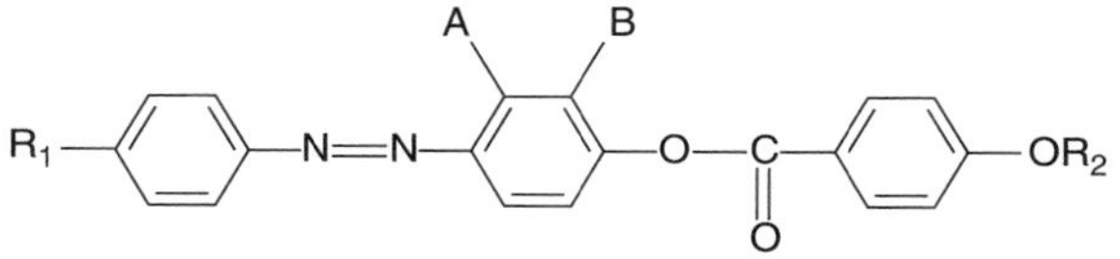

$R_1 = CH_3$
$A = Cl$
$B = H$
$R_2 = n-C_4H_9$

Thermophysical Properties:

solid   → nematic 112 °C

nematic   → isotropic 165 °C

Analytical Properties: Separation of close-boiling disubstituted benzenes

Reference 1

Name: 2-chloro-4′-n-methyl-4-(4-ethoxybenzoyloxy) azobenzene

Structure:

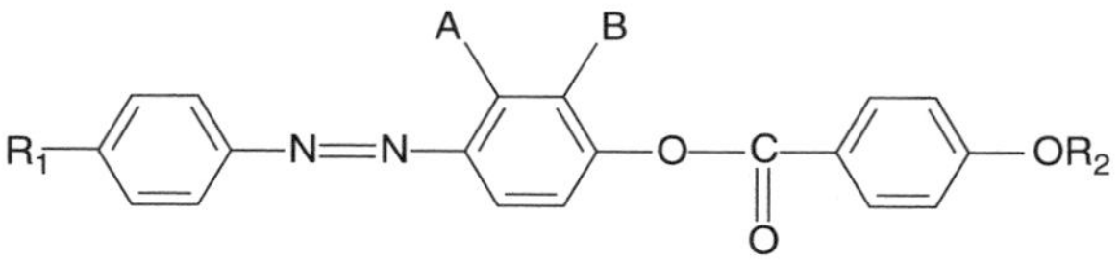

$R_1 = CH_3$
$A = Cl$
$B = H$
$R_2 = C_2H_5$

Thermophysical Properties:

solid   → nematic 128.3 °C

nematic   → isotropic 185 °C

Analytical Properties: Separation of close-boiling disubstituted benzenes

Reference 1

(Continued)

Mesogenic Stationary Phases (Continued)

Name: p-cyano-p′-pentoxyazoxybenzene
Structure:

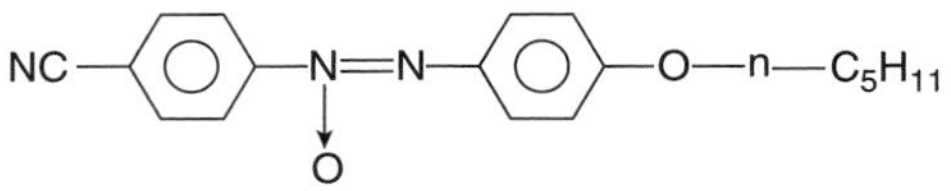

Thermophysical Properties:
solid   → nematic 124 °C
nematic   → isotropic 153 °C

Analytical Properties: Complete separation of ethyltoluenes, chlorotoluenes, bromotoluenes and dichlorobenzenes. Also: ethylbenzene from xylenes and propylbenzene from ethyltoluenes
Reference 3

Name: p-cyano-p′-pentoxyazobenzene
Structure:

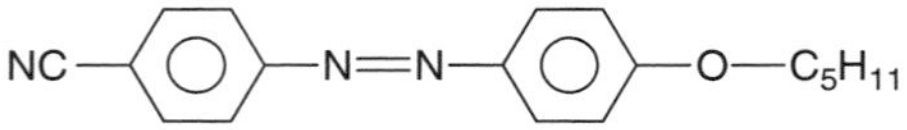

Thermophysical Properties:
solid   → nematic 106 °C
nematic   → isotropic 116.5 °C

Analytical Properties: Separation of ethyltoluenes, chlorotoluenes, bromotoluenes, dichlorobenzenes. Also: ethylbenzenes from xylenes and propylbenzenes from ethylbenzenes
Reference 3

Name: p-cyano-p′-pentoxyazoxybenzene (mixed isomers)
Structure:

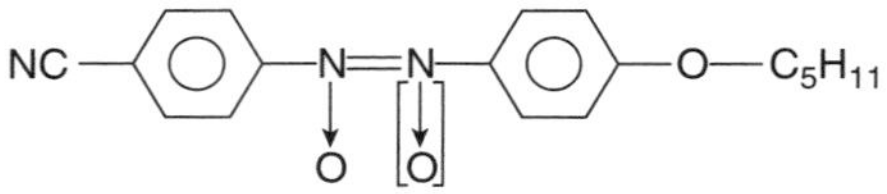

Thermophysical Properties:
solid   → nematic 93.5 °C
nematic   → isotropic 146.5 °C

Analytical Properties: Complete separation of ethyltoluenes, chlorotoluenes, bromotoluenes and dichlorobenzenes. Also: ethylbenzene from xylenes and propylbenzene from ethyltoluenes
Reference 30

Name: p-cyano-p′-octoxyazoxybenzene (mixed isomers)
Structure:

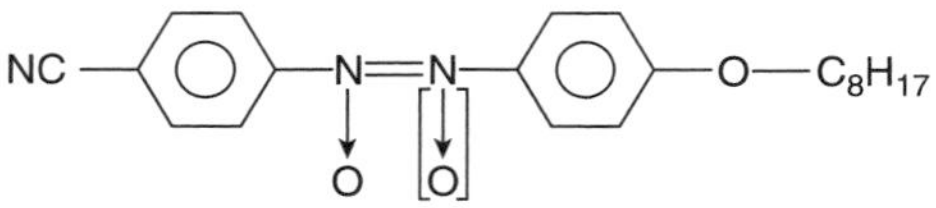

Mesogenic Stationary Phases (Continued)

Thermophysical Properties:

solid   → smectic 71 °C

smectic   → nematic 117 °C

nematic   → isotropic 135 °C

Analytical Properties: Separation of ethyltoluenes, chlorotoluenes, bromotoluenes, and dichlorobenzenes. Also: ethylbenzene from xylenes and propylbenzene from ethylbenzenes

Reference 3

Name: p-cyano-p′-octoxyazoxybenzene

Structure:

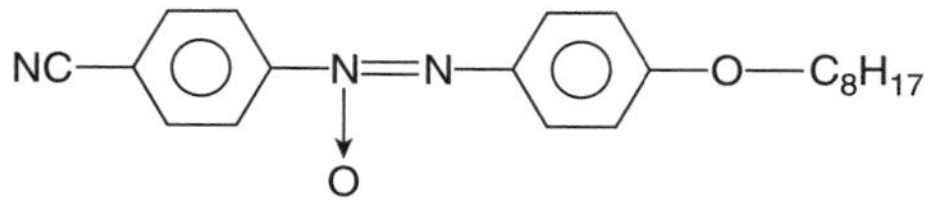

Thermophysical Properties:

solid   → smectic 100.5 °C

smectic   → nematic 138.5 °C

nematic   → isotropic 148.5 °C

Analytical Properties: Separation of ethyltoluenes, chlorotoluenes, bromotoluenes, and dichlorobenzenes. Also: ethylbenzene from xylenes and propylbenzene from ethylbenzenes

Reference 3

Name: 4′-n-butyl-4(4-n-butoxybenzoyloxy) azobenzene

Structure:

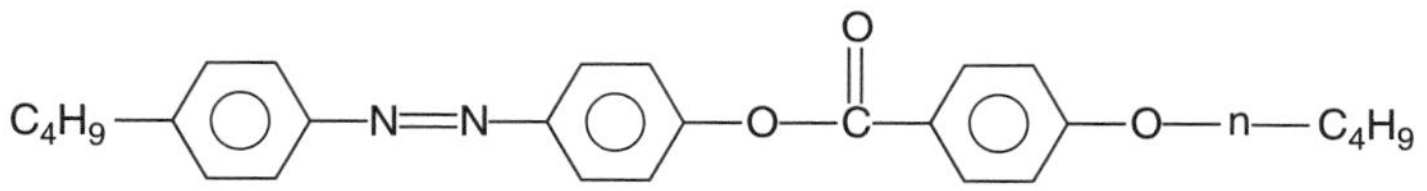

Thermophysical Properties:

solid   → nematic 94 °C

nematic   → isotropic 234 °C

Analytical Properties: Separation of chlorinated biphenyls

Reference 4

Name: 4-4′-di-n-heptyloxyazoxybenzene

Structure:

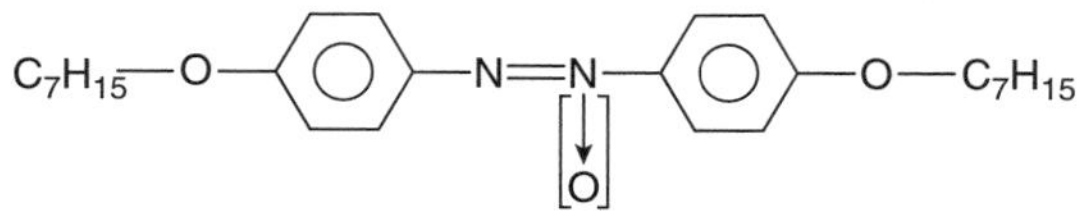

Thermophysical Properties:

solid   → nematic 95 °C

nematic   → isotropic 127 °C

Analytical Properties: Separation of meta and para-xylene in nematic region

Reference 5

(*Continued*)

Mesogenic Stationary Phases (Continued)

Name: 4,4'-di-n-hexyloxyazoxybenzene
Structure:

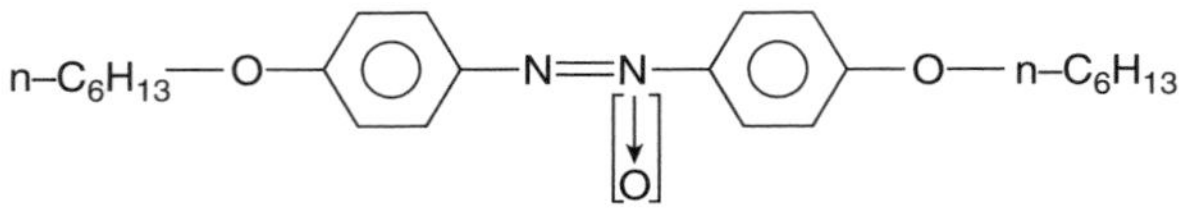

Thermophysical Properties:
solid   → nematic 81 °C
nematic   → isotropic 129 °C

Analytical Properties: Separation of meta and para-xylene using gas chromatography
Reference 5, 6

Name: 4'-methoxy-4-(-4-n-butoxybenzoyloxy) azobenzene
Structure:

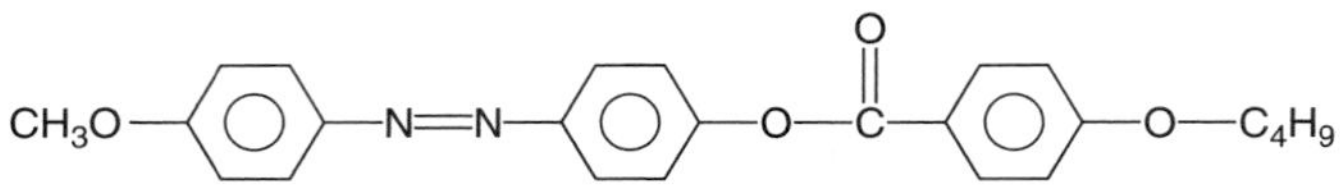

Thermophysical Properties:
solid   → nematic 116 °C
nematic   → isotropic 280 °C

Analytical Properties: Separation of chlorinated biphenyls
Reference 4

Name: 2-methyl-4'-n-butyl-4-(4-n-butoxybenzoyloxy) azobenzene
Structure:

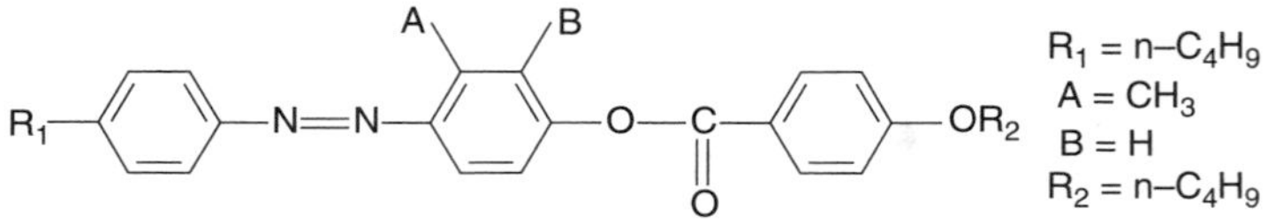

$R_1 = n–C_4H_9$
$A = CH_3$
$B = H$
$R_2 = n–C_4H_9$

Thermophysical Properties:
solid   → nematic 90 °C
nematic   → isotropic 175 °C

Analytical Properties: Separation of close-boiling disubstituted benzenes
Reference 1

Name: 2-methyl-4'-n-butyl-4-(p-methoxycinnamoyloxy) azobenzene
Structure:

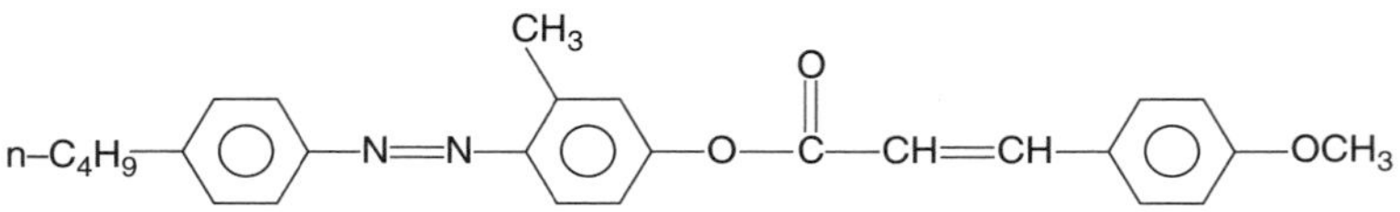

Mesogenic Stationary Phases (Continued)

Thermophysical Properties:

solid　　　→ nematic 109 °C

nematic　 → isotropic 253 °C

Analytical Properties: Separation of positional isomers of aromatic hydrocarbons

Reference 7

Name: 2-methyl-4′-methoxy-4-(4-ethoxybenzoyloxy) azobenzene

Structure:

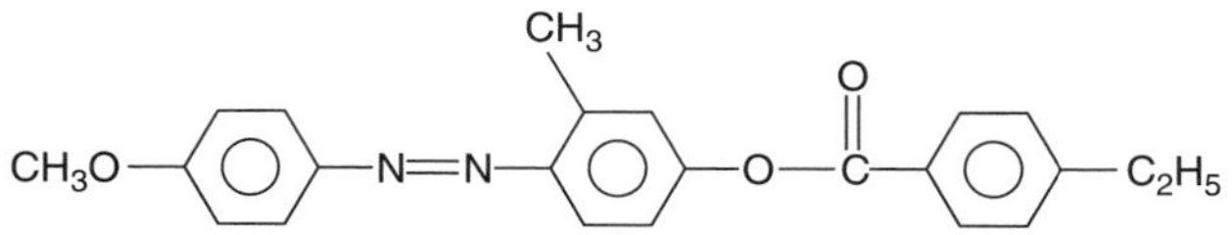

Thermophysical Properties:

solid　　　→ nematic 125 °C

nematic　 → isotropic 244 °C

Analytical Properties: Separation of chlorinated biphenyls

Reference 4

Name: 2-methyl-4′-methoxy-4-(4-methoxybenzoyloxy) azobenzene

Structure:

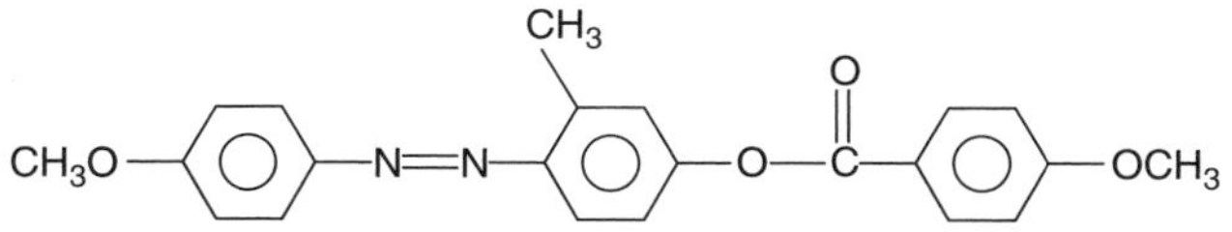

Thermophysical Properties:

solid　　　→ nematic 160 °C

nematic　 → isotropic 253 °C

Analytical Properties: Separation of chlorinated biphenyls

Reference 4

Name: 2-methyl-4′-ethyl-4-(4′-methoxycinnamyloxy) azobenzene

Structure:

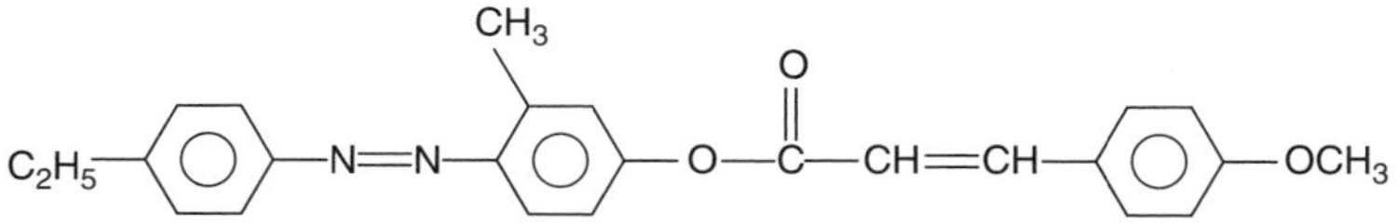

Thermophysical Properties:

solid　　　→ nematic 126 °C

nematic　 → isotropic 262 °C

Analytical Properties: Separation of polyaromatic hydrocarbons and insect sex pheromones

Reference 5

(Continued)

Mesogenic Stationary Phases (Continued)

Name: 2-methyl-4′-methoxy-4-(p-methoxycinnamoyloxy) azobenzene
Structure:

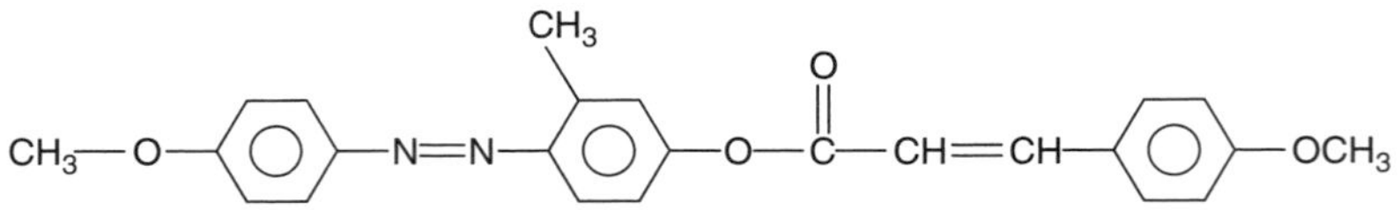

Thermophysical Properties:

solid   → nematic 149 °C
nematic   → isotropic 298 °C

Analytical Properties: Separation of positional isomers of aromatic compounds and geometrical isomers of sex pheromones

Reference 7

Name: 2-methyl-4′-methyl-4-(4-ethoxybenzoyloxy) azobenzene
Structure:

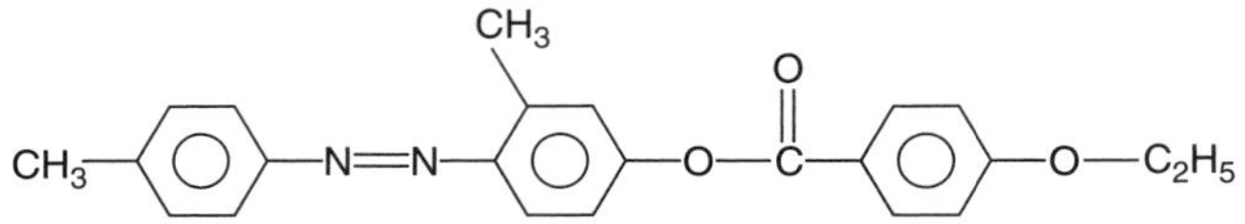

Thermophysical Properties:

solid   → nematic 125 °C
nematic   → isotropic 220 °C

Analytical Properties: Separation of chlorinated biphenyls
Reference 4

Name: 4,4′-azoxyphenetole
Structure:

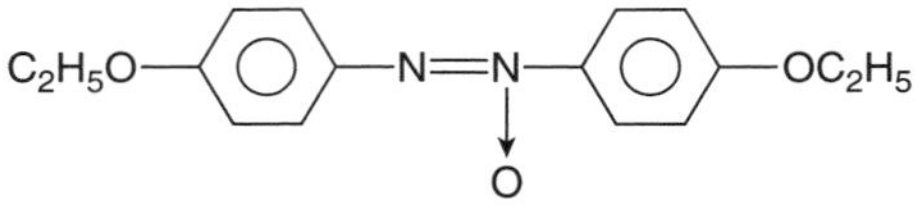

Thermophysical Properties:

solid   → nematic 138 °C
nematic   → isotropic 168 °C

Analytical Properties: Separation of meta and para isomers of disubstituted benzenes
Reference 8

Name: 4,4-biphenylene-bis-[p-(heptyloxy) benzoate]
Structure:

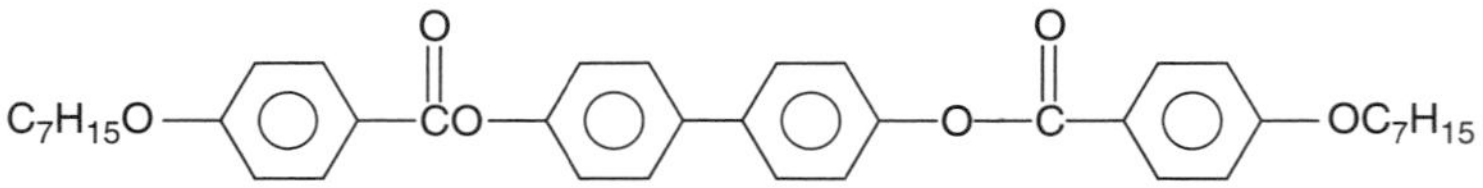

Thermophysical Properties:

solid   → smectic 150 °C
smectic   → nematic 211 °C
nematic   → isotropic 316 °C

Analytical Properties: Separation of meta and para isomers of disubstituted benzenes
Reference 8

Mesogenic Stationary Phases (Continued)

Name: p'-ethylazoxybenzene p-cyanobenzoate (mixed isomers)
Structure:

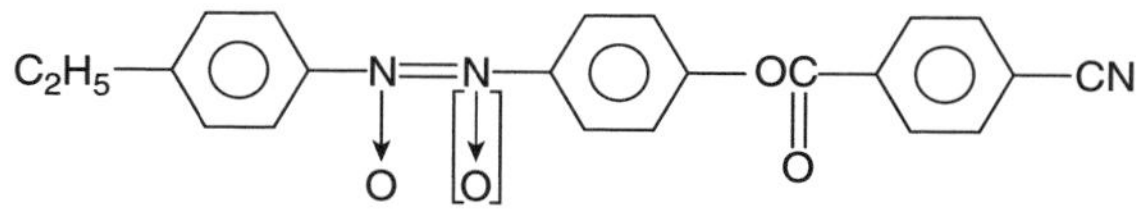

Thermophysical Properties:

melting range   → 114–136 °C

nematic   → isotropic > 306 °C

Analytical Properties: Separation of substituted xylenes

Reference 9

Name: p'-ethylazoxybenzene p-cyanobenzoate (pure isomer)
Structure:

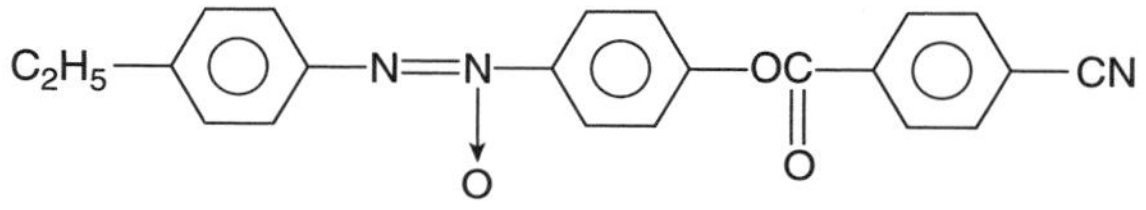

Thermophysical Properties:

solid   → nematic 115 °C

nematic   → isotropic 294 °C

Analytical Properties: Separation of nitronaphthalenes

Reference 10

Name: p'-ethylazobenzene p'-cyanobenzoate
Structure:

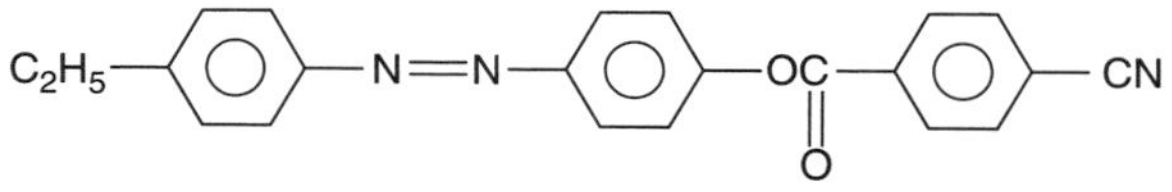

Thermophysical Properties:

solid   → nematic 138–140 °C

nematic   → isotropic 292 °C

Analytical Properties: Separation of substituted xylenes

Reference 9

Name: p'-ethylazobenzene p-methylbenzoate
Structure:

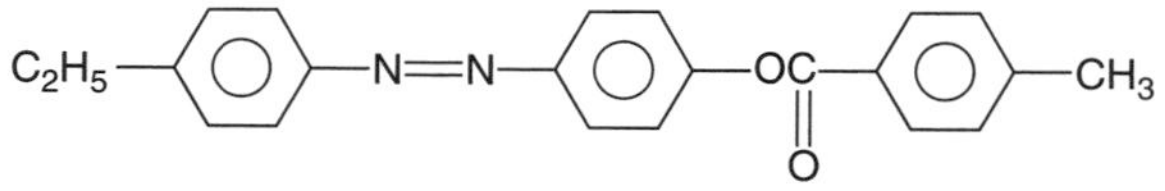

Thermophysical Properties:

solid   → nematic 108 °C

nematic   → isotropic 230 °C

Analytical Properties: Separation of nitronaphthalenes

Reference 10

(Continued)

Mesogenic Stationary Phases (Continued)

Name: p-ethylazoxybenzene p'-methylbenzoate (mixed isomers)
Structure:

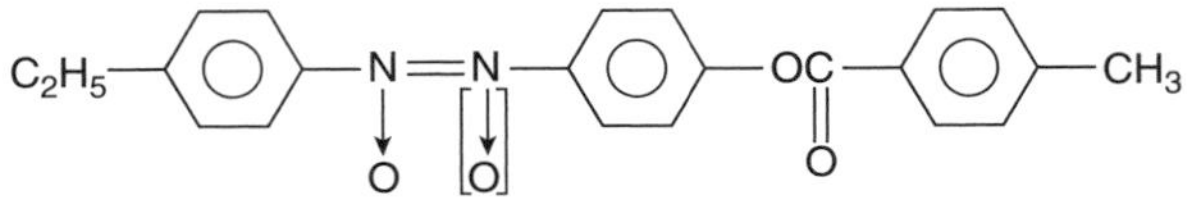

Thermophysical Properties:

(directly after crystallization)	(after melting and cooling)
crystal → nematic 97.5 °C	crystal → nematic 87.5–97.5 °C
nematic → isotropic 250.5 °C	nematic → isotropic 250.5 °C

Analytical Properties: Separation of substituted xylenes

Reference 9

Name: 4'-methoxybiphenyl-4,4-[-(allyloxy)phenyl] benzoate
Structure:

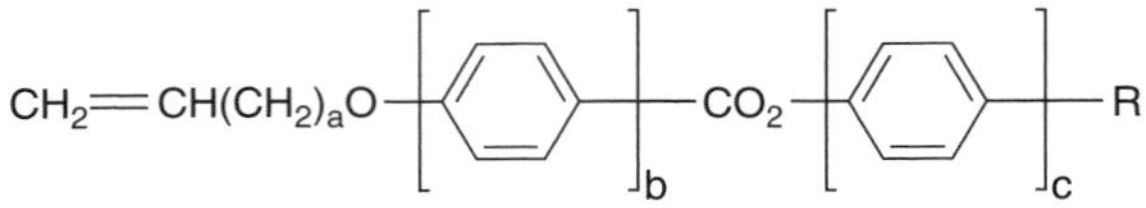

a = 1
b = 2
c = 2
R = OCH$_3$

Thermophysical Properties:
solid   → nematic 214 °C
nematic   → isotropic 290 °C

Analytical Properties: Suggested for separation of polycyclic aromatic compounds

Reference 11

Name: (S)-4-[(2-methyl-1-butoxy)carbonyl]phenyl 4-[4-(4-pentenyloxy)phenyl] benzoate
Structure:

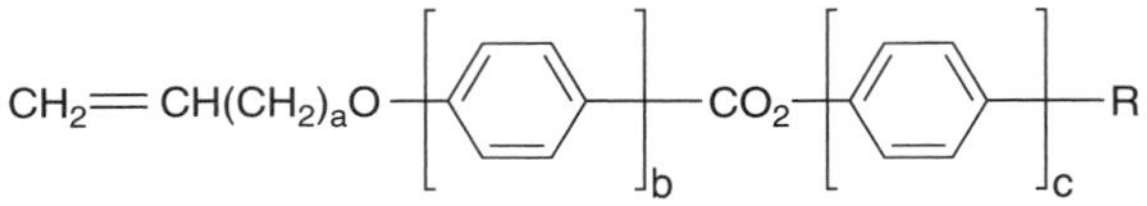

a = 3
b = 2
c = 1
R = COOCH$_2$C*H(CH$_3$)CH$_2$CH$_3$

Thermophysical Properties:
solid   → smectic 105 °C
smectic   → isotropic 198 °C

Analytical Properties: Suggested for separation of polycyclic aromatic compounds

Reference 11

Name: 4-methoxyphenyl 4-[4-(allyloxy) phenyl] benzoate
Structure:

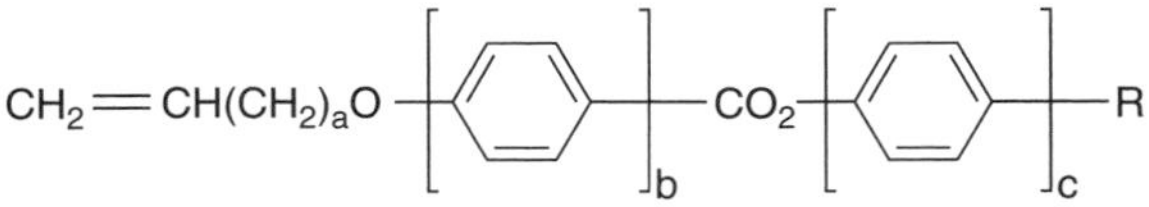

a = 1
b = 2
c = 1
R = OCH$_3$

Mesogenic Stationary Phases (Continued)

Thermophysical Properties:
solid   → nematic 137 °C
nematic   → isotropic 243 °C

Analytical Properties: Suggested for the separation of polycyclic aromatic compounds

Reference 11

Name: 4-methoxyphenyl 4-[4-(4-pentenyloxy) phenyl] benzoate
Structure:

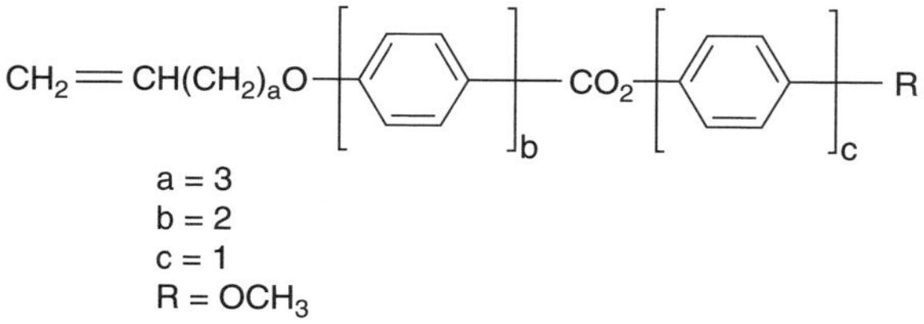

$$CH_2 = CH(CH_2)_aO - \left[\langle\rangle \right]_b - CO_2 - \left[\langle\rangle \right]_c - R$$

a = 3
b = 2
c = 1
R = OCH$_3$

Thermophysical Properties:
solid   → smectic 133 °C
smectic   → nematic 172 °C
nematic   → isotropic 253 °C

Analytical Properties: Suggested for separation of polycyclic aromatic compounds

Reference 11

Name: p-phenylene-bis-4-n-heptyloxybenzoate
Structure:

$$C_7H_{15}O - \langle\rangle - COO - \langle\rangle - OOC - \langle\rangle - OH_{15}C_7$$

Thermophysical Properties:
solid   → smectic 83 °C
smectic   → nematic 125 °C
nematic   → isotropic 204 °C

Analytical Properties: Separation of 1- and 2-ethylnapthalene; baseline separation of pyrazines

Reference 12

Name: 4-[(4-dodecyloxyphenyl)azoxy]-benzonitrile
Structure:

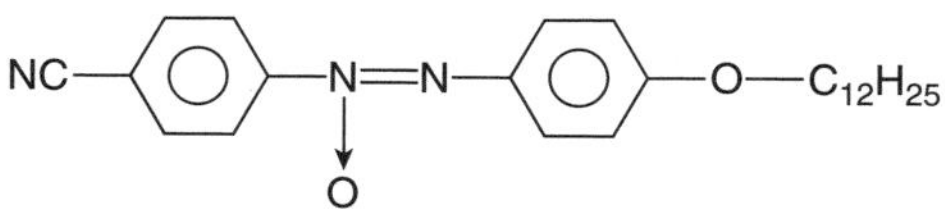

$$NC - \langle\rangle - N = N - \langle\rangle - O - C_{12}H_{25}$$
$$\downarrow$$
$$O$$

Thermophysical Properties:
solid   → smectic 106 °C
smectic   → isotropic 147 °C

Analytical Properties: Marginal effectiveness in separating disubstituted benzene isomers

Reference 13

(Continued)

Mesogenic Stationary Phases (Continued)

Name: 4-[(4-pentyloxyphenyl)azoxy]-benzonitrile (mixed isomers)
Structure:

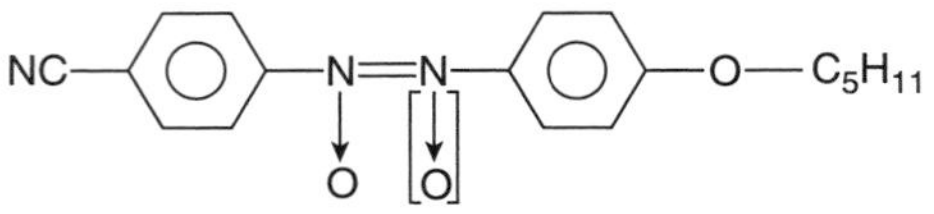

Thermophysical Properties:
solid   → nematic 94 °C
nematic   → isotropic 141.5 °C

Analytical Properties: —does not separate diethylbenzene (DEB) isomers
                           —good separation of disubstituted benzene isomers
Reference 13

Name: 4-[(4-octyloxyphenyl)azoxy]-benzonitrile
Structure:

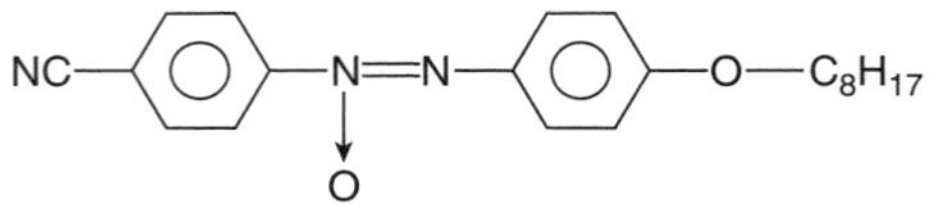

Thermophysical Properties:
solid   → smectic 101.5 °C
smectic   → nematic 137 °C
nematic   → isotropic 151.5 °C

Analytical Properties: Separates diethylbenzene isomers
Reference 13

Name: 4-[(4-pentyloxyphenyl) azoxy]-benzonitrile (pure isomers)
Structure:

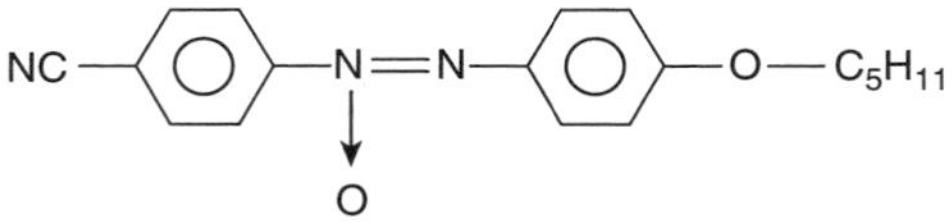

Thermophysical Properties:
solid   → nematic 124 °C
nematic   → isotropic 153 °C

Analytical Properties: Complete separation of dichlorobenzene or bromotoluene isomers at 126 °C. Complete separation of chlorotoluene isomers at 87 °C; partial separation of m- and p-xylenes at 87 °C
Reference 13

Name: 4,4′-bis(p-methoxybenzylidene amino)-3,3′-dichloro biphenyl
Structure:

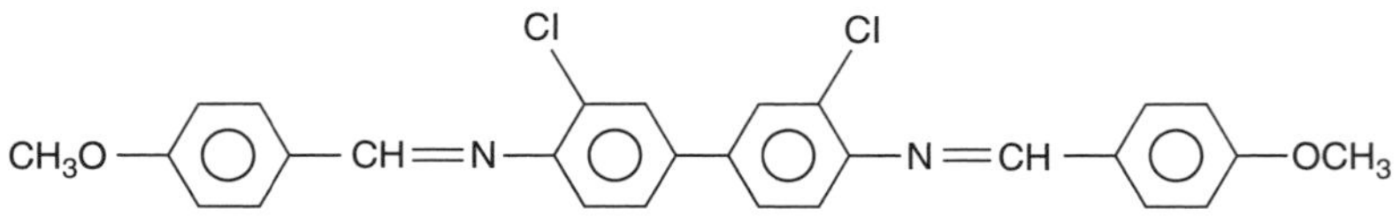

Mesogenic Stationary Phases (Continued)

Thermophysical Properties:

solid   → nematic 154 °C

nematic   → isotropic 344 °C

Analytical Properties: Separation of dimethyl benzene isomers, dihalo benzene isomers (Cl, Br), halo-ketone benzene isomers, dimethoxy benzene isomers

Reference 14

Name: azoxybenzene p-cyano-p′-heptyl carbonate (mixed isomers)

Structure:

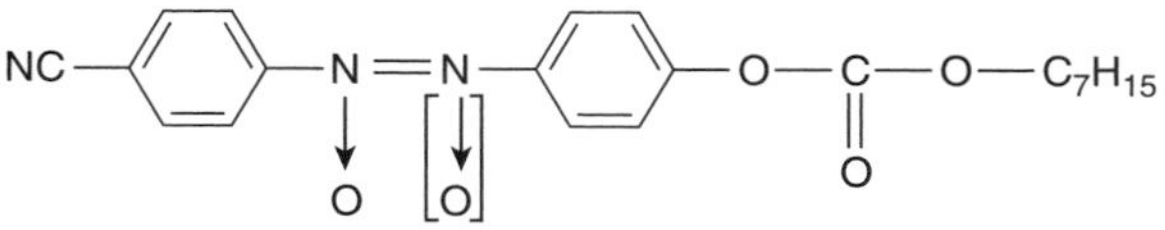

Thermophysical Properties:

solid   → nematic 66 °C

Analytical Properties: Separation of disubstituted benzene isomers

Reference 15

Name: azoxybenzene p-cyano-p′-octyl carbonate (mixed isomers)

Structure:

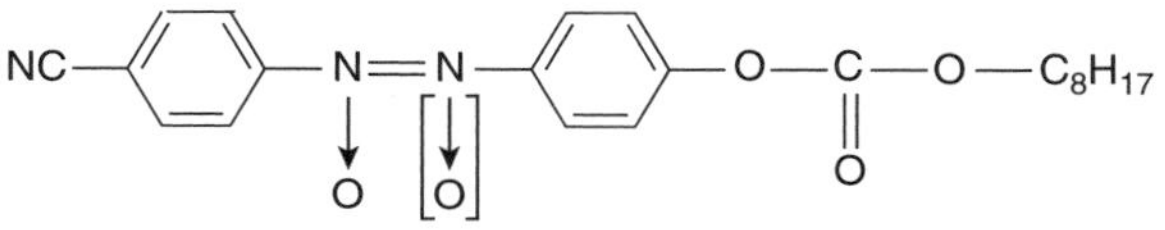

Thermophysical Properties:

solid   → smectic 60.5 °C

smectic   → nematic 119.5 °C

Analytical Properties: Separation of ethyltoluenes, chlorotoluenes, bromotoluenes, dichlorobenzenes. Also: ethylbenzenes from xylenes and propylbenzene from ethylbenzenes

Reference 15

Name: azoxybenzene p-cyano-p′-pentyl carbonate (pure isomer)

Structure:

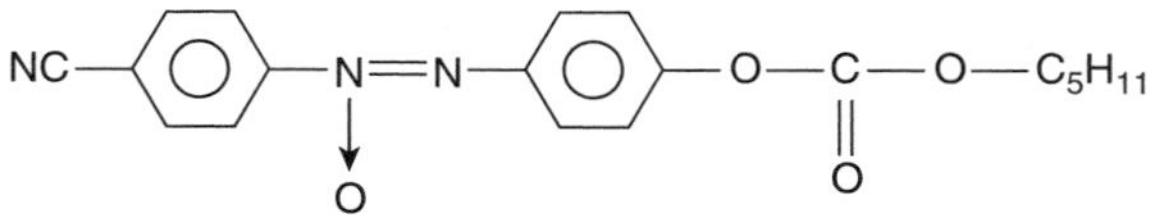

Thermophysical Properties:

solid   → nematic 60.5 °C

nematic   → isotropic 132 °C

Analytical Properties: Separation of ethyltoluenes, chlorotoluenes, bromotoluenes, dichlorobenzenes. Also: ethylbenzenes from xylenes and propylbenzene from ethylbenzenes

Reference 3

(Continued)

Mesogenic Stationary Phases (Continued)

Name: azoxybenzene p-cyano-p′-pentyl carbonate (mixed isomers)
Structure:

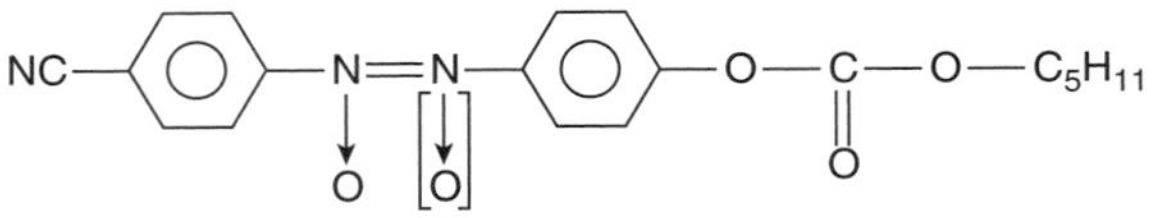

$$NC-\bigcirc-N=N-\bigcirc-O-C-O-C_5H_{11}$$

Thermophysical Properties:

solid $\rightarrow$ nematic 96–100 °C

Analytical Properties: Separation of disubstituted benzene isomers

Reference 15

Name: cyanoazoxybenzene decyl carbonate (mixed isomers)
Structure:

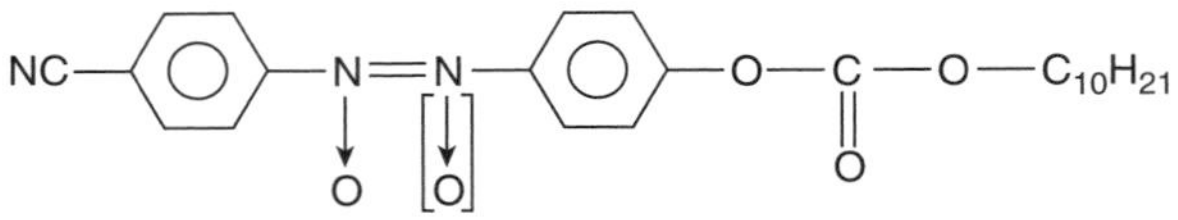

$$NC-\bigcirc-N=N-\bigcirc-O-C-O-C_{10}H_{21}$$

Thermophysical Properties:

solid $\rightarrow$ smectic 74 °C
smectic $\rightarrow$ isotropic 125.5 °C

Analytical Properties: Separation of polycyclic hydrocarbons

Reference 16

Name: cyanoazoxybenzene hexyl carbonate (mixed isomers)
Structure:

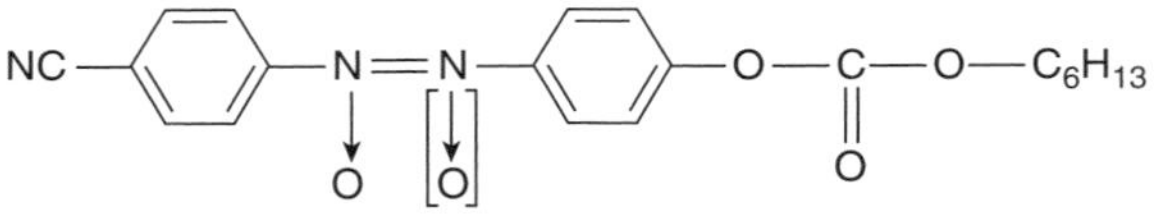

$$NC-\bigcirc-N=N-\bigcirc-O-C-O-C_6H_{13}$$

Thermophysical Properties:

solid $\rightarrow$ nematic 73–76 °C
nematic $\rightarrow$ isotropic 137 °C

Analytical Properties: Separation of xylene and ethyltoluene isomers

Reference 16

Name: cyanoazoxybenzene nonyl carbonate (mixed isomers)
Structure:

$$NC-\bigcirc-N=N-\bigcirc-O-C-O-C_9H_{19}$$

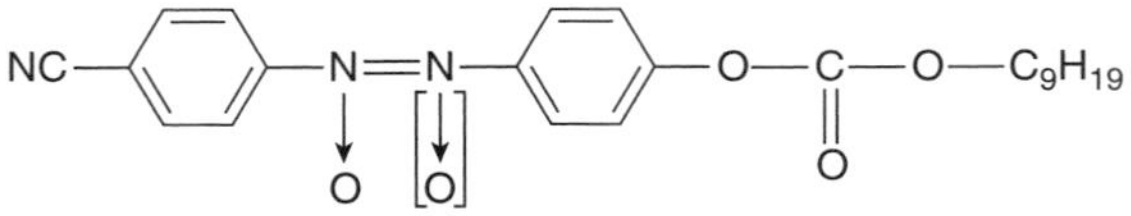

Mesogenic Stationary Phases (Continued)

Thermophysical Properties:

solid   → smectic 61 °C

smectic   → nematic 124 °C

nematic   → isotropic 127 °C

Analytical Properties: Separation of polycyclic hydrocarbons

Reference 16

Name: p-(p-ethoxyphenylazo) phenyl crotonate

Structure:

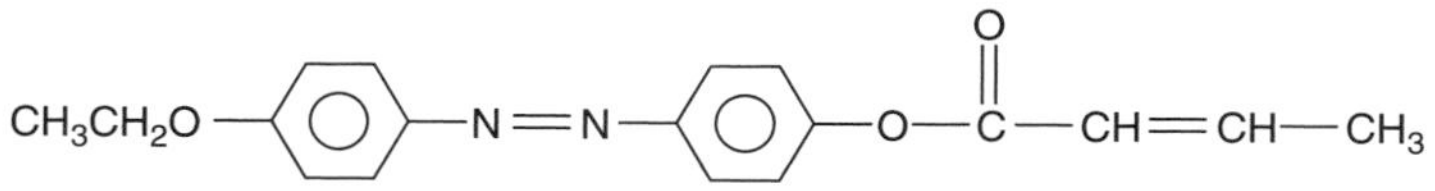

Thermophysical Properties:

solid   → nematic 110 °C

nematic   → isotropic 197 °C

Analytical Properties: Separation of aromatic isomers

Reference 12

Name: carbonyl-bis-(D-leucine isopropyl ester)

Structure:

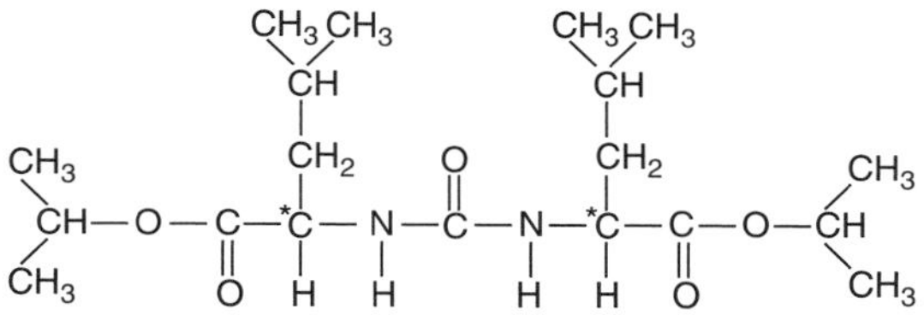

Thermophysical Properties:

solid   → smectic 55 °C   *Note:* The asterisk indicates a chiral center.

smectic   → isotropic 110 °C

Analytical Properties: Baseline and near-baseline separations of racemic mixtures of N-perfluoroacyl-2-aminoethyl benzenes, trifluoroacetyl (TFA), pentafluoropropionyl (PFP), heptafluorobutyl (HFB)

Reference 17

Name: carbonyl-bis-(L-valine isopropyl ester)

Structure:

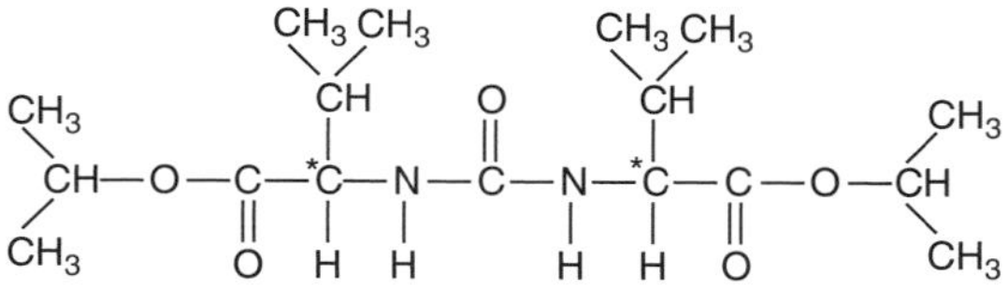

Thermophysical Properties:

solid   → smectic[1] 91 °C   *Note:* This compound exhibits 2 stable smectic states prior

smectic1   → smectic[2] 99 °C   to melting; the asterisk indicates a chiral center.

smectic2   → isotropic 109 °C

Analytical Properties: Separation of enantiomers

Reference 17

(Continued)

Mesogenic Stationary Phases (Continued)

Name: carbonyl-bis-(L-valine t-butyl ester)
Structure:

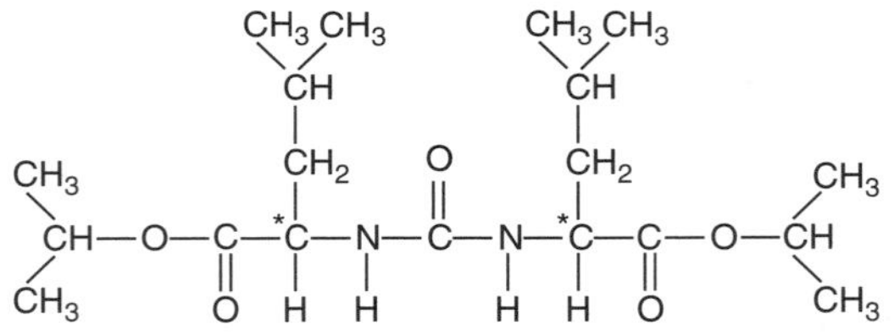

Thermophysical Properties:

solid   → smectic 98 °C   *Note:* The asterisk indicates a chiral center.
smectic   → isotropic 402 °C

Analytical Properties: Separation of enantiomers

Reference 17

Name: carbonyl-bis-(L-valine ethyl ester)
Structure:

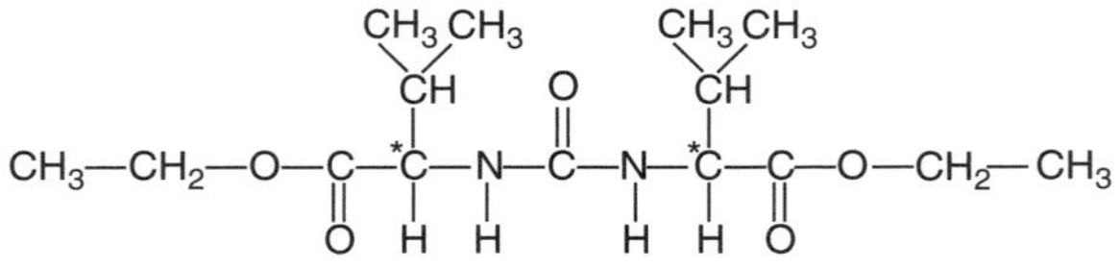

Thermophysical Properties:

solid   → smectic 88 °C   *Note:* The asterisk indicates a chiral center.
smectic   → isotropic 388 °C

Analytical Properties: Separation of enantiomers

Reference 17

Name: carbonyl-bis-(L-valine methylester)
Structure:

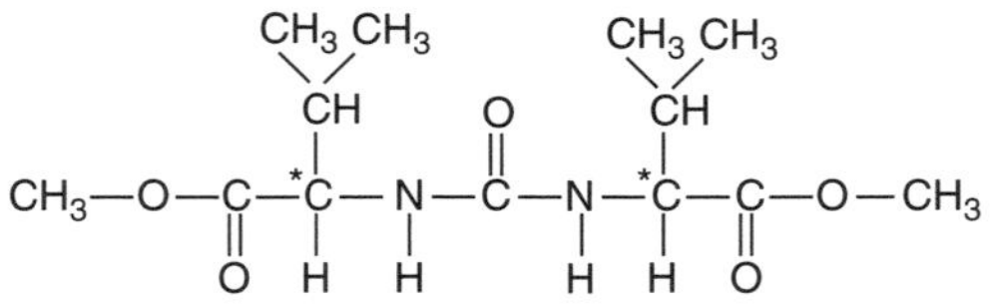

Thermophysical Properties:

solid   → smectic 382 °C   *Note:* The asterisk indicates a chiral center.
smectic   → isotropic 415 °C

Analytical Properties: Separation of enantiomers

Reference 17

Mesogenic Stationary Phases (Continued)

Name: phenylcarboxylate ester (systematic name not available)
Structure:

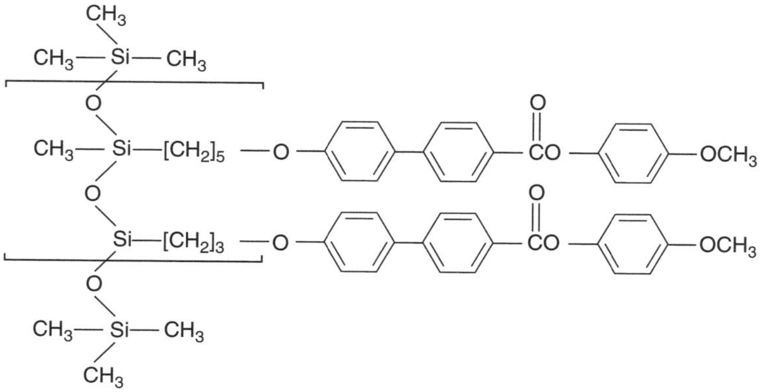

Thermophysical Properties:

solid   → smectic 118 °C
smectic   → isotropic 300 °C

Analytical Properties: Separation of 3 and 4 member methylated polycyclic aromatic hydrocarbons (PAH), on basis of length-to-breadth ratio (l/b); as l/b increases, retention time decreases, cross-linking increases retention times, separation of methylcrypene isomers

Reference 18

Name: p-cyano-p′-octoxyazobenzene
Structure:

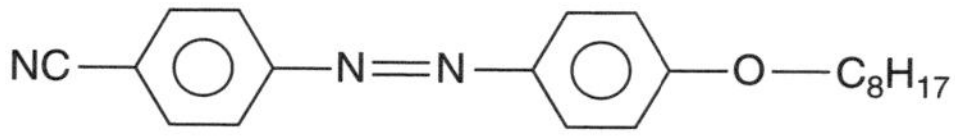

Thermophysical Properties:

solid   → nematic 101 °C
nematic   → isotropic 111 °C

Analytical Properties: Separation of ethyltoluenes, chlorotoluenes, bromotoluenes, dichlorobenzenes. Also: ethylbenzenes from xylenes and propylbenzene from ethylbenzenes
Reference 3

Name: p-n-butoxy benzoic acid
Structure:

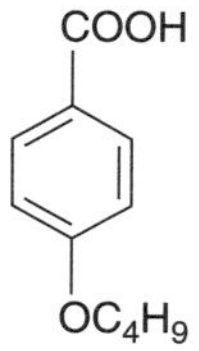

(Continued)

Mesogenic Stationary Phases (Continued)

Thermophysical Properties:

solid 100 °C

mesomorphous 150 °C (not well characterized)

isotropic 160 °C

Analytical Properties: Separation of methyl and monoalkyl substituted benzenes as well as organoelemental compounds (for example, dimethyl mercury)

Reference 19

Name: p-[(p-methoxybenzylidene)-amino]phenylacetate
Structure:

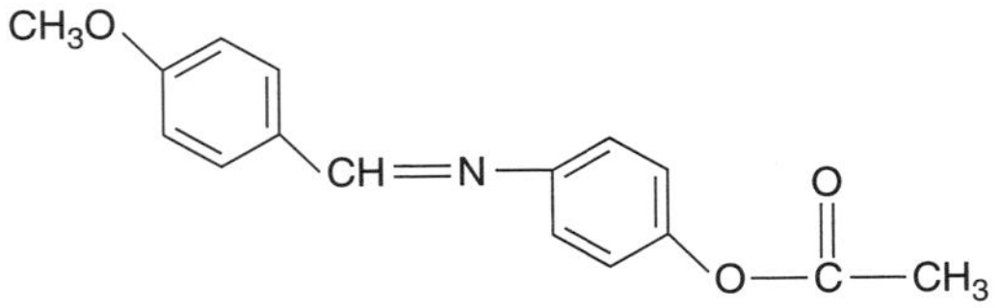

Thermophysical Properties:

solid   → nematic 80 °C

nematic   → isotropic 108 °C

Analytical Properties: Separation of substituted phenols, selectivity is best at the lower end of the nematic range

Reference 20

Name: poly (mesogen/methyl) siloxane (PMMS)—compound has not been named
Structure:

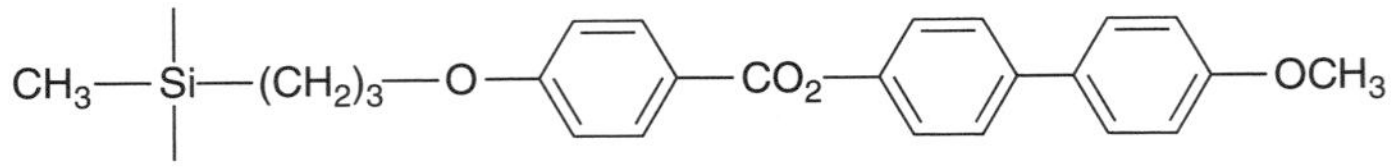

Thermophysical Properties:

solid   → nematic 70 °C

nematic   → isotropic 300 °C

high thermal stability

Analytical Properties: Separation of methylchrysene isomers
Reference 21, 22

Name: N,N′-bis-(p-butoxybenzylidene)-bis-p-toluidine (BBBT)
Structure:

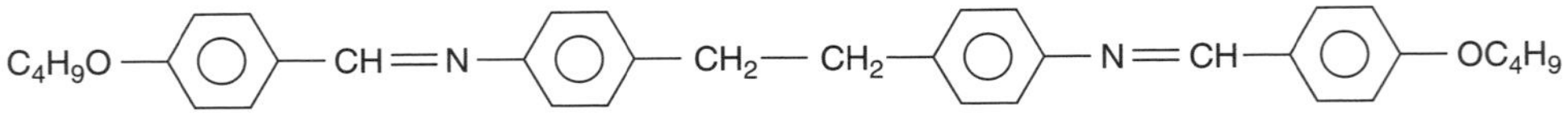

Thermophysical Properties:

solid   → smectic 159 °C

smectic   → nematic 188 °C

nematic   → isotropic 303 °C

Analytical Properties: Separation of polycyclic aromatic hydrocarbons on the basis of length-to-breadth ratio
Reference 23

Mesogenic Stationary Phases (Continued)

Name: N,N'-bis(ethoxy-benzylidene)-α,α''-bi-p-toluidine (BEBT)
Structure:

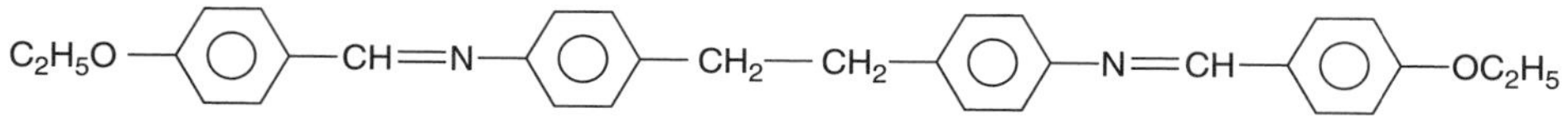

$$C_2H_5O \longleftrightarrow CH = N \longleftrightarrow CH_2 \longrightarrow CH_2 \longleftrightarrow N = CH \longleftrightarrow OC_2H_5$$

Thermophysical Properties:

solid $\rightarrow$ nematic 173 °C
nematic $\rightarrow$ isotropic 341 °C

Analytical Properties: Separation of polynuclear aromatic hydrocarbons
Reference 24

Name: N,N'-bis(n-heptoxy-benzylidene)-α,α''-bi-p-toluidine (BHpBT)
Structure:

$$C_7H_{15}O - \phi - CH = N - \phi - CH_2 - CH_2 - \phi - N = CH - \phi - OC_7H_{15}$$

$$\phi = \langle \bigcirc \rangle$$

Thermophysical Properties:

solid $\rightarrow$ smectic 119 °C
smectic $\rightarrow$ nematic 238 °C
nematic $\rightarrow$ isotropic 262 °C

Analytical Properties: Separation of polynuclear aromatic hydrocarbons
Reference 24

Name: N,N'-bis(n-hexoxy-benzylidene)-α,α'-bi-p-toluidine (BHxBT)
Structure:

$$H_{13}C_6O - \phi - CH = N - \phi - CH_2 - CH_2 - \phi - N = CH - \phi - OC_6H_{13}$$

$$\phi = \langle \bigcirc \rangle$$

Thermophysical Properties:

solid $\rightarrow$ smectic 127 °C
smectic $\rightarrow$ nematic 229 °C
nematic $\rightarrow$ isotropic 276 °C

Analytical Properties: Separation of methyl and nitro derivatives of naphthalene; separation of higher hydrocarbons
Reference 25

Name: N,N'-bis (p-methoxybenzylidene)-α,α'-bi-p-toluidine (BMBT)
Structure:

$$CH_3O - \phi - CH = N - \phi - CH_2 - CH_2 - \phi - N = CH - \phi - OCH_3$$

$$\phi = \langle \bigcirc \rangle$$

(Continued)

Mesogenic Stationary Phases (Continued)

Thermophysical Properties:

solid   → nematic 181 °C

nematic   → isotropic 320 °C

Analytical Properties: Separation of androstane and cholestane alcohols and ketones, good separation of azaheterocyclic compounds column bleed of BMBT can occur during prolonged periods of operation of elevated temperatures

Reference 26

Name: N,N-bis(n-octoxy-benzylidene)-α,α'-bi-p-toluidine (BoBT)
Structure:

$$C_8H_{17}O —\phi— CH{=}N—\phi— CH_2— CH_2—\phi—N{\equiv}CH—\phi— OC_8H_{17}$$

$$\phi = \langle\bigcirc\rangle$$

Thermophysical Properties:

solid   → smectic 118 °C

smectic   → nematic 244 °C

nematic   → isotropic 255 °C

Analytical Properties: Separation of polynuclear aromatic hydrocarbons

Reference 24

Name: N,N-bis(n-pentoxy-benzylidene)-α,α'-bi-p-toluidine (BPeBT)
Structure:

$$C_5H_{11}O —\phi—CH{=}N—\phi— CH_2— CH_2—\phi—N{=}CH—\phi—OC_5H_{11}$$

$$\phi = \langle\bigcirc\rangle$$

Thermophysical Properties:

solid   → smectic 139 °C

smectic   → nematic 208 °C

nematic   → isotropic 283 °C

Analytical Properties: Separation of polynuclear aromatic hydrocarbons

Reference 24

Name: N,N'-bis (p-phenylbenzylidene)-α,α'-bi-p-toluidine (BphBT)
Structure:

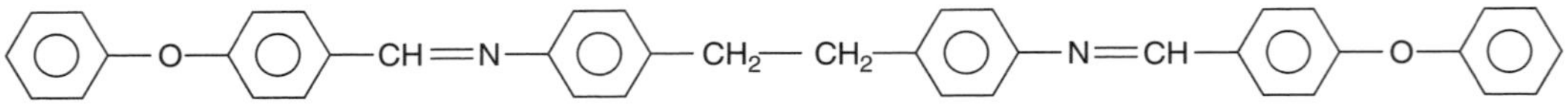

Thermophysical Properties:

solid   → nematic 257 °C

nematic   → isotropic 403 °C

Analytical Properties: Separation of unadulterated steroids; used chromatographically in the temperature range of 260–270 °C.

Reference 27

Mesogenic Stationary Phases (Continued)

Name: N,N′-bis(n-propoxy-benzylidene)-α,α′-bi-p-toluidine (BPrBT)
Structure:

$$C_3H_7O — \phi — CH = N — \phi — CH_2 — CH_2 — \phi — N = CH — \phi — OC_3H_7$$

$$\phi = \langle \bigcirc \rangle$$

Thermophysical Properties:

solid   → smectic 169 °C
smectic   → nematic 176 °C
nematic   → isotropic 311 °C

Analytical Properties: Separation of polynuclear aromatic hydrocarbons
Reference 24

Cholesteric Phases

Name: cholesteryl acetate
Structure:

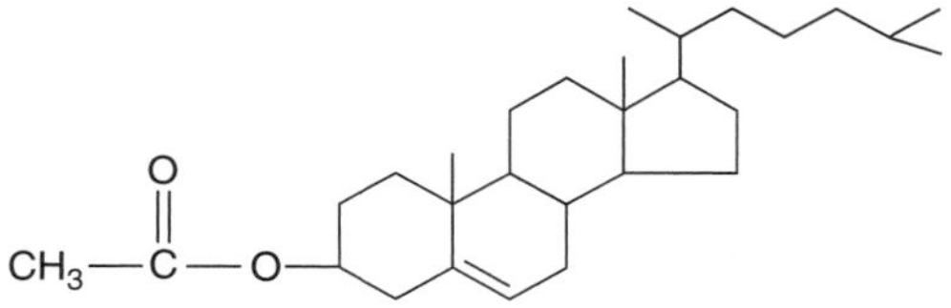

Thermophysical Properties:

solid   → cholesteric 94.5 °C
cholesteric → isotropic 116.5 °C

Analytical Properties: Separation of aromatics and paraffins
Reference 28

Name: (S)-4′-[(2-methyl-1-butoxy)carbonyl] biphenyl-4-yl 4(allyloxy) benzoate
Structure:

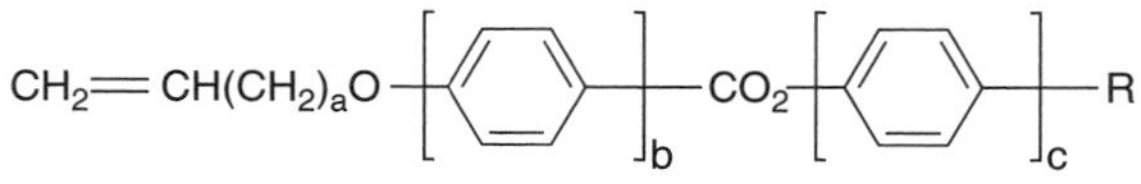

$$CH_2 = CH(CH_2)_aO \left[\bigcirc \right]_b CO_2 \left[\bigcirc \right]_c R$$

$$a = 1$$
$$b = 1$$
$$c = 2$$
$$R = COOCH_2C^*H(CH_3)CH_2CH_3$$

Thermophysical Properties:

solid   → smectic 100 °C
smectic   → cholesteric 150 °C
cholesteric → isotropic 188 °C

Analytical Properties: Suggested for separation of polycyclic aromatic compounds.
Reference 11

(Continued)

Cholesteric Phases (Continued)

Name: (S)-4′-[(2-methyl-1-butoxy)carbonyl] biphenyl-4-yl 4-[4-(allyloxy)phenyl] benzoate

Structure:

$$CH_2 = CH(CH_2)_aO \left[\right]_b CO_2 \left[\right]_c R$$

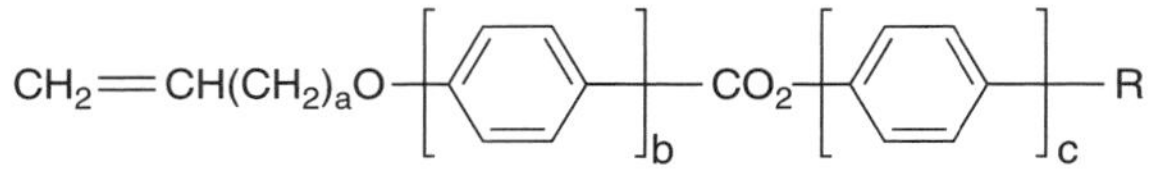

a = 1
b = 2
c = 2
R = COOCH_2C*H(CH_3)CH_2CH_3

Thermophysical Properties:

solid   → smectic 152 °C
smectic   → cholesteric 240 °C
cholesteric   → isotropic 278 °C

Analytical Properties: Suggested for separation of polycyclic aromatic compounds.

Reference 11

Name: (S)-4′-[(2-methyl-1-butoxy)biphenyl-4-yl 4-[4-4-pentenyloxy) phenyl] benzoate

Structure:

$$CH_2 = CH(CH_2)_aO \left[\right]_b CO_2 \left[\right]_c R$$

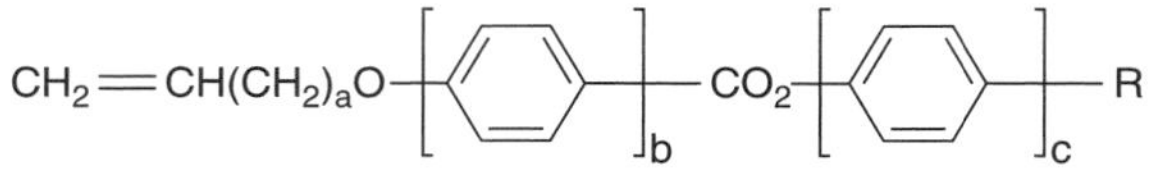

a = 3
b = 2
c = 2
R = COOCH_2C*H(CH_3)CH_2CH_3

Thermophysical Properties:

solid   → smectic 135 °C
smectic   → cholesteric 295 °C
cholesteric   → isotropic 315 °C

Analytical Properties: Suggested for separation of polycyclic aromatic compounds

Reference 11

Name: (S)-4-[(2-methyl-1-butoxy)carbonyl]phenyl 4-[4-(allyloxy)phenyl] benzoate

Structure:

$$CH_2 = CH(CH_2)_aO \left[\right]_b CO_2 \left[\right]_c R$$

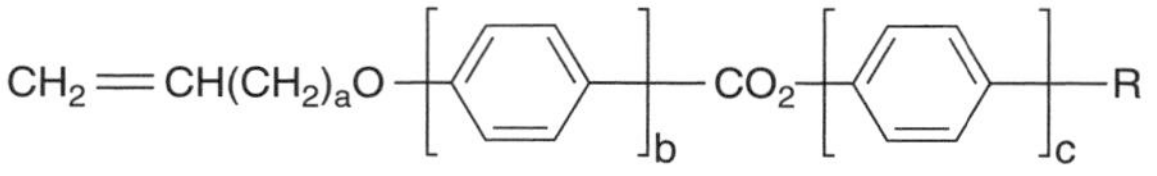

a = 1
b = 2
c = 1
R = COOCH_2C*H(CH_3)CH_2CH_3

Thermophysical Properties:

solid   → smectic 118 °C
smectic   → cholesteric 198 °C
cholesteric   → isotropic 213 °C

Analytical Properties: Suggested for separation of polycyclic aromatic compounds

Reference 11

Cholesteric Phases (Continued)

Name: cholesterol cinnamate
Structure:

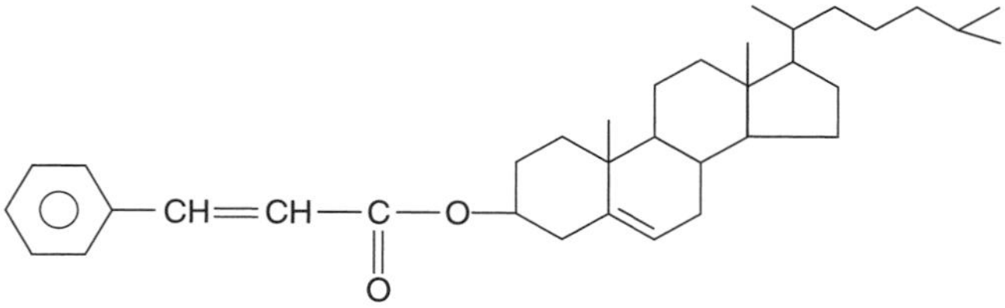

Thermophysical Properties:

solid $\rightarrow$ cholesteric 160 °C

cholesteric $\rightarrow$ isotropic 210 °C

Analytical Properties: Separation of olefinic positional isomers

Reference 12, 29

Name: cholesterol-p-chlorocinnamate (CpCC)
Structure:

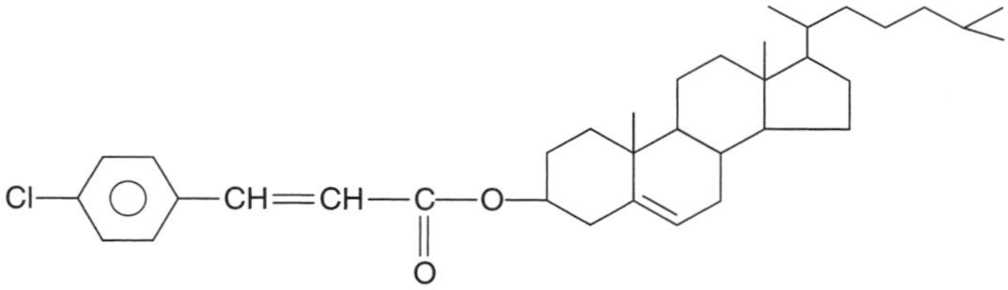

Thermophysical Properties:

solid $\rightarrow$ cholesteric 144 °C

cholesteric $\rightarrow$ isotropic 268 °C

Analytical Properties: Separation of diastereomeric amides and carbamates; the separation of olefinic geometrical isomers is dependent upon the position of the double bond

Reference 29, 30

Name: cholesterol-p-methylcinnamate
Structure:

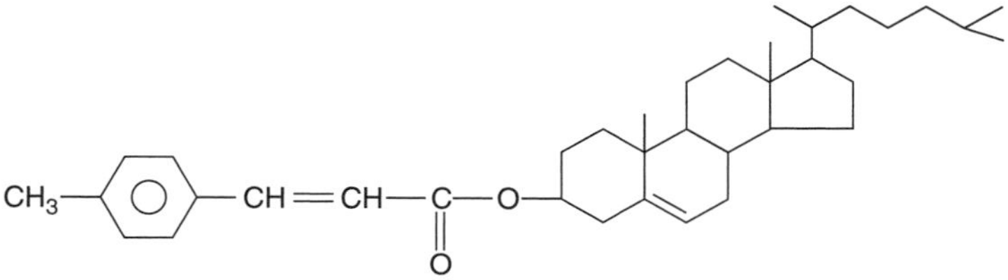

Thermophysical Properties:

solid $\rightarrow$ cholesteric 157 °C

cholesteric $\rightarrow$ isotropic 254 °C

Analytical Properties: Separation of olefinic positional isomers

Reference 29

(Continued)

Cholesteric Phases (Continued)

Name: cholesterol-p-methoxycinnamate
Structure:

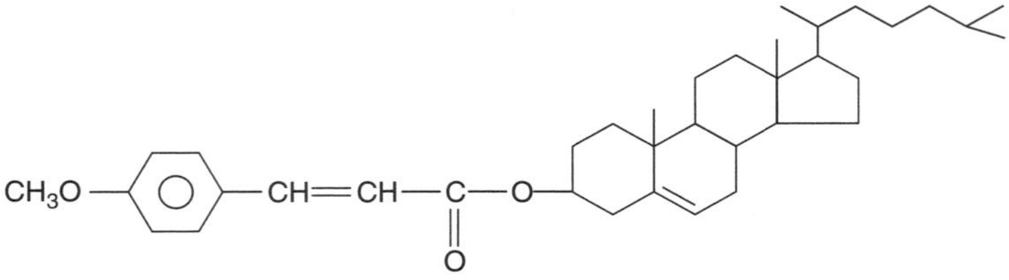

Thermophysical Properties:
solid   → cholesteric 165 °C
cholesteric → isotropic 255 °C

Analytical Properties: Separation of olefinic positional isomers

Reference 29

Name: cholesterol p-nitro cinnamate
Structure:

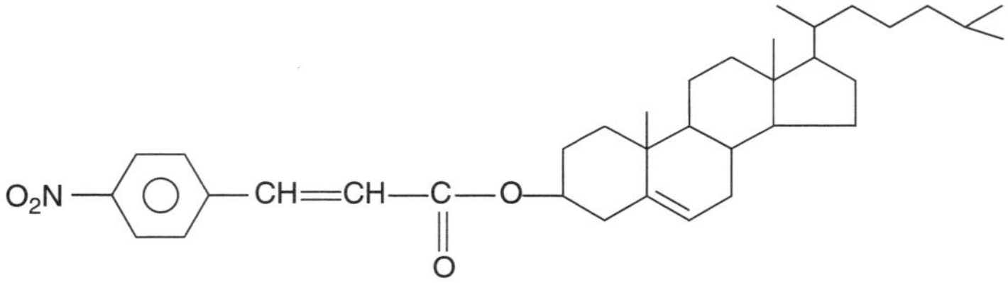

Thermophysical Properties:
solid   → cholesteric 167 °C
cholesteric → isotropic 265 °C

Analytical Properties: Separation of geometrical isomers (2 and 3 octadecene) using p-substituted cholesterols. (Best separation) p-NO$_2$ > p-MeO > cholesterol cinnamate > p-Me > p-Cl (worst separation) for unsaturation occurring within 4 carbon atoms from the terminal methyl the above order holds for separations of tetradecen-1-ol acetates; for unsaturation on carbons 5-12 from the terminal methyl of the tetradecen-1-ol of acetates the best separation is the reverse of the above.

Reference 29

Name: cholesteryl nonanoate
Structure:

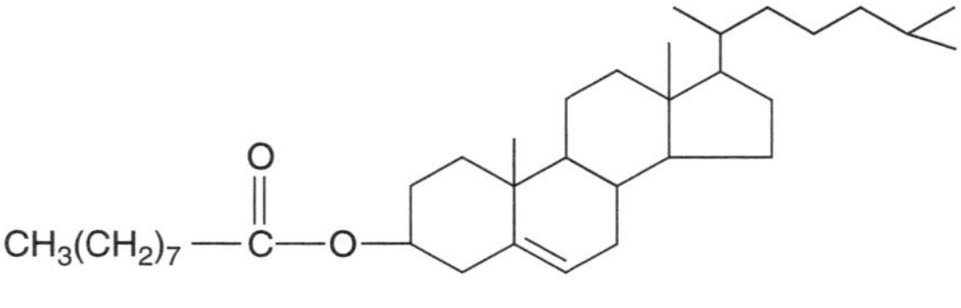

Cholesteric Phases (Continued)

Thermophysical Properties:

solid $\rightarrow$ smectic 77.5 °C

smectic $\rightarrow$ cholesteric 80.5 °C

cholesteric $\rightarrow$ isotropic 92 °C

Analytical Properties: Separation of aromatics and paraffins

Reference 28

Name: cholesteryl valerate

Structure:

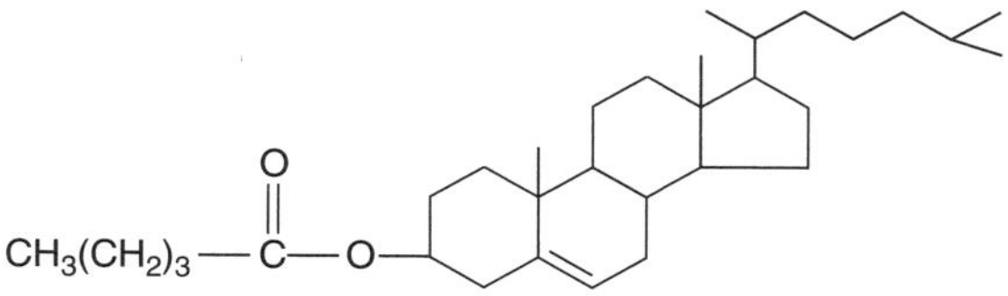

$$CH_3(CH_2)_3 - \overset{O}{\underset{\|}{C}} - O-$$

Thermophysical Properties:

solid $\rightarrow$ cholesteric 93 °C

cholesteric $\rightarrow$ isotropic 101.5 °C

Analytical Properties: Separation of aromatics and paraffins

Reference 28

TRAPPING SORBENTS

The following table provides a listing of the major types of sorbents used in sampling, concentrating, odor profiling, and air and water pollution research [1–6]. These materials are useful in a wide variety of research and control applications. Many can be obtained commercially in different sizes, depending upon the application involved. The purpose of this table is to aid in the choice of a sorbent for a given analysis. Information that is specific for solid phase microextraction (SPME) is provided elsewhere in this chapter.

REFERENCES

1. Borgstedt, H. U., H. W. Emmel, E. Koglin, R. G. Melcher, A. Peters, and J. M. L. Sequaris. *Analytical Problems*. Berlin: Springer-Verlag, 1986.
2. Averill, W., and J. E. Purcell. "Concentration and GC Determination of Organic Compounds from Air and Water." *Gas Chromatography Newsletter* 6, no. 2 (1978): 30.
3. Gallant, R. F., J. W. King, P. L. Levins, and J. F. Piecewicz. Characterization of Sorbent Resins for Use in Environmental Sampling, Report EPA-600/7-78-054, March 1978.
4. Chladek, E., and R. S. Marano. "Use of Bonded Phase Silica Sorbents for the Sampling of Priority Pollutants in Waste Waters." *Journal of Chromatographic Science* 22 (1984): 313.
5. Good, T. J. "Applications of Bonded-Phase Materials." *American Laboratory,* July 1981, 36.
6. Beyermann, K. *Organic Trace Analysis.* New York: Halsted Press (of John Wiley and Sons), 1984.

Trapping Sorbents

Sorbent	Desorption Solvents	Applications
Activated carbon	carbon disulfide, methylene chloride, diethyl ether, diethyl ether with 1 % methanol, diethyl ether with 5 % 2-propanol (caution: CS_2 and CH_3OH can react in the presence of charcoal)	Used for common volatile organics; examples include methylene chloride, vinyl chloride, chlorinated aliphatics, aromatics, acetates; more data is provided in the table entitled "Adsorbents for Gas Chromatography"

Notes: Metallic or salt impurities in the sorbent can sometimes cause the irreversible adsorption of electron-rich oxygen functionalities; examples include 1-butanol, 2-butanone, and 2-ethoxyacetate; recovery rate is often poor for polar compounds.

Sorbent	Desorption Solvents	Applications
Graphitized carbon-black	carbon disulfide, methylene chloride, diethyl ether (or thermal desorption can be used)	Used for common volatile aliphatic and aromatic compounds, organic acids and alcohols, and chlorinated aliphatics; more data is provided in the table entitled "Adsorbents for Gas Chromatography"

Notes: These sorbents are hydrophobic and are not very sensitive to moisture; the possibility of thermal desorption makes them of value for "trace-level" analyses.

Sorbent	Desorption Solvents	Applications
Silica gel	methanol, ethanol, water, diethyl ether	Used for polar compound collection and concentration; examples include alcohols, phenols, chlorophenols, chlorinated aromatics, aliphatic and aromatic amines, nitrogen dioxide; more data is provided in the table entitled "Adsorbents for Gas Chromatography"

Notes: Useful for compounds that can't be recovered from the charcoal sorbents; the most serious problem with silica is the effect of water, which can cause desorption of the analytes of interest, and the heating effect involved can sometimes initiate reactions such as polymerization of the analyte.

Sorbent	Desorption Solvents	Applications
Activated alumina	water, diethyl ether, methanol	Used for polar compounds such as alcohols, glycols, ketones, aldehydes; has also been used for polychlorinated biphenyls and phthalates; more data is provided in the table entitled "Adsorbents for Gas Chromatography"

Notes: Similar in application to silica gel.

Sorbent	Desorption Solvents	Applications
Porous polymers	hexane, diethyl ether, alcohols (thermal desorption also possible in some cases)	Used for a wide range of compounds that include phenols, acidic and basic organics, pesticides, priority pollutants; more data is provided in the table entitled "Porous Polymer Phases"

Notes: The most commonly used porous polymer sorbent is Tenax-GC, although the Porapak and Chromosorb Century series have also been used; Tenax-GC has been used with thermal desorption methods, but can release toluene, benzene, and trichloroethylene residues at the higher temperatures; in addition to Tenax-GC, XAD 2-8, Porapak-N, Chromosorbs 101, 102, 103, and 106 have found applications, sometimes in "stacked" sampling devices (for example, a sorbent column of Tenax-GC and Chromosorb 106 in tandem); Chromosorb 106, a very low polarity polymer, has the lowest retention of water with respect to organic materials, and is well suited for use as a back-up sorbent.

(Continued)

Trapping Sorbents (Continued)

Sorbent	Desorption Solvents	Applications
Bonded phases	methanol, hexane, diethyl ether	Used for specialized applications in pesticides, herbicides, and polynuclear aromatic hydrocarbons

Notes: Most expensive of the common sorbents; useful for the collection of organic samples from water.

Sorbent	Desorption Solvents	Applications
Molecular sieves	carbon disulfide, hexane diethyl ether	Have been used for the collection of aldehydes, alcohols, and for acrolein

Notes: Molecular sieve 13-X is the main molecular sieve to be used as a trapping adsorbent; the sorbents will also retain water.

COOLANTS FOR CRYOTRAPPING

The following table provides fluids (in some cases mixtures) that can be used to chill trapping sorbents (or any cold trap) for the purge and trap sampling of solutes [1–3]. In each case the ratio is mass/mass.

REFERENCES

1. Gordon, A. J., and R. A. Ford. *The Chemists Companion: A Handbook of Practical Data, Techniques and References*, New York: Wiley Interscience, 1972.
2. Bruno, T. J. "Chromatographic Cryofocusing and Cryotrapping with the Vortex Tube." *Journal of Chromatographic Science* 32, no. 3 (1994): 112.
3. Bruno, T. J. "Simple, Quantitative Headspace Analysis by Cryoadsorption on a Short Alumina PLOT Column." *Journal of Chromatographic Science* 47, no. 8 (2009): 1.

Coolant	Temperature, °C
Crushed ice + sodium chloride (3:1)	−21
Crushed ice + calcium chloride (1.2:2)	−39
Vortex tube	−40
Liquid nitrogen + n-butyl amine (slush)	−50
Crushed ice + calcium chloride (1.4:2)	−55
Liquid nitrogen + chloroform (1:1)	−63
Liquid nitrogen + t-butyl amine (slush)	−68
Dry ice	−78
Liquid nitrogen + acetone (slush)	−95
Liquid nitrogen + ethanol (slush)	−120
Liquid nitrogen + methyl cyclohexane (slush)	−126
Liquid nitrogen + n-pentane (slush)	−131
Liquid nitrogen + 1,5-hexadiene (slush)	−141
Liquid nitrogen + isopentane (slush)	−160
Liquid argon	−186
Liquid nitrogen	−196

SORBENTS FOR THE SEPARATION OF VOLATILE INORGANIC SPECIES

The following sorbents have proven useful for the adsorptive separation of volatile inorganic species [1]. This material is used with permission of John Wiley and Sons.

REFERENCE

1. MacDonald, J. C. *Inorganic Chromatographic Analysis: Chemical Analysis Series*, Vol. 78. New York: John Wiley and Sons, 1985.

Separation Material	Typical Separations
Alumina	O_2, N_2, CO_2
Beryllium oxide	H_2S, H_2O, NH_3
Silica gel	O_2/N_2, CO_2, O_3, H_2S, SO_2
Chromium(III) oxide	O_2, N_2, Ar, He
Clay minerals (Attapulgite, Sepiolite)	O_2, N_2, CO, CO_2
Kaolin	He, O_2, N_2, CO, CO_2
Sodium-, lithium fluoride, alumina	MoF_6, SbF_5, UF_6, F
Quartz granules	Ta, Re, Ru, Os, Ir: oxides, hydroxides
Chromosorb 102	Element hydrates
Graphite	NH_3, N_2, H_2
Synthetic diamond	CF_2O, CO_2
Molecular sieve	Hydrogen isotopes
Carbon molecular sieve	O_2, N_2, CO, CO_2, N_2O, SO_2, H_2S
XAD resins	NH_3, SO_2, H_2S, CO, CO_2, H_2O
Porapak Q	GeH_4, SnH_4, AsH_3, SbH_3, $Sn(CH_3)_4$
Porapak QS polymers	H_2S, CH_3SH, $(CH_3)_2S$, $(CH_3)_2S_x$, SO_2
Porapak P	Chlorides of Si, Sn, Ge, P, As, Ti, V, Sb
Teflon	F, MoF_6, SbF_6, SbF_3

Source: From MacDonald, J. C., *Inorganic Chromatographic Analysis: Chemical Analysis Series*, Vol. 78, John Wiley & Sons, New York, 1985. With permission.

ACTIVATED CARBON AS A TRAPPING SORBENT FOR TRACE METALS

Activated carbon, which is a common trapping sorbent for organic species, can also be used for trace metals [1]. This material is typically used by passing the samples through a thin layer (50–150 mg) of the activated carbon that is supported on a filter disk. It can also be used by shaking 50–150 mg of activated carbon in the solution containing the heavy metal, and then filtering the sorbent out of the solution.

REFERENCE

1. Alfasi, Z. B., and C. M. Wai. *Preconcentration Techniques for Trace Elements.* Boca Raton, FL: CRC Press, 1992.

Matrices	Trace Metals	Complexing Agents
Water	Ag, Bi, Cd, Co, Cu, Fe, In, Mg, Mn, Ni, Pb, Zn	(NaOH; pH 7–8)
Water	Ag, As, Ca, Cd, Ce, Co, Cu, Dy, Fe, La, Mg, Mn, Nb, Nd, Ni, Pb, Pr, Sb, Sc, Sn, U, V, Y, Zn	8-quinolinol
Water	Ba, Co, Cs, Eu, Mn, Zn	APDC, DDTC, PAN, 8-quinolinol
Water	Hg, methyl mercury	—
Water	Hg (halide)	—
Water	Hg (halide)	—
Water	U	L-ascorbic acid
HNO_3, water, Al, KCl	Ag, Bi, Cd, Cu, Hg, Pb, Zn	dithizone
Mn, MnO_3, Mn salts	Bi, Cd, Co, Cu, Fe, In, Ni, Pb, Tl, Zn	ethyl xanthate
Co, $Co(NO_3)_2$	Ag, Bi	APDC
Ni, $Ni(NO_3)_2$	Ag, Bi	APDC
Mg, $Mg(NO_3)_2$	Ag, Cu, Fe, Hg, In, Mn, Pb, Zn	(pH 8.1–9)
Al	Cd, Co, Cu, Ni, Pb	thioacetamide
Ag, $TlNO_3$	Bi, Co, CU, Fe, In, Pb	xenol orange
Cr salts	Ag, Bi, Cd, Co, Cu, In, Ni, Pb, Tl, Zn	HAHDTC
Co, In, Pb, Ni, Zn	Ag, Bi, Cu, Tl	DDTC
Se	Cd, Co, Cu, Fe, Ni, Pb, Zn	DDTC
$NaClO_4$	Ag, Bi, Cd, Co, Cu, Fe, Hg, In, Mn, Ni, Pb	(pH 6)

Abbreviations:
APDC: ammonium pyrrolidinecarbodithioate
DDTC: diethyldithiocarbamate
HAHDTC: hexamethyleneammonium hexaethylenedithiocarbamate;
PAN: 1-(2-pyridylazo)-2-naphthol

REAGENT IMPREGNATED RESINS AS TRAPPING
SORBENTS FOR HEAVY METALS

Reagent impregnated resins can be used as trapping sorbents for the preconcentration of heavy metals [1]. These materials can be used in the same way as activated carbons.

REFERENCE

1. Alfasi, Z. B., and C. M. Wai. *Preconcentration Techniques for Trace Elements*. Boca Raton, FL: CRC Press, 1992.

Reagents	Adsorbents	Metals
TBP	porous polystyrene DVB resins	U
TBP	Levextrel (polystyrene DVB resins)	U
DEHPA	Levextrel	Zn
DEHPA	XAD-2	Zn
Alamine 336	XAD-2	U
LIX-63	XAD-2	Co, Cu, Fe, Ni, etc.
LIX-64N, -65N	XAD-2	Cu
Hydroxyoximes	XAD-2	Cu
Kelex 100	XAD-2	Co, Cu, Fe, Ni
Kelex 100	XAD-2,4,7,8,11	Cu
Dithizone, STTA	polystyrene DVB resins	Hg
Dithizone (acetone)	XAD-1,2,4,7,8	Hg, methyl mercury
DMABR	XAD-4	Au
Pyrocatechol violet	XAD-2	In, Pb
TPTZ	XAD-2	Co, Cu, Fe, Ni, Zn

Abbreviations:
  TBP: tributyl phosphate
  DEHPA: di-ethylhexyl phosphoric acid
  STTA: monothiothenolytrifluoroacetone
  DMABR: 5-(4-dimethylaminobenzylidene)-rhodanine
  TPTZ: 2,4,6-tri(2-pyridyl)-1,3,5-triazine
  LIX 63: aliphatic α-hydroxyoxime
  LIX 65N: 2-hydroxy-5-nonylbensophenoneoxime
  LIX 64N: a mixture of LIX 65N with approximately 1 % (vol/vol) of LIX-63

REAGENT IMPREGNATED FOAMS AS TRAPPING
SORBENTS FOR INORGANIC SPECIES

Reagent impregnated foams can be used as trapping sorbents for the preconcentration of heavy metals [1]. These materials can be used in the same way as activated carbons.

REFERENCE

1. Alfasi, Z. B., and C. M. Wai. *Preconcentration Techniques for Trace Elements.* Boca Raton, FL: CRC Press, 1992.

Matrices	Elements	Conc.	Foam Type	Reagents
Water	^{131}I, ^{203}Hg	Traces	Polyether	Alamine 336
Natural water				
Water	Bi, Cd, Co, Cu, Fe, Hg, Ni, Pb, Sn, Zn	Traces	Polyether	Amberlite LA-2
Water	Co, Fe, Mn	Traces to µg/1	Polyether Polyether	PAN
Natural water	Cd	µg/1	Polyether	PAN
Water	Au, Hg	µg/1	Polyether	PAN
Water	Ni	Traces to µg/1	—	DMG, α-benzyldioxime
Water	Cr	µg/1	Polyether	DPC
Water	Hg, methyl-Hg, phenyl-Hg	µg/1	Polyether	DADTC
Natural water	Sn	Traces	Polyether	toluene-3,4-dithiol
Water	Cd, Co, Fe, Ni	Traces	Polyether	Aliquot
Water	Th	Traces	Polyether	PMBP HDEHP-TBP
Water	PO_4^{3-}	Traces		Amine-molybdate-TBP

Abbreviations:
PAN: 1-(2-pyridylazo)-2-naphthol
DMG: dimethylglyoxime
DPC: 1,5-diphenylcarbazide
DADTC: diethylammonium diethyldithiocarbamate
PMBP: 1-phenyl-3-methyl-4-bensoyl-pyrazolone-5
HDEHP: bis-[2-ethylhexyl]phosphate
TBP: tributyl phosphate

CHELATING AGENTS FOR THE ANALYSIS OF INORGANICS BY GAS CHROMATOGRAPHY

The following table provides guidance in choosing a chelating agent for the analysis of inorganic species by gas chromatography [1–3]. The key to the abbreviation list is provided below.

REFERENCES

1. Guiochon, G., and C. Pommier. *Gas Chromatography of Inorganics and Organometallics.* Ann Arbor, MI: Ann Arbor Science Publishers, 1973.
2. Robards, K., E. Patsalides, and S. Dilli. "Review: Gas Chromatography of Metal Beta-Diketonates and Their Analogues." *Journal of Chromatography* 41 (1987): 1–41.
3. Robards, K., and E. Patsalides. "Comparison of the Liquid and Gas Chromatography of Five Classes of Metal Complexes." *Journal of Chromatography A*, 844 (1999): 181–90.

acac =	acetylacetonate
dibm =	2,6-dimethyl-3,5-heptanedionate
fod =	1,1,1,2,2,3,3,-heptafluoro-7,7-dimethyl-4,6-octanedionate
hfa =	hexafluoroacetylacetonate
tacac =	monothioacetylacetonate
tfa =	trifluoroacetylacetonate
thd =	2,2,6,6-tetramethyl-3,5-heptadionate
tpm =	1,1,1-trifluoro-5,5-dimethyl-2,4-hexanedionate

Aluminum

In Mixture With	Complex
Be, Sc	acac
Be	acac
Cr	acac
Be, Cr	acac
Be, Sc	tfa
Be, Rh	tfa
Cr, Rh	tfa
Cr, Rh	tfa
Cu, Fe	tfa
Ga, In	tfa
Fe	tfa
Cr, Rh, Zr	tfa
Be, Ga, In, Tl	tfa
Be, Cr	hfa
Be, Cr, Cu	hfa
Be, Cr, Fe	hfa
Be, Cu, Cr, Fe, Pd, Y	fod
Cr, Fe	tpm
Cr, Fe, Cu	tpm
Be, Cr, Fe, Ni	dibm
Traces on U	tfa
Traces	tfa, hfa

Beryllium

In Mixture With	Complex
Al, Sc	acac
Cu	acac
Al, Cr	acac
Al, Sc	tfa
Al, Ga, Tl, In	tfa
Al, Cr	hfa
Al, Cr, Cu	hfa
Al. Cr, Fe	hfa
Al, Cu, Cr, Fe, Pd, Y	fod
Al, Cr, Fe, Ni	dibm
Traces	tfa

Chromium

In Mixture With	Complex
Al, Be	acac
Al	acac
Al, Rh	tfa
Al, Rh, Zr	tfa
Al, Be	hfa
Al, Be, Cu	hfa
Al, Be, Fe	hfa
Fe, Rh	hfa
Ru	hfa, tpm
Al, Fe	tpm
Al, Fe, Cu	tpm
Al, Be, Cu, Fe, Pd, Y	fod
Al, Be, Fe, Ni	dibm
Traces in Fe	tfa
Traces	tfa, hfa

Cobalt

In Mixture With	Complex
Ru	tfa, hfa
Ni, Pd	tacac
Traces	fod

Copper

In Mixture With	Complex
Be	acac
Al, Fe	tfa
Fe	tfa
Al, Be, Cr	hfa
Fe	hfa
Al, Cr, Fe	tpm
Al, Be, Cr, Fe, Pd, Y	fod

Gallium

In Mixture With	Complex
Al, In	tfa
Al, Be, In, Tl	tfa

Indium

In Mixture With	Complex
Al, Ga	tfa
Al, Be, Ga, Tl	tfa

Iron

In Mixture With	Complex
Al, Cu	tfa
Al	tfa
Cr	tfa
Cu	tfa
Al, Be, Cr	hfa
Cu	hfa
Cr, Rh	hfa
Al, Cr	tpm
Al, Cr, Cu	tpm
Al, Be, Cr, Cu, Pd, Y	fod
Al, Be, Cr, Ni	dibm

Nickel

In Mixture With	Complex
Co, Pd	tacac
Al, Be, Cr, Fe	dibm

Paladium

In Mixture With	Complex
A, Be, Cr, Cu, Fe, Y	fod
Co, Ni	tacac

Rare Earths

In Mixture With	Complex
Sc	thd
Sc, Y	tpm
Sc, Y	fod
Other rare earths	*

Rhodium

In Mixture With	Complex
Al, Cr	tfa
Al, Cr	tfa
Al, Cr, Zr	tfa
Cr, Fe	hfa
Traces	tfa

Ruthenium

In Mixture With	Complex
Co	tfa, hfa
Cr	hfa

Scandium

In Mixture With	Complex
Al, Be	acac
Al, Be	tfa
Rare earths	thd
Rare earths, Y	tpm
Rare earths, Y	fod

Thallium

In Mixture With	Complex
Al, Be, Ga, In	tfa

Thorium

In Mixture With	Complex
U	fod

Uranium

In Mixture With	Complex
Th	fod

Yttrium

In Mixture With	Complex
Sc, rare earths	tpm
Sc, rare earths	fod
Al, Be, Cr, Cu, Fe, Pd	fod

Zirconium

In Mixture With	Complex
Al, Cr, Rh	tfa

BONDED PHASE MODIFIED SILICA SUBSTRATES
FOR SOLID PHASE EXTRACTION

The following table provides the most commonly used bonded phase modified silica substrates used in solid phase extraction, reproduced with permission from Fritz 1999 [1]. Additional information on many of these materials can be found in the table entitled "More Common HPLC Stationary Phases" in the HPLC chapter in this book.

REFERENCE

1. Fritz, J. S. *Solid Phase Extraction*. New York: Wiley-VCH, 1999.

Phase	Polarity of Phase	Designation
Octadecyl, endcapped	Strongly apolar	C18ec
Octadecyl	Strongly apolar	C18
Octyl	Apolar	C8
Ethyl	Slightly polar	C2
Cyclohexyl	Slightly polar	CH
Phenyl	Slightly polar	PH
Cyanopropyl	Polar	CN
Diol	Polar	2OH
Silica gel	Polar	SiOH
Carboxymethyl	Weak cation exchanger	CBA
Aminopropyl	Weak anion exchanger	NH_2
Propylbenzene sulfonic acid	Strong cation exchanger	SCX
Trimethylaminopropyl	Strong anion exchanger	SAX

Source: From Fritz, J. S., *Solid Phase Extraction*, Wiley-VCH, New York, 1999. With permission.

SOLID PHASE MICROEXTRACTION SORBENTS

The following tables provide information on the selection and optimization of solid phase microextraction fibers [1]. The reader is also advised to consult the tables for headspace analysis in this chapter.

REFERENCE

1. Shirey, R. Supelco Corp., Bellefonte, PA, Private Communication, 2009.

FIBER SELECTION CRITERIA

The main fiber selection parameters are polarity and relative molecular mass. This table provides general guidelines on the applicability of available fibers relative to these two parameters. The fibers are characterized by the extraction mechanism, either adsorption or absorption. Adsorbent fibers contain particles suspended in Polydimethylsiloxane (PDMS) or Carbowax.

Fiber	Type of Fiber	Polarity	RMM Range
7 µm PDMS	Absorbent	Nonpolar	150–700
30 µm PDMS	Absorbent	Nonpolar	80–600
85 µm Polyacrylate	Absorbent	Moderately Polar	60–450
100 µm PDMS	Absorbent	Nonpolar	55–400
50 µm Carbowax (PEG)	Adsorbent	Polar	50–400
PDMS-DVB	Adsorbent	Bipolar	50–350
Carbowax-DVB	Adsorbent	Polar	50–350
PDMS-DVB-Carboxen	Adsorbent	Bipolar	40–270
PDMS-Carboxen	Adsorbent	Bipolar	35–180
Carbopak Z-PDMS	Adsorbent	Nonpolar	50–500

Abbreviations:
  PDMS: Polydimethylsiloxane
  DVB: Divinylbenzene (3–5 µm particles)
  PEG: Polyethylene glycol
  Carboxen: Carboxen 1006 (contains micro, meso, and macro tapered pores)
    (3–5 µm particles)
  RMM Range: Relative molecular mass range that is the ideal range for opti-
    mum extraction. Ranges can be extended by varying extraction times, but
    results will not be optimized.

Phase Material Characteristics:

Polydimethylsiloxane (PDMS)

Similar in properties to the OV-1 or SE-30 phases discussed in the tables on silicone liquid phases; nonpolar fluid suitable for nonpolar or slightly polar analytes; thicker coatings extract more analyte, but require longer extraction times.

Polyacrylate

Rigid solid material; moderate polarity; diffusion of analytes through bulk is relatively slow because of rigidity of material; relatively higher desorption temperatures required because of rigidity of material; can be oxidized easily at higher temperatures; must use oxygen-free carrier gas and ensure GC system is leak-free; fibers are very solvent resistant; darkens to a brown color upon exposure to temperatures in excess of 280 °C, but fiber is generally still usable until color becomes black.

Carbowax (polyethylene glycol, PEG)

Similar in properties to the PEG coatings used extensively in chromatography and described elsewhere in this book; moderately polar; highly cross-linked to counteract water solubility; sensitive to attack by oxygen at temperatures in excess of 220 °C, at which point the fiber will darken and become powdery; requires use of high purity carrier gas (typically Heat 99.999 % purity) treated for oxygen contamination.

Divinlybenzene (DVB)

Similar to the properties of divinylbenzene porous polymer phases described elsewhere in this book; higher polarity than carbowax, and when combined with carbowax results in a more polar phase; like polyacrylate, it is a solid particle that must be carried in a liquid to coat on a fiber.

Carboxen

Similar to the material used in carboxen porous layer open tubular (PLOT) columns; structure has an approximately even distribution of macro-, meso-, and micropores, making it valuable for smaller analytes; larger analytes can show hysteresis that must be addressed by desorption at 280 °C.

EXTRACTION CAPABILITY OF SOLID PHASE MICROEXTRACTION SORBENTS

This table shows the extraction capability of the fibers for acetone, a small moderately polar analyte, for 4-nitrophenol, a medium size polar analyte, and benzo(GHI)perylene, a large nonpolar analyte. This provides a general guideline for fiber selection.

Fiber	Approx. Linear Conc. Range Acetone 10 min Ext[a] (FID)	Approx. Linear Conc. Range 4- Nitrophenol 20 min Ext[b] (GC/MS)	Approx. Linear Conc. Range Benzo perylene 20 min Ext (GHI)
7 μm PDMS	100 ppm and up	Not extracted	100 ppt to 500 ppb
30 μm PDMS	10 ppm and up	10 ppm and up	100 ppt to 10 ppm
85 μm Polyacrylate	1 ppm to 1000 ppm	5 ppb to 100 ppm	500 ppt to 10 ppm
100 μm PDMS	500 ppb to 1000 ppm	500 ppb to 500 ppm	500 ppt to 10 ppm
50 μm Carbowax (PEG)	1 ppm to 1000 ppm	5 ppb to 50 ppm	25 ppb to 10 ppm
PDMS-DVB	50 ppb to 100 ppm	25 ppb to 10 ppm	10 ppb to 1 ppm
Carbowax-DVB	100 ppb to 100 ppm	5 ppb to 10 ppm	50 ppb to 5 ppm
PDMS-DVB-Carboxen	25 ppb to 10 ppm	50 ppb to 10 ppm	100 ppb to 1 ppm poorly desorbed
PDMS-Carboxen	5 ppb to 5 ppm	100 ppb to 10 ppm	Not desorbed
Carbopak Z-PDMS	10 ppm to 500 ppm	5 ppm to 100 ppm	500 ppt to 100 ppb

Note: In each case, the concentration is expressed on a mass basis (e.g., ppm mass/mass).
[a] Water sample contains 25 % NaCl (mass/mass).
[b] Water sample contains 25 % NaCl (mass/mass) acidified to pH = 2 with 0.05 M phosphoric acid.
1 ppm = 1 part in 1×10^{6}
1 ppb = 1 part in 1×10^{9}
1 ppt = 1 part in 1×10^{12}

Typical Phase Volumes of SPME Fiber Coatings

Fiber Coating Thickness/ Type	Type of Fiber Core	Fiber Core Diameter (mm)	Phase volume (m^3 or μl)
100 μm PDMS	fused silica	0.110	0.612
100 μm PDMS	metal	0.130	0.598
30 μm PDMS	fused silica	0.110	0.132
30 μm PDMS	metal	0.130	0.136
7 μm PDMS	fused silica	0.110	0.028
7 μm PDMS	metal	0.130	0.030
85 μm PA	fused silica	0.110	0.543
60 μm PEG	metal	0.130	0.358
15 μm Carbopack Z/PDMS	metal	0.130	0.068
65 μm PDMS/DVB	fused silica	0.120	0.418
65 μm PDMS/DVB	proprietary	0.130	0.440
65 μm PDMS/DVB	metal	0.130	0.440
75 μm Carboxen-PDMS	fused silica	0.120	0.502
85 μm Carboxen-PDMS	proprietary	0.130	0.528
85 μm Carboxen-PDMS	metal	0.130	0.528
50/30μm DVB/Carboxen	metal		
Carboxen layer		0.130	0.151
DVB layer		0.190	0.377
50/30 μm DVB/Carboxen	metal		
Carboxen layer		0.130	0.151
DVB layer		0.190	0.377
60 μm PDMS-DVB HPLC	proprietary	0.160	0.459

SALTING OUT REAGENTS FOR HEADSPACE ANALYSIS

The following table provides data on the common salts used for salting out in chromatographic headspace analysis, as applied to direct injection methods and to solid phase microextraction [1,2]. Data are provided for the most commonly available salts, although others are possible. Sodium citrate, for example, occurs as the dihydrate and the pentahydrate. The pentahydrate is not as stable as the dihydrate, however, and dries out on exposure to air, forming cakes. Potassium carbonate occurs as the dihydrate, trihydrate and sesquihydrate, however data is provided only for the anhydrous material. The solubility is provided as the number of grams that can dissolve in 100 mL of water at the indicated temperature. The vapor enhancement cited is the degree of increase of the concentration of vapor over the solution of a 2 % (mass/mass) ethanol solution in water at 60 °C [3,4].

REFERENCES

1. Lide, D. R., ed. *CRC Handbook for Chemistry and Physics*. 90th ed. Boca Raton, FL: CRC Press, 2010.
2. NIST Chemistry Web Book, (www.webbook.nist.gov/chemistry/), 2009.
3. Machata, G. "Determination of Alcohol in Blood by Gas Chromatography." *Clinical Chemistry Newsletter* 4, no. 2 (1972): 29.
4. Ioffe, B. V., and A. G. Vitenberg. *Head Space Analysis and Related Methods in Gas Chromatography*. New York: Wiley Interscience, 1983.

Salt	Formula	Rel. Mol. Mass	Density	Solubility		Vapor Enhancement
				Cold Water	Hot Water	
Potassium carbonate	K_2CO_3	138.21	2.428 at 14 °C	112[a]	156[b]	8
Ammonium sulfate	$(NH_4)_2SO_4$	132.13	1.769 at 50 °C	70.6[c]	103.8[b]	5
Sodium citrate (dihydrate)	$Na_3C_6H_5O_7 \cdot 2H_2O$	294.10		72[d]	167[b]	5
Sodium chloride	$NaCl$	58.44	2.165[e]	37.5[a]	39.12[b]	3
Ammonium chloride	NH_4Cl	53.49	1.527	29.7[c]	75.8[b]	2

[a] 20 °C
[b] 100 °C
[c] 0 °C
[d] 25 °C
[e] specific gravity, 25/4 °C

PARTITION COEFFICIENTS OF COMMON FLUIDS IN AIR-WATER SYSTEMS

The following table provides the partition coefficients (or distribution coefficients), $K = C_s/C_v$, (solid/vapor) at various temperatures, for application in gas chromatographic headspace analysis [1,2]. The values marked with an asterisk were determined from a linear regression of experimental data.

REFERENCES

1. Ioffe, B. V., and A. G. Vitenberg. *Head Space Analysis and Related Methods in Gas Chromatography.* New York: Wiley Interscience, 1983.
2. Kolb, B., and L. S. Ettre. *Static Headspace Gas Chromatography: Theory and Practice.* New York: Wiley-VCH, 1996.

Partition Coefficient, K

Fluid	20 °C	25 °C	30 °C	40 °C	50 °C	60 °C
Cyclohexane				0.077	0.055*	0.040
n-Hexane				0.14	0.068*	0.043
Tetrachloroethylene				1.48	1.28*	1.27
1,1,1-Trichloromethane				1.65	1.53*	1.47
o-Xylene				2.44	1.79*	1.31
Toluene	4.6	3.6	2.9	2.82	2.23*	1.77
Benzene	4.8	4.0	3.4	2.90	3.18*	2.27
Dichloromethane				5.65	4.29*	3.31
n-Butylacetate	126	87	59	31.4	20.6*	13.6
Ethylacetate	210	150	108	62.4	42.7*	29.3
Methyl ethylketone	600	380	283	139.5	109*	68.8
n-Butanol	4660	3600	2710	647	384*	238
Ethanol	7020	5260	4440	1355	820*	511
1,4-Dioxane	8000	5750	4330	1618	1002*	624
m-Xylene	5.9	4.0	3.9			
1-Propanol	5480	4090	3210		479*	
Acetone	752	551	484			

VAPOR PRESSURE AND DENSITY OF SATURATED WATER VAPOR

The following table provides the temperature dependence of the saturated vapor pressure and vapor density of water. This information is useful in gas chromatographic headspace analysis and for SPME sampling [1,2].

REFERENCES

1. Kolb, B., and L. S. Ettre. *Static Headspace Gas Chromatography: Theory and Practice*. New York: Wiley-VCH, 1997.
2. Lide, D. R., ed. *CRC Handbook for Chemistry and Physics*. 90th ed. Boca Raton, FL: CRC Press, 2010.

°C	p° (kPa)	p° (torr)	d (µg/mL)
10	1.2	9.2	9.4
20	2.3	17.5	17.3
30	4.2	31.8	30.3
40	7.4	55.3	51.1
50	12.3	92.5	83.2
60	19.9	149.4	130.5
70	31.1	233.7	198.4
80	47.2	355.1	293.8
90	69.9	525.8	424.1
100	101.1	760.0	598.0
110	142.9	1074.5	826.5
120	198.1	1489.1	1122.0

SOLVENTS FOR SAMPLE PREPARATION FOR
GAS CHROMATOGRAPHIC ANALYSIS

Many different solvents are used to prepare samples for analysis by gas chromatography, and it would be impossible to list all of them in one place. In this table, the most common solvents are provided, along with relevant properties [1–5]. The solubility parameter and the dielectric constant for each solvent is used to choose the best match for the solutes present in the sample, based on polarity considerations [6,7]. Unless otherwise indicated, the dielectric constant is provided at 20 °C. The solvent viscosity is most commonly used to optimize the operation of automatic samplers. A delay must be programmed in the filling sequence for highly viscous solvents. Unless otherwise indicated, the viscosity is provided at 20 °C. The normal boiling temperature is provided to guide the selection of injector temperatures and pressure programs (along with sample decomposition considerations). It is also used in optimization of automatic sampler programs, since highly volatile solvents can form bubbles in the syringe barrel if repeat pumps are not programmed. The recommended starting oven temperatures are provided to guide the development of temperature programs for splitless injector operations, to take advantage of solvent focusing at the head of the column. Temperature ranges marked with an asterisk indicate that subambient starting temperature will be needed, most easily obtained with a vortex tube or a cryoblast valve installed on the gas chromatograph.

REFERENCES

1. Willard, H. H., L. L. Merritt, J. A. Dean, and F. A. Settle. *Instrumental Methods of Analysis.* 6th ed. Belmont: Wadsworth Publishing Co., 1981.
2. Dreisbach, R. R. *Physical Properties of Chemical Compounds, Number 22 of the Advances in Chemistry Series.* Washington, DC: American Chemical Society, 1959.
3. Krstulovic, A. M., and P. R. Brown. *Reverse Phase High Performance Liquid Chromatography.* New York: John Wiley and Sons (Interscience), 1982.
4. Lide, D. R., ed. *CRC Handbook for Chemistry and Physics.* 90th ed. Boca Raton, FL: CRC Press, 2010.
5. Poole, C. F. *The Essence of Chromatography.* Amsterdam: Elsevier, 2003.
6. Hoy, K. L. "Tables of Experimental Dipole Moments." *Journal of Paint Technology* 42 (1970): 541.
7. Barton, A. F. M. *Handbook of Solubility Parameters and Other Cohesion Parameters.* 2nd ed. Boca Raton, FL: CRC Press, 1991.

Solvent	Solubility Parameter, δ	Viscosity mPa•s (20 °C)	Dielectric Constant (20 °C)	Normal Boiling Temperature (°C)	Recommended Starting Oven Temperature (°C)
Acetone	9.62	0.30(25)	20.7(25)	56.3	35–45*
Acetonitrile	12.11	0.34(25)	37.5	81.6	55–65
Benzene	9.16	0.65	2.284	80.1	55–65
1-Butanol	11.60	2.95	17.8	117.7	85–100
2-Butanol	11.08	4.21	15.8(25)	99.6	75–85
n-Butyl acetate	8.69	0.73		126.1	100–115
n-Butyl chloride	8.37	0.47(15)		78.4	55–65
Carbon tetrachloride	8.55	0.97	2.238	76.8	55–65
Chlorobenzene	9.67	0.80	2.708	131.7	100–120
Chloroform	9.16	0.58	4.806	61.2	30–50*
Cyclohexane	8.19	0.98	2.023	80.7	55–65
Cyclopentane	8.10	0.44	1.965	49.3	20–35*
o-Dichlorobenzene	10.04	1.32(25)	9.93(25)	180.5	155–165
N,N-Dimethylacetamide		2.14	37.8	166.1	135–145
Dimethylformamide	11.79	0.92	36.7	153.0	125–140
Dimethyl sulfoxide	12.8	2.20	4.7	189.0	165–175
1,4-Dioxane	10.13	1.44(15)	2.209(25)	101.3	75–85
2-Ethoxyethanol		2.05		135.6	100–120
Ethyl acetate	8.91	0.46	6.02(25)	77.1	50–65
Diethyl ether	7.53	0.24	4.335	34.6	10–25
Glyme (ethylene glycol dimethyl ether)		0.46(25)		93.0	65–75
n-Heptane	7.50	0.42	1.92	98.4	60–80
n-Hexadecane		3.34		287.0	250–270
n-Hexane	7.27	0.31	1.890	68.7	40–60
Isobutyl alcohol	11.24	4.70(15)	15.8(25)	107.7	70–90
Methanol	14.50	0.55	32.63(25)	64.7	35–55
2-Methoxyethanol	11.68	1.72	16.9	124.6	95–110
2-Methoxyethyl acetate				144.5	120–135
Methylene chloride	9.88	0.45(15)	9.08	39.8	10–35
Methylethylketone	9.45	0.42(15)	18.5	79.6	50–70
n-Nonane	7.64	0.72	1.972	150.8	125–140
n-Pentane	7.02	0.24	1.84	36.1	10–25*
Petroleum ether		0.30		30–60	10–30*
1-Propanol	12.18	2.26	20.1(25)	97.2	70–85
2-Propanol	11.44	2.86(15)	18.3(25)	82.3	55–70
Pyridine	10.62	0.95	12.3(25)	115.3	95–110
Tetrachloroethylene	9.3	0.93(15)		121.2	90–110
Tetrahydrofuran	9.1	0.55	7.6	66.0	35–55*
Toluene	8.93	0.59	2.379(25)	110.6	80–100
Trichloroethylene	9.16	0.57	3.4(16)	87.2	60–75
1,2,2-Trichloro-1,1,2-trifluoroethane		0.71		47.6	25–35*
2,2,4-Trimethylpentane	6.86	0.50	1.94	99.2	70–85
o-Xylene	9.06	0.81	2.568	144.4	120–135
p-Xylene			2.270	138.5	120–135

DERIVATIZING REAGENTS FOR GAS CHROMATOGRAPHY

The following table lists some of the more common derivatizing reagents used in gas chromatography for the purposes of (1) increasing sample volatility, (2) increasing sample thermal stability, (3) reducing sample-support interactions, and (4) increasing sensitivity toward a particular detector. The table is divided into reagents for acylation, alkylation, esterification, pentafluorophe-nylation, and silylation. The conditions and concentrations used in derivatization must be carefully considered, since one can often cause more problems than one cures using these methods. Such problems include poor peak resolution, incomplete reactions and side products, and less than stoichiometric yields of products. The reader is referred to the citation list for more detail on the reagents, conditions, and difficulties.

REFERENCES

General References

1. Blau, K., and G. S. King, eds. *Handbook of Derivatives for Chromatography*. London: Heyden, 1978.
2. Knapp, D. R. *Handbook of Analytical Derivatization Reactions*. New York: John Wiley and Sons, 1979.
3. Drozd, J. *Chemical Derivatization in Gas Chromatography*. Amsterdam: Elsevier, 1981.
4. Poole, C. F., and S. A. Schutte. *Contemporary Practice of Chromatography*. Amsterdam: Elsevier, 1984.
5. Grob, R. L. *Modern Practice of Gas Chromatography*. New York: John Wiley and Sons, 1985.
6. Braithwaite, A., and F. J. Smith. *Chromatographic Methods*. London: Chapman and Hall, 1985.
7. Merritt, C. In *Ancillary Techniques of Gas Chromatography*. Edited by L. S. Ettre and W. H. McFadder. New York: Wiley Interscience, 1969.
8. Hammarstrand, K., and E. J. Bonelli. *Derivative Formation in Gas Chromatography*. Walnut Creek, CA: Varian Aerograph, 1968.
9. Vanden Heuvel, W. J. A. *Gas Chromatography of Steroids in Biological Fluids*. New York: Plenum Press, 1965.

Acylating Reagents

1. Brooks, C. J. W., and E. C. Horning. "Gas Chromatographic Studies of Catecholamines, Tryptamines, and Other Biological Amines." *Analytical Chemistry* 36, no. 8 (1964): 1540.
2. Imai, K., M. Sugiura, and Z. Tamura. "Catecholamines in Rat Tissues and Serum Determined by Gas Chromatographic Method." *Chemical & Pharmaceutical Bulletin* 19 (1971): 409–11.
3. Scoggins, M. W., L. Skurcenski, and D. S. Weinberg. "Gas Chromatographic Analysis of Geometric Diamine Isomers as Tetramethyl Derivatives." *Journal of Chromatographic Science* 10 (1972): 678.

Esterification Reagents

1. Shulgin, A. T. "Separation and Analysis of Methylated Phenols as Their Trifluoroacetate Ester Derivatives." *Analytical Chemistry* 36, no. 4 (1964): 920.
2. Argauer, R. J. "Rapid Procedure for the Chloroacetylation of Microgram Quantities of Phenols and Detection by Electron: Capture Gas Chromatography." *Analytical Chemistry* 40, no. 1 (1968): 122.
3. Vanden Heuvel, W. J. A., W. L. Gardiner, and E. C. Horning. "Characterization and Separation of Amines by Gas Chromatography." *Analytical Chemistry* 36, no. 8 (1964): 1550.
4. Änggård, E., and S. Göran. "Gas Chromatography of Catecholamine Metabolites Using Electron Capture Detection and Mass Spectrometry." *Analytical Chemistry* 41, no. 10 (1969): 1250.
5. Alley, C. C., J. B. Brooks, and G. Choudhary. "Electron Capture Gas-Liquid Chromatography of Short Chain Acids as Their 2,2,2-Trichloroethyl Esters." *Analytical Chemistry* 48, no. 2 (1976): 387.
6. Godse, D. D., J. J. Warsh, and H. C. Stancer. "Analysis of Acidic Monoamine Metabolites by Gas Chromatography-Mass Spectrometry." *Analytical Chemistry* 49, no. 7 (1977): 915.

7. Matin, S. B., and M. Rowland. "Electron-Capture Sensitivity Comparison of Various Derivatives of Primary and Secondary Amines." *Journal of Pharmaceutical Sciences* 61, no. 8 (1972): 1235.
8. Bertani, L. M., S. W. Dziedzic, D. D. Clarke, and S. E. Gitlow. "A Gas-Liquid Chromatographic Method for the Separation and Quantitation of Nomethanephrine and Methanephrine in Human Urine." *Clinica Chimica Acta* 30, no. 2 (1970): 227–33.
9. Kawai, S., and Z. Tamura. "Gas Chromatography of Catecholamines as Their Trifluoroacetates." *Chemical & Pharmaceutical Bulletin* 16, no. 4 (1968): 699.
10. Moffat, A. C., and E. C. Horning. "A New Derivative for the Gas-Liquid Chromatography of Picogram Quantities of Primary Amines of the Catecholamine Series." *Biochimica et Biophysica Acta* 222, no. 1 (1970): 248–50.
11. Lamparski, L. I., and T. J. Nestrick. "Determination of Trace Phenols in Water by Gas Chromatographic Analysis of Heptafluorobutyl Derivatives." *Journal of Chromatography* 156 (1978): 143.
12. Mierzwa, S., and S. Witek. "Gas-Liquid Chromatographic Method with Electron-Capture Detection for the Determination of Residues of Some Phenoxyacetic Acid Herbicides in Water as Their 2,2-Trichloro-ethyl Esters." *Journal of Chromatography* 136 (1977): 105.
13. Hoshika, Y. "Gas Chromatographic Separation of Lower Aliphatic Primary Amines as Their Sulphur-Containing Schiff Bases Using a Glass Capillary Column." *Journal of Chromatography* 136 (1977): 253.
14. Brooks, J. B., C. C. Alley, and J. A. Liddle. "Simultaneous Esterification of Carboxyl and Hydroxyl Groups with Alcohols and Heptafluorobutyric Anhydride for Analysis by Gas Chromatography." *Analytical Chemistry* 46, no. 13 (1974): 1930.
15. Deyrup, C. L., S. M. Chang, R. A. Weintraub, and H. A. Moye. "Simultaneous Esterification and Acylation of Pesticides for Analysis by Gas Chromatography. 1. Derivatization of Glyphosate and (Aminomethyl) Phasphonic Acid with Fluorinated Alcohols-Perfluoronated Anhydrides." *Journal of Agricultural and Food Chemistry* 33, no. 5 (1985): 944.
16. Samar, A. M., J. L. Andrieu, A. Bacconin, J. C. Fugier, H. Herilier, and G. Faucon. "Assay of Lipids in Dog Myocardium Using Capillary Gas Chromatography and Derivatization with Boron Trifluoride and Methanols." *Journal of Chromatography* 339, no. 1 (1985): 25–34.

Pentafluoro Benzoyl Reagents

1. Mosier, A. R., C. E. Andre, and F. G. Viets, Jr. "Identification of Aliphatic Amines Volatilized From Cattle Feedyard." *Environmental Science & Technology* 7, no. 7 (1973): 642.
2. DeBeer, J., C. Van Petegham, and Al. Heyndridex. "Electron Capture-Gas-Liquid Chromatography (EC-GLC) Determination of the Herbicidal Monohalogenated Phenoxyalkyl Acid Mecoprop in Tissues, Urine and Plasma After Derivatization with Pentafluorobenzylbromide." *Veterinary & Human Toxicology* 21 (1979): 172.
3. Davis, B. "Crown Ether Catalyzed Derivatization of Carboxylic Acids and Phenols with Pentafluorobenzyl Bromide for Electron Capture Gas Chromatography." *Analytical Chemistry* 49, no. 6 (1977): 832.
4. Avery, M. J., and G. A. Junk. "Gas Chromatography/Mass Spectrometry Determination of Water-Soluble Primary Amines as Their Pentafluorobenzaldehyde Imines." *Analytical Chemistry* 57, no. 4 (1985): 790.

Silyating Reagents

1. Metcalfe, L. D., and R. J. Martin. "Gas Chromatography of Positional Isomers of Long Chain Amines and Related Compounds." *Analytical Chemistry* 44, no. 2 (1972): 403.
2. Sen, H. P., and P. L. McGeer. "Gas Chromatography of Phenolic and Catecholic Amines as the Trimethylsilyl Ethers." *Biochemical and Biophysical Research Communications* 13, no. 5 (1963): 390.
3. Fogelgvist, E., B. Josefsson, and C. Roos. "Determination of Carboxylic Acids and Phenols in Water by Extractive Alkylation Using Pentafluorobenzylation, Glass Capillary G.C. and Electron Capture Detection." *Journal of High Resolution Chromatography & Chromatography Communications* 3 (1980): 568.
4. Poole, C. F. C., W. F. Sye, S. Singhawangcha, F. Hsu, A. Zlatkis, A. Arfwidsson, and J. Vessman. "New Electron-Capturing Pentafluorophenyldialkylchlorosilanes as Versatile Derivatizing Reagents for Gas Chromatography." *Journal of Chromatography* 199 (1980): 123.

5. Quilliam, M. A., K. K. Ogilvie, K. L. Sadana, and J. B. Westmore. "Study of Rearrangement Reactions Occurring During Gas Chromatography of Tert-butyl-dimethylsilyl Ether Derivatives of Uridine." *Journal of Chromatography* 194 (1980): 379.
6. Poole, C. F., and A. Zlatkis. "Trialkylsilyl Ether Derivatives (Other Than TMS) for Gas Chromatography and Mass Spectrometry." *Journal of Chromatographic Science* 17, no. 3 (1979): 115.
7. Francis, A. J., E. D. Morgan, and C. F. Poole. "Flophemesyl Derivatives of Alcohols, Phenols, Amines and Carboxylic Acids and Their Use in Gas Chromatography with Electron-Capture Detection." *Journal of Chromatography* 161 (1978): 111.
8. Harvey, D. J. "Comparison of Fourteen Substituted Silyl Derivatives for the Characterization of Alcohols, Steroids and Cannabinoids by Combined Gas-Liquid Chromatography and Mass Spectrometry." *Journal of Chromatography* 147 (1978): 291.
9. Quilliam, M. A., and J. M. Yaraskavitch. "Tertbutyldiphenylsilyl Derivatization for Liquid Chromatography and Mass Spectrometry." *Journal of Liquid Chromatography & Related Technologies* 8, no. 3 (1985): 449.

Derivatizing Reagents for Gas Chromatography

Acylating Reagents

Derivatizing Reagent	Structure/Formula	Notes
Acetic anhydride	$(CH_3CO)_2O$	Used for amino acids, steroids, urinary sugars, pesticides and herbicides, and narcotics
Chloracetic anhydride	$(CH_2ClCO)_2O$	Useful for electron capture detection of lower aliphatic primary amines
2,4'-Dibromoaceto-phenone	$BrCH_2\text{—}C\text{=}O$	Used for short and medium chain aliphatic carboxylic acids
Heptafluorobutyric anhydride	$(CF_3CF_2CF_2CO)_2O$	Used in basic solution for alcohols, amines, nitrosamines, amino acids, and steroids; heptafluorobutylimidazole is used in a similar fashion in the analysis of phenols
Pentafluorobenz-aldehyde		Useful for electron capture detection of several primary amines
Pentafluorobenzoyl chloride		Useful for electron capture detection of several primary amines
Pentafluoropropionic anhydride	$(CF_3CF_2CO)_2O$	Used for aromatic monoamines and their metabolites
Propionic anhydride	$(CH_3CH_2CO)_2O$	Used for amines, amino acids, narcotics
Pivalic anhydride	$[(CH_3)_3CCO]_2O$	Used for hormone analysis

Derivatizing Reagents for Gas Chromatography (Continued)

Acylating Reagents

Derivatizing Reagent	Structure/Formula	Notes
2-Thiophene aldehyde		Used for electron capture detection of lower aliphatic primary amines
Trifluoroacetic anhydride	$(CF_3CO)_2O$	Used for phenols, amines, amino acids, amino phosphoric acids, saccharides, vitamins
N-Trifluoroacetyl-imidazole		Useful for the relatively straightforward acylation of hydroxyl groups, secondary or tertiary amines
Diazomethane	$CH_2 = N = N$ + -	Used as a common alkylating agent; acts on acidic and enolic groups rapidly, and more slowly on other groups with replaceable hydrogens (the use of a Lewis acid catalyst such as BF_3 is sometimes helpful). All diazoalkanes are toxic and sometimes explosive, and are used in microscale operations only.
Trimethylanilinium hydroxide (TMAH) (in methanol)		Useful for methylation of amines
Pentafluorobenzyl bromide		Useful for the derivatization of acids, amides and phenols, providing great increase in sensitivity toward electron capture detection

Esterification Reagents

Derivatizing Reagent	Structure/Formula	Notes
Boron trifluoride + methanol	$BF_3 + CH_3OH$	Useful for carboxylic acids (aromatic and aliphatic), fatty acids, fatty acid esters, Krebs cycle acids
Boron trifluoride + n-propanol	$BF_3 + CH_3(CH_2)_2OH$	Useful for fatty acid, lactic acid, and succinic acid
N,N-Dimethyl-formamide dimethyl acetal		Useful in the formation of fatty acid esters, and for N-protected amino acids, sulfonamides, barbiturates
2-Bromopropane	$(CH_3)_2CHBr$	Used for amino acids and amides
n-Butanol	$CH_3(CH_2)_3OH$	Used for carboxylic acids and amino acids
Hydrogen chloride + methanol	$HCl + CH_3OH$	Useful for carboxylic acids, branched chain fatty acids, oxalic acid, amino acids, lipids; HCl serves as a catalytic agent

(Continued)

Derivatizing Reagents for Gas Chromatography (Continued)

Esterification Reagents

Derivatizing Reagent	Structure/Formula	Notes
Sodium methoxide	CH_3ONa in CH_3OH	Used for the transesterification of lipids
Sulfuric acid + methanol	$H_2SO_4 + CH_3OH$	Useful for carboxylic and fatty acids
Tetramethyl ammonium hydroxide	$(CH_3)_4NOH$ in CH_3OH	Useful for carboxylic acids, fatty acids, alkyd and polyester resins
Thionyl chloride	$SOCl_2$	Useful in the formation of esters of carboxylic acids and other acidic functional groups
2,2,2-Trichloroethanol	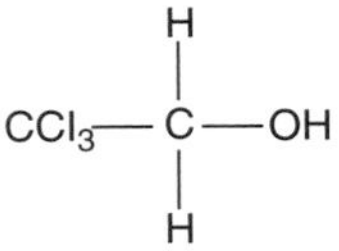	Useful in the esterification of short chain acids following electron capture detection; sometimes used with trifluoroacetic anhydride in the presence of H_2SO_4
Triethyl orthoformate	$HC(OC_2H_5)_3$	Used for aminophosphoric acids
Trimethylphenyl-ammonium hydroxide	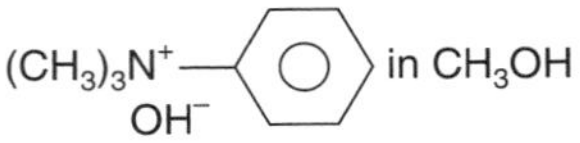 in CH_3OH	Used for fatty acids, aromatic acids, herbicides, pesticides

Pentafluorophenyl Reagents

α-Bromopentafluoro-toluene	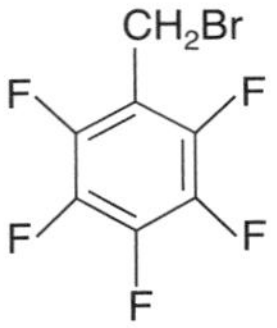	Used to etherify sterols and phenols, in diethyl ether with the presence of potassium t-butanolate
Pentafluorobenz-aldehyde	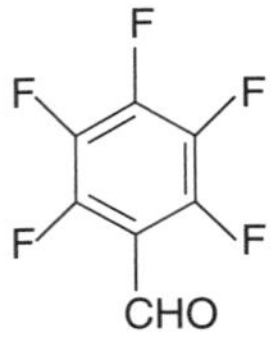	Used in derivatizing primary amines; greatly enhances electron capture detector response (to the picogram level)
Pentafluorobenzyl alcohol	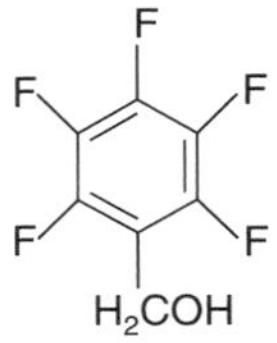	Used in derivatizing carboxylic acids
Pentafluorobenzyl bromide	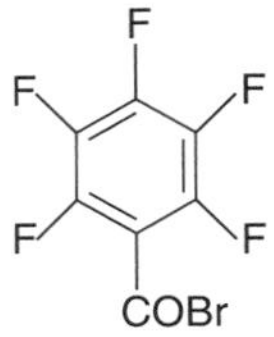	Used in the derivatization of carboxylic acids, phenols, mercaptans, and sulfamides; lachrymator; potentially unstable; high sensitivity for electron capture detection; not usable for formic acid

Derivatizing Reagents for Gas Chromatography (Continued)

Pentafluorophenyl Reagents

Derivatizing Reagent	Structure/Formula	Notes
Pentafluorobenzyl chloride	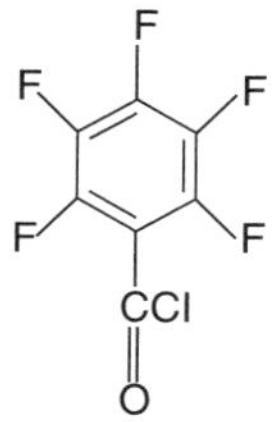	Used in the derivatization of amines, phenols, and alcohols; used in a solution of NaOH
Pentafluorobenzyl chloroformate	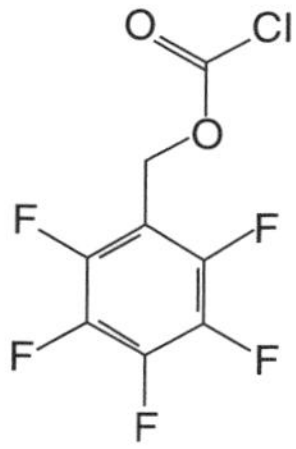	Used in derivatization of tertiary amines
Pentafluorobenzyl hydroxylamine	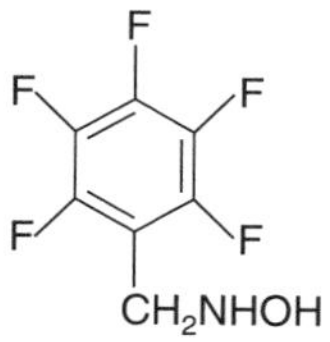	Used in derivitization of ketones; can form both syn- and anti-isomers (two peaks)
Pentafluorophenacetyl chloride	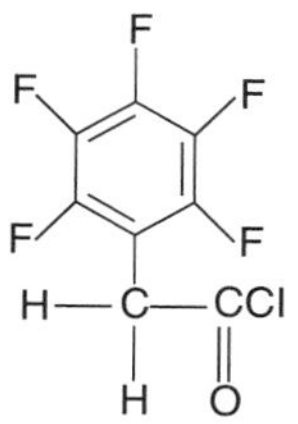	Used in derivatization of alcohols, phenols, and amines
Pentafluorophenyl-hydrazine	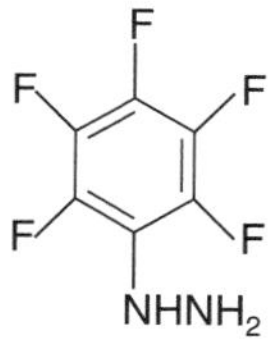	Used in derivatization of ketones; can form both the syn- and anti-isomers, resulting in two peaks
Pentafluorophenoxy-acetyl chloride	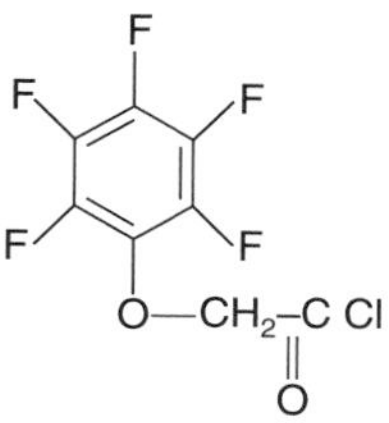	Used in derivatization of alcohols, phenols, and amines

(Continued)

Derivatizing Reagents for Gas Chromatography (Continued)

Silyating Reagents

Derivatizing Reagent	Structure/Formula	Notes
Bis(dimethylsilyl) acetamide (BSDA)	CH_3—C=N—$Si(CH_3)_2$ with O and H below, and H—$Si(CH_3)_2$	Similar in use and application to DMCS (see below)
N,N-Bis (trimethyl-silyl)-acetamide (BSA)	$Si(CH_3)_3$ —O— CH_3 C=N$Si(CH_3)_3$	More reactive than HMDS (see below) or TMCS, but forming essentially similar derivatives; useful for alcohols, amines, amino acids, carboxylic acids, penicillic acid, purine, and pyrimidene bases
Bis(trimethylsilyl) trifluoroacetamide (BSTFA)	$Si(CH_3)_3$ —O— CF_3—C=N$Si(CH_3)_3$	Similar in use and application to BSA, but the derivatives are more volatile; by-products often elute with the solvent front; reacts more strongly than HMDS or TMCS; may promote enol-TMS formation unless ketone groups are protected
Dimethylchlorosilane (DMCS)	H, $(CH_3)_2Si$—Cl	Similar in use and application to TMCS and HMDS, but usually forming more volatile and less thermally stable derivatives; also finds use in surface deactivation of chromatographic columns and injectors
1,1,1,3,3,3-Hexamethyl disilizane (HMDS)	$(CH_3)_3$-Si-NH-Si$(CH_3)_3$	Useful for such compounds as sugars, phenols, alcohols, amines, thiols, steroids; especially recommended for citric acid cycle compounds and amino acids; reaction is often carried out in pyridine or dimethyl formamide (the latter being preferred for 17-keto steroids); care must be taken to eliminate moisture; lowest silyl donating strength of all common silating reagents
1,1,1,3,3,3-Hexamethyl disiloxane (HMDSO)	$(CH_3)_3$Si-O-Si-$(CH_3)_3$	Similar in use and application to HMDS (see above)
N-Methyl-N-(trimethylsilyl)-acetamide (MSTA)	CH_3—C—N—$Si(CH_3)_3$ with CH_3 above N and O (double bond) below C	Similar in use and application to HMDS, but somewhat higher "silyl donating" strength
N-Methyl-N-(trimethylsilyl) tri-fluoroacetamide (MSTFA)	CF_3—C—N—$Si(CH_3)_3$ with CH_3 above N and O (double bond) below C	Similar to MSTA, but produces the most volatile derivatives of all common silylating agents; particularly useful with low molecular mass derivatives

Derivatizing Reagents for Gas Chromatography (Continued)

Silyating Reagents

Derivatizing Reagent	Structure/Formula	Notes
Tetramethyldisilazane (TMDS)	$(CH_3)_2\,Si-NH-Si\,(CH_3)_2$ with H on each Si	Similar in use and application to DMCS
N-Trimethylsilyl diethylamine (TMSDEA)	$(CH_3)_3\text{-}Si\text{-}N\text{-}(C_2H_5)_2$	Similar in use and application to DMCS
N-Trimethylsilyl imidazole (TMSIM)	$(CH_3)_3\,Si-N$ (imidazole ring)	Generally useful reagent with a high silyl donor ability; will not react with amino groups; will not cause formation enol-ether on unprotected ketone groups; especially useful for ecdysones, norepinephrine, dopamine, steroids, sugars, sugar phosphates, and ketose isomers

Pentafluorophenyl Reagents

Derivatizing Reagent	Structure/Formula	Notes
Trimethylchlorosilane (TMCS)	$(CH_3)_3SiCl$	Similar properties and applications as for HMDS; useful for amino acid analyses; provides good response for electron capture detection; has relatively low silyl donating ability, and is usually used in the presence of a base such as pyridine; may cause enol-ether formation with unprotected ketone groups; often used as a catalyst with other silylating reagents
Halomethylflophemesyl reagents	C_6F_5-Si-Y with CH_3 and R substituents; $R = CH_2Cl$, $Y = Cl$	Similar in use and applications to the flophemesyl and alkylflophemesyl reagents
Halomethyldimethyl silyl reagents	XCH_2-Si-Y with two CH_3; $X = Cl, Br, I$; $Y = Cl, N(C_2H_5)_2, NHSi(CH_3)_2CH_2X$	Family of derivatizing agents that improve sensitivity of analyte to the electron capture detector; the response enhancement is in the order expected: $I > Br > Cl > > F$, reverse order of the volatility of these compounds. The iodomethyldimethylsilyl reagents are unstable, and these derivatives are usually prepared in situ.

(Continued)

Derivatizing Reagents for Gas Chromatography (Continued)

Pentafluorophenyl Reagents

Derivatizing Reagent	Structure/Formula	Notes
Flophemesyl reagents	$C_6F_5-\underset{\underset{R}{|}}{\overset{\overset{CH_3}{|}}{Si}}-Y$ $R = CH_3$ $Y = Cl, NH_2, N(C_2H_5)_2$	Family of reagents forming derivatives that have stabilities similar to those produced by TMSIM, BSA, MSTFA, BSTFA, with additional electron capture detection sensitivity enhancement; usually used in pyridine as a solvent; reactions subject to steric considerations
Alkylflophemesyl reagents	$C_6F_5-\underset{\underset{R}{|}}{\overset{\overset{CH_3}{|}}{Si}}-Y$ $R = CH(CH_3)_2, C(CH_3)_3$ $Y = Cl$	Family of reagents forming derivatives of somewhat higher stability than the flophemesyl reagents; reactions subject to steric considerations

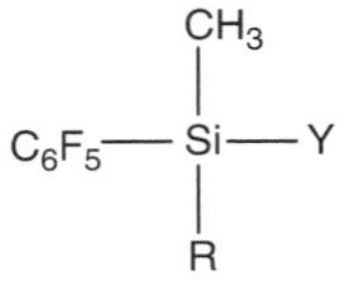

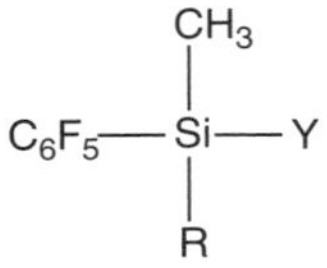

Miscellaneous Reagents

Derivatizing Reagent	Structure/Formula	Notes
Boronation reagents	$(OH)_2B\text{-}R$ $R = CH_3, \text{-}C(CH_3)_3$	Used to block two vicinal hydroxy groups, derivatives have very distinctive mass spectra that are easily identified
Carbon disulfide	CS_2	Used to derivatize primary amines to yield isothiocyanates
Dansyl chloride	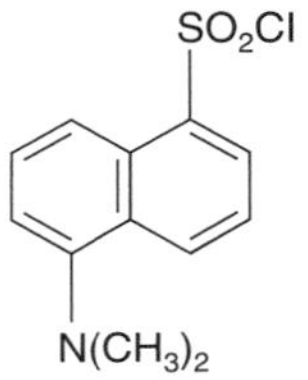	Used for derivatization of tripeptides; provides high sensitivity toward spectrofluorimetric detection
Dimethyldiacetoxy-silane	$(Cl)_2Si(CH_3)_2$	Used in similar applications as the boronation reagents in pyridene or trimethylamine solvent
2,4-Dinitrophenyl-hydrazine	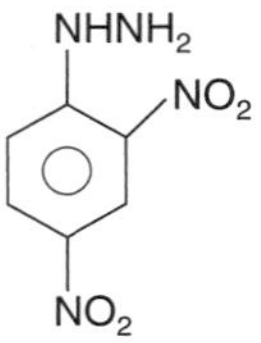	Useful in derivatizing carbonyl compounds, and also provides a "spot test" for these compounds
l-Fluoro-2,4-dinitro-fluorobenzene	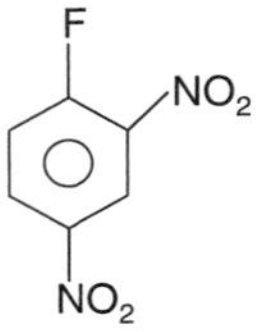	Useful for derivatizing C_1-C_4 primary and secondary amines, providing high electron capture detector response; this reagent is also useful for primary alicyclic amines
Girard reagent T	$(CH_3)_3N\text{-}Cl\text{-}CH_2CONHNH_2$ 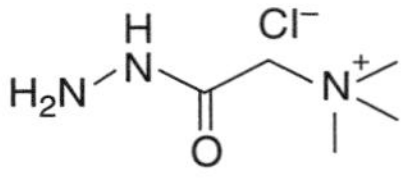	Useful for derivatization of saturated aldehydes

Derivatizing Reagents for Gas Chromatography (Continued)

Miscellaneous Reagents

Derivatizing Reagent	Structure/Formula	Notes
Hydrazine	NH_2NH_2	Used for the analysis of C-terminal peptide residue species
Methyl iodide + silver oxide	$CH_3I + Ag_2O$ (in di-methylformamide)	Used to convert polyhydroxy compounds to the methyl ethers
Methyloxamine hydrochloride	$CH_3\text{-}O\text{-}NH \cdot HCl$	Used in derivatization of steroids and carbohydrates
2-Methylthioaniline	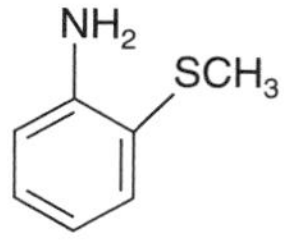	Used to form sulfur bearing derivatives of benzaldehydes
Phenyl isocyanate	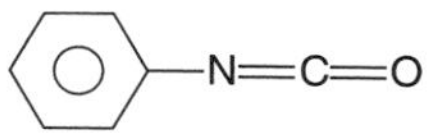	Used for derivatization of N-terminal peptide residue
2,4,6-Trichloro-phenylhydrazine	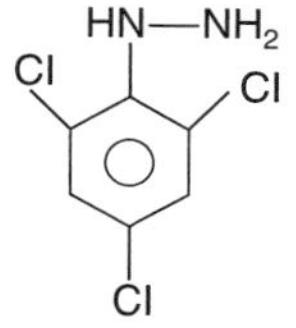	Used for derivatization of carbonyl compounds

DETECTORS FOR GAS CHROMATOGRAPHY

The following table provides some comparative data for the selection and operation of the more common detectors applied to capillary and packed column gas chromatography [1–7].

REFERENCES

1. Hill, H. H., and D. McMinn, eds. *Detectors for Capillary Chromatography*. New York: Wiley-Interscience, John Wiley & Sons, Inc., 1992.
2. Buffington, R., and M. K. Wilson. *Detectors for Gas Chromatography: A Practical Primer*. Avondale, PA: Hewlett Packard Corp, 1987.
3. Buffington, R. *GC-Atomic Emission Spectroscopy using Microwave Plasmas*. Avondale, PA: Hewlett Packard Corp, 1988.
4. Liebrand, R. J., ed. *Basics of GC/IRD and GC/IRD/MS*. Avondale, PA: Hewlett Packard Corp, 1993.
5. Bruno, T. J. "A Review of Hyphenated Chromatographic Instrumentation." *Separation and Purification Methods* 29, no. 1 (2000): 63–89.
6. Bruno, T. J. "A Review of Capillary and Packed Column Chromatographs." *Separation and Purification Methods* 29, no. 1 (2000): 27–61.
7. Sevcik, J. *Detectors in Gas Chromatography, Journal of Chromatography Library*. Vol. 4. Amsterdam: Elsevier, 1976.

Detectors for Gas Chromatography

Detector	Limit of Detection	Linearity	Selectivity	Comments
Thermal conductivity detector (TCD, katharometer)	1×10^{-10} g propane (in helium carrier gas)	1×10^6	Universal response, concentration detector	• Ultimate sensitivity depends on analyte thermal conductivity difference with carrier gas. • Since thermal conductivity is temperature dependent, response depends on cell temperature. • Wire selection depends on chemical nature of analyte. • Helium is recommended as carrier and make-up gas. When analyzing mixtures containing hydrogen, one can use a mixture of 8.5 % (mass/mass) hydrogen in helium. See the table entitled "Carrier Gas Properties" for information on this mixture.
Gas density balance detector (GADE)	1×10^{-9} g, H_2 with SF_6 as carrier gas	1×10^6	Universal response, concentration detector	• Response and sensitivity is based on difference in relative molecular mass of analyte with that of the carrier gas; approximate calibration can be done on the basis of relative density. • The sensing elements (hot wires) never touch sample, thus making GADE suitable for the analysis of corrosive analytes such as acid gases; gold sheathed tungsten wires are most common. • Best used with SF_6 as a carrier gas, switched between nitrogen when analyses are required. • Detector can be sensitive to vibrations and should be isolated on a cushioned base
Flame ionization detector (FID)	$1 \times 10^{-11} - 1 \times 10^{-10}$ g	1×10^7	Organic compounds with C-H bonds	• Ultimate sensitivity depends on the number of C-H bonds on analyte. • Nitrogen is recommended as carrier gas and make-up gas to enhance sensitivity. • Sensitivity depends on carrier, make-up, and jet gas flow-rates. • Column must be positioned 1–2 mm below the base of the flame tip. • Jet gases must be of high purity.

(Continued)

Detectors for Gas Chromatography (Continued)

Detector	Limit of Detection	Linearity	Selectivity	Comments
Nitrogen-phosphorous detector (NPD, thermionic detector, alkali flame ionization detector)	4×10^{-13}–1×10^{-11} g of nitrogen compounds; 1×10^{-13}–1×10^{-12} g of phosphorous compounds	1×10^4	10^5–10^6 by mass selectivity of N or P over carbon	• Does not respond to inorganic nitrogen such as N_2 or NH_3. • Jet gas flow rates are critical to optimization. • Response is temperature dependent. • Used for trace analysis only, and is very sensitive to contamination. • Avoid use of phosphate detergents or leak detectors. • Avoid tobacco use nearby. • Solvent-quenching is often a problem.
Electron capture detector (ECD)	5×10^{-14}–1×10^{-12} g	1×10^4	Selective for compounds with high electron affinity, such as chlorinated organics; concentration detector	• Sensitivity depends on number of halogen atoms on analyte. • Used with nitrogen or argon/methane (95/5, mass/mass) carrier and make-up gases. See the table entitled "Carrier Gas Properties" for information on this mixture. • Carrier and make-up gases must be pure and dry. • The radioactive ^{63}Ni source is subject to regulation and periodic inspection.
Flame photometric detector (FPD)	2×10^{-11} g of sulfur compounds, 9×10^{-13} g of phosphorous compounds	1×10^3 for sulfur compounds; 1×10^4 for phosphorous compounds	10^5 to 1 by mass selectivity of S or P over carbon	• Hydrocarbon quenching can result from high levels of CO_2 in the flame. • Self-quenching of S and P analytes can occur with large samples. • Gas flows are critical to optimization. • Response is temperature dependent. • Condensed water can be a source of window fogging and corrosion.

Detector	Minimum detectable amount	Linear dynamic range	Selectivity	Comments
Photoionization detector (PID)	$1 \times 10^{-12} - 1 \times 10^{-11}$ g	1×10^7	Depends on ionization potentials of analytes	• Used with lamps with energies of 10.0–10.2 eV. • Detector will have response to ionizable compounds such as aromatics and unsaturated organics, some carboxylic acids, aldehydes, esters, ketones, silanes, iodo- and bromoalkanes, alkylamines and amides, and some thiocyanates.
Sulfur chemiluminescence detector (SCD)	1×10^{-12} g of sulfur in sulfur compounds	1×10^4	10^7 by mass selectivity of S over carbon	• Equimolar response to all sulfur compounds to within 10 %. • Requires pure hydrogen and oxygen combustion gases. • Instrument generates ozone in-situ, which must be catalytically destroyed at detector outlet. • Catalyst operates at 950–975 °C. • Detector operated at reduced pressure (10^3 Pa). • Only high purity solvents should be used.
Electrolytic conductivity detector (ECD, Hall detector)	$10 \times 10^{-13} - 1 \times 10^{-12}$ g of chlorinated compounds, 2×10^{-12} g of sulfur compounds, 4×10^{-12} g of nitrogen compounds	1×10^6 for chlorinated compounds; 10^4 for sulfur and nitrogen compounds	10^6 by mass selectivity of Cl over carbon. 10^5-10^6 by mass selectivity of S and N over carbon	• Carbon particles in conductivity chamber can be problematic. • Frequent cleaning and maintenance is required. • Often used in conjunction with a photoionization detector. • For chlorine, use hydrogen as the reactant gas and n-propanol as the electrolyte. • For nitrogen or sulfur, hydrogen or oxygen can be used as reactant gas, and water of methanol as the electrolyte. • Ultrahigh purity reactant gases are required.
Ion mobility detector (IMD)	1×10^{-12} g	$1 \times 10^3 - 1 \times 10^4$	10^3	• Amenable to use in handheld instruments. • Linear dynamic range of 10^3 for radioactive sources and 10^5 for photo-ionization sources. • Selectivity depends on mobility differences of ions. • Has been used for a wide variety of compounds including amino acids, halogenated organics, explosives.

(Continued)

Detectors for Gas Chromatography (Continued)

Detector	Limit of Detection	Linearity	Selectivity	Comments
Mass selective detector (MSD, mass spectrometer, MS)	1×10^{-11} g (single ion monitoring) 1×10^{-8} g (scan mode)	1×10^5	Universal	• Quadrupole and magnetic sector instruments available. • Must operate under moderate vacuum (1×10^{-4} Pa). • Requires a molecular jet separator to operate with packed columns. • Amenable to library searching for qualitative identification. • Requires tuning of electronic optics over the entire m/e range of interest. • See tables for mass spectrometry for structure elucidation and identification.
Infrared detector (IRD)	1×10^{-9} g of a strong infrared absorber	1×10^3	Universal for compounds with mid-infrared active functionality	• A costly and temperamental instrument that requires high purity carrier gas, a nitrogen purge of optical components (purified air will, in general, not be adequate). • Must be isolated from vibrations. • Presence of carbon dioxide is a typical impurity band at 2200–2300 cm⁻¹. • Requires frequent cleaning and optics maintenance. • Amenable to library searching for qualitative identification. • See tables for infrared functionalities for structure elucidation and identification.
Atomic emission detector (AED)	1×10^{-13} –2×10^{-11} g of each element	1×10^3 –1×10^4	10^3 –10^5, element to element	• Requires the use of ultra high purity carrier and plasma gases. • Plasma produced in a microwave cavity operated at 2450 MHz. • Scavenger gases (H_2, O_2) are used as dopants. • Photodiode array is used to detect emitted radiation.

RECOMMENDED OPERATING RANGES FOR HOT WIRE
THERMAL CONDUCTIVITY DETECTORS

The following table provides guidance in the operation of hot wire thermal conductivity detectors. The operating trances are provided in mA dc for detector cells operated between 25 and 200 °C [1]. The current ranges and the cold resistances provided are for typical wire lengths and configurations.

REFERENCE

1. Gow-Mac Instrument Company Manual SB-13. *Thermal Conductivity Detector Elements for Gas Analysis*, Bethlehem, PA, 1995.

Substance	Carrier Gas				
	H_2mA-dc	HemA-dc	N_2mA-dc	CO_2, ArmA-dc	Cold ResistanceOhms, 25 °C
Tungsten, W	250–500	250–400	100–175	90–130	18
Tungsten-Rhenium, WX (97 %–3 %)	250–400	230–375	100–150	90–130	26–32
Nickel, Ni 99.8 %	300–500	300–450	125–150	100–130	12.5
Gold Sheathed Tungsten AuW	250–400	250–375	100–150	75–120	24

CHEMICAL COMPATIBILITY OF THERMAL CONDUCTIVITY DETECTOR WIRES

The following table provides guidance in the selection of hot wires for use in thermal conductivity detectors (TCD) [1–3]. This information is applicable to the operation of packed and open tubular columns. Some of the entries in this table deal with analytes and others deal with solutions that might be used to clean the TCD cell.

REFERENCES

1. Gow-Mac Instrument Company Manual SB-13. *Thermal Conductivity Detector Elements for Gas Analysis*, Bethlehem, PA, 1995.
2. Seveik, J. *Detectors in Gas Chromatography.* Amsterdam: Elsevier Scientific Publishing Co., 1976.
3. Lawson, A. E., and J. M. Miller. *Journal of Gas Chromatography* 4, no. 8 (1966): 273–84.

Substance	Tungsten (W)	Rhenium-Tungsten (WX)	Nickel (Ni)	Gold-Sheathed Tungsten (AuW)
Air/oxygen	Good	Good	Good	Very Good
Water	Good	Good	Good	Good
Steam	Good below 700 °C	Good below 700 °C	Good	Good
Ammonia/amines	Good	Good	Poor in presence of water	Poor[a]
Carbon monoxide/ carbon dioxide	Good	Good	Good	Good
Hydrogen	Good	Good	Good	Good
Nitrogen	Good	Good	Good	Fair
Fluorine	Poor (fluoride forms at 20 °C)	Poor (fluoride forms at 20 °C)	Good	Poor
Chlorine	Fair	Fair	Good	Fair
Bromine	Fair	Fair	Good	Fair
Iodine	Fair	Fair	Good	Fair
Sulfur	Fair	Good	Poor	Good
Hydrogen sulfide/sulfur dioxide (sulfuric acid)	Fair	Fair	Poor	Good
Hydrogen chloride	Fair	Fair	Good	Fair
Aqua regia	Fair	Fair	Poor	Poor
Hydrogen fluoride	Fair	Fair	Good	Fair
Hydrogen fluoride/nitric acid	Poor	Poor	Good	Poor

[a] Gold sheathed tungsten filaments are attacked by amines, but the process is somewhat reversible. The baseline departure will recover, but the peak will develop a significant tail.

DATA FOR THE OPERATION OF GAS DENSITY DETECTORS

The following data provide useful guidance in the operation and optimization of procedures with the gas density balance detector in gas chromatography [1]. The property values were calculated with the Refprop database [2].

REFERENCES

1. Nerheim, A. G. *Analytical Chemistry* 35 (1963): 1640.
2. Lemmon, E. W., M. L. Huber, and M. O. McLinden. REFPROP, Reference Fluid Thermodynamic and Transport Properties, NIST Standard Reference Database 23, Version 8. National Institute of Standards and Technology, Gaithersburg, MD, 2007.

Argon, Ar, 24 psia:

Temp. °C	Density g/L	Cp/Cv	Viscosity μPa-s
30	2.6251	1.6712	22.887
60	2.3877	1.6703	24.735
90	2.1899	1.6697	26.521
120	2.0224	1.6692	28.249
150	1.8787	1.6688	29.925

Carbon Dioxide, CO_2, 24 psia:

Temp. °C	Density g/L	Cp/Cv	Viscosity μPa-s
30	2.9120	1.2950	15.179
60	2.6441	1.2802	16.614
90	2.4221	1.2679	18.018
120	2.2350	1.2576	19.391
150	2.0749	1.2487	20.731

Helium, He, 24 psia:

Temp. °C	Density g/L	Cp/Cv	Viscosity μPa-s
30	0.26258	1.6665	20.075
60	0.23895	1.6665	21.417
90	0.21923	1.6665	22.726
120	0.20251	1.6665	24.006
150	0.18816	1.6665	25.259

Hydrogen, H_2, 24 psia:

Temp. °C	Density g/L	Cp/Cv	Viscosity μPa-s
30	0.13222	1.4047	9.0188
60	0.12032	1.4015	9.6186
90	0.11039	1.3997	10.200
120	0.10197	1.3987	10.766
150	0.094745	1.3982	11.317

Nitrogen, N$_2$, 24 psia:

Temp. °C	Density g/L	Cp/Cv	Viscosity µPa-s
30	1.8396	1.4022	18.052
60	1.6734	1.4013	19.399
90	1.5348	1.4002	20.695
120	1.4175	1.3989	21.945
150	1.3169	1.3973	23.153

Sulfur Hexafluoride, SF$_6$, 24 psia:

Temp. °C	Density g/L	Cp/Cv	Viscosity µPa-s
30	9.7615	1.0997	15.646
60	8.8401	1.0913	17.105
90	8.0832	1.0851	18.514
120	7.4489	1.0804	9.869
150	6.9090	1.0768	21.177

1,1,1,2-tetrafluoroethane, R134a, CF$_3$CFH$_2$, 24 psia:

Temp. °C	Density g/L	Cp/Cv	Viscosity µPa-s
30	6.9186	1.1247	12.013
60	6.2347	1.1116	13.209
90	5.6842	1.1021	14.365
120	5.2285	1.0949	15.486
150	4.8436	1.0890	16.575

COMMON SPURIOUS SIGNALS OBSERVED IN GC-MS

The following table provides guidance in the recognition of spurious signals (m/z peaks) that will sometimes be observed in measured mass spectra [1]. Often, the occurrence of these signals can be predicted by the recent history of the instrument or the method being used. This is especially true if the mass spectrometer is interfaced to a gas chromatograph.

REFERENCES

1. Maintaining Your GC-MS System Agilent Technologies. Applications Manual, 2001. Available online at www.agilent.com/chem.

Ions Observed, m/z	Possible Compound	Possible Source
13, 14, 15, 16	methane[a]	Chlorine reagent gas
18	water[a]	Residual impurity, outgasing of ferrules, septa and seals
14, 28	nitrogen[a]	Residual impurity, outgasing of ferrules, septa and seals; leaking seal
16, 32	oxygen[a]	Residual impurity, outgasing of ferrules, septa and seals; leaking seal
44	carbon dioxide[a]	Residual impurity, outgasing of ferrules, septa and seals; leaking seal; note it may be mistaken for propane in a sample
31, 51, 69, 100, 119, 131, 169, 181, 214, 219, 264, 376, 414, 426, 464, 502, 576, 614	perfluorotributyl amine (PFTBA), and related ions	This is a common tuning compound; may indicate a leaking valve
31	methanol	Solvent; can be used as a leak detector
43, 58	acetone	Solvent; can be used as a leak detector
78	benzene	Solvent; can be used as a leak detector
91, 92	toluene	Solvent; can be used as a leak detector
105, 106	xylenes	Solvent; can be used as a leak detector
151, 153	trichloroethane	Solvent; can be used as a leak detector
69	fore pump fluid, PFTBA	Back diffusion of fore pump fluid, possible leaking valve of tuning compound vial
73, 147, 207, 221, 281, 295, 355, 429	dimethylpolysiloxane	Bleed from a column or septum, often during high temperature program methods in GC-MS
77, 94, 115, 141, 168, 170, 262, 354, 446	diffusion pump fluid	Back diffusion from diffusion pump, if present
149	phthalates	Plasticizer in vacuum seals, gloves
X−14 peaks	hydrocarbons	Loss of a methylene group indicates a hydrocarbon sample

[a] It is possible to operate the analyzer to ignore these common background impurities. They will be present to contribute to poor vacuum if these impurities result from a significant leak.

PHASE RATIO FOR CAPILLARY COLUMNS

The phase ratio is an important parameter used in the design of capillary (open tubular) column separations [1]. This quantity relates the partition coefficient (K) to the partition ratio (k):

$$K = k\beta,$$

where β is the phase ratio, defined as the ratio of the volume occupied by the gas or mobile phase (V_m) relative to that occupied by the liquid or stationary phase (V_s). For wall coated open tubular columns, the phase ratio can be found from:

$$\beta = r/2d_f,$$

where r is the internal radius of the column, and d_f is the thickness of the stationary phase film. The following table provides the phase ratio for common combinations of column internal diameter and stationary phase film thickness. These values are given to the nearest whole number, since only an approximate value is needed for most analytical applications.

REFERENCE

1. Sandra, P. *High Resolution Gas Chromatography*. Avondale, PA: Hewlett Packard Corporation, 1989.

Phase Ratio for Capillary Columns

Film Thickness, μm	Column Inside Diameter, mm							
	0.05	0.10	0.20	0.30	0.32	0.40	0.50	0.53
0.03	417	833	1667	2500	2667	3333	4167	4417
0.06	208	417	833	1250	1333	1667	2083	2208
0.1	125	250	500	750	800	1000	1250	1325
0.2	63	125	250	375	400	500	625	663
0.3	42	83	167	250	267	333	417	442
0.4	31	63	125	188	200	250	313	331
0.5	25	50	100	150	160	200	250	265
0.6	21	42	83	125	133	167	208	221
0.7	18	36	71	107	114	143	179	189
0.8	16	31	63	94	100	125	156	166
0.9	14	28	56	83	89	111	139	147
1.0	13	25	50	75	80	100	125	133
1.5	8	17	34	50	53	67	83	88
2.0	6.3	13	25	38	40	50	63	66
2.5	5	10	20	30	34	40	50	53
3.0	4	8	17	25	27	33	42	44
3.5	4	7	14	21	23	29	18	38
4.0	3	6	13	19	20	25	32	33
4.5	3	6	11	17	18	22	29	29
5.0	2.5	5	10	15	16	20	25	27
5.5	2	5	9	14	15	18	23	24
6.0	2	4	8	13	13	17	21	22
6.5	2	4	8	12	12	15	19	20
7.0	2	4	7	11	11	14	18	19
7.5	2	3	7	10	11	13	17	18
8.0	2	3	6	9	10	13	16	17
8.5	1	3	6	9	9	12	15	16
9.0	1	3	6	8	9	11	14	15

PRESSURE DROP IN OPEN TUBULAR COLUMNS

The pressure drop across an open tubular or capillary column is often important for optimization of chromatographic analyses [1]. Column performance is typically assessed by the height equivalent to a theoretical plate (HETP), which is based on the average linear carrier gas velocity. As the average linear velocity increases, the head pressure and carrier gas flow rate increases as well. One may express the pressure drop across the column as:

$$\Delta p = p_i - p_o,$$

where Δp is the pressure drop, p_i is the inlet or head pressure, and p_o is the outlet pressure. The head pressure is typically a gauge pressure measured electronically, while the outlet pressure is the barometric pressure that can be measured electronically with a mercury barometer or with an aneroid barometer. Concern for the spillage of mercury has caused an increase in the number of laboratories employing an electronic measure for this. In relation to the average carrier gas velocity:

$$\Delta p = 8\eta Lu/r_c^2,$$

where η is the carrier gas viscosity, L is the column length, and r_c is the column internal radius. For helium carrier gas at 100 °C, the following tables provide the pressure drop in units of psig and kPa.

REFERENCES

1. Hinshaw, J. V. "Open Tubular Column Pressures and Flows, GC Troubleshooting," *LC-GC* 7, no. 3 (1989): 237–39.

For 10 m columns:

Diameter, $2r_c$, mm	0.750	0.530	0.320	0.200	0.100
Carrier Gas Velocity, u, cm/sec	Pressure Drop, psig				
10	0.19	0.38	1.0	2.7	10.0
20	0.38	0.75	2.1	5.3	21.2
30	0.56	1.1	3.1	7.9	31.8
40	0.75	1.5	4.1	10.6	42.3
60	1.1	2.3	6.2	15.9	63.5
80	1.5	3.0	8.3	21.3	84.7

For 25 m columns:

Diameter, $2r_c$, mm	0.750	0.530	0.320	0.200	0.100
Carrier Gas Velocity, u, cm/sec	Pressure Drop, psig				
10	0.47	0.94	2.6	6.6	26.5
20	0.94	1.9	5.2	13.3	52.9
30	1.4	2.8	7.8	19.8	79.4
40	1.9	3.8	10.3	26.5	
60	2.8	5.7	15.5	39.7	
80	3.8	7.5	20.7	52.9	

For 50 m columns:

Diameter, 2r_c, mm	0.750	0.530	0.320	0.200	0.100
Carrier Gas Velocity, u, cm/sec	**Pressure Drop, psig**				
10	0.94	1.9	5.2	13.2	52.9
20	1.9	3.8	10.3	26.5	
30	2.8	5.7	15.5	39.7	
40	3.8	7.5	20.7	52.9	
60	5.6	11.3	31.0	79.4	
80	7.5	15.1	41.3		

For 10 m columns:

Diameter, 2r_c, mm	0.750	0.530	0.320	0.200	0.100
Carrier Gas Velocity, u, cm/sec	**Pressure Drop, kPa, Gauge**				
10	1.3	2.6	6.9	18.6	69.0
20	2.6	5.2	14.5	36.5	146.2
30	3.9	7.6	21.4	54.5	219.3
40	5.2	10.3	28.3	73.1	291.7
60	7.6	15.9	42.7	109.6	437.8
80	10.3	20.7	57.2	146.9	584.0

For 25 m columns:

Diameter, 2r_c, mm	0.750	0.530	0.320	0.200	0.100
Carrier Gas Velocity, u, cm/sec	**Pressure Drop, kPa, Gauge**				
10	3.2	6.5	17.9	45.5	182.7
20	6.5	13.1	35.9	91.7	364.7
30	9.7	19.3	53.8	136.5	547.5
40	13.1	26.2	71.0	182.7	
60	19.3	39.3	106.9	273.7	
80	26.2	51.7	142.7	364.7	

For 50 m columns:

Diameter, 2r_c, mm	0.750	0.530	0.320	0.200	0.100
Carrier Gas Velocity, u, cm/sec	**Pressure Drop, kPa, Gauge**				
10	6.5	13.1	35.9	91.0	364.7
20	13.1	26.2	71.0	182.7	
30	19.3	39.3	106.9	273.7	
40	26.2	51.7	142.7	364.7	
60	38.6	77.9	213.7	547.5	
80	51.7	104.1	284.8	0.0	

MINIMUM RECOMMENDED INJECTOR SPLIT RATIOS
FOR CAPILLARY COLUMNS

In order to avoid overloading high efficiency open tubular or capillary columns (with theoretical plate counts between 400,000 and 600,000), it is necessary to split the flow in the injector. Split ratios that are too low will result in distorted peak shapes and poor analyses. As a first approximation, the lowest split ratio that can be used is dependent upon the column internal diameter. Secondary factors then include the solute properties (polarity, etc.), column temperature (or temperature program), liner volume, injector volume, and stationary phase properties. The following table provides the minimum split ratios that should be considered for typical capillary columns [1].

REFERENCE

1. Rood, D. "Gas Chromatography Problem Solving and Troubleshooting" *Journal of Chromatographic Science* 37, no. 3 (1999): 88.

Column Diameter, mm	Minimum Split Ratio
0.18	1:25
0.20	1:20
0.25	1:15–1.20
0.32	1:10–1:12
0.53	1:3–1:5

MARTIN–JAMES COMPRESSIBILITY FACTOR AND GIDDINGS PLATE HEIGHT CORRECTION FACTOR

The following table provides the Martin–James compressibility factor, j [1], and the Giddings plate height correction factor, f [2], for chromatographically useful pressures. These quantities are defined as:

$$j = 3/2 \left[\frac{\left[\left(P_i^{abs} / P_o \right)^2 - 1 \right]}{\left[\left(P_i^{abs} / P_o \right)^3 - 1 \right]} \right]$$

$$f = 9/8 \left[\frac{\left[\left(P_i^{abs} / P_o \right)^4 - 1 \right] \left[\left(P_i^{abs} / P_o \right)^2 - 1 \right]}{\left[\left(P_i^{abs} / P_o \right)^3 - 1 \right]^2} \right],$$

where P_i is the absolute inlet pressure, and P_o is the outlet pressure.

The inlet pressures listed in the table are gauge pressures; the pressures used in the calculations of j and f are absolute pressures. Thus, atmospheric pressure had already been accounted for in the inlet pressure. The outlet pressure is taken as standard atmospheric pressure. As an example, for a measured gauge pressure of 137.9 kPa (20 psig), the ratio P_i^{abs}/P_o is 2.361. The actual value of the atmospheric pressure will vary day to day and with altitude, thus if an exact value for j or f is desired, local pressure measurements must be made.

REFERENCES

1. Grob, R. L. *Modern Practice of Gas Chromatography*. 2nd ed. New York: John Wiley and Sons (Wiley Interscience), 1985.
2. Lee, M. L., F. J. Yang, and K. D. Bartle. *Open Tubular Column Gas Chromatography*. New York: John Wiley and Sons (Wiley Interscience), 1984.

Pressure	j	f
15.0	0.638	1.034
16.0	0.622	1.037
17.0	0.606	1.039
18.0	0.592	1.042
19.0	0.578	1.044
20.0	0.564	1.046
25.0	0.505	1.057
30.0	0.456	1.066
35.0	0.416	1.074
40.0	0.381	1.080
45.0	0.352	1.085
50.0	0.327	1.090
55.0	0.305	1.093
60.0	0.286	1.096

Gas Hold-Up Volume

There are a few instances in which it is important to determine the gas hold-up volume of the chromatographic system consisting of the injector, column and detector swept volumes. Noxious volumes are by definition unswept and are generally minimized by design. The most common application of the gas hold-up volume or measurement is in the determination of the average column flow rate with the following equation:

$$F_{ave} = \pi r^2 L / t_m,$$

where F_{ave} is the average flow rate, L is the length of the column in cm, and t_m is the average retention time of a marker compound that is minimally retained. The following table provides potential, minimally retained markers for various detectors:

Detector	Minimally Retained Marker Compound
FID	methane, n-butane
TCD	methane, n-butane, air
MSD	methane, n-butane, air
ECD	sulfur hexafluoride, methylene chloride
NPD	acetonitrile

Methane is usually easily obtainable from a natural gas line and n-butane is easily obtained from a disposable lighter. For the liquids, it is important to only use an aliquot of the headspace or to use a permeation vial [1].

Another application in which the gas hold-up volume is needed is in the use of chromatographic retention parameters for solute identification. Chromatographic parameters include net retention volumes, relative retentions, specific retention volumes, and retention indices. Here, it is important to evaluate the applicability of a minimally retained marker in each case, since even a very light solute such as methane can show retentive behavior. It is usually best to use an extrapolative method to estimate the hold-up, although the chromatographic behavior of methane is often used in these procedures as well [2]. A convenient way to dispense the methane is with a permeation tube methanizer [3].

REFERENCES

1. Bruno, T. J. "Permeation Tube Approach to Long-Term Use of Automatic Sampler Retention Index Standards." *Journal of Chromatography, A* 704, no. 1 (1995): 157–62.
2. Miller, K. E., and T. J. Bruno. "Isothermal Kovats Retention Indices of Sulfur Compounds on a Poly(5 % phenyl–95 % dimethyl siloxane) Stationary Phase." *Journal of Chromatography, A* 1007 (2003): 117–25.
3. Bruno, T. J. "Simple and Efficient Methane-Marker Devices for Chromatographic Samples." *Journal of Chromatography, A* 721, no. 1 (1996): 157–64.

CRYOGENS FOR SUBAMBIENT TEMPERATURE GAS CHROMATOGRAPHY

The following table lists properties of common cryogenic fluids used to produce subambient temperatures for gas chromatographic columns [1–5]. These properties are of value in designing low temperature chromatographic experiments efficiently and safely. Due to the potential dangers in handling extremely low temperatures and high pressures, appropriate precautions must be observed. These precautions must include protective clothing and shielding to prevent frostbite. Most cryogenic fluids can create a health hazard if they are vaporized in an inhabited area. Even small quantities can contaminate and displace air in a relatively short period of time. It may be advisable to locate a self-contained breathing apparatus immediately outside the laboratory in which the cryogens are being used. The effect of low temperatures on construction materials (of G.C. ovens and columns, for example) should also be considered. In this respect, differential expansion and tensile strength changes are pertinent issues. A dew point versus moisture content table is also provided to allow the user to estimate the effects of ambient and impurity water. The viscosity data are provided in cP, which is equivalent to mPa·s, the appropriate SI unit. The freezing points are reported at 0.101325 MPa (1 atm), and the expansion ratios are reported at STP.

If temperatures no lower than approximately −40 °C are required, the use of the Ranque-Hilsch vortex tube should be considered [6–8]. This device requires a source of clean, dry compressed air at a pressure of approximately 0.70 MPa (100 psi) for proper operation. The flow-rate of air that is required depends on the volume of space to be cooled.

REFERENCES

1. Zabetakis, M. G. *Safety with Cryogenic Fluids*. New York: Plenum Press, 1967.
2. Cook, G. A., ed. *Argon, Helium and the Rare Gases*. New York: John Wiley and Sons (Interscience), 1961.
3. Brettell, T. A., and R. L. Grob. *American Laboratory* 17, no. 10 (1985): 19.
4. Cowper, C. J., and A. J. DeRose. *The Analysis of Gases by Chromatography*. Oxford: Pergamon Press, 1983.
5. Matheson Gas Data Book. 4th ed. East Rutherford: The Matheson Company, 1966.
6. Bruno, T. J. "Vortex Cooling for Subambient Temperature Gas Chromatography." *Analytical Chemistry* 58, no. 7 (1986): 1596.
7. Bruno, T. J. "Vortex Refrigeration of HPLC Components, LC." *Liquid Chromatography HPLC Magazine* 4, no. 2 (1986): 134.
8. Bruno, T. J. "Laboratory Applications of the Vortex Tube." *Journal of Chemical Education* 64, no. 11 (1987): 987.

Cryogen Name	Relative Molecular Mass	Freezing Point °C (K)	Heat of Fusion J/g	Normal Boiling Point °C (K)	Heat of Vaporization J/g
Argon Ar	39.948	−189.4 (83.8)	27.6	−185.9 (87.3)	163.2

Critical Temperature °C (K)	Critical Pressure MPa	Critical Density g/L	Vapor Pressure MPa	Gas Density g/L	Liquid/Gas Expansion Ratio
−122.3 (150.9)	4.89	530.5	a	1.63	860

Heat Capacity C_p J/(kg·K)	Heat Capacity C_v J/(kg·K)	Thermal Conductivity $\times 10^{-2}$ w/(m·K)	Viscosity Pa·s $\times 10^5$ (cP)	Solubility in Water 0 °C, V/V
523.8 (21 °C)	313.8 (15.6 °C)	1.44 (233 K)	2.21 (21 °C)	0.056

Cryogen Name	Relative Molecular Mass	Freezing Point °C (K)	Heat of Fusion J/g	Normal Boiling Point °C (K)	Heat of Vaporization J/g
Carbon Dioxide CO_2	44.01	−78.5[b] (194.7)	198.7	−56.6 (216.6)	151.5

Critical Temperature °C (K)	Critical Pressure MPa	Critical Density g/L	Vapor Pressure MPa	Gas Density g/L	Liquid/Gas Expansion Ratio
31.1 (304.2)	7.38	468	5.72 (21 °C)	1.98	790

Heat Capacity C_p J/(kg·K)	Heat Capacity C_v J/(kg·K)	Thermal Conductivity $\times 10^{-2}$ w/(m·K)	Viscosity Pa·s $\times 10^5$ (cP)	Solubility in Water 0 °C, V/V
831.8 (15.6 °C)	638.8 (15.6 °C)	1.17 (233 K)	1.48 (21 °C)	0.90

Cryogen Name	Relative Molecular Mass	Freezing Point °C (K)	Heat of Fusion J/g	Normal Boiling Point °C (K)	Heat of Vaporization J/g
Helium He	4.003	−272[c] (1)	c	−269.0 (4.2)	23.0 (15 °C)

Critical Temperature °C (K)	Critical Pressure MPa	Critical Density g/L	Vapor Pressure MPa	Gas Density g/L	Liquid/Gas Expansion Ratio
−268.0 (5.2)	0.23	69.3	a	0.16	780

Heat Capacity C_p J/(kg·K)	Heat Capacity C_v J/(kg·K)	Thermal Conductivity × 10^{-2} w/(m·K)	Viscosity Pa·s × 10^5 (cP)	Solubility in Water 0 °C, V/V
5221.6 (21 °C)	3146.4 (15.6 °C)	12.76 (233 K)	1.96 (21 °C)	0.0086

Cryogen Name	Relative Molecular Mass	Freezing Point °C (K)	Heat of Fusion J/g	Normal Boiling Point °C (K)	Heat of Vaporization J/g
Methane CH_4	16.04	−182.6 (90.6)	58.6	−161.5 (87.3)	510.0

Critical Temperature °C (K)	Critical Pressure MPa	Critical Density g/L	Vapor Pressure MPa	Gas Density g/L	Liquid/Gas Expansion Ratio
−82.1 (190.1)	4.64	162.5	a	0.7174	650

Heat Capacity C_p J/(kg·K)	Heat Capacity C_v J/(kg·K)	Thermal Conductivity × 10^{-2} w/(m·K)	Viscosity Pa·s × 10^5 (cP)	Solubility in Water 0 °C, V/V
2205.4 (15.6 °C)	1687.0 (15.6 °C)	2.57 (233 K)	1.20 (21 °C)	

Cryogen Name	Relative Molecular Mass	Freezing Point °C (K)	Heat of Fusion J/g	Normal Boiling Point °C (K)	Heat of Vaporization J/g
Nitrogen N_2	28.013	−210.1 (63.1)	25.5	−195.81 (77.3)	199.6

Critical Temperature °C (K)	Critical Pressure MPa	Critical Density g/L	Vapor Pressure MPa	Gas Density g/L	Liquid/Gas Expansion Ratio
−146.9 (150.9)	3.4	311	a	1.14	710

Heat Capacity C_p J/(kg·K)	Heat Capacity C_v J/(kg·K)	Thermal Conductivity × 10^{-2} w/(m·K)	Viscosity Pa·s × 10^5 (cP)	Solubility in Water 0 °C, V/V
1030.6 (21 °C)	738.6 (21 °C)	2.11 (233 K)	1.744 (15 °C)	0.023

Cryogen Name	Relative Molecular Mass	Freezing Point °C (K)	Heat of Fusion J/g	Normal Boiling Point °C (K)	Heat of Vaporization J/g
Oxygen O_2	31.999	−218.8 (54.4)	13.8	−183.0 (90.2)	213.0

Critical Temperature °C (K)	Critical Pressure MPa	Critical Density g/L	Vapor Pressure MPa	Gas Density g/L	Liquid/Gas Expansion Ratio
−118.4 (154.8)	5.04	410	a	1.3	875

Heat Capacity C_p J/(kg·K)	Heat Capacity C_v J/(kg·K)	Thermal Conductivity $\times 10^{-2}$ w/(m·K)	Viscosity Pa·s $\times 10^5$ (cP)	Solubility in Water 0 °C, V/V
910.9 (15 °C)	650.2 (15 °C)	2.11 (233 K)	2.06 (20 °C)	0.0489

[a] Fluid is supercritical at ambient temperature.
[b] Solid sublimes at atmospheric pressure.
[c] Helium will not solidify at 1 atmosphere pressure (0.101325 MPa). The approximate pressure at which solidification can occur is calculated to be 2535 kPa.

Dew Point: Moisture Content

Dew Point°F	Dew Point°C	Moisture ppm (vol/vol)	Dew Point°F	Dew Point°C	Moisture ppm (vol/vol)
−130	−90.0	0.1	−83	−63.9	6.20
−120	−84.4	0.25	−82	−63.3	6.60
−110	−78.9	0.63	−81	−62.8	7.20
−105	−76.1	1.00	−80	−62.2	7.80
−104	−75.6	1.08	−79	−61.7	8.40
−103	−75.0	1.18	−78	−61.1	9.10
−102	−74.4	1.29	−77	−60.6	9.80
−101	−73.9	1.40	−76	−60	10.50
−100	−73.3	1.53	−75	−59.4	11.40
−99	−72.8	1.66	−74	−58.9	12.30
−98	−72.2	1.81	−73	−58.3	13.30
−97	−71.7	1.96	−72	−57.8	14.30
−96	−71.7	2.15	−71	−57.2	15.40
−95	−70.6	2.35	−70	−56.7	16.60
−94	−70.0	2.54	−69	−56.1	17.90
−93	−69.4	2.76	−68	−55.6	19.20
−92	−68.9	3.00	−67	−55.0	20.60
−91	−68.3	3.28	−66	−54.4	22.10
−90	−67.8	3.53	−65	−53.9	23.60
−89	−67.2	3.84	−64	−53.3	25.60
−88	−66.7	4.15	−63	−52.8	27.50
−87	−66.1	4.50	−62	−52.2	29.40
−86	−65.6	4.78	−61	−51.7	31.70
−85	−65.0	5.30	−60	−51.1	34.00
−84	−64.4	5.70			

CHAPTER 2

High-Performance Liquid Chromatography

CONTENTS

MODES OF LIQUID CHROMATOGRAPHY

The following flowchart provides a rough guide among the various liquid chromatographic techniques, based on sample properties [1].

REFERENCE

1. Courtesy of Millipore Corporation, Waters Chromatography Division.

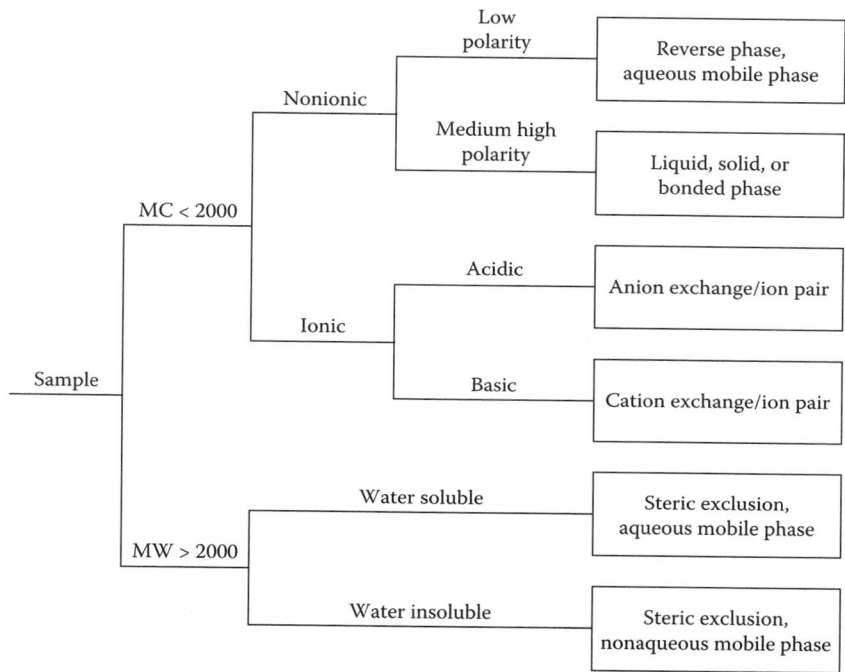

SOLVENTS FOR LIQUID CHROMATOGRAPHY

The following table provides the important physical properties for the selection of solvent systems for high performance liquid chromatography (HPLC) [1–7]. These properties are required for proper detector selection and the prediction of expected column pressure gradients. The values of the dielectric constant aid in estimating the relative solubilities of solutes and other solvents. Data on adsorption energies of useful HPLC solvents on silica and alumina (the eluotropic series) can be found in the chapter on thin layer chromatography. Here we present the values for alumina, ε°, not because this is a common surface encountered in HPLC, but because there is more data on this surface than for silica. These numbers should be used for trend analysis. The data presented were measured at 20 °C, unless otherwise indicated (in parentheses). The solubility parameters, δ, defined fundamentally as the cohesive energy per unit volume, were calculated from vapor pressure data [8] or estimated from group contribution methods [9]. Those values obtained by group contribution are indicated by an asterisk. Solubility parameters are presented in units of $(cal^{1/2}cm^{-3/2})$.

REFERENCES

1. Willard, H. H., L. L. Merritt, J. A. Dean, and F. A. Settle. *Instrumental Methods of Analysis.* 6th ed. Belmont: Wadsworth Publishing Co., 1981.
2. Snyder, L. R., and J. J. Kirkland. *Introduction to Modern Liquid Chromatography.* 2nd ed. New York: John Wiley and Sons (Interscience), 1979.
3. Dreisbach, R. R. *Physical Properties of Chemical Compounds, Number 22 of the Advances in Chemistry Series.* Washington, DC: American Chemical Society, 1959.
4. Krstulovic, A. M., and P. R. Brown. *Reverse Phase High Performance Liquid Chromatography.* New York: John Wiley and Sons (Interscience), 1982.
5. Lide, D. R., ed. *CRC Handbook for Chemistry and Physics.* 90th ed. Boca Raton, FL: CRC Press, 2010.
6. Poole, C. F., and S. A. Shuttle. *Contemporary Practice of Chromatography.* Amsterdam: Elsevier, 1984.
7. Braithwaite, A., and F. J. Smith. *Chromatographic Methods.* 4th ed. London: Chapman and Hale, 1985.
8. Hoy, K. L. "New Values of the Solubility Parameters from Vapor Pressure Data." *Journal of Paint Technology* 42 (1970): 541.
9. Barton, A. F. M. *Handbook of Solubility Parameters and Other Cohesion Parameters.* 2nd ed. Boca Raton, FL: CRC Press, 1991.

Solvents for Liquid Chromatography

Solvent	$\varepsilon°$	δ	Viscosity mPa•s (20 °C)	UV Cutoff nm	Refractive Index (20 °C)	Normal Boiling Temperature (°C)	Dielectric Constant (20 °C)
Acetic acid	1.0	13.01	1.31(15 °C)		1.372	117.9	6.2
Acetone	0.56	9.62	0.30(25)	330	1.359	56.3	20.7(25)
Acetonitrile	0.65	12.11	0.34(25)	190	1.344	81.6	37.5
Benzene	0.32	9.16	0.65	278	1.501	80.1	2.284
1-Butanol		11.60	2.95	215	1.399	117.7	17.8
2-Butanol		11.08	4.21	260	1.397	99.6	15.8(25)
n-Butyl acetate		8.69	0.73	254	1.394	126.1	
n-Butylchloride		8.37	0.47(15)	220	1.402	78.4	
Carbon tetrachloride	0.18	8.55	0.97	263	1.460	76.8	2.238
Chlorobenzene	0.30	9.67	0.80	287	1.525	131.7	2.708
Chloroform	0.40	9.16	0.58	245	1.446	61.2	4.806
Cyclohexane	0.04	8.19	0.98	200	1.426	80.7	2.023
Cyclopentane	0.05	8.10	0.44	200	1.406	49.3	1.965
o-Dichlorobenzene		10.04	1.32(25)	295	1.551	180.5	9.93(25)
N,N-Dimethylacetamide			2.14	268	1.438	166.1	37.8
Dimethylformamide		11.79	0.92	268	1.430	153.0	36.7
Dimethyl sulfoxide	0.62	12.8	2.20	286	1.478	189.0	4.7
1,4-Dioxane	0.56	10.13	1.44(15)	215	1.422	101.3	2.209(25)
2-Ethoxyethanol			2.05	210	1.408	135.6	
Ethyl acetate	0.58	8.91	0.46	256	1.372	77.1	6.02(25)
Ethyl ether	0.38	7.53	0.24	218	1.352	34.6	4.335
Glyme (ethylene glycol dimethyl ether)			0.46(25)	220	1.380	93.0	
n-Heptane	0.01	7.50	0.42	200	1.388	98.4	1.92
n-Hexadecane			3.34	200	1.434	287.0	
n-Hexane	0.01	7.27	0.31	200	1.375	68.7	1.890
Isobutyl alcohol		11.24	4.70(15)	220	1.396	107.7	15.8(25)
Methanol	0.95	14.50	0.55	205	1.328	64.7	32.63(25)
2-Methoxyethanol		11.68	1.72	210	1.402	124.6	16.9
2-Methoxyethyl acetate				254	1.402	144.5	
Methylene chloride	0.42	9.88	0.45(15)	233	1.424	39.8	9.08
Methylethylketone	0.51	9.45	0.42(15)	329	1.379	79.6	18.5
Methylisoamylketone		8.65		330	1.406	−144.0	
Methylisobutylketone	0.43	8.58	0.54(25)	334	1.396	116.5	
N-Methyl-2-pyrrolidone			1.67(25)	285	1.488	202.0	32.0
n-Nonane		7.64	0.72	200	1.405	150.8	1.972
n-Pentane	0.00	7.02	0.24	200	1.357	36.1	1.84
Petroleum ether	0.01		0.30	226		30-60	
β-Phenethylamine				285	1.529(25)	197–198	
1-Propanol	0.82	12.18	2.26	210	1.386	97.2	20.1(25)
2-Propanol	0.82	11.44	2.86(15)	205	1.377	82.3	18.3(25)
Propylene carbonate		13.3			1.419	240.0	
Pyridine	0.71	10.62	0.95	330	1.510	115.3	12.3(25)
Tetrachloroethylene		9.3	0.93(15)	295	1.506	121.2	
Tetrahydrofuran	0.45	9.1	0.55	212	1.407	66.0	7.6

(Continued)

Solvents for Liquid Chromatography (Continued)

Solvent	ε°	δ	Viscosity mPa•s (20 °C)	UV Cutoff nm	Refractive Index (20 °C)	Normal Boiling Temperature (°C)	Dielectric Constant (20 °C)
Tetramethyl urea				265	1.449(25)	175.2	23.0
Toluene	0.29	8.93	0.59	284	1.497	110.6	2.379(25)
Trichloroethylene		9.16	0.57	273	1.477	87.2	3.4(16)
1,2,2-Trichloro-1,2,2-trifluoroethane			0.71	231	1.356(25)	47.6	
2,2,4-Trimethylpentane	0.01	6.86	0.50	215	1.391	99.2	1.94
Water	large	23.53	1.00	< 190	1.333	100.0	80.0
o-Xylene	0.26	9.06	0.81	288	1.505	144.4	2.568
p-Xylene				290	1.5004	138.5	2.270

INSTABILITY OF HPLC SOLVENTS

Solvents that are commonly used in high performance liquid chromatography frequently have inherent chemical instabilities that must be considered when designing an analysis or in the interpretation of results [1,2]. In many cases, such solvents are obtainable with stabilizers added to control the instability or to slow the reaction. Reactive solvents that do not have stabilizers must be used quickly or be given proper treatment. In either case, it is important to understand that the solvents (as they may be used in an analysis) are not necessarily pure materials.

REFERENCES

1. Sadek, P. C. *The HPLC Solvent Guide*. 2nd ed. New York: Wiley Interscience, 2002.
2. Bruno, T. J., and G. C. Straty. *Journal of Research of the National Bureau of Standards* (U.S.) 91, no. 3 (1986): 135–38.

Instability of HPLC Solvents

Solvent	Contaminants, Reaction Products	Stabilizers
Ethers		
Diethyl ether	peroxides[a]	2–3 % (vol/vol) ethanol[b] 1–10 ppm (mass/mass) BHT (1.5–3.5 % ethanol) + (0.2–0.5 % water) + (5–10 ppm (mass/mass) BHT)
Isopropyl ether	peroxides[a]	0.01 % (mass/mass) hydroquinone 5–100 ppm (mass/mass) BHT
1,4-Dioxane	peroxides[a]	25–1500 ppm (mass/mass) BHT
Tetrahydrofuran	peroxides[a]	25–250 ppm (mass/mass) BHT
Chlorinated Alkanes		
Chloroform	hydrochloric acid, chlorine, phosgene (CCl_2O)	0.5–1 % (vol/vol) ethanol 50–150 ppm (mass/mass) amylene[c] various ethanol amylene blends
Dichloromethane	hydrochloric acid, chlorine, phosgene (CCl_2O)	25 ppm (mass/mass) amtlene 25 ppm (mass/mass) cyclohexene 400–600 ppm (mass/mass) methanol various amylene methanol blends
Alcohols		
Ethanol	water; numerous denaturants are commonly added	
Methanol	water; formal dehydrate (at elevated temperature)	
Acetone:	diacetone alcohol, and higher oligomers	

[a] The peroxide concentration that is usually considered hazardous is 250 ppm (mass/mass).
[b] Ethanol does not actually stabilize diethyl ether, nor is it a peroxide scavenger, although it was thought to be so in the past. It is still available in chromatographic solvents to preserve the utility of retention relationships and analytical methods.
[c] Amylene is a generic name for 2-methyl-2-butene.
Abbreviations:
   BHT: 2,6-di-*t*-butyl-p-cresol

ULTRAVIOLET ABSORBANCE OF REVERSE PHASE MOBILE PHASES

The following table provides guidance in the selection of mobile phases that are to be used in conjunction with ultraviolet spectrophotometric detection [1,2]. The data in this table differ from the other solvent tables in this volume in that the wavelength dependence of absorbance is provided here. Moreover, common mixed mobile phases are considered here. The percentages that are given are on the basis of (vol/vol). This material is used by permission of John Wiley and Sons, Inc.

REFERENCES

1. Snyder, L. R., J. J. Kirkland, and J. Glajch. *Practical HPLC Method Development.* New York: John Wiley and Sons, 1997.
2. Li, J. "Signal to Noise Optimization in HPLC UV detection." B. *LC/GC* 10 (1992): 856.

Ultraviolet Absorbance of Reverse Phase Mobile Phases

	Absorbance (AU) at Wavelength (nm) Specified									
	200	205	210	215	220	230	240	250	260	280
Solvents										
Acetonitrile	0.05	0.03	0.02	0.01	0.01	<0.01				
Methanol	2.06	1.00	0.53	0.37	0.24	0.11	0.05	0.02	<0.01	
Degassed	1.91	0.76	0.35	0.21	0.15	0.06	0.02	<0.01		
Isopropanol	1.80	0.68	0.34	0.24	0.19	0.08	0.04	0.03	0.02	0.02
Tetrahydrofuran										
Fresh	2.44	2.57	2.31	1.80	1.54	0.94	0.42	0.21	0.09	0.05
Old[a]	>2.5	>2.5	>2.5	>2.5	>2.5	>2.5	>2.5	>2.5	2.5	1.45
Acids and Bases										
Acetic acid, 1%	2.61	2.63	2.61	2.43	2.17	0.87	0.14	0.01	<0.01	
Hydrochloric acid, 6 mM (0.02%)	0.11	0.02	<0.01							
Phosphoric acid, 0.1%	<0.01									
Trifluoroacetic acid										
0.1% in water	1.20	0.78	0.54	0.34	0.20	0.06	0.02	<0.01		
0.1% in acetonitrile	0.29	0.33	0.37	0.38	0.37	0.25	0.12	0.04	0.01	<0.01
Ammonium phosphate, dibasic, 50 mM	1.85	0.67	0.15	0.02	<0.01					
Triethylamine, 1%	2.33	2.42	2.50	2.45	2.37	1.96	0.50	0.12	0.04	<0.01
Buffers and Salts										
Ammonium acetate, 10 mM	1.88	0.94	0.53	0.29	0.15	0.02	<0.01			
Ammonium bicarbonate, 10 mM	0.41	0.10	0.01	<0.01						
EDTA (ethylenediaminetetra-acetic acid), disodium, 1 mM	0.11	0.07	0.06	0.04	0.03	0.03	0.02	0.02	0.02	0.02
HEPES [N-(2-hydroxyethyl) piperazine-N′-2-ethanesulfonic acid], 10 mM pH 7.6	2.45	2.50	2.37	2.08	1.50	0.29	0.03	<0.01		

(Continued)

Ultraviolet Absorbance of Reverse Phase Mobile Phases (Continued)

	Absorbance (AU) at Wavelength (nm) Specified									
	200	205	210	215	220	230	240	250	260	280
MES [2-(N- morpholino) ethanesulfonic acid], 10 mM, pH 6.0	2.42	2.38	1.89	0.90	0.45	0.06	<0.01			
Potassium phosphate										
Monobasic, 10 mM	0.03	<0.01								
Dibasic, 10 mM	0.53	0.16	0.05	0.01	<0.01					
Sodium acetate, 10 mM	1.85	0.96	0.52	0.30	0.15	0.03	<0.01			
Sodium chloride, 1 mM	2.00	1.67	0.40	0.10	<0.01					
Sodium citrate, 10 mM	2.48	2.84	2.31	2.02	1.49	0.54	0.12	0.03	0.02	0.01
Sodium formate, 10 mM	1.00	0.73	0.53	0.33	0.20	0.03	<0.01			
Sodium phosphate, 100 mM, pH 6.8	1.99	0.75	0.19	0.06	0.02	0.01	0.01	0.01	0.01	<0.01
Tris-hydrochloric acid, 20 mM										
pH 7.0	1.40	0.77	0.28	0.10	0.04	<0.01				
pH 8.0	1.80	1.90	1.11	0.43	0.13	<0.01				

[a] For additional information, see the table entitled "Instability of HPLC Solvents" in this book.

Source: Practical HPLC Method Development, 2nd ed., John Wiley & Sons, New York, 1997. With permission.

ULTRAVIOLET ABSORBANCE OF NORMAL PHASE MOBILE PHASES

The following table provides guidance in the selection of mobile phases that are to be used in conjunction with ultraviolet spectrophotometric detection [1].

REFERENCE

1. Snyder, L. R., J. J. Kirkland, and J. Glajch. *Practical HPLC Method Development.* New York: John Wiley and Sons, 1997.

Ultraviolet Absorbance of Normal Phase Mobile Phases

Solvent	Absorbance (A) at Wavelength (nm) Indicated						
	200	210	220	230	240	250	260
Ethyl acetate	> 1.0	> 1.0	> 1.0	> 1.0	> 1.0	> 1.0	0.10
Ethyl ether	> 1.0	> 1.0	0.46	0.27	0.18	0.10	0.05
Hexane	0.54	0.20	0.07	0.03	0.02	0.01	0.00
Methylene chloride	> 1.0	> 1.0	> 1.0	1.4	0.09	0.00	0.00
Methyl-*t*-butyl ether	> 1.0	0.69	0.54	0.45	0.26	0.11	0.05
n-Propanol	> 1.0	0.65	0.35	0.15	0.07	0.03	0.01
i-Propanol	> 1.0	0.44	0.20	0.11	0.05	0.03	0.02
Tetrahydrofuran	> 1.0	> 1.0	0.70	0.50	0.30	0.16	0.09

SOME USEFUL ION PAIRING AGENTS

The following table provides a short list of ion pair chromatographic modifiers, for use in the separation ionic or ionizable species [1–4]. The use of these modifiers can often greatly improve the chromatographic performance of both normal and reverse phase systems. In many cases new column technology has superseded the use of ion pairing agents, especially when mass spectrometry is used with HPLC. Ion pairing agents can cause numerous difficulties when the column is interfaced with a mass spectrometer.

REFERENCES

1. Poole, C. F., and S. A. Schuette. *Contemporary Practice of Chromatography*. Amsterdam: Elsevier Science Publishers, 1984.
2. Snyder, L. R., and J. J. Kirkland. *Introduction to Modern Liquid Chromatography*, New York: John Wiley and Sons, 1979.
3. Krstulovic, A. M., and P. R. Brown. *Reversed-Phase High Performance Liquid Chromatography*. New York: John Wiley and Sons, 1982.
4. Basic Principles in Reversed Phase Chromatography. Amersham Biosciences Online Education Centre, 2002.

Some Useful Ion Pairing Agents

Ion Type/Examples	Applications/Notes
Perchloric acid	Used for a wide range of basic analytes; typically used at 0.1 M concentration in reverse phase solvent system, and at approximately the same concentration in a water buffer system on the stationary phase in normal mode. See the compatibilities information presented in the next table for cautions.
Trifluoroacetic acid (TFA)	One of the most common ion pairing agents used in HPLC; used for solutes that form positive ions; it is volatile and is therefore often easily removed; it has low absorption within detection wavelengths.
Heptafluorobutyric acid (HFBA)	Used with analytes that form positive ions.
Pentafluoropropionic acid (PFPA)	Used with analytes that form positive ions.
Bis-(2-ethylhexyl) phosphate	Used for cationic species of intermediate polarity, such as phenols; typically used in reverse phase, on bis-(2-ethylhexyl) phosphoric acid/chloroform stationary phase at a pH $\approx$ 3.8.
N,N-Dimethyl protriptyline	Used for carboxylic acids; typically used in normal phase, with a basic (pH $\approx$ 9) buffered stationary phase, and an organic mobile phase.
Quarternary amines: tetramethyl, tetrabutyl, palmityltrimethyl-ammonium salts, usually in chloride or phosphate forms.	Used for strong and weak acids, sulfonated dyes, carboxylic acids, in normal phase applications, typical buffer pH values are between 6 and 8.5, with an organic mobile phase; in reverse phase, the mobile phase is typically aqueous plus a polar organic modifier at nearly neutral pH values.
Tertiary amines: tri-n-octyl amine	Used for carboxylic acids and sulfonates; used in reverse phase mode with a water + buffer + approximately 0.05 M perchloric acid mobile phase.
Sulfonates, alkyl and aromatic: methane or heptanesulfonate, camphorsulfonic acid	Used for strong and weak bases, benzalkonium salts, and catecholamines.
Alkyl sulfates: octyl, decyl, dodecyl, and lauryl sulfates	Used in similar applications as the sulfonates, but provide a different selectivity; typically used in reverse phase mode, often using a water + methanol + sulfuric acid mobile phase.

MATERIALS COMPATIBLE WITH AND RESISTANT TO 72 % PERCHLORIC ACID

The perchloric acid mentioned in the previous table on ion pairing agents must be handled with great care since it can be a very powerful oxidizing agent. Cold perchloric acid at a concentration of 70 % (mass/mass) or less is not considered a very strong oxidizing agent. At concentrations of 73 % or higher, or at lower concentrations but at higher temperatures, perchloric acid is a powerful oxidant. The following table provides some guidance in handling this material in the laboratory [1].

REFERENCE

1. Furr, A. K., ed. *CRC Handbook of Laboratory Safety.* 5th ed. Boca Raton, FL: CRC Press, 2000.

Material	Comments
Elastomers	
Gum Rubber	each batch must be tested to determine compatibility
Vitons	slight swelling only
Metals and Alloys	
Tantalum	excellent
Titanium (chemically pure grade)	excellent
Zirconium	excellent
Columbium (Niobium)	excellent
Hastelloy	slight corrosion rate
Plastics	
Polyvinyl chloride	
Teflon	
Polyethylene	
Polypropylene	
Kel-F	
Vinylidine fluoride	
Saran	
Epoxies	
Others	
Glass	
Glass-lined steel	
Alumina	
Fluorolube	

Incompatible

Plastics

Polyamide (nylon)
Modacrylic ester, Dynel (35–85 %) acrylonitrile
Polyester (dacron)
Bakelite
Lucite
Micarta
Cellulose-based lacquers, metals
Copper
Copper alloys (brass, bronze, etc.) for very shock-sensitive perchlorate salts
Aluminum (dissolves at room temperature)
High nickel alloys (dissolves), others
Cotton
Wood
Glycerin-lead oxide (letharge)

MORE COMMON HPLC STATIONARY PHASES

The following table provides a summary of the general characteristics of the most popular stationary phases used in modern high performance liquid chromatography [1–7]. The most commonly used phases are the bonded reverse phase materials, in which separation control is a function of the mobile (liquid) phase. The selection of a particular phase and solvent system is an empirical procedure involving survey analyses. The references provided below will assist the reader in this procedure.

REFERENCES

1. Snyder, L. R., and J. J. Kirkland. *Introduction to Modern Liquid Chromatography*. 2nd ed. New York: John Wiley and Sons, 1979.
2. Poole, C. F., and S. A. Schuette. *Contemporary Practice of Chromatography*. Amsterdam: Elsevier, 1984.
3. Krstulovic, A. M., and P. R. Brown. *Reverse-Phase High Performance Liquid Chromatography*. New York: John Wiley and Sons (Interscience), 1982.
4. Berridge, J. C. *Techniques for the Automated Optimization of HPLC Separations*. Chichester: John Wiley and Sons, 1985.
5. Braithwaite, A., and F. J. Smith. *Chromatographic Methods*. 4th ed. London: Chapman and Hall, 1985.
6. Sander, L. C., K. E. Sharpless, and M. Pursch. *Journal of Chromatography A*. 880 (2000): 189–202.
7. Snyder, L. R., J. J. Kirkland, and J. Glajch. *Practical HPLC Method Development*. New York: John Wiley and Sons, 1997.

More Common HPLC Stationary Phases

Phase Type	Bond Type	Functional Group	Separation Mode	Notes and Applications
			Solid Sorbents	
Silica (pure)	SiO_2	Si–OH	adsorption	Usually used with nonpolar mobile phase, since it is the most polar sorbent; selectivity is based on differences in number and location of polar groups; results can be unpredictable due to changes in the surface due to adsorption; water or acetic acid is often added (in low concentrations) to the mobile phase to better control surface characteristics; usually the best choice for normal phase and preparative scale separations.
Controlled pore glass	deglassed bosilicate		adsorption	Made by deglassing borosilicate glass and subsequent removal of B_2O_3; pressure stable; can be used in acid and alkali, but not strong alkali; can be sterilized; can be derivatized; narrow pore sizes, rigidity, high pore volume are advantages, used for macromolecular samples, especially biologicals, used mainly in preparatory or industrial scales, rather than analytical scales.
Alumina, acidic	Al_2O_3–A	—	adsorption-normal phase	Similar in characteristics and application to silica; a classic Lewis acid, lacking two electrons in the Al center, having an approximate pH of 4.5; this phase can be treated to make it more retentive to electron rich species.
Alumina, neutral	Al_2O_3		adsorption-normal phase	Prepared as a neutral surface, approximate pH = 7.5, used for separation of aromatics and moieties that contain electronegative groups such as oxygen; should not be heated above 150 °C; more prone to chemisorption than silica; somewhat lesser efficiency (plate height) than silica.
Alumina, base treated	Al_2O_3–B		adsorption-normal phase, weak cation exchange	Base treatment makes the phase suitable for separation of hydrogen bonding or cationic species; approximate pH = 10; weak cation exchanger, should not be heated above 150 °C.
Zirconia (pure)	ZrO_2	—	normal phase, ion exchange	Can be used over the entire pH range, 1–14; can be heated to 200 °C; has a stable particle size that will not shrink; surface is free of silanol groups; can function in ion exchange mode since it is a Lewis acid.
Titania	TiO_2		normal and reverse phase	Has basic –OH groups on the surface; stable at high pH; has been used for the separation of phosphopeptides; has been modified to sol–gel phases modified with poly(dimethylsiloxane).

Name	Composition/Structure	Type	Description
Porous graphitic carbon	intertwined ribbons of porous graphite	carbon network	primarily reverse phase, can be used in normal phase	Spherical macrostructure with a crystalline graphitic surface; stable over wide pH range; thermally stable to high temperatures; lower capacity and efficiency than silica phases; unique selectivity; can be used to separate very polar compounds; can be modified.
Magnesium silicate (Florisil)	MgO_3Si	—	polar adsorbent	Florisil is a magnesium substituted silica with a highly polar surface; typically has a relatively large particle size (approximately 200 μm), and therefore high flow rates are possible even with viscous samples; used with many official methods, and in cases where the Lewis acidity of alumina would be problematic; must be used with caution since aromatics, amines esters and other compounds can be chemisorbed.
Hydroxyapatite	$Ca_{10}(PO_4)_6(OH)_2$	—	polar adsorbent; specific interactions	Hexagonally crystallized calcium phosphate; pressure stable to 15 MPa; typically used with a linear gradient of a potassium or sodium phosphate buffer at a pH of approximately 6.8; useful for the separation of proteins and other biopolymers, nucleic acids, viruses; see the entry under specialized HPLC phases.

Polymeric Phases

Name	Structure	Type	Description
Styrene-divinyl benzene		reverse phase	Polymer must have at least 8 % divinyl benzene to be suitable for high pressure; bed volume will change with solvent or ionic strength of mobile phase, once a solvent is chosen, it generally cannot be changed; structure can be micro- and macroporous, allowing larger molecules to enter structure; stable at pH 1–13; chromatographic behavior similar to ODS but with specific interactions (π–π) for aromatics; can be modified for ion exchange.
Agarose			Cross-linked polysaccharide stable over pH = 1–14; can be derivatized for affinity chromatography.

(Continued)

More Common HPLC Stationary Phases (Continued)

Phase Type	Bond Type	Functional Group	Separation Mode	Notes and Applications
Bonded Phases, Straight Chain				
ODS	Si–O–Si–C	octadecyl, n–C_{18}–$(CH_2)_{17}$–CH_3, hydrocarbon chain	bonded, reverse phase	Octadecylsilane; most common material used in HPLC; high resolution possible; pH must be maintained between 2 and 7.
C2	Si–O–Si–C	–CH_2CH_3	moderately polar bonded, reverse phase	A moderately polar phase that is used for aqueous samples, blood and urine samples; moderate polarity derives from the polar substrate, silica; has a polarity similar to a cyclohexyl bonded phase.
OS	Si–O–Si–C	octyl, n–C_8 hydrocarbon chain	bonded, reverse phase	Octylsilane; lower resolution and retention than the octadecyl bonded phase; useful when separations involve species of greatly different polarity.
C30	Si–O–Si–C	triacontyl, C_{30} hydrocarbon chain	bonded, reverse phase	A polymeric phase useful for the separation of caratenoid compounds, fullerenes.
TMS	Si–O–Si–C	methyl, CH_3	bonded, reverse phase	Tetramethylsilane; lowest resolution of reverse-phase packings; useful for "survey" separations and for large molecules.
ODA	Al–O–Si–R	octadecyl, n–C_{18} $(CH_2)_{17}CH_3$, hydrocarbon chain	bonded, reverse phase	Far less used than the silica bonded phases, although alumina chemistry can be more facile than silica chemistry; has been used for separation of small and larger peptide molecules.
OA	Al–O–Si–R	octyl, n–C_8 hydrocarbon chain	bonded, reverse phase	Far less used than the silica bonded phases, although alumina chemistry can be more facile than silica chemistry; has been used for separation of small and larger peptide molecules.
Bonded Phases, Functionalized[a]				
Bonded diol	Si–O–Si–C	OH OH (–C–C– hydrocarbon chain with hydroxyl groups)	polar bonded phase	A polar phase that has a hydrogen bonding capability similar to that of unbonded silica; useful in size-exclusion chromatography, and in the analysis of glycols and glycerol, oils, lipids, and related compounds.

Name	Linkage	Structure	Phase type	Description
Carboxyl acid, CBA	Si–O–Si–C	–CH$_2$CH$_2$COOH	polar bonded phase	Medium polarity phase that has a weak cation exchange capability useful for strong cations; above pH = 4.8, most of the functional groups are negatively charged, and therefore the phase can be used for cationic compounds; lowering pH to 2.8 elutes retained analytes.
Cyclyhexyl, CH	Si–O–Si–C	–C$_6$H$_9$	moderately polar bonded phase	A moderately polar phase that is used for aqueous samples; moderate polarity derives from the polar substrate, silica; has a polarity similar to a C2 bonded phase.
Bonded nitrile	Si–O–Si–C	–CH$_2$CH$_2$CH$_2$–C≡N	moderately polar bonded phase	Moderate polar phase, but with selectivity modified with respect to silica; less sensitive to mobile phase impurities than silica; less retentive than OS; many nitrile phases are less stable than OS; also called cyanopropyl phase.
Bonded nitro	Si–O–Si–C	–NO$_2$	polar bonded phase	
Bonded amine	Si–O–Si–C	CH$_2$CH$_2$CH$_2$–NH$_2$	polar bonded phase	Selectivity is modified with respect to silica through the aminopropyl functionality; the propyl linkage can interact with nonpolar interactions; highly polar phase overall; phase is less stable than cyano or diol phases; can utilize hydrogen bonding and ion exchange mechanisms; protonates below pH = 9.8; useful for sugar and carbohydrate separations; not recommended for samples that contain aldehydes and ketones.
Phenyl	Si–O–Si–C	(benzene ring) often represented as φ	normal or reverse phase	Lower efficiency than other bonded phases; more polar than ODS, OS, and TMS phases; used with both normal and reverse phase solvent systems.
Polybutadiene	Zr–C	[CH$_2$–C=C–CH$_2$]$_n$	reverse phase	Similar to ODS in separation characteristics; can be used up to 150 °C.
Carbon on zirconia	Zr–C		reverse phase	This elemental carbon on zirconia is useful in the separation of diastereomers.
Polystyrene on zirconia	Zr–C		reverse phase	Separations are similar to those obtained with phenyl-bonded silica; can be used up to 150 °C.

a In this context, functionalized refers to functionalization beyond straight chain hydrocarbons.

(Continued)

More Common HPLC Stationary Phases (Continued)

Phase Type	Bond Type	Functional Group	Separation Mode[a]	Notes and Applications
		Bonded Phases, Ion Exchange[a]		
Bonded amine	Si–O–Si–C	$CH_2CH_2CH_2–NH_2$	polar bonded phase	Selectivity is modified with respect to silica through the aminopropyl functionality; the propyl linkage can interact via nonpolar interactions; highly polar phase overall; phase is less stable than cyano or diol phases; can utilize hydrogen bonding and ion exchange mechanisms, and as such is a weak anion exchanger; protonates below pH = 9.8; useful for sugar and carbohydrate separations; not recommended for samples that contain aldehydes and ketones.
Benzene sulfonic acid	Si–O–Si–C	$CH_2CH_2\phi–SO_3^-H^+$	ion exchange	Separates cations, with divalent ions more strongly retained than monovalent ions; phosphate buffer systems are often used, sometimes with low concentrations of polar nonaqueous modifiers added; the presence of the benzene group on the benzenesulfonic acid moiety gives this phase a dual nature, and the ability to separate based upon nonpolar interactions.
Propyl, ethylene diamine	Si–O–Si–C	$–CH_2CH_2CH_2– NHCH_2CH_2NH_2$	ion exchange	Weak anion exchange phase for aqueous and biological samples; incorporates a bidentate ligand to form chelate complexes useful for metal separations; less polar than the propyl amine bonded phase.
Propyl sulfonic acid	Si–O–Si–C	$–CH_2CH_2CH_2– SO_3^-Na +$	ion exchange	Strong cation exchange substrate for aqueous and biological samples; effective for the separation of weaker cations such as pyridinium compounds
Propyl, trimethylamino	Si–O–Si–C	$–CH_2CH_2CH_2–N^+ Cl^-(CH_3)_2$	ion exchange	Strong anion exchange phase for aqueous and biological samples suitable for weaker anions such as carboxylic acids; properties may be modified or conditioned by proper formulation of buffer mobile phases (see the appropriate table in the Solution Properties chapter).

[a] Note that while the principal separation mechanism is ion exchange, the organic moieties on many of these phases can interact through nonpolar interactions as well. Thus, many phases are mixed mode.

Note: ϕ denotes a phenyl group.

ELUOTROPIC VALUES OF SOLVENTS ON OCTADECYLSILANE

The following table provides, for comparative purposes, eluotropic values on octadecyl silane (ODS) and octyl silane (OS) for common solvents [1,2].

REFERENCES

1. Krieger, P. A. *High Purity Solvent Guide*. McGaw Park: Burdick and Jackson Laboratories, 1984.
2. Ahuja, S. *Trace and Ultratrace Analysis by HPLC*. New York: John Wiley and Sons, 1992.

Eluotropic Values of Solvents on Octadecylsilane

Solvent	Eluotropic Value, ODS	Eluotropic Value, OS
Acetic acid	—	2.7
Acetone	8.8	9.3
Acetonitrile	3.1	3.3
1,4-Dioxane	11.7	13.5
Dimethylformamide	7.6	9.4
Methanol	1.0	1.0
Ethanol	3.1	3.2
n-Propanol	10.1	10.8
2-Propanol	8.3	8.4
Tetrahydrofuran	3.7	—

MESH-SIZE RELATIONSHIPS

The following table provides the relationship between particle sizes and standard sieve mesh sizes. However, it should be noted that the trend in HPLC has been toward shorter columns containing much finer particles than the standard sieves will separate. These values will be of use when packing relatively large diameter columns for bench-top elutions.

Mesh-Size Relationships

Mesh Range	Top Screen Opening, μm	Bottom Screen Opening, μm	Micron Screen, μm	Range Ratio
80/100	177	149	28	1.19
100/120	149	125	24	1.19
100/140	149	105	44	1.42
120/140	125	105	20	1.19
140/170	105	88	17	1.19
170/200	88	74	14	1.19
200/230	74	63	11	1.19
230/270	63	53	10	1.19
270/325	53	44	9	1.20
325/400	44	37	7	1.19

EFFICIENCY OF HPLC COLUMNS

The efficiency of a column used for HPLC describes the ability of the column to produce sharp narrow peaks. Typically, the efficiency is represented at the plate number, N. The plate number can be estimated by:

$$N = 3500 \ L/d_p,$$

where L is the column length in cm, and d_p is the particle diameter in (μm). The following table provides the plate number for optimized test conditions for various combinations of column length and particle diameter. It therefore represents the upper limit of efficiency, and can be used as a column diagnostic measurement [1].

REFERENCE

1. Snyder, L. R., J. J. Kirkland, and J. L. Glajch. *Practical HPLC Method Development*. 2nd ed. New York: John Wiley and Sons, 1997. Reproduced with permission.

Efficiency of HPLC Columns

Particle Diameter, μm	Column Length, cm	Plate Number
10	15	6000–7000
10	25	8000–10,000
5	10	7000–9000
5	15	10,000–12,000
5	25	17,000–20,000
3	5	6000–7000
3	7.5	9000–11,000
3	10	12,000–14,000
3	15	17,000–20,000

COLUMN FAILURE PARAMETERS

The point at which a column used for HPLC will fail depends largely upon how the operator uses it. Eventually, however, all HPLC columns will fail. The onset of column failure can be monitored by two common failure parameters, the peak asymmetry factor, A_s, and the peak tailing factor. These parameters are defined according to the figure below:

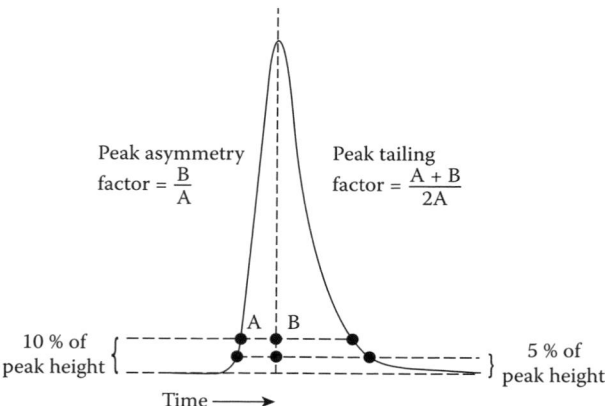

The two parameters are related, and the following table provides the interconversion [1].

REFERENCE

1. Snyder, L. R., J. J. Kirkland, and J. L. Glajch. *Practical HPLC Method Development.* 2nd ed. New York: John Wiley and Sons, 1997. Reproduced with permission.

Peak Asymmetry Factor (A_s, 10 %)	Peak Tailing Factor (5 %)
1.0	1.0
1.3	1.2
1.6	1.4
1.9	1.6
2.2	1.8
2.5	2.0

To put these factors into context, column performance can be described as in the following table:

Peak Asymmetry Factor (A_s, 10 %)	Column Performance
1.0–1.05	excellent, new
1.2	acceptable
2.0	degraded, approaching poor
4.0	not usable

COLUMN REGENERATION SOLVENT SCHEMES

When HPLC columns become fouled or inefficient, it is sometimes possible to regenerate performance to some degree. The following solvent schemes have been found helpful [1,2]. Some workers recommend daily flushing of columns to maintain performance and avoid problems. Another cause of efficiency loss can be settling of the packing, especially if the column has been in service for several months. In some cases, it is possible to top off the columns with packing material. If packing material is not available, glass beads of the appropriate size can be used.

REFERENCES

1. Meyer, V. R. *Practical High Performance Liquid Chromatography*. 4th ed. Chichester: John Wiley and Sons, 2004.
2. Majors, R. E. *The Cleaning and Regeneration of Reverse Phase HPLC Columns*. Europe: LC-GC, July 2003.

Silica Adsorbent Columns:

Pump the following solvents sequentially at the rate of 1–3 mL/min.
75 mL of tetrahydrofuran
75 mL of methanol

If acidic impurities are suspected:
75 mL of 1–5 % (vol/vol) of pyridine in water

If basic impurities are suspected:
75 mL of 1–5 % (vol/vol) of acetic acid in water
75 mL of tetrahydrofuran
75 mL of *t*-butylmethyl ether
75 mL of n-hexane or hexanes

Bonded Phase Columns (OS, ODS, phenyl, and nitrile phases):

Pump the following solvents sequentially at the rate of 0.5–2 mL/min.
75 mL of water, while injecting 100 μL of dimethyl sulfoxide four times
75 mL of methanol
75 mL of chloroform
75 mL of methanol

Another sequence that is possible, usually offline from the HPLC instrument, is the following:

water
0.1 M sulfuric acid
water

It is recommended that this be done with a spare pump used only for this solvent sequence.

When proteins have been separated on a reverse phase column, the following solvent systems can be used for regeneration and cleaning:

1 % (vol/vol) acetic acid in water
1 % trifluoroacetic acid in water
0.1 % trifluoroacetic acid + propanol, 40:60 (vol/vol)

Note: this mixture is relatively viscous, so a low flow rate should be used.

triethylamine + propanol 40:60 (vol/vol)
    (adjust triethylamine to pH = 2.5 with 0.25 N phosphoric acid before mixing)
aqueous urea or guanidine, 5–8 M (adjusted to pH 6–8)
aqueous sodium chloride, sodium phosphate or sodium sulphate at 0.5–1.0 M
dimethyl sulfoxide + water, 50:50 (vol/vol)
dimethylformamide + water, 50:50 (vol/vol)

When metal ions have been introduced into a reverse phase column, the organic solvents listed above are often ineffective. In those cases, the following mixture may be useful:

0.05 M ethylenediaminetetraacetic acid (EDTA)
water flush

Anion Exchange Columns:

Pump the following solvents sequentially at the rate of 0.5–2 mL/min.
    75 mL of water
    75 mL of methanol
    75 mL of chloroform, then, apply a methanol to water gradient

Cation Exchange Columns:

Pump the following solvents sequentially at the rate of 0.5–2 mL/min.
    75 mL of water with 100 μL dimethyl sulfoxide injected four times
    75 mL of tetrahydrofuran
    water flush

Styrene-Divinylbenzene Polymer Columns:

Pump the following solvents sequentially at the rate of 0.5–2 mL/min.
    40 mL toluene, or 40 mL tetrahydrofuran (peroxide free)

SPECIALIZED STATIONARY PHASES FOR LIQUID CHROMATOGRAPHY

The following table provides information on the properties and application of some of the more specialized bonded, adsorbed, and polymeric phases used in modern high performance liquid chromatography. In many cases the phases are not commercially available, and the reader is referred to the appropriate literature citation for details on the synthesis.

REFERENCES

1. Pietrzyk, D. J., and W. J. Cahill. "Amberlite XAD-4 as a Stationary Phase for Preparative Liquid Chromatography in a Radially Compressed Column." *Journal of Liquid Chromatography & Related Technologies* 5, no. 4 (1982): 781–95.
2. Nikolov, Z., M. Meagher, and P. Reilly. "High-Performance Liquid Chromatography of Trisaccharides on Amine-Bonded Silica Columns." *Journal of Chromatography* 321 (1985): 393–99.
3. Ascalone, V., and L. Dal Bo. "Determination of Ceftriaxone, A Novel Cephalosporin, in Plasma, Urine and Saliva by High-Performance Liquid Chromatography on an NH_2 Bonded-Phase Column." *Journal of Chromatography* 273 (1983): 357.
4. Pharr, D. Y., P. C. Uden, and S. Siggia. "A 3-(p-Acetylphenoxy) Propylsilane Bonded Phase for Liquid Chromatography of Basic Amines and Other Nitrogen Compounds." *Journal of Chromatographic Science* 23 (1985): 391.
5. Felix, G., and C. Bertrand. "Separation of Polyaromatic Hydrocarbons on Caffeine-Bonded Silica Gel." *Journal of Chromatography* 319 (1985): 432.
6. Felix, G., C. Bertrand, and F. Van Gastel. "A New Caffeine Bonded Phase for Separation of Polyaromatic Hydrocarbons and Petroleum Asphaltenes by High-Performance Liquid Chromatography." *Chromatographia* 20, no. 3 (1985): 155.
7. Bruner, F., G. Bertoni, and P. Ciccioli. "Comparison of Physical and Gas Chromatographic Properties of Sterling FT and Carbopack-C Graphitized Carbon Blacks." *Journal of Chromatography* 120 (1976): 307–19.
8. Bruner, F., P. Ciccioli, G. Crescentini, and M. T. Pistolesi. "Role of the Liquid Phase in Gas-Liquid-Solid Chromatography and Its Influence on Column Performance: An Experimental Approach." *Analytical Chemistry* 45, no. 11, (1973): 1851.
9. Ciccioli, P., and Liberti, A. "Microbore Columns Packed with Graphitized Carbon Black for High-Performance Liquid Chromatography." *Journal of Chromatography* 290, 173 (1984).
10. DiCorcia, A., Liberti, A., and Samperi, R. "Gas-Liquid-Solid Chromatography-Theoretical Aspects and Analysis of Polar Compounds." *Analytical Chemistry* 45, no. 7 (1973): 1228.
11. Hatada, K., T. Kitayama, S.-I. Shimizu, H. Yuki, W. Harris, and O. Vogl. "High-Performance Liquid Chromatography of Aromatic Compounds on Polychloral." *Journal of Chromatography* 248 (1982): 63–68.
12. Abe, A., K. Tasaki, K. Inomata, and O. Vogl. "Conformational Rigidity of Polychloral: Effect of Bulky Substituents on the Polymerization Mechanism." *Macromolecules* 19 (1986): 2707.
13. Kubisa, P., L. S. Corley, T. Kondo, M. Jacovic, and O. Vogl. "Haloaldehyde Polymers. XXIII: Thermal and Mechanical Properties of Chloral Polymers." *Polymer Engineering and Science* 21, no. 13 (1981): 829.
14. Veuthey, J.-L., M.-A. Bagnoud, and W. Haerdi. "Enrichment of Amino and Carboxylic Acids Using Copper-Loaded Silica Pre-Columns Coupled On-Line with HPLC." *International Journal of Environmental Analytical Chemistry* 26 (1986): 157–66.
15. Guyon, F., A. Foucault, M. Caude, and R. Rosset. "Separation of Sugars by HPLC on Copper Silicate Gel." *Carbohydrate Research* 140 (1985): 135–38.
16. Leonard, J. L., F. Guyon, and P. Fabiani. "High-Performance Liquid Chromatography of Sugars on Copper (11) Modified Silica Gel." *Chromatographia* 18 (1984): 600.
17. Yamazaki, S., H. Omori, and C. Eon Oh. "High Performance Liquid Chromatography of Alkaline-Earth Metal Ions Using Reversed-Phase Column Coated with N-n-Dodecylimminodiacetic Acid." *Journal of High Resolution Chromatography & Chromatography Communications* 9 (1986): 765.

18. Miller, N. T., and C. H. Shieh. "Preparative Hydrophobic Interaction Chromatography of Proteins Using Ether Based Chemically Bonded Phases." *Journal of Liquid Chromatography & Related Technologies* 9, no. 15 (1986): 3269.

19. Williams, R. C., J. F. Vasta-Russell, J. L. Glajch, and K. Golebiowski. "Separation of Proteins on a Polymeric Fluorocarbon High-Performance Liquid Chromatography Column Packing." *Journal of Chromatography* 371 (1986): 63–70.

20. Hirayama, C., H. Ihara, T. Yoshinga, H. Hirayama, and Y. Motozato. "Novel Packing for High Pressure Liquid Chromatography. Partially Alkylated and Cross-Linked PMLG Spherical Particles." *Journal of Liquid Chromatography & Related Technologies* 9, no. 5 (1986): 945–54.

21. Kawasaki, T., W. Kobayashi, K. Ikeda, S. Takahashi, and H. Monma. "High-Performance Liquid Chromatography Using Spherical Aggregates of Hydroxyapatite Micro-Crystals as Adsorbent." *European Journal of Biochemistry* 157 (1986): 291–95.

22. Kawasaki, T., and W. Kobayashi. "High-Performance Liquid Chromatography Using Novel Square Tile-Shaped Hydroxyapatite Crystals as Adsorbent." *Biochemistry International* 14, no. 1 (1987): 55–62.

23. Funae, Y., S. Wada, S. Imaoka, S. Hirotsune, M. Tominaga, S. Tanaka, T. Kishimoto, and M. Maekawa. "Chromatographic Separation of α-Acid Glyco-Protein from α-Antitrypsin by High-Performance Liquid Chromatography Using a Hydroxyapatite Column." *Journal of Chromatography* 381 (1986): 149–52.

24. Kadoya, T., T. Isobe, M. Ebihara, T. Ogawa, M. Sumita, H. Kuwahara, A. Kobayashi, T. Ishikawa, and T. Okuyama. "A New Spherical Hydroxyapatite for High Performance Liquid Chromatography of Proteins." *Journal of Liquid Chromatography & Related Technologies* 9, no. 16 (1986): 3543–57.

25. Kawasaki, T., M. Niikura, S. Takahashi, and W. Kobayashi. "High-Performance Liquid Chromatography Using Improved Spherical Hydroxyapatite Particles as Adsorbent: Efficiency and Durability of the Column." *Biochemistry International* 13, no. 6 (1986): 969–82.

26. Bernardi, G. "Chromatography of Nucleic Acids on Hydroxyapatite Columns." *Methods in Enzymology* 21D (1971): 95–139.

27. Bernardi, G. "Chromatography of Proteins on Hydroxyapatite." *Methods in Enzymology* 27 (1973): 471–79.

28. Bernardi, G. "Chromatography of Proteins on Hydroxyapatite." *Methods in Enzymology* 22 (1971): 325–39.

29. Figueroa, A., C. Corradini, B. Feibush, and B. Karger. "High-Performance Immobilized-Metal Affinity Chromatography of Proteins on Iminodiacetic Acid Silica-Based Bonded Phases." *Journal of Chromatography* 371 (1986): 335–52.

30. Danielson, N. D., S. Ahmed, J. A. Huth, and M. A. Targrove. "Characterization of Organomagnesium Modified Kel-f Polymers as Column Packings." *Journal of Liquid Chromatography & Related Technologies* 9, no. 4 (1986): 727–43.

31. Taylor, P. J., and P. L. Sherman. "Liquid Crystals as Stationary Phases for High Performance Liquid Chromatography." *Journal of Liquid Chromatography & Related Technologies* 3, no. 1 (1980): 21–40.

32. Taylor, P. J., and P. L. Sherman. "Liquid Crystals as Stationary Phases for High Performance Liquid Chromatography." *Journal of Liquid Chromatography & Related Technologies* 2, no. 9 (1979): 1271–90.

33. Felix, G., and C. Bertrand. "HPLC on Pentafluorobenzamidopropyl Silica Gel." *Journal of High Resolution Chromatography & Chromatography Communications* 8 (1985): 362.

34. Kurosu, Y., H. Kawasaki, X.-C. Chen, Y. Amano, Y.-I. Fang, T. Isobe, and T. Okuyama. "Comparison of Retention Times of Polypeptides in Reversed Phase High Performance Liquid Chromatography on Polystyrene Resin and on Alkyl Bonded Silica." *Bunseki Kagaku* 33 (1984): E301–E308.

35. Yang, Y.-B., and M. Verzele. "New Water-Compatible Modified Polystyrene as a Stationary Phase for High-Performance Liquid Chromatography." *Journal of Chromatography* 387 (1987): 197.

36. Nieminen, N., and P. Heikkila. "Simultaneous Determination of Phenol, Cresols and Xylenols in Workplace Air, Using a Polystyrene-Divinylbenzene Column and Electrochemical Detection." *Journal of Chromatography* 360 (1986): 271.

37. Yang, Y. B., F. Nevejans, and M. Verzele. "Reversed-Phase and Cation-Exchange Chromatography on a New Poly(Styrenedivinylbenzene) High Capacity, Weak Cation-Exchanger." *Chromatographia* 20, no. 12 (1985): 735.

38. Tweeten, K. A., and T. N. Tweeten. "Reversed-Phase Chromatography of Proteins on Resin-Based Wide-Pore Packings." *Journal of Chromatography* 359 (1986): 111.

39. Lee, D. P., and A. D. Lord. "A High Performance Phase for the Organic Acids." *LC-GC* 5, no. 3 (1987): 261.

40. Miyake, K., F. Kitaura, N. Mizuno, and H. Terada. "Determination of Partition Coefficient and Acid Dissociation Constant by High-Performance Liquid Chromatography on Porous Polymer Gel as a Stationary Phase." *Chemical & Pharmaceutical Bulletin* 35, no. 1 (1987): 377.

41. Joseph, J. M. "Selectivity of Poly(Styrene-divinylbenzene) Columns." *ACS Symposium Series* 297 (1986): 83–100.

42. Cope, M. J., and I. E. Davidson. "Use of Macroporous Polymeric High-Performance Liquid Chromatographic Columns in Pharmaceutical Analysis." *Analyst* 112 (1987): 417.

43. Werkhoven-Goewie, C. E., W. M. Boon, A. J. J. Praat, R. W. Frei, U. A. Th. Brinkman, and C. J. Little. "Preconcentration and LC Analysis of Chlorophenols, Using a Styrene-Divinyl-Benzene Copolymeric Sorbent and Photochemical Reaction Detection." *Chromatographia* 16 (1982): 53.

44. Smith, R. M. "Selectivity Comparisons of Polystyrenedivinylbenzene Columns." *Journal of Chromatography* 291 (1984): 372–76.

45. Köhler, J. "Poly (Vinylpyrrolidone)-Coated Silicia: A Versatile, Polar Stationary Phase for H.P.L.C." *Chromatographia* 21 (1986): 573.45.

46. Murphy, L. J., S. Siggia, and P. C. Uden. "High-Performance Liquid Chromatography of Nitroaromatic Compounds on an N-Propylaniline Bonded Stationary Phase." *Journal of Chromatography* 366 (1986): 161.

47. Felix, G., and C. Bertrand. "HPLC on n-Propyl Picryl Ether Silica Gel." *Journal of High Resolution Chromatography & Chromatography Communications* 7 (1984): 714.

48. Risner, C. H., and J. R. Jezorek. "The Chromatographic Interaction and Separation of Metal Ions with 8-Quinolinol Stationary Phases in Several Aqueous Eluents." *Analytica Chimica Acta* 186 (1986): 233.

49. Shahwan, G. J., and J. R. Jezorek. "Liquid Chromatography of Phenols on an 8-Quinolinol Silica Gel-Iron (III) Stationary Phase." *Journal of Chromatography* 256 (1983): 39–48.

50. Krauss, G.-J. "Ligand-Exchange H.P.L.C. of Uracil Derivatives on 8-Hydroxyquinoline-Silica-Polyol." *Journal of High Resolution Chromatography & Chromatography Communications* 9 (1986): 419–20.

51. Hansen, S. A., P. Helboe, and M. Thomsen. "High-Performance Liquid Chromatography on Dynamically Modified Silica." *Journal of Chromatography* 360 (1986): 53–62.

52. Helboe, P. "Separation of Corticosteroids by High-Performance Liquid Chromatography on Dynamically Modified Silica." *Journal of Chromatography* 366 (1986): 191–96.

53. Hansen, D. H., P. Helboe, and M. Thomsen. "Dynamically Modified Silica: The Use of Bare Silica in Reverse-Phase High-Performance Liquid Chromatography." *Trends in Analytical Chemistry* 4, no. 9 (1985): 233.

54. Flanagan, R. J. "High-Performance Liquid Chromatographic Analysis of Basic Drugs on Silica Columns Using Non-Aqueous Ionic Eluents." *Journal of Chromatography* 323 (1985): 173–89.

55. Vespalec, R., M. Ciganková, and J. Viska. "Effect of Hydrothermal Treatment in the Presence of Salts on the Chromatographic Properties of Silica Gel." *Journal of Chromatography* 354 (1986): 129.

56. Unger, K. K., G. Jilge, J. N. Kinkel, and M. T. W. Hearn. "Evaluation of Advanced Silica Packings for the Separation on Biopolymers by High-Performance Liquid Chromatography." *Journal of Chromatography* 359 (1986): 61–72.

57. Lullmann, C., H.-G. Genieser, and B. Jastorff. "Structural Investigations on Reversed-Phase Silicas." *Journal of Chromatography* 354 (1986): 434–37.

58. Schou, O., and P. Larsen. "Preparation of 6,9,12-Trioxatridecylmethylsilyl Substituted Silica, A New Stationary Phase for Liquid Chromatography." *Acta Chemica Scandinavica* B35 (1981): 337.

59. Desideri, P. G., L. Lepri, L. Merlini, and L. Checchini. "High-Performance Liquid Chromatography of Amino Acids and Peptides on Silica Coated with Ammonium Tungstophosphate." *Journal of Chromatography* 370 (1986): 75.

Specialized Stationary Phases for Liquid Chromatography

Name: Amberlite XAD-4

Structure: Macroporous polystyrene-divinylbenzene nonpolar adsorbent, 62–177 μm particle size.

Analytical Properties: Used mainly in preparative scale HPLC; stable over entire pH range; (1–13) sometimes difficult to achieve column to column reproducibility due to packing the irregular particles. Relatively lower efficiency than alkyl bonded phases; particles tend to swell as the organic content of the mobile phase increases.

Reference: 1

NAME: amine bonded phase

Structure: NH_2 functionality with a Si–O–Si–C or Si–C linkage.

Analytical Properties: Polar phase useful for sugar and carbohydrate separation; not recommended for samples that contain aldehydes and ketones.

Reference: 2,3

Name: 3-(p-acetophenoxy) propyl bonded phase

Structure:

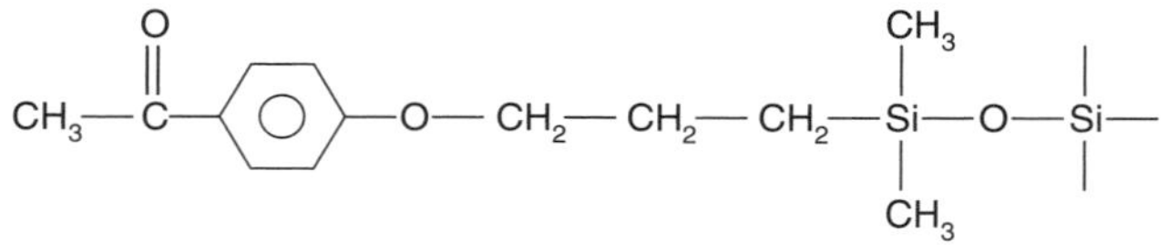

Analytical Properties: Selective for aromatic amines, with the selectivity being determined by the interactions with the carbonyl group.

Reference: 4

Name: caffeine bonded phase

Structure:

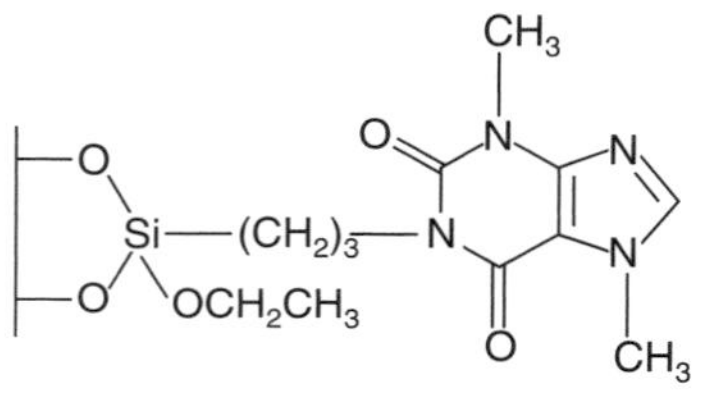

Analytical Properties: Separation of polynuclear aromatic hydrocarbons (of the type often encountered in petroleum residue work) by donor-acceptor complex formation.

Reference: 5,6

Name: graphitized carbon black

Structure: Carbon subjected to + 1300 °C in helium atmosphere, resulting in a graphite-like structure in the form of polyhedra, with virtually no unsaturated bonds, ions, lone electron pairs or free radicals.

Analytical Properties: Especially for use in microbore columns; suggested for lower aromatics but with some potential for higher molecular mass compound separations.

Reference: 7–10

Name: polychloral (polytrichloroacetaldehyde)

Structure:

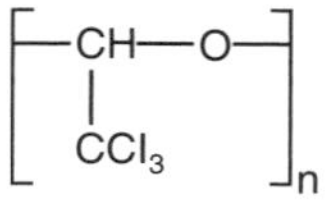

Specialized Stationary Phases for Liquid Chromatography (Continued)

Analytical Properties: Separation of lower aromatic hydrocarbons and small fused ring systems using toluene and hexane methanol as the stationary phases; the relatively low pressure rating on the polymeric phase limits solvent flow rate.
Reference: 11,12,13

Name: cyclam-copper-silica
Structure:

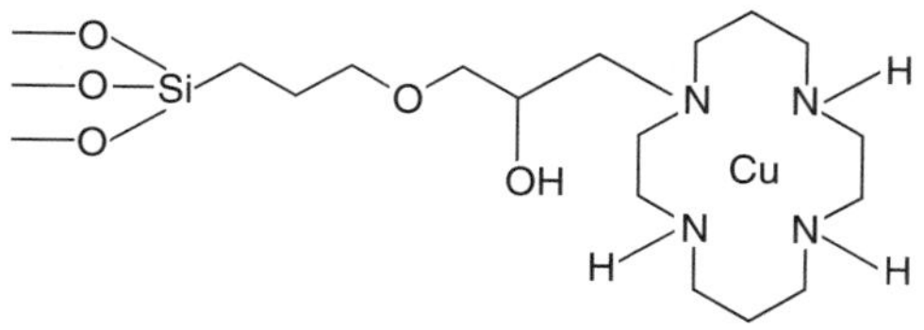

Analytical Properties: This phase has found use in preconcentrating carboxylic acids on precolumns.
Reference: 14

Name: bis-dithiocarbamate-copper-silica
Structure:

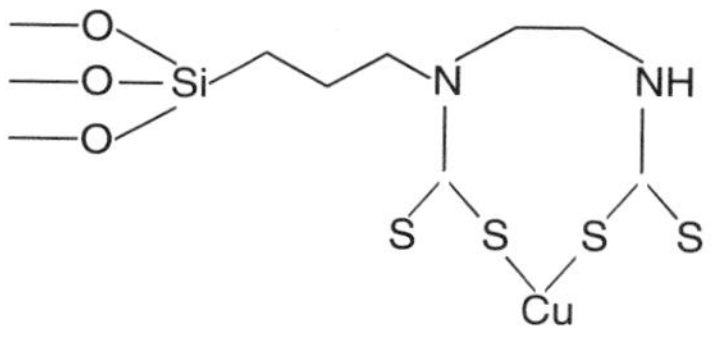

Analytical Properties: This phase has found use in preconcentrating amino acids on precolumns.
Reference: 14

Name: copper (II) coated silica gel
Structure: $(\equiv Si\text{–}O)_2 Cu(NH_3)_x (H_2O)_y$
  $x = 1$ or 2

Analytical Properties: Separation of sugars and amino sugars by ligand exchange or partitioning interactions using water + acetonitrile + ammonia liquid phases. The phase is usually prepared by treating silica gel with ammoniacal copper sulfate solution prior to packing.
Reference: 15,16

Name: N-n-dodecyliminodiacetic acid (coated on silica gel)
Structure:

$$CH_3(CH_2)_{11}N \diagup\diagdown \begin{matrix} CH_2COOH \\ CH_2COOH \end{matrix}$$

Analytical Properties: Separation of alkaline-earth metal ions.
Reference: 17

Name: ether bonded phase
Structure:

$$CH_3(O\text{—}CH_2\text{—}CH_2)_2\text{—}O\text{—}(CH_2)_3\text{—}Si\text{—}$$

on 15–20 μm wide-pore silica

Analytical Properties: Separation by hydrophobic interaction chromatography, using aqueous salt solutions near pH = 7; used primarily in protein work.
Reference: 18

(Continued)

Specialized Stationary Phases for Liquid Chromatography (Continued)

Name: fluorocarbon polymer phase

Structure: Proprietary information of E.I. duPont de Nemours Corp.

Analytical Properties: Similar separations as obtained using C_3 bonded silica, with a much larger pH stability range than silica based phases; useful for protein and peptide separations using TFA (trifluoroacetic acid) as a mobile phase modifier; less mechanical stability than silica based phases.

Reference: 19

Name: poly (-methyl-L-glutamate) (PMLG)

Structure: Partially cross-linked, with long chain alkyl branches:

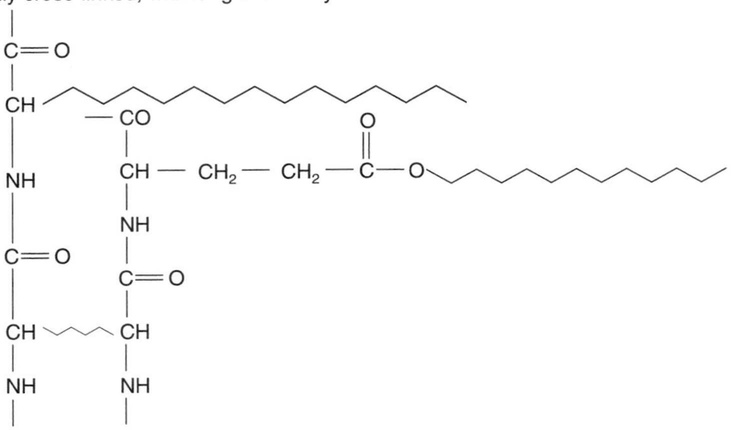

Analytical Properties: Separation similar to ODS, but with somewhat higher stability in alkaline solutions; particles are spherical and macroporous.

Reference: 20

Name: hydroxyapatite adsorbent

Structure: $Ca_{10}(PO_4)_6(OH)_2$ crystalline, nonstoichiometric mineral rich in surface ions (primarily carbonate).

Analytical Properties: Separation of proteins; overcomes some difficulties associated with ion exchange; selectivity and efficiency depend to some extent on particle geometry (i.e., sphere, plate, etc.)

Reference: 21–29

Name: iminodiacetic acid bonded phase

Structure:

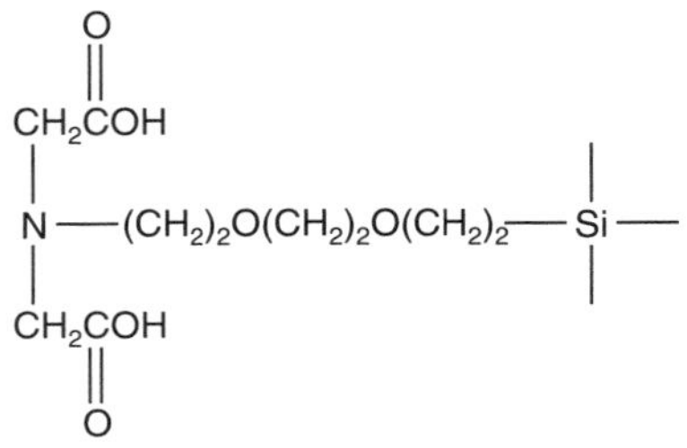

Analytical Properties: Separation of proteins by immobilized-metal affinity chromatography (HPIMAC) with Cu(II) or Zn(II) present in the mobile phase.

Reference: 29

Name: Kel-F (polychlorotrifluoroethylene)

Structure: (exact structure is proprietary, 3M Company)

Analytical Properties: Highly inert, even more nonpolar than hydrocarbon phases, with sufficient mechanical integrity to withstand high pressures; can be functionalized with $-CH_3$, $CH_3(CH_2)^-_3$, and phenyl (using Grignard reactions) to increase selectivity.

Reference: 30

Specialized Stationary Phases for Liquid Chromatography (Continued)

Name: 2-ethylhexyl carbonate coated or bonded on cholesteryl silica (room temperature liquid crystal)

Structure:

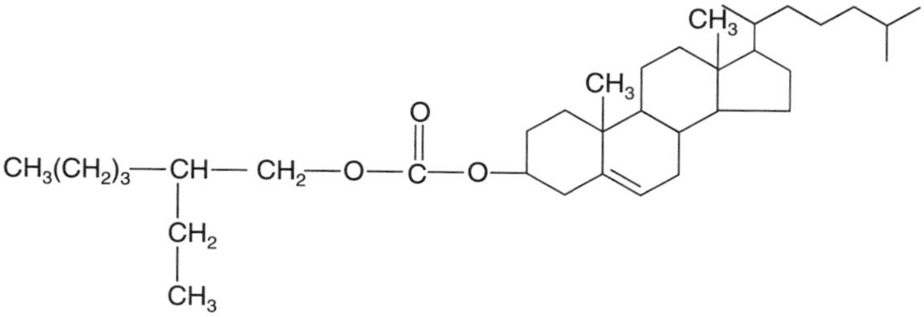

Analytical Properties: Has been used for the separation of estrogens and corticoid steroids; liquid crystal phase retains some order when coated on an active substrate.

Reference: 31,32

Name: cholesteryl-2-ethylhexanoate (room temperature liquid crystal) on silica

Structure:

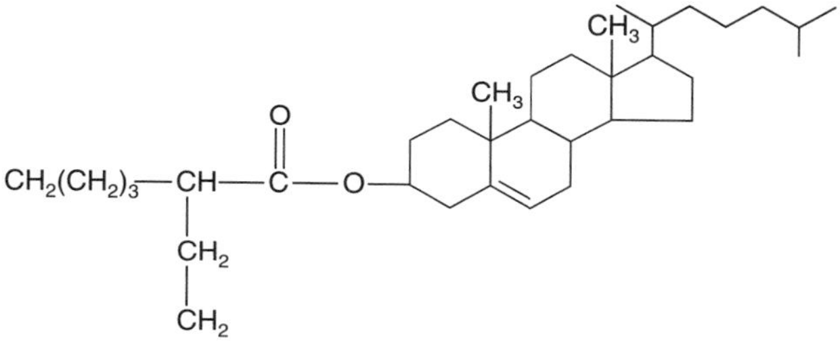

Analytical Properties: Has been used for the separation of androstenediones and testosterone.

Reference: 31,32

Name: pentafluorobenzamidopropyl silica gel

Structure:

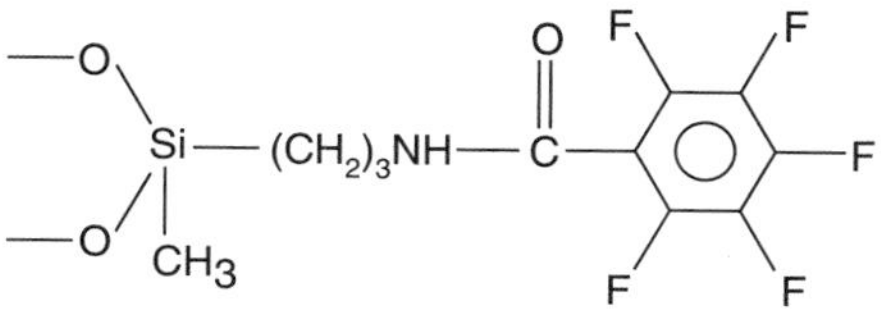

Analytical Properties: Separations via interactions with B-electrons of solutes; can be used in both normal and reverse phase for such B-donor systems as polynuclear aromatic hydrocarbons.

Reference: 33

Name: hydroxymethyl polystyrene

Structure:

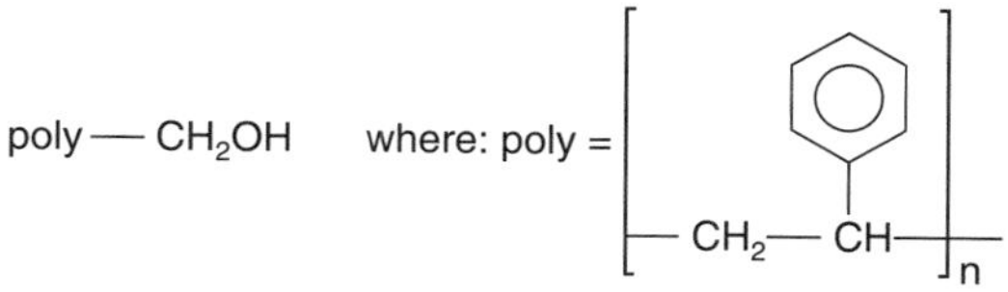

Analytical Properties: Separation of polypeptides; usually gives shorter retention times than ODS; hydrophobic interactions not as strong as with ODS.

Reference: 34

(Continued)

Specialized Stationary Phases for Liquid Chromatography (Continued)

Name: polystyrene

Structure:

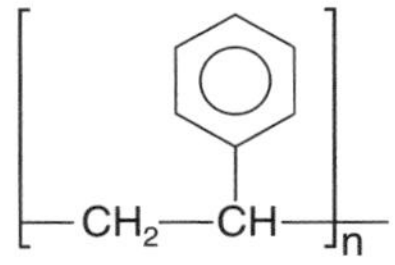

Analytical Properties: Separation of polypeptides with results similar to those obtainable with ODS; higher stability at high pH levels (to allow the phase to be washed); stronger hydrophobic interactions than ODS in reverse phase mode.

Reference: 35

Name: polystyrene-divinylbenzene (PS-DVB)

Structure:

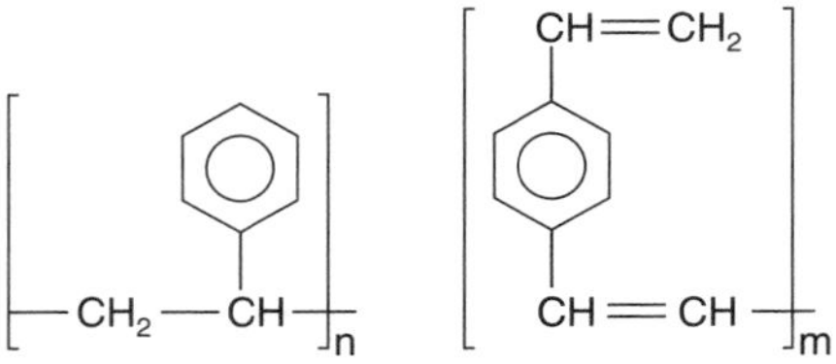

(exact structure is proprietary)

Analytical Properties: Useful for the separation of relatively polar compounds such as phenols, carboxylic acids, organic anions, nucleosides, alkylarylketones, chlorophenols, barbiturates, thimine derivatives; good stability under high and low pH; reasonable mechanical integrity at high carrier pressure; compatible with buffered liquid phases.

Reference: 36–44

Name: poly(vinylpyrrolidone) or (PVP) on silica

Structure:

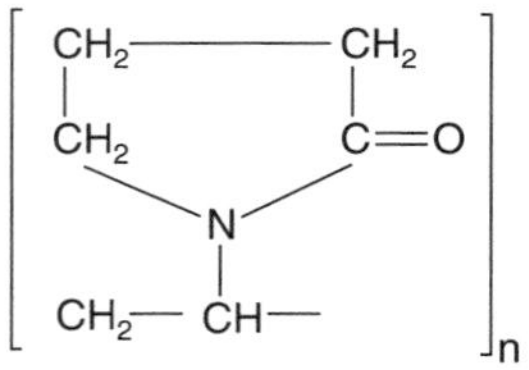

Analytical Properties: Separates aromatic and polynuclear aromatic hydrocarbons; can be used in normal phase mode (commonly using n-heptane or n-heptane + dichloromethane liquid phases) or reverse phase mode (commonly using methanol + water, acetonitrile + water, or phosphate buffered liquid phases).

Reference: 45

Name: bonded n-propylaniline

Structure:

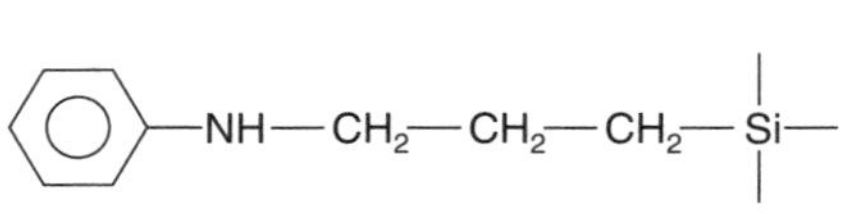

Analytical Properties: Selectivity is based on charge transfer interactions; nitroaromatic compounds are separated essentially according to the number of nitro groups, the higher number compounds being most strongly retained when using methanol/water mobile phases.

Reference: 46

Specialized Stationary Phases for Liquid Chromatography (Continued)

Name: n-propylpicrylether bonded phase

Structure:

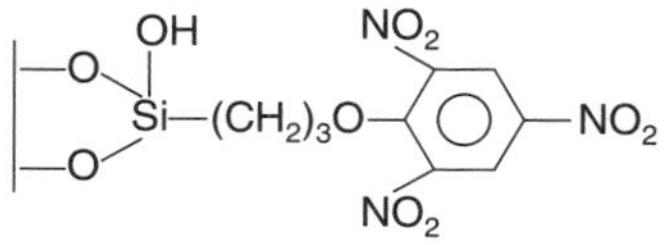

Analytical Properties: Separation of aromatic species, including polynuclear aromatic species, by charge transfer interactions.

Reference: 47

Name: 8-quinolinol bonded phases

Structure:

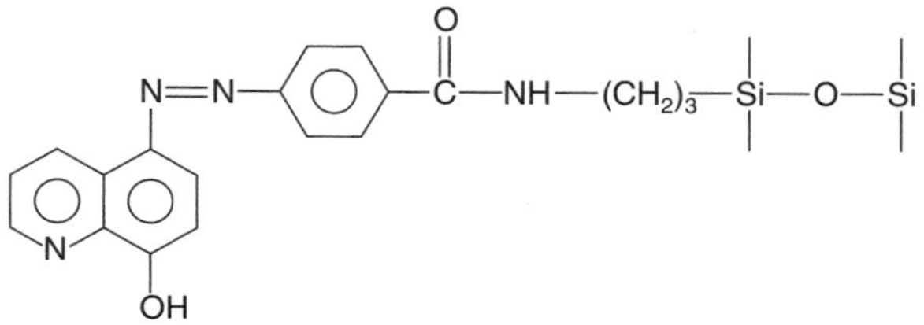

Analytical Properties: Separates phenols and EPA priority pollutants; often used with metal ions (such as iron (III)) as chelate ligands; 8-quinolinol has a high affinity for oxygen moieties, and will form complexes with upward of 60 metal ions; often with an acidic aqueous mobile phase.

Reference: 48–50

Name: cetyltrimethyl ammonium bromide adsorbed on silica

Structure:

$$(CH_3)_3N^+(n\text{–}C_{16}H_{33})Br^-$$

Analytical Properties: Has been used to separate aromatic hydrocarbons, heterocyclic compounds, phenols and aryl amines using methanol/water/phosphate buffer; extent of adsorption affects retention times; also used as a mobile phase modifier to provide a dynamically modified silica.

Reference: 51–57

Name: 6,9,12-trioxatridecylmethyl bonded phase

Structure:

$$CH_3\text{—}(OCH_2CH_2)_3(CH_2)_3Si(CH_3)(OH)OSi\text{—}$$

Analytical Properties: Phase is very well wetted by water, allowing mobile phases with high water concentration to be used. Somewhat higher efficiency and selectivity than ODS, but with similar separation properties.

Reference: 58

Name: ammonium tungstophosphate on silica

Structure: tungstophosphoric acid with ammonium nitrate.

Analytical Properties: Separation of compounds containing the NH_4^+ group, such as amino acids and peptides; the coated silica also behaves as a reversed phase for the separation of aliphatic and aromatic acids; high selectivity for glycine and tyrosine oligomers.

Reference: 59

CHIRAL STATIONARY PHASES FOR LIQUID CHROMATOGRAPHY

The following table provides information on the properties and application of some of the more specialized stationary phases used to carry out the separation enantiomeric mixtures. In many cases the phases are not commercially available, and the reader is referred to the appropriate literature citation for details on the synthesis.

REFERENCES

General References

1. Armstrong, D. W. "Chiral Stationary Phases for High Performance Liquid Chromatographic Separation of Enantiomers: A Mini-Review." *Journal of Liquid Chromatography & Related Technologies* 7, no. S-2 (1984): 353.
2. Dappen, R., H. Arm, and V. R. Meyer. "Applications and Limitations of Commercially Available Chiral Stationary Phases for High-Performance Liquid Chromatography." *Chromatography Reviews* (CHREV 200), 1, 1986.
3. Armstrong, D. W., S. Chen, C. Chang, and S. Chang. "A New Approach for the Direct Resolution of Racemic Beta Adrenergic Blocking Agents by HPLC." *Journal of Liquid Chromatography & Related Technologies* 15, no. 3 (1992): 545–56.
4. Armstrong, D. W. "The Evolution of Chrial Stationary Phases for Liquid Chromatography." *Journal of the Chinese Chemical Society* 45 (1998): 581–90.
5. *Cyclobond Handbook: A Guide to Understanding Cyclodextrin Bonded Phases for Chiral LC Separations.* 6th ed. Wippany, NJ: Advanced Separation Technologies, 2002.
6. *Chirobiotic Handbook: A Guide to Using Macrocyclic Glycopeptide Bonded Phases for Chiral LC Separations.* 4th ed. Wippany, NJ: Advanced Separation Technologies, 2002.
7. Busch, K. W., and M. A. Busch. eds. *Chiral Analysis.* Amsterdam: Elsevier, 2006.

Cited References

1. Erlandsson, P., L. Hansson, and R. Isaksson. "Direct Analytical and Preparative Resolution of Enantiomers Using Albumin Adsorbed to Silica as a Stationary Phase." *Journal of Chromatography* 370 (1986): 475.
2. Kuesters, E., and D. Giron. "Enantiomeric Separation of the Beta-Blocking Drugs Pindolol and Bipindolol Using a Chiral Immobilized Protein Stationary Phase, HRC&CC." *Journal of High Resolution Chromatography, Chromatography Communication* 9, no. 9 (1986): 531.
3. Hermannson, J., and M. Eriksson. "Direct Liquid Chromatographic Resolution of Acidic Drugs Using a Chiral 1-Acid Glycoprotein Column (Enantiopac)." *Journal of Liquid Chromatography & Related Technologies* 9, no. 2&3 (1986): 621.
4. Naobumi, O., and K. Hajimu. "HPLC Separation of Amino Acid Enantiomers on Urea Derivatives of L-Valine Bonded to Silica Gel." *Journal of Chromatography* 285 (1984): 198.
5. Okamoto, Y., H. Sakamoto, K. Hatada, and M. Irie. "Resolution of Enantiomers by HPLC on Cellulose Trans- and Cis-tris (4-Phenylazophenylcarbamate)." *Chemistry Letters* 983 (1986).
6. Ichid, A., T. Shibata, I. Okamoto, Y. Yuki, H. Namikoshi, and Y. Toga. "Resolution of Enantiomers by HPLC on Cellulose Derivatives." *Chromatographia* 19 (1984): 280.
7. Tagahara, K., J. Koyama, T. Okatani, and Y. Suzuta. "Chromatographic Resolution of Racemic Tetrahydroberbeine Alkaloids by Using Cellulose Tris (Phenylcarbamate) Stationary Phase." *Chemical & Pharmaceutical Bulletin* 34 (1986): 5166.
8. Klemisch, W., and A. von Hodenberg. "Separation on Crosslinked Acetylcellulose." *Journal of High Resolution Chromatography & Chromatography Communications* 9, no. 12 (1986): 765–67.
9. Rimboock, K., F. Kastner, and A. Mannschreck. "Microcrystallinetribenzoyl Cellulose: A High-Performance Liquid Chromatographic Sorbent for the Separation of Enantiomers." *Journal of Chromatography* 351 (1986): 346.

10. Lindner, K., and A. Mannschreck. "Separation of Enantiomers by High-Performance Liquid Chromatography on Triacetylcellulose." *Journal of Chromatography* 193 (1980): 308–10.

11. Gubitz, G., W. Jellenz, and D. Schonleber. "High Performance Liquid Chromatographic Resolution of the Optical Isomers of D,L-Tryptophane, D,L-5-Hydroxytryptophan and D,L-Dopa on Cellulose Columns." *Journal of High Resolution Chromatography & Chromatography Communications* 3 (1980): 31.

12. Takayanagi, H., O. Hatano, K. Fujimura, and T. Ando. "Ligand-Exchange High-Performance Liquid Chromatography of Dialkyl Sulfides." *Analytical Chemistry* 57 (1985): 1840.

13. Armstrong, D. W. U.S. Patent #4,539,399. Whippany, NJ: Assigned to Advanced Separation Technologies Inc., 1985.

14. Armstrong, D. W., and W. Demond. "Cyclodextrin Bonded Phases for the Liquid Chromatographic Separation of Optical, Geometrical, and Structural Isomers." *Journal of Chromatographic Science* 22 (1984): 411.

15. Armstrong, D. W., T. J. Ward, R. D. Armstrong, and T. J. Beesley. "Separation of Drug Stereoisomers by the Formation of β-Cyclodextrin Inclusion Complexes." *Science* 232 (1986): 1132.

16. Armstrong, D. W., W. DeMond, and B. P. Czech. "Separation of Metallocene Enantiomers by Liquid Chromatography: Chiral Recognition via Cyclodextrin Bonded Phases." *Analytical Chemistry* 57 (1985): 481.

17. Armstrong, D. W., W. DeMond, A. Alak, W. L. Hinze, T. E. Riehl, and K. H. Bui. "Liquid Chromatographic Separation of Diastereomers and Structural Isomers on Cyclodextrin-Bonded Phases." *Analytical Chemistry* 57 (1985): 234.

18. Weaver, D. E., and R. van Lier. "Coupled β-Cyclodextrin and Reverse-Phase High-Performance Liquid Chromatography for Assessing Biphenyl Hydroxylase Activity in Hepatic 9000 g Supernatant." *Analytical Biochemistry* 154 (1986): 590.

19. Armstrong, D. W. "Optical Isomer Separation by Liquid Chromatography." *Analytical Chemistry* 59 (1987): 84A.

20. Chang, C. A., Q. Wu, and L. Tan. "Normal-Phase High-Performance Liquid Chromatographic Separations of Positional Isomers of Substituted Benzoic Acids with Amine and b-Cyclodextrin Bonded-Phase Columns." *Journal of Chromatography* 361 (1986): 199.

21. Cline-Love, L., and M. Arunyanart. "Cyclodextrin Mobile-Phase and Stationary-Phase Liquid Chromatography." *ACS Symposium Series* 297 (1986): 226.

22. Feitsma, K., J. Bosman, B. Drenth, and R. DeZeeuw. "A Study of the Separation of Enantiomers of Some Aromatic Carboxylic Acids by High-Performance Liquid Chromatography on a β-Cyclodextrin-Bonded Stationary Phase." *Journal of Chromatography* 333 (1985): 59.

23. Fujimura, K., T. Ueda, and T. Ando. "Retention Behavior of Some Aromatic Compounds on Chemically Bonded Cyclodextrin Silica Stationary Phase in Liquid Chromatography." *Analytical Chemistry* 55 (1983): 446.

24. Hattori, K., K. Takahashi, M. Mikami, and H. Watanabe. "Novel High-Performance Liquid Chromatographic Adsorbents Prepared by Immobilization of Modified Cyclodextrins." *Journal of Chromatography* 355 (1986): 383.

25. Ridlon, C. D., and H. J. Issaq. "Effect of Column Type and Experimental Parameters on the HPLC Separation of Dipeptides." *Journal of Liquid Chromatography & Related Technologies* 9, no. 15 (1986): 3377.

26. Sybilska, D., J. Debowski, J. Jurczak, and J. Zukowski. "The α- and β-Cyclodextrin Complexation as a Tool for the Separation of o-, m- and p-Nitro- Cis- and Trans-Cinnamic Acids by Reversed-Phase High-Performance Liquid Chromatography." *Journal of Chromatography* 286 (1984): 163.

27. Chang, C. A., Q. Wu, and M. P. Eastman. "Mobile Phase Effects on the Separations of Substituted Anilines with β-Cyclodextrin-Bonded Column." *Journal of Chromatography* 371 (1986): 269.

28. Maguire, J. H. "Some Structural Requirements for Resolution of Hydantoin Enantiomers with β-Cyclodextrin Liquid Chromatography Column." *Journal of Chromatography* 387 (1987): 453.

29. Sinibaldi, M., V. Carunchio, C. Coradini, and A. M. Girelli. "High-Performance Liquid Chromatographic Resolution of Enantiomers on Chiral Amine Bonded Silica Gel." *Chromatographia* 18, no. 81 (1984): 459.

30. Gasparrini, D., D. Misti, and C. Villani. "Chromatographic Optical Resolution on Trans-1,2-Diaminocyclohexane Derivatives: Theory and Applications." *Chirality* 4 (1992): 447–58.

31. Zhong, Q., X. Han, L. He, T. Beesley, W. Trahanovsky, and D. Armstrong. "Chromatographic Evaluation of Poly (Trans-1,2-Cyclohexanediyl-Bis Acrylamide) as a Chiral Stationary Phase for HPLC." *Journal of Chromatography A* 1066 (2005): 55–70.

32. Pettersson, C., and H. W. Stuurman. "Direct Separation of Enantiomer of Ephedrine and Some Analogues by Reversed-Phase Liquid Chromatography Using (+)-di-n-Butytartrate as the Liquid Stationary Phase." *Journal of Chromatographic Science* 22 (1984): 441.

33. Weems, H. B., M. Mushtaq, and S. K. Yang. "Resolution of Epoxide Enantiomers of Polycyclic Aromatic Hydrocarbons by Chiral Stationary-Phase High-Performance Liquid Chromatography." *Analytical Biochemistry* 148 (1985): 328.

34. Weems, H., M. Mushtaq, P. Fu, and S. Yang. "Direct Separation of Non-k-Region Monool and Diol Enantiomers of Phenanthrene, Benz[a]anthracene, and Chrysene by High-Performance Liquid Chromatography with Chiral Stationary Phases." *Journal of Chromatography* 371 (1986): 211.

35. Yang, S., M. Mushtaq, and P. Fu. "Elution Order-Absolute Configuration of k-Region Dihydrodiol Enantiomers of Benz[a]anthracene Derivatives in Chiral Stationary Phase High Performance Liquid Chromatography." *Journal of Chromatography* 371 (1986): 195–209.

36. Wainer, I. "Applicability of HPLC Chiral Stationary Phases to Pharamacokinetic and Disposition Studies on Enantiomeric Drugs." *Methodological Surveys in Biochemistry and Analysis, Subseries A* 16 (1986): 243.

37. Vaughan, G. T., and B. V. Millborrow. "The Resolution by HPLC of RS-[2-^{14}C] Me 1',4'-Cis-Diol of Abscisic Acid and the Metabolism of (−)-R-and-S-Abscisic Acid." *Journal of Experimental Botany* 35, no. 150 (1984): 110–120.

38. Tambute, A., P. Gareil, M. Caude, and P. Rosset. "Preparative Separation of Racemic Tertiary Phosphine Oxides by Chiral High-Performance Liquid Chromatography." *Journal of Chromatography* 363 (1986): 81–93.

39. Wainer, I., and T. Doyle. "The Direct Enantiomeric Determination of (−) and (+)-Propranolol in Human Serum by High-Performance Liquid Chromatography on a Chiral Stationary Phase." *Journal of Chromatography* 306 (1984): 405–11.

40. Okamoto, Y., H. Mohri, M. Ishikura, K. Hatada, and H. Yuki. "Optically Active Poly (Diphenyl-2-pyridylmethyl methacrylate): Asymmetric Synthesis, Stability of Helix, and Chiral Recognition Ability." *Journal of Polymer Science: Polymer Symposium* 74 (1986): 125–39.

41. Gubitz, G., and S. Mihellyes. "Direct Separation of 2-Hydroxy Acid Enantiomers by High-Performance Liquid Chromatography on Chemically Bonded Chiral Phases." *Chromatographia* 19 (1984): 257.

42. Schulze, J., and W. Konig. "Enantiomer Separation by High-Performance Liquid Chromatography on Silica Gel with Covalently Bound Mono-Saccharides." *Journal of Chromatography* 355 (1986): 165.

43. Kip, J., P. Van Haperen, and J. C. Kraak. R-N-(Pentafluorobenzoyl) Phenylglycine as a Chiral Stationary Phase for the Separation of Enantiomers by High-Performance Liquid Chromatography." *Journal of Chromatography* 356 (1986): 423.

44. Gelber, L. R., B. L. Karger, J. L. Neumeyer, and B. Feibush. "Ligand Exchange Chromatography of Amino Alcohols. Use of Schiff Bases in Enantiomer Resolution." *Journal of the American Chemical Society* 106 (1984): 7729.

45. Dabashi, Y., and S. Hara. "Direct Resolution of Enantiomers by Liquid Chromatography with the Novel Chiral Stationary Phase Derived from (R,R)-Tartamide." *Tetrahedron Letters* 26, no. 35 (1985): 4217.

46. Facklam, C., H. Pracejus, G. Oehme, and H. Much. "Resolution of Enantiomers of Amino Acid Derivatives by High-Performance Liquid Chromatography in a Silica Gel Bonded Chiral Amide Phase." *Journal of Chromatography* 257 (1983): 118.

47. Okamoto, Y., S. Honda, K. Hatada, and H. Yuki. "IX. High-Performance Liquid Chromatographic Resolution of Enantiomers on Optically Active Poly (Tri-Phenylmethyl Methacrylate)." *Journal of Chromatography* 350 (1985): 127.

48. Okamoto, Y., and K. Hatada. "Resolution of Enantiomers by HPLC on Optically Active Poly (Triphenylmethyl Methacrylate)." *Journal of Liquid Chromatography & Related Technologies* 9, nos. 2&3 (1986): 369.

49. Armstrong, D. W., Y. Tang, S. Chen, C. Zhou, J. R. Bagwill, and J. R. Chen. "Macrocyclic Antibiotics as a New Class of Chiral Selectors for Liquid Chromatography." *Analytical Chemistry* 66, no. 9 (1994): 1473–84.

50. Ekborg-Ott, K. H., Y. Liu, and D. W. Armstrong. "Highly Enantioselective HPLC Separations Using the Covalently Bonded Macrocyclic Antibiotic Ristocetin A Chiral Stationary Phase." *Chirality* 10 (1998): 434–83.

51. Armstrong, D. W., Y. Liu, and K. H. Ekborg-Ott. "A Covalently Bonded Teicoplanin Chiral Stationary Phase for HPLC Enantioseparations." *Chirality* 7, no. 6 (1995): 474–97.

52. Berthod, A., Y. Liu, J. R. Bagwill, and D. W. Armstrong. "Facile RPLC Enantioresolution of Native Amino Acids and Peptides Using a Teicoplanin Chiral Stationary Phase." *Journal of Chromatography A* 731 (1996): 123–37.

53. Berthod, A., X. Chen, J. P. Kullman, D. W. Armstrong, F. Gasparrini, I. D'Acquarica, C. Villani, and A. Carotti. "Role of Carbohydrate Moieties in Chiral Recognition on Teicoplanin Based LC Stationary Phases." *Analytical Chemistry* 72 (2000): 1736–39.

54. Aboul-Enein, H. Y., and V. Serignese. "Enantiomeric Separation of Several Cyclic Imides on a Macrocyclic Antibiotic (Vancomycin) Chiral Stationary Phase Under Normal and Reverse Phase Conditions." *Chirality* 10 (1998): 358–61.

55. Lehotay, J., J. Hrobonova, J. Krupcik, and J. Cizmarik. "Chiral Separation of Enantiomers of Amino Acid Derivatives by HPLC on Vancomycin and Teichoplanin Chiral Stationary Phases." *Pharmazie* 53 (1998): 863–65.

Chiral Stationary Phases for Liquid Chromatography

Name: bovine serum albumin (covalently fixed to silica gel)

Structure: Prolate ellipsoid 14×4 nm, with a molecular mass of 66,500. Amount absorbed is dependent on buffer pH, with the maximum at pH = 4.9.

Analytical Properties: Separation of bopindolol and also separation of pindolol after derivatization with isopropyl isocyanate. Separation of DL mixtures of enantiomers; can be used on both the analytical and preparative scales. Changes in pH will cause this phase to leach from the column. Storage at 4 °C is recommended.

Reference: 1,2

Name: α-acid glycoprotein

Structure: Structure is proprietary (Enantiopac, LKB Co.)

Analytical Properties: Separation of the drugs ibuprofen, ketoprofen, naproxen, 2-phenoxypropionic acid, bendroflumethiazide, ethotoin, hexobarbital, disopyramide and RAC 109.
Retention and selectivity of the solutes can be regulated by addition of the tertiary amine N,N,-dimethyloctylamine (DMOA) to the mobile phase.
DMOA decreases retention time and the enantioselectivity of the weaker acids but has opposite effects on the stronger acids.

Reference: 3

Name: N-(t-butylamino carbonyl-L-valine) bonded silica

Structure:

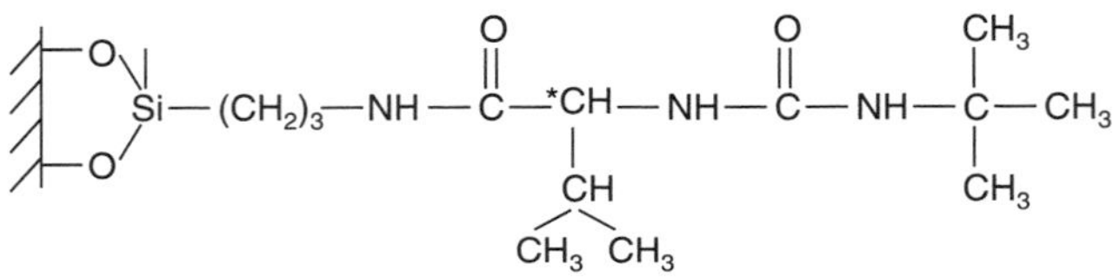

Analytical Properties: Separation of amino acid enantiomers; most effective of the L-valine urea derivatives; depends on hydrogen bond interactions usually prepared on LiChro-sorb (10 μm); hexane plus isopropanol modifier has been used as the liquid phase.

Reference: 4

Name: cellulose cis and trans tris (4-phenylazophenyl carbamate) (CPAPC)

Structure:

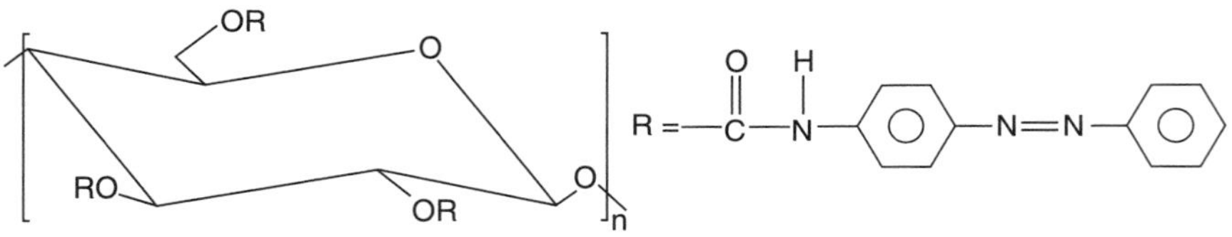

Analytical Properties: Trans isomer provides excellent resolution of racemic mixtures such as atropine, pindolol, flavanone; resolution decreases quickly with increasing cis isomer concentration; the cis/trans equilibrium is controlled by UV radiation, and the phase is adsorbed to silica gel; liquid phase of hexane with 10 % 2-propanol has been found useful.

Reference: 5

Name: cellulose triacetate

Structure:

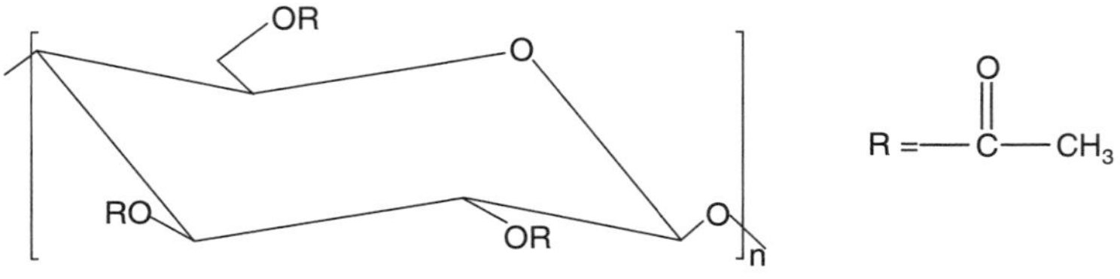

Chiral Stationary Phases for Liquid Chromatography (Continued)

Analytical Properties: Shows chiral recognition for many racemates and is especially effective for substrates with a phosphorous atom at an asymmetric center. However, the degree of chiral recognition is not so high in general.
Reference: 6

Name: cellulose tribenzoate
Structure:

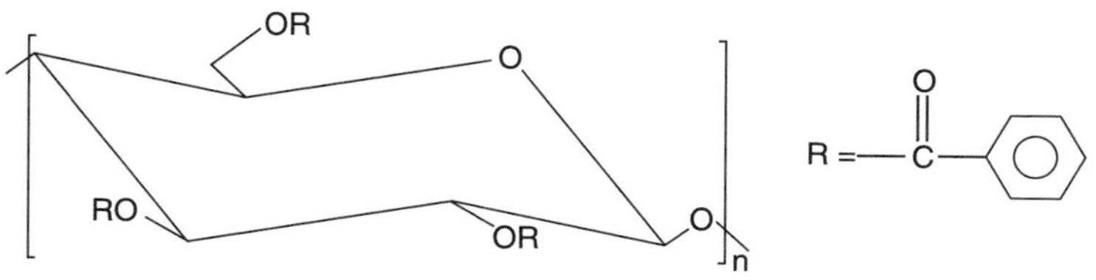

Analytical Properties: Demonstrates good chiral recognition for the racemates with carbonyl group(s) in the neighborhood of an asymmetric center.
Reference: 6

Name: cellulose tribenzyl ether
Structure:

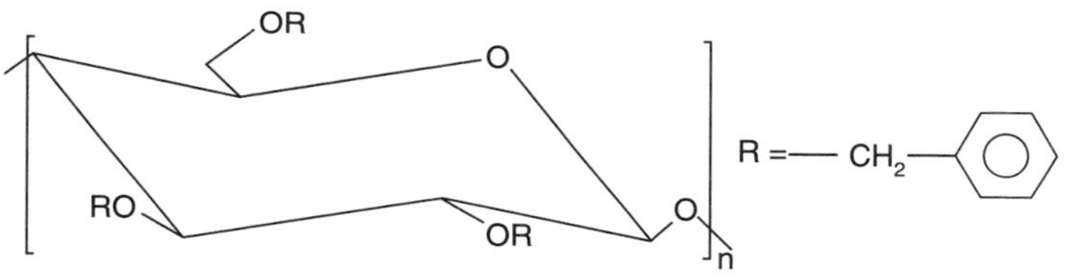

Analytical Properties: Effective with protic solvents used as mobile phases.
Reference: 6

Name: cellulose tricinnamate
Structure:

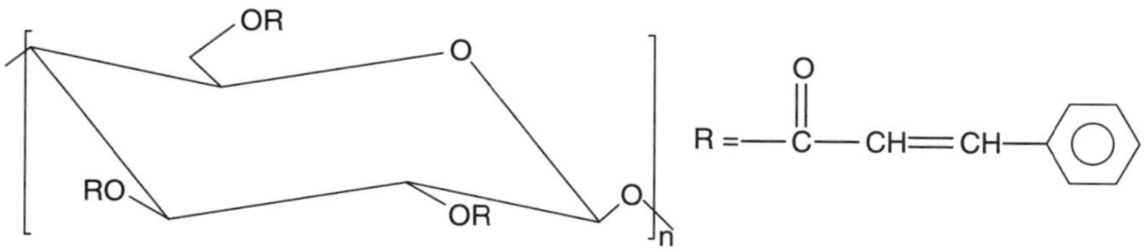

Analytical Properties: Shows high chromatographic retention times and a good chiral recognition for many aromatic racemates and barbiturates.
Reference: 6

Name: cellulose tris (phenylcarbamate)
Structure:

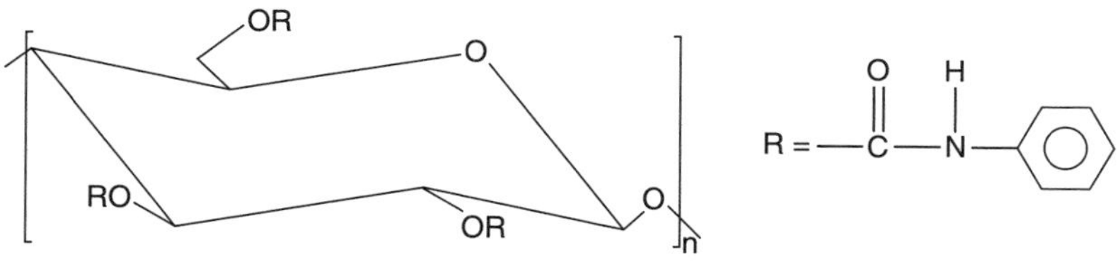

(Continued)

Chiral Stationary Phases for Liquid Chromatography (Continued)

Analytical Properties: Separation of racemic mixtures of alkaloids, using ethanol as the eluent; 1-isomers tend to be more strongly retained than d-isomers.
Good chiral recognition of sulfoxides and high affinity for racemates having an –OH or –NH group, through hydrogen bonding.
Reference: 6,7

Name: cross-linked acetylcellulose

Structure: Cellulose with one of the –OH groups acylated.

ANALYTICAL PROPERTIES: Separation of enantiomers (such as etozolin, piprozolin, ozolinon, and bunolol) using an ethanol/water, 95/5 (V/V%) liquid phase.
Reference: 8

Name: microcrystalline tribenzoylcellulose

Structure: Same structure as cellulose trobenzoate (coated on macroporous silica gel).

Analytical Properties: Resolution of trans-1,2, diphenyloxirane, 2-methyl-3-(2′-methylphenyl)-4(3H)-quinazolinone and some aromatic hydrocarbons.
Reference: 9

Name: triacetylcellulose

Structure: Same structure as cellulose triacetate.

Analytical Properties: Microcrystalline triacetylcellulose swells in organic solvents.
Separation of racemic thioamides, sulfoxides, organophosphorus compounds, drugs, and amino acids derivatives.
Separations of these racemates were achieved at pressures at or above 4.9 MPa.
Reference: 10

Name: untreated cellulose (average particle size of 7 μm)
Structure:

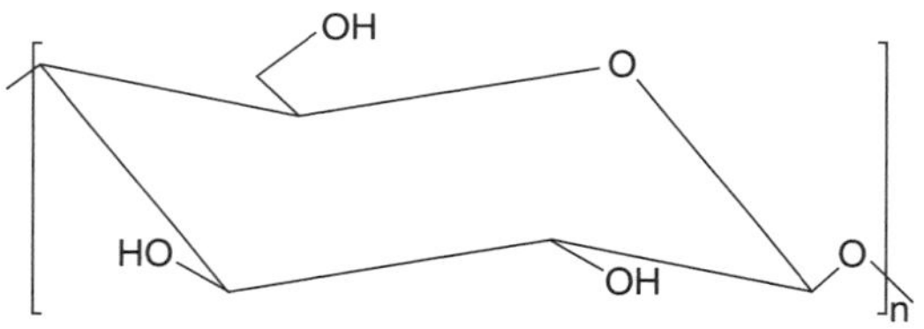

Analytical Properties: Complete resolution of D,L-tryptophane, and D,L-5-hydroxytryptophane.
Reference: 11

Name: copper (II) 2-amino-1-cyclopentene-1-dithio carboxylate
Structure:

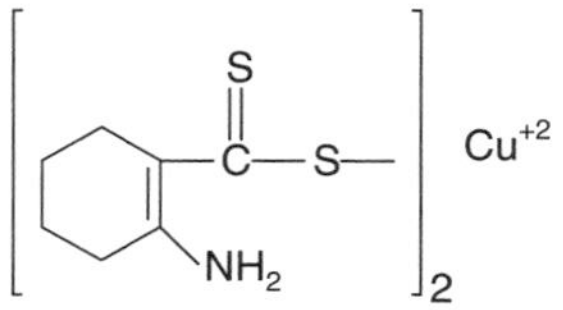

(bonded to silica)

Analytical Properties: Separation of dialkyl sulfides when hexane containing methanol or acetonitrile was used as the mobile phase.
Reference: 12

Chiral Stationary Phases for Liquid Chromatography (Continued)

Name: α-cyclodextrin bonded phase
Structure:
subunit:

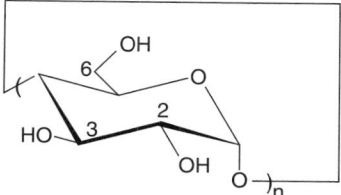

cavity:

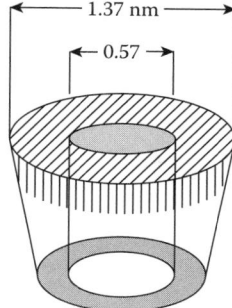

Analytical Properties: α-cyclodextrin (cyclohexamolyose); reversed phase separation of barbiturates and other drugs, and aromatic amino acids. The substrate is composed of six glucose units and has a relative molecular mass of 972. The cavity diameter is 0.57 nm and the substrate has a water solubility of 14.5 g/mL.

Reference: 13–28

Name: β-cyclodextrin bonded phase
Structure:
subunit:

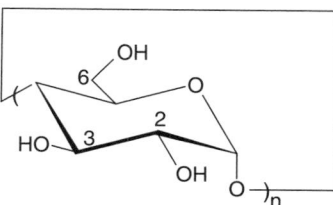

cavity:

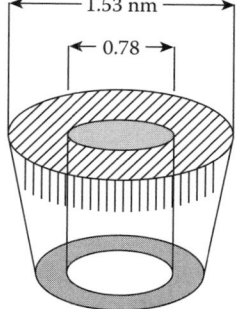

(Continued)

Chiral Stationary Phases for Liquid Chromatography (Continued)

Ligand:

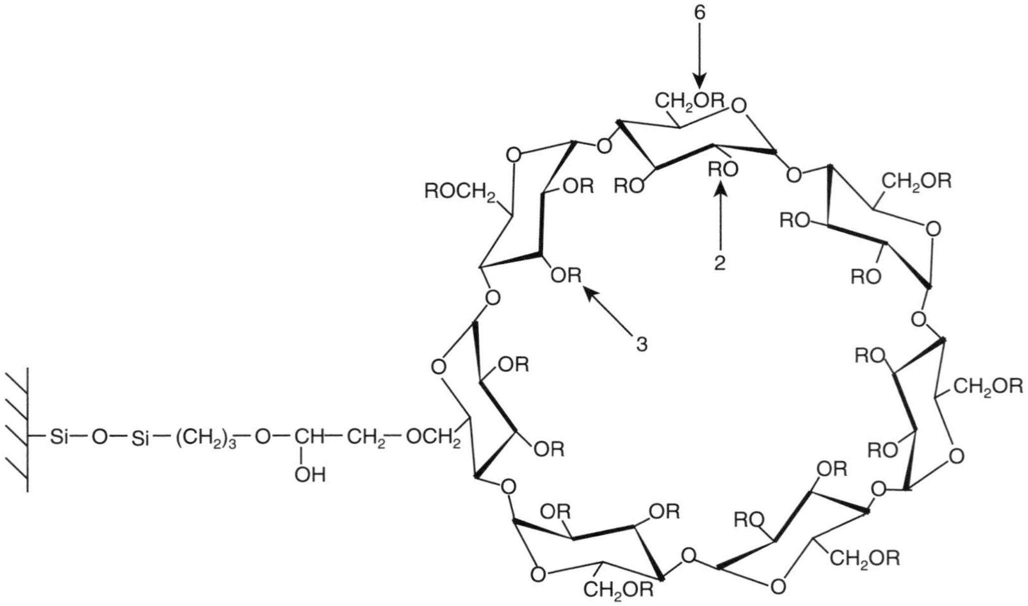

Note that this structure also illustrates the linkage to silica through one primary –OH group. Either one or two such linkages usually attach the substrate to the silica.

Analytical Properties: β-cyclodextrin (cycloheptamylose); normal phase separation of positional isomers of substituted benzoic acids; reverse phase separation of dansyl and napthyl amino acids, several aromatic drugs, steroids, alkaloids, metallocenes, binapthyl crown ethers, aromatics acids, aromatic amines, and aromatic sulfoxides. This substrate has seven glucose units and has a relative molecular mass of 1135. The inside cavity has a diameter of 0.78 nm, and the substrate has a water solubility of 1.85 g/mL, although this can be increased by derivatization.

Reference: 13–28

Name: β-cyclodextrin, dimethylated bonded phase

Structure:

cavity:

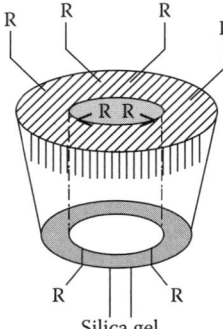

R-ligand:

–CH₃

Analytical Properties: β-cyclodextrin DM (cycloheptamylose-DM); reversed phase separation of a variety of structural and geometrical isomers; useful for the separation of analytes that have a carbonyl group off of the stereogenic center.

Reference: 13–28

Chiral Stationary Phases for Liquid Chromatography (Continued)

Name: β-cyclodextrin, acylated bonded phase
Structure:
cavity:

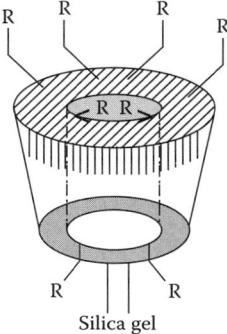

R-ligand:

$$-COCH_3$$

Analytical Properties: β-cyclodextrin AC (cycloheptamylose-AC); reversed phase separation of steroids and polycyclic compounds.
Reference: 13–28

Name: β-cyclodextrin, hydroxypropyl ether modified bonded phase
Structure:
cavity:

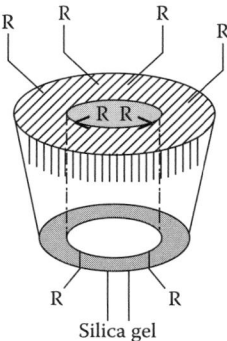

R-ligand:

$$\overset{\displaystyle OH}{\underset{\displaystyle *}{-\!\!-\!\!-CH_2CHCH_3}}$$

Analytical Properties: β-cyclodextrin SP or RSP (cycloheptamylose-SP, RSP); note that the modifying ligand has a stereogenic center; useful for reversed phase separation of a variety of analytes, especially for enantiomers that have bulky substituents that are beta to the stereogenic center; can be used for cyclic hydrocarbons and for t-boc amino acids.
Reference: 13–28

(Continued)

Chiral Stationary Phases for Liquid Chromatography (Continued)

Name: β-cyclodextrin, napthylethyl carbamate modified bonded phase

Structure:
cavity:

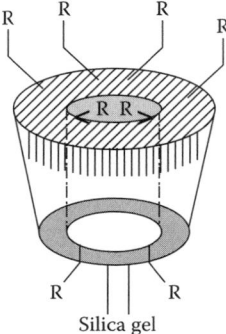

R-ligand:

$$-CONHCH\underset{*}{-}\overset{\displaystyle CH_3}{|}$$

Analytical Properties: β-cyclodextrin SN or RN (cycloheptamylose-SN, RN); note that the modifying ligand has a stereogenic center; useful for normal phase, reversed phase and polar organic phase separation under specific circumstances; the substrate performs best with normal phase and polar organic mobile phases; the SN modification has shown the highest selectivity.

Reference: 13–28

Name: β-cyclodextrin, 3,5-dimethylphenyl carbamate modified bonded phase

Structure:
cavity:

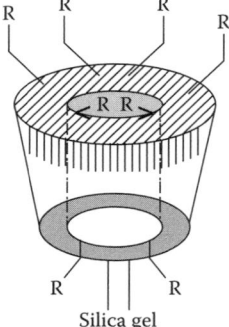

Chiral Stationary Phases for Liquid Chromatography (Continued)

R-ligand:

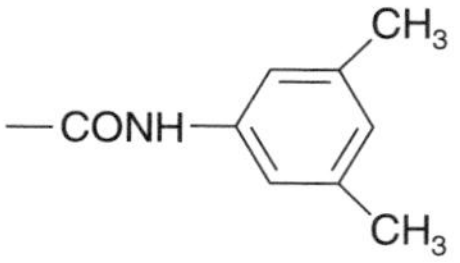

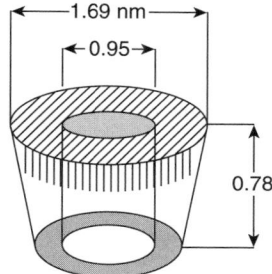

Analytical Properties: β-cyclodextrin DMP (cycloheptamylose-DMP); useful for normal phase, reversed phase and polar organic phase separation under specific circumstances; the substrate performs best with normal phase and polar organic mobile phases.

Reference: 13–28

Name: γ-cyclodextrin bonded phase
Structure:
subunit:

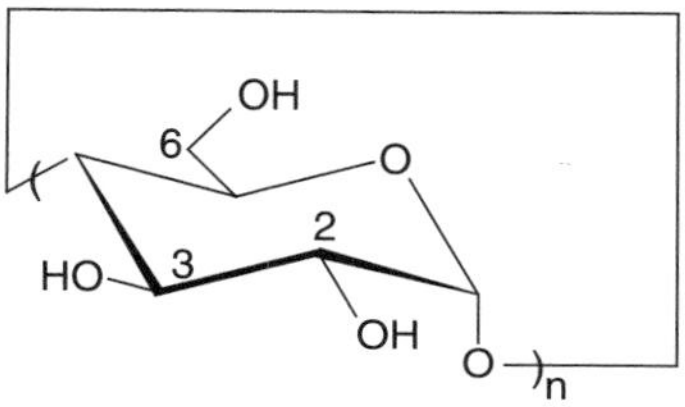

cavity:

Analytical Properties: γ-cyclodextrin (cyclooctylamalyose), reverse phase separation of stereoisomers of polycyclic aromatic hydrocarbons. The substrate has eight glucose units and has a relative molecular mass of 1297. The cavity has a diameter of 0.59 nm, and the substrate has a water solubility of 23.2 g/mL.

Reference: 13–28

Name: (−)trans-1,2-cyclohexanediamine
Structure:

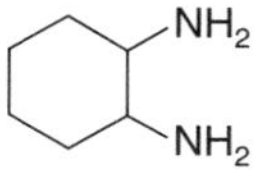

(bonded to silica gel)

Analytical Properties: Resolution of such enantiomeric compounds as 2,2′-dihydroxy-1,1′-binapthyl and trans-1,2-cyclo hexandiol

Reference: 29

(Continued)

Chiral Stationary Phases for Liquid Chromatography (Continued)

Name: poly (tert-1,2-cyclohexanediyl-bis acrylamide)
Structure:

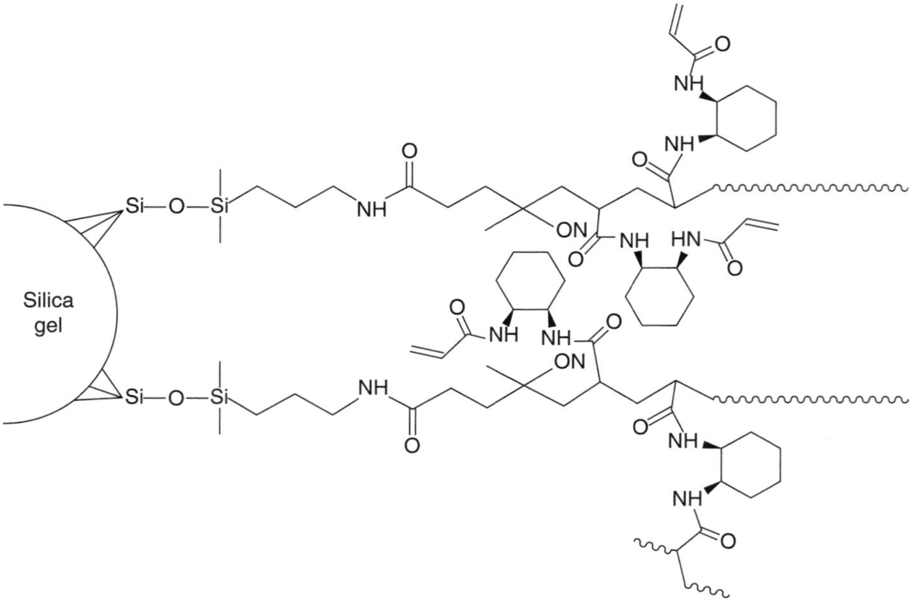

Analytical Properties: Synthetic, covalently bonded chiral stationary phase consisting of a thin, ordered selector layer bonded to the silica surface; high stability and loadability, relatively easy to scale up; reversal of elution order is possible (R,R) to (S,S) configuration; used for resolution of–aryloxyacetic acids, alcohols, sulfoxides, selenoxides, phosphinates, t-phospine oxides, benzodiazepines without derivatization; amines, amino acids, amino alcohols, nonsteriod anti-inflamatory drugs with derivatization.

Reference: 30,31

Name: (+)-di-n-butyltartate
Structure:

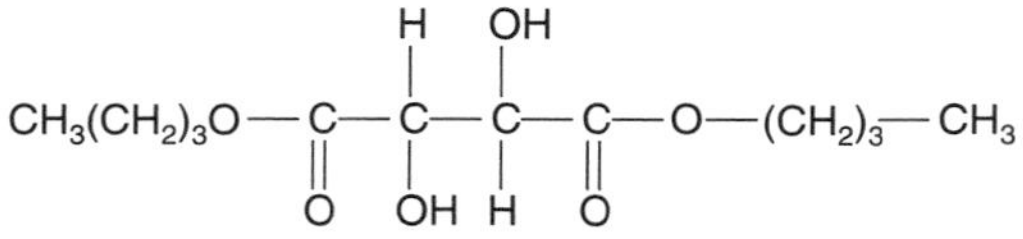

(adsorbed on phenyl bonded silica)

Analytical Properties: Resolution of ephedrine and nonephedrine.

Reference: 32

Name: (S)-N-(3,5-dinitrobenzoyl) leucine or (S)-DNBL
Structure:

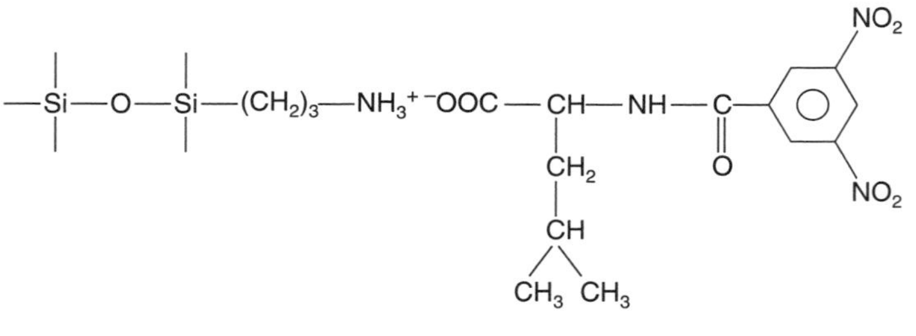

Chiral Stationary Phases for Liquid Chromatography (Continued)

Analytical Properties: Resolution of several enantiomers of polycyclic aromatic hydrocarbons, for example, chrysene 5,6-epoxide, dibenz[a,h]anthracene 5,6 epoxide, 7-methyl benz[a]anthracene 5,6-epoxide. Resolution of barbiturates, mephenytoin, benzodiazepinones, and succinimides. Direct separation of some mono-ol and diol enantiomers of phenanthrene, benz[a]anthrene and chrysene. Ionically bonded to silica gel, this phase provides resolution of enantiomers of cis-dihydroidiols of unsubstituted and methyl- and bromo-substituted benz[a]anthracene derivatives having hydroxyl groups that adopt quasiequatorial-quasiaxial and/or quasiaxial-quasiequatorial conformation.

Reference: 33–37

Name: (R)-N-(3,5-dinitrobenzyl) phenylglycine

Structure:

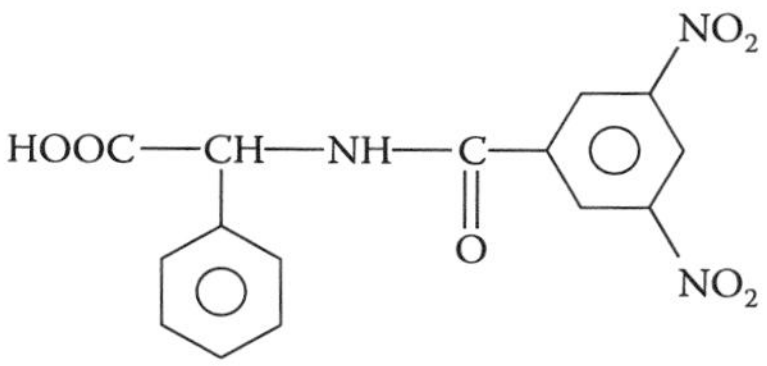

Analytical Properties: Ionically bonded to silica, this phase provides good resolution of enantiomeric quasiequatorial trans-dehydriols of unsubstituted and methyl and bromo-substituted benz[a]anthracene derivatives. Covalently bonded to silica, this phase provides good resolution of enantiomeric pairs of quasidiaxial trans-dihydrodiols of unsubstituted and methyl- and bromo-substituted benz[a]anthracene derivatives. By addition of a third solvent (chloroform) to the classical binary mixture (hexane-alcohol) of the mobile phase, resolution of enantiomers of tertiary phosphine oxides is possible.

Reference: 33–35,38,39

Name: poly(diphenyl-2-pyridylmethyl methacrylate) or PD2PyMa

Structure:

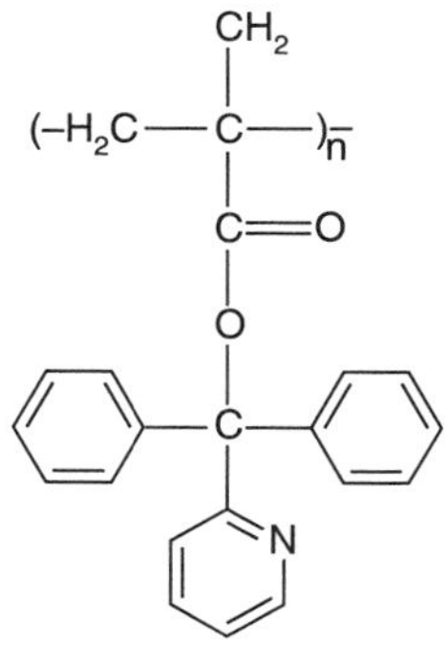

(coated on macroporous silica gel)

Analytical Properties: Resolution of such compounds as racemic 1,2-diphenol-ethanol, 2-2′-dihydroxyl-1,1′-dinapthyl, 2,3-diphenyloxirane, and phenyl-2-pyrid-o-toly-1-methanol.
Reference: 40

(Continued)

Chiral Stationary Phases for Liquid Chromatography (Continued)

Name: poly(1,2-diphenylethylenediamine acrylamide) or P-CAP DP

Structure:

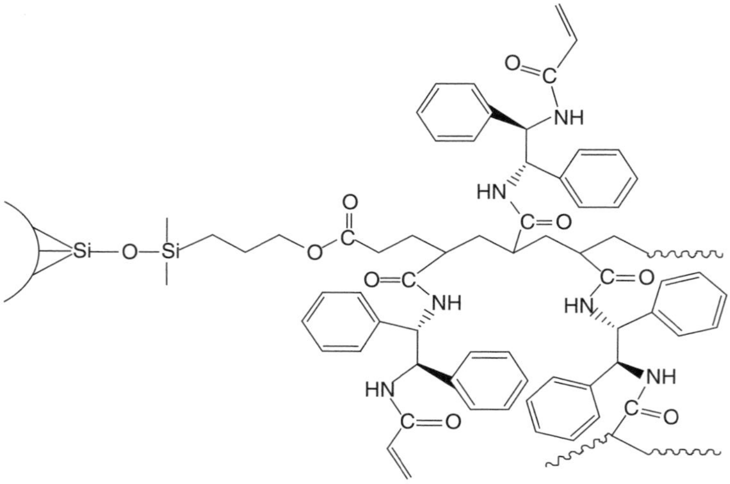

Analytical Properties: Synthetic, covalently bonded chiral stationary phase consisting of a thin, ordered selector layer bonded to the silica surface; high stability and loadability, relatively easy to scale up; reversal of elution order is possible (R,R) to (S,S) configuration.

Reference: 30,31

Name: L-hydroxyproline

Structure:

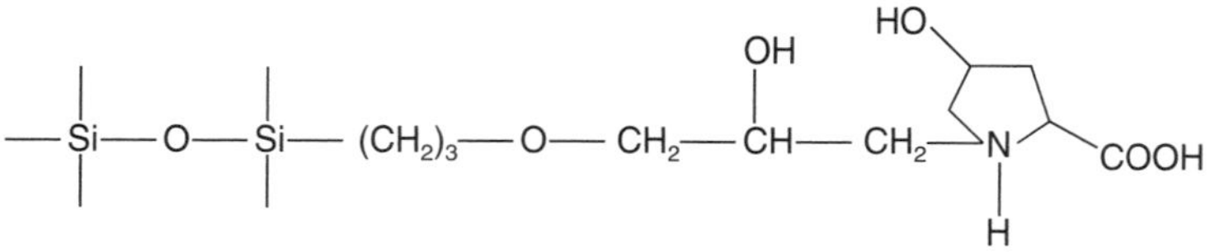

(as a fixed ligand on a silica gel; Cu (II) used as complexing agent)

Analytical Properties: Chiral phases containing L-hydroxy-proline as a fixed ligand show high enantioselectivity for 2-hydroxy acids; these phases have resolved some aromatic as well as aliphatic 2-hydroxy acids.

Reference: 41

Name: 1-isothiocyanato-D-glucopyranosides

Structure:

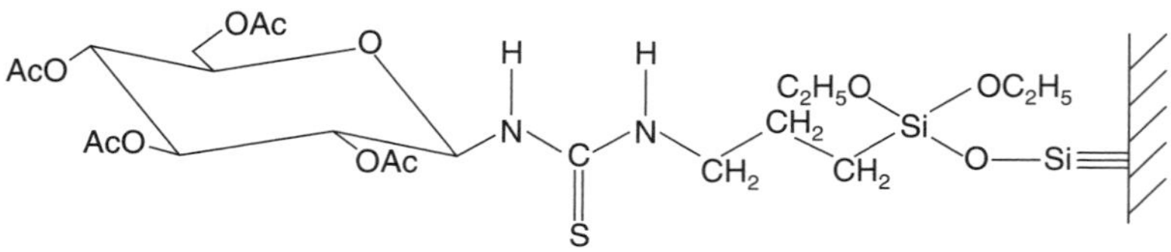

Chiral Stationary Phases for Liquid Chromatography (Continued)

CSP 2

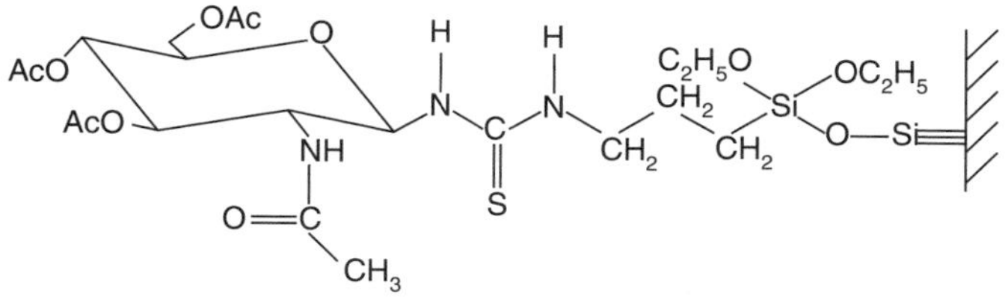

CSP 3

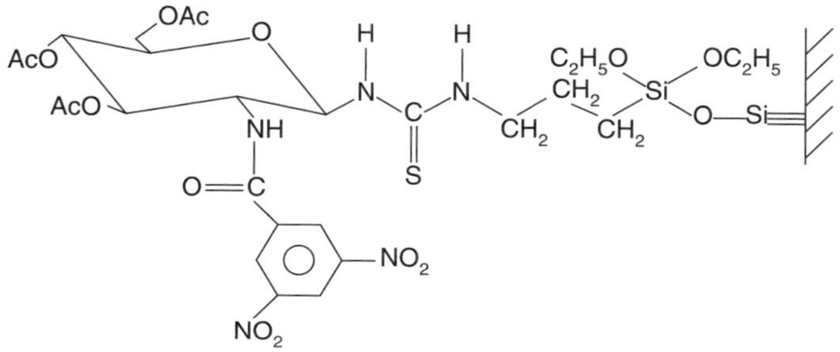

Analytical Properties:

CSP 1: (chiral stationary phase) Separates some chiral binapthyl derivatives when mixtures of hexane diethyl ether, dichloromethane or dioxane are used as the mobile phase.

CSP 2: Separates compounds with carbamate or amide functions. (Mixtures of n-hexane and 2-propanol can be used as mobile phase.)

CSP 3: Separation of compounds separated by CSP 2. In addition, separation of compounds with carbanoyl or amide functions and some amino alcohols which have pharmaceutical relevance (β-blockers).

Reference: 42

Name: N-isopropylamino carbonyl-L-valine bonded silica

Structure:

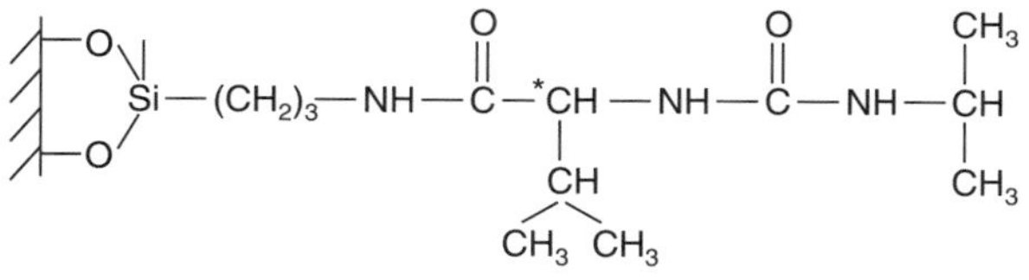

Analytical Properties: Separation of amino acid enantiomers; good chiral recognition of N-acetyl amino acid methyl esters; depends on hydrogen bond interactions; hexane with isopropanol modifier has been used as the liquid phase; usually prepared on LiChrosorb (10 μm).

Reference: 4

Name: (R)- or (S)-N-(2-napthyl)alanine

Structure:

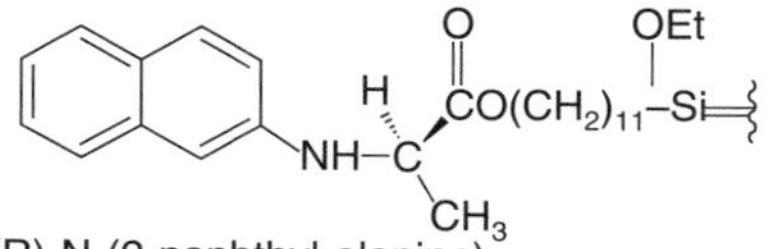

(R)-N-(2-naphthyl-alanine)

Analytical Properties: High selectivities for a variety of dinitrobenzoyl derivatized compounds.

Reference: 19

(Continued)

Chiral Stationary Phases for Liquid Chromatography (Continued)

Name: (S)-1-(α-napthyl)ethylamine
Structure:

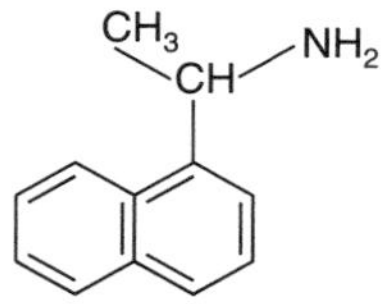

Analytical Properties: Separation of 3,5-dinitrobenzoyl derivatives of amino acids; 3,5-dinitroanilide derivatives of carboxylic acids.
Reference: 19

Name: R-N-(pentafluorobenzoyl) phenylglycine
Structure:

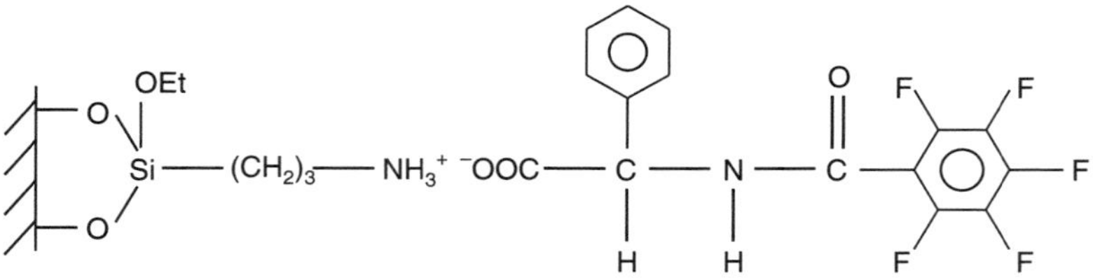

Analytical Properties: Higher selectivity for nitrogen-containing racemates than R-N-(3,5-dinitrobenzoyl) phenylglycine. Examples of nitrogen-containing racemates include succinimides, hydantoins, and mandelates.
Reference: 43

Name: N-phenylaminocarbonyl-L-valine bonded silica
Structure:

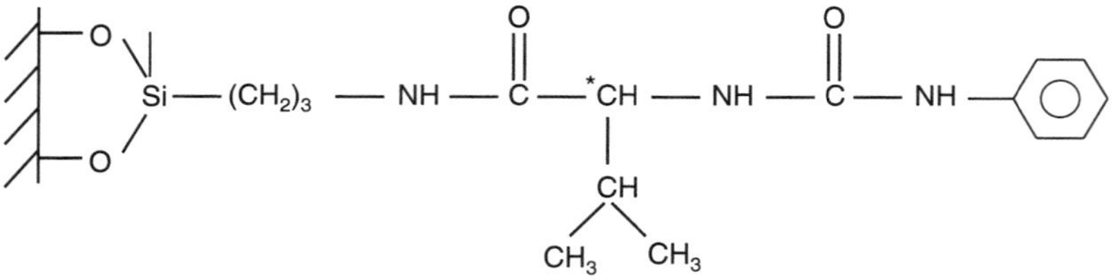

Analytical Properties: Separation of amino acid enantiomers by hydrogen bond interactions; usually prepared on LiCrosorb (10 μ:m); hexane plus isopropanol modifier commonly used as liquid phase.
Reference: 4

Name: (L-proline) copper (II)
Structure:

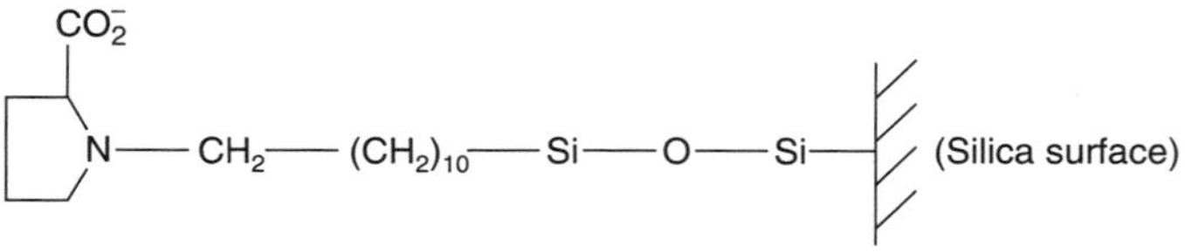

Chiral Stationary Phases for Liquid Chromatography (Continued)

Analytical Properties: Separation of primary α-amino alcohols, for example β-hydroxyphenethylamines and catecholamines.

Reference: 44

Name: derivative of (R,R)-tartramide

Structure:

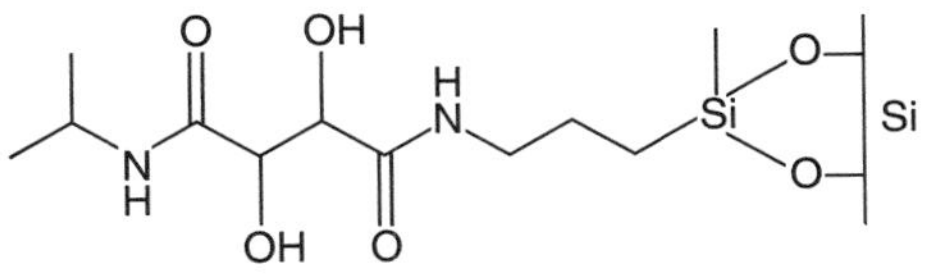

Analytical Properties: Resolution of a series of β-hydroxycarboxylic acids as tert-butylamide derivatives.

Reference: 45

Name: tert-butylvalinamide

Structure:

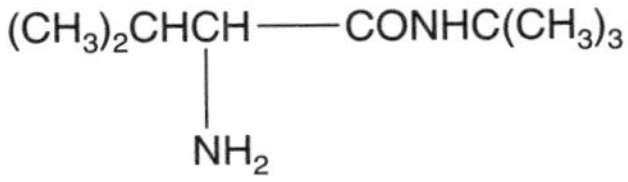

$$(CH_3)_2CHCH\text{——}CONHC(CH_3)_3$$
$$|$$
$$NH_2$$

Analytical Properties: Resolution of heavier amino acid derivatives.

Reference: 46

Name: (+)-poly(triphenylmethyl methacrylate) or (+)-PTrMA

Structure:

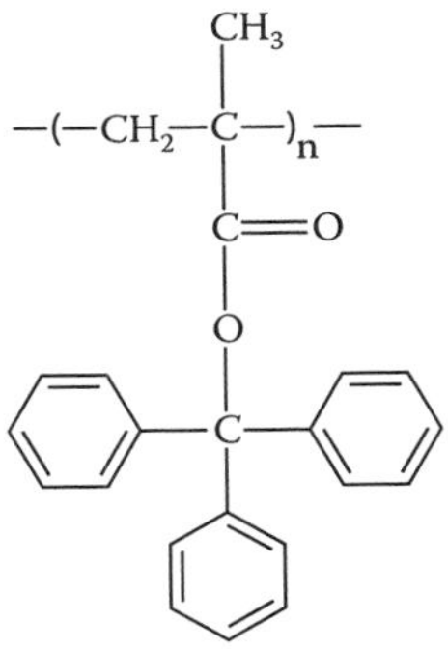

Analytical Properties: Resolution of enantiomers such as trans-1,3-cyclohexene dibenzoate, 3,5-pentylene dibenzoate, 3,5-dichlorobenzoate, triacetylacetonates, and racemic compounds having phosphorous as a chiral center. This phase will also resolve achiral compounds.

Reference: 45–48

(*Continued*)

Chiral Stationary Phases for Liquid Chromatography (Continued)

Name: ristocetin-A bonded phase

Structure:

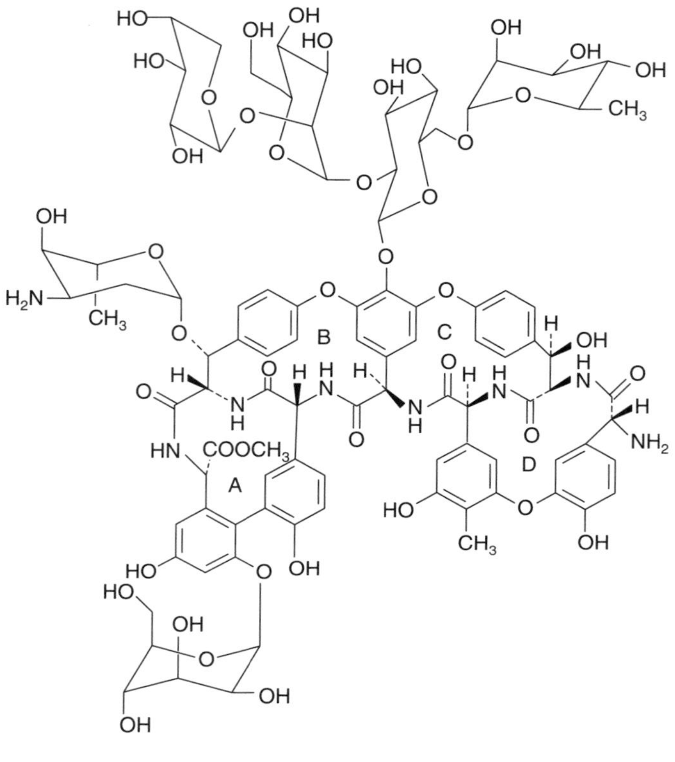

Analytical Properties: Substrate has 38 chiral centers and seven aromatic rings surrounding four cavities, making this the most structurally complex of the macrocyclic glycopeptides. Substrate has a relative molecular mass of 2066. This phase can be used in normal, reverse and polar organic phase separations; selective for anionic chiral species. With polar organic mobile phases, it can be used for α-hydroxy acids, profens, and N-blocked amino acids; in normal phase mode, it can be used for imides, hydantoins, and N-blocked amino acids, and in reverse phase it can be used for α-hydroxy and halogenated acids, substituted aliphatic acids, profens, N-blocked amino acids, hydantoins, and peptides.

Reference: 49,50

Chiral Stationary Phases for Liquid Chromatography (Continued)

Name: teicoplanin bonded phase
Structure:

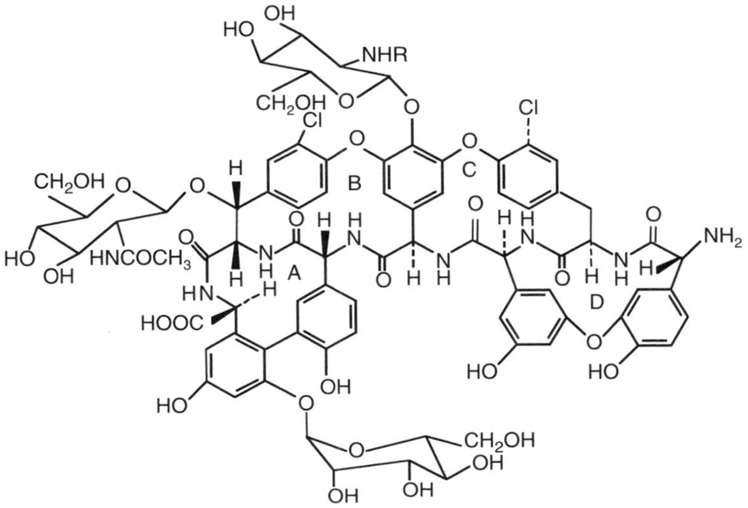

Analytical Properties: Substrate has 20 chiral centers and seven aromatic rings surrounding four cavities. Substrate has a relative molecular mass of 1885. Separation occurs through chiral hydrogen bonding sites, B–B interactions and inclusion complexation in polar organic, normal and reverse mobile phases. Useful for the resolution of α, β, γ, or cyclic amino acids, small peptides, N-derivatized amino acids.
Reference: 51,52

Name: teicoplanin aglycone bonded phase
Structure:

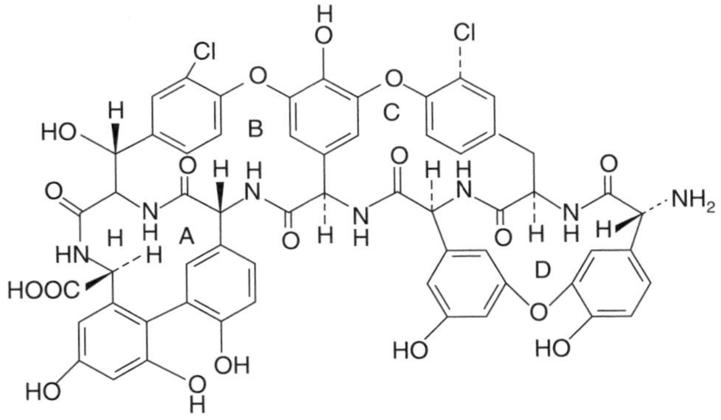

Analytical Properties: Substrate has eight chiral centers and seven aromatic rings surrounding four cavities. Substrate has a relative molecular mass of 1197. Separation occurs through chiral hydrogen bonding sites, π-π interactions and inclusion complexation in polar organic, normal and reverse mobile phases. Highly selective for amino acids (α, β, γ, or cyclic), some N-blocked amines, many neutral cyclic compounds, peptides, diazepines, hydantoins, oxazolidinones, and sulfoxides.
Reference: 53

(Continued)

Chiral Stationary Phases for Liquid Chromatography (Continued)

Name: vancomycin bonded phase

Structure:

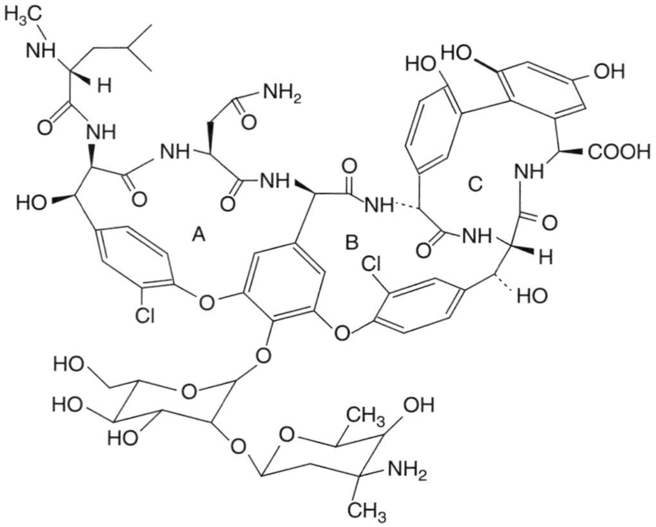

Analytical Properties: Substrate has 18 chiral centers and five aromatic rings surrounding three cavities. Substrate has a relative molecular mass of 1449, an isoelectric point of 7.2, with pKs of 2.9, 7.2, 8.6, 9.6, 10.4, and 11.7. Separation occurs through chiral hydrogen bonding sites, π-π interactions, a peptide binding site and inclusion complexation in polar organic, normal and reverse mobile phases. Selective for cyclic amines, amides, acids, esters and neutral molecules; high sample capacity.

Reference: 54,55

DETECTORS FOR LIQUID CHROMATOGRAPHY

The following table provides some comparative data for the selection and operation of the more common detectors applied to high performance liquid chromatography [1–5]. In general, the operational parameters provided are for optimized systems, and represent the maximum obtainable in terms of sensitivity and linearity. In this table, the molar extinction coefficient is represented by ε.

REFERENCES

1. Pryde, A., and M. T. Gilbert. *Applications of High Performance Liquid Chromatography*. London: Chapman and Hall, 1979.
2. Hamilton, R. J., and P. A. Sewell. *Introduction to High Performance Liquid Chromatography*. London: Chapman and Hall, 1977.
3. Ahuja, S. *Trace and Ultratrace Analysis by HPLC, Chemical Analysis Series*. New York: John Wiley and Sons, 1991.
4. Snyder, L. R., J. J. Kirkland, and J. Glajch. *Practical HPLC Method Development*. New York: John Wiley and Sons, 1997.
5. Bruno, T. J. "A Review of Hyphenated Chromatographic Instrumentation." *Separation and Purification Methods* 29, no. 1 (2000): 63–89.

Detectors for Liquid Chromatography

Detector	Sensitivity	Linearity	Selectivity	Comments
Ultraviolet spectrophotometer	1×10^{-9} g (for compounds of $\varepsilon = 10,000–20,000$)	1×10^4	For UV-active functionalities, on the basis of absorptivity.	Relatively insensitive to flow and temperature fluctuations; nondestructive, useful with gradient elution; use mercury lamp for 254 nm, and quartz-iodine lamp for 350–700 nm; often a diode-array instrument is used to obtain entire UV-vis. Spectrum.
Refractive index detector (RID)	1×10^{-7} g	1×10^4	Universal, dependent on refractive index difference with mobile phase.	Relatively insensitive to flow fluctuations, but sensitive to temperature fluctuations; nondestructive, cannot be used with gradient elution; solvents must be degassed to avoid bubble formations; laser-based RI detectors offer higher sensitivity.
Fluorometric detector	1×10^{-11} g	1×10^5	For fluorescent species with conjugated bonding and/or aromaticity.	Relatively insensitive to temperature and flow fluctuations; nondestructive; can be used with gradient elution; often, chemical derivatization is done on analytes to form fluorescent species; uses deuterium lamp for 190–400 nm, or tungsten lamp for 350–600 nm.
Electrochemical detectors Amperometric	1×10^{-9} g		Responds to –OH functionalities	Used for aliphatic and aromatic –OH compounds, amines, and indoles; pulsed potential units are most sensitive, can be used with gradient elution and organic mobile phases; senses compounds in oxidatitive or reductive modes; mobile phases must be highly pure and purged of O_2.
Conductivity detector	1×10^{-9} g	2×10^4	Specific to ionizable compounds	Uses postcolumn derivatization to produce ionic species; especially useful for certain halogen, sulfur and nitrogen compounds.
Mass spectrometers	Interface dependent	Interface dependent	Universal, within limits imposed by interface.	Complex, expensive devices highly dependent on an efficient interface; electrospray and thermospray interfaces are most common; linear response is difficult to achieve.

ULTRAVIOLET DETECTION OF CHROMOPHORIC GROUPS

The following table is provided to aid the use in the application of ultraviolet spectrophotometric detectors. The data here are used to evaluate the potential of detection of individual chromophoric moities on analytes [1–3].

REFERENCES

1. Willard, H. H, L. L. Merritt, J. A. Dean, and F. A. Settle. *Instrumental Methods of Analysis*. 7th ed. Belmont: Wadsworth Publishing Co., 1988.
2. Silverstein, R. M., G. C. Bassler, and T. C. Morrill. *Spectrometric Identification of Organic Compounds*. 4th ed. New York: John Wiley and Sons, 1981.
3. Lambert, J. B., H. F. Shuruell, L. Verbit, R. G. Cooks, and G. H. Stout. *Organic Structural Analysis*. New York: MacMillan Publishing Co., 1976.

Ultraviolet Detection of Chromophoric Groups

Chromophore	Functional Group	λ_{max} nm	ε_{max}	λ_{max} nm	ε_{max}	λ_{max} nm	ε_{max}
Ether	-O-	185	1000				
Thioether	-S-	194	4600	215	1600		
Amine	-NH$_2$-	195	2800				
Amide	-CONH$_2$	<210	—				
Thiol	-SH	195	1400				
Disulfide	-S-S-	194	5500	255	400		
Bromide	-Br	208	300				
Iodide	-I	260	400				
Nitrile	-C≡N	160	—				
Acetylide (alkyne)	-C≡C-	175–180	6000				
Sulfone	-SO$_2^-$	180	—				
Oxime	-NOH	190	5000				
Azido	>C=N-	190	5000				
Alkene	-C=C-	190	8000				
Ketone	>C=O	195	1000	270–285	18–30		
Thioketone	>C=S	205	strong				
Esters	-COOR	205	50				
Aldehyde	-CHO	210	strong	280–300	11–18		
Carboxyl	-COOH	200–210	50–70				
Sulfoxide	>S→O	210	1500				
Nitro	-NO$_2$	210	strong				
Nitrite	-ONO	220–230	1000–2000	300–4000	10		
Azo	-N=N-	285–400	3–25				
Nitroso	-N=O	302	100				
Nitrate	-ONO$_2$	270 (shoulder)	12				
Conjugated hydrocarbon	-(C=C)$_2$-(acyclic)	210–230	21,000				
Conjugated hydrocarbon	-(C=C)$_3$-	260	35,000				
Conjugated hydrocarbon	-(C=C)$_4$-	300	52,000				
Conjugated hydrocarbon	-(C=C)$_5$-	330	118,000				
Conjugated hydrocarbon	-(C=C)$_2$-(alicyclic)	230–260	3000–8000				

		λ	ε	λ	ε	λ	ε
Conjugated hydrocarbon	C=C—C≡C	219	6500				
Conjugated system	C=C—C=N	220	23,000				
Conjugated system	C=C—C=O	210–250	10,000–20,000			300–350	weak
Conjugated system	C=C—NO$_2$	229	9500				
Benzene		184	46,700	202	6900	255	170
Diphenyl	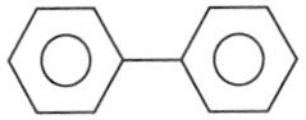			246	20,000		
Naphthalene	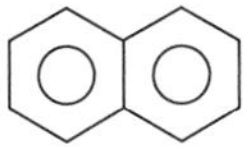	220	112,000	275	5600	312	175
Anthracene		252	199,000	375	7900		
Pyridine		174	80,000	195	6000	251	1700
Quinoline	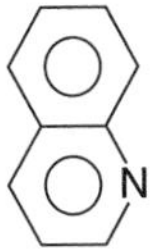	227	37,000	270	3600	314	2750
Isoquinoline	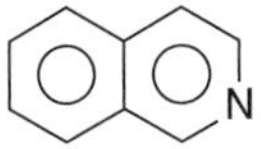	218	80,000	266	4000	317	3500

Note: φ also typically denotes a phenyl group.

DERIVATIZING REAGENTS FOR HPLC

The following table provides a listing of the common reagents used in HPLC. Most of these reagents are used to impart a chromophoric or fluorescent group in the sample to enhance the detectability. Occasionally, a derivatization procedure is done in order to enhance selectivity, but this is the exception.

REFERENCES

1. Umagat, H., P. Kucera, and L.-F. Wen. "Total Amino Acid Analysis Using Pre-Column Fluorescence Derivatization." *Journal of Chromatography* 239 (1982): 463.
2. Nimura, N., and T. Kinoshita. "Fluorescent Labeling of Fatty Acids with 9-Anthryldiazomethane (ADAM) for High Performance Liquid Chromatography." *Analytical Letters* 13, no. A3 (1980): 191.
3. Lehrfeld, J. "Separation of Some Perbenzoylated Carbohydrates by High Performance Liquid Chromatography." *Journal of Chromatography* 120 (1976): 141.
4. Durst, H. D., M. Milano, E. J. Kikta, S. A. Connelly, and E. Grushka. "Phenacyl Esters of Fatty Acids via Crown Ether Catalysts for Enhanced Ultraviolet Detection in Liquid Chromatography." *Analytical Chemistry* 47, no. 11 (1975): 1797.
5. Farinotti, R., Ph. Siard, J. Bourson, S. Kirkiacharian, B. Valeur, and G. Mohuzier. "4-Bromomethyl-6,7-Dimethoxycoumarin as a Fluorescent Label for Carboxylic-Acids in Chromatographic Detection." *Journal of Chromatography* 269, no. 2 (1983): 81.
6. Lam, S., and E. Grushka. "Labeling of Fatty Acids with 4-Bromomethyl-7-Methoxycoumarin via Crown Ether Catalyst for Fluorimetric Detection in High-Performance Liquid Chromatography." *Journal of Chromatography* 158 (1978): 207–14.
7. Korte, W. D. "9-(Chloromethyl) Anthracene: A Useful Derivatizing Reagent for Enhanced Ultraviolet and Fluorescence Detection of Carboxylic Acids with Liquid Chromatography." *Journal of Chromatography* 243 (1982): 153.
8. Ahnoff, M., I. Grundevik, A. Arfwidsson, J. Fonselius, and B.-A. Persson. "Derivatization with 4-Chloro-7-Nitrobenzofurazan for Liquid Chromatographic Determination of Hydroxyproline in Collagen Hydrolysate." *Analytical Chemistry* 53 (1981): 485.
9. Linder, W. "N-Chloromethyl-4-Nitro-Phthalimide as a Derivatizing Reagent for HPLC." *Journal of Chromatography* 198 (1980): 367.
10. Lawrence, J. F. *Organic Trace Analysis by Liquid Chromatography.* New York: Academic Press, 1981.
11. Avigad, G. "Dansyl Hydrazine as a Fluorimetric Reagent for Thin-Layer Chromatographic Analysis of Reducing Sugars." *Journal of Chromatography* 139 (1977): 343–47.
12. Lloyd, J. B. F. "Phenanthramidazoles as Fluorescent Derivatives in the Analysis of Fatty Acids by High-Performance Liquid Chromatography." *Journal of Chromatography* 189 (1980): 359–73.
13. Frei, R. W., and J. F. Lawrence, eds. *Chemical Derivatization in Analytical Chemistry.* Vol. 1. New York: Plenum Press, 1981.
14. Goto, J., S. Komatsu, N. Goto, and T. Nambara. "A New Sensitive Derivatization Reagent for Liquid Chromatographic Separation of Hydroxyl Compounds." *Chemical & Pharmaceutical Bulletin* 29, no. 3 (1981): 899.
15. Musson, D. G., and L. A. Sternson. "Conversion of Arylhydroxylamines to Electrochemically Active Derivatives Suitable for High-Performance Liquid Chromatographic Analysis with Amperometric Detection." *Journal of Chromatography* 188 (1980): 159.
16. Lankmayr, E. P., K. W. Budna, K. Müller, F. Nachtmann, and F. Rainer. "Determination of d-Penicillamine in Serum by Fluorescence Derivatization and Liquid Column Chromatography." *Journal of Chromatography* 222 (1981): 249.
17. Mopper, K., W. L. Stahovec, and L. Johnson. "Trace Analysis of Aldehydes by Reversed-Phase High-Performance Liquid Chromatography and Precolumn Fluorigenic Labeling with 5,5-Dimethyl-1,3-Cyclohexanedione." *Journal of Chromatography* 256 (1983): 243.

18. Lawrence, J. F., and R. W. Frei. *Chemical Derivatization in Liquid Chromatography.* Amsterdam: Elsevier, 1976.

19. Carey, M. A., and H. E. Persinger. "Liquid Chromatographic Determination of Traces of Aliphatic Carbonyl Compounds and Glycols as Derivatives that Contain the Dinitrophenyl Group." *Journal of Chromatographic Science* 10 (1972): 537.

20. Fitzpatrick, F. A., S. Siggia, and J. Dingman Sr. "High Speed Liquid Chromatography of Derivatized Urinary 17-Keto Steroids." *Analytical Chemistry* 44, no. 13 (1972): 2211.

21. Pietrzyk, D. J., and E. P. Chan. "Determination of Carbonyl Compounds by 2-Diphenylacetyl-1,3-Indandione-1-Hydrazone." *Analytical Chemistry* 42, no. 1 (1970): 37.

22. Braun, R. A., and W. A. Mosher. "2-Diphenylacetyl-1,3-Indandione 1-Hydrazone: A New Reagent for Carbonyl Compounds." *Journal of the American Chemical Society* 80 (1958): 3048.

23. Schäfer, M., and E. Mutschuler. "Fluorimetric Determination of Oxprenolol in Plasma by Direct Evaluation of Thin-Layer Chromatograms." *Journal of Chromatography* 164 (1979): 247.

24. Moye, H. A., and A. J. Boning, Jr. "A Versatile Fluorogenic Labelling Reagent for Primary and Secondary Amines: 9-Fluorenylmethyl Chloroformate." *Analytical Letters* 12, no. B1 (1979): 25.

25. Lehninger, A. L. *Biochemistry.* 2nd ed. New York: Worth Publishers, 1978.

26. Roos, R. W. "Determination of Conjugated and Esterified Estrogens in Pharmaceutical Tablet Dosage Forms by High-Pressure, Normal-Phase Partition Chromatography." *Journal of Chromatographic Science* 14 (1976): 505.

27. DeLeenheer, A., J. E. Sinsheimer, and J. H. Burckhalter. "Fluorometric Determination of Primary and Secondary Aliphatic Amines by Reaction with 9-Isothiocyanatoacridine." *Journal of Pharmaceutical Sciences* 62, no. 8 (1973): 1370.

28. Clark, C. R., and M. M. Wells. "Precolumn Derivatization of Amines for Enhanced Detectability in Liquid Chromatography." *Journal of Chromatographic Science* 16 (1978): 332.

29. Hulshoff, A., H. Roseboom, and J. Renema. "Improved Detectability of Barbiturates in High-Performance Liquid Chromatography by Pre-Column Labelling and Ultraviolet Detection." *Journal of Chromatography* 186 (1979): 535.

30. Matthees, D. P., and W. C. Purdy. "Napthyldiazomethane as a Derivatizing Agent for the High-Performance Liquid Chromatography Detection of Bile Acids." *Analytica Chimica Acta* 109 (1979): 161.

31. Kuwata, K., M. Uebori, and Y. Yamazaki. "Determination of Phenol in Polluted Air as p-Nitrobenzeneazophenol Derivative by Reversed Phase High Performance Liquid Chromatography." *Analytical Chemistry* 52 (1980): 857.

32. Nachtmann, F., H. Spitzy, and R. W. Frei. "Rapid and Sensitive High-Resolution Procedure for Digitalis Glycoside Analysis by Derivatization Liquid Chromatography." *Journal of Chromatography* 122 (1976): 293.

33. Knapp, D. R., and S. Krueger. "Use of o-p-Nitrobenzyl-N,N-Diisopropylisourea as a Chromogenic Reagent for Liquid Chromatographic Analysis of Carboxylic Acids." *Analytical Letters* 8, no. 9 (1975): 603–10.

34. Jupille, T. "UV-Visible Absorption Derivatization in Liquid Chromatography." *Journal of Chromatographic Science* 17 (1979): 160–67.

35. Dunlap, K. L., R. L. Sandridge, and Jürgen Keller. "Determination of Isocyanates in Working Atmospheres by High Speed Liquid Chromatography." *Analytical Chemistry* 48, no. 3 (1976): 497–99.

36. Politzer, I. R., G. W. Griffin, B. J. Dowty, and J. L. Laseter. "Enhancement of Ultraviolet Detectability of Fatty Acids for Purposes of Liquid Chromatographic-Mass Spectrometric Analyses." *Analytical Chemistry* 6, no. 6 (1973): 539–46.

37. Cox, G. B. "Estimation of Volatile N-Nitrosamines by High-Performance Liquid Chromatography." *Journal of Chromatography* 83 (1973): 471–81.

38. Borch, R. F. "Separation of Long Chain Fatty Acids as Phenacyl Esters by High Pressure Liquid Chromatography." *Analytical Chemistry* 47, no. 14 (1975): 2437–39.

39. Poole, C. F., S. Singhawangcha, A. Zlatkis, and E. D. Morgan. "Polynuclear Aromatic Boronic Acids as Selective Fluorescent Reagents for HPTLC and HPLC." *Journal of High Resolution Chromatography & Chromatography Communications* 1 (1978): 96–97.

40. Björkqvist, B., and H. Toivonen. "Separation and Determination of Aliphatic Alcohols by High-Performance Liquid Chromatography with U.V. Detection." *Journal of Chromatography* 153 (1978): 265–70.

41. Munger, D., L. A. Sternson, A. J. Repta, and T. Higuchi. "High-Performance Liquid Chromatographic Analysis of Dianhydrogalactitol in Plasma by Derivatization with Sodium Diethyldithiocarbamate." *Journal of Chromatography* 143 (1977): 375–82.

42. Sugiura, T., T. Hayashi, S. Kawai, and T. Ohno. "High Speed Liquid Chromatographic Determination of Putrescine, Spermidine and Spermine." *Journal of Chromatography* 110 (1975): 385–88.

43. Suzuki, Y., and K. Tani. "High-Speed Liquid-Chromatography of the Aliphatic Alcohols as Their Trityl Ether Derivatives." *Buneseki Kagaku* 28 (1979): 610.

Derivatizing Reagents for HPLC

Derivatizing Reagent	Structure/Formula	Notes
N-(9-acridinyl) malemide	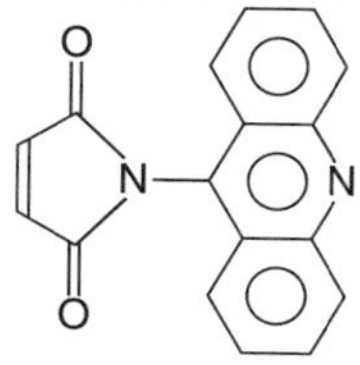	Used for the precolumn preparation of fluorescent derivatives of thiols. Reference 1
9-Anthryldiazomethane	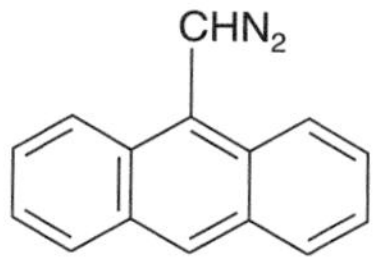	Used for the precolumn preparation of fluorescent derivatives of carboxylic acids; reagent reacts well with fatty acids at room temperature to give intensely fluorescent esters. Reference 2
Benzoyl chloride	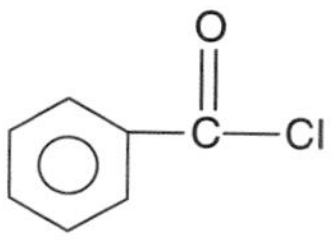	Used to introduce chromophores into alcohols and amines using pyridine as a solvent; efficient means for the isolation of carbohydrates in complex mixtures. Reference 3
Benzyl bromide	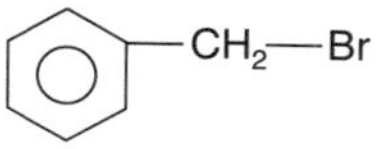	Used to introduce chromophores into carboxylic acids. Reference 4
4-Bromomethyl-6,7-dimethoxycoumarin	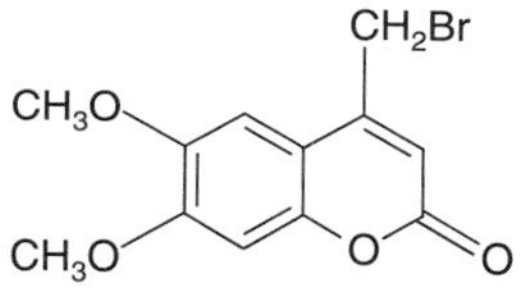	Used for the precolumn preparation of fluorescent derivatives of carboxylic acids using acetone as solvent and with crown ether and alkali as catalysts. Reference 5
4-Bromomethyl-7-methoxycoumarin (Br-Mmc)	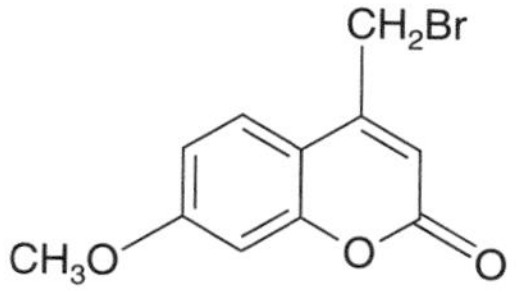	Used for the precolumn preparation of fluorescent derivatives of carboxylic acids, using a crown ether (18-crown-6) as a catalyst. Reference 6
p-Bromophenacyl bromide	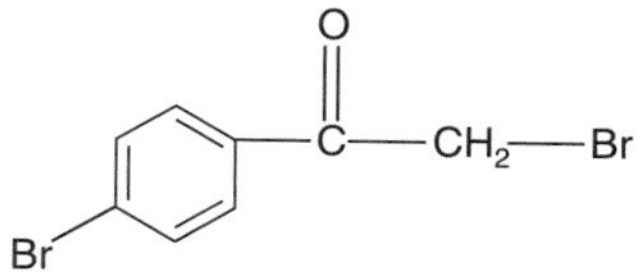	Used to introduce chromophores into carboxylic acids; crown ethers are used as phase transfer agents (for example, 18-crown-6 and dicyclohexyl-18-crown-6). Reference 4
9-(chloromethyl) Anthracene (9-CIMA)	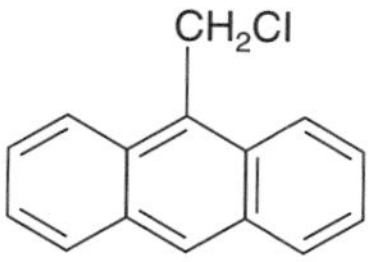	Used for the precolumn preparation of fluorescent derivatives of carboxylic acids, using cyclohexane as a solvent. Reference 7

(Continued)

Derivatizing Reagents for HPLC (Continued)

Derivatizing Reagent	Structure/Formula	Notes
4-Chloro-7-nitrobenzo-2-oxa-1,3-diazole (NBD-Cl)	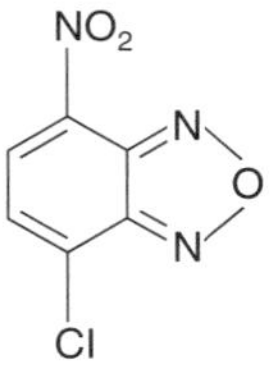	Used for the precolumn preparation of fluorescent derivatives of primary and secondary amines, phenols, and thiols (4-chloro-7-nitrobenzofuran). Reference 8
N-Chloromethyl-4-nitrophthalimide	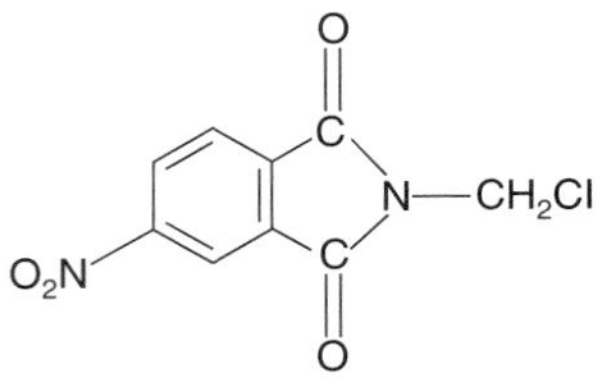	Used to introduce chromophores into carboxylic acids. Reference 9
Dansyl chloride (DnS-Cl)	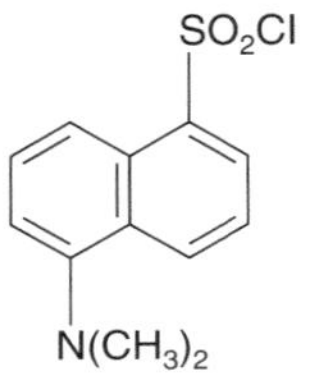	Used for the precolumn preparation of fluorescent derivatives of primary and secondary amines, phenols, amino acids, and imidazoles. Reference 10
Dansyl hydrazine (DnS-H)	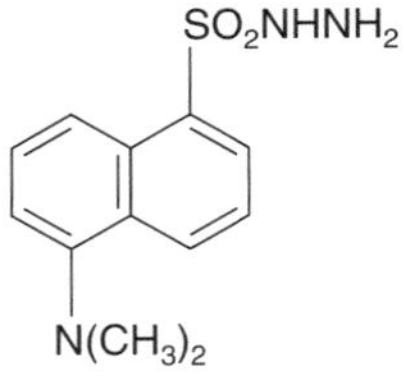	Used for the precolumn preparation of fluorescent derivatives of aldehydes and ketones; optimal derivatization of glucose and other sugars occurs at pH 2-3. Reference 11
9,10-Diaminophen-anthrene	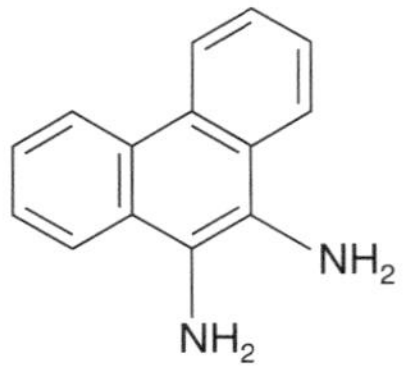	Used for the precolumn preparation of fluorescent derivatives of carboxylic acids. Reference 12
Diazo-4-aminobenzonitrile	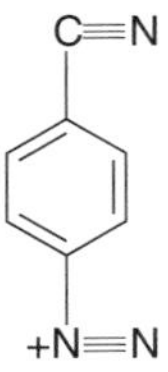	Used to introduce chromophores in phenols. Reference 13

Derivatizing Reagents for HPLC (Continued)

Derivatizing Reagent	Structure/Formula	Notes
2,5-di-n-Butylamino-naphthalene-1-sulfonyl chloride (BnS-Cl)	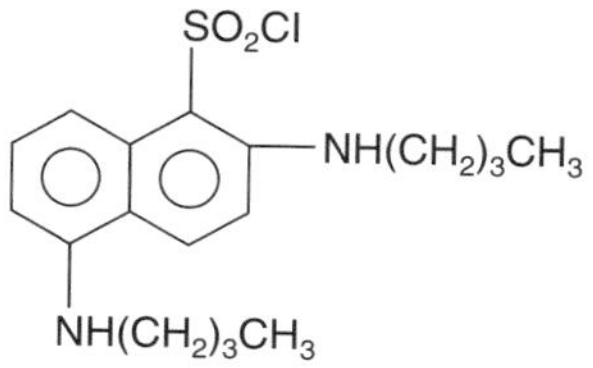	Used for the precolumn preparation of fluorescent derivatives of primary and secondary amines, phenols, amino acids, and imidazoles. Reference 10
4-Dimethylamino-1-naphthoyl nitrile	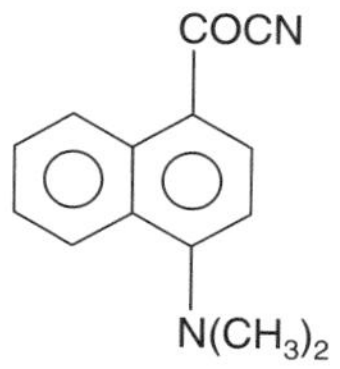	Used for the precolumn preparation of fluorescent derivatives of primary and secondary (but not tertiary) alcohols. Reference 14
p-Dimethylamino-phenyl isocyanate	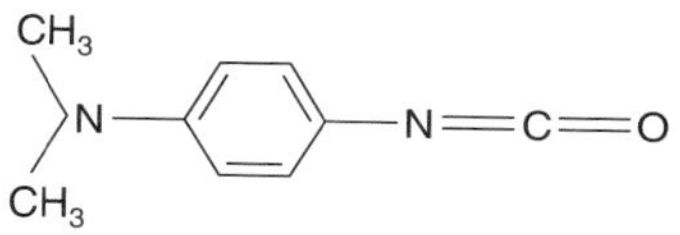	Used to introduce chromophores into alcohols. After reaction, excess reagent must be removed as it interferes with ensuing analysis. Reference 15
5-Dimethylamino-naphthalene 1-sulfonylaziridine (dansylaziridine)	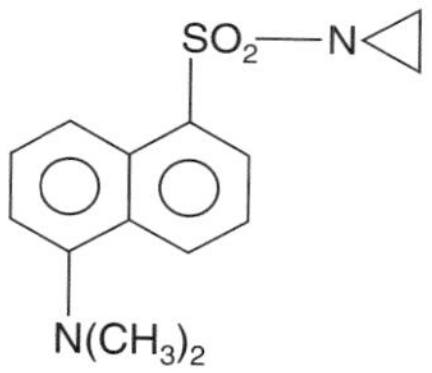	Used for the precolumn preparation of fluorescent derivatives of thiols; optimum derivatization conditions were found to be pH 8.2 with a minimum of a 2.7 fold molar reagent excess using a reaction time of 1 h at 60 °C. Under these conditions only free sulfhydryl groups are derivatized. Reference 16
5,5-Dimethyl-1,3-cyclohexanedione (dimedone)	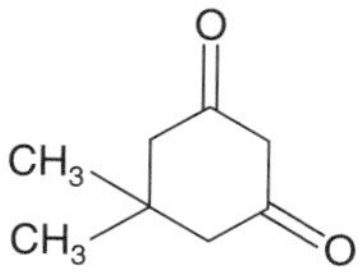	Used for the precolumn preparation of fluorescent derivatives of aldehydes, using isopropanol as a solvent in the presence of ammonium acetate. Reference 17
3,5-Dinitrobenzyl chloride	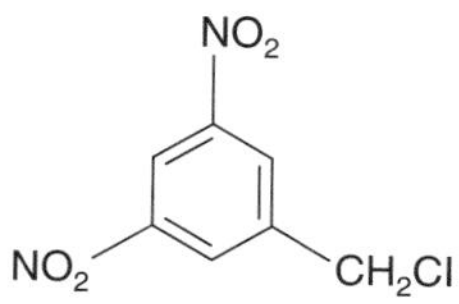	Used to introduce chromophores into amines (forming phenyl substituted amines), alcohols, glycols, and phenols. References 18,19
2,4-Dinitro phenyl-hydrazine	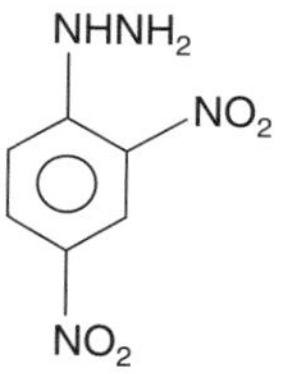	Used to introduce chromophores into aldehydes and ketones in a solution of carbonyl-free methanol; detection of the more common 17-keto steroids as their 2,4-dinitrophenyl derivatives from urine and plasma; suggested potential for clinical use. Reference 20

(Continued)

Derivatizing Reagents for HPLC (Continued)

Derivatizing Reagent	Structure/Formula	Notes
2-Diphenylacetyl-1,3-indandione-1-hydrazone		Used for the precolumn preparation of fluorescent derivatives of aldehydes and ketones; reagent suggested to be especially useful because of its application on the micro level, for the analysis and identification of carbonyl compounds in smog, polluted air, biochemical and pharmaceutical mixtures. Reagent does not appear to be useful for analysis of sugars. Derivatives are fluorescent in the UV_2 as solids and in solution. References 21, 22
1-Ethoxy-4-(dichloro-s-triazinyl) naphthalene (EDTN)		Used for the precolumn preparation of fluorescent derivatives of primary and secondary alcohols and phenols. Reference 23
9-Fluorenylmethyl chloroformate (FMOCCl)		Used for the precolumn preparation of fluorescent derivatives of primary and secondary amines in acetone solvent; in the presence of sodium borate, derivation proceeds rapidly under alkaline conditions. Reference 24
Fluorescamine (fluram)		Used for the precolumn preparation of fluorescent derivatives of primary amines and amino acids by HPLC. Reference 10
1-Fluoro-2,4-dinitrobenzene		Used to introduce chromophores into amines using Sanger's procedure. Reference 25

Derivatizing Reagents for HPLC (Continued)

Derivatizing Reagent	Structure/Formula	Notes
p-Iodobenzensulfonyl chloride	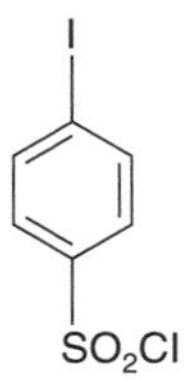 I—C6H4—SO2Cl	Used to introduce chromophores into alcohols and phenols; aids in separation of estrogen derivatives. Reference 26
9-Isothiocyanato-acridine	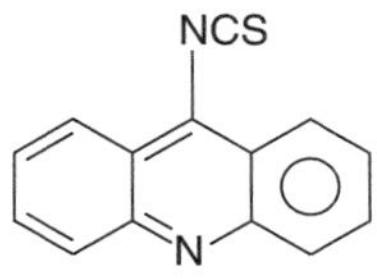 NCS (acridine ring)	Used for the precolumn preparation of fluorescent derivatives of some primary and secondary amines using toluene as a solvent; only amines with $PK_a \geq 9.33$ have been successfully determined. Reference 27
p-Methoxybenzoyl chloride	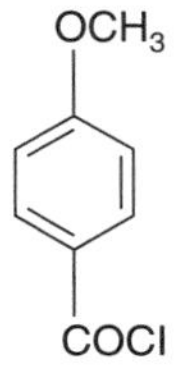 OCH_3—C6H4—COCl	Used to introduce chromophores into amines using the reagent in tetrahydrofuran. Reference 28
2-Naphthacyl bromide (NPB)	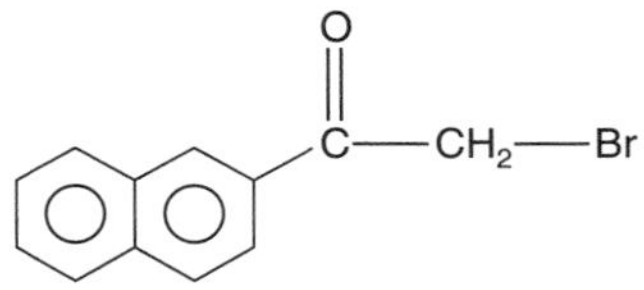 naphthalene—C(=O)—CH2—Br	Used to introduce chromophores into amines in acetone as a solvent, with cesium bromide as a catalyst; it is suggested that elevated temperatures (up to 80 °C) are necessary for the complete derivatization of compounds containing diisopropylamines. Reference 29
1-Naphthyldiazo-methane	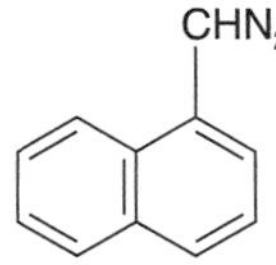 naphthalene—CHN_2	Used to introduce chromophores into carboxylic acids; reagent is prepared from 1-naphthaldehyde hydrazone by oxidation with Hg(II) oxide, with diethyl ether as a solvent; acetic acid will destroy excess reagent. Reference 30
p-Nitrobenzenediazo-niumtetrafluoro-borate	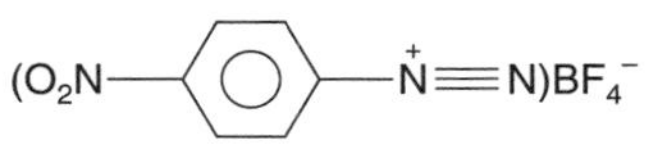 $(O_2N$—C6H4—$N^+{\equiv}N)BF_4^-$	Used to introduce chromophores into phenols, suggested derivation takes place in aqueous medium at pH 11.5 Reference 31
p-Nitrobenzoyl chloride (4-NBCl)	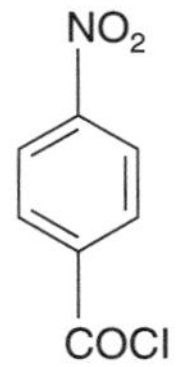 NO_2—C6H4—COCl	Used to introduce chromophores into alcohols and amines, using pyridine as the solvent; with silica gel as the stationary phase, relatively low-viscosity, low-polarity solvents can be used for detection of digitalis glycosides by HPLC following derivatization with p-nitrobenzoyl chloride. Reference 32

(Continued)

Derivatizing Reagents for HPLC (Continued)

Derivatizing Reagent	Structure/Formula	Notes
p-Nitrobenzyl-N,N'-diisopropylisourea	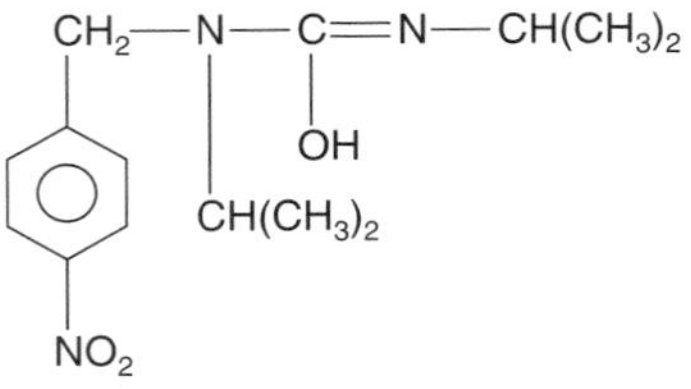	Used to introduce chromophores into carboxylic acids, without the need for a base catalyst, and under mild conditions; picomolar concentrations are rendered detectable. Reference 33
p-Nitrobenzyl-hydroxylamine hydrochloride	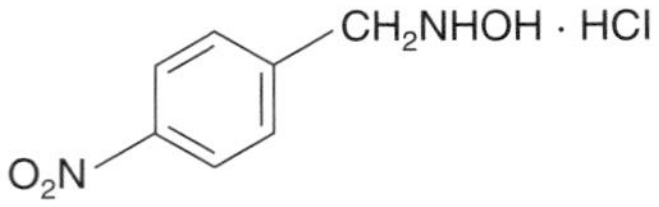	Used to introduce chromophores into ketones and aldehydes. Reference 34
N-p-Nitrobenzyl-N-n-propylamine	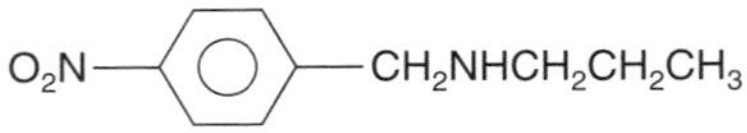	Used to introduce chromophores into isocyanates; suggested for use in determining isocyanate levels in air down to 0.2 ppm in a 20 L air sample. Reference 35
1-(p-nitro)Benzyl-3-p-tolyltriazine	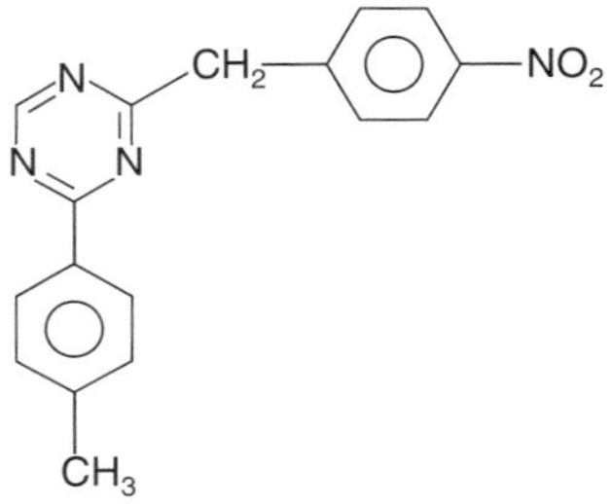	Used to introduce chromophores into carboxylic acids. Reference 36
p-Nitrophenyl chloroformate	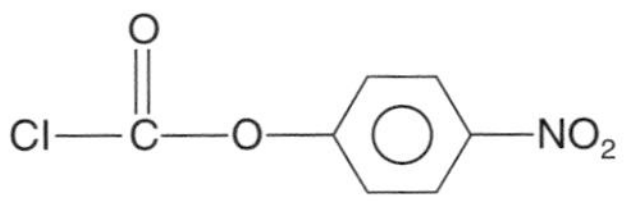	Used to introduce chromophores into alcohols. Reference 37
Phenacyl bromide	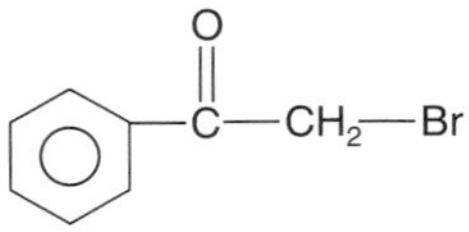	Used to introduce chromophores into carboxylic acids; provides for the subsequent analysis of fatty acid mixtures on the microgram scale using HPLC. Reference 38
Phenanthreneboronic acid	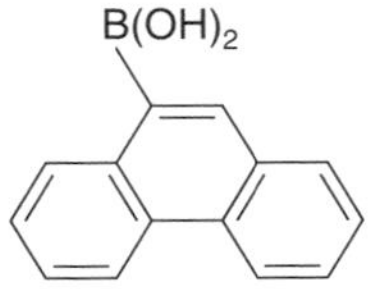	Used for the precolumn preparation of fluorescent derivatives of bifunctional compounds. Reference 39

Derivatizing Reagents for HPLC (Continued)

Derivatizing Reagent	Structure/Formula	Notes
Phenyl isocyanate	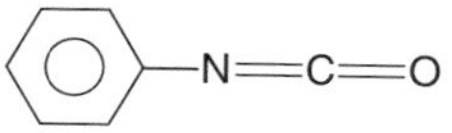 C$_6$H$_5$—N=C=O	Used to introduce chromophores into alcohols; thermal lability of the derivatives can cause problems; can also be used in the presence of water, but more reagent is required in this case. Reference 40
o-Phthaldialdehyde (OPT)	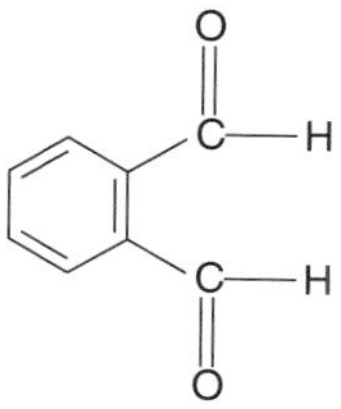 benzene ring with two adjacent CHO groups	Used for the precolumn preparation of fluorescent derivatives of amines and amino acids, in the presence of mercaptoethanol (or ethanethiol) and borate buffer. Reference 1
Pyridoxal	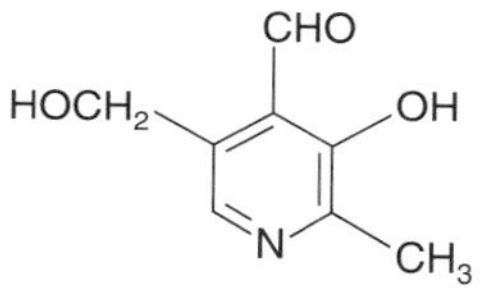 pyridine ring with CHO, OH, HOCH$_2$, and CH$_3$ substituents	Used for the precolumn preparation of fluorescent derivatives of amino acids. Reference 18
Pyruvoyl chloride (2,4-dinitrophenyl hydrazone)	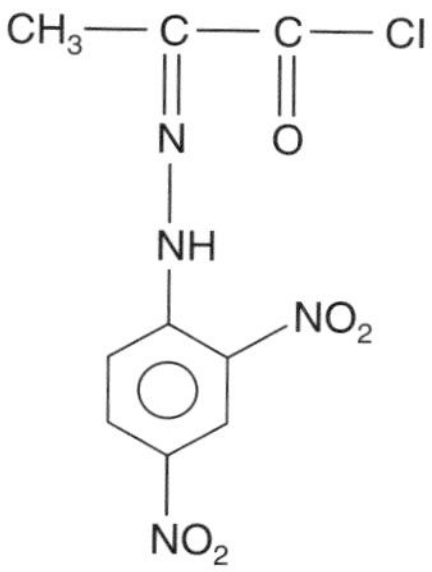 CH$_3$—C(=N—NH—C$_6$H$_3$(NO$_2$)$_2$)—C(=O)—Cl	Used to introduce chromophores into alcohols, amines, ketones, aldehydes, mercaptans, and phenols; aids in separation of estrogen derivatives. Reference 26
Sodium diethyldithio-carbamate (DDTC)	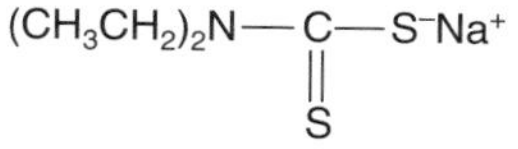 (CH$_3$CH$_2$)$_2$N—C(=S)—S$^-$Na$^+$	Used to introduce chromophores into epoxides in the presence of phosphate buffer; dithiocarbamates retain high nucleophilicity, and are often water soluble. Reference 41
n-Succinimidyl-p-nitrophenylacetate	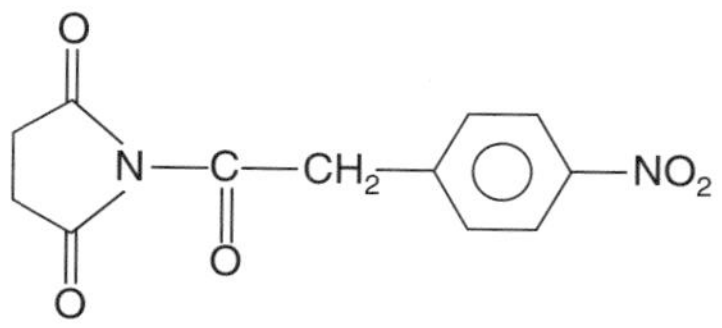 succinimidyl—N—C(=O)—CH$_2$—C$_6$H$_4$—NO$_2$	Used to introduce chromophores into amines; reacts under mild conditions without the need for catalysis. Reference 34

(Continued)

Derivatizing Reagents for HPLC (Continued)

Derivatizing Reagent	Structure/Formula	Notes
p-Toluenesulfonyl chloride (TsCl)	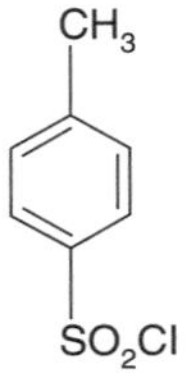 CH₃ — SO_2Cl (ring structure)	Used to introduce chromophores into amines; aids in resolution of putrescine, spermidine and spermine by HPLC; excess TsCl must be removed (by extraction with hexane, for example) before analysis. Reference 41
Trityl chloride	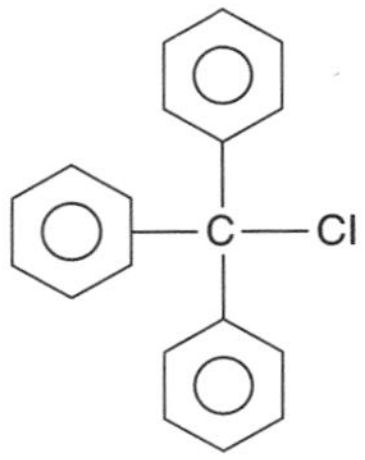 (three phenyl rings) C—Cl	Used to introduce chromophores into alcohols. Reference 42,43

CHAPTER 3

Thin Layer Chromatography

CONTENTS

STRENGTH OF COMMON TLC SOLVENTS

The following table contains the common solvents used in thin layer chromatography, with a measure of their "strengths" on silica gel and alumina. The solvent strength parameter, $\varepsilon°$, is defined as a relative energy of adsorption per unit area of standard adsorbent [1–3]. It is defined as zero on alumina when pentane is used as the solvent. This series is what was called the eluotropic series in the older literature. For convenience, the solvent viscosity is also provided. Note that the viscosity is tabulated in cP for the convenience of most users. This is equivalent to mPa·s in the SI convention. Additional data on these solvents may be found in the tables on high performance liquid chromatography.

REFERENCES

1. Snyder, L. R. *Principles of Adsorption Chromatography*. New York: Marcel Dekker, 1968.
2. Willard, H. H., L. L. Merritt, J. A. Dean, and F. A. Settle. *Instrumental Methods of Analysis*. 7th ed. New York, Belmont: Van Nostrand, 1988.
3. Hamilton, R., and S. Hamilton. *Thin Layer Chromatography*. Chichester: John Wiley and Sons (on behalf of "Analytical Chemistry by Open Learning," London), 1987.

Strength of Common TLC Solvents

Solvent	$\varepsilon°$ (Al$_2$O$_3$)	Viscosity cP, 20°	$\varepsilon°$ (SiO$_2$)
Fluoroalkanes	−0.25	—	
n-Hexane	0.00	0.23	0.00
n-Pentane	0.001	0.23	0.00
2,2,4-Trimethylpentane (isooctane)	0.01	0.54	
n-Heptane	0.01	0.41	
n-Decane	0.04	0.92	
Cyclohexane	0.04	1.00	−0.05
Cyclopentane	0.05	0.47	
Carbon disulfide	0.15	0.37	0.14
Tetrachloromethane (carbon tetrachloride)	0.18	0.97	
1-Chloropentane (n-pentylchloride)	0.26	0.43	
Diisopropyl ether	0.28	0.37	
2-Chloropropane (isopropyl chloride)	0.29	0.33	
Methylbenzene (toluene)	0.29	0.59	
1-Chloropropane (n-propyl chloride)	0.30	0.35	
Chlorobenzene	0.30	0.80	
Benzene	0.32	0.65	0.25
Bromoethane (ethyl bromide)	0.37	0.41	
Diethyl ether (ether)	0.38	0.23	0.38
Trichloromethane (chloroform)	0.40	0.57	
Dichloromethane (methylene chloride)	0.42	0.44	
Tetrahydrofuran	0.45	0.55	
1,2-Dichloroethane	0.49	0.79	
Butanone (methyl ethyl ketone)	0.51	0.43	
Propanone (acetone)	0.56	0.32	0.47
1,4-Dioxane	0.56	1.54	0.49
Ethyl ethanoate (ethyl acetate)	0.58	0.45	0.38
Methyl ethanoate (methyl acetate)	0.60	0.37	
1-Pentanol (n-pentanol)	0.61	4.1	
Dimethyl sulfoxide (DMSO)	0.62	2.24	
Aminobenzene (aniline)	0.62	4.4	
Nitromethane	0.64	0.67	
Cyanomethane (acetonitrile)	0.65	0.37	0.50
Pyridine	0.71	0.94	
2-Propanol (isopropanol)	0.82	2.3	
Ethanol	0.88	1.20	
Methanol	0.95	0.60	
Ethylene glycol	1.11	19.9	
Ethanoic acid (acetic acid)	large	1.26	
Water	large	1.00	

MODIFICATION OF THE ACTIVITY OF ALUMINA BY ADDITION OF WATER

The following table describes five different activity grades of commercial alumina used in chromatography [1–3]. The activity grades are defined by the degree of adsorption of azobenzene (called azobenzene number) on the types of hydrated alumina. Those types are prepared by heating commercial alumina to redness, giving grade I, and then adding controlled amounts of water and allowing equilibration in a closed vessel. The azobenzene number decreases with the amount of water added. The R_f value is the ratio of distance traveled by the solute spot to that traveled by the solvent.

REFERENCES

1. Randerath, K. *Thin Layer Chromatography.* New York: Verlag Chemie-Academic Press, Weinheim Bergstr., 1968.
2. Gordon, A. J., and R. A. Ford. *The Chemist's Companion: A Handbook of Practical Data, Techniques, and References.* New York: John Wiley and Sons, 1972.
3. Brockmann, H., and H. Schodder. "Aluminum Oxide with Buffered Adsorptive Properties for Purposes of Chromatographic Adsorption." *Berichte der Deutschen Chemischen Gesellschaft* 74B (1941): 73.

Modification of the Activity of Alumina by Addition of Water

Water Added (wt/wt%)	Activity Grade	Azobenzene Number [maximum adsorption of azobenzene (10^{-5} mol/g)]	R_f (p-Amino-azobenzene)
0	I	26	0.00
3	II	21	0.13
6	III	18	0.25
10	IV	13	0.45
15	V	0	0.55

STATIONARY AND MOBILE PHASES

The following table provides a comprehensive guide to the selection of thin layer chromatography media and solvents for a given chemical family. Mixed mobile phases are denoted with a slash, /, between components, and where available, the proportions are given. Among the references are several excellent texts [1–3,60], review articles [4–24], and original research papers and reports [25–59,61–98]. A table of abbreviations follows this section.

REFERENCES

1. Krebs, K. G., D. Heusser, and H. Wimmer. *Thin Layer Chromatography, A Laboratory Handbook,* edited by E. Stahl. New York: Springer-Verlag, 1969.
2. Bobbitt, J. B. *Thin Layer Chromatography.* New York: Reinhold, 1963.
3. Touchstone, J. C. *Techniques and Application of Thin Layer Chromatography.* New York: John Wiley, 1985, 1972.
4. Pataki, G. "Paper, Thin-Layer, and Electrochromatography of Amino Acids in Biological Material." *Z. Klin. Chem.* 2 (1964): 129; *Chemical Abstracts* 64 (1966): 5425c.
5. Padley, F. B. Thin-Layer Chromatography of Lipids, Thin-layer Chromatography, Proceedings Symposium, Rome 1963, 87 (Pub. 1964).
6. Honjo, M. "Thin-Layer Chromatography of Nucleic Acid Derivatives." *Kagaku No Ryoiki, Zokan* 64 (1964): 1.
7. Kazumo, T. "Thin-Layer Chromatography of Bile Acids." *Kagaku No Ryoiki, Zokan* 64 (1964): 19.
8. Nakazawa, Y. "Thin-Layer Chromatography of Compound Lipids." *Kagaku No Ryoiki, Zokan* 64 (1964): 31.
9. Nishikaze, O. "Separation and Quantitative Analysis of Adrenocortical Hormone and Its Metabolite (C_{21}) by Thin-Layer Chromatography." *Kagaku No Ryoiki, Zokan* 64 (1964): 37.
10. Shikita, M., H. Kakizazi, and B. Tamaoki. "Thin-Layer Chromatography of Radioactive Substances." *Kagaku No Ryoiki, Zokan* 64 (1964): 45.
11. Mo, I., and Y. Hashimoto. "Method of Thin-Layer Zone Electrophoresis." *Kagaku No Ryoiki, Zokan* 64 (1964): 61.
12. Kinoshita, S. "Thin-Layer Chromatography of Sugar Esters." *Kagaku No Ryoiki, Zokan* 64 (1964): 79.
13. Okada, M. "Thin-Layer Chromatography of Cardiotonic Glycosides." *Kagaku No Ryoiki, Zokan* 64 (1964): 103.
14. Omoto, T. "Thin-Layer Chromatography of Toad Toxin." *Kagaku No Ryoiki, Zokan* 64 (1964): 115.
15. Furuya, C., and H. Itokawa. "Thin-Layer Chromatography of Triterpenoids." *Kagaku No Ryoiki, Zokan* 64 (1964): 123.
16. Zenda, H. "Thin-Layer Chromatography of Aconitine-Type Alkaloids." *Kagaku No Ryoiki, Zokan* 64 (1964): 133.
17. Hara, S., and H. Tanaka. "Thin-Layer Chromatography of Mixed Pharmaceutical Preparations." *Kagaku No Ryoiki, Zokan* 64 (1964): 141.
18. Katsui, G. "Thin-Layer Chromatography of Vitamins." *Kagaku No Ryoiki, Zokan* 64 (1964): 157.
19. Fujii, S., and M. Kamikura. "Thin-Layer Chromatography of Pigments." *Kagaku No Ryoiki, Zokan* 64 (1964): 173.
20. Hosogai, Y. "Thin-Layer Chromatography of Organic Chlorine Compounds." *Kagaku No Ryoiki, Zokan* 64 (1964): 185.
21. Takeuchi, T. "Thin-Layer Chromatography of Metal Complex Salts." *Kagaku No Ryoiki, Zokan* 64 (1964): 197.
22. Yamakawa, H., and K. Tanigawa. "Thin-Layer Chromatography of Organic Metal Compounds." *Kagaku No Ryoiki, Zokan* 64 (1964): 209.
23. Takitani, S., and K. Kawanabe. "Thin-Layer Chromatography of Inorganic Ions (Anions)." *Kagaku No Ryoiki, Zokan* 64 (1964): 221.
24. Ibayashi, H. "Thin-Layer Chromatography of Steroid Hormones and Its Clinical Application." *Kagaku No Ryoiki, Zokan* 64 (1964): 227.

25. Chilingarov, A. O., and N. M. Sobchinskaya. "Quantitative Ultramicroanalysis of Monoamine Dansyl Derivatives in Biological Material." *Lab. Delo* (1980): 333; *Chemical Abstracts* 93 (1980): 109910t.

26. Heacock, R. A., C. Nerenberg, and A. N. Payza. "The Chemistry of the 'Aminochromes': Part I. The Preparation and Paper Chromatography of Pure Adrenochrome." *Canadian Journal of Chemistry* 36 (1958): 853.

27. Heacock, R. A., and W. S. Powell. "Adrenochrome and Related Compounds." *Progress in Medicinal Chemistry* 9 (1972): 275.

28. Heacock, R. A., and B. D. Scott. "The Chemistry of the 'Aminochromes': Part IV. Some New Aminochromes and Their Derivatives." *Canadian Journal of Chemistry* 38 (1960): 516.

29. Heacock, R. A. "The Chemistry of Adrenochrome and Related Compounds." *Chemical Reviews* 59 (1959): 181.

30. Suryaraman, M. G., and W. T. Cave. "Detection of Some Aliphatic Saturated Long Chain Hydrocarbon Derivatives by Thin-Layer Chromatography." *Analytica Chimica Acta* 30 (1964): 96; *Chemical Abstracts* 60 (1964): 7463e.

31. Knappe, E., D. Peteri, and I. Rohdewald. "Thin-Layer Chromatographic Identification of Technically Important Polyhydric Alcohols." *Z. Anal. Chem.* 199 (1964): 270; *Chemical Abstracts* 60 (1964): 7464f.

32. Horak, V., and R. F. X. Klein. "Microscale Group Test for Carbonyl Compounds." *Journal of Chemical Education* 62 (1985): 806.

33. Jaminet, F. "Paper Microchromatography in Phytochemical Analysis. Application to Congolian Strychnos." *J. Pharm. Belg.* 8 (1953): 339, 449; *Chemical Abstracts* 48 (1954): 8482c.

34. Neu, R. "A New Color Method for Determining Alkaloids and Organic Bases with Sodium Tetraphenylborate." *Journal of Chromatography* 11 (1963): 364; *Chemical Abstracts* 59 (1963): 12181d.

35. Marini-Bettolo, B. G., and E. Caggiano. "Paper Chromatography and Electrophoresis of Tertiary Bases." Liblice, Czech. 91 (1961); *Chemical Abstracts* 60 (1964): 838d.

36. Knappe, E., and I. Rohdewald. "Impregnation of Chromatographic Thin Layers with Polyesters. III. Thin-Layer Chromatographic Identification of Acetoacetic Acid Amides." *Z. Anal. Chem.* 208 (1965): 195; *Chemical Abstracts* 62 (1965): 12424f.

37. Lane, E. S. "Thin-Layer Chromatography of Long-Chain Tertiary Amines and Related Compounds." *Journal of Chromatography* 18 (1965): 426; *Chemical Abstracts* 63 (1965): 7630f.

38. Ashworth, M. R. F., and G. Bohnstedt. "Reagent for the Detection and Determination of N-Active Hydrogen." *Talanta* 13 (1966): 1631; *Chemical Abstracts* 66 (1967): 25889x.

39. Heacock, R. A. "The Aminochromes." *Advances in Heterocyclic Chemistry.* Edited by A. R. Katritsky. London: Academic Press, 1965; *Chemical Abstracts* 65 (1966): 5432d.

40. Knappe, E., D. Peteri, and I. Rohdewald. "Impregnation of Chromatographic Thin Layers with Polyesters for the Separation and Identification of Substituted 2-Hydroxybenzophenones and Other Ultraviolet Absorbers." *Z. Anal. Chem.* 197 (1963): 364; *Chemical Abstracts* 60 (1964): 762g.

41. Hara, S., and M. Takeuchi. "Systematic Analysis of Bile Acids and Their Derivatives by Thin Layer Chromatography." *Journal of Chromatography* 11 (1963): 565; *Chemical Abstracts* 60 (1964): 838f.

42. Hauck, A. "Detection of Caffeine by Paper Chromatography." *Deut. Z. Gerichtl. Med.* 54 (1963): 98; *Chemical Abstracts* 60 (1964): 838b.

43. Knappe, E., and I. Rohdewald. "Thin-Layer Chromatography of Dicarboxylic Acids. IV. Combination of Thin-Layer Chromatographic Systems for the Identification of Individual Components in Dicarboxylic Acid Mixtures." *Z. Anal. Chem.* 210 (1965): 183; *Chemical Abstracts* 63 (1965): 3600f.

44. Passera, C., A. Pedrotti, and G. Ferrari. "Thin-Layer Chromatography of Carboxylic Acids and Ketoacids of Biological Interest." *Journal of Chromatography* 14 (1964): 289; *Chemical Abstracts* 60 (1964): 16191f.

45. Knappe, E., and D. Peteri. "Thin-Layer Chromatography of Dicarboxylic Acids. I. Separations in the Homologous Series Oxalic to Sebacic Acids." *Z. Anal. Chem.* 188 (1962): 184; *Chemical Abstracts* 57 (1962): 11836a.

46. Peteri, D. "Thin-Layer Chromatography of Dicarboxylic Acids. II. Separation of Carbocyclic Dicarboxylic Acids." *Z. Anal. Chem.* 158 (1962): 352; *Chemical Abstracts* 57 (1962): 11836b.

47. Dutta, S. P., and A. K. Baruta. "Separation of Cis- and Trans-Isomers of α,β-Unsaturated Acids by Thin-Layer Chromatography." *Journal of Chromatography* 29 (1967): 263; *Chemical Abstracts* 67 (1967): 96616n.

48. Dalmaz, Y., and L. Peyrin. "Rapid Procedure for Chromatographic Isolation of DOPA, DOPAC, Epinephrine, Norepinephrine and Dopamine from a Single Urinary Sample at Endogenous Levels." *Journal of Chromatography* 145 (1978): 11; *Chemical Abstracts* 88 (1978): 59809c.

49. Baumgartner, H., W. Ridl, G. Klein, and S. Preindl. "Improved Radioenzymic Assay for the Determination of Catecholamines in Plasma." *Clinica Chimica Acta* 132 (1983): 111; *Chemical Abstracts* 99 (1983): 99459k.

50. Hansson, C., G. Agrup, H. Rorsman, A. M. Rosengren, and E. Rosengren. "Chromatographic Separation of Catecholic Amino Acids and Catecholamines on Immobilized Phenylboronic Acid." *Journal of Chromatography* 161 (1978): 352; *Chemical Abstracts* 90 (1979): 50771d.

51. Endo, Y., and Y. Ogura. "Separation of Catecholamines on the Phosphocellulose Column." *Japanese Journal of Pharmacology* 23 (1973): 491; *Chemical Abstracts* 80 (1974): 12002s.

52. Wada, H., A. Yamatodani, and T. Seki. "Systematic Determination of Amino Acids, Amines and Some Nucleotides Using Dansylchloride." *Kagaku No Ryoiki, Zokan* 114 (1976): 1; *Chemical Abstracts* 87 (1977): 1904f.

53. Head, R. J., R. J. Irvine, and J. A. Kennedy. "The Use of Sodium Borate Impregnated Silica Gel Plates for the Separation of 3-0-Methyl Catecholamines from Their Corresponding Catecholamines." *Journal of Chromatography Science* 14 (1976): 578; *Chemical Abstracts* 86 (1977): 39601x.

54. Adamec, O., J. Matis, and M. Galvanek. "Fractionation and Quantitative Determination of Urinary 17-Hydroxycorticosteroids by Thin Layer Chromatography on Silica Gel." *Steroids* 1 (1963): 495.

55. Adamec, O., J. Matis, and M. Galvanek. "Chromatographic Separation of Corticoids on a Thin-Layer of Silica Gel." *Lancet* 279, no. 7220 (1962): 81–82; *Chemical Abstracts* 56 (1962): 9034d.

56. Knappe, E., and I. Rohdewald. "Thin-Layer Chromatography of Dicarboxylic Acids. V. Separation and Identification of Hydroxy Dicarboxylic Acids, of Di- and Tricarboxylic Acids of the Citrate Cycle, and Some Other Dicarboxylic Acids of Plant Origin." *Z. Anal. Chem.* 211 (1965): 49; *Chemical Abstracts* 63 (1965): 7333c.

57. Snegotskii, V. I., and V. A. Snegotskaya. "Thin-Layer Chromatography of Sulfur Compounds." *Zavodskaya Laboratoriya* 35 (1969): 429; *Chemical Abstracts* 71 (1969): 23436b.

58. Borecky, J., J. Gasparic, and M. Vecera. "Identification of Organic Compounds. XXV. Identification and Separation of Aliphatic C_1–C_{18} Alcohols by Paper Chromatography." *Chem. Listy* 52 (1958): 1283; *Chemical Abstracts* 53 (1958): 8039h.

59. Hörhammer, L., H. Wagner, and H. Hein. "Thin-Layer Chromatography of Flavonoids on Silica Gel." *Journal of Chromatography* 13 (1964): 235; *Chemical Abstracts* 60 (1964): 13856c.

60. Mikes, O., ed. *Laboratory Handbook of Chromatographic Methods.* London: D. Van Nostrand Co., Ltd., 1966.

61. Wright, J. Detection of Humectants in Tobacco by Thin Layer Chromatography. London: Chem. & Ind., 1963.

62. Korte, F., and J. Vogel. "Thin-Layer Chromatography of Lactones, Lactams and Thiolactones." *Journal of Chromatography* 9 (1962): 381; *Chemical Abstracts* 58 (1963): 9609c.

63. Heacock, R. A., and M. E. Mahon. "Paper Chromatography of Some Indole: Derivatives on Acetylated Paper." *Journal of Chromatography* 6 (1961): 91.

64. Hackman, R. H., and M. Goldberg. "Microchemical Detection of Melanins." *Analytical Biochemistry* 41 (1971): 279; *Chemical Abstracts* 74 (1971): 136114a.

65. Preussmann, R., G. Neurath, G. Wulf-Lorentzen, D. Daiber, and H. Hengy. "Color Formation and Thin-Layer Chromatography for N-Nitrosocompounds." *Z. Anal. Chem.* 202 (1964): 187.

66. Preussmann, R., D. Daiber, and H. Hengy. "Sensitive Color Reaction for Nitrosamines on Thin-Layer Chromatography." *Nature* 201 (1964): 502; *Chemical Abstracts* 60 (1964): 12663e.

67. Hranisavljevic-Jakovljevic, M., I. Pejkovic-Tadic, and A. Stojiljkovic. "Thin-Layer Chromatography of Isomeric Oximes." *Journal of Chromatography* 12 (1963): 70; *Chemical Abstracts* 60 (1964): 7d.

68. Abraham, M. H., A. G. Davies, D. R. Llewellyn, and E. M. Thain. "The Chromatographic Analysis of Organic Peroxides." *Analytica Chimica Acta* 17 (1957): 499; *Chemical Abstracts* 53 (1959): 120b.

69. Seeboth, H. "Thin-Layer Chromatography Analysis of Phenols." *Monatsber. Deut. Akad. Wiss. Berlin* 5 (1963): 693; *Chemical Abstracts* 61 (1964): 2489c.

70. Knappe, E., and I. Rohdewald. "Thin-Layer Chromatographic Identification of Simple Phenols Using the Coupling Products with Fast Red Salt AL." *Z. Anal. Chem.* 200 (1964): 9; *Chemical Abstracts* 60 (1964): 9913g.

71. Donner, R., and K. Lohs. "Cobalt Chloride in the Detection of Organic Phosphate Ester by Paper and Especially Thin-Layer Chromatography." *Journal of Chromatography* 17 (1965): 349; *Chemical Abstracts* 62 (1965): 13842d.

72. Engel, J. F., and J. E. Barney. "Chromatographic Separation of Hydrogenation Products of Dibenz[a,h] anthracene." *Journal of Chromatography* 29 (1967): 232; *Chemical Abstracts* 57 (1967): 96617p.

73. Kucharczyk, N., J. Fohl, and J. Vymetal. "Thin-Layer Chromatography of Aromatic Hydrocarbons and Some Heterocyclic Compounds." *Journal of Chromatography* 11 (1963): 55; *Chemical Abstracts* 59 (1963): 9295g.

74. Perifoy, P. V., S. C. Slaymaker, and M. Nager. "Tetracyanoethylene as a Color-Developing Reagent for Aromatic Hydrocarbons." *Analytical Chemistry* 31 (1959): 1740; *Chemical Abstracts* 54 (1970): 5343e.

75. Kodicek, E., and K. K. Reddi. "Chromatography of Nicotinic Acid Derivatives." *Nature* 168 (1951): 475; *Chemical Abstracts* 46 (1952): 3601g.

76. Heacock, R. A., and M. E. Mahon. "The Color Reactions of the Hydroxyskatoles." *Journal of Chromatography* 17 (1965): 338; *Chemical Abstracts* 62 (1965): 13824g.

77. Martin, H. P. "Reversed Phase Paper Chromatography and Detection of Steroids of the Cholesterol Class." *Biochimica et Biophysica Acta* 25 (1957): 408.

78. Lisboa, B. P. "Application of Thin-Layer Chromatography to the Steroids of the Androstane Series." *Journal of Chromatography* 13 (1964): 391; *Chemical Abstracts* 60 (1964): 13890b.

79. Lisboa, B. P. "Separation and Characterization of Δ^5-3-Hydroxy-C_{19}-Steroids by Thin-Layer Chromatography." *Journal of Chromatography* 19 (1965): 333; *Chemical Abstracts* 63 (1965): 16403h.

80. Lisboa, B. P. "Thin-Layer Chromatography of Δ^4-3-Oxosteroids of the Androstane Series." *Journal of Chromatography* 19 (1965): 81; *Chemical Abstracts* 63 (1965): 13619e.

81. Lisboa, B. P. "Thin-Layer Chromatography of Steroids." *J. Pharm. Belg.* 20 (1965): 435; *Chemical Abstracts* 65 (1966): 570c.

82. Partridge, S. M. "Aniline Hydrogen Phthalate as a Spraying Reagent for Chromatography of Sugars." *Nature* 164 (1949): 443.

83. Grossert, J. S., and R. F. Langler. "A New Spray Reagent for Organosulfur Compounds." *Journal of Chromatography* 97 (1974): 83; *Chemical Abstracts* 82 (1976): 25473n.

84. Petranek, J., and M. Vecera. "Identification of Organic Compounds. XXIV. Separation and Identification of Sulfides by Paper Chromatography." *Chem. Listy* 52 (1958): 1279; *Chemical Abstracts* 53 (1958): 8039d.

85. Bican-Fister, T., and V. Kajganovic. "Quantitative Analysis of Sulfonamide Mixtures by Thin-Layer Chromatography." *Journal of Chromatography* 16 (1964): 503; *Chemical Abstracts* 62 (1965): 8943d.

86. Bican-Fister, T., and V. Kajganovic. "Separation and Identification of Sulfonamides by Thin-Layer Chromatography." *Journal of Chromatography* 11 (1963): 492; *Chemical Abstracts* 60 (1964): 372f.

87. Reisch, J., H. Bornfleth, and J. Rheinbay. "Thin-Layer Chromatography of Some Useful Sulfonamides." *Pharm. Ztg., Ver. Apotheker-Ztg.* 107 (1962): 920; *Chemical Abstracts* 60 (1964): 372e.

88. Prinzler, H. W., H. Tauchmann, and C. Tzcharnke. "Thin-Layer Chromatographic Separation of Organic Sulfoxides and Dinitrothioethers. Some Observations on Reproducibility and Structural Influence. II. Separation of Sulfoxide Mixtures by One and Two-Dimensional Thin Layer Chromatography." *Journal of Chromatography* 29 (1967): 151; *Chemical Abstracts* 67 (1967): 96615m.

89. Wolski, T. "Color Reactions for the Detection of Sulfoxides." *Chem. Anal.* (Warsaw) 14 (1969): 1319; *Chemical Abstracts* 72 (1970): 106867q.

90. Bergstrom, G., and C. Lagercrantz. "Diphenylpicrylhydrazyl as a Reagent for Terpenes and Other Substances in Thin-Layer Chromatography." *Acta Chemica Scandinavica* 18 (1964): 560; *Chemical Abstracts* 61 (1964): 2491h.

91. Dietz, W., and K. Soehring. "Identification of Thiobarbituric Acids in Urine by Paper Chromatography." *Archives of Pharmacology* 290 (1957): 80; *Chemical Abstracts* 52 (1958): 4736d.

92. Curtis, R. F., and C. T. Philips. "Thin-Layer Chromatography of Thiophene Derivatives." *Journal of Chromatography* 9 (1962): 366; *Chemical Abstracts* 58 (1963): 10705c.

93. Salame, M. "Detection and Separation of the Most Important Organophosphorus Pesticides by Thin-Layer Chromatography." *Journal of Chromatography* 16 (1964): 476; *Chemical Abstracts* 62 (1965): 11090b.

94. Knappe, E., and I. Rohdewald. "Thin-Layer Chromatography of Substituted Ureas and Simple Urethanes." *Z. Anal. Chem.* 217 (1966): 110; *Chemical Abstracts* 64 (1966): 16601g.

95. Fishbein, L., and J. Fawkes. "Detection and Thin-Layer Chromatography of Sulfur Compounds. I. Sulfoxides, Sulfones and Sulfides." *Journal of Chromatography* 22 (1966): 323; *Chemical Abstracts* 65 (1966): 6281e.

96. Prinzler, H. W., D. Pape, H. Tauchmann, M. Teppke, and C. Tzcharnke. "Thin-Layer Chromatography of Organic Sulfur Compounds." *Ropa Uhlie* 8 (1966): 13; *Chemical Abstracts* 65 (1966): 9710h.

97. Karaulova, E. N., T. S. Bobruiskaya, and G. D. Gal'pern. "Thin-Layer Chromatography of Sulfoxides." *Zh. Analit. Khim.* 21 (1966): 893; *Chemical Abstracts* 65 (1966): 16046f.

98. Knappe, E., and K. G. Yekundi. "Impregnation of Chromatographic Thin Layers with Polyesters. II. Separation and Identification of Lower and Middle Fatty Acids via the Hydroxamic Acid." *Z. Anal. Chem.* 203 (1964): 87; *Chemical Abstracts* 61 (1964): 5915e.

Stationary and Mobile Phases

Abbreviations/Solvent Table

Abbreviation	Solvent Name	Abbreviation	Solvent Name
Ac	acetone	Et_2O	diethylether
Ace	acetate	Foram	amylformate
AcOH	acetic acid	HCl	hydrochloric acid
n-AmOH	n-amyl alcohol	H_3BO_3	boric acid
t-AmOH	t-amyl alcohol	Hex	hexane
$AmSO_4$	ammonium sulfate	HForm	formic acid
i-BuAc	isobutylacetate	MeCl	methylene chloride
BuFor	n-butylformate	MeCN	acetonitrile
i-BuOH	isobutanol	MEK	methylethylketone
n-BuOH	n-butanol	MeOH	methanol
$i-Bu_2O$	diisobutylether	NaAc	sodium acetate
CCl_4	carbon tetrachloride	NH_3	ammonia, aqueous
C_2HCl_3	trichloroethene	Petet	petroleum ether
$CHCl_3$	chloroform	Ph	phosphate
$(CH_2)_6$	cyclohexane	PhOH	phenol
C_6H_6	benzene	PrAc	propylacetate
$n-C_6H_{14}$	n-hexane	PrFor	propylformate
$n-C_7H_{16}$	n-heptane	Progl	propylene glycol
$i-C_8H_{18}$	isooctane	i-PrOH	isopropanol
$(ClCH_2)_2$	dichloroethane	n-PrOH	n-propanol
DEAE	diethyl aminoethyl	$i-Pr_2NH$	diisopropylamine
Diox	dioxane	$i-Pr_2O$	diisopropylether
DMF	dimethylformamide	Py	pyridine
EtFor	ethylformate	THF	tetrahydrofuran
EtOAc	ethylacetate	Tol	toluene
EtOH	ethanol	w	water
Et_2NH	diethylamine	m-X	m-xylene

Stationary and Mobile Phases

Family	Stationary Phase	Mobile Phase	Ref.
Adrenaline and derivatives	alumina (two dimensional)	C_6H_6/EtOAc (60:40)	25
		$CHCl_3$/EtOH/Tol (90:6.5:3.1)	
Adrenochromes	cellulose	AcOH(2 %)/w	26,27
	Whatman #1 (descending)	AcOH(2 %)/w	26–29
Alcohols	silica gel (G-coated)	EtOAc/Hex	30
Alcohols, polyhydric	alumina or Kieselguhr (impregnated with polyamide) or silica gel	$CHCl_3$/Tol/HForm or n-BuOH/NH_3 or $CHCl_3$	31
Aldehydes	silica gel (G-coated)	EtOAc/Hex	30
Aldehydes, 2,4-dinitro-phenylhydrazones	alumina	C_6H_6 or $CHCl_3$ or Et_2O or C_6H_6/Hex	3
	alumina IB	MeCl or Tol/THF (4:1)	32
	silica gel	Hex/EtOAc (4:1 or 3:2)	3
	silica gel IB	MeCl or Tol/THF (4:1)	32
Alkaloids	alumina	i-BuOH/AcOH or i-BuOH/NH_3 or i-PrOH/AcOH	33
	alumina		3
	cellulose (impregnated with formamide)	$CHCl_3$ or EtOH or $(CH_2)_6$/$CHCl_3$ (3:7)	3
	paper (S&S #2043b)	C_6H_6/n-C_7H_{16}/$CHCl_3$/Et_2NH (6:5:1:0.02)	34
	paper electrophoresis	n-BuOH/HCl(25 %)/w (100:26:39)	35
	silica gel	i-BuOH/AcOH or i-BuOH/NH_3 or i-PrOH/HOAc	3
		C_6H_6/EtOH (9:1) or $CHCl_3$/Ac/Et_2NH (5:4:1)	
Amides	Kieselguhr (adipic acid impregnated)	i-Pr_2O/Petet/CCl_4/HForm/w	36
	silica gel	i-Pr_2O/Petet/CCl_4/HForm/w	36
Amines	alumina	Ac/n-C_7H_{16} (1:1)	3
	alumina G	i-BuAc or i-BuAc/AcOH	37
	Keiselguhr G	Ac/w (99:1)	3
	silica gel	EtOH (95 %)/NH_3 (25 %) (4:1)	3
	silica gel (aromatic only)		38
Amino acids	alumina	n-BuOH/AcOH/w (3:1:1) or Py/w	3
	cellulose	n-BuOH/AcOH/w (4:1:1)	3
	cellulose (two-dimensional)	n-BuOH/Ac/NH_3/w (10:10:5:2) followed by i-PrOH/HForm/w (20:1:5)	3
	silica gel		3
		n-BuOH/AcOH/w (3 or 4:1:1) or PhOH/w (3:1) or n-PrOH/NH_3 (34 %) (2:1)	
Aminochromes	Whatman #1 (acid washed)	w or AcOH/w or MeOH/w or EtOH/w or n-BuOH/AcOH/w or i-PrOH/w	26,39
Barbiturates	silica gel	$CHCl_3$/n-BuOH/NH_3 (25 %) (14:8:1)	3
Benzophenones, hydroxy	alumina or cellulose or Kieselguhr (impregnated with adipic acid triethylene glycol polyester) or silica gel	HForm/m-X	40

(Continued)

Stationary and Mobile Phases (Continued)

Family	Stationary Phase	Mobile Phase	Ref.
Bile acids	silica gel	C_6H_6/Et_2O (4:1) or Et_2O/AcOH (99.6:0.4) or $CHCl_3$/MeOH (9:1)	41
Caffeine	chromatography paper	n-BuOH/NH_3 or n-BuOH/HForm	42
Carboxylic acids	Keiselguhr/polyethylene glycol	i-Bu_2O/HForm/w (90:7:3)	43
	Polyamide powder	i-Pr_2O/Petet/CCl_4/HForm/w (50:20:20:8:1) or MeCN/EtOAc/ HForm or BuForm/EtOAc/HForm	43
	silica gel (G-coated)	EtOH/NH_3/THF	30
	silica gel ($CaSO_4$ impregnated)	n-PrOH/NH_3 or EtOH/$CHCl_3$/NH_3	44
	silica gel/polyethylene glycol M-1000	i-Pr_2O/HForm/w (90:7:3)	45,46 43
Carboxylic acids, unsaturated	silica gel	$CHCl_3$/MeOH	47
Catecholamines	alumina		48
	boric acid gel (neutral pH)	HCl (0.025N)	48
	Kieselguhr		
	phenylboronate		49
	phosphocellulose	Dilute acids	50
		Ph buffer (pH = 6.2)/EDTA	51
Catecholamines, dansyl derivatives	alumina (two dimensional)	C_6H_6/EtOAc (60:40) or $CHCl_3$/EtOH/ Tol (90:6.5:3.5)	25
	Amberlite IRC50		52
Catecholamines, o-methyl derivatives	silica gel (sodium borate impregnated)		53
Corticosteroids	silica gel	EtOH(5 %)/MeCl or EtOH/$CHCl_3$	54,55
Coumarins	polyamide	MeOH/w (4:1 or 3:2)	3
	silica gel G	Petet/EtOAc (2:1)	3
	silica gel G (impregnated with NaAc)	Tol/EtFor/HForm (5:4:1)	3
		EtOAc/Skellysolve B	3
	silicic acid (starch bound)		
Dicarboxylic acids	Kieselguhr/polyethylene glycol	i-Pr_2O/HForm/w (90:7:3)	43
	polyamide powder	i-Pr_2O/Petet/CCl_4/HForm/w (50:20:20:8:1) or MeCN/EtOAc/ HForm (9:1:1) or BuFor/EtOAc/ HForm (9:1:1)	43
	polyamide	MeCN/PrFor/PrAc/HForm (45:45:10:10) or i-Pr_2O/Petet/ CCl_4/ HForm/w (50:20:20:8:1) or n-AmOH/CCl_4/HFor (3:2:1)	56
	Woelm DC powder		
	silica gel	i-Pr_2O/HForm/w (90:7:3)	43
	silica gel (G-coated)	EtOH/NH_3/THF	30
Diols (see Alcohols, polyhydric)			
Disulfides	alumina	Hex	57
Disulfides, 3,5-dinitro-benzoates	Whatman #3 (impregnated with 10 % paraffin oil in cyclohexane), $(CH_2)_6$	DMF/MeOH/w or Foram/MeOH/w	58

Stationary and Mobile Phases (Continued)

Family	Stationary Phase	Mobile Phase	Ref.
Flavinoids	paper	n-BuOH/AcOH/w or EtOAc/w or AcOH/w or C_6H_6/AcOH/w	60
	polyamide		3
	silica gel	MeOH/H_2O	59
	silica gel (impregnated with NaAc)	C_6H_6/Py/AcOH (36:9:5) Petet/EtOAc (2:1)	3
	silicic acid (starch bound)	Tol/EtForm/HFor (5:4:1)	3
		EtOAc/Skellysolve B	
Glycerides	silica gel G	$CHCl_3$/C_6H_6 (7:3)	3
	silica gel G (impregnated with silver nitrate)	$CHCl_3$/AcOH (99.5:0.5)	3
Glycolipids	silica gel G	n-PrOH/NH_3 (12 %) (4:1)	3
Glycols, polyethylene	paper	n-PrOH/EtOAc/w (7:1:2) or n-BuOH/ AcOH/w (4:1:5) or t-AmOH/n-PrOH/w (8:2:3) or EtOAc/AcOH/w (9:2:2)	60
	silica gel	Ac or n-BuOH/AcOH/w	61
Hydroxamates	silica gel	i-Pr_2O or i-Pr_2O/EtOAc (1:4) or i-Pr_2O/i-C_8H_{18}	62
Hydroxamic acids	Kieselguhr G (impregnated with diethy-lene glycol or triethylene glycol adipate polyesters)	i-Pr_2O/Petet/CCl_4/HForm/w (50:20:20:8:1)	98
Indoles	acetylated (ascending)	$CHCl_3$/MeOH/w (10:10:6)	63
	cellulose (thin-layer)	w or HCl (0.005N) or n-BuOH/ AcOH/w (12:3:5) or C_6H_6/AcOH/w (125:72:3)	64
α-Ketoacids	silica gel ($CaSO_4$ impregnated)	EtOH/$CHCl_3$/NH_3	44
Ketones, 2,4-dinitro-phenyl hydrazones	alumina IB	MeCl or Tol/THF (4:1)	32
	silica Gel IB	MeCl or Tol/THF (4:1)	32
Lactams	silica gel	i-Pr_2O or i-Pr_2O/EtOAc (1:4) or i-Bu_2O/i-C_8H_{18}	62
Lactones	silica gel	i-Pr_2O or i-Pr_2O/EtOAc (1:4) or i-Bu_2O/i-C_8C_{18}	62
Lipids	alumina	Petet/Et_2O (95:5)	3
	silica gel G	Petet/Et_2O/AcOH (90:10:1)	3
	silicic acid	$CHCl_3$/MeOH/w (80:25:3)	3
Mercaptans (see Thiols)			
Nitrosamines	silica gel	Hex/Et_2O/MeCl	65
	Kieselgel	MeCl/Hex/Et_2O (2:3:4) (aliphatic, aromatic); MeCl/Hex/Et_2O (5:7:10) (cyclic)	66
Nucleotides	cellulose	$AmSO_4$ (sat'd)/NaAc(1 M)/ i-PrOH (80:18:2)	3
	cellulose (on DEAE)	HCl (aq)	3
Oximes	silica gel G	C_6H_6/EtOAc or C_6H_6/MeOH (abs)	67
Peroxides	silicone filter paper	w/EtOH/$CHCl_3$	68
Phenols	alumina	Et_2O	3
	alumina/AcOH	C_6H_6	3

(Continued)

Stationary and Mobile Phases (Continued)

Family	Stationary Phase	Mobile Phase	Ref.
	silica gel A	$CHCl_3$/AcOH (5:1) or $CHCl_3$/Ac/AcOH (10:2:1) or C_6H_6/AcOH (5:1) or Petet (80 °)/CCl_4/AcOH (4:6:1) or $CHCl_3$/Ac/Et_2NH (4:2:0.2)	69
	silica gel G	C_6H_6/Diox/AcOH (90:25:4)	3
	silica gel/oxalic acid	C_6H_6	70
	silica gel/potassium carbonate	MeCl/EtOAc/Et_2NH (92:5:3 or 93:5:2)	70
Phosphates, esters	alumina	Hex/C_6H_6/MeOH (2:1:1) or Hex/MeOH/Et_2O	71
	Kieselgel	Hex/C_6H_6/MeOH (2:1:1) or Hex/MeOH/Et_2O	71
Phospholipids	silica gel G	$CHCl_3$/MeOH/w	3
Polynuclear aromatics	alumina	CCl_4	3
	alumina	C_6H_6/$(CH_2)_6$ (15:85)	72
	silica gel	Hex or CH_3CHCl_2 or C_2HCl_3 or CCl_4	73,74
Polypeptides	Sephadex G-25	w or NH_3 (0.05 M)	3
	silica gel G	$CHCl_3$/MeOH (9:1) or $CHCl_3$/Ac (9:1)	3
Pyridines	Whatman #1 (descending)	n-BuOH/w or n-BuOH/w/NH_3 or Ac or i-BuOH/w or MEK/AcOH/w	75
Pyridines, quaternary salts (descending)	Whatman #1	Ac/w or $AmSO_4$/Ph buffer (pH = 6.8)/n-PrOH(2 %) or n-PrOH	75
Purines	silica gel	Ac/$CHCl_3$/n-BuOH/NH_3 (25 %) (3:3:4:1)	3
Pyrrole, tri-carboxylic acid	silica gel	n-BuOH/EtOH/NH_3/w (10:10:1:1)	64
Skatoles, hydroxy	silica gel G	i-Pr_2O or $(ClCH_2)_2$/i-Pr_2NH (6:1)	76
Steroids	alumina	$CHCl_3$EtOH (96:4)	3
	paper	Petet/Tol/MeOH/w or Petet/C_6H_6/MeOH/w	60
	paper (impregnated with kerosene)	n-PrOH/w	77
			78,79
	silica gel G	EtOAc/$(CH_2)_6$/EtOH(abs) or EtOAc/$(CH_2)_6$ or $CHCl_3$/EtOH (abs) or C_6H_6/EtOH or n-C_6H_{14}/EtOAc or EtOAc/n-C_6H_{14}/EtOH(abs)/AcOH or EtOAc/n-C_6H_{14}/AcOH	80,81
Sugars	cellulose	n-BuOH/Py/w (6:4:3) or EtOAc/Py/w (2:1:2)	3
	Kieselguhr G (buffered with 0.02N NaAc)	EtOAc/i-PrOH/w	3
	silica gel (buffered with H_3BO_3)	C_6H_6/AcOH/MeOH (1:1:3)	3
	silica gel (impregnated with sodium bisulfite)	EtOAc/AcOH/MeOH/w (6:1:5:1) or n-PrOH/w (85:15) or i-PrOH/EtOAc/w (7:1:2) or MEK/AcOH/w (6:1:3)	3
	silica gel G	n-PrOH/conc NH_3/w (6:2:1)	3
	Whatman #1 (descending-two dimensional)	PhOH or n-BuOH/AcOH	82

Stationary and Mobile Phases (Continued)

Family	Stationary Phase	Mobile Phase	Ref.
Sugars, aldoses	paper	EtOAc/Py/w (2:1:2) or n-BuOH/ AcOH/w (4:1:5) or n-BuOH/EtOH/ H_2O (5:1:4) or EtOAc/AcOH/w (9:2:2) or EtOAc/AcOH/HForm/w or EtOAc/Py/NaAc (sat'd)	60
	Whatman #1	PhOH or n-BuOH/AcOH	82
Sugars, carbamates	silica gel	n-BuOH/H_3BO_3 (0.03 M) (9:1)	
Sugars, deoxy	Whatman #1	PhOH or n-BuOH/AcOH	82
Sugars, ketoses	paper	EtOAc/Py/w (2:1:2) or n-BuOH/ AcOH/w (4:1:5) or n-BuOH/EtOH/ H_2O (5:1:4) or EtOAc/AcOH/w (9:2:2) or w/PhOH (pH = 5.5)	60
	Whatman #1	PhOH or n-BuOH/AcOH	82
Sulfides	alumina	Hex	75
	alumina	$CHCl_3$/MeOH	96
	silica gel	CCl_4 or C_6H_6	83
	silica gel DF-5	Ac/C_6H_6 or Tol/EtOAc	95
Sulfilimines, p-nitrosobenzene sulfonyl	Whatman #4 (impregnated with formamide)	C_6H_6 or C_6H_6/$(CH_2)_6$	84
Sulfonamides	Kieselguhr	$CHCl_3$/MeOH (9:1) or $CHCl_3$/MeOH/ NH_3	85
	silica gel	Et_2O or $CHCl_3$/MeOH (10:1)	86
	silica gel (neutral)	n-BuOH/MeOH/Ac/Et_2NH (9:1:1:1)	87
	silica gel (G)	$CHCl_3$/EtOH/n-C_7H_{16}	3
Sulfones	alumina	Et_2O or Hex/Ac (1:1)	57
	silica gel DF-5	Ac/C_6H_6 or Tol/EtOAc	95
Sulfones, esters	alumina	Et_2O or Hex/Ac (1:1)	57
Sulfones, hydroxy-ethyl	alumina	Hex/w (1:3)	57
Sulfoxides	alumina	C_6H_6/Py (20:1) and Diox	88
	alumina	Ac/CCl_4 (1:4)	97
	silica gel	Ac or EtOAc or $CHCl_3$/Et_2O	83
	silica gel DF-5	Ac/C_6H_6 or Tol/EtOAc	95
	Whatman #1	PhOH/w (8:3) or n-BuOH/AcOH/w (9:1:2.5)	89
Sulfoxides, hydroxy-ethyl	alumina	Et_2O or Hex/Ac (1:1) or Hex/Et_2O (1:3)	57
Terpenes	alumina	C_6H_6 or C_6H_6/Petet or C_6H_6/EtOH	3
	silica gel G	i-Pr_2O or i-Pr_2O/Ac	3
	silica gel/gypsum	$CHCl_3$/C_6H_6 (1:1)	90
	silicic acid (starch bond)	n-C_6H_{14}/EtOAc (85:15)	3
Thiobarbiturates	paper	n-AmOH/n-BuOH/25 % NH_3 (2:2:1)	91
Thiolactones	silica gel	i-Pr_2O or i-Pr_2O/EtOAc (1:4) or i-Bu_2O/i-C_8H_{18}	62

(Continued)

Stationary and Mobile Phases (Continued)

Family	Stationary Phase	Mobile Phase	Ref.
Thiols	alumina	Hex	57
	alumina (activated)	AcOH/MeCN (3:1)	96
	alumina (5 % cetane impregnated)	AcOH/MeCN (3:1)	96
	silica gel	EtOAc or $CHCl_3$	83
Thiophenes	alumina G	Petet (40–60 °C)	92
	silica gel	MeOH or C_6H_6/$CHCl_3$ (9:1)	92
Thiophosphate, esters		Petet or C_6H_6/$CHCl_3$ or Ac or EtOH or EtOAc or MeOH	93
Ureas	acetylated plates	CCl_4/EtOAc/EtOH (100:5:2)	94
	silica gel	CCl_4/MeCl/EtOAc/HOAc (70:50:15:10)	94
Urethanes (See ureas)			

TYPICAL STATIONARY AND MOBILE PHASE SYSTEMS USED IN THE SEPARATION OF VARIOUS INORGANIC IONS

The following table lists a series of stationary and mobile systems that are used in the separation of various inorganic ions [1–8]. The list is far from detailed and the reader is advised to consult the given references for details.

REFERENCES

1. Kirchner, J. G. *Thin Layer Chromatography.* 2nd ed. New York: Wiley-Interscience, 1978.
2. Bobbitt, J. M. *Thin Layer Chromatography.* New York: Reinhold, 1963.
3. Randerath, K. *Thin Layer Chromatography.* New York: Academic Press, 1963.
4. Randerath, K. *Thin Layer Chromatography.* 2nd ed. Weinheim: Verlag, Chemie, 1975.
5. Gagliardi, E., and B. Brodar. *Chromatographia* 2 (1969): 267.
6. Gagliardi, E., and B. Brodar. *Chromatographia* 3 (1970): 7.
7. Gagliardi, E., and B. Brodar. *Chromatographia* 3 (1970): 320.
8. MacDonald, J. C., ed. *Inorganic Chromatographic Analysis.* New York: John Wiley and Sons, 1985.

Typical Stationary and Mobile Phase Systems Used in the Separation of Various Inorganic Ions

Stationary Phase	Mobile Phase	Solvent Ratio	Separated Ions
Silica Gel G	butanol/1.5N HCl/2,5-hehanedione	100:20:0.5	hydrogen sulfide group
Silica gel G	acetone/conc. HCl/2,5-hexanedione	100:1:0.5	ammonium sulfide group
Silica gel G	water sat'd ethyl acetate/tributyl phosphate	100:4	U, Ga, Al
Silica gel G	ethanol/acetic acid	100:1	alkali metals
Silica gel G	acetone/1-butanol/conc. NH$_4$OH/ water	65:25:10:5	halogens
Silica gel G	methanol/conc. NH$_4$OH/10 % trichloroacetic acid/water	50:15:5:30	phosphates
Dowex 1-cellulose (1:1)	1 M aqueous sodium nitrate		halogens
Cellulose	HCl (or HBr)/alcohol mixtures	variable	Groups IA, IIA, IIIB, IVB, VB, VIB, transition metals
Cellulose	1-butanol/water/HCl	8:1:1	Fe, Al, Ga, Ti, In
Cellulose	acetic acid/pyridine/conc. HCl	80:6:20	ammonium sulfide group
DEAE cellulose	sodium azide/HCl	variable	Cd, Cu, Hg
Amberlite CG 400 and CG 120	HCl/HNO$_3$	variable	Pb, Bi, Sn, Sb, Cu, Cr, Hg

Sat'd = Saturated.
Conc. = Concentrated.

SPRAY REAGENTS IN THIN LAYER CHROMATOGRAPHY

The following table lists the most popular spray reagents needed to identify organic compounds on chromatographic plates. These reagents have been thoroughly covered in several books [1–3], and reviews [4–23]. Due to the aerosol nature of the spray and the chemical hazards associated with several of these chemicals, the use of a fume hood is highly recommended. The original references of the spray reagents are given in order to provide information about their results with individual compounds [24–138]. A list and description of some complicated protocols follows this section of the chapter.

Note: $1\gamma = 1\ \mu g/cm^2$ on a TLC plate.

REFERENCES

1. Krebs, K. G., D. Heusser, and H. Wimmer. "Spray Reagents." In *Thin Layer Chromatography, A Laboratory Handbook,* edited by E. Stahl. New York: Springer-Verlag, 1969.
2. Bobbitt, J. M. "Visualization." In *Thin Layer Chromatography.* New York: Reinhold, 1963.
3. Touchstone, J. C. "Visualization Procedures." In *Techniques and Application of Thin Layer Chromatography.* New York: John Wiley and Sons, 1985.
4. Pataki, G. "Paper, Thin-Layer, and Electrochromatography of Aminoacids in Biological Material." *Z. Klin. Chem.* 2 (1964): 129; *Chemical Abstracts* 64 (1966): 5425c.
5. Padley, F. B. Thin-Layer Chromatography of Lipids, Thin-Layer Chromatography, Proceedings Symposium, Rome 1963, 87 (Pub. 1964).
6. Honjo, M. "Thin-Layer Chromatography of Nucleic Acid Derivatives." *Kagaku No Ryoiki, Zokan* 64 (1964): 1.
7. Kazumo, T. "Thin-Layer Chromatography of Bile Acids." *Kagaku No Ryoiki, Zokan* 64 (1964): 19.
8. Nakazawa, Y. "Thin-Layer Chromatography of Compound Lipids." *Kagaku No Ryoiki, Zokan* 64 (1964): 31.
9. Nishikaze, O. "Separation and Quantitative Analysis of Adrenocortical Hormone and Its Metabolite (C_{21}) by Thin-Layer Chromatography." *Kagaku No Ryoiki, Zokan* 64 (1964): 37.
10. Shikita, M., H. Kazikazi, and B. Tamaoki. "Thin-Layer Chromatography of Radioactive Substances." *Kagaku No Ryoiki, Zokan* 64 (1964): 45.
11. Mo, I., and Y. Hashimoto. "Method of Thin-Layer Zone Electrophoresis." *Kagaku No Ryoiki, Zokan* 64 (1964): 61.
12. Kinoshita, S. "Thin-Layer Chromatography of Sugar Esters." *Kagaku No Ryoiki, Zokan* 64 (1964): 79.
13. Okada, M. "Thin-Layer Chromatography of Cardiotonic Glycosides." *Kagaku No Ryoiki, Zokan* 64 (1964): 103.
14. Omoto, T. "Thin-Layer Chromatography of Toad Toxin." *Kagaku No Ryoiki, Zokan* 64 (1964): 115.
15. Furnya, C., and H. Itokawa. "Thin-Layer Chromatography of Triterpenoids." *Kagaku No Ryoiki, Zokan* 64 (1964): 123.
16. Zenda, H. "Thin-Layer Chromatography of Aconitine-Type Alkaloids." *Kagaku No Ryoiki, Zokan* 64 (1964): 133.
17. Hara, S., and H. Tanaka. "Thin-Layer Chromatography of Mixed Pharmaceutical Preparations." *Kagaku No Ryoiki, Zokan* 64 (1964): 141.
18. Katsui, G. "Thin-Layer Chromatography of Vitamins." *Kagaku No Ryoiki, Zokan* 64 (1964): 157.
19. Fujii, S., and M. Kamikura. "Thin-Layer Chromatography of Pigments." *Kagaku No Ryoiki, Zokan* 64 (1964): 173.
20. Hosogai, Y. "Thin-Layer Chromatography of Organic Chlorine Compounds." *Kagaku No Ryoiki, Zokan* 64 (1964): 185.
21. Takeuchi, T. "Thin-Layer Chromatography of Metal Complex Salts." *Kagaku No Ryoiki, Zokan* 64 (1964): 197.

22. Yamakawa, H., and K. Tanigawa. "Thin-Layer Chromatography of Organic Metal Compounds." *Kagaku No Ryoiki, Zokan* 64 (1964): 209.

23. Ibayashi, H. "Thin-Layer Chromatography of Steroid Hormones and Its Clinical Application." *Kagaku No Ryoiki, Zokan* 64 (1964): 227.

24. Beckett, A. H., M. A. Beavan, and A. E. Robinson. "Paper Chromatography: Multiple Spot Formation by Sympathomimetic Amines in the Presence of Acids." *Journal of Pharmacy and Pharmacology* 12 (1960): 203T; *Chemical Abstracts* 55 (1961): 9785c.

25. Heacock, R. A., and B. D. Scott. "The Chemistry of the 'Aminochromes': Part IV. Some New Aminochromes and Their Derivatives." *Canadian Journal of Chemistry* 38 (1960): 516.

26. Matthews, J. S. "Steroids (CCXXIII) Color Reagent for Steroids in Thin-Layer Chromatography." *Biochimica et Biophysica Acta* 69 (1963): 163; *Chemical Abstracts* 58 (1963): 14043d.

27. Wasicky, R., and O. Frehden. "Spot-Plate Tests in the Examination of Drugs (I) Aldehyde and Amine Tests for the Recognition of Ethereal Oils." *Mikrochimica Acta* 1 (1937): 55; *Chemical Abstracts* 31 (1937): 5944.

28. Lane, E. S. "Thin-Layer Chromatography of Long-Chain Tertiary Amines and Related Compounds." *Journal of Chromatography* 18 (1965): 426; *Chemical Abstracts* 63 (1965): 7630f.

29. Neu, R. "A New Color Method for Determining Alkaloids and Organic Bases with Sodium Tetraphenylborate." *Journal of Chromatography* 11 (1963): 364; *Chemical Abstracts* 59 (1963), 12181d.

30. Zinser, M., and C. Baumgartel. "Thin-Layer Chromatography of Ergot Alkaloids." *Arch. Pharm.* 297 (1964): 158; *Chemical Abstracts* 60 (1964): 13095f.

31. Ashworth, M. R. F., and G. Bohnstedt. "Reagent for the Detection and Determination of N-Active Hydrogen." *Talanta* 13 (1966): 1631.

32. Whittaker, V. P., and S. Wijesundera. "Separation of Esters of Choline." *Biochemistry Journal* 51 (1952): 348; *Chemical Abstracts* 46 (1952): 7940g.

33. Heacock, R. A., and M. E. Mahon. "The Color Reactions of the Hydroxyskatoles." *Journal of Chromatography* 17 (1965): 338; *Chemical Abstracts* 62 (1965): 13824g.

34. Micheel, F., and H. Schweppe. "Paper chromatographic separation of hydrophobic compounds with acetylated cellulose paper." *Mikrochimica Acta* 53 (1954); *Chemical Abstracts* 48 (1954): 4354i.

35. Smyth, R. B., and G. G. Mckeown. "Analysis of Arylamines and Phenols in Oxidation-Type Hair Dyes by Paper Chromatography." *Journal of Chromatography* 16 (1964): 454; *Chemical Abstracts* 62 (1963): 8930e.

36. Kawerau, E., and T. Wieland. "Aminoacids Chromatograms." *Nature* 168 (1951): 77; *Chemical Abstracts* 46 (1952): 382h.

37. Sturm, A., and H. W. Scheja. "Separation of Phenolic Acids by High Voltage Electrophoresis." *Journal of Chromatography* 16 (1964): 194; *Chemical Abstracts* 62 (1965): 6788b.

38. Feigl, F. *Spot Tests in Organic Analysis.* 7th ed. Amsterdam: Elsevier Publishing Co., 1966.

39. Curzon, G., and J. Giltrow. "A Chromatographic Color Reagent for a Group of Aminoacids." *Nature* 172 (1953): 356.

40. Heacock, R. A., C. Nerenberg, and A. N. Payza. "The Chemistry of the 'aminochromes': Part I. The Preparation and Paper Chromatography of Pure Adrenochrome." *Canadian Journal of Chemistry* 36 (1958): 853.

41. Heacock, R. A. "The Aminochromes." In *Advances in Heterocyclic Chemistry,* edited by A. R. Katritsky. London: Academic Press, 1965; *Chemical Abstracts* 65 (1966): 5432d.

42. Wieland, T., and L. Bauer. "Separation of Purines and Aminoacids." *Angewandte Chemie* 63 (1951): 511; *Chemical Abstracts* 46 (1952): 1082h.

43. Hara, S., and M. Takeuchi. "Systematic Analysis of Bile Acids and Their Derivatives by Thin Layer Chromatography." *Journal of Chromatography* 11 (1963): 565; *Chemical Abstracts* 60 (1964): 838f.

44. Anthony, W. L., and W. T. Beher. "Color Detection of Bile Acids Using Thin Layer Chromatography." *Journal of Chromatography* 13 (1964): 570; *Chemical Abstracts* 60 (1964): 13546c.

45. Hauck, A. "Detection of Caffeine by Paper Chromatography." *Deut. Z. Gerichtl. Med.,* 54 (1963): 98; *Chemical Abstracts* 60 (1964): 838b.

46. Suryaraman, M. G., and W. T. Cave. "Detection of Some Aliphatic Saturated Long Chain Hydrocarbon Derivatives by Thin-Layer Chromatography." *Analytica Chimica Acta* 30 (1964): 96; *Chemical Abstracts* 60 (1964): 7463e.

47. Passera, C., A. Pedrotti, and G. Ferrari. "Thin-Layer Chromatography of Carboxylic Acids and Ketoacids of Biological Interest." *Journal of Chromatography* 14 (1964): 289; *Chemical Abstracts* 60 (1964): 16191f.

48. Grant, D. W. "Detection of Some Aromatic Acids." *Journal of Chromatography* 10 (1963): 511; *Chemical Abstracts* 59 (1963): 5772a.

49. Roux, D. G. "Some Recent Advances in the Identification of Leucoanthocyanins and the Chemistry of Condensed Tanins." *Nature* 180 (1957): 973; *Chemical Abstracts* 52 (1958): 5212f.

50. Abbott, D. C., H. Egan, and J. Thompson. "Thin-Layer Chromatography of Organochlorine Pesticides." *Journal of Chromatography* 16 (1964): 481; *Chemical Abstracts* 62 (1965): 11090c.

51. Adamec, O., J. Matis, and M. Galvanek. "Fractionation and Quantitative Determination of Urinary 17-Hydroxycorticosteroids by Thin Layer Chromatography on Silica Gel." *Steroids* 1 (1963): 495.

52. French, D., M. L. Levine, J. H. Pazur, and E. Norberg. "Studies on the Schardinger Dextrins. The Preparation and Solubility Characteristics of Alpha, Beta and Gamma Dextrins." *Journal of the American Chemical Society* 71 (1949): 353.

53. Knappe, E., and I. Rohdewald. "Thin-Layer Chromatography of Dicarboxylic Acids. V. Separation and Identification of Hydroxy Dicarboxylic Acids, of Di- and Tricarboxylic Acids of the Citrate Cycle, and Some Other Dicarboxylic Acids of Plant Origin." *Z. Anal. Chem.* 211 (1965): 49; *Chemical Abstracts* 63 (1965): 7333c.

54. Wright, J. "Detection of Humectants in Tobacco by Thin Layer Chromatography." London: Society of Chemical Industry, 1963.

55. Toennies, G., and J. J. Kolb. "Techniques and Reagents for Paper Chromatography." *Analytical Chemistry* 23 (1951): 823; *Chemical Abstracts* 45 (1951): 8392i.

56. Kaufmann, H. P., and A. K. Sen Gupta. "Terpenes as Constituents of the Unsaponifiables of Fats." *Chemische Berichte* 97 (1964): 2652; *Chemical Abstracts* 61 (1964): 14723b.

57. Gage, T. B., C. D. Douglass, and S. H. Wender. "Identification of Flavonoid Compounds by Filter Paper Chromatography." *Analytical Chemistry* 23 (1951): 1582; *Chemical Abstracts* 46 (1952): 2449c.

58. Hörhammer, L., H. Wagner, and K. Hein. "Thin Layer Chromatography of Flavonoids on Silica Gel." *Journal of Chromatography* 13 (1964): 235; *Chemical Abstracts* 60 (1964): 13856c.

59. Nakamura, H., and J. J. Pisano. "Specific Detection of Primary Catecholamines and Their 3-0-Methyl Derivatives on Thin-Layer Plates Using a Fluorigenic Reaction with Fluorescamine." *Journal of Chromatography* 154 (1978): 51; *Chemical Abstracts* 89 (1978): 117958x.

60. Neu, R. "Analyses of Washing and Cleaning Agents. XVIII. A New Test for Polyethylene Glycols and Their Esters." *Chemical Abstracts* 49 (1955): 16475c; *Ibid.* 54 (1960): 2665e.

61. Korte, F., and J. Vogel. "Thin-Layer Chromatography of Lactones, Lactams and Thiolactones." *Journal of Chromatography* 9 (1962): 381; *Chemical Abstracts* 58 (1963): 9609c.

62. Harley-Mason, J., and A. A. P. G. Archer. "p-Dimethylamino-Cinnamaldehyde as a Spray Reagent for Indole Derivatives on Paper Chromatograms." *Biochemistry Journal* 69 (1958): 60; *Chemical Abstracts* 52 (1958): 18600g.

63. Heacock, R. A., and M. E. Mahon. "Paper Chromatography of Some Indole Derivatives on Acetylated Paper." *Journal of Chromatography* 6 (1961): 91.

64. Adams, C. W. M. "A Perchloric Acid-Naphthoquinone Method for the Histochemical Localization of Cholesterol." *Nature* 192 (1961): 331.

65. Bennet-Clark, T. A., M. S. Tamblah, and N. P. Kefford. "Estimation of Plant Growth Substances by Partition Chromatography." *Nature* 169 (1951): 452; *Chemical Abstracts* 46 (1952): 6181c.

66. Gordon, S. A., and R. P. Weber. "Estimation of Indole Acetic Acid." *Plant Physiology* 26 (1951): 192; *Chemical Abstracts* 45 (1951): 4605c.

67. Dickmann, S. R., and A. L. Crockett. "Reactions of Xanthydrol: (IV) Determination of Tryptophan in Blood Plasma and Proteins." *Journal of Biological Chemistry* 220 (1956): 957; *Chemical Abstracts* 49 (1956): 7028h.

68. Mangold, H. K., B. G. Lamp, and H. Schlenk. "Indicators for the Paper Chromatography of Lipids." *Journal of the American Chemical Society* 77 (1953): 6070; *Chemical Abstracts* 50 (1956): 5074f.

69. Witter, R. F., G. V. Marinetti, A. Morrison, and L. Heicklin. "Paper Chromatography of Phospholipids with Solvent Mixtures of Ketones and Acetic Acid." *Archives of Biochemistry and Biophysics* 68 (1957): 15; *Chemical Abstracts* 51 (1957): 12200a.

70. Martin, H. P. "Reversed Phase Paper Chromatography and Detection of Steroids of the Cholesterol Class." *Biochimica et Biophysica Acta* 25 (1957): 408.

71. Preussmann, R., D. Daiber, and H. Hengy. "Sensitive Color Reaction for Nitrosamines on Thin-Layer Chromatography." *Nature* 201 (1964): 502; *Chemical Abstracts* 60 (1964): 12663e.

72. Preussmann, R., G. Neurath, G. Wulf-Lorentzen, D. Daiber, and H. Hengy. "Color Formation and Thin-Layer Chromatography of N-Nitrosocompounds." *Z. Anal. Chem.* 202 (1964): 187.

73. Hranisavljevic-Jakovljevic, M., I. Pejkovic-Tadic, and A. Stojiljkovic. "Thin-Layer Chromatography of Isomeric Oximes." *Journal of Chromatography* 12 (1963): 70; *Chemical Abstracts* 60 (1964): 7d.

74. Abraham, M. H., A. G. Davies, D. R. Llewellyn, and E. M. Thain. "Chromatographic Analysis of Organic Peroxides." *Analytica Chimica Acta* 17 (1957): 499; *Chemical Abstracts* 53 (1959): 120b.

75. Knappe, E., and D. Peteri. "Thin-Layer Chromatographic Identification of Organic Peroxides." *Z. Anal. Chem.*, 190 (1962): 386; *Chemical Abstracts* 58 (1963): 5021a.

76. Servigne, Y., and C. Duval. "Paper Chromatographic Separation of Mineral Anions Containing Sulfur." *Compt. Rend.* 245 (1957): 1803; *Chemical Abstracts* 52 (1958): 5207b.

77. Lisboa, B. P. "Characterization of Δ^4-3-Oxo-C_{21}-Steroids on Thin-Layer Chromatography." *Journal of Chromatography* 16 (1964): 136; *Chemical Abstracts* 62 (1965): 3409.

78. Sherma, J., and L. V. S. Hood. "Thin-Layer Solubilization Chromatography: (I) Phenols." *Journal of Chromatography* 17 (1965): 307; *Chemical Abstracts* 62 (1965): 13819b.

79. Gumprecht, D. L. "Paper Chromatography of Some Isomeric Monosubstituted Phenols." *Journal of Chromatography* 18 (1965): 336; *Chemical Abstracts* 63 (1965): 7630h.

80. Barton, G. M. "α,α-Dipyridyl as a Phenol-Detecting Reagent." *Journal of Chromatography* 20 (1965): 189; *Chemical Abstracts* 64 (1966): 2724a.

81. Sajid, H. "Separation of Chlorinated Cresols and Chlorinated Xylenols by Thin-Layer Chromatography." *Journal of Chromatography* 18 (1965): 419; *Chemical Abstracts* 63 (1965): 7630d.

82. Seeboth, H. "Thin-Layer Chromatography Analysis of Phenols." *Monatsber. Deut. Akad. Wiss. Berlin* 5 (1963): 693; *Chemical Abstracts* 61 (1964): 2489c.

83. Burke, W. J., A. D. Potter, and R. M. Parkhurst. "Neutral Silver Nitrate as a Reagent in the Chromatographic Characterization of Phenolic Compounds." *Analytical Chemistry* 32 (1960): 727; *Chemical Abstracts* 54 (1960): 13990d.

84. Perifoy, P. V., S. C. Slaymaker, and M. Nager. "Tetracyanoethylene as a Color-Developing Reagent for Aromatic Hydrocarbons." *Analytical Chemistry* 31 (1959): 1740; *Chemical Abstracts* 54 (1960): 5343e.

85. Bate-Smith, E. C., and R. G. Westall. "Chromatographic Behavior and Chemical Structure (I) Naturally Occurring Phenolic Substances." *Biochimica et Biophysica Acta* 4 (1950): 427; *Chemical Abstracts* 44 (1950): 5677a.

86. Noirfalise, A., and M. H. Grosjean. "Detection of Phenothiazine Derivatives by Thin-Layer Chromatography." *Journal of Chromatography* 16 (1964): 236; *Chemical Abstracts* 62 (1965): 10295f.

87. Schreiber, K., O. Aurich, and G. Osske. "Solanum Alkaloids (XVIII): Thin-Layer Chromatography of Solanum Steroid Alkaloids and Steroidal Sapogenins." *Journal of Chromatography* 12 (1963): 63; *Chemical Abstracts* 60 (1964): 4442h.

88. Clarke, E. G. C. "Identification of Solanine." *Nature* 181 (1958): 1152; *Chemical Abstracts* 53 (1959): 7298h.

89. Donner, R., and K. Lohs. "Cobalt Chloride in the Detection of Organic Phosphate Ester by Paper and Especially Thin-Layer Chromatography." *Journal of Chromatography* 17 (1965): 349; *Chemical Abstracts* 62 (1965): 13842d.

90. Kucharczyk, N., J. Fohl, and J. Vymetal. "Thin-Layer Chromatography of Aromatic Hydrocarbons and Some Heterocyclic Compounds." *Journal of Chromatography* 11 (1963): 55; *Chemical Abstracts* 59 (1963): 9295g.

91. Kodicek, E., and K. K. Reddi. "Chromatography of Nicotinic Acid Derivatives." *Nature* 168 (1951): 475; *Chemical Abstracts* 46 (1952): 3601g.

92. Hodgson, E., E. Smith, and F. E. Guthrie. "Two-Dimensional Thin-Layer Chromatography of Tobacco Alkaloids and Related Compounds." *Journal of Chromatography* 20 (1965): 176; *Chemical Abstracts* 64 (1966): 3960b.

93. Stevens, P. J. "Thin-Layer Chromatography of Steroids. Specificity of Two Location Reagents." *Journal of Chromatography* 14 (1964): 269; *Chemical Abstracts* 61 (1964): 2491b.

94. Lisboa, B. P. "Application of Thin-Layer Chromatography to the Steroids of the Androstane Series." *Journal of Chromatography* 13 (1964): 391; *Chemical Abstracts* 60 (1964): 13890b.

95. Lisboa, B. P. "Separation and Characterization of Δ^5-3-Hydroxy-C_{19}-Steroids by Thin-Layer Chromatography." *Journal of Chromatography* 19 (1965): 333; *Chemical Abstracts* 63 (1965): 16403h.

96. Lisboa, B. P. "Thin-Layer Chromatography of Δ^4-3-Oxosteroids of the Androstane Series." *Journal of Chromatography* 19 (1965): 81; *Chemical Abstracts* 63 (1965): 13619e.

97. Neher, R., and A. Wettstein. "Steroids (CVII) Color Reactions; Corticosteroids in the Paper Chromatogram." *Helvetica Chimica Acta* 34 (1951): 2278; *Chemical Abstracts* 46 (1952): 3110d.

98. Michalec, C. "Paper Chromatography of Cholesterol and Cholesterol Esters." *Naturwissenschaften* 42 (1955): 509; *Chemical Abstracts* 51 (1957): 5884a.

99. Scheidegger, J. J., and E. Cherbuliez. "Hederacoside A, A Heteroside Extracted from English Ivy." *Helvetica Chimica Acta* 38 (1955): 547; *Chemical Abstracts* 50 (1956): 1685g.

100. Richter, E. "Detection of Sterols with Naphthoquinone-Perchloric Acid on Silica Gel Layers." *Journal of Chromatography* 18 (1965): 164; *Chemical Abstracts* 63 (1965): 7653a.

101. Lisboa, B. P. "Thin-Layer Chromatography of Steroids." *J. Pharm. Belg.* 20 (1965): 435; *Chemical Abstracts* 65 (1966): 570c.

102. Adachi, S. "Thin-Layer Chromatography of Carbohydrates in the Presence of Bisulfite." *Journal of Chromatography* 17 (1965): 295; *Chemical Abstracts* 62 (1965): 13818g.

103. Bryson, J. L., and T. J. Mitchell. "Spraying Reagents for the Detection of Sugar." *Nature* 167 (1951): 864; *Chemical Abstracts* 45 (1951): 8408b.

104. Sattler, L., and F. W. Zerban. "Limitations of the Anthrone Test for Carbohydrates." *Journal of the American Chemical Society* 72 (1950): 3814; *Chemical Abstracts* 45 (1951): 1039b.

105. Bacon, J. S. D., and J. Edelmann. "Carbohydrates of the Jerusalem Artichoke and Other Compositae." *Biochemistry Journal* 48 (1951): 114; *Chemical Abstracts* 45 (1951): 5242b.

106. Timell, T. E., C. P. J. Glaudemans, and A. L. Currie. "Spectrophotometric Method for Determination of Sugars." *Analytical Chemistry* 28 (1956): 1916.

107. Hay, G. W., B. A. Lewis, and F. Smith. "Thin-Film Chromatography in the Study of Carbohydrates." *Journal of Chromatography* 11 (1963): 479; *Chemical Abstracts* 60 (1964): 839b.

108. Edward, J. T., and D. M. Waldron. "Detection of Deoxy Sugars, Glycols and Methyl Pentoses." *Journal of Chemical Society* (1952): 3631; *Chemical Abstracts* 47 (1953): 1009h.

109. Johanson, R. "New Specific Reagent for Keto-Sugars." *Nature* 172 (1953): 956.

110. Adachi, S. "Use of Dimedon for the Detection of Keto Sugars by Paper Chromatography." *Analytical Biochemistry* 9 (1964): 224; *Chemical Abstracts* 61 (1964): 13616g.

111. Sattler, L., and F. W. Zerban. "New Spray Reagents for Paper Chromatography of Reducing Sugars." *Analytical Chemistry* 24 (1952): 1862; *Chemical Abstracts* 47 (1953): 1543d.

112. Bailey, R. W., and E. J. Bourne. "Color Reactions Given by Sugars and Diphenylamine-Aniline Spray Reagents on Paper Chromatograms." *Journal of Chromatography* 4 (1960): 206; *Chemical Abstracts* 55 (1961): 4251c.

113. Buchan, J. L., and R. J. Savage. "Paper Chromatography of Starch-Conversion Products." *Analyst* 77 (1952): 401; *Chemical Abstracts* 48 (1954): 8568c.

114. Schwimmer, S., and A. Bevenue. "Reagent for Differentiation of 1,4- and 1,6-Linked Glucosaccharides." *Science* 123 (1956): 543; *Chemical Abstracts* 50 (1956): 8376a.

115. Partridge, S. M. "Aniline Hydrogen Phthalate as a Spraying Reagent for Chromatography of Sugars." *Nature* 164 (1949): 443.

116. Grossert, J. S., and R. F. Langler. "A New Spray Reagent for Organosulfur Compounds." *Journal of Chromatography* 97 (1974): 83; *Chemical Abstracts* 82 (1976): 25473n.

117. Snegotskii, V. I., and V. A. Snegotskaya. "Thin-Layer Chromatography of Sulfur Compounds." *Zavodskaya Laboratoriya* 35 (1969): 429; *Chemical Abstracts* 71 (1969): 23436b.

118. Fishbein, L., and J. Fawkes. "Detection and Thin-Layer Chromatography of Sulfur Compounds. I. Sulfoxides, Sulfones and Sulfides." *Journal of Chromatography* 22 (1966): 323; *Chemical Abstracts* 65 (1966): 6281e.

119. Svoronos, P. D. N. On the Synthesis and Characteristics of Sulfonyl Sulfilimines Derived from Aromatic Sulfides, Dissertation, Washington, DC: Georgetown University, 1980. (Available At University Microfilms, Order No. 8021272)

120. Petranek, J. and Vecera, M., "Identification of Organic Compounds. XXIV. Separation and Identification of Sulfides by Paper Chromatography." *Chem. Listy* 52 (1958): 1279; *Chemical Abstracts* 53 (1958): 8039d.

121. Bican-Fister, T., and V. Kajganovic. "Quantitative Analysis of Sulfonamide Mixtures by Thin-Layer Chromatography." *Journal of Chromatography* 16 (1964): 503; *Chemical Abstracts* 62 (1965): 8943d.

122. Bratton, A. C., and E. K. Marshall, Jr. "A New Coupling Component for Sulfanilamide Determination." *Journal of Biological Chemistry* 128 (1939): 537.

123. Borecky, J. "Pinakryptol Yellow, Reagent for the Identification of Arenesulfonic Acids." *Journal of Chromatography* 2 (1959): 612; *Chemical Abstracts* 54 (1960): 16255a.

124. Pollard, F. H., G. Nickless, and K. W. C. Burton. "A Spraying Reagent for Anions." *Journal of Chromatography* 8 (1962): 507; *Chemical Abstracts* 58 (1963): 3873b.

125. Coyne, C. M., and G. A. Maw. "Paper Chromatography for Aliphatic Sulfonates." *Journal of Chromatography* 14 (1964): 552; *Chemical Abstracts* 61 (1964): 7679d.

126. Wolski, T. "Color Reactions for the Detection of Sulfoxides." *Chemical Analysis* (Warsaw) 14 (1969): 1319; *Chemical Abstracts* 72 (1970): 106867q.

127. Suchomelova, L., V. Horak, and J. Zyka. "The Detection of Sulfoxides." *Microchemical Journal* 9 (1965): 196; *Chemical Abstracts* 63 (1965): 9062a.

128. Thompson, J. F., W. N. Arnold, and C. J. Morris. "A Sensitive Qualitative Test for Sulfoxides on Paper Chromatograms." *Nature* 197 (1963): 380; *Chemical Abstracts* 58 (1963): 7351d.

129. Karaulova, E. N., T. S. Bobruiskaya, and G. D. Gal'pern. "Thin-Layer Chromatography of Sulfoxides." *Zhurnal Analiticheskoi Khimii* 21 (1966): 893; *Chemical Abstracts* 65 (1966): 16046f.

130. Bergstrom, G., and C. Lagercrantz. "Diphenylpicrylhydrazyl as a Reagent for Terpenes and Other Substances in Thin-Layer Chromatography." *Acta Chemica Scandinavica* 18 (1964): 560; *Chemical Abstracts* 61 (1964): 2491h.

131. Urx, M., J. Vondrackova, L. Kovarik, O. Horsky, and M. Herold. "Paper Chromatography of Tetracyclines." *Journal of Chromatography* 11 (1963): 62; *Chemical Abstracts* 59 (1963): 9736g.

132. Dietz, W., and K. Soehring. "Identification of Thiobarbituric Acids in Urine by Paper Chromatography." *Arch. Pharm.* 290 (1957): 80; *Chemical Abstracts* 52 (1958): 4736d.

133. Prinzler, H. W., D. Pape, H. Tauchmann, M. Teppke, and C. Tzcharnke. "Thin-Layer Chromatography of Organic Sulfur Compound." *Ropa Uhlie* 8 (1966): 13; *Chemical Abstracts* 65 (1966): 9710h.

134. Curtis, R. F., and G. T. Philips. "Thin-Layer Chromatography of Thiophene Derivatives." *Journal of Chromatography* 9 (1962): 366; *Chemical Abstracts* 58 (1963): 10705c.

135. Salame, M. "Detection and Separation of the Most Important Organo-Phosphorus Pesticides by Thin-Layer Chromatography." *Journal of Chromatography* 16 (1964): 476; *Chemical Abstracts* 62 (1965): 11090b.

136. Siliprandi, D., and M. Siliprandi. "Separation and Determination of Phosphate Esters of Thiamine." *Biochimica et Biophysica Acta* 14 (1954): 52; *Chemical Abstracts* 49 (1955): 6036f.

137. Nuernberg, E. "Thin-Layer Chromatography of Vitamins." *Deut., Apotheker-Zfg.* 101 (1961): 268; *Chemical Abstracts* 60 (1964): 372.

138. Mariani, A., and C. Vicari. "Determination of Vitamin D in the Presence of Interfering Substances." *Chemical Abstracts* 60 (1964): 373a.

Spray Reagents in Thin Layer Chromatography

Family/Functional Group	Test	Result	Ref.
Adrenaline (and derivatives)	2,6-dichloroquinonechloroimide (0.5 % in absolute ethanol)	Variety of colors	1
	potassium ferricyanide (0.6 % in 0.5 % sodium hydroxide	Red spots	25
Adrenochromes	4-N,N-diemthylaminocinnamaldehyde	Blue-green to grey-green spots	26
	Ehrlich reagent	Blue-violet to red-violet spots	26
	zinc acetate (20 %)	Blue or yellow fluorescent spots	26
Alcohols	ceric ammonium sulfate (or nitrate)	Yellow/green spots on red background	1,3
	2,2-Diphenylpicrylhydrazyl (0.06 % in chloroform)	Yellow spots on purple background after heating (110 °C, 5 min)	3
	vanillin (1 % in conc. sulfuric acid)	Variety of spots after heating (120 °C); good only for higher alcohols	27
Aldehydes	o-dianisidine (saturated solution in acetic acid)	Variety of spots	28
	2,4-dinitrophenylhydrazine	Blue colors (saturated ketones); olive green colors (saturated aldehydes); slow developing colors (unsaturated carbonyl compounds)	1
	2,4-diphenylpicrylhydrazyl (0.06 % in chloroform)	Yellow spots on a purple background after heating (110 °C, 5 min)	3
	hydrazine sulfate (1 % in 1N hydrochloric acid)	Spots under UV (especially after heating)	3
Aldehydes, carotenoids	Tollens reagent	Dark spots	1
	Rhodamine (1–5 % in ethanol)	Variety of spots after treatment with strong alkali (sensitivity 0.03 µg)	1
Alkaloids	Bromcresol green (0.05 % in ethanol)	Green spots, especially after exposure to ammonia	3
	Chloramine-T (10 % aqueous)	Rose spots after exposure to hydrochloric acid and heat	4
	cobalt (II) thiocyanate	Blue spots on a light pink background	29
	p-N,N-dimethylaminobenzaldehyde (4 % in 1:3 hydrochloric acid/methanol)	Characteristic spots for individual alkaloids	4
	iodine/potassium iodide (in 2N acetic acid)	Variety of spots	3
	Kalignost test	Orange/red spots fluorescing under long-wave UV	30
	Sonnenschein test	Variety of spots	1
Alkaloids (ergot or fungal)	p-N,N-dimethylaminobenzaldehyde/ sulfuric acid	Blue spots	31
Amides	chlorine/pyrazolinone/cyanide	Red spots turning blue (detection limit 0.5 µg)	32
	hydroxylamine/ferric chloride	Variety of spots	33

Amines (all types unless specified)	alizarin (0.1 % in ethanol)	Violet spots on yellow background	3
	chlorine/pyrazolinon/cyanide	Red spots turning blue (aromatic only)	32
	cobalt (II) thiocyanate	Blue spots on white/pink background	29
	diazotization and α-naphthol coupling	Variety of spots (1 ° aromatic amines only)	1
	Ehrlich reagent	Yellow spots for aromatic amines	34
	Fast Blue B Salt	Variety of spots (only for amines that can couple)	1
	Glucose/phosphoric acid (4 %)	Variety of spots (aromatic amines only) especially after heating	35
Amines (all types unless specified; cont.)	malonic acid (0.2 %)/salicylaldehyde (0.1 %) (in ethanol)	Yellow spots after heating (120 °C, 15 min)	3,4
	1,2-naphthoquinone-4-sulfonic acid, sodium salt (0.5 % in 1N acetic acid)	Variety of colors after 30 min (aromatic amines only)	36
	ninhydrin	Red colors when exposed to ammonium hydroxide	37
	p-nitroaniline, diazotized	Variety of colored spots	38
	nitroprusside (2.5 %)/acetaldehyde (5 %)/sodium carbonate (1 %)	Variety of spots (2 ° aliphatic only)	39
	picric acid (3 % in ethanol)/sodium hydroxide (10 %) (5:1)	Orange spots	4
	potassium iodate (1 %)	Variety of spots for phenylethylamines (after heating)	3
	vanillin-potassium hydroxide	Variety of colors	40
Amino acids	dehydroascorbic acid (0.1 % in 95 % n-butanol)	Variety of colored spots	3
	2,4-dinitrofluorobenzene	Variety of spots	1
	Isatin-zinc acetate	Variety of colors	1
	Folin reagent	Variety of colors	1
	ninhydrin	Red colors when exposed to ammonium hydroxide	36
	vanillin/potassium hydroxide	Variety of colors	40
Amino alcohols	alizarin (0.1 % in ethanol)	Violet on yellow background	3
Aminochromes	p-N,N-dimethylaminocinnamaldehyde	Variety of colors	41
	Ehrlich reagent	Violet spots	26,41
	ferric chloride (3 %)	Gray-brown spots	41
	p-nitroaniline, diazotized	Red/brown spots	26,41
	sodium bisulfite, aqueous	Yellow fluorescence under UV	41,42
Aminosugars	ninhydrin	Red colors when exposed to ammonium hydroxide	37
Ammonium salts, quaternary	cobalt (II) thiocyanate	Variety of spots	29

(Continued)

Spray Reagents in Thin Layer Chromatography (Continued)

Family/Functional Group	Test	Result	Ref.
Anhydrides	hydroxylamine/ferric chloride	Variety of spots	33
Arginine	Sakaguchi reagent	Orange/red spots	1
Azulenes	EP reagent	Blue spots (room temperature) that fade to green/yellow shades and can be regenerated with steam	1
Barbiturates	cobalt (II) nitrate (2 %)/lithium hydroxide (0.5 %)	Variety of colors	1
	cupric sulfate/quinine/pyridine	Variety of colors (white, yellow, violet)	1
	s-diphenylcarbazone (0.1 % in ethanol)	Purple spots	3
	ferrocyanide/hydrogen peroxide	Yellow/red colors	1
	fluorescein (0.005 % in 0.5 M ammonia)	Variety of spots under long or short-wave UV	43
	mercurous nitrate (1 %)	Variety of spots	1
	Zwikker reagent	Variety of spots	1
Bile acids	anisaldehyde/sulfuric acid	Variety of spots	4
	antimony trichloride (in chloroform)	Variety of spots	44
	perchloric acid (60 %)	Fluorescent spots (long wave UV) after heating (150 °C, 10 min)	44
	sulfuric acid	Variety of spots	44,45
Bromides	fluorescein/hydrogen peroxide	Nonfluorescent spots	1
Caffeine	chloramine-T	Pink–red spots	1
	silver nitrate (2 % in 10 % sulfuric acid)	Carmine-red spots (limit 2γ)	46
Carboxylic acids	Bromcresol blue (0.5 % in 0.2 % citric acid)	Yellow spots on blue background	3
	Bromothymol blue (0.2 % in ethanol, pH = 7)	Yellow spots upon exposure to ammonia	47
	2,6-dichlorophenol/indophenol (0.1 % in ethanol)	Red spots on blue background after heating	48
	hydrogen peroxide (0.3 %)	Blue fluorescence under long-wave UV	49
Carboxylic acids, ammonium salts	Schweppe reagent	Dark brown spots	1
Catechins	p-toluenesulfonic acid (20 % in chloroform)	Fluorescent spots under long-wave UV	50
Catecholamines	ethylenediamine (50 %)	Spots under short/long wave UV after heating (50 °C, 20 min)	1

Compound class	Reagent	Observation	Ref.
Chlorides, alkyl	2,6-dichlorophenol indophenol (0.2 %)/silver nitrate (3 %) in ethanol	Variety of spots	1
	silver nitrate (0.5 % in ethanol)	Dark spots upon UV irradiation	51
	silver nitrate/formaldehyde	Dark grey spots	1
	silver nitrate/hydrogen peroxide	Dark spots	1
Chlorinated insecticides and pesticides	diphenylamine (0.5 %)/zinc chloride (0.5 %) in acetone	Variety of colors upon heating (200 °C)	1
	2-phenoxyethanol (5 %) in 0.05 % silver nitrate	Variety of spots	4
	silver nitrate/formaldehyde	Dark grey spots	1
	o-toluidine (0.5 %) in ethanol	Green spots under UV (sensitivity 0.5 µg)	4
Choline derivatives	dipicrylamine (0.2 % in 50 % aqueous acetone)	Red spots on yellow background	1
Corticosteroids	Blue Tetrazolium (0.05 %)/sodium hydroxide (2.5 M)	Violet spots (limit 1γ/cm²)	1,52
	2,3,5-Triphenyl-H-tetrazolium chloride (2 % in 0.5 NaOH)	Red spots after heating (100 °C, 5 min)	1
Coumarins	Benedict reagent	Fluorescent spots under long-wave UV	1
	potassium hydroxide (5 % in methanol)	Variety of spots under long-wave UV	1
Dextrins	iodine/potassium iodide	Blue–black spots (α–dextrins); brown–yellow spots (β- or γ-dextrins)	53
Dicarboxylic acids	bromocresol purple (0.04 % in basic 50 % ethanol, pH = 10)	Yellow spots on blue background	4,54
Diols (1,2-)	lead tetraacetate (1 % in benzene)	White spots after heating (110 °C, 5 min; limit 2 µg)	55
Disulfides	iodine (1.3 % in ethanol)/sodium azide (3.3 % in ethanol)	White spots on brown iodine background	3
	nitroprusside (sodium)	Red spots	56
Diterpenes	antimony (III) chloride/acetic acid	Reddish yellow to blue-violet	57
Esters	hydroxylamine/ferric chloride	Variety of spots	33
Flavonoids	aluminum chloride	Yellow fluorescence on long-wave UV	58
	antimony (III) chloride (10 % in chloroform)	Fluorescence on long-wave UV	59
	Benedict's reagent	Fluorescence on long-wave UV (only for o-dihydroxy compounds)	59
	lead acetate (basic, 25 %)	Fluorescent spots	4,50
	p-toluenesulfonic acid (20 % in chloroform)	Fluorescent spots under long-wave UV after heating (100 °C, 10 min)	
Fluorescamines	perchloric acid (70 %)	Blue fluorescent spots	60
Glycols, polyethylene	quercetin/sodium tetraphenylborate	Orange–red spots	61
Glycolipids	diphenylamine (5 % in ethanol) dissolved in 1:1 hydrochloric acid/acetic acid	Blue–grey spots	1

(Continued)

Spray Reagents in Thin Layer Chromatography (Continued)

Family/Functional Group	Test	Result	Ref.
Glycosides, triterpene	Liebermann–Burchard reagent	Fluorescence under long-wave UV	1
Hydroxamates	ferric chloride (10 % in acetic acid)	Brown spots	62
Hydroxamic acids	ferric chloride (1–5 % in 0.5N hydrochloric acid)	Red spots	1
Imidazoles	p-anisidine/amyl nitrite	Red/brown spots	3
Indoles	chlorine/pyrazolinone/cyanide	Red spots turning blue after a few minutes (limit 0.5 μg)	32
	cinnamaldehyde/hydrochloric acid	Red spots	1
	p-N,N-dimethylaminocinnamaldehyde	Variety of colored spots	63
	Ehrlich reagent	Purple for indoles; blue for hydroxyindoles	9,34,64
	ferric chloride (0.001 M) in 5 % perchloric acid	Red spots	3
	naphthoquinone/perchloric acid	Orange spots	65
	perchloric acid (5 %)/ferric chloride (0.001 M)	Variety of colored spots	66
	Prochazka reagent	Fluorescent (yellow/orange/green) spots under long wave UV	1
	Salkowski reagent	Variety of colored spots	67
	van Urk (or Stahl) reagent	Variety of colored spots	1
	xyanthydrol (0.1 % in acidified ethanol)	Variety of colored spots after heating (100 °C)	68
Iodides	Sonnenschein test	Variety of spots	1
α-Ketoacids	2,6-dichlorophenol/indophenol (0.1 % in ethanol)	Pink spots upon heating	4,48
	o-phenylenediamine (0.05 % in 10 % trichloroacetic acid or 0.2 % in 0.1N H$_2$SO$_4$/ethanol)	Green fluorescence under long wave UV after heating (100 °C, 2 min)	1
Ketones	o-dianisidine (saturated solution in acetic acid)	Characteristic spots	28
	2,4-dinitrophenylhydrazine	Yellow–red spots	3
Lactones	hydroxylamine/ferric chloride	Variety of colors	33
Lipids	α-cyclodextrin	Variety of spots (for straight chain lipids)	53
	2′,7′-dichlorofluorescein (0.2 %) in ethanol	Spots under long-wave UV	1,69
	fluorescein	Spots after treatment with steam	1
	Rhodamine 6G (1 % in acetone)	Spots under long-wave UV	70
	tungstophosphoric acid (20 % in ethanol)	Variety of colored spots after heating	71

	Reagent	Result	Ref.
Mercaptans (see Thiols)			
Nitrocompounds	p-N,N-dimethylaminobenzaldehyde/ stannous chloride/ hydrochloric acid	Yellow spots	3
Nitrosamines	diphenylamine/palladium chloride	Violet spots after exposure to short-wave UV (limit 0.5γ)	1,72
	sulfanilic acid (0.5 %)/α-naphthylamine (0.05 %) in 30 % acetic acid	Spraying is preceded by short-wave UV irradiation (3 min); aliphatic nitrosamines yield red/violet spots, while aromatic ones green/blue spots (limit 0.2–0.5γ)	1,72,73
Oximes	cupric chloride (0.5 %)	Immediate green spots (β-oximes); green-brown spots after 10 min (α-oximes)	74
Peroxides	ammonium thiocyanate (1.2 %)/ferrous sulfate (4 %)	Brown-red spots	74
	N,N-dimethyl-p-phenylene diammonium dichloride	Purple spots	76
	ferrous thiocyanate	Red-brown spots	1,75
	iodide (potassium)/starch	Blue spots	1
Persulfates	benzidine (0.05 % in 1N acetic acid)	Blue spots	77
Phenols	anisaldehyde/sulfuric acid	Variety of colors	1,78
	p-anisidine/ammonium vanadate	Variety of spots on pink background	3
	benzidine, diazotized	Variety of colors	79
	ceric ammonium nitrate (46 % in 2 M nitric acid)	Variety of spots	80
	α,α'-dipyridyl (0.5 %)/ferric chloride (0.5 %) in ethanol	Variety of spots	4,81
	emerson	Red-orange to pink spots	1
	fast Blue B salt	Variety of spots	1
	ferric chloride (1–5 % in 0.5N HCl)	Blue-greenish spots	1
	Folin–Denis reagent	Variety of spots	82
	Gibbs reagent	Variety of colors	1
	Millon reagent	Variety of colors after heating	1
	naphthoquinone/perchloric acid	Yellow spots (phenol, catechol); dark blue spots (resorcinol)	65
	p-nitroaniline, diazotized	Variety of colored spots	38
	p-nitrobenzenediazonium fluoroborate	Variety of spots	84
	silver nitrate (saturated in acetone)	Pink to deep green colors	84
	stannic chloride (5 %) in equal volumes of chloroform/ acetic acid	Variety of spots after heating (100 °C, 5 min)	1
	tetracyanoethylene (10 % in benzene)	Variety of colors	85
	Tollen's (or Zaffaroni) reagent	Dark spots	86
	vanillin (1 % in sulfuric acid)	Variety of colors after heating	27

(Continued)

Spray Reagents in Thin Layer Chromatography (Continued)

Family/Functional Group	Test	Result	Ref.
Phenols, chlorinated	Folin–Denis reagent	Variety of spots	82
Phenothiazines	ferric chloride (5 %)/perchloric acid (20 %)/nitric acid (50 %; 1:9:10)	Variety of colors	4,87
	formaldehyde (0.03 % in phosphoric acid)	Variety of spots	88,89
	palladium (II) chloride (0.5 % pH < 7)	Variety of spots	1
Phosphates, esters	cobalt (II) chloride (1 % in acetone or acetic acid)	Blue spots upon warming the plate at 40 °C	90
Polynuclear aromatics	formaldehyde (2 %) in conc. sulfuric acid	Variety of colors	91
	tetracyanoethylene (10 % in benzene)	Variety of colors	85
Purines	fluorescein (0.005 % in 0.5 M ammonia)	Variety of spots under long- or short-wave UV	43
Pyrazolones	ferric chloride (5 %)/acetic acid (2N; 1:11)	Variety of colors	4
Pyridines	König reagent	Variety of spots (for free α-position pyridines)	92,93
Pyridines, quaternary	König reagent	Blue-white fluorescence under UV	93
Pyrimidines	Fluorescein (0.005 % in 0.5 M ammonia)	Variety of spots under long- or short-wave UV	43
Pyrones (α- and γ-)	Neu reagent	Fluorescent spots under long-wave UV	1
Quinine derivatives	formic acid vapors	Fluorescent blue spots	3
Sapogenins	Komarowsky reagent	Yellow/pink spots	94
	paraformaldehyde (0.03 % in 85 % phosphoric acid)	Variety of spots	88
	zinc chloride (30 % in methanol)	Fluorescent spots after heating (105 °C, 1 h) in a moisture-free atmosphere	94
Steroids	anisaldehyde/sulfuric acid	Variety of colors	95,96,97
	antimony (III) chloride (in acetic acid)	Variety of colors	57,96
	Carr-Price reagent	Variety of colors	1
	chlorosulfonic acid/acetic acid	Fluorescence under long-wave UV	78,95
	Dragendorff reagent	Variety of spots	88,89
	formaldehyde (0.03 % in phosphoric acid)	Variety of spots	96
	Hanes and Isherwood reagent	Variety of spots (only for 3-hydroxy-Δ⁵-steroids)	1
	Liebermann—Burchard reagent	Fluorescence under long-wave UV	1,44
	perchloric acid (20 %)	Fluorescent spots (long-wave UV) after heating (150 °C, 10 min)	96
	phosphomolybdic acid	Blue color	95,96,98
	phosphoric acid (50 %)	Fluorescent spots after heating (120 °C) (limit 0.005γ)	99

Category	Reagent	Description	Ref
	phosphotungstic acid (10 % in ethanol)	Variety of spots	1, 100
	stannic chloride (5 %) in equal volumes of chloroform/acetic acid (1:1)	Variety of spots after heating (100 °C, 5 min)	1
	sulfuric acid	Variety of spots	50
	p-toluenesulfonic acid (20 % in chloroform)	Fluorescent spots under long-wave UV	96
	trichloroacetic acid (50 % aqueous)	Variety of colors	95,96
Sterols	Zimmerman reagent	Variety of colors	
	antimony (III) chloride (50 % in acetic acid)	Variety of spots	99
	bismuth (III) chloride	Fluorescence under long-wave UV	1
	chlorosulfonic acid/acetic acid	Fluorescence under long-wave UV	1
	Liebermann–Burchard reagent	Fluorescence under long-wave UV	1
	1,2-naphthoquinone-4-sulfonic acid/perchloric acid	Pink spots that change to blue upon prolonged heating (cholesterol limit 0.03γ)	65,101
	phosphoric acid (50 %)	Fluorescent spots after heating (120 °C, 15 min)	98,102
	phosphotungstic acid (10 % in ethanol)	Variety of spots	99
	stannic chloride (5 %) in equal volumes of chloroform/acetic acid	Variety of spots after heating (100 °C, 5 min)	1
	sulfuric acid	Variety of spots	102
Sugars	o-aminodiphenyl (0.3 %)/orthophosphoric acid (5 %)	Brown spots after heating	103
	aniline/phosphoric acid	Variety of colors	104
	anisaldehyde/sulfuric acid	Variety of colors	1,78
	Anthrone test	Yellow spots	105
	benzidine/trichloroacetic acid	Red-brown/dark spots	106
	carbazole/sulfuric acid	Violet spots on blue background	103
	Lewis-Smith reagent	Brown spots	107
	naphthoquinone/perchloric acid	Pink–brown spots (glucose, mannose, lactose, sucrose)	65
	naphthoresorcinol (0.2 % in ethanol)/phosphoric acid (10:1)	Variety of spots after heating (100 °C, 5–10 min)	1
	naphthoresorcinol (0.1 %)/sulfuric acid (10 %)	Variety of spots after heating (100 °C, 5–10 min)	1
	orcinol reagent	Variety of spots	1
	permanganate, potassium (0.5 % in 1N sodium hydroxide)	Variety of spots after heating (100 °C)	108
	phenol (3 %)/sulfuric acid (5 % in ethanol)	Brown spots after heating (100 °C, 10 min)	103

(Continued)

Spray Reagents in Thin Layer Chromatography (Continued)

Family/Functional Group	Test	Result	Ref.
	silver nitrate (0.2 % in methanol)/ammonia (saturated)/sodium methoxide (2 % in methanol)	Variety of spots after heating (110 °C, 10 min)	1
	silver nitrate/sodium hydroxide	Variety of spots	1
	sulfuric acid	Variety of spots	108
	thymol (0.5 %) in sulfuric acid (5 %)	Pink spots after heating (120 °C, 20 min)	103
Sugars, deoxy	metaperiodate/p-nitroaniline	Fluorescent (long-wave UV) yellow spots	109
Sugars, ketoses	Anthrone test	Bright purple (pentoses); orange-yellow (heptoses); blue fluorescence (aldoses)	110
	dimedone (0.3 %)/phosphoric acid (10 % in ethanol)	Dark-grey spots (white light); dark-pink fluorescing spots (UV) after heating (110 °C, 15 min)	1, 111
Sugars, reducing	4-aminohippuric acid	Fluorescence under long-wave UV	112
	aniline/diphenylamine/phosphoric acid	Variety of colors	113,114,115
	aniline hydrogen phthalate	Variety of colors (limit 1 µg)	116
	p-anisidine phthalate	Variety of colors	1
	3,5-dinitrosalicylic acid (0.5 % in 4 % sodium hydroxide)	Brown spots (sensitivity 1 µg)	3
Sulfides	ceric ammonium nitrate (in 2 M HNO_3)	Colorless spots (limit < 100 µg/spot)	117
	chloranil (1 %) in benzene	Yellow–brown spots	119
	2,3-dichloro-5,6-dicyano-1,4-benzoquinone (2 %) in benzene	Purple–blue spots changing to orange upon ammonia exposure	119
	Gibbs reagent	Yellow–brown spots changing to blue-orange upon exposure to ammonia	119
	iodine vapors	Brown spots	118
	tetracyanoethylene (2 %) in benzene	Orange spots	119
	N,2,6-trichloro p-benzoquinoneimine (2 %) in ethanol	Brown spots	119
Sulfilimines	potassium permanganate	Colorless spots	120
Sulfilimines, p-nitro-benzene-sulfonyl	tin chloride/4-N,N-dimethylaminobenz-aldehyde	Yellow spots	121
Sulfites	malachite green oxalate	White spots on blue background	4
Sulfonamides	chlorine/pyrazolinone/cyanide	Red spots changing to blue	32
	diazotization and coupling	Variety of spots (limit 0.25γ)	122,123
	Ehrlich	Variety of colors	124

Compound	Reagent	Result	Reference
	chloranil (1 %) in benzene	Pink turning to violet or green after heating	119
	2,3-dichloro-5,6-dicyano-1,4-benzoquinone (2 %) in benzene	Lilac-violet turning to yellow-green upon ammonia exposure	119
Sulfones	Gibbs reagent	Violet turning to tan upon exposure to ammonia and heat	119
	iodine vapors	Brown spots	118
	tetracyanoethylene (2 %) in benzene	Pink to yellow upon exposure to ammonia and heat	119
Sulfonic acids	Pinacryptol yellow (0.1 %)	Yellow–orange spots under long-wave UV	124
	silver nitrate/fluorescein	Yellow spots under long-wave UV	125,126
Sulfoxides	acetyl bromide	Yellow–orange spots	127
	ceric ammonium nitrate (40 %) in 2 M nitric acid	Brown spots after heating (especially good for α-polychlorosulfoxides); limit 80 μg/spot	117
	chloranil (1 %) in benzene	Yellow–blue spots	119
	2,3-dichloro-5,6-dicyano-1,4-benzoquinone (2 %) in benzene	Orange–crimson spots	119
	Dragendorff reagent	Orange–brown–red spots (limit 30–150γ)	128
	Gibbs reagent	Yellow turning to brown upon ammonia exposure	168
	Iodide (sodium)/starch	Brown spots (limits 0.01 μmol/20 μl solution)	129
	iodine vapors	Brown spots	118,130
	tetracyanoethylene (2 %) in benzene	Yellow or crimson turning to white or tan upon exposure to ammonia	119
	N,2,6-trichloro-p-benzoquinoneimine (2 %) in ethanol	Yellow spots	119
Terpenes	anisaldehyde/sulfuric acid	Variety of colors	1
	antimony (V) chloride	Variety of colors	1
	Carr-Price reagent	Variety of colors	1
	diphenylpicrylhydrazyl in chloroform	Yellow spots on purple background after heating (110 °C) (limit 1γ/0.5 cm diameter)	131
	phenol (50 % in carbon tetrachloride)	Variety of spots upon exposure to bromine vapors	3
	vanillin (1 % in 50 % H_3PO_4)	Variety of spots after heating (120 °C, 20 min)	4
Tetracyclines	ammonium hydroxide	Yellow fluorescence under long-wave UV	132
Thioacids	silver nitrate/ammonium hydroxide/sodium chloride	Yellow-brown spots	1
Thiobarbiturates	cupric sulfate (0.5 %)/diethylamine (3 % in methanol)	Green spots (limit 15γ)	3,133
Thiolactones	nitroprusside (sodium), basic	Red spots	62

(Continued)

Spray Reagents in Thin Layer Chromatography (Continued)

Family/Functional Group	Test	Result	Ref.
Thiols (Mercaptans)	ceric ammonium nitrate (in 2 M nitric acid)	Colorless spots on yellow background (limit < 100 μg/spot)	117
	iodine (1.3 % in ethanol)/ethanol	White spots in brown iodine background	3
	nitroprusside (sodium; 3 %)	Red spots	134
Thiophenes	Isatin (0.4 % in conc. sulfuric acid)	Variety of colors	135
Thiophosphates, esters	ferric chloride/sulfosalicylic acid	White spots on violet background	136
	palladium (II) chloride (0.5 % in acidified water)	Variety of spots	1,136
	periodic acid (10 % in 70 % perchloric acid)	Variety of spots	3
Unsaturated compounds	fluorescein (0.1 % in ethanol)/bromine	Yellow spots on a pink background upon exposure to bromine vapors	1
	osmium tetroxide vapors	Brown/black spots	3,95
Ureas	p-N,N-dimethylaminobenzaldehyde (1 % in ethanol)	Characteristic spots after exposure to hydrochloric acid	4
Vitamin A	antimony (V) chloride	Variety of colors	1
	Carr-Price reagent	Variety of colors	1
	sulfuric (50 % in methanol) followed by heating	Blue spots that turn brown	1
Vitamin B1	dipicrylamine	Characteristic spots	3
	Thiochrome	Variety of spots under long-wave UV	137
Vitamin B6	N,2,6-trichloro-p-benzoquinoneimine (0.1 % in ethanol)	Blue spots after exposure to ammonia	3
Vitamin B6, acetal	2,6-dibromo-p-benzoquinone-4-chlorimine (0.4 % in methanol)	Characteristic spots	138
Vitamin C	cacotheline (2 % aqueous)	Purple spot after heating (100 °C)	3
	iodine (0.005 %) in starch (0.4 %)	White spot on blue background	3
	methoxynitroaniline/sodium nitrite	Blue spots on orange background	3
Vitamin D	antimony (V) chloride	Variety of colors	139
	Carr-Price reagent	Variety of colors	1
	trichloroacetic (1 % in chloroform)	Variety of spots after heating (120 °C, 5 min)	1
Vitamin E	2′,7′-dichlorofluorescein (0.01 % in ethanol)	Spots under long-wave UV light	1
	α,α′-dipyridyl (0.5 %)/ferric chloride (0.5 % in ethanol)	Variety of colors	1

PROTOCOL FOR REAGENT PREPARATION

The following section gives a summary for the preparation of the major spray reagents listed in the previous section (Spray Reagents in Thin Layer Chromatography). Reference to the original literature is recommended for any reagents not listed here [1–4].

REFERENCES

1. Krebs, K. G., D. Heusser, and H. Wimmer. *Thin Layer Chromatography, A Laboratory Handbook,* edited by E. Shahl. New York: Springer-Verlag, 1969.
2. Bobbitt, J. B. *Thin Layer Chromatography.* New York: Reinhold, 1963.
3. Touchstone, J. C., and M. F. Dobbins. *Practice of Thin Layer Chromatography.* New York: John Wiley and Sons, 1983.
4. Randerath, K. *Thin-Layer Chromatography.* 2nd ed. Verlag Chemie, GmbH. (in the United States, Academic Press, New York, 1968).

acetic anhydride-sulfuric acid

See Liebermann–Burchard reagent.

alizarin

A saturated solution of alizarin in ethanol is sprayed on the moist plate, which is then placed in a chamber containing 25 % ammonium hydroxide solution to yield a variety of colors.

aluminum chloride

A 1 % aluminum chloride solution in ethanol is sprayed on the plate, which is then observed under long-wave UV light.

4-aminoantipyrine-potassium ferricyanide

See Emerson reagent.

4-aminobiphenyl-phosphoric acid

See Lewis-Smith reagent.

4-aminohippuric acid

A 0.3 % 4-aminohippuric acid solution in ethanol is sprayed on the plate, which is then heated at 140 °C (8 min) and observed under long-wave UV light.

ammonium hydroxide

The chromatogram is placed in a chamber containing 25 % ammonium hydroxide, dried, and then observed under long-wave UV light.

aniline-diphenylamine-phosphoric acid

An aniline (1 g)/diphenylamine (1 g)/phosphoric acid (5 mL) solution in acetone (50 mL) is sprayed on the plate, which is then heated at 85 °C (10 min) yielding a variety of colors.

aniline-phosphoric acid

A 20 % aniline solution in n-butanol, saturated with an aqueous (2N) orthophosphoric acid solution is sprayed on the plate, which is then heated at 105 °C (10 min) yielding a variety of colors.

aniline phthalate

An aniline (1 g)/o-phthalic acid (1.5 g) solution in n-butanol (100 mL; saturated with water) is sprayed on the plate, which is then heated at 105 °C (10 min) yielding a variety of colors.

anisaldehyde-sulfuric acid

A 1 % anisaldehyde solution in acetic acid (acidified by conc. sulfuric acid) is sprayed on the plate, which is then heated at 105 °C to yield a variety of colors.

p-anisidine phthalate

A 0.1 M solution of p-anisidine and phthalic acid in ethanol is sprayed on the plate, which is then heated at 100 °C (10 min) to yield a variety of colors.

anthrone

A 1 % anthrone solution in 60 % aqueous ethanol solution acidified with 10 mL 60 percent phosphoric acid is sprayed on the plate, which is then heated at 110 °C (5 min) to yield yellow spots.

antimony (III) chloride

See Carr–Price reagent.

antimony (III) chloride-acetic acid

A 20 % antimony (III) chloride solution in 75 % chloroform-acetic acid solution is sprayed on the plate, which upon heating at 100 °C (5 min) yields a variety of colors.

antimony (V) chloride

A 20 % antimony (V) chloride solution in chloroform or carbon tetrachloride is sprayed on the plate yielding a variety of colors upon heating.

Benedict's reagent

A solution that is 0.1 M in cupric sulfate, 1.0 M in sodium citrate and 1.0 M in sodium carbonate is sprayed on the plate, which is then observed under long-wave UV light.

benzidine diazotized

A 0.5 % benzidine solution in 0.005 % hydrochloric acid is mixed with an equal volume of 10 % sodium nitrite solution in water; the mixture is sprayed on the plate to yield a variety of colors.

benzidine-trichloroacetic acid

A 0.5 % benzidine in (1:1:8) acetic acid/trichloroacetic acid/ethanol is sprayed on the plate to yield red-brown spots upon heating (110 °C) or exposure to unfiltered UV light (15 min).

bismuth (III) chloride

A 33 % ethanol solution of bismuth (III) chloride is sprayed on the plate, which upon heating (110 °C) yields fluorescent spots under long-wave UV light.

carbazole-sulfuric acid

A 0.5 % carbazole in ethanol/sulfuric acid (95:5) is sprayed on the plate, which yields violet spots (on blue background) after heating at 120 °C (10 min).

Carr–Price reagent

A 25 % antimony (III) chloride solution in chloroform or carbon tetrachloride is sprayed on the plate, which is heated at 100 °C (10 min) to yield a variety of colors.

ceric ammonium sulfate

A 1 % solution of ceric ammonium sulfate in strong acids (phosphoric, nitric) is sprayed on the plate to yield yellow/green spots on a red background, after heating at 105 °C (10 min).

chloramine-T

A 10 % chloramine-T solution is sprayed on the plate, followed by 1 N hydrochloric acid. The chromatogram is dried and exposed to 25 % ammonium hydroxide and warmed.

chlorine-pyrazolinone-cyanide

An equal volume mixture of 0.2 M 1-phenyl-3-methyl-2-pyrazolin-5-one solution (in pyridine) and 1 M aqueous potassium cyanide solution is sprayed on the plate that has been previously exposed to chlorine vapors. The resulting red spots turn blue after a few minutes.

chlorosulfonic acid-acetic acid

A 35 % chlorosulfonic acid solution in acetic acid is sprayed on the plate, which is then heated at 130 °C (5 min) to produce fluorescence under long-wave UV.

cinnamaldehyde-hydrochloric acid

A 5 % cinnamaldehyde solution in ethanol (acidified with hydrochloric acid) is sprayed on the plate, which is then placed in a hydrochloric acid chamber to yield red spots.

cobalt (II) thiocyanate

An ammonium thiocyanate (15 %)/cobalt (II) chloride (5 %) solution in water is sprayed on the plate yielding blue spots.

cupric sulfate-quinine-pyridine

A solution that is 0.4 % in cupric sulfate, 0.04 % in quinine hydrochloride and 4 % in pyridine in water is sprayed on the plate followed by a 0.5 % aqueous potassium permanganate solution. A variety of colors (white, yellow, violet) is detected on the chromatogram.

α-cyclodextrin

A 30 % α-cyclodextrin solution in ethanol is sprayed on the plate, which is further developed in an iodine chamber.

diazonium

See Fast Blue B salt.

diazotization and coupling reagent

A 1 % sodium nitrite solution (in 1 M hydrochloric acid) is sprayed on the plate, followed by a 0.2 % α-naphthol solution in 1 M potassium hydroxide and drying.

4-N,N-dimethylaminobenzaldehyde-sulfuric acid

A 0.125 % solution of 4-N,N-dimethylaminobenzaldehyde in 65 % sulfuric acid mixed with 5 % ferric chloride (0.05 mL per 100 mL solution) is sprayed on the plate giving a variety of spots.

4-N,N-dimethylaminocinnamaldehyde

A 0.2 % solution of 4-N,N-dimethylaminocinnamaldehyde in 6N HCl/ethanol (1:4) is sprayed on the plate, which is then heated at 105 °C (5 min) revealing a variety of colored spots. Vapors of aqua regia tend to intensify the spots.

2,4-dinitrofluorobenzene

A 1 % sodium bicarbonate solution in 0.025 M sodium hydroxide is sprayed on the plate followed by a 2,4-dinitrofluorobenzene (10 %) solution in methanol. Heating the plate in the dark (40 °C, one hour) and further spraying it with diethyl ether yields a variety of spots.

2,4-dinitrophenylhydrazine

A 0.4 % solution of 2,4-dinitrophenylhydrazine in 2N hydrochloric acid is sprayed on the plate followed by a 0.2 % solution of potassium ferricyanide in 2N hydrochloric acid yielding orange/yellow spots.

Dragendorff reagent

A 1.7 % aqueous solution of basic bismuth nitrate in weak acids (tartaric, acetic) mixed with an aqueous potassium iodide or barium chloride solution is sprayed on the plate to yield a variety of spots.

Ehrlich reagent

A 1 % 4-N,N-dimethylaminobenzaldehyde solution in ethanol is sprayed on the plate, which is dried and then placed in a hydrochloric acid chamber to yield various spots.

Emerson reagent

A 2 % 4-aminoantipyrine solution in ethanol is sprayed on the plate, followed by an 8 % aqueous potassium ferricyanide solution. The chromatogram is then placed in a chamber containing 25 % ammonium hydroxide.

EP

A 0.3 % solution of 4-N,N-dimethylaminobenzaldelyde in acetic acid/phosphoric acid/water (10:1:4) is sprayed on the plate to yield a variety of spots.

Fast Blue B Salt (diazonium)

A 0.5 % aqueous solution of Fast Blue B Salt is sprayed on the plate followed by a 0.1 M sodium hydroxide.

ferric chloride–perchloric acid

A solution made out of 5 mL 5 % aqueous ferric chloride, 45 mL 20 % perchloric acid and 50 mL 50 % nitric acid is sprayed on the plate to yield a variety of spots.

ferric chloride–sulfosalicylic acids

The plate is first exposed to a bromine atmosphere then sprayed with a 0.1 % ethanolic solution of ferric chloride. After air drying (15 min) the chromatogram is sprayed with a 1 % ethanolic solution of sulfosalicylic acid to yield a variety of spots.

ferrocyanide–hydrogen peroxide

0.5 g ammonium chloride is added to a 0.1 % potassium ferrocyanide solution in 0.2 % hydrochloric acid and the resulting solution is sprayed on the plate, which is then dried (100 °C). The chromatogram is further sprayed with 30 % hydrogen peroxide, heated (150 °C, 30 min) and sprayed with 10 % potassium carbonate to yield yellow/red spots.

ferrous thiocyanate

A 2:3 mixture of a 4 % aqueous ferrous sulfate and 1.3 % acetone solution of ammonium thiocyanate is sprayed on the plate yielding red-brown spots.

fluorescein-hydrogen peroxide

A 0.1 % fluorescein solution in 50 % aqueous ethanol is sprayed on the plate followed by a 15 % hydrogen peroxide in glacial acetic acid and heated (90 °C, 20 min) yielding nonfluorescent spots.

Folin reagent

A 0.02 % sodium 1,2-naphthoquinone-4-sulfonate in 5 % sodium carbonate is sprayed on the plate, which is then dried to yield a variety of colors.

Folin–Denis reagent

A tungstomolybdophosphoric acid solution is sprayed on the plate, which is then exposed to ammonia vapors.

Gibbs reagent

A 0.4 % methanolic solution of 2,6-dibromoquinonechloroimide is sprayed on the plate followed by a 10 % aqueous sodium carbonate yielding a variety of spots.

glucose–phosphoric acid

A 2 % glucose solution in phosphoric acid/water/ethanol/n-butanol (1:4:3:3) is sprayed on the plate followed by heating (115 °C, 10 min) to yield a variety of spots.

hydroxylamine-ferric chloride

A 1:2 mixture of a 10 % hydroxylammonium chloride/10 % potassium hydroxide in aqueous ethanol is sprayed on the plate followed by drying. The chromatogram is then sprayed with an ether solution of ferric chloride in hydrochloric acid to yield a variety of spots.

iodide (potassium) starch

A 1 % potassium iodide solution in 80 % aqueous acetic acid is sprayed on the plate followed by a 1 % aqueous starch solution. A pinch of zinc dust is recommended as an addition to the potassium iodide solution.

iodide (sodium) starch

A solution made by mixing a 5 % starch/0.5 % sodium iodide solution with an equal volume of concentrated hydrochloric acid is sprayed on the plate, which is then exposed to dry sodium hydroxide (desiccator) and evacuated (30–60 min) to yield brown spots.

isatin-zinc acetate

An isatin (1 %)/zinc acetate (1.5 %) solution in isopropanol acidified with acetic acid is sprayed on the plate, which is then heated to yield a variety of spots.

Kalignost reagent

A 1 % solution of sodium tetraphenylborate in aqueous butanone is sprayed on the plate, followed by a 0.015 % methanolic solution of fischtin or quercetin to yield orange-red spots that fluoresce under long-wave UV.

Komarowski reagent

A 2 % methanolic solution of p-hydroxybenzaldehyde that is 5 % in sulfuric acid is sprayed on the plate, which is then heated (105 °C, 3 min) to yield yellow or pink spots.

König reagent

A 2 % p-aminobenzoic acid in ethanolic hydrochloric acid (0.6 M) is sprayed on the plate that has been exposed (1 hr) to vapors of cyanogen bromide.

Lewis–Smith reagent

o-Aminobiphenyl (0.3 g dissolved in 100 ml of a 19:1 ethanol/phosphoric acid mixture) is sprayed on the plate, which is then heated at 110 °C (15 min).

Liebermann-Burchard reagent

A freshly prepared mixture of 5 mL acetic anhydride/5 mL conc. sulfuric acid in 50 mL cold absolute ethanol is sprayed on the plate, which is heated at 100 °C (10 min) and observed under long-wave UV light.

malachite green oxalate

A 1 % ethanolic potassium hydroxide solution is sprayed, the plate heated (150 °C, 5 min), and further sprayed with a buffered (pH = 7) water/acetone solution of malachite green oxalate to yield white spots on blue background.

metaperiodate (sodium)-p-nitroaniline

A 35 % saturated solution of sodium metaperiodate is sprayed on the plate, which is left to dry (10 min). The chromatogram is then sprayed with a 0.2 % p-nitroaniline solution in ethanol/hydrochloric acid (4:1) to yield fluorescing (long-wave UV) yellow spots.

methoxynitroaniline - sodium nitrite

A 0.02 M 4-methoxy-2-nitroaniline solution in 50 % aqueous acetic acid/5N sulfuric acid is sprayed on the plate, which is dried and re-sprayed with 0.2 % sodium nitrite to yield blue spots on an orange background.

Millon reagent

A solution of mercury (5 g) in fuming nitric acid (10 mL) diluted with water (10 mL) is sprayed on the plate to yield yellow/orange spots that are intensified by heat (100 °C).

1,2-naphthoquinone-4-sulfonic acid/perchloric acid

A 0.1 % 1,2-naphthoquinone-4-sulfonic acid solution in ethanol/perchloric acid/40 % formaldehyde/water (20:10:1:9) is sprayed on the plate, which is then heated (70 °C) to yield pink spots that turn to blue on prolonged heating.

Neu reagent

A 1 % methanolic solution of the β-aminoethylester of diphenylboric acid is sprayed on the plate to yield fluorescent spots under long wave UV light.

ninhydrin

A ninhydrin solution (0.3 % in acidified n-butanol or 0.2 % in ethanol) is sprayed on the plate, which is then heated (110 °C). The resulting spots are stabilized by spraying with a solution made of 1 mL saturated aqueous cupric nitrate, 0.2 mL 10 % nitric acid and 100 mL 95 % ethanol, to yield red spots when exposed to ammonium hydroxide (25 %).

p-nitroaniline, diazotized

A solution made by mixing 0.1 % aqueous p-nitroaniline/0.2 % aqueous sodium nitrite/10 % aqueous potassium carbonate (1:1:2) is sprayed on the plate to yield colored spots.

p-nitroaniline, diazotized (buffered)

A solution of 0.5 % p-nitroaniline (in 2N hydrochloric acid), 5 % aqueous sodium nitrite and 20 % aqueous sodium acetate (10:1:30) is sprayed on the plate to yield a variety of colored spots.

nitroprusside (sodium)

A solution made by mixing sodium nitroprusside (1.5 g), 2N hydrochloric acid (5 mL), methanol (95 mL), and 25 % ammonium hydroxide (10 mL) is sprayed on the plate to yield a variety of colors.

nitroprusside (sodium), basic

A 2 % sodium nitroprusside solution in 75 % ethanol is sprayed on the plate, which has already been treated with 1N sodium hydroxide to yield red spots.

orcinol

A mixture consisting of 0.6 % ethanolic orcinol and 1 % ferric chloride in dilute sulfuric acid is sprayed on the plate, which is further heated (100 °C, 10 min) to yield characteristic spots.

Prochazka reagent

A 10 % formaldehyde solution in 5 % hydrochloric acid solution in ethanol is sprayed on the plate, which is then heated to yield fluorescent spots (yellow/orange/green) under long-wave UV.

quercetin-sodium tetraphenylborate

A mixture of quercetin (0.015 % in methanol) and sodium tetraphenyl-borate (1 % in n-butanol saturated with water) is sprayed on the plate to yield orange/red spots.

quinaldine

A 1–1.5 % solution of 3,5-diaminobenzoic acid dihydrochloride in 30 % phosphoric acid is sprayed on the plate, which is then heated (100 °C, 15 min) to yield fluorescent (green/yellow) spots under long-wave UV or (in case of high concentrations) brown spots in daylight.

Sakaguchi reagent

A 0.1 % acetone solution of 8-hydroxyquinoline is sprayed on the plate followed by a 0.2 % 0.5N sodium hydroxide solution to yield orange/red spots.

Salkowski reagent

A 0.01 M aqueous ferric chloride/35 % perchloric acid solution is sprayed on the plate, which is then heated (60 °C, 5 min) to yield a variety of colors intensified when exposed to aqua regia.

Schweppe reagent

A mixture of 2 % aqueous glucose/2 % ethanolic aniline in n-butanol is sprayed on the plate, which is heated (125 °C, 5 min) to yield a variety of spots.

silver nitrate-ammonium hydroxide-sodium chloride

A mixture of silver nitrate (0.05 M)/ammonium hydroxide (5 %) is sprayed on the plate, followed by drying and further spraying with 10 % aqueous sodium chloride to yield yellow/brown spots.

silver nitrate-fluorescein

A mixture of silver nitrate (2 %)/sodium-fluorescein (0.2 %) in 80 % ethanol is sprayed on the plate to yield yellow spots on pink background.

silver nitrate-formaldehyde

The plate is consecutively sprayed with 0.05 M ethanolic silver nitrate, 35 % aqueous formaldehyde, 2 M potassium hydroxide and, finally, a solution made of equal volumes of hydrogen peroxide (30 %) and nitric acid (65 %). Each spraying is preceded by a 30 min. drying and at the end the plate is kept in the dark for 12 hr before exposing to sunlight to yield dark grey spots.

silver nitrate-hydrogen peroxide

A 0.05 % silver nitrate solution in water/cellosolve/acetone (1:10:190; to which a drop of 30 % hydrogen peroxide has been added) is spayed on the plate, which is then treated under unfiltered UV to yield dark spots.

silver nitrate-sodium hydroxide

A saturated silver nitrate solution is sprayed on the plate followed by a 0.5 M aqueous/methanol solution. Subsequent drying (100 °C, 2 min) yields a variety of spots.

Sonnenschein reagent

A 2 % ceric sulfate solution in 20 % aqueous trichloroacetic acid (that has been acidified with sulfuric acid) is sprayed on the plate. A variety of colors appears upon heating (110 °C, 5 min).

Stahl

See van Urk reagent.

sulfanilic acid-1-naphthylamine

A mixture of 2 % sulfanilic acid/1-naphthylamine in 30 % acetic acid is sprayed on the plate to yield a variety (violet/green/blue) of colors.

thiochrome

A 0.3 M aqueous potassium ferricyanide solution that is 15 % in sodium hydroxide is sprayed on the plate yielding a variety of spots under long-wave UV.

Tollen's reagent

See Zaffaroni reagent.

vanillin-potassium hydroxide

A 2 % solution of vanillin in n-propanol is sprayed on the plate, which is heated (100 °C, 10 min) and sprayed again with 1 % ethanolic potassium hydroxide. Reheating yields a variety of colors observed under daylight.

van Urk (Stahl) reagent

A 0.5 % solution of 4-N,N-dimethylaminobenzaldehyde in concentrated hydrochloric acid/ethanol (1:1) is sprayed on the preheated plate, which is then subjected to aqua regia vapors to yield a variety of colors.

Zaffaroni (Tollen's) reagent

A mixture of silver nitrate (0.02 M)/ammonium hydroxide (5 M) is sprayed on the plate, which is then heated (105 °C, 10 min) to yield black spots.

Zwikker reagent

A 1 % cobaltous nitrate in absolute ethanol is sprayed on the plate, which is dried (at room temperature) and exposed to a wet chamber containing 25 % ammonium hydroxide.

CHAPTER 4

Supercritical Fluid Extraction and Chromatography

CONTENTS

SOME USEFUL FLUIDS FOR SUPERCRITICAL FLUID EXTRACTION AND CHROMATOGRAPHY

The following table lists some useful carrier and modifier fluids for supercritical fluid extraction and chromatography, along with relevant properties [1–3]. The critical properties are needed to determine successful fluid operating ranges. Where possible, experimental values are provided. In some cases, however, values calculated with a group contribution approach are presented [2]. These entries are marked with an asterisk. The dipole moment is provided to assess fluid polarity, although these values can be temperature dependent, especially with the more complex fluids. Occasionally, conformations change with temperature, resulting in a change in the dipole moment. Data on ultraviolet cutoff are provided to allow the application of UV-vis monitoring instrumentation. Data are not provided if the only electronic transition is in the very low wavelength range, and the spectrum is largely flat. With respect to the halocarbon fluids, if a commonly used refrigerant designator is available, it is presented with the chemical name. The fluids listed here have either been used or proposed for use in supercritical fluid chromatography or supercritical fluid extraction. The reader should also note that some of these fluids (for example, methanol and toluene) will undergo serious chemical degradation under near critical conditions while in contact with stainless steels and other common materials [4–7].

REFERENCES

1. Bruno, T. J., and P. D. N. Svoronos. *CRC Handbook of Basic Tables for Chemical Analysis.* 2nd ed. Boca Raton, FL: CRC Press, 2003.
2. Bruno, T. J., and J. F. Ely. *Supercritical Fluid Technology, Reviews in Modern Theory and Applications.* Boca Raton, FL: CRC Press, 1991.
3. Bruno, T. J. *CRC Handbook for the Identification and Analysis of Alternative Refrigerants.* Boca Raton, FL: CRC Press, 1995.
4. Bruno, T. J., and G. C. Straty. "Thermophysical Property Measurement on Chemically Reactive Systems: A Case Study." *Journal of Research of the National Bureau of Standards* 91, no. 3 (1986): 135.
5. Straty, G. C., M. J. Ball, and T. J. Bruno. "Experimental Determination of the PVT Surface for Benzene." *Journal of Chemical & Engineering Data* 32 (1987): 163.
6. Ashe, W. "Mobile Phases for Supercritical Fluid Chromatography." *Chromatographia* 11, no. 7 (1978): 411.
7. Goodwin, A. R. H. and Mehl, J. B. "Measurement of the Dipole Moments of Seven Partially Fluorinated Hydrocarbons with a Radiofrequency Reentrant Cavity Resonator." *International Journal of Thermophysics* 18, (1997): 795–806.

Inorganic Fluids

Fluid	T_c (°C)	ρ_c (g/mL)	P_c (MPa)	μ (D)	UV Cutoff, nm
Carbon dioxide	31.413	0.460	7.4	0.0	
Ammonia	132.5	0.235	11.4	1.47	
Nitrous oxide	36.5	0.450	7.3	0.167	
Sulfur dioxide	157.8	0.520	7.9	1.63	
Sulfur hexafluoride	45.6	0.730	3.76	0.0	
Water	374.1	0.40	22.1	1.85	
Xenon	16.6	1.155	5.9		

Hydrocarbon Fluids

Fluid	T_c (°C)	ρ_c (g/mL)	P_c (MPa)	μ (D)	UV Cutoff, nm
Ethane	32.19	0.203	4.871	0.0	
Propane	96.8	0.220	4.26	≥ 0.05	
n-Butane	152.05	1.225	3.80		
n-Pentane	196.6	0.232	3.4	NA	
n-Hexane	234.2	0.234	3.0	0.08	250
Benzene	288.9	0.304	4.9	0.0	325
Toluene	320.8	0.29	4.2	0.084	325

Alcohols, Ethers, and Ketones

Fluid	T_c (°C)	ρ_c (g/mL)	P_c (MPa)	μ (D)	UV Cutoff, nm
Methanol	239.5	0.272	8.0	1.70	255
Ethanol	240.8	0.275	6.14	1.7	255
Isopropanol	235.3	0.273	4.8	1.66	255
Dimethyl ether	126.9	0.271	5.24	1.3	
Diethyl ether	193.4	0.265	3.651	1.3	225
Acetone	235	0.278	4.7	2.9	350

Halocarbons

Fluid	T_c (°C)	ρ_c (g/mL)	P_c (MPa)	μ (D)	UV Cutoff, nm
Fluoromethane, R-41	41.9	0.301	5.6	1.8	285
Difluoromethane, R-32	78.41	0.430	5.83	1.978	240
Trifluoromethane, R-23	25.83	0.526	4.82	1.65	300
Chloromethane, R-40	143.15	0.364	6.7	1.87	240
Chlorodifluoromethane, R-22	96.15	0.521	4.97	1.44	220
Dichlorofluoromethane, R-21	178.5	0.522	5.2	1.24	235
Trichlorofluoromethane	198	0.554	4.4	NA	
Chlorotrifluoromethane, R-13	28.9	0.578	3.9	0.5	220
Dichlorodifluoromethane, R-12	111.5	0.558	4.0	0.51	245
Fluoroethane, R-161	102.16	0.288	4.70	1.94	210
1,1-Difluoroethane, R-152a	113.5	0.365	4.49	2.262	
1,1,1-Trifluoroethane, R-143a	73.1	0.434	3.76	2.32	
1,1,2-Trifluoroethane, R-143	156.75	0.466	4.52	1.68 (35.9 °C) 1.75 (136.9 °C)	
1,1,1,2-Tetrafluoroethane, R-134a	101.06	0.515	4.06	2.058	
1,1,2,2-Tetrafluoroethane, R-134	118.95	0.542	4.56	0.991 (36 °C) 0.250 (140 °C)	300
Pentafluoroethane, R-125	68.3	0.572	3.631	1.54	
1,1-Dichlorotetrafluoro ethane, R-114a	145.7	0.582	3.6		
1-Chloro-1,1-difluoroethane	137.1	0.435	4.12	2.14	265
1,2-Dichloro tetrafluoroethane	145.7	0.582	3.26	0.668 (35 °C) 0.699 (137 °C)	240
2-Chloro-1,1,1,2-tetrafluoroethane, R-124	122.5	0.554	3.63	1.469	230
Chloropentafluoroethane, R-115	79.9	0.596	3.12	0.52	245
1,1,1,2,3,3-Hexafluoropropane, R-236ea	141.1	0.571	3.533	NA	
1,1,1,2,3,3,3-Heptafluoropropane, R-227ea	102.8	0.580	2.94	NA	
2-Chloroheptafluoro propane, R-217ba	127.5*	0.592*	3.12*	NA	210
Bis(difluoromethyl) ether, E-134	147.6	0.522	4.302	1.739 (36 °C) 1.840 (173 °C)	

P-ρ-T TABLE FOR CARBON DIOXIDE

The following table provides a numerical listing of the P-ρ-T surface for carbon dioxide in the region of interest for supercritical fluid extraction and chromatography. These data were calculated using an empirical equation of state (the Schmidt-Wagner equation) [1,2], the parameters of which were determined from a fit of experimental P-V-T measurements [3–7]. Note that the pressures are tabulated in bars for convenience. The appropriate SI unit of pressure is the megapascal (1 bar = 0.1 MPa).

As an alternative to this or any density table, we recommend the application of reliable databases that have equations of state implemented [8–10].

REFERENCES

1. Schmidt, R., and W. Wagner. "A New Form of the Equation of State for Pure Substances and Its Application to Oxygen." *Fluid Phase Equilibria* 19 (1985): 175.
2. Ely, J. F. National Bureau of Standards, Boulder, CO, private communication (coefficients for carbon dioxide), 1986.
3. Poling, B. E., J. M. Prausnitz, and J. P. O'Connell. *The Properties of Gases and Liquids.* 5th ed. New York: McGraw-Hill, 2000.
4. Prausnitz, J. M. *Molecular Thermodynamics of Fluid Phase Equilibria.* Englewood Cliffs: Prentice-Hall, 1969.
5. Chao, K. C., and R. A. Greenkorn. *The Thermodynamics of Fluids.* New York: Marcel Dekker, 1975.
6. Jacobson, R. T., and R. B. Stewart. "Thermodynamic Properties of Nitrogen Including Liquid and Vapor Phases from 63K to 2000K with Pressures to 10,000 Bar." *Journal of Physical and Chemical Reference Data* 2, no. 4 (1973): 757.
7. Ely, J. F. Proceedings of the 63rd Gas Processors Association Annual Convention, pg. 9, 1984.
8. Angus, S., B. Armstrong, and K. M. deReuck. *Carbon Dioxide, International Thermodynamic Tables of the Fluid State.* Oxford: Pergamon Press, 1976.
9. Rowley, R. L., W. V. Wilding, J. L. Oscarson, N. A. Zundel, T. L. Marshall, T. E. Daubert, and R. P. Danner. *DIPPR Data Compilation of Pure Compound Properties.* New York: Design Institute for Physical Properties AIChE, 2008.
10. Lemmon, E. W., M. O. McLinden, and M. L. Huber. *REFPROP, Reference Fluid Thermodynamic and Transport Properties, NIST Standard Reference Database 23.* Gaithersburg, MD: National Institute of Standards and Technology, 2005.

TEMPERATURE = 308.15 K

Density ρ (mol/L)	Pressure P (bar)	Density ρ (mol/L)	Pressure P (bar)	Density ρ (mol/L)	Pressure P (bar)
0.20	5.009	4.80	68.878	9.40	79.865
0.30	7.429	4.90	69.450	9.50	79.935
0.40	9.792	5.00	69.998	9.60	80.003
0.50	12.100	5.10	70.523	9.70	80.069
0.60	14.353	5.20	71.027	9.80	80.135
0.70	16.552	5.30	71.509	9.90	80.200
0.80	18.698	5.40	71.970	10.00	80.264
0.90	20.791	5.50	72.410	10.10	80.328
1.00	22.832	5.60	72.831	10.20	80.391
1.10	24.821	5.70	73.233	10.30	80.454
1.20	26.760	5.80	73.617	10.40	80.517
1.30	28.648	5.90	73.983	10.50	80.580
1.40	30.487	6.00	74.332	10.60	80.643
1.50	32.278	6.10	74.665	10.70	80.706
1.60	34.020	6.20	74.981	10.80	80.771
1.70	35.716	6.30	75.282	10.90	80.835
1.80	37.365	6.40	75.569	11.00	80.901
1.90	38.968	6.50	75.841	11.10	80.968
2.00	40.526	6.60	76.100	11.20	81.037
2.10	42.040	6.70	76.345	11.30	81.107
2.20	43.510	6.80	76.578	11.40	81.178
2.30	44.938	6.90	76.800	11.50	81.252
2.40	46.323	7.00	77.010	11.60	81.328
2.50	47.668	7.10	77.209	11.70	81.407
2.60	48.971	7.20	77.398	11.80	81.489
2.70	50.235	7.30	77.577	11.90	81.574
2.80	51.460	7.40	77.746	12.00	81.663
2.90	52.646	7.50	77.907	12.10	81.756
3.00	53.795	7.60	78.060	12.20	81.854
3.10	54.907	7.70	78.205	12.30	81.957
3.20	55.983	7.80	78.342	12.40	82.065
3.30	57.024	7.90	78.473	12.50	82.180
3.40	58.030	8.00	78.597	12.60	82.301
3.50	59.002	8.10	78.515	12.70	82.430
3.60	59.941	8.20	78.828	12.80	82.567
3.70	60.847	8.30	78.935	12.90	82.713
3.80	61.722	8.40	79.037	13.00	82.868
3.90	62.565	8.50	79.135	13.10	83.033
4.00	63.379	8.60	79.228	13.20	83.210
4.10	64.162	8.70	79.318	13.30	83.399
4.20	64.917	8.80	79.405	13.40	83.601
4.30	65.644	8.90	79.488	13.50	83.817
4.40	66.343	9.00	79.568	13.60	84.047
4.50	67.015	9.10	79.646	13.70	84.294
4.60	67.661	9.20	79.721	13.80	84.558
4.70	68.282	9.30	79.794	13.90	84.841

(Continued)

TEMPERATURE = 308.15 K (Continued)

Density ρ (mol/L)	Pressure P (bar)	Density ρ (mol/L)	Pressure P (bar)	Density ρ (mol/L)	Pressure P (bar)
14.00	85.143	17.70	125.850	21.40	324.601
14.10	85.466	17.80	128.308	21.50	334.300
14.20	85.811	17.90	130.874	21.60	344.307
14.30	86.179	18.00	133.554	21.70	354.627
14.40	86.573	18.10	136.350	21.80	365.268
14.50	86.993	18.20	139.267	21.90	376.236
14.60	87.441	18.30	142.309	22.00	387.540
14.70	87.919	18.40	145.478	22.10	399.185
14.80	88.427	18.50	148.781	22.20	411.179
14.90	88.969	18.60	152.221	22.30	423.529
15.00	89.545	18.70	155.802	22.40	436.242
15.10	90.158	18.80	159.530	22.50	449.326
15.20	90.808	18.90	163.407	22.60	462.787
15.30	91.499	19.00	167.440	22.70	476.634
15.40	92.231	19.10	171.633	22.80	490.874
15.50	93.007	19.20	175.991	22.90	505.514
15.60	93.829	19.30	180.519	23.00	520.562
15.70	94.700	19.40	185.222	23.10	536.026
15.80	95.620	19.50	190.105	23.20	551.914
15.90	96.592	19.60	195.173	23.30	568.233
16.00	97.619	19.70	200.432	23.40	584.991
16.10	98.703	19.80	205.887	23.50	602.197
16.20	99.845	19.90	211.544	23.60	619.859
16.30	101.049	20.00	217.408	23.70	637.984
16.40	102.317	20.10	223.485	23.80	656.582
16.50	103.652	20.20	229.781	23.90	675.660
16.60	105.055	20.30	236.302	24.00	695.227
16.70	106.530	20.40	243.053	24.10	715.292
16.80	108.079	20.50	250.040	24.20	735.864
16.90	109.706	20.60	257.271	24.30	756.952
17.00	111.412	20.70	264.750	24.40	778.564
17.10	113.201	20.80	272.484	24.50	800.709
17.20	115.076	20.90	280.480	24.60	823.398
17.30	117.040	21.00	288.743	24.70	846.639
17.40	119.096	21.10	297.280	24.80	870.441
17.50	121.428	21.20	306.098	24.90	894.816
17.60	123.498	21.30	315.203	25.00	919.772

TEMPERATURE = 313.15 K

Density ρ (mol/L)	Pressure P (bar)	Density ρ (mol/L)	Pressure P (bar)	Density ρ (mol/L)	Pressure P (bar)
0.20	5.095	4.80	72.148	9.40	87.401
0.30	7.559	4.90	72.180	9.50	87.558
0.40	9.968	5.00	73.449	9.60	87.714
0.50	12.323	5.10	74.006	9.70	87.869
0.60	14.625	5.20	74.661	9.80	88.023
0.70	16.874	5.30	75.235	9.90	88.176
0.80	19.070	5.40	75.788	10.00	88.329
0.90	21.215	5.50	76.322	10.10	88.481
1.00	23.309	5.60	76.836	10.20	88.634
1.10	25.352	5.70	77.332	10.30	88.787
1.20	27.346	5.80	77.810	10.40	88.940
1.30	29.291	5.90	78.271	10.50	89.095
1.40	31.188	6.00	78.714	10.60	89.250
1.50	33.038	6.10	79.142	10.70	89.407
1.60	34.840	6.20	79.553	10.80	89.566
1.70	36.597	6.30	79.950	10.90	89.727
1.80	38.308	6.40	80.331	11.00	89.890
1.90	39.975	6.50	80.699	11.10	90.056
2.00	41.598	6.60	81.053	11.20	90.224
2.10	43.177	6.70	81.395	11.30	90.397
2.20	44.714	6.80	81.723	11.40	90.572
2.30	46.209	6.90	82.040	11.50	90.753
2.40	47.664	7.00	82.345	11.60	90.937
2.50	49.078	7.10	82.640	11.70	91.127
2.60	50.453	7.20	82.923	11.80	91.322
2.70	51.788	7.30	83.197	11.90	91.523
2.80	53.086	7.40	83.462	12.00	91.730
2.90	54.346	7.50	83.717	12.10	91.944
3.00	55.570	7.60	83.964	12.20	92.166
3.10	56.758	7.70	84.202	12.30	92.396
3.20	57.911	7.80	84.433	12.40	92.635
3.30	59.029	7.90	84.657	12.50	92.883
3.40	60.114	8.00	84.874	12.60	93.141
3.50	61.165	8.10	85.084	12.70	93.410
3.60	62.185	8.20	85.289	12.80	93.691
3.70	63.172	8.30	85.487	12.90	93.984
3.80	64.129	8.40	85.681	13.00	94.290
3.90	65.056	8.50	85.869	13.10	94.610
4.00	65.953	8.60	86.053	13.20	94.945
4.10	66.821	8.70	86.233	13.30	95.296
4.20	67.661	8.80	86.409	13.40	95.664
4.30	68.474	8.90	86.581	13.50	96.049
4.40	69.260	9.00	86.750	13.60	96.453
4.50	70.019	9.10	86.917	13.70	96.878
4.60	70.754	9.20	87.080	13.80	97.323
4.70	71.463	9.30	87.241	13.90	97.791

(Continued)

TEMPERATURE = 313.15 K (Continued)

Density ρ (mol/L)	Pressure P (bar)	Density ρ (mol/L)	Pressure P (bar)	Density ρ (mol/L)	Pressure P (bar)
14.00	98.283	17.70	148.798	21.40	363.790
14.10	98.799	17.80	151.603	21.50	374.021
14.20	99.341	17.90	154.522	21.60	384.564
14.30	99.911	18.00	157.558	21.70	395.426
14.40	100.510	18.10	160.717	21.80	406.613
14.50	101.139	18.20	164.001	21.90	418.132
14.60	101.800	18.30	167.414	22.00	429.990
14.70	102.495	18.40	170.962	22.10	442.195
14.80	103.224	18.50	174.647	22.20	454.753
14.90	103.991	18.60	178.474	22.30	467.672
15.00	104.796	18.70	182.449	22.40	480.958
15.10	105.640	18.80	186.574	22.50	494.620
15.20	106.527	18.90	190.855	22.60	508.664
15.30	107.458	19.00	195.297	22.70	523.099
15.40	108.435	19.10	199.904	22.80	537.931
15.50	109.459	19.20	204.681	22.90	553.168
15.60	110.533	19.30	209.633	23.00	568.818
15.70	111.659	19.40	214.766	23.10	601.389
15.80	112.839	19.50	220.084	23.20	601.389
15.90	114.075	19.60	225.592	23.30	618.325
16.00	115.369	19.70	231.296	23.40	635.705
16.10	116.725	19.80	237.201	23.50	653.539
16.20	118.143	19.90	243.314	23.60	671.833
16.30	119.627	20.00	249.638	23.70	690.597
16.40	121.179	20.10	256.181	23.80	709.839
16.50	122.801	20.20	262.947	23.90	729.567
16.60	124.496	20.30	269.943	24.00	749.790
16.70	126.268	20.40	277.175	24.10	770.517
16.80	128.118	20.50	284.648	24.20	791.756
16.90	130.049	20.60	292.368	24.30	813.518
17.00	132.065	20.70	300.342	24.40	835.810
17.10	134.168	20.80	308.576	24.50	858.643
17.20	136.362	20.90	317.076	24.60	882.025
17.30	138.649	21.00	325.848	24.70	905.967
17.40	141.033	21.10	334.899	24.80	930.477
17.50	143.517	21.20	344.235	24.90	955.567
17.60	146.104	21.30	353.863	25.00	981.245

TEMPERATURE = 318.15 K

Density ρ (mol/L)	Pressure P (bar)	Density ρ (mol/L)	Pressure P (bar)	Density ρ (mol/L)	Pressure P (bar)
0.20	5.180	4.90	76.115	9.70	95.636
0.30	7.689	5.00	76.842	9.80	95.886
0.40	10.144	5.10	77.547	9.90	96.435
0.50	12.546	5.20	78.231	10.00	96.385
0.60	14.896	5.30	78.895	10.10	96.635
0.70	17.194	5.50	80.163	10.20	96.886
0.80	19.440	5.60	80.769	10.30	97.138
0.90	21.637	5.70	81.356	10.40	97.392
1.00	23.783	5.80	81.926	10.50	97.648
1.10	25.881	5.90	82.479	10.60	97.906
1.20	27.930	6.00	83.015	10.70	98.167
1.30	29.931	6.10	83.535	10.80	98.431
1.40	31.885	6.20	84.039	10.90	98.669
1.50	33.793	6.30	84.529	11.00	98.970
1.60	35.655	6.40	85.004	11.10	99.246
1.70	37.472	6.50	85.466	11.20	99.527
1.80	39.245	6.60	85.914	11.30	99.813
1.90	40.974	6.70	86.349	11.40	100.104
2.10	44.305	6.80	86.771	11.50	100.402
2.20	45.908	6.90	87.182	11.60	100.707
2.30	47.470	7.00	87.582	11.70	101.018
2.40	48.992	7.10	87.970	11.80	101.338
2.50	50.475	7.20	88.348	11.90	101.666
2.60	51.919	7.30	88.716	12.00	102.003
2.70	53.326	7.40	89.074	12.10	102.349
2.80	54.695	7.50	89.424	12.20	102.705
2.90	56.028	7.60	89.764	12.30	103.073
3.00	57.325	7.70	90.097	12.40	103.452
3.10	58.588	7.80	90.422	12.50	103.843
3.20	59.816	7.90	90.739	12.60	104.247
3.30	61.110	8.00	91.409	12.70	104.665
3.40	62.172	8.10	91.953	12.80	105.097
3.50	63.301	8.20	91.651	12.90	105.545
3.60	64.399	8.30	91.943	13.00	106.009
3.70	65.466	8.40	92.230	13.10	106.490
3.80	66.503	8.50	92.512	13.20	106.990
3.90	67.511	8.60	92.789	13.30	107.508
4.00	68.490	8.70	93.062	13.40	108.046
4.10	69.441	8.90	93.596	13.50	108.606
4.20	70.364	9.00	93.859	13.60	109.187
4.30	71.261	9.10	94.118	13.70	109.792
4.40	72.131	9.20	94.375	13.80	110.422
4.50	72.976	9.30	94.631	13.90	111.077
4.60	73.797	9.40	94.884	14.00	111.759
4.70	74.593	9.50	95.136	14.10	112.469
4.80	75.365	9.60	95.386	14.20	113.209

(*Continued*)

TEMPERATURE = 318.15 K (Continued)

Density ρ (mol/L)	Pressure P (bar)	Density ρ (mol/L)	Pressure P (bar)	Density ρ (mol/L)	Pressure P (bar)
14.30	113.980	17.90	178.346	21.50	413.691
14.40	114.783	18.00	181.735	21.60	424.764
14.50	115.619	18.10	185.251	21.70	436.159
14.60	116.492	18.20	188.897	21.80	447.885
14.70	117.401	18.30	192.678	21.90	459.948
14.80	118.348	18.40	196.598	22.00	472.354
14.90	119.336	18.50	200.661	22.10	485.111
15.00	120.366	18.60	204.871	22.20	498.227
15.10	121.440	18.70	209.233	22.30	511.708
15.20	122.559	18.80	213.751	22.40	525.561
15.30	123.725	18.90	218.430	22.50	539.794
15.40	124.942	19.00	223.275	22.60	554.415
15.50	126.209	19.10	228.291	22.70	569.431
15.60	127.530	19.20	233.481	22.80	584.849
15.70	128.907	19.30	238.852	22.90	600.677
15.80	130.341	19.40	244.408	23.00	616.924
15.90	131.836	19.50	250.154	23.10	633.596
16.00	133.392	19.60	256.096	23.20	650.702
16.10	135.014	19.70	262.239	23.30	668.250
16.20	136.702	19.80	268.588	23.40	686.248
16.30	138.461	19.90	275.149	23.50	704.704
16.40	140.291	20.00	281.927	23.60	723.627
16.50	142.196	20.10	288.928	23.70	743.025
16.60	144.178	20.20	296.158	23.80	762.907
16.70	146.241	20.30	303.623	23.90	783.280
16.80	148.387	20.40	311.327	24.00	804.155
16.90	150.618	20.50	319.278	24.10	825.540
17.00	152.938	20.60	327.482	24.20	847.444
17.10	155.351	20.70	335.943	24.30	869.877
17.20	157.858	20.80	344.669	24.40	892.847
17.30	160.463	20.90	353.666	24.50	916.364
17.40	163.170	21.00	362.940	24.60	940.437
17.50	165.982	21.10	372.497	24.70	965.077
17.60	168.902	21.20	382.345	24.80	990.294
17.70	171.933	21.30	392.488	24.90	1016.096
17.80	175.080	21.40	402.935	25.00	1042.495

TEMPERATURE = 323.15 K

Density ρ (mol/L)	Pressure P (bar)	Density ρ (mol/L)	Pressure P (bar)	Density ρ (mol/L)	Pressure P (bar)
0.20	5.266	4.80	78.537	9.40	102.340
0.30	7.819	4.90	79.372	9.50	102.691
0.40	10.320	5.00	80.186	9.60	103.042
0.50	12.768	5.10	80.978	9.70	103.392
0.60	15.166	5.20	81.750	9.80	103.743
0.70	17.513	5.30	82.502	9.90	104.094
0.80	19.810	5.40	83.235	10.00	104.447
0.90	22.057	5.50	83.948	10.10	104.801
1.00	24.256	5.60	84.643	10.20	105.157
1.10	26.407	5.70	85.320	10.30	105.515
1.20	28.511	5.80	85.980	10.40	105.876
1.30	30.568	5.90	86.624	10.50	106.240
1.40	32.579	6.00	87.251	10.60	106.608
1.50	34.544	6.10	87.862	10.70	106.980
1.60	36.465	6.20	88.458	10.80	107.357
1.70	38.342	6.30	89.040	10.90	107.738
1.80	40.176	6.40	89.607	11.00	108.125
1.90	41.967	6.50	90.161	11.10	108.519
2.00	43.716	6.60	90.702	11.20	108.918
2.10	45.424	6.70	91.230	11.30	109.325
2.20	47.092	6.80	91.746	11.40	109.740
2.30	48.720	6.90	92.250	11.50	110.162
2.40	50.309	7.00	92.743	11.60	110.593
2.50	51.860	7.10	93.225	11.70	111.034
2.60	53.373	7.20	93.697	11.80	111.485
2.70	54.849	7.30	94.159	11.90	111.946
2.80	56.289	7.40	94.612	12.00	112.419
2.90	57.693	7.50	95.055	12.10	112.903
3.00	59.063	7.60	95.491	12.20	113.400
3.10	60.398	7.70	95.918	12.30	113.911
3.20	61.700	7.80	96.337	12.40	114.436
3.30	62.970	7.90	96.749	12.50	114.975
3.40	64.207	8.00	97.155	12.60	115.530
3.50	65.413	8.10	97.554	12.70	116.101
3.60	66.588	8.20	97.947	12.80	116.690
3.70	67.733	8.30	98.334	12.90	117.297
3.80	68.849	8.40	98.716	13.00	117.923
3.90	69.936	8.50	99.094	13.10	118.569
4.00	70.996	8.60	99.467	13.20	119.236
4.10	72.027	8.70	99.836	13.30	119.924
4.20	73.032	8.80	100.201	13.40	120.636
4.30	74.011	8.90	100.564	13.50	121.372
4.40	74.965	9.00	100.923	13.60	122.133
4.50	75.894	9.10	101.280	13.70	122.920
4.60	76.798	9.20	101.635	13.80	123.735
4.70	77.679	9.30	101.988	13.90	124.578

(Continued)

TEMPERATURE = 323.15 K (Continued)

Density ρ (mol/L)	Pressure P (bar)	Density ρ (mol/L)	Pressure P (bar)	Density ρ (mol/L)	Pressure P (bar)
14.00	125.452	17.70	195.213	21.40	442.025
14.10	126.356	17.80	198.699	21.50	453.300
14.20	127.294	17.90	202.309	21.60	464.895
14.30	128.265	18.00	206.046	21.70	476.819
14.40	129.272	18.10	209.916	21.80	489.077
14.50	130.316	18.20	213.920	21.90	501.676
14.60	131.399	18.30	218.065	22.00	514.624
14.70	132.521	18.40	222.353	22.10	527.928
14.80	133.686	18.50	226.789	22.20	541.595
14.90	134.894	18.60	231.378	22.30	555.632
15.00	136.148	18.70	236.123	22.40	570.046
15.10	137.448	18.80	241.030	22.50	584.846
15.20	138.798	18.90	246.102	22.60	600.037
15.30	140.199	19.00	251.345	22.70	615.629
15.40	141.652	19.10	256.764	22.80	631.628
15.50	143.161	19.20	262.363	22.90	648.043
15.60	144.727	19.30	268.147	23.00	664.880
15.70	146.352	19.40	274.121	23.10	682.149
15.80	148.039	19.50	280.290	23.20	699.857
15.90	149.789	19.60	286.660	23.30	718.013
16.00	151.606	19.70	293.236	23.40	736.623
16.10	153.492	19.80	300.023	23.50	755.698
16.20	155.448	19.90	307.027	23.60	775.246
16.30	157.478	20.00	314.253	23.70	795.274
16.40	159.584	20.10	321.706	23.80	815.791
16.50	161.770	20.20	329.393	23.90	836.808
16.60	164.036	20.30	337.320	24.00	858.331
16.70	166.388	20.40	345.491	24.10	880.371
16.80	168.826	20.50	353.914	24.20	902.937
16.90	171.355	20.60	362.593	24.30	926.037
17.00	173.977	20.70	371.536	24.40	949.682
17.10	176.696	20.80	380.747	24.50	973.881
17.20	179.514	20.90	390.235	24.60	998.644
17.30	182.435	21.00	400.004	24.70	1023.980
17.40	185.461	21.10	410.061	24.80	1049.901
17.50	188.598	21.20	420.412	24.90	1076.415
17.60	191.847	21.30	431.065	25.00	1103.533

TEMPERATURE = 328.15 K

Density ρ (mol/L)	Pressure P (bar)	Density ρ (mol/L)	Pressure P (bar)	Density ρ (mol/L)	Pressure P (bar)
0.20	5.351	4.80	81.670	9.40	109.781
0.30	7.949	4.90	82.590	9.50	110.235
0.40	10.495	5.00	83.489	9.60	110.690
0.50	12.990	5.10	84.368	9.70	111.145
0.60	15.435	5.20	85.226	9.80	111.602
0.70	17.831	5.30	86.064	9.90	112.059
0.80	20.178	5.40	86.884	10.00	112.519
0.90	22.477	5.50	87.685	10.10	112.982
1.00	24.728	5.60	88.468	10.20	113.447
1.10	26.932	5.70	89.234	10.30	113.916
1.20	29.089	5.80	89.984	10.40	114.389
1.30	31.201	5.90	90.716	10.50	114.867
1.40	33.269	6.00	91.434	10.60	115.349
1.50	35.291	6.10	92.135	10.70	115.837
1.60	37.271	6.20	92.822	10.80	116.331
1.70	39.207	6.30	93.495	10.90	116.831
1.80	41.101	6.40	94.154	11.00	117.339
1.90	72.953	6.50	94.800	11.10	117.854
2.00	44.765	6.60	95.433	11.20	118.378
2.10	46.536	6.70	96.053	11.30	118.910
2.20	48.268	6.80	96.662	11.40	119.452
2.30	49.961	6.90	97.259	11.50	120.004
2.40	51.616	7.00	97.846	11.60	120.566
2.50	53.234	7.10	98.421	11.70	121.141
2.60	54.815	7.20	98.987	11.80	121.727
2.70	56.360	7.30	99.544	11.90	122.326
2.80	57.869	7.40	100.091	12.00	122.938
2.90	59.344	7.50	100.629	12.10	123.564
3.00	60.785	7.60	101.160	12.20	124.206
3.10	62.192	7.70	101.682	12.30	124.863
3.20	63.567	7.80	102.197	12.40	125.536
3.30	64.911	7.90	102.706	12.50	126.227
3.40	66.222	8.00	103.207	12.60	126.396
3.50	67.504	8.10	103.703	12.70	127.664
3.60	68.755	8.20	104.193	12.80	128.411
3.70	69.977	8.30	104.677	12.90	129.180
3.80	71.170	8.40	105.157	13.00	129.970
3.90	72.336	8.50	105.633	13.10	130.782
4.00	73.474	8.60	106.104	13.20	131.619
4.10	74.585	8.70	106.572	13.30	132.480
4.20	75.671	8.80	107.036	13.40	133.366
4.30	76.731	8.90	107.498	13.50	134.280
4.40	77.766	9.00	107.958	13.60	135.221
4.50	78.777	9.10	108.415	13.70	136.192
4.60	79.764	9.20	108.872	13.80	137.193
4.70	80.728	9.30	109.327	13.90	138.225

(*Continued*)

TEMPERATURE = 328.15 K (Continued)

Density ρ (mol/L)	Pressure P (bar)	Density ρ (mol/L)	Pressure P (bar)	Density ρ (mol/L)	Pressure P (bar)
14.00	139.291	17.70	218.605	21.40	481.052
14.10	140.390	17.80	222.428	21.50	492.839
14.20	141.526	17.90	226.379	21.60	504.951
14.30	142.698	18.00	230.462	21.70	517.397
14.40	142.909	18.10	234.682	21.80	530.181
14.50	145.160	18.20	239.042	21.90	543.312
14.60	146.453	18.30	243.546	22.00	556.797
14.70	147.789	18.40	248.199	22.10	570.642
14.80	149.171	18.50	253.005	22.20	584.855
14.90	150.599	18.60	257.696	22.30	599.442
15.00	152.076	18.70	263.093	22.40	614.413
15.10	153.603	18.80	268.384	22.50	629.773
15.20	155.183	18.90	273.846	22.60	645.530
15.30	156.817	19.00	279.483	22.70	661.693
15.40	158.508	19.10	285.301	22.80	678.268
15.50	160.257	19.20	291.303	22.90	695.264
15.60	162.067	19.30	297.496	23.00	712.689
15.70	163.940	19.40	303.883	23.10	730.550
15.80	165.878	19.50	310.471	23.20	748.856
15.90	167.884	19.60	317.264	23.30	767.615
16.00	169.960	19.70	324.267	23.40	786.835
16.10	172.108	19.80	331.487	23.50	806.525
16.20	174.332	19.90	338.928	23.60	826.693
16.30	176.633	20.00	346.596	23.70	847.348
16.40	179.014	20.10	354.496	23.80	868.499
16.50	181.478	20.20	362.635	23.90	890.155
16.60	184.028	20.30	371.017	24.00	912.324
16.70	186.666	20.40	379.650	24.10	935.017
16.80	189.396	20.50	388.538	24.20	958.241
16.90	192.221	20.60	397.688	24.30	982.008
17.00	195.143	20.70	407.106	24.40	1006.326
17.10	198.166	20.80	416.797	24.50	1031.205
17.20	201.293	20.90	426.769	24.60	1056.655
17.30	204.527	21.00	437.027	24.70	1082.686
17.40	207.872	21.10	447.578	24.80	1109.308
17.50	211.331	21.20	458.428	24.90	1136.533
17.60	214.907	21.30	469.584	25.00	1164.369

TEMPERATURE = 333.15 K

Density ρ (mol/L)	Pressure P (bar)	Density ρ (mol/L)	Pressure P (bar)	Density ρ (mol/L)	Pressure P (bar)
0.20	5.437	4.80	84.771	9.40	117.211
0.30	8.078	4.90	85.775	9.50	117.772
0.40	10.670	5.00	86.758	9.60	118.333
0.50	13.212	5.10	87.720	9.70	118.897
0.60	15.704	5.20	88.664	9.80	119.462
0.70	18.149	5.30	89.588	9.90	120.030
0.80	20.545	5.40	90.494	10.00	120.601
0.90	22.895	5.50	91.382	10.10	121.175
1.00	25.197	5.60	92.253	10.20	121.754
1.10	27.454	5.70	93.106	10.30	122.337
1.20	29.666	5.80	93.944	10.40	122.925
1.30	31.832	5.90	94.765	10.50	123.519
1.40	33.955	6.00	95.572	10.60	124.120
1.50	36.035	6.10	96.363	10.70	124.727
1.60	38.072	6.20	97.140	10.80	125.342
1.70	40.067	6.30	97.903	10.90	125.965
1.80	42.021	6.40	98.653	11.00	126.596
1.90	43.934	6.50	99.391	11.10	127.237
2.00	45.807	6.60	100.115	11.20	127.888
2.10	47.641	6.70	100.828	11.30	128.549
2.20	49.437	6.80	101.529	11.40	129.222
2.30	51.194	6.90	102.220	11.50	129.906
2.40	52.914	7.00	102.900	11.60	130.603
2.50	54.598	7.10	103.569	11.70	131.314
2.60	56.246	7.20	104.229	11.80	132.038
2.70	57.859	7.30	104.880	11.90	132.777
2.80	59.437	7.40	105.523	12.00	133.532
2.90	60.982	7.50	106.157	12.10	134.303
3.00	62.493	7.60	106.783	12.20	135.091
3.10	63.972	7.70	107.402	12.30	135.897
3.20	65.419	7.80	108.013	12.40	136.721
3.30	66.835	7.90	108.619	12.50	137.566
3.40	68.220	8.00	109.218	12.60	138.430
3.50	69.575	8.10	109.811	12.70	139.316
3.60	70.902	8.20	110.400	12.80	140.224
3.70	72.200	8.30	110.983	12.90	141.156
3.80	73.469	8.40	111.562	13.00	142.111
3.90	74.712	8.50	112.137	13.10	143.092
4.00	75.928	8.60	112.709	13.20	144.099
4.10	77.118	8.70	113.278	13.30	145.133
4.20	78.283	8.80	113.844	13.40	146.196
4.30	79.422	8.90	114.408	13.50	147.288
4.40	80.538	9.00	114.970	13.60	148.411
4.50	81.630	9.10	115.531	13.70	149.566
4.60	82.699	9.20	116.091	13.80	150.753
4.70	83.746	9.30	116.651	13.90	151.976

(Continued)

TEMPERATURE = 333.15 K (Continued)

Density ρ (mol/L)	Pressure P (bar)	Density ρ (mol/L)	Pressure P (bar)	Density ρ (mol/L)	Pressure P (bar)
14.00	153.234	17.70	242.085	21.40	520.006
14.10	154.529	17.80	246.242	21.50	532.301
14.20	155.863	17.90	250.531	21.60	544.926
14.30	157.237	18.00	254.958	21.70	557.888
14.40	158.652	18.10	259.526	21.80	571.194
14.50	160.111	18.20	264.238	21.90	584.851
14.60	161.614	18.30	269.100	22.00	598.867
14.70	163.164	18.40	274.115	22.10	613.248
14.80	164.763	18.50	279.288	22.20	628.002
14.90	166.411	18.60	284.622	22.30	643.136
15.00	168.111	18.70	290.123	22.40	658.658
15.10	169.866	18.80	295.795	22.50	674.574
15.20	171.676	18.90	301.642	22.60	690.894
15.30	173.543	19.00	307.669	22.70	707.623
15.40	175.471	19.10	313.881	22.80	724.771
15.50	177.461	19.20	320.283	22.90	742.344
15.60	179.515	19.30	326.879	23.00	760.352
15.70	181.636	19.40	333.676	23.10	778.802
15.80	183.825	19.50	340.677	23.20	797.702
15.90	186.086	19.60	347.888	23.30	814.061
16.00	188.421	19.70	355.315	23.40	836.887
16.10	190.832	19.80	362.962	23.50	857.189
16.20	193.321	19.90	370.836	23.60	877.975
16.30	195.892	20.00	378.941	23.70	899.254
16.40	198.548	20.10	387.283	23.80	921.036
16.50	201.290	20.20	395.868	23.90	943.328
16.60	204.122	20.30	404.702	24.00	966.141
16.70	207.047	20.40	413.790	24.10	989.484
16.80	210.068	20.50	423.139	24.20	1013.365
16.90	213.187	20.60	432.754	24.30	1037.796
17.00	216.409	20.70	442.641	24.40	1062.785
17.10	219.735	20.80	452.807	24.50	1088.342
17.20	223.169	20.90	463.258	24.60	1114.478
17.30	226.715	21.00	474.000	24.70	1141.202
17.40	230.376	21.10	485.039	24.80	1168.526
17.50	234.156	21.20	496.382	24.90	1196.459
17.60	238.058	21.30	508.036	25.00	1225.013

TEMPERATURE = 338.15 K

Density ρ (mol/L)	Pressure P (bar)	Density ρ (mol/L)	Pressure P (bar)	Density ρ (mol/L)	Pressure P (bar)
0.20	5.522	4.80	87.842	9.40	124.630
0.30	8.208	4.90	88.929	9.50	125.300
0.40	10.845	5.00	89.995	9.60	125.971
0.50	13.433	5.10	91.042	9.70	126.645
0.60	15.973	5.20	92.069	9.80	127.322
0.70	18.466	5.30	93.078	9.90	128.002
0.80	20.912	5.40	94.070	10.00	128.687
0.90	23.311	5.50	95.044	10.10	129.376
1.00	25.665	5.60	96.001	10.20	130.070
1.10	27.975	5.70	96.942	10.30	130.771
1.20	30.240	5.80	97.867	10.40	131.477
1.30	32.461	5.90	98.776	10.50	132.191
1.40	34.639	6.00	99.671	10.60	132.912
1.50	36.775	6.10	100.552	10.70	133.642
1.60	38.870	6.20	101.418	10.80	134.380
1.70	40.923	6.30	102.272	10.90	135.128
1.80	42.936	6.40	103.112	11.00	135.886
1.90	44.910	6.50	103.941	11.10	136.655
2.00	46.844	6.60	104.757	11.20	137.436
2.10	48.740	6.70	105.562	11.30	138.229
2.20	50.598	6.80	106.356	11.40	139.034
2.30	52.419	6.90	107.139	11.50	139.854
2.40	54.204	7.00	107.912	11.60	140.687
2.50	55.954	7.10	108.676	11.70	141.536
2.60	57.668	7.20	109.431	11.80	142.104
2.70	59.348	7.30	110.177	11.90	143.282
2.80	60.994	7.40	110.914	12.00	144.181
2.90	62.607	7.50	111.644	12.10	145.099
3.00	64.188	7.60	112.367	12.20	146.035
3.10	65.737	7.70	113.082	12.30	146.992
3.20	67.256	7.80	113.792	12.40	147.969
3.30	68.743	7.90	114.495	12.50	148.968
3.40	70.201	8.00	115.193	12.60	149.990
3.50	71.630	8.10	115.885	12.70	151.036
3.60	73.031	8.20	116.573	12.80	152.106
3.70	74.404	8.30	117.257	12.90	153.201
3.80	75.749	8.40	117.937	13.00	154.324
3.90	77.068	8.50	118.613	13.10	155.473
4.00	78.361	8.60	119.287	13.20	156.652
4.10	79.628	8.70	119.958	13.30	157.860
4.20	80.871	8.80	120.628	13.40	159.100
4.30	82.090	8.90	121.296	13.50	160.371
4.40	83.285	9.00	121.963	13.60	161.676
4.50	84.458	9.10	122.629	13.70	163.016
4.60	85.607	9.20	123.296	13.80	164.391
4.70	86.736	9.30	123.963	13.90	165.804

(*Continued*)

TEMPERATURE = 338.15 K (Continued)

Density ρ (mol/L)	Pressure P (bar)	Density ρ (mol/L)	Pressure P (bar)	Density ρ (mol/L)	Pressure P (bar)
14.00	167.255	17.70	265.632	21.40	558.884
14.10	168.746	17.80	270.121	21.50	571.681
14.20	170.279	17.90	274.748	21.60	584.813
14.30	171.855	18.00	279.515	21.70	598.287
14.40	173.476	18.10	284.429	21.80	612.111
14.50	175.142	18.20	289.491	21.90	626.290
14.60	176.857	18.30	294.707	22.00	640.833
14.70	178.621	18.40	300.082	22.10	655.746
14.80	180.437	18.50	305.618	22.20	671.037
14.90	182.306	18.60	311.320	22.30	686.713
15.00	184.230	18.70	317.194	22.40	702.782
15.10	186.211	18.80	323.242	22.50	719.251
15.20	188.252	18.90	329.471	22.60	736.128
15.30	190.353	19.00	335.885	22.70	753.421
15.40	192.518	19.10	342.488	22.80	771.137
15.50	194.749	19.20	349.286	22.90	789.285
15.60	197.047	19.30	356.283	23.00	807.872
15.70	199.415	19.40	363.484	23.10	826.907
15.80	201.856	19.50	370.895	23.20	846.399
15.90	204.372	19.60	378.520	23.30	866.355
16.00	206.965	19.70	386.365	23.40	886.784
16.10	209.639	19.80	394.436	23.50	907.695
16.20	212.394	19.90	402.737	23.60	929.096
16.30	215.236	20.00	411.275	23.70	950.997
16.40	218.165	20.10	420.054	23.80	973.407
16.50	221.185	20.20	429.081	23.90	996.334
16.60	224.299	20.30	438.362	24.00	1019.788
16.70	227.509	20.40	447.901	24.10	1043.779
16.80	230.820	20.50	457.705	24.20	1068.315
16.90	234.233	20.60	467.781	24.30	1093.408
17.00	237.752	20.70	478.133	24.40	1119.066
17.10	241.380	20.80	488.769	24.50	1145.301
17.20	245.121	20.90	499.694	24.60	1172.121
17.30	248.978	21.00	510.914	24.70	1199.537
17.40	252.954	21.10	522.437	24.80	1227.561
17.50	257.052	21.20	534.269	24.90	1256.202
17.60	261.277	21.30	546.416	25.00	1285.472

TEMPERATURE = 343.15 K

Density ρ (mol/L)	Pressure P (bar)	Density ρ (mol/L)	Pressure P (bar)	Density ρ (mol/L)	Pressure P (bar)
0.20	5.607	4.80	90.888	9.40	132.038
0.30	8.337	4.90	92.056	9.50	132.819
0.40	11.019	5.00	93.205	9.60	133.602
0.50	13.654	5.10	94.335	9.70	134.388
0.60	16.241	5.20	95.446	9.80	135.179
0.70	18.782	5.30	96.539	9.90	135.974
0.80	21.277	5.40	97.615	10.00	136.774
0.90	23.727	5.50	98.675	10.10	137.580
1.00	26.132	5.60	99.718	10.20	138.393
1.10	28.494	5.70	100.745	10.30	139.212
1.20	30.812	5.80	101.757	10.40	140.039
1.30	33.087	5.90	102.754	10.50	140.875
1.40	35.321	6.00	103.737	10.60	141.719
1.50	37.513	6.10	104.706	10.70	142.573
1.60	39.664	6.20	105.662	10.80	143.437
1.70	41.775	6.30	106.605	10.90	144.313
1.80	43.847	6.40	107.536	11.00	145.199
1.90	45.880	6.50	108.445	11.10	146.099
2.00	47.875	6.60	109.363	11.20	147.011
2.10	49.833	6.70	110.259	11.30	147.937
2.20	51.753	6.80	111.146	11.40	148.878
2.30	53.638	6.90	112.022	11.50	149.834
2.40	55.487	7.00	112.889	11.60	150.806
2.50	57.301	7.10	113.747	11.70	151.795
2.60	59.081	7.20	114.596	11.80	152.802
2.70	60.827	7.30	115.437	11.90	153.828
2.80	62.541	7.40	116.271	12.00	154.873
2.90	64.222	7.50	117.097	12.10	155.938
3.00	65.872	7.60	117.916	12.20	157.024
3.10	67.491	7.70	118.730	12.30	158.133
3.20	69.080	7.80	119.537	12.40	159.264
3.30	70.639	7.90	120.339	12.50	160.419
3.40	72.169	8.00	121.136	12.60	161.600
3.50	73.670	8.10	121.928	12.70	162.806
3.60	75.144	8.20	122.717	12.80	164.039
3.70	76.591	8.30	123.502	12.90	165.300
3.80	78.011	8.40	124.284	13.00	166.589
3.90	79.405	8.50	125.063	13.10	167.909
4.00	80.774	8.60	125.840	13.20	169.261
4.10	82.119	8.70	126.616	13.30	170.644
4.20	83.439	8.80	127.390	13.40	172.061
4.30	84.736	8.90	128.163	13.50	173.513
4.40	86.010	9.00	128.937	13.60	175.000
4.50	87.261	9.10	129.710	13.70	176.525
4.60	88.491	9.20	130.485	13.80	178.089
4.70	89.700	9.30	131.261	13.90	179.693

(Continued)

TEMPERATURE = 343.15 K (Continued)

Density ρ (mol/L)	Pressure P (bar)	Density ρ (mol/L)	Pressure P (bar)	Density ρ (mol/L)	Pressure P (bar)
14.00	181.338	17.70	289.231	21.40	597.680
14.10	183.026	17.80	294.050	21.50	610.975
14.20	184.758	17.90	299.011	21.60	624.611
14.30	186.536	18.00	304.118	21.70	638.593
14.40	188.362	18.10	309.374	21.80	652.929
14.50	190.238	18.20	314.784	21.90	667.627
14.60	192.164	18.30	320.352	22.00	682.693
14.70	194.143	18.40	326.083	22.10	698.134
14.80	196.176	18.50	331.980	22.20	713.958
14.90	198.266	18.60	338.047	22.30	730.173
15.00	200.414	18.70	344.290	22.40	746.785
15.10	202.623	18.80	350.713	22.50	763.803
15.20	204.894	18.90	357.321	22.60	781.235
15.30	207.229	19.00	364.118	22.70	799.087
15.40	209.632	19.10	371.108	22.80	817.369
15.50	212.103	19.20	378.298	22.90	836.088
15.60	214.646	19.30	385.692	23.00	855.252
15.70	217.262	19.40	393.294	23.10	874.869
15.80	219.954	19.50	401.110	23.20	894.949
15.90	222.725	19.60	409.146	23.30	915.500
16.00	225.577	19.70	417.406	23.40	936.529
16.10	228.512	19.80	425.896	23.50	958.047
16.20	231.534	19.90	434.621	23.60	980.061
16.30	234.645	20.00	443.587	23.70	1002.582
16.40	237.848	20.10	452.800	23.80	1025.617
16.50	241.145	20.20	462.264	23.90	1049.177
16.60	244.540	20.30	471.987	24.00	1073.271
16.70	248.036	20.40	481.973	24.10	1097.908
16.80	251.635	20.50	492.229	24.20	1123.098
16.90	255.341	20.60	502.760	24.30	1148.851
17.00	259.157	20.70	513.573	24.40	1175.178
17.10	263.086	20.80	524.674	24.50	1202.088
17.20	267.132	20.90	536.069	24.60	1229.591
17.30	271.298	21.00	547.764	24.70	1257.699
17.40	275.587	21.10	559.767	24.80	1286.421
17.50	280.003	21.20	572.082	24.90	1315.770
17.60	284.550	21.30	584.718	25.00	1345.755

TEMPERATURE = 348.15 K

Density ρ (mol/L)	Pressure P (bar)	Density ρ (mol/L)	Pressure P (bar)	Density ρ (mol/L)	Pressure P (bar)
0.20	5.692	4.80	93.911	9.40	139.434
0.30	8.467	4.90	95.160	9.50	140.326
0.40	11.194	5.00	96.391	9.60	141.223
0.50	13.874	5.10	97.603	9.70	142.123
0.60	16.508	5.20	98.797	9.80	143.029
0.70	19.097	5.30	99.974	9.90	143.941
0.80	21.642	5.40	101.134	10.00	144.859
0.90	24.142	5.50	102.278	10.10	145.784
1.00	26.598	5.60	103.406	10.20	146.716
1.10	29.011	5.70	104.519	10.30	147.657
1.20	31.382	5.80	105.618	10.40	148.606
1.30	33.712	5.90	106.702	10.50	149.565
1.40	36.000	6.00	107.773	10.60	150.534
1.50	38.247	6.10	108.830	10.70	151.514
1.60	40.455	6.20	109.875	10.80	152.506
1.70	42.624	6.30	110.907	10.90	153.510
1.80	44.754	6.40	111.928	11.00	154.528
1.90	46.847	6.50	112.938	11.10	155.559
2.00	48.902	6.60	113.936	11.20	156.605
2.10	50.920	6.70	114.925	11.30	157.666
2.20	52.902	6.80	115.904	11.40	158.743
2.30	54.850	6.90	116.873	11.50	159.838
2.40	56.762	7.00	117.833	11.60	160.950
2.50	58.641	7.10	118.785	11.70	162.081
2.60	60.486	7.20	119.729	11.80	163.231
2.70	62.298	7.30	120.666	11.90	164.402
2.80	64.078	7.40	121.595	12.00	165.595
2.90	65.827	7.50	122.518	12.10	166.809
3.00	67.545	7.60	123.435	12.20	168.047
3.10	69.233	7.70	124.346	12.30	169.308
3.20	70.892	7.80	125.252	12.40	170.595
3.30	72.521	7.90	126.154	12.50	171.908
3.40	74.122	8.00	127.051	12.60	173.247
3.50	75.696	8.10	127.944	12.70	174.615
3.60	77.243	8.20	128.834	12.80	176.012
3.70	78.763	8.30	129.721	12.90	177.439
3.80	80.257	8.40	130.606	13.00	178.898
3.90	81.726	8.50	131.489	13.10	180.389
4.00	83.171	8.60	132.370	13.20	181.913
4.10	84.591	8.70	133.251	13.30	183.472
4.20	85.988	8.80	134.131	13.40	185.068
4.30	87.362	8.90	135.011	13.50	186.700
4.40	88.714	9.00	135.893	13.60	188.371
4.50	90.045	9.10	136.775	13.70	190.083
4.60	91.354	9.20	137.659	13.80	191.835
4.70	92.642	9.30	138.545	13.90	193.630

(Continued)

TEMPERATURE = 348.15 K (Continued)

Density ρ (mol/L)	Pressure P (bar)	Density ρ (mol/L)	Pressure P (bar)	Density ρ (mol/L)	Pressure P (bar)
14.00	195.470	17.70	312.867	21.40	639.390
14.10	197.355	17.80	318.014	21.50	650.180
14.20	199.287	17.90	323.307	21.60	664.315
14.30	201.269	18.00	328.751	21.70	678.802
14.40	203.300	18.10	334.348	21.80	693.648
14.50	205.384	18.20	340.104	21.90	708.860
14.60	207.522	18.30	346.021	22.00	724.445
14.70	209.716	18.40	352.105	22.10	740.411
14.80	211.968	18.50	358.360	22.20	756.765
14.90	214.279	18.60	364.791	22.30	773.515
15.00	216.652	18.70	371.401	22.40	790.668
15.10	219.088	18.80	378.195	22.50	808.232
15.20	221.590	18.90	385.178	22.60	826.215
15.30	224.160	19.00	392.355	22.70	844.624
15.40	226.800	19.10	399.730	22.80	863.468
15.50	229.512	19.20	407.308	22.90	882.755
15.60	232.299	19.30	415.095	23.00	902.493
15.70	235.163	19.40	423.095	23.10	922.691
15.80	238.106	19.50	431.314	23.20	943.357
15.90	241.132	19.60	439.756	23.30	964.499
16.00	244.242	19.70	448.427	23.40	986.127
16.10	247.439	19.80	457.333	23.50	1008.249
16.20	250.727	19.90	466.478	23.60	1030.875
16.30	254.107	20.00	475.869	23.70	1054.012
16.40	257.582	20.10	485.511	23.80	1077.672
16.50	261.156	20.20	495.409	23.90	1101.863
16.60	264.832	20.30	505.570	24.00	1126.595
16.70	268.612	20.40	515.999	24.10	1151.877
16.80	272.499	20.50	526.702	24.20	1177.719
16.90	276.497	20.60	537.685	24.30	1204.132
17.00	280.609	20.70	548.955	24.40	1231.125
17.10	284.838	20.80	560.517	24.50	1258.710
17.20	289.188	20.90	572.378	24.60	1286.896
17.30	293.662	21.00	584.545	24.70	1315.693
17.40	298.263	21.10	597.022	24.80	1345.114
17.50	302.995	21.20	609.818	24.90	1375.169
17.60	307.862	21.30	622.938	25.00	1405.870

TEMPERATURE = 353.15 K

Density ρ (mol/L)	Pressure P (bar)	Density ρ (mol/L)	Pressure P (bar)	Density ρ (mol/L)	Pressure P (bar)
0.20	5.777	4.80	96.913	9.40	146.816
0.30	8.596	4.90	98.243	9.50	147.822
0.40	11.368	5.00	99.554	9.60	148.832
0.50	14.094	5.10	100.848	9.70	149.849
0.60	16.776	5.20	102.125	9.80	150.872
0.70	19.412	5.30	103.385	9.90	151.901
0.80	22.005	5.40	104.628	10.00	152.938
0.90	24.555	5.50	105.857	10.10	153.983
1.00	27.062	5.60	107.070	10.20	155.036
1.10	29.527	5.70	108.268	10.30	156.100
1.20	31.951	5.80	109.453	10.40	157.716
1.30	34.334	5.90	110.624	10.50	158.257
1.40	36.676	6.00	111.781	10.60	159.353
1.50	38.979	6.10	112.927	10.70	160.460
1.60	41.244	6.20	114.060	10.80	161.581
1.70	43.469	6.30	115.182	10.90	162.716
1.80	45.658	6.40	116.292	11.00	163.865
1.90	47.809	6.50	117.392	11.10	165.030
2.00	49.923	6.60	118.482	11.20	166.211
2.10	52.002	6.70	119.562	11.30	167.408
2.20	54.046	6.80	120.633	11.40	168.624
2.30	56.056	6.90	121.695	11.50	169.858
2.40	58.031	7.00	122.749	11.60	171.112
2.50	59.973	7.10	123.795	11.70	172.386
2.60	61.883	7.20	124.834	11.80	173.681
2.70	63.761	7.30	125.866	11.90	174.999
2.80	65.607	7.40	126.892	12.00	176.340
2.90	67.423	7.50	127.911	12.10	177.704
3.00	69.209	7.60	128.926	12.20	179.094
3.10	70.966	7.70	129.935	12.30	180.510
3.20	72.693	7.80	130.940	12.40	181.953
3.30	74.393	7.90	131.941	12.50	183.424
3.40	76.064	8.00	132.939	12.60	184.925
3.50	77.709	8.10	133.933	12.70	186.455
3.60	79.328	8.20	134.925	12.80	188.017
3.70	80.921	8.30	135.915	12.90	189.611
3.80	82.488	8.40	136.904	13.00	191.239
3.90	84.032	8.50	137.891	13.10	192.902
4.00	85.551	8.60	138.878	13.20	194.600
4.10	87.047	8.70	139.864	13.30	196.336
4.20	88.520	8.80	140.852	13.40	198.111
4.30	89.971	8.90	141.840	13.50	199.925
4.40	91.400	9.00	142.830	13.60	201.780
4.50	92.809	9.10	143.822	13.70	203.678
4.60	94.196	9.20	144.816	13.80	205.620
4.70	95.564	9.30	145.814	13.90	207.607

(Continued)

TEMPERATURE = 353.15 K (Continued)

Density ρ (mol/L)	Pressure P (bar)	Density ρ (mol/L)	Pressure P (bar)	Density ρ (mol/L)	Pressure P (bar)
14.00	209.642	17.70	336.528	21.40	675.013
14.10	211.724	17.80	342.001	21.50	689.294
14.20	213.857	17.90	347.625	21.60	703.925
14.30	216.042	18.00	353.403	21.70	718.913
14.40	218.280	18.10	359.340	21.80	734.265
14.50	220.573	18.20	365.438	21.90	749.989
14.60	222.924	18.30	371.703	22.00	766.090
14.70	225.333	18.40	378.139	22.10	782.578
14.80	227.803	18.50	384.749	22.20	799.459
14.90	230.335	18.60	391.539	22.30	816.741
15.00	232.933	18.70	398.513	22.40	834.431
15.10	235.597	18.80	405.676	22.50	852.538
15.20	238.330	18.90	413.032	22.60	871.070
15.30	241.134	19.00	420.586	22.70	890.033
15.40	244.011	19.10	428.342	22.80	909.437
15.50	246.964	19.20	436.306	22.90	929.290
15.60	249.996	19.30	444.483	23.00	949.600
15.70	253.107	19.40	452.878	23.10	970.375
15.80	256.302	19.50	461.496	23.20	991.625
15.90	259.582	19.60	470.341	23.30	1013.357
16.00	262.950	19.70	479.420	23.40	1035.581
16.10	266.409	19.80	488.738	23.50	1058.306
16.20	269.961	19.90	498.300	23.60	1081.541
16.30	273.610	20.00	508.112	23.70	1105.295
16.40	277.358	20.10	518.180	23.80	1129.577
16.50	281.208	20.20	528.508	23.90	1154.397
16.60	285.163	20.30	539.104	24.00	1179.765
16.70	289.277	20.40	549.972	24.10	1205.691
16.80	293.401	20.50	561.119	24.20	1232.185
16.90	297.690	20.60	572.551	24.30	1259.256
17.00	302.097	20.70	584.274	24.40	1286.915
17.10	306.625	20.80	596.294	24.50	1315.173
17.20	311.277	20.90	608.617	24.60	1344.040
17.30	316.057	21.00	621.251	24.70	1373.528
17.40	320.968	21.10	634.200	24.80	1403.646
17.50	326.015	21.20	647.473	24.90	1434.407
17.60	331.200	21.30	661.075	25.00	1465.822

TEMPERATURE = 363.15 K

Density ρ (mol/L)	Pressure P (bar)	Density ρ (mol/L)	Pressure P (bar)	Density ρ (mol/L)	Pressure P (bar)
0.20	5.947	4.80	102.862	9.40	161.534
0.30	8.854	4.90	104.351	9.50	162.770
0.40	11.716	5.00	105.823	9.60	164.013
0.50	14.534	5.10	107.279	9.70	165.264
0.60	17.309	5.20	108.719	9.80	166.523
0.70	20.041	5.30	110.144	9.90	167.792
0.80	22.731	5.40	111.553	10.00	169.070
0.90	25.380	5.50	112.949	10.10	170.358
1.00	27.988	5.60	114.330	10.20	171.658
1.10	30.556	5.70	115.698	10.30	172.970
1.20	33.084	5.80	117.054	10.40	174.294
1.30	35.573	5.90	118.397	10.50	175.632
1.40	38.024	6.00	119.728	10.60	176.984
1.50	40.436	6.10	121.048	10.70	178.351
1.60	42.812	6.20	122.357	10.80	179.734
1.70	45.151	6.30	123.656	10.90	181.133
1.80	47.454	6.40	124.945	11.00	182.550
1.90	49.722	6.50	126.225	11.10	183.985
2.00	51.955	6.60	127.497	11.20	185.439
2.10	54.154	6.70	128.759	11.30	186.913
2.20	56.319	6.80	130.014	11.40	188.409
2.30	58.452	6.90	131.262	11.50	189.926
2.40	60.552	7.00	132.503	11.60	191.466
2.50	62.620	7.10	133.737	11.70	193.030
2.60	64.658	7.20	134.966	11.80	194.618
2.70	66.665	7.30	136.189	11.90	196.233
2.80	68.643	7.40	137.408	12.00	197.874
2.90	70.591	7.50	138.622	12.10	199.542
3.00	72.510	7.60	139.832	12.20	201.240
3.10	74.402	7.70	141.038	12.30	202.968
3.20	76.267	7.80	142.242	12.40	204.727
3.30	78.104	7.90	143.444	12.50	206.517
3.40	79.916	8.00	144.643	12.60	208.341
3.50	81.702	8.10	145.841	12.70	210.200
3.60	83.463	8.20	147.039	12.80	212.094
3.70	85.200	8.30	148.236	12.90	214.025
3.80	86.912	8.40	149.433	13.00	215.994
3.90	88.602	8.50	150.631	13.10	218.003
4.00	90.269	8.60	151.830	13.20	220.052
4.10	91.914	8.70	153.030	13.30	222.143
4.20	93.538	8.80	154.234	13.40	224.278
4.30	95.141	8.90	155.440	13.50	226.457
4.40	96.723	9.00	156.649	13.60	228.682
4.50	98.286	9.10	157.863	13.70	230.956
4.60	99.830	9.20	159.081	13.80	233.278
4.70	101.355	9.30	160.305	13.90	235.651

(Continued)

TEMPERATURE = 363.15 K (Continued)

Density ρ (mol/L)	Pressure P (bar)	Density ρ (mol/L)	Pressure P (bar)	Density ρ (mol/L)	Pressure P (bar)
14.00	238.077	17.70	383.886	21.40	751.988
14.10	240.556	17.80	390.007	21.50	767.243
14.20	243.091	17.90	396.287	21.60	782.857
14.30	245.684	18.00	402.728	21.70	798.839
14.40	248.335	18.10	409.336	21.80	815.195
14.50	251.048	18.20	416.114	21.90	831.932
14.60	253.823	18.30	423.067	22.00	849.059
14.70	256.664	18.40	430.199	22.10	866.581
14.80	259.571	18.50	437.513	22.20	884.508
14.90	262.546	18.60	445.016	22.30	902.847
15.00	265.593	18.70	452.711	22.40	921.605
15.10	268.713	18.80	460.603	22.50	940.791
15.20	271.908	18.90	468.697	22.60	960.413
15.30	275.180	19.00	476.997	22.70	980.478
15.40	278.532	19.10	485.508	22.80	1000.995
15.50	281.967	19.20	494.236	22.90	1021.973
15.60	285.486	19.30	503.184	23.00	1043.421
15.70	289.092	19.40	512.360	23.10	1065.345
15.80	292.787	19.50	521.766	23.20	1087.757
15.90	296.575	19.60	531.410	23.30	1110.664
16.00	300.458	19.70	541.295	23.40	1134.075
16.10	304.438	19.80	551.429	23.50	1158.001
16.20	308.519	19.90	561.815	23.60	1182.449
16.30	312.703	20.00	572.460	23.70	1207.430
16.40	316.994	20.10	583.370	23.80	1232.954
16.50	321.393	20.20	594.549	23.90	1259.029
16.60	325.905	20.30	606.005	24.00	1285.667
16.70	330.533	20.40	617.742	24.10	1312.876
16.80	335.279	20.50	629.768	24.20	1340.669
16.90	340.146	20.60	642.087	24.30	1369.054
17.00	345.139	20.70	654.707	24.40	1398.042
17.10	350.261	20.80	667.633	24.50	1427.645
17.20	355.514	20.90	680.872	24.60	1457.873
17.30	360.903	21.00	694.430	24.70	1488.738
17.40	366.431	21.10	708.314	24.80	1520.250
17.50	372.101	21.20	722.531	24.90	1552.421
17.60	377.918	21.30	737.086	25.00	1585.264

TEMPERATURE = 373.15 K

Density ρ (mol/L)	Pressure P (bar)	Density ρ (mol/L)	Pressure P (bar)	Density ρ (mol/L)	Pressure P (bar)
0.20	6.117	4.80	108.746	9.40	176.191
0.30	9.111	4.90	110.393	9.50	177.660
0.40	12.063	5.00	112.025	9.60	179.138
0.50	14.972	5.10	113.641	9.70	180.626
0.60	17.840	5.20	115.243	9.80	182.125
0.70	20.667	5.30	116.830	9.90	183.635
0.80	23.454	5.40	118.404	10.00	185.157
0.90	26.202	5.50	119.965	10.10	186.693
1.00	28.910	5.60	121.514	10.20	188.242
1.10	31.579	5.70	123.050	10.30	189.806
1.20	34.211	5.80	124.575	10.40	191.385
1.30	36.806	5.90	126.089	10.50	192.980
1.40	39.363	6.00	127.593	10.60	194.592
1.50	41.885	6.10	129.086	10.70	196.221
1.60	44.371	6.20	130.571	10.80	197.869
1.70	46.822	6.30	132.046	10.90	199.537
1.80	49.239	6.40	133.513	11.00	201.225
1.90	51.622	6.50	134.972	11.10	202.935
2.00	53.971	6.60	136.424	11.20	204.666
2.10	56.289	6.70	137.869	11.30	206.421
2.20	58.574	6.80	139.307	11.40	208.200
2.30	60.828	6.90	140.740	11.50	210.004
2.40	63.052	7.00	142.167	11.60	211.834
2.50	65.245	7.10	143.590	11.70	213.691
2.60	67.409	7.20	145.008	11.80	215.577
2.70	69.544	7.30	146.423	11.90	217.491
2.80	71.651	7.40	147.834	12.00	219.436
2.90	73.729	7.50	149.242	12.10	221.413
3.00	75.781	7.60	150.648	12.20	223.422
3.10	77.807	7.70	152.053	12.30	225.465
3.20	79.806	7.80	153.456	12.40	227.543
3.30	81.780	7.90	154.858	12.50	229.656
3.40	83.729	8.00	156.261	12.60	231.807
3.50	85.655	8.10	157.664	12.70	233.997
3.60	87.556	8.20	159.067	12.80	236.227
3.70	89.435	8.30	160.472	12.90	238.497
3.80	91.291	8.40	161.880	13.00	240.811
3.90	93.126	8.50	163.289	13.10	243.168
4.00	94.939	8.60	164.702	13.20	245.570
4.10	96.731	8.70	166.119	13.30	248.019
4.20	98.503	8.80	167.540	13.40	250.516
4.30	100.256	8.90	168.966	13.50	253.062
4.40	101.990	9.00	170.398	13.60	255.660
4.50	103.705	9.10	171.835	13.70	258.310
4.60	105.403	9.20	173.280	13.80	261.015
4.70	107.083	9.30	174.731	13.90	263.775

(Continued)

TEMPERATURE = 373.15 K (Continued)

Density ρ (mol/L)	Pressure P (bar)	Density ρ (mol/L)	Pressure P (bar)	Density ρ (mol/L)	Pressure P (bar)
14.00	266.593	17.70	431.239	21.40	828.597
14.10	269.471	17.80	438.001	21.50	844.815
14.20	272.409	17.90	444.929	21.60	861.404
14.30	275.410	18.00	452.027	21.70	878.369
14.40	278.476	18.10	459.299	21.80	895.719
14.50	281.609	18.20	466.749	21.90	913.461
14.60	284.810	18.30	474.382	22.00	931.602
14.70	288.081	18.40	482.201	22.10	950.151
14.80	291.426	18.50	490.212	22.20	969.115
14.90	294.845	18.60	498.419	22.30	988.502
15.00	298.340	18.70	506.826	22.40	1008.320
15.10	301.915	18.80	515.438	22.50	1028.576
15.20	305.571	18.90	524.260	22.60	1049.280
15.30	309.311	19.00	533.296	22.70	1070.440
15.40	313.137	19.10	542.553	22.80	1092.063
15.50	317.051	19.20	552.034	22.90	1114.159
15.60	321.056	19.30	561.744	23.00	1136.737
15.70	325.155	19.40	571.690	23.10	1159.805
15.80	329.350	19.50	581.875	23.20	1183.372
15.90	333.643	19.60	592.306	23.30	1207.447
16.00	338.038	19.70	602.988	23.40	1232.041
16.10	342.537	19.80	613.926	23.50	1257.161
16.20	347.144	19.90	625.126	23.60	1282.818
16.30	351.861	20.00	636.593	23.70	1309.022
16.40	356.690	20.10	648.334	23.80	1335.782
16.50	361.636	20.20	660.354	23.90	1363.108
16.60	366.701	20.30	672.658	24.00	1391.012
16.70	371.888	20.40	685.254	24.10	1419.502
16.80	377.201	20.50	698.147	24.20	1448.589
16.90	382.643	20.60	711.343	24.30	1478.285
17.00	388.217	20.70	724.848	24.40	1508.600
17.10	393.927	20.80	738.669	24.50	1539.545
17.20	399.776	20.90	752.813	24.60	1571.132
17.30	405.768	21.00	767.285	24.70	1603.371
17.40	411.907	21.10	782.093	24.80	1636.275
17.50	418.196	21.20	797.243	24.90	1669.856
17.60	424.638	21.30	812.742	25.00	1704.124

TEMPERATURE = 383.15 K

Density ρ (mol/L)	Pressure P (bar)	Density ρ (mol/L)	Pressure P (bar)	Density ρ (mol/L)	Pressure P (bar)
0.20	6.287	4.80	114.577	9.40	190.786
0.30	9.369	4.90	116.380	9.50	192.489
0.40	12.409	5.00	118.169	9.60	194.203
0.50	15.410	5.10	119.944	9.70	195.931
0.60	18.370	5.20	121.706	9.80	197.671
0.70	21.292	5.30	123.455	9.90	199.425
0.80	24.175	5.40	125.192	10.00	201.194
0.90	27.020	5.50	126.918	10.10	202.979
1.00	29.828	5.60	128.632	10.20	204.780
1.10	32.599	5.70	130.335	10.30	206.598
1.20	35.334	5.80	132.029	10.40	208.434
1.30	38.033	5.90	133.713	10.50	210.288
1.40	40.696	6.00	135.388	10.60	212.163
1.50	43.326	6.10	137.054	10.70	214.058
1.60	45.921	6.20	138.712	10.80	215.974
1.70	48.483	6.30	140.363	10.90	217.913
1.80	51.013	6.40	142.007	11.00	219.875
1.90	53.510	6.50	143.645	11.10	221.862
2.00	55.975	6.60	145.276	11.20	223.873
2.10	58.410	6.70	146.093	11.30	225.911
2.20	60.814	6.80	148.524	11.40	227.977
2.30	63.189	6.90	150.142	11.50	230.070
2.40	65.534	7.00	151.755	11.60	232.193
2.50	67.851	7.10	153.365	11.70	234.347
2.60	70.139	7.20	154.973	11.80	236.532
2.70	72.401	7.30	156.578	11.90	238.750
2.80	74.635	7.40	158.182	12.00	241.002
2.90	76.843	7.50	159.785	12.10	243.289
3.00	79.025	7.60	161.387	12.20	245.612
3.10	81.183	7.70	162.989	12.30	247.972
3.20	83.316	7.80	164.592	12.40	250.372
3.30	85.425	7.90	166.196	12.50	252.811
3.40	87.510	8.00	167.802	12.60	255.292
3.50	89.573	8.10	169.409	12.70	257.815
3.60	91.614	8.20	171.020	12.80	260.382
3.70	93.633	8.30	172.634	12.90	262.995
3.80	95.631	8.40	174.252	13.00	265.655
3.90	97.608	8.50	175.875	13.10	268.362
4.00	99.566	8.60	177.502	13.20	271.119
4.10	101.504	8.70	179.136	13.30	273.928
4.20	103.423	8.80	180.776	13.40	276.789
4.30	105.325	8.90	182.423	13.50	279.704
4.40	107.208	9.00	184.077	13.60	282.676
4.50	109.074	9.10	185.740	13.70	285.705
4.60	110.924	9.20	187.412	13.80	288.793
4.70	112.758	9.30	189.094	13.90	291.941

(Continued)

TEMPERATURE = 383.15 K (Continued)

Density ρ (mol/L)	Pressure P (bar)	Density ρ (mol/L)	Pressure P (bar)	Density ρ (mol/L)	Pressure P (bar)
14.00	295.153	17.70	478.539	21.40	904.840
14.10	298.429	17.80	485.935	21.50	922.013
14.20	301.771	17.90	493.505	21.60	939.566
14.30	305.181	18.00	501.253	21.70	957.507
14.40	308.662	18.10	509.182	21.80	975.843
14.50	312.214	18.20	517.297	21.90	994.582
14.60	315.841	18.30	525.602	22.00	1013.731
14.70	319.544	18.40	534.102	22.10	1033.299
14.80	323.324	18.50	542.801	22.20	1053.292
14.90	327.186	18.60	551.704	22.30	1073.720
15.00	331.130	18.70	560.815	22.40	1094.590
15.10	335.159	18.80	570.139	22.50	1115.911
15.20	339.275	18.90	579.681	22.60	1137.690
15.30	343.480	19.00	589.446	22.70	1159.938
15.40	347.778	19.10	599.439	22.80	1182.661
15.50	352.171	19.20	609.665	22.90	1205.870
15.60	356.660	19.30	620.128	23.00	1229.572
15.70	361.249	19.40	630.835	23.10	1253.778
15.80	365.941	19.50	641.791	23.20	1278.495
15.90	370.737	19.60	653.000	23.30	1303.735
16.00	375.642	19.70	664.469	23.40	1329.505
16.10	380.657	19.80	676.202	23.50	1355.816
16.20	685.786	19.90	688.207	23.60	1382.678
16.30	391.032	20.00	700.487	23.70	1410.101
16.40	396.397	20.10	713.050	23.80	1438.094
16.50	401.885	20.20	725.900	23.90	1455.668
16.60	407.499	20.30	739.045	24.00	1495.833
16.70	413.242	20.40	752.489	24.10	1525.601
16.80	419.118	20.50	766.240	24.20	1555.981
16.90	425.129	20.60	780.303	24.30	1586.985
17.00	431.280	20.70	794.685	24.40	1618.624
17.10	437.574	20.80	809.393	24.50	1650.910
17.20	444.014	20.90	824.432	24.60	1683.853
17.30	450.604	21.00	839.809	24.70	1717.466
17.40	457.347	21.10	855.532	24.80	1751.761
17.50	464.249	21.20	871.607	24.90	1786.749
17.60	471.311	21.30	888.041	25.00	1822.442

TEMPERATURE = 393.15 K

Density ρ (mol/L)	Pressure P (bar)	Density ρ (mol/L)	Pressure P (bar)	Density ρ (mol/L)	Pressure P (bar)
0.20	6.457	4.80	120.359	9.40	205.320
0.30	9.626	4.90	122.318	9.50	207.258
0.40	12.755	5.00	124.263	9.60	209.211
0.50	15.847	5.10	126.196	9.70	211.178
0.60	18.900	5.20	128.118	9.80	213.162
0.70	21.915	5.30	130.027	9.90	215.161
0.80	24.894	5.40	131.926	10.00	217.178
0.90	27.837	5.50	133.815	10.10	219.213
1.00	30.743	5.60	135.693	10.20	221.267
1.10	33.615	5.70	137.563	10.30	223.341
1.20	36.452	5.80	139.424	10.40	225.436
1.30	39.254	5.90	141.277	10.50	227.552
1.40	42.024	6.00	143.122	10.60	229.690
1.50	44.760	6.10	144.960	10.70	231.852
1.60	47.464	6.20	146.792	10.80	234.039
1.70	50.136	6.30	148.617	10.90	236.250
1.80	52.777	6.40	150.438	11.00	238.489
1.90	55.388	6.50	152.253	11.10	240.754
2.00	57.968	6.60	154.064	11.20	243.048
2.10	60.519	6.70	155.871	11.30	245.371
2.20	63.041	6.80	157.675	11.40	247.725
2.30	65.534	6.90	159.477	11.50	250.110
2.40	68.000	7.00	161.276	11.60	252.528
2.50	70.439	7.10	163.073	11.70	254.980
2.60	72.851	7.20	164.870	11.80	257.467
2.70	75.238	7.30	166.666	11.90	259.990
2.80	77.599	7.40	168.462	12.00	262.551
2.90	79.935	7.50	170.259	12.10	265.150
3.00	82.246	7.60	172.057	12.20	267.789
3.10	84.534	7.70	173.857	12.30	270.470
3.20	86.799	7.80	175.660	12.40	273.193
3.30	89.042	7.90	177.465	12.50	275.959
3.40	91.262	8.00	179.274	12.60	278.771
3.50	93.462	8.10	181.087	12.70	281.630
3.60	95.640	8.20	182.904	12.80	284.537
3.70	97.798	8.30	184.728	12.90	287.493
3.80	99.936	8.40	186.557	13.00	290.500
3.90	102.055	8.50	188.393	13.10	293.560
4.00	104.156	8.60	190.236	13.20	296.674
4.10	106.239	8.70	192.086	13.30	299.843
4.20	108.304	8.80	193.946	13.40	303.069
4.30	110.352	8.90	195.815	13.50	306.355
4.40	112.384	9.00	197.693	13.60	309.700
4.50	114.400	9.10	199.582	13.70	313.108
4.60	116.401	9.20	201.483	13.80	316.580
4.70	118.387	9.30	203.395	13.90	320.118

(Continued)

TEMPERATURE = 393.15 K (Continued)

Density ρ (mol/L)	Pressure P (bar)	Density ρ (mol/L)	Pressure P (bar)	Density ρ (mol/L)	Pressure P (bar)
14.00	323.723	17.70	525.748	21.40	980.720
14.10	327.398	17.80	533.774	21.50	998.841
14.20	331.144	17.90	541.980	21.60	1017.352
14.30	334.964	18.00	550.371	21.70	1036.262
14.40	338.858	18.10	558.950	21.80	1055.577
14.50	342.830	18.20	567.724	21.90	1075.306
14.60	346.882	18.30	576.695	22.00	1095.456
14.70	351.015	18.40	585.869	22.10	1116.036
14.80	355.232	18.50	595.249	22.20	1137.053
14.90	359.534	18.60	604.841	22.30	1158.516
15.00	363.925	18.70	614.649	22.40	1180.433
15.10	368.407	18.80	624.678	22.50	1202.812
15.20	372.981	18.90	634.933	22.60	1225.662
15.30	377.651	19.00	645.419	22.70	1248.922
15.40	382.419	19.10	656.141	22.80	1272.810
15.50	387.288	19.20	667.104	22.90	1297.127
15.60	392.259	19.30	678.313	23.00	1321.949
15.70	397.336	19.40	689.774	23.10	1347.288
15.80	402.522	19.50	701.491	23.20	1373.152
15.90	407.819	19.60	713.471	23.30	1399.551
16.00	413.230	19.70	725.719	23.40	1426.495
16.10	418.758	19.80	738.240	23.50	1453.994
16.20	424.406	19.90	751.041	23.60	1482.057
16.30	430.177	20.00	764.126	23.70	1510.695
16.40	436.075	20.10	777.503	23.80	1539.918
16.50	442.101	20.20	791.176	23.90	1569.737
16.60	448.261	20.30	805.153	24.00	1600.162
16.70	454.556	20.40	819.438	24.10	1631.204
16.80	460.990	20.50	834.039	24.20	1662.875
16.90	467.567	20.60	848.962	24.30	1695.186
17.00	474.290	20.70	864.213	24.40	1728.147
17.10	481.162	20.80	879.798	24.50	1761.772
17.20	488.188	20.90	895.725	24.60	1796.070
17.30	495.371	21.00	912.000	24.70	1831.055
17.40	502.714	21.10	928.630	24.80	1866.739
17.50	510.222	21.20	945.622	24.90	1903.133
17.60	517.899	21.30	962.983	25.00	1940.252

TEMPERATURE = 403.15 K

Density ρ (mol/L)	Pressure P (bar)	Density ρ (mol/L)	Pressure P (bar)	Density ρ (mol/L)	Pressure P (bar)
0.20	6.627	4.80	126.101	9.40	219.796
0.30	9.883	4.90	128.213	9.50	221.971
0.40	13.101	5.00	130.314	9.60	224.162
0.50	16.283	5.10	132.404	9.70	226.370
0.60	19.428	5.20	134.483	9.80	228.597
0.70	22.537	5.30	136.552	9.90	230.843
0.80	25.612	5.40	138.612	10.00	233.109
0.90	28.651	5.50	140.663	10.10	235.395
1.00	31.656	5.60	142.705	10.20	237.703
1.10	34.628	5.70	144.740	10.30	240.034
1.20	37.566	5.80	146.768	10.40	242.388
1.30	40.472	5.90	148.789	10.50	244.767
1.40	43.346	6.00	150.803	10.60	247.170
1.50	46.189	6.10	152.813	10.70	249.601
1.60	49.000	6.20	154.817	10.80	252.058
1.70	51.782	6.30	156.816	10.90	254.544
1.80	54.534	6.40	158.812	11.00	257.059
1.90	57.256	6.50	160.805	11.10	259.605
2.00	59.591	6.60	162.794	11.20	262.182
2.10	62.617	6.70	164.782	11.30	264.792
2.20	65.255	6.80	166.768	11.40	267.435
2.30	67.867	6.90	168.753	11.50	270.113
2.40	70.453	7.00	170.737	11.60	272.828
2.50	73.013	7.10	172.722	11.70	275.579
2.60	75.548	7.20	174.707	11.80	278.370
2.70	78.058	7.30	176.693	11.90	281.199
2.80	80.544	7.40	178.682	12.00	284.070
2.90	83.007	7.50	180.672	12.10	286.983
3.00	85.447	7.60	182.666	12.20	289.939
3.10	87.864	7.70	184.664	12.30	292.941
3.20	90.260	7.80	186.665	12.40	295.989
3.30	92.635	7.90	188.672	12.50	299.084
3.40	94.989	8.00	190.684	12.60	302.229
3.50	97.324	8.10	192.702	12.70	305.424
3.60	99.638	8.20	194.727	12.80	308.671
3.70	101.934	8.30	196.760	12.90	311.972
3.80	104.212	8.40	198.800	13.00	315.327
3.90	106.471	8.50	200.849	13.10	318.740
4.00	108.714	8.60	202.908	13.20	322.211
4.10	110.940	8.70	204.976	13.30	325.742
4.20	113.149	8.80	207.056	13.40	329.334
4.30	115.343	8.90	209.146	13.50	332.990
4.40	117.523	9.00	211.249	13.60	336.710
4.50	119.687	9.10	213.365	13.70	340.498
4.60	121.838	9.20	215.494	13.80	344.354
4.70	123.976	9.30	217.638	13.90	348.281

(Continued)

CRC HANDBOOK OF BASIC TABLES FOR CHEMICAL ANALYSIS

TEMPERATURE = 403.15 K (Continued)

Density ρ (mol/L)	Pressure P (bar)	Density ρ (mol/L)	Pressure P (bar)	Density ρ (mol/L)	Pressure P (bar)
14.00	352.280	17.70	572.841	21.40	1056.245
14.10	356.353	17.80	581.489	21.50	1075.307
14.20	360.503	17.90	590.326	21.60	1094.770
14.30	364.731	18.00	599.355	21.70	1114.642
14.40	369.039	18.10	608.580	21.80	1134.931
14.50	373.430	18.20	618.006	21.90	1155.645
14.60	377.905	18.30	627.637	22.00	1176.791
14.70	382.468	18.40	637.479	22.10	1198.378
14.80	387.119	18.50	647.534	22.20	1220.413
14.90	391.862	18.60	657.809	22.30	1242.906
15.00	396.698	18.70	668.308	22.40	1265.865
15.10	401.631	18.80	679.036	22.50	1289.298
15.20	406.662	18.90	689.997	22.60	1313.214
15.30	411.794	19.00	701.198	22.70	1337.622
15.40	417.030	19.10	712.642	22.80	1362.531
15.50	422.372	19.20	724.335	22.90	1387.951
15.60	427.823	19.30	736.283	23.00	1413.890
15.70	433.386	19.40	748.491	23.10	1440.359
15.80	439.063	19.50	760.964	23.20	1467.366
15.90	444.858	19.60	773.707	23.30	1494.921
16.00	450.773	19.70	786.727	23.40	1523.035
16.10	456.811	19.80	800.029	23.50	1551.718
16.20	462.975	19.90	813.619	23.60	1580.979
16.30	469.268	20.00	827.503	23.70	1610.830
16.40	475.694	20.10	841.687	23.80	1641.281
16.50	482.256	20.20	856.176	23.90	1672.342
16.60	488.956	20.30	870.997	24.00	1704.025
16.70	495.799	20.40	886.097	24.10	1736.340
16.80	502.788	20.50	901.541	24.20	1769.300
16.90	509.926	20.60	917.316	24.30	1802.915
17.00	517.216	20.70	933.429	24.40	1837.198
17.10	524.663	20.80	949.886	24.50	1872.159
17.20	532.270	20.90	966.695	24.60	1907.812
17.30	240.041	21.00	983.861	24.70	1944.168
17.40	547.980	21.10	1001.392	24.80	1981.239
17.50	556.090	21.20	1019.294	24.90	2019.040
17.60	564.375	21.30	1037.576	25.00	2057.581

TEMPERATURE = 413.15 K

Density ρ (mol/L)	Pressure P (bar)	Density ρ (mol/L)	Pressure P (bar)	Density ρ (mol/L)	Pressure P (bar)
0.20	6.796	4.80	131.805	9.40	234.217
0.30	10.139	4.90	134.070	9.50	236.628
0.40	13.446	5.00	136.326	9.60	239.058
0.50	16.718	5.10	138.571	9.70	541.508
0.60	19.955	5.20	140.808	9.80	243.979
0.70	23.158	5.30	143.036	9.90	246.471
0.80	26.327	5.40	145.256	10.00	248.986
0.90	29.463	5.50	147.468	10.10	251.524
1.00	32.566	5.60	149.673	10.20	254.087
1.10	35.638	5.70	151.873	10.30	256.675
1.20	38.677	5.80	154.066	10.40	259.290
1.30	41.686	5.90	156.254	10.50	261.931
1.40	44.664	6.00	158.438	10.60	264.601
1.50	47.612	6.10	160.617	10.70	267.300
1.60	50.531	6.20	162.793	10.80	270.029
1.70	53.421	6.30	164.966	10.90	272.790
1.80	56.283	6.40	167.137	11.00	275.583
1.90	59.117	6.50	169.306	11.10	278.410
2.00	61.924	6.60	171.474	11.20	281.271
2.10	64.705	6.70	173.641	11.30	284.168
2.20	67.459	6.80	175.809	11.40	287.102
2.30	70.189	6.90	177.977	11.50	290.074
2.40	72.893	7.00	180.146	11.60	293.085
2.50	75.573	7.10	182.317	11.70	296.137
2.60	78.229	7.20	184.490	11.80	299.231
2.70	80.863	7.30	186.667	11.90	302.368
2.80	83.473	7.40	188.847	12.00	305.550
2.90	86.062	7.50	191.031	12.10	308.777
3.00	88.629	7.60	193.220	12.20	312.052
3.10	91.175	7.70	195.415	12.30	315.375
3.20	93.701	7.80	197.616	12.40	318.748
3.30	96.207	7.90	199.824	12.50	322.173
3.40	98.694	8.00	202.039	12.60	325.650
3.50	101.162	8.10	204.262	12.70	329.182
3.60	103.612	8.20	206.494	12.80	332.770
3.70	106.045	8.30	208.736	12.90	336.416
3.80	108.460	8.40	210.987	13.00	340.121
3.90	110.860	8.50	213.250	13.10	343.887
4.00	113.243	8.60	215.524	13.20	347.715
4.10	115.611	8.70	217.810	13.30	351.607
4.20	117.964	8.80	220.109	13.40	355.566
4.30	120.303	8.90	222.422	13.50	359.592
4.40	122.628	9.00	224.749	13.60	363.687
4.50	124.941	9.10	227.091	13.70	637.854
4.60	127.241	9.20	229.450	13.80	372.094
4.70	129.528	9.30	231.825	13.90	376.409

(Continued)

TEMPERATURE = 413.15 K (Continued)

Density ρ (mol/L)	Pressure P (bar)	Density ρ (mol/L)	Pressure P (bar)	Density ρ (mol/L)	Pressure P (bar)
14.00	380.801	17.70	619.798	21.40	1131.422
14.10	385.273	17.80	629.065	21.50	1151.421
14.20	389.825	17.90	638.527	21.60	1171.831
14.30	394.461	18.00	648.189	21.70	1192.661
14.40	399.182	18.10	658.054	21.80	1213.919
14.50	403.991	18.20	668.127	21.90	1235.613
14.60	408.889	18.30	678.413	22.00	1257.750
14.70	413.880	18.40	688.917	22.10	1280.339
14.80	418.964	18.50	699.642	22.20	1303.389
14.90	424.145	18.60	710.594	22.30	1326.907
15.00	429.426	18.70	721.778	22.40	1350.903
15.10	434.808	18.80	733.199	22.50	1375.386
15.20	440.294	18.90	744.861	22.60	1400.365
15.30	445.886	19.00	756.770	22.70	1425.847
15.40	451.588	19.10	768.931	22.80	1451.844
15.50	457.402	19.20	781.349	22.90	1478.363
15.60	463.330	19.30	794.030	23.00	1505.416
15.70	469.376	19.40	806.978	23.10	1533.011
15.80	475.542	19.50	820.200	23.20	1561.157
15.90	481.831	19.60	833.702	23.30	1589.867
16.00	488.247	19.70	847.488	23.40	1619.148
16.10	494.791	19.80	861.565	23.50	1649.012
16.20	501.468	19.90	875.938	23.60	1679.470
16.30	508.281	20.00	890.614	23.70	1710.531
16.40	515.232	20.10	905.598	23.80	1742.207
16.50	522.325	20.20	920.898	23.90	1774.509
16.60	529.563	20.30	936.518	24.00	1807.447
16.70	536.950	20.40	952.466	24.10	1841.034
16.80	544.489	20.50	968.747	24.20	1875.281
16.90	552.184	20.60	985.369	24.30	1910.199
17.00	560.038	20.70	1002.338	24.40	1945.801
17.10	568.056	20.80	1019.661	24.50	1982.099
17.20	576.240	20.90	1037.345	24.60	2019.104
17.30	584.594	21.00	1055.396	24.70	2056.830
17.40	593.124	21.10	1073.822	24.80	2095.289
17.50	601.831	21.20	1092.631	24.90	2134.494
17.60	610.721	21.30	1111.828	25.00	2174.459

TEMPERATURE = 423.15 K

Density ρ (mol/L)	Pressure P (bar)	Density ρ (mol/L)	Pressure P (bar)	Density ρ (mol/L)	Pressure P (bar)
0.20	6.965	4.80	137.476	9.40	248.586
0.30	10.396	4.90	139.894	9.50	251.233
0.40	13.791	5.00	142.303	9.60	253.902
0.50	17.153	5.10	144.704	9.70	256.594
0.60	20.482	5.20	147.096	9.80	259.308
0.70	23.778	5.30	149.482	9.90	262.048
0.80	27.042	5.40	151.861	10.00	264.812
0.90	30.274	5.50	154.234	10.10	267.602
1.00	33.475	5.60	156.602	10.20	270.420
1.10	36.645	5.70	158.965	10.30	273.265
1.20	39.785	5.80	161.323	10.40	276.140
1.30	42.896	5.90	163.678	10.50	279.045
1.40	45.977	6.00	166.030	10.60	281.981
1.50	49.031	6.10	168.379	10.70	284.949
1.60	52.056	6.20	170.726	10.80	287.950
1.70	55.054	6.30	173.072	10.90	290.986
1.80	58.025	6.40	175.417	11.00	294.057
1.90	60.971	6.50	177.762	11.10	297.165
2.00	63.890	6.60	180.108	11.20	300.311
2.10	66.784	6.70	182.454	11.30	303.495
2.20	69.654	6.80	184.802	11.40	306.720
2.30	72.500	6.90	187.153	11.50	309.986
2.40	75.322	7.00	189.506	11.60	313.295
2.50	78.121	7.10	191.864	11.70	316.648
2.60	80.899	7.20	194.225	11.80	320.046
2.70	83.654	7.30	196.591	11.90	323.490
2.80	86.388	7.40	198.962	12.00	326.983
2.90	89.101	7.50	201.340	12.10	330.525
3.00	91.795	7.60	203.724	12.20	334.118
3.10	94.469	7.70	206.116	12.30	337.763
3.20	97.124	7.80	208.516	12.40	341.462
3.30	99.760	7.90	210.924	12.50	345.216
3.40	102.379	8.00	213.342	12.60	349.027
3.50	104.980	8.10	215.770	12.70	352.896
3.60	107.564	8.20	218.209	12.80	356.825
3.70	110.133	8.30	220.660	12.90	360.815
3.80	112.685	8.40	223.123	13.00	364.869
3.90	115.223	8.50	225.598	13.10	368.988
4.00	117.746	8.60	228.088	13.20	373.173
4.10	120.255	8.70	230.591	13.30	377.427
4.20	122.751	8.80	233.110	13.40	381.751
4.30	125.234	8.90	235.645	13.50	386.147
4.40	127.705	9.00	238.197	13.60	390.617
4.50	130.164	9.10	240.766	13.70	395.163
4.60	132.612	9.20	243.353	13.80	399.786
4.70	135.049	9.30	245.960	13.90	404.489

(*Continued*)

TEMPERATURE = 423.15 K (Continued)

Density ρ (mol/L)	Pressure P (bar)	Density ρ (mol/L)	Pressure P (bar)	Density ρ (mol/L)	Pressure P (bar)
14.00	409.274	17.70	666.606	21.40	1206.264
14.10	414.143	17.80	676.487	21.50	1227.195
14.20	419.097	17.90	686.570	21.60	1248.548
14.30	424.139	18.00	696.860	21.70	1270.331
14.40	429.272	18.10	707.360	21.80	1292.553
14.50	434.497	18.20	718.076	21.90	1315.222
14.60	439.817	18.30	729.012	22.00	1338.347
14.70	445.234	18.40	740.173	22.10	1361.934
14.80	450.750	18.50	751.563	22.20	1385.994
14.90	456.368	18.60	763.188	22.30	1410.535
15.00	462.091	18.70	775.052	22.40	1435.565
15.10	467.920	18.80	787.160	22.50	1461.095
15.20	473.859	18.90	799.517	22.60	1487.132
15.30	479.910	19.00	812.130	22.70	1513.686
15.40	486.075	19.10	825.002	22.80	1540.766
15.50	492.358	19.20	838.139	22.90	1568.383
15.60	498.761	19.30	851.548	23.00	1596.546
15.70	505.288	19.40	865.232	23.10	1625.264
15.80	511.940	19.50	879.198	23.20	1654.548
15.90	518.721	19.60	893.452	23.30	1684.409
16.00	525.635	19.70	907.999	23.40	1714.855
16.10	532.683	19.80	922.845	23.50	1745.899
16.20	539.870	19.90	937.996	23.60	1777.550
16.30	547.198	20.00	953.459	23.70	1809.820
16.40	554.671	20.10	969.239	23.80	1842.720
16.50	562.292	20.20	985.343	23.90	1876.260
16.60	570.065	20.30	1001.777	24.00	1910.453
16.70	577.993	20.40	1018.548	24.10	1945.309
16.80	586.079	20.50	1035.662	24.20	1980.842
16.90	594.327	20.60	1053.125	24.30	2017.062
17.00	602.741	20.70	1070.945	24.40	2053.982
17.10	611.325	20.80	1089.129	24.50	2091.614
17.20	620.082	20.90	1107.683	24.60	2129.972
17.30	629.016	21.00	1126.615	24.70	2169.066
17.40	638.131	21.10	1145.932	24.80	2208.912
17.50	647.432	21.20	1165.640	24.90	2249.522
17.60	656.922	21.30	1185.749	25.00	2290.909

TEMPERATURE = 433.15 K

Density ρ (mol/L)	Pressure P (bar)	Density ρ (mol/L)	Pressure P (bar)	Density ρ (mol/L)	Pressure P (bar)
0.20	7.135	4.80	143.118	9.40	262.905
0.30	10.652	4.90	145.687	9.50	265.789
0.40	14.136	5.00	148.249	9.60	268.696
0.50	17.588	5.10	150.804	9.70	271.629
0.60	21.008	5.20	153.352	9.80	274.588
0.70	24.397	5.30	155.895	9.90	277.573
0.80	27.755	5.40	158.433	10.00	280.587
0.90	31.083	5.50	160.966	10.10	283.629
1.00	34.381	5.60	163.495	10.20	286.802
1.10	37.650	5.70	166.021	10.30	289.805
1.20	40.890	5.80	168.544	10.40	292.940
1.30	44.103	5.90	171.064	10.50	296.108
1.40	47.287	6.00	173.584	10.60	299.310
1.50	50.445	6.10	176.102	10.70	302.547
1.60	53.577	6.20	178.620	10.80	305.821
1.70	56.682	6.30	181.138	10.90	309.131
1.80	59.762	6.40	183.657	11.00	312.481
1.90	62.817	6.50	186.177	11.10	315.870
2.00	65.848	6.60	188.700	11.20	319.300
2.10	68.856	6.70	191.225	11.30	322.772
2.20	71.840	6.80	193.754	11.40	326.287
2.30	74.802	6.90	196.286	11.50	329.847
2.40	77.741	7.00	198.824	11.60	333.453
2.50	80.659	7.10	201.366	11.70	337.107
2.60	83.556	7.20	203.915	11.80	340.809
2.70	86.433	7.30	206.470	11.90	344.561
2.80	89.290	7.40	209.033	12.00	348.365
2.90	92.127	7.50	211.603	12.10	352.221
3.00	94.946	7.60	214.182	12.20	356.132
3.10	97.747	7.70	216.771	12.30	360.099
3.20	100.530	7.80	219.369	12.40	364.124
3.30	103.296	7.90	221.978	12.50	368.207
3.40	106.045	8.00	224.599	12.60	372.350
3.50	108.779	8.10	227.231	12.70	376.556
3.60	111.497	8.20	229.877	12.80	380.826
3.70	114.200	8.30	232.536	12.90	385.161
3.80	116.889	8.40	235.210	13.00	389.564
3.90	119.564	8.50	237.898	13.10	394.035
4.00	122.227	8.60	240.603	13.20	398.577
4.10	124.876	8.70	243.324	13.30	403.192
4.20	127.514	8.80	246.062	13.40	407.881
4.30	130.140	8.90	248.819	13.50	412.646
4.40	132.755	9.00	251.595	13.60	417.490
4.50	135.360	9.10	254.391	13.70	422.414
4.60	137.955	9.20	257.207	13.80	427.419
4.70	140.541	9.30	260.045	13.90	432.509

(Continued)

TEMPERATURE = 433.15 K (Continued)

Density ρ (mol/L)	Pressure P (bar)	Density ρ (mol/L)	Pressure P (bar)	Density ρ (mol/L)	Pressure P (bar)
14.00	437.686	17.70	713.256	21.40	1280.781
14.10	442.950	17.80	723.747	21.50	1302.640
14.20	448.306	17.90	734.447	21.60	1324.931
14.30	453.754	18.00	745.360	21.70	1347.665
14.40	459.296	18.10	756.492	21.80	1370.848
14.50	464.937	18.20	767.846	21.90	1394.489
14.60	470.677	18.30	779.428	22.00	1418.597
14.70	476.519	18.40	791.241	22.10	1443.179
14.80	482.465	18.50	803.292	22.20	1468.246
14.90	488.518	18.60	815.584	22.30	1493.806
15.00	494.681	18.70	828.123	22.40	1519.867
15.10	500.956	18.80	840.915	22.50	1546.440
15.20	507.345	18.90	853.963	22.60	1573.533
15.30	513.852	19.00	867.274	22.70	1601.155
15.40	520.479	19.10	880.853	22.80	1629.317
15.50	527.229	19.20	894.705	22.90	1658.028
15.60	534.105	19.30	908.836	23.00	1687.299
15.70	541.109	19.40	923.251	23.10	1717.139
15.80	548.245	19.50	937.957	23.20	1747.558
15.90	555.516	19.60	952.958	23.30	1778.567
16.00	562.924	19.70	968.261	23.40	1810.177
16.10	570.474	19.80	983.872	23.50	1842.398
16.20	578.168	19.90	999.797	23.60	1875.241
16.30	586.009	20.00	1016.042	23.70	1908.718
16.40	594.001	20.10	1032.613	23.80	1942.839
16.50	602.147	20.20	1049.517	23.90	1977.617
16.60	610.450	20.30	1066.760	24.00	2013.062
16.70	618.915	20.40	1084.349	24.10	2049.187
16.80	627.545	20.50	1102.290	24.20	2086.004
16.90	636.343	20.60	1120.591	24.30	2123.524
17.00	645.313	20.70	1139.257	24.40	2161.762
17.10	654.460	20.80	1158.297	24.50	2200.728
17.20	663.786	20.90	1177.718	24.60	2240.436
17.30	673.296	21.00	1197.526	24.70	2280.899
17.40	682.994	21.10	1217.728	24.80	2322.131
17.50	692.884	21.20	1238.333	24.90	2364.144
17.60	702.970	21.30	1259.349	25.00	2406.952

TEMPERATURE = 443.15 K

Density ρ (mol/L)	Pressure P (bar)	Density ρ (mol/L)	Pressure P (bar)	Density ρ (mol/L)	Pressure P (bar)
0.20	7.304	4.80	148.733	9.40	277.176
0.30	10.908	4.90	151.453	9.50	280.296
0.40	14.480	5.00	154.166	9.60	283.442
0.50	18.022	5.10	156.875	9.70	286.616
0.60	21.533	5.20	159.578	9.80	289.818
0.70	25.015	5.30	162.277	9.90	293.050
0.80	28.467	5.40	164.973	10.00	296.313
0.90	31.890	5.50	167.665	10.10	299.607
1.00	35.285	5.60	170.355	10.20	302.934
1.10	38.653	5.70	173.044	10.30	306.294
1.20	41.993	5.80	175.730	10.40	309.690
1.30	45.307	5.90	178.417	10.50	313.121
1.40	48.594	6.00	181.103	10.60	316.589
1.50	51.856	6.10	183.789	10.70	320.094
1.60	55.093	6.20	186.477	10.80	323.640
1.70	58.305	6.30	189.167	10.90	327.225
1.80	61.494	6.40	191.859	11.00	330.852
1.90	64.659	6.50	194.555	11.10	334.522
2.00	67.801	6.60	197.254	11.20	338.236
2.10	70.920	6.70	199.957	11.30	341.996
2.20	74.018	6.80	202.666	11.40	345.802
2.30	77.095	6.90	205.380	11.50	349.655
2.40	80.151	7.00	208.101	11.60	353.559
2.50	83.187	7.10	210.829	11.70	357.512
2.60	86.203	7.20	213.564	11.80	361.518
2.70	89.201	7.30	216.308	11.90	365.578
2.80	92.180	7.40	219.061	12.00	369.692
2.90	95.141	7.50	221.824	12.10	373.863
3.00	98.084	7.60	224.598	12.20	378.091
3.10	101.011	7.70	227.382	12.30	382.380
3.20	103.921	7.80	230.179	12.40	386.729
3.30	106.816	7.90	232.988	12.50	391.141
3.40	109.695	8.00	235.811	12.60	395.617
3.50	112.560	8.10	238.648	12.70	400.159
3.60	115.411	8.20	241.500	12.80	404.769
3.70	118.249	8.30	244.367	12.90	409.448
3.80	121.073	8.40	247.252	13.00	414.199
3.90	123.886	8.50	250.153	13.10	419.022
4.00	126.686	8.60	253.072	13.20	423.920
4.10	129.475	8.70	256.010	13.30	428.895
4.20	132.254	8.80	258.968	13.40	433.948
4.30	135.022	8.90	261.947	13.50	439.082
4.40	137.781	9.00	264.946	13.60	444.298
4.50	140.531	9.10	267.968	13.70	449.599
4.60	143.272	9.20	271.013	13.80	454.986
4.70	146.006	9.30	274.082	13.90	460.462

(Continued)

TEMPERATURE = 443.15 K (Continued)

Density ρ (mol/L)	Pressure P (bar)	Density ρ (mol/L)	Pressure P (bar)	Density ρ (mol/L)	Pressure P (bar)
14.00	466.028	17.70	759.744	21.40	1354.986
14.10	471.688	17.80	770.841	21.50	1377.768
14.20	447.443	17.90	782.154	21.60	1400.995
14.30	483.295	18.00	793.687	21.70	1424.675
14.40	489.247	18.10	805.446	21.80	1448.816
14.50	495.300	18.20	817.434	21.90	1473.426
14.60	501.459	18.30	829.657	22.00	1498.514
14.70	507.724	18.40	842.119	22.10	1524.089
14.80	514.099	18.50	854.826	22.20	1550.159
14.90	520.585	18.60	867.782	22.30	1576.735
15.00	527.187	18.70	880.993	22.40	1603.825
15.10	533.905	18.80	894.463	22.50	1631.438
15.20	540.743	18.90	908.193	22.60	1659.584
15.30	547.704	19.00	922.204	22.70	1688.273
15.40	554.790	19.10	936.485	22.80	1717.514
15.50	562.005	19.20	951.047	22.90	1747.317
15.60	569.351	19.30	965.897	23.00	1777.693
15.70	576.831	19.40	981.039	23.10	1808.652
15.80	584.448	19.50	996.479	23.20	1840.204
15.90	592.206	19.60	1012.223	23.30	1872.360
16.00	600.107	19.70	1028.278	23.40	1905.131
16.10	608.155	19.80	1044.650	23.50	1938.528
16.20	616.353	19.90	1061.343	23.60	1972.561
16.30	624.704	20.00	1078.366	23.70	2007.243
16.40	633.211	20.10	1095.724	23.80	2042.585
16.50	641.879	20.20	1113.424	23.90	2078.598
16.60	650.711	20.30	1131.472	24.00	2115.295
16.70	659.710	20.40	1149.875	24.10	2152.687
16.80	668.880	20.50	1168.640	24.20	2190.787
16.90	678.225	20.60	1187.773	24.30	2229.607
17.00	687.748	20.70	1207.283	24.40	2269.161
17.10	697.454	20.80	1227.175	24.50	2309.460
17.20	707.346	20.90	1247.458	24.60	2350.518
17.30	717.428	21.00	1268.138	24.70	2392.349
17.40	727.705	21.10	1289.223	24.80	2434.965
17.50	738.180	21.20	1310.721	24.90	2478.381
17.60	748.859	21.30	1332.639	25.00	2522.610

TEMPERATURE = 453.15 K

Density ρ (mol/L)	Pressure P (bar)	Density ρ (mol/L)	Pressure P (bar)	Density ρ (mol/L)	Pressure P (bar)
0.20	7.473	4.80	154.323	9.40	291.041
0.30	11.164	4.90	157.193	9.50	294.756
0.40	14.824	5.00	160.058	9.60	298.141
0.50	18.456	5.10	162.919	9.70	301.556
0.60	22.058	5.20	165.777	9.80	305.001
0.70	25.632	5.30	168.632	9.90	308.479
0.80	29.178	5.40	171.485	10.00	311.991
0.90	32.697	5.50	174.336	10.10	215.536
1.00	36.188	5.60	177.186	10.20	319.117
1.10	39.654	5.70	180.036	10.30	322.735
1.20	43.094	5.80	182.886	10.40	326.390
1.30	46.508	5.90	185.737	10.50	330.083
1.40	49.898	6.00	188.590	10.60	333.817
1.50	53.263	6.10	191.444	10.70	337.591
1.60	56.605	6.20	194.302	10.80	341.407
1.70	59.924	6.30	197.163	10.90	345.267
1.80	63.220	6.40	200.028	11.00	349.172
1.90	66.494	6.50	202.898	11.10	353.122
2.00	69.747	6.60	205.773	11.20	357.120
2.10	72.978	6.70	208.654	11.30	361.166
2.20	76.190	6.80	211.542	11.40	365.262
2.30	79.381	6.90	214.438	11.50	369.409
2.40	82.553	7.00	217.341	11.60	373.609
2.50	85.706	7.10	220.254	11.70	377.863
2.60	88.841	7.20	223.176	11.80	382.172
2.70	91.059	7.30	226.108	11.90	386.538
2.80	95.059	7.40	229.051	12.00	390.962
2.90	98.142	7.50	232.006	12.10	395.446
3.00	101.210	7.60	234.974	12.20	399.992
3.10	104.262	7.70	237.954	12.30	404.601
3.20	107.299	7.80	240.949	12.40	409.275
3.30	110.322	7.90	243.958	12.50	414.015
3.40	113.331	8.00	246.982	12.60	418.823
3.50	116.327	8.10	250.023	12.70	423.700
3.60	119.310	8.20	253.081	12.80	428.649
3.70	122.281	8.30	253.157	12.90	433.672
3.80	125.240	8.40	259.251	13.00	438.769
3.90	128.188	8.50	262.364	13.10	443.944
4.00	131.126	8.60	265.498	13.20	449.197
4.10	134.054	8.70	268.653	13.30	454.531
4.20	136.973	8.80	271.830	13.40	459.947
4.30	139.883	8.90	275.030	13.50	465.448
4.40	142.785	9.00	278.253	13.60	471.036
4.50	145.679	9.10	281.501	13.70	476.713
4.60	148.567	9.20	284.774	13.80	482.480
4.70	151.448	9.30	288.074	13.90	488.341

(Continued)

TEMPERATURE = 453.15 K (Continued)

Density ρ (mol/L)	Pressure P (bar)	Density ρ (mol/L)	Pressure P (bar)	Density ρ (mol/L)	Pressure P (bar)
14.00	494.296	17.70	806.067	21.40	1428.889
14.10	500.349	17.80	817.766	21.50	1452.593
14.20	506.502	17.90	829.688	21.60	1476.752
14.30	512.757	18.00	841.838	21.70	1501.375
14.40	519.116	18.10	854.220	21.80	1526.470
14.50	525.582	18.20	866.839	21.90	1552.046
14.60	532.157	18.30	879.699	22.00	1578.112
14.70	538.844	18.40	892.807	22.10	1604.677
14.80	545.646	18.50	906.166	22.20	1631.749
14.90	552.564	18.60	919.782	22.30	1659.338
15.00	559.602	18.70	933.660	22.40	1687.453
15.10	566.762	18.80	947.805	22.50	1716.104
15.20	574.047	18.90	962.223	22.60	1745.301
15.30	581.459	19.00	976.919	22.70	1775.054
15.40	589.003	19.10	991.899	22.80	1805.372
15.50	596.680	19.20	1007.168	22.90	1836.265
15.60	604.494	19.30	1022.732	23.00	1867.745
15.70	612.447	19.40	1038.597	23.10	1899.821
15.80	620.543	19.50	1054.768	23.20	1932.504
15.90	628.785	19.60	1071.252	23.30	1965.806
16.00	637.176	19.70	1088.055	23.40	1999.736
16.10	645.720	19.80	1105.183	23.50	2034.307
16.20	654.419	19.90	1122.642	23.60	2069.529
16.30	663.277	20.00	1140.439	23.70	2105.415
16.40	672.298	20.10	1158.580	23.80	2141.976
16.50	681.485	20.20	1177.071	23.90	2179.223
16.60	690.842	20.30	1195.921	24.00	2217.170
16.70	700.372	20.40	1215.134	24.10	2255.828
16.80	710.080	20.50	1234.719	24.20	2295.210
16.90	719.968	20.60	1254.682	24.30	2335.329
17.00	730.041	20.70	1275.031	24.40	2376.197
17.10	740.303	20.80	1295.772	24.50	2417.829
17.20	750.757	20.90	1316.913	24.60	2460.236
17.30	761.409	21.00	1338.462	24.70	2503.433
17.40	772.261	21.10	1360.427	24.80	2547.434
17.50	783.319	21.20	1382.814	24.90	2592.252
17.60	794.586	21.30	1405.632	25.00	2637.902

TEMPERATURE = 463.15 K

Density ρ (mol/L)	Pressure P (bar)	Density ρ (mol/L)	Pressure P (bar)	Density ρ (mol/L)	Pressure P (bar)
0.20	7.642	4.80	159.890	9.40	305.582
0.30	11.419	4.90	162.910	9.50	309.172
0.40	15.168	5.00	165.926	9.60	312.795
0.50	18.889	5.10	168.939	9.70	316.450
0.60	22.582	5.20	171.951	9.80	320.138
0.70	26.248	5.30	174.961	9.90	323.862
0.80	29.888	5.40	177.971	10.00	327.622
0.90	33.502	5.50	180.980	10.10	331.418
1.00	37.090	5.60	183.990	10.20	335.253
1.10	40.653	5.70	187.001	10.30	339.127
1.20	44.192	5.80	190.014	10.40	343.041
1.30	47.707	5.90	193.029	10.50	346.996
1.40	51.199	6.00	196.047	10.60	350.995
1.50	54.668	6.10	199.069	10.70	355.037
1.60	58.114	6.20	202.096	10.80	359.124
1.70	61.539	6.30	205.127	10.90	363.258
1.80	64.942	6.40	208.165	11.00	367.440
1.90	68.325	6.50	211.208	11.10	371.670
2.00	71.688	6.60	214.259	11.20	375.951
2.10	75.031	6.70	217.318	11.30	380.283
2.20	78.355	6.80	220.385	11.40	384.668
2.30	81.660	6.90	223.461	11.50	389.108
2.40	84.948	7.00	226.547	11.60	393.604
2.50	88.218	7.10	229.644	11.70	398.157
2.60	91.471	7.20	232.752	11.80	402.768
2.70	94.707	7.30	235.872	11.90	407.440
2.80	97.928	7.40	239.005	12.00	412.174
2.90	101.134	7.50	242.152	12.10	416.971
3.00	104.325	7.60	245.313	12.20	421.833
3.10	107.502	7.70	248.489	12.30	426.762
3.20	110.665	7.80	251.681	12.40	431.759
3.30	113.815	7.90	254.889	12.50	436.826
3.40	116.953	8.00	258.115	12.60	441.965
3.50	120.079	8.10	261.359	12.70	447.177
3.60	123.193	8.20	264.623	12.80	452.465
3.70	126.297	8.30	267.906	12.90	457.830
3.80	129.391	8.40	271.210	13.00	463.273
3.90	132.475	8.50	274.535	13.10	468.798
4.00	135.549	8.60	277.883	13.20	474.405
4.10	138.616	8.70	281.255	13.30	480.097
4.20	141.674	8.80	284.650	13.40	485.875
4.30	144.725	8.90	288.071	13.50	491.743
4.40	147.769	9.00	291.517	13.60	497.701
4.50	150.807	9.10	294.991	13.70	503.752
4.60	153.840	9.20	298.492	13.80	509.899
4.70	156.867	9.30	302.022	13.90	516.142

(Continued)

TEMPERATURE = 463.15 K (Continued)

Density ρ (mol/L)	Pressure P (bar)	Density ρ (mol/L)	Pressure P (bar)	Density ρ (mol/L)	Pressure P (bar)
14.00	522.486	17.70	852.224	21.40	1502.503
14.10	528.931	17.80	864.522	21.50	1527.126
14.20	535.480	17.90	877.050	21.60	1552.214
14.30	542.136	18.00	889.813	21.70	1577.778
14.40	548.901	18.10	902.815	21.80	1603.825
14.50	555.778	18.20	916.061	21.90	1630.365
14.60	562.768	18.30	929.556	22.00	1657.406
14.70	569.875	18.40	943.305	22.10	1684.957
14.80	577.101	18.50	957.313	22.20	1713.028
14.90	584.449	18.60	971.586	22.30	1741.628
15.00	591.921	18.70	986.128	22.40	1770.767
15.10	599.521	18.80	1000.945	22.50	1800.455
15.20	607.251	18.90	1016.042	22.60	1830.700
15.30	615.114	19.00	1031.425	22.70	1861.515
15.40	623.112	19.10	1047.100	22.80	1892.907
15.50	631.250	19.20	1063.072	22.90	1924.889
15.60	639.529	19.30	1079.347	23.00	1957.471
15.70	647.954	19.40	1095.931	23.10	1990.662
15.80	656.526	19.50	1112.830	23.20	2024.475
15.90	665.250	19.60	1130.050	23.30	2058.920
16.00	674.129	19.70	1147.597	23.40	2094.008
16.10	683.166	19.80	1165.478	23.50	2129.752
16.20	692.364	19.90	1183.699	23.60	2166.162
16.30	701.727	20.00	1202.267	23.70	2203.250
16.40	711.258	20.10	1221.187	23.80	2241.028
16.50	720.962	20.20	1240.467	23.90	2279.509
16.60	730.841	20.30	1260.115	24.00	2318.704
16.70	740.899	20.40	1280.135	24.10	2358.627
16.80	751.141	20.50	1300.537	24.20	2399.291
16.90	761.570	20.60	1321.326	24.30	2440.707
17.00	772.190	20.70	1342.510	24.40	2482.890
17.10	783.005	20.80	1364.097	24.50	2525.852
17.20	794.019	20.90	1386.095	24.60	2569.608
17.30	805.236	21.00	1408.509	24.70	2614.171
17.40	816.661	21.10	1431.350	24.80	2659.555
17.50	828.297	21.20	1454.623	24.90	2705.775
17.60	840.150	21.30	1478.338	25.00	2752.844

TEMPERATURE = 473.15 K

Density ρ (mol/L)	Pressure P (bar)	Density ρ (mol/L)	Pressure P (bar)	Density ρ (mol/L)	Pressure P (bar)
0.20	7.811	4.80	165.437	9.40	319.720
0.30	11.675	4.90	168.605	9.50	323.545
0.40	15.512	5.00	171.772	9.60	327.405
0.50	19.322	5.10	174.937	9.70	331.299
0.60	23.106	5.20	178.102	9.80	335.231
0.70	26.864	5.30	181.267	9.90	339.199
0.80	30.597	5.40	184.432	10.00	343.207
0.90	34.306	5.50	187.599	10.10	347.254
1.00	37.990	5.60	190.768	10.20	351.341
1.10	41.651	5.70	193.940	10.30	355.471
1.20	45.289	5.80	197.115	10.40	359.644
1.30	48.904	5.90	200.294	10.50	363.861
1.40	52.497	6.00	203.477	10.60	368.124
1.50	56.069	6.10	206.666	10.70	372.433
1.60	59.620	6.20	209.861	10.80	376.791
1.70	63.150	6.30	213.063	10.90	381.198
1.80	66.660	6.40	216.272	11.00	385.656
1.90	70.152	6.50	219.489	11.10	390.165
2.00	73.624	6.60	222.715	11.20	394.728
2.10	77.078	6.70	225.950	11.30	399.346
2.20	80.514	6.80	229.196	11.40	404.020
2.30	83.933	6.90	232.452	11.50	408.752
2.40	87.335	7.00	235.721	11.60	413.543
2.50	90.721	7.10	239.001	11.70	418.394
2.60	94.092	7.20	242.295	11.80	423.307
2.70	97.448	7.30	245.603	11.90	428.284
2.80	100.789	7.40	248.925	12.00	433.327
2.90	104.116	7.50	252.263	12.10	438.436
3.00	107.430	7.60	255.617	12.20	443.614
3.10	110.731	7.70	258.988	12.30	448.861
3.20	114.020	7.80	262.377	12.40	454.181
3.30	117.297	7.90	265.784	12.50	459.574
3.40	120.563	8.00	269.211	12.60	465.043
3.50	123.818	8.10	272.658	12.70	470.589
3.60	127.064	8.20	276.127	12.80	476.214
3.70	130.299	8.30	279.617	12.90	481.920
3.80	133.527	8.40	283.131	13.00	487.708
3.90	136.745	8.50	286.668	13.10	493.582
4.00	139.956	8.60	290.229	13.20	499.542
4.10	143.160	8.70	293.816	13.30	505.591
4.20	146.358	8.80	297.430	13.40	511.730
4.30	149.549	8.90	301.071	13.50	517.963
4.40	152.735	9.00	304.740	13.60	524.290
4.50	155.916	9.10	308.439	13.70	530.715
4.60	159.093	9.20	312.167	13.80	537.239
4.70	162.267	9.30	315.927	13.90	543.864

(Continued)

TEMPERATURE = 473.15 K (Continued)

Density ρ (mol/L)	Pressure P (bar)	Density ρ (mol/L)	Pressure P (bar)	Density ρ (mol/L)	Pressure P (bar)
14.00	550.594	17.70	898.216	21.40	1575.840
14.10	557.430	17.80	911.110	21.50	1601.378
14.20	564.374	17.90	924.242	21.60	1627.394
14.30	571.430	18.00	937.614	21.70	1653.896
14.40	578.599	18.10	951.233	21.80	1680.893
14.50	585.885	18.20	965.103	21.90	1708.393
14.60	593.288	18.30	979.230	22.00	1736.407
14.70	600.814	18.40	993.617	22.10	1764.943
14.80	608.463	18.50	1008.271	22.20	1794.011
14.90	616.239	18.60	1023.197	22.30	1823.620
15.00	624.144	18.70	1038.399	22.40	1853.781
15.10	632.181	18.80	1053.885	22.50	1884.502
15.20	640.354	18.90	1069.658	22.60	1915.795
15.30	648.664	19.00	1085.725	22.70	1947.669
15.40	657.116	19.10	1102.092	22.80	1980.135
15.50	665.712	19.20	1118.763	22.90	2013.204
15.60	674.455	19.30	1135.746	23.00	2046.885
15.70	683.348	19.40	1153.046	23.10	2081.191
15.80	692.395	19.50	1170.670	23.20	2116.131
15.90	701.599	19.60	1188.623	23.30	2151.718
16.00	710.963	19.70	1206.911	23.40	2187.963
16.10	720.490	19.80	1225.542	23.50	2224.878
16.20	730.185	19.90	1244.522	23.60	2262.474
16.30	740.050	20.00	1263.857	23.70	2300.763
16.40	750.090	20.10	1283.554	23.80	2339.758
16.50	760.307	20.20	1303.620	23.90	2379.471
16.60	770.706	20.30	1324.062	24.00	2419.914
16.70	781.290	20.40	1344.887	24.10	2461.101
16.80	792.064	20.50	1366.102	24.20	2503.044
16.90	803.030	20.60	1387.714	24.30	2545.757
17.00	814.194	20.70	1409.732	24.40	2589.254
17.10	825.560	20.80	1432.162	24.50	2633.546
17.20	837.130	20.90	1455.012	24.60	2678.650
17.30	848.911	21.00	1478.289	24.70	2724.578
17.40	860.905	21.10	1502.003	24.80	2771.345
17.50	873.118	21.20	1526.160	24.90	2818.965
17.60	885.553	21.30	1550.770	25.00	2867.454

TEMPERATURE = 493.15 K

Density ρ (mol/L)	Pressure P (bar)	Density ρ (mol/L)	Pressure P (bar)	Density ρ (mol/L)	Pressure P (bar)
0.20	8.149	4.80	176.475	9.40	347.872
0.30	12.186	4.90	179.939	9.50	352.166
0.40	16.199	5.00	183.404	9.60	356.498
0.50	20.187	5.10	186.871	9.70	360.871
0.60	24.152	5.20	190.341	9.80	365.286
0.70	28.094	5.30	193.813	9.90	369.743
0.80	32.014	5.40	197.289	10.00	374.244
0.90	35.911	5.50	200.770	10.10	378.790
1.00	39.787	5.60	204.256	10.20	383.382
1.10	43.643	5.70	207.747	10.30	383.022
1.20	47.477	5.80	211.245	10.40	392.710
1.30	51.292	5.90	214.750	10.50	397.448
1.40	55.088	6.00	218.262	10.60	402.238
1.50	58.864	6.10	221.784	10.70	407.080
1.60	62.622	6.20	225.314	10.80	411.976
1.70	66.363	6.30	228.855	10.90	416.927
1.80	70.086	6.40	232.406	11.00	421.935
1.90	73.793	6.50	235.968	11.10	427.001
2.00	77.483	6.60	239.543	11.20	432.127
2.10	81.157	6.70	243.130	11.30	437.313
2.20	84.817	6.80	246.732	11.40	442.562
2.30	88.462	6.90	250.347	11.50	447.875
2.40	92.092	7.00	253.978	11.60	453.254
2.50	95.710	7.10	257.625	11.70	458.699
2.60	99.314	7.20	261.289	11.80	464.213
2.70	102.906	7.30	264.970	11.90	469.798
2.80	106.486	7.40	268.669	12.00	475.455
2.90	110.055	7.50	272.388	12.10	481.185
3.00	113.613	7.60	276.127	12.20	486.991
3.10	117.161	7.70	279.887	12.30	492.874
3.20	120.699	7.80	283.668	12.40	498.836
3.30	124.228	7.90	287.473	12.50	504.879
3.40	127.749	8.00	291.300	12.60	511.004
3.50	131.262	8.10	295.152	12.70	517.214
3.60	134.767	8.20	299.029	12.80	523.511
3.70	138.266	8.30	302.933	12.90	529.895
3.80	141.759	8.40	306.864	13.00	536.371
3.90	145.245	8.50	310.822	13.10	542.938
4.00	148.727	8.60	314.810	13.20	549.601
4.10	152.205	8.70	318.827	13.30	556.360
4.20	155.679	8.80	322.876	13.40	563.217
4.30	159.149	8.90	326.956	13.50	570.176
4.40	162.617	9.00	331.069	13.60	577.238
4.50	166.083	9.10	335.217	13.70	584.405
4.60	169.547	9.20	339.399	13.80	591.680
4.70	173.011	9.30	343.617	13.90	599.065

(Continued)

TEMPERATURE = 493.15 K (Continued)

Density ρ (mol/L)	Pressure P (bar)	Density ρ (mol/L)	Pressure P (bar)	Density ρ (mol/L)	Pressure P (bar)
14.00	606.563	17.70	989.715	21.40	1721.725
14.10	614.176	17.80	1003.793	21.50	1749.089
14.20	621.906	17.90	1018.122	21.60	1776.952
14.30	629.757	18.00	1032.707	21.70	1805.324
14.40	637.729	18.10	1047.551	21.80	1834.214
14.50	645.827	18.20	1062.661	21.90	1863.631
14.60	654.053	18.30	1078.041	22.00	1893.584
14.70	662.410	18.40	1093.696	22.10	1924.084
14.80	670.900	18.50	1109.633	22.20	1955.140
14.90	679.526	18.60	1125.857	22.30	1986.762
15.00	688.292	18.70	1142.373	22.40	2018.960
15.10	697.199	18.80	1159.186	22.50	2051.744
15.20	706.252	18.90	1176.303	22.60	2085.126
15.30	715.453	19.00	1193.729	22.70	2119.115
15.40	724.805	19.10	1211.471	22.80	2153.722
15.50	734.311	19.20	1229.533	22.90	2188.959
15.60	743.975	19.30	1247.924	23.00	2224.836
15.70	753.800	19.40	1266.647	23.10	2261.365
15.80	763.789	19.50	1285.711	23.20	2298.558
15.90	773.946	19.60	1305.121	23.30	2336.425
16.00	784.273	19.70	1324.884	23.40	2374.979
16.10	794.776	19.80	1345.006	23.50	2414.233
16.20	805.457	19.90	1365.495	23.60	2454.197
16.30	816.319	20.00	1386.356	23.70	2494.885
16.40	827.368	20.10	1407.598	23.80	2536.310
16.50	838.606	20.20	1429.228	23.90	2578.484
16.60	850.037	20.30	1451.251	24.00	2621.420
16.70	861.665	20.40	1473.676	24.10	2665.131
16.80	873.495	20.50	1496.511	24.20	2709.632
16.90	885.530	20.60	1519.762	24.30	2754.936
17.00	897.774	20.70	1543.438	24.40	2801.056
17.10	910.232	20.80	1567.546	24.50	2848.007
17.20	922.908	20.90	1592.094	24.60	2895.804
17.30	935.807	21.00	1617.090	24.70	2944.460
17.40	948.932	21.10	1642.543	24.80	2993.990
17.50	962.289	21.20	1668.461	24.90	3044.411
17.60	975.882	21.30	1694.852	25.00	3095.736

TEMPERATURE = 513.15 K

Density ρ (mol/L)	Pressure P (bar)	Density ρ (mol/L)	Pressure P (bar)	Density ρ (mol/L)	Pressure P (bar)
0.20	8.487	4.80	187.446	9.40	375.869
0.30	12.697	4.90	191.204	9.50	380.628
0.40	16.885	5.00	194.966	9.60	385.431
0.50	21.051	5.10	198.733	9.70	390.280
0.60	25.197	5.20	202.505	9.80	395.175
0.70	29.322	5.30	206.283	9.90	400.119
0.80	33.427	5.40	210.068	10.00	405.111
0.90	37.513	5.50	213.860	10.10	410.153
1.00	41.580	5.60	217.660	10.20	415.247
1.10	45.629	5.70	221.470	10.30	420.394
1.20	49.660	5.80	225.288	10.40	425.595
1.30	53.673	5.90	229.117	10.50	430.852
1.40	57.670	6.00	232.957	10.60	436.165
1.50	61.650	6.10	236.809	10.70	441.537
1.60	65.615	6.20	240.673	10.80	446.968
1.70	69.565	6.30	244.550	10.90	452.461
1.80	73.499	6.40	248.442	11.00	458.016
1.90	77.420	6.50	252.348	11.10	463.635
2.00	81.327	6.60	256.269	11.20	469.320
2.10	85.220	6.70	260.207	11.30	475.071
2.20	89.102	6.80	264.162	11.40	480.892
2.30	92.971	6.90	268.135	11.50	486.783
2.40	96.828	7.00	272.127	11.60	492.746
2.50	100.675	7.10	276.138	11.70	498.782
2.60	104.512	7.20	280.170	11.80	504.894
2.70	108.338	7.30	284.223	11.90	511.082
2.80	112.155	7.40	288.298	12.00	517.349
2.90	115.964	7.50	292.396	12.10	523.697
3.00	119.764	7.60	296.518	12.20	530.127
3.10	123.557	7.70	300.664	12.30	536.641
3.20	127.343	7.80	304.837	12.40	543.241
3.30	131.123	7.90	309.035	12.50	549.929
3.40	134.896	8.00	313.262	12.60	556.707
3.50	138.665	8.10	317.517	12.70	563.576
3.60	142.428	8.20	321.801	12.80	570.540
3.70	146.188	8.30	326.116	12.90	577.599
3.80	149.944	8.40	330.461	13.00	584.756
3.90	153.697	8.50	334.840	13.10	592.013
4.00	157.448	8.60	339.251	13.20	599.373
4.10	161.197	8.70	343.697	13.30	606.837
4.20	164.945	8.80	348.178	13.40	614.407
4.30	168.693	8.90	352.696	13.50	622.087
4.40	172.440	9.00	357.251	13.60	629.878
4.50	176.189	9.10	361.548	13.70	637.783
4.60	179.939	9.20	366.479	13.80	645.803
4.70	183.691	9.30	371.153	13.90	653.942

(Continued)

TEMPERATURE = 513.15 K (Continued)

Density ρ (mol/L)	Pressure P (bar)	Density ρ (mol/L)	Pressure P (bar)	Density ρ (mol/L)	Pressure P (bar)
14.00	662.203	17.70	1080.589	21.40	1866.632
14.10	670.587	17.80	1095.842	21.50	1895.814
14.20	679.097	17.90	1111.358	21.60	1925.517
14.30	687.735	18.00	1127.144	21.70	1955.751
14.40	696.506	18.10	1143.204	21.80	1986.526
14.50	705.410	18.20	1159.543	21.90	2017.852
14.60	714.452	18.30	1176.167	22.00	2049.738
14.70	723.633	18.40	1193.081	22.10	2082.195
14.80	732.958	18.50	1210.291	22.20	2115.232
14.90	742.428	18.60	1227.802	22.30	2148.860
15.00	752.046	18.70	1245.620	22.40	2183.089
15.10	761.817	18.80	1263.752	22.50	2217.931
15.20	771.742	18.90	1282.202	22.60	2253.395
15.30	781.826	19.00	1300.978	22.70	2289.493
15.40	792.070	19.10	1320.084	22.80	2326.236
15.50	802.480	19.20	1339.527	22.90	2363.636
15.60	813.057	19.30	1359.315	23.00	2401.704
15.70	823.805	19.40	1379.452	23.10	2440.451
15.80	834.729	19.50	1399.946	23.20	2479.890
15.90	845.830	19.60	1420.803	23.30	2520.033
16.00	857.114	19.70	1442.030	23.40	2560.892
16.10	868.583	19.80	1463.634	23.50	2602.479
16.20	880.242	19.90	1485.622	23.60	2644.808
16.30	892.093	20.00	1508.001	23.70	2687.891
16.40	904.142	20.10	1530.778	23.80	2731.741
16.50	916.392	20.20	1553.961	23.90	2776.372
16.60	928.846	20.30	1577.557	24.00	2821.797
16.70	941.510	20.40	1601.573	24.10	2868.030
16.80	954.386	20.50	1626.018	24.20	2915.085
16.90	967.480	20.60	1650.899	24.30	2962.976
17.00	980.796	20.70	1676.224	24.40	3011.717
17.10	994.338	20.80	1702.001	24.50	3061.323
17.20	1008.110	20.90	1728.239	24.60	3111.810
17.30	1022.117	21.00	1754.946	24.70	3163.191
17.40	1036.364	21.10	1782.130	24.80	3215.483
17.50	1050.855	21.20	1809.800	24.90	3268.701
17.60	1065.595	21.30	1837.964	25.00	3322.860

TEMPERATURE = 533.15 K

Density ρ (mol/L)	Pressure P (bar)	Density ρ (mol/L)	Pressure P (bar)	Density ρ (mol/L)	Pressure P (bar)
0.20	8.825	4.80	198.358	9.40	403.719
0.30	13.207	4.90	202.408	9.50	408.941
0.40	17.570	5.00	206.465	9.60	414.213
0.50	21.914	5.10	210.530	9.70	419.535
0.60	26.240	5.20	214.603	9.80	424.909
0.70	30.548	5.30	218.685	9.90	430.335
0.80	34.839	5.40	222.777	10.00	435.816
0.90	39.112	5.50	226.879	10.10	441.352
1.00	43.370	5.60	230.992	10.20	445.945
1.10	47.611	5.70	235.117	10.30	452.596
1.20	51.837	5.80	239.255	10.40	458.307
1.30	56.049	5.90	243.406	10.50	464.079
1.40	60.246	6.00	247.572	10.60	469.913
1.50	64.429	6.10	251.752	10.70	475.812
1.60	68.599	6.20	255.948	10.80	481.775
1.70	72.756	6.30	260.160	10.90	487.806
1.80	76.902	6.40	264.390	11.00	493.905
1.90	81.035	6.50	268.638	11.10	500.074
2.00	85.157	6.60	272.904	11.20	506.315
2.10	89.269	6.70	277.191	11.30	512.629
2.20	93.371	6.80	281.498	11.40	519.017
2.30	97.463	6.90	285.826	11.50	525.482
2.40	101.546	7.00	290.177	11.60	532.026
2.50	105.621	7.10	294.551	11.70	538.026
2.60	109.688	7.20	298.949	11.80	545.354
2.70	113.747	7.30	303.371	11.90	552.143
2.80	117.800	7.40	307.820	12.00	559.016
2.90	121.847	7.50	312.295	12.10	565.977
3.00	125.889	7.60	316.799	12.20	573.027
3.10	129.925	7.70	321.330	12.30	580.168
3.20	133.957	7.80	325.891	12.40	587.402
3.30	137.985	7.90	330.483	12.50	594.731
3.40	142.010	8.00	335.106	12.60	602.157
3.50	146.032	8.10	339.762	12.70	609.681
3.60	150.052	8.20	344.451	12.80	617.307
3.70	154.071	8.30	349.175	12.90	625.036
3.80	158.088	8.40	353.934	13.00	632.871
3.90	162.106	8.50	358.730	13.10	640.813
4.00	166.124	8.60	363.564	13.20	648.865
4.10	170.143	8.70	368.436	13.30	657.029
4.20	174.164	8.80	373.348	13.40	665.307
4.30	178.186	8.90	378.301	13.50	673.703
4.40	182.212	9.00	383.296	13.60	682.217
4.50	186.242	9.10	388.334	13.70	690.854
4.60	190.275	9.20	393.416	13.80	699.615
4.70	194.314	9.30	398.544	13.90	708.503

(Continued)

TEMPERATURE = 533.15 K (Continued)

Density ρ (mol/L)	Pressure P (bar)	Density ρ (mol/L)	Pressure P (bar)	Density ρ (mol/L)	Pressure P (bar)
14.00	717.520	17.70	1170.870	21.40	2010.642
14.10	726.669	17.80	1187.288	21.50	2041.633
14.20	735.953	17.90	1203.984	21.60	2073.169
14.30	745.374	18.00	1220.962	21.70	2105.260
14.40	754.936	18.10	1238.229	21.80	2137.914
14.50	764.641	18.20	1255.788	21.90	2171.143
14.60	774.492	18.30	1273.647	22.00	2204.956
14.70	784.492	18.40	1291.811	22.10	2239.363
14.80	794.645	18.50	1310.285	22.20	2274.376
14.90	804.952	18.60	1329.075	22.30	2310.004
15.00	815.417	18.70	1348.188	22.40	2346.259
15.10	826.044	18.80	1367.628	22.50	2383.152
15.20	836.836	18.90	1387.403	22.60	2420.694
15.30	847.795	19.00	1407.519	22.70	2458.896
15.40	858.926	19.10	1427.981	22.80	2497.770
15.50	870.231	19.20	1448.797	22.90	2537.328
15.60	881.714	19.30	1469.973	23.00	2577.581
15.70	893.379	19.40	1491.515	23.10	2618.542
15.80	905.230	19.50	1513.431	23.20	2660.223
15.90	917.269	19.60	1535.726	23.30	2702.637
16.00	929.501	19.70	1558.410	23.40	2745.796
16.10	941.929	19.80	1581.487	23.50	2789.714
16.20	954.558	19.90	1604.966	23.60	2834.403
16.30	967.390	20.00	1628.854	23.70	2879.877
16.40	980.432	20.10	1653.158	23.80	2926.149
16.50	993.685	20.20	1677.886	23.90	2973.233
16.60	1007.155	20.30	1703.046	24.00	3021.144
16.70	1020.845	20.40	1728.646	24.10	3069.895
16.80	1034.761	20.50	1754.693	24.20	3119.501
16.90	1048.906	20.60	1781.196	24.30	3169.976
17.00	1063.284	20.70	1808.163	24.40	3221.335
17.10	1077.901	20.80	1835.602	24.50	3273.594
17.20	1092.761	20.90	1863.522	24.60	3326.768
17.30	1107.868	21.00	1891.931	24.70	3380.872
17.40	1123.228	21.10	1920.839	24.80	3435.922
17.50	1138.845	21.20	1950.254	24.90	3491.935
17.60	1154.724	21.30	1980.185	25.00	3548.926

TEMPERATURE = 553.15 K

Density ρ (mol/L)	Pressure P (bar)	Density ρ (mol/L)	Pressure P (bar)	Density ρ (mol/L)	Pressure P (bar)
0.20	9.162	4.80	209.218	9.40	431.431
0.30	13.717	4.90	213.559	9.50	437.114
0.40	18.255	5.00	217.909	9.60	442.851
0.50	22.777	5.10	222.270	9.70	448.644
0.60	27.282	5.20	226.643	9.80	454.494
0.70	31.773	5.30	231.027	9.90	460.401
0.80	36.248	5.40	235.424	10.00	466.367
0.90	40.709	5.50	239.834	10.10	472.395
1.00	45.156	5.60	244.258	10.20	478.484
1.10	49.590	5.70	248.698	10.30	484.637
1.20	54.011	5.80	253.153	10.40	490.855
1.30	58.149	5.90	257.625	10.50	497.139
1.40	62.816	6.00	262.114	10.60	503.492
1.50	67.201	6.10	266.621	10.70	509.913
1.60	71.576	6.20	271.147	10.80	516.406
1.70	75.940	6.30	275.693	10.90	522.972
1.80	80.294	6.40	280.259	11.00	529.612
1.90	84.640	6.50	284.847	11.10	536.327
2.00	88.976	6.60	289.457	11.20	543.120
2.10	93.305	6.70	294.090	11.30	549.993
2.20	97.626	6.80	298.747	11.40	556.946
2.30	101.940	6.90	303.429	11.50	563.982
2.40	106.248	7.00	308.137	11.60	571.102
2.50	110.549	7.10	312.872	11.70	578.309
2.60	114.846	7.20	317.634	11.80	585.604
2.70	119.137	7.30	322.425	11.90	592.988
2.80	123.425	7.40	327.245	12.00	600.465
2.90	127.708	7.50	332.096	12.10	608.035
3.00	131.989	7.60	336.979	12.20	615.701
3.10	136.267	7.70	341.893	12.30	623.465
3.20	140.544	7.80	346.842	12.40	631.328
3.30	144.819	7.90	351.824	12.50	639.294
3.40	149.094	8.00	356.842	12.60	647.363
3.50	153.368	8.10	361.897	12.70	655.539
3.60	157.643	8.20	366.989	12.80	663.823
3.70	161.919	8.30	372.120	12.90	672.217
3.80	166.197	8.40	377.291	13.00	680.724
3.90	170.478	8.50	382.502	13.10	689.346
4.00	174.761	8.60	387.756	13.20	698.086
4.10	179.049	8.70	393.052	13.30	706.945
4.20	183.340	8.80	398.393	13.40	715.927
4.30	187.637	8.90	403.779	13.50	725.033
4.40	191.939	9.00	409.211	13.60	734.267
4.50	196.248	9.10	414.691	13.70	743.630
4.60	200.563	9.20	420.220	13.80	753.126
4.70	204.886	9.30	425.800	13.90	762.757

(*Continued*)

TEMPERATURE = 553.15 K (Continued)

Density ρ (mol/L)	Pressure P (bar)	Density ρ (mol/L)	Pressure P (bar)	Density ρ (mol/L)	Pressure P (bar)
14.00	772.526	17.70	1260.593	21.40	2153.826
14.10	782.435	17.80	1278.169	21.50	2186.623
14.20	792.487	17.90	1296.036	21.60	2219.986
14.30	802.686	18.00	1314.199	21.70	2253.927
14.40	813.034	18.10	1332.665	21.80	2288.455
14.50	823.534	18.20	1351.438	21.90	2323.581
14.60	834.188	18.30	1370.524	22.00	2359.315
14.70	845.001	18.40	1389.929	22.10	2395.668
14.80	855.975	18.50	1409.659	22.20	2432.651
14.90	867.114	18.60	1429.721	22.30	2470.275
15.00	878.420	18.70	1450.120	22.40	2508.551
15.10	889.897	18.80	1470.862	22.50	2547.490
15.20	901.548	18.90	1491.954	22.60	2587.105
15.30	913.377	19.00	1513.402	22.70	2627.406
15.40	925.387	19.10	1535.213	22.80	2668.407
15.50	937.582	19.20	1557.393	22.90	2710.118
15.60	949.965	19.30	1579.950	23.00	2752.553
15.70	962.540	19.40	1602.890	23.10	2795.723
15.80	975.311	19.50	1626.220	23.20	2839.643
15.90	988.281	19.60	1649.947	23.30	2884.324
16.00	1001.454	19.70	1674.078	23.40	2929.780
16.10	1014.835	19.80	1698.622	23.50	2976.024
16.20	1028.426	19.90	1723.584	23.60	3023.070
16.30	1042.233	20.00	1748.974	23.70	3070.931
16.40	1056.260	20.10	1774.798	23.80	3119.622
16.50	1070.510	20.20	1801.065	23.90	3169.156
16.60	1084.988	20.30	1827.782	24.00	3219.549
16.70	1099.698	20.40	1854.958	24.10	3270.815
16.80	1114.645	20.50	1882.601	24.20	3322.969
16.90	1129.833	20.60	1920.719	24.30	3376.025
17.00	1145.268	20.70	1939.321	24.40	3430.000
17.10	1160.952	20.80	1968.415	24.50	3484.909
17.20	1176.892	20.90	1998.011	24.60	3540.767
17.30	1193.092	21.00	2028.116	24.70	3597.592
17.40	1209.557	21.10	2058.742	24.80	3655.398
17.50	1226.292	21.20	2089.895	24.90	3714.203
17.60	1243.302	21.30	2121.587	25.00	3774.024

TEMPERATURE = 573.15 K

Density ρ (mol/L)	Pressure P (bar)	Density ρ (mol/L)	Pressure P (bar)	Density ρ (mol/L)	Pressure P (bar)
0.20	9.500	4.80	220.032	9.40	459.012
0.30	14.227	4.90	224.662	9.50	465.154
0.40	18.940	5.00	229.304	9.60	471.355
0.50	23.638	5.10	233.960	9.70	477.616
0.60	28.324	5.20	238.630	9.80	483.938
0.70	32.996	5.30	243.314	9.90	490.324
0.80	37.656	5.40	248.015	10.00	496.773
0.90	42.304	5.50	252.732	10.10	503.289
1.00	46.940	5.60	257.466	10.20	509.872
1.10	51.565	5.70	262.218	10.30	516.524
1.20	56.180	5.80	266.989	10.40	523.246
1.30	60.785	5.90	271.779	10.50	530.040
1.40	65.381	6.00	276.590	10.60	536.907
1.50	69.967	6.10	281.423	10.70	543.850
1.60	74.546	6.20	286.277	10.80	550.869
1.70	79.116	6.30	291.154	10.90	557.966
1.80	83.679	6.40	296.056	11.00	565.144
1.90	88.236	6.50	300.982	11.10	572.403
2.00	92.786	6.60	305.934	11.20	579.745
2.10	97.330	6.70	310.912	11.30	587.173
2.20	101.870	6.80	315.918	11.40	594.687
2.30	106.404	6.90	320.952	11.50	602.291
2.40	110.935	7.00	326.015	11.60	609.985
2.50	115.463	7.10	331.109	11.70	617.771
2.60	119.987	7.20	336.234	11.80	625.652
2.70	124.509	7.30	341.391	11.90	633.629
2.80	129.030	7.40	346.581	12.00	641.705
2.90	133.549	7.50	351.806	12.10	649.880
3.00	138.068	7.60	357.066	12.20	658.159
3.10	142.587	7.70	362.362	12.30	666.541
3.20	147.107	7.80	367.695	12.40	675.030
3.30	151.628	7.90	373.067	12.50	683.628
3.40	156.151	8.00	378.478	12.60	692.337
3.50	160.677	8.10	383.930	12.70	701.160
3.60	165.205	8.20	389.423	12.80	710.097
3.70	169.738	8.30	394.959	12.90	719.152
3.80	174.275	8.40	400.539	13.00	728.328
3.90	178.816	8.50	406.164	13.10	737.626
4.00	183.364	8.60	411.835	13.20	747.048
4.10	187.918	8.70	417.553	13.30	756.599
4.20	192.479	8.80	423.320	13.40	766.279
4.30	197.048	8.90	429.137	13.50	776.091
4.40	201.625	9.00	435.005	13.60	786.039
4.50	206.212	9.10	440.925	13.70	796.125
4.60	210.808	9.20	446.899	13.80	806.351
4.70	215.414	9.30	452.927	13.90	816.720

(Continued)

TEMPERATURE = 573.15 K (Continued)

Density ρ (mol/L)	Pressure P (bar)	Density ρ (mol/L)	Pressure P (bar)	Density ρ (mol/L)	Pressure P (bar)
14.00	827.235	17.70	1349.794	21.40	2296.253
14.10	837.899	17.80	1368.521	21.50	2330.849
14.20	848.715	17.90	1387.552	21.60	2366.035
14.30	859.686	18.00	1406.894	21.70	2401.821
14.40	870.814	18.10	1426.551	21.80	2438.218
14.50	882.103	18.20	1446.530	21.90	2475.237
14.60	893.556	18.30	1466.837	22.00	2512.887
14.70	905.177	18.40	1487.477	22.10	2551.181
14.80	916.967	18.50	1508.457	22.20	2590.130
14.90	928.931	18.60	1529.783	22.30	2629.745
15.00	941.072	18.70	1551.461	22.40	2670.037
15.10	953.394	18.80	1573.498	22.50	2711.019
15.20	965.899	18.90	1595.900	22.60	2752.702
15.30	978.592	19.00	1618.674	22.70	2795.099
15.40	991.476	19.10	1641.827	22.80	2838.221
15.50	1004.554	19.20	1665.365	22.90	2882.083
15.60	1017.831	19.30	1689.296	23.00	2926.695
15.70	1031.309	19.40	1713.627	23.10	2972.072
15.80	1044.994	19.50	1738.364	23.20	3018.226
15.90	1058.889	19.60	1763.516	23.30	3065.170
16.00	1072.997	19.70	1789.090	23.40	3112.919
16.10	1087.324	19.80	1815.093	23.50	3161.486
16.20	1101.872	19.90	1841.533	23.60	3210.886
16.30	1116.647	20.00	1868.418	23.70	3261.131
16.40	1131.653	20.10	1895.756	23.80	3312.238
16.50	1146.893	20.20	1923.555	23.90	3364.220
16.60	1162.373	20.30	1951.824	24.00	3417.092
16.70	1178.096	20.40	1980.570	24.10	3470.870
16.80	1194.068	20.50	2009.802	24.20	3525.569
16.90	1210.293	20.60	2039.530	24.30	3581.204
17.00	1226.776	20.70	2069.761	24.40	3637.792
17.10	1243.522	20.80	2100.504	24.50	3695.348
17.20	1260.535	20.90	2131.770	24.60	3753.888
17.30	1277.821	21.00	2163.566	24.70	3813.431
17.40	1295.385	21.10	2195.903	24.80	3873.991
17.50	1313.231	21.20	2228.790	24.90	3935.587
17.60	1331.366	21.30	2262.237	25.00	3998.235

SOLUBILITY PARAMETERS OF THE MOST COMMON FLUIDS FOR SUPERCRITICAL FLUID EXTRACTION AND CHROMATOGRAPHY

The following table provides the solubility parameters, δ^*, for the most common fluids and modifiers used in supercritical fluid extraction and chromatography. The data presented in the first table are for carrier or solvent supercritical fluids at a reduced temperature T_r of 1.02 and a reduced pressure P_r of 2. These values were calculated with the equation of Lee and Kesler [1,2]. The data presented in the second table are for liquid solvents that are potential modifiers [3].

The solubility parameter is defined as the square root of the cohesive energy density. The most common presentation of the solubility parameter is in units of $(cal^{1/2} cm^{-3/2})$. Here, the cohesive energy is expressed per unit volume. A more modern format that is found in some of the literature after 1990 utilizes SI units derived from cohesive pressures. It is possible to convert between the two scales with the following equations:

$$\delta^*(cal^{1/2}\ cm^{-3/2}) = 0.48888 \times \delta^*(MPa^{1/2}) \tag{4.1}$$

$$\delta^*(MPa^{1/2}) = 2.0455 \times \delta^*(cal^{1/2}\ cm^{-3/2}) \tag{4.2}$$

Thus, as a rough guide, the solubility parameters expressed in the SI system of cohesive pressures are numerically approximately double the values expressed in the older system.

REFERENCES

1. Lee, B. I., and M. G. Kesler. "Generalized Thermodynamic Correlation Based on 3-Parameter Corresponding States." *AIChE Journal* 21 (1975): 510.
2. Schoenmakers, P. J., and L. G. M. Vunk. "Mobile and Stationary Phases for Supercritical Fluid Chromatography". *Advances in Chromatography*, edited by J. C. Giddings, E. Grushka, and P. R. Brown, Vol. 30. New York: Marcel Dekker, 1989.
3. Barton, A. *CRC Handbook of Solubility and Cohesive Energy Parameters.* Boca Raton, FL: CRC Press, 1983.

Solubility Parameters of Supercritical Fluids

In this table, the solubility parameter, δ^* (in $cal^{1/2} cm^{-3/2}$), was obtained by the methods outlined in Lee and Kesler (1975) and Schoenmakers and Vunk (1989), and the conversion to the pressure scale was done by applying Equation 4.2.

Supercritical Fluid	δ^* ($cal^{1/2} cm^{-3/2}$)	δ^* ($MPa^{1/2}$)
Carbon dioxide	7.5	15.3
Nitrous oxide	7.2	14.7
Sulfur hexafluoride	5.5	11.3
Ammonia	9.3	19.0
Xenon	6.1	12.5
Ethane	5.8	11.9
Propane	5.5	11.3
n-Butane	5.3	10.8
Diethyl ether	5.4	11.0

Solubility Parameters of Liquid Solvents

In this table, we provide solubility parameters for some liquid solvents that can be used as modifiers in supercritical fluid extraction and chromatography. The solubility parameters (in $MPa^{1/2}$) were obtained from Barton (1983), and those in $cal^{1/2}$ $cm^{-3/2}$ were obtained by application of Equation 4.2, for consistency. It should be noted that other tabulations exist in which these values are slightly different, since they were calculated from different measured data or models. The reader is therefore cautioned that these numbers are for trend analysis and separation design only. For other applications of cohesive parameter calculations, it may be more advisable to consult a specific compilation. This table should be used along with the table on modifier decomposition, since many of these liquids show chemical instability, especially in contact with active surfaces.

Liquid	δ^* ($cal^{1/2}$ $cm^{-3/2}$)	δ^* ($MPa^{1/2}$)
n-Pentane	7.0	14.4
n-Hexane	7.3	14.9
n-Heptane	7.5	15.3
Cyclohexane	8.2	16.8
Benzene	9.1	18.7
Toluene	8.9	18.3
Acetone	9.6	19.7
Methyl ethyl ketone	9.4	19.3
Chloroform	9.1	18.7
Dichloromethane	9.9	20.2
Trichloroethene	9.1	18.7
Methanol	14.5	29.7
Ethanol	12.8	26.2
Diethyl ether	7.5	15.4
Tetrahydrofuran	9.0	18.5
1,4-Dioxane	10	20.5
Water	23.5	48.0

INSTABILITY OF MODIFIERS USED WITH SUPERCRITICAL FLUIDS

Liquid modifiers that are commonly used to increase the effective polarity of supercritical fluids such as carbon dioxide frequently have inherent chemical instabilities that must be considered when designing an analysis, or in the interpretation of results [1–3]. In many cases, such solvents are obtainable with stabilizers added to control the instability or to slow the reaction. Reactive solvents that do not have stabilizers must be used quickly or be given proper treatment. In either case, it is important to understand that the solvents (as they may be used in an analysis) are not necessarily pure materials. The reader is cautioned that many of the other fluids listed earlier in this section are thermally unstable; this table only treats chemical instabilities that are considerable at typical laboratory ambient temperature.

REFERENCES

1. Sadek, P. C. *The HPLC Solvent Guide*. 2nd ed. New York: Wiley Interscience, 2002.
2. Bruno, T. J., and G. C. Straty. *Journal of Research of the National Bureau of Standards (U.S.)* 91, no. 3 (1986):135–38.
3. Ashe, W. "Mobile Phases for Supercritical Fluid Chromatography." *Chromatographia* 11 no. 7 (1978): 411.

Solvent	Contaminants, Reaction Products	Stabilizers
Ethers		
Diethyl ether	peroxides[1]	2–3 % (vol/vol) ethanol[2] 1–10 ppm (mass/mass) BHT (1.5–3.5 % ethanol) + (0.2–0.5 % water) + (5–10 ppm (mass/mass) BHT)
Isopropyl ether	peroxides[1]	0.01 % (mass/mass) hydroquinone 5–100 ppm (mass/mass) BHT
1,4-Dioxane	peroxides[1]	25–1500 ppm (mass/mass) BHT
tetrahydrofuran	peroxides[1]	25–250 PPM (mass/mass) BHT
Chlorinated Alkanes		
Chloroform	Hydrochloric acid, chlorine, phosgene (CCl_2O)	0.5–1 % (vol/vol) ethanol 50–150 ppm (mass/mass) amylene[3] various ethanol amylene blends
Dichloromethane	Hydrochloric acid, chlorine, phosgene (CCl_2O)	25 ppm (mass/mass) amylene 25 ppm (mass/mass) cyclohexene 400–600 ppm (mass/mass) methanol various amylene methanol blends
Alcohols		
Ethanol	water, numerous denaturants are commonly added	
Methanol	water; formaldehyde (at elevated temperature)	
Acetone	diacetone alcohol and higher oligomers	

Notes:
1. The peroxide concentration that is usually considered hazardous is 250 ppm (mass/mass). See the treatment of peroxide hazards in the Laboratory Safety chapter of this book (Chapter 14).
2. Ethanol does not actually stabilize diethyl ether, nor is it a peroxide scavenger, although it was thought to be so in the past. It is still available in chromatographic solvents to preserve the utility of retention relationships and analytical methods.
3. Amylene is a generic name for 2-methyl-2-butene.

Abbreviations:
BHT: 2,6-di-t-butyl-p-cresol.

CHAPTER 5

Electrophoresis

CONTENTS

SEPARATION RANGES OF POLYACRYLAMIDE GELS

The following table provides a rough guide to the separation ranges of polyacrylamide gels that have varying gel concentrations, T, expressed in a percentage, as a function of relative molecular mass [1,2].

REFERENCES

1. Andrews, A. T. *Electrophoresis: Theory Techniques and Biochemical and Clinical Applications.* 2nd ed. Oxford: Oxford University Press, 1986, reproduced with permission.
2. Hames, B. D., ed. *Gel Electrophoresis of proteins: A Practical Approach*, 3rd ed. The Practical Approach Series. New York: Oxford University Press, 1998.

T (percentage)	Optimum Relative Molecular Mass Range
3–5	Above 100,000
5–12	20,000–150,000
10–15	10,000–80,000
15+	Below 15,000

PREPARATION OF POLYACRYLAMIDE GELS

The following table provides (in recipe format) the typical proportions of reagents needed to prepare 100 mL of the starting material for polyacrylamide gels [1,2]. The factor T is the gel concentration and is related to the ability to separate a given relative molecular mass range. Typically, the tertiary aliphatic amines N,N,N,N-tetramethylethylenediamine (TEMED) or 3-dimethylamino-propionitrile (DMAPN) are used to catalyze the reaction. Note that gelation does not occur readily below T = 2.5 %.

REFERENCES

1. Andrews, A. T. *Electrophoresis: Theory Techniques and Biochemical and Clinical Applications.* 2nd ed. Oxford: Oxford University Press, 1986, reproduced with permission.
2. www.protocol-online.org/prot/Molecular_Biology/Electrophoresis/Polyacrylamide_Gel_ Electrophoresis__PAGE_/index.html, accessed 2009.

Constituent	Amounts Required for Gels with		
	T = 5 %	T = 7.5 %	T = 10 %
Acrylamide	4.75 g	7.125 g	9.50 g
Biscrylamide	0.25 g	0.375 g	0.50 g
TEMED or DMAPN	0.05 mL	0.05 mL	0.05 mL
Ammonium persulfate	0.05 g	0.05 g	0.05 g

BUFFER MIXTURES COMMONLY USED FOR POLYACRYLAMIDE GEL ELECTROPHORESIS

The following table provides suggested buffers used for polyacrylamide gel electrophoresis. This list is by no means exhaustive, however these buffers are the most common [1,2].

REFERENCES

1. Andrews, A. T. *Electrophoresis: Theory Techniques and Biochemical and Clinical Applications.* 2nd ed. Oxford: Oxford University Press, 1986, reproduced with permission.
2. McClellan, T.A. "Electrophoresis Buffers for Polyacrylamide Gels at Various pH." *Analytical Biochemistry* 126 (2004): 94–99.

Approximate pH Range	Primary Buffer Constituent	pH Adjusted to the Desired Value with
2.4–6.0	0.1 M citric acid	1 M NaOH
2.8–3.8	0.05 M formic acid	1 M NaOH
4.0–5.5	0.05 M acetic acid	1 M NaOH or tris
5.2–7.0	0.05 M maleic acid	1 M NaOH or tris
6.0–8.0	0.05 M KH_2PO_4 or NaH_2PO_4	1 M NaOH
7.0–8.5	0.05 M sodium diethyl barbiturate (veronal)	1 M HCl
7.2–9.0	0.05 M tris	1 M HCl or glycine
8.5–10.0	0.015 M $Na_2B_4O_7$	1 M HCl or NaOH
9.0–10.5	0.05 M glycine	1 M NaOH
9.0–11.0	0.025 M $NaHCO_3$	1 M NaOH

Note that tris is a buffer made using tris(hydroxymethyl)aminomethane, also abbreviated as THAM.

PROTEINS FOR INTERNAL STANDARDIZATION OF
POLYACRYLAMIDE GEL ELECTROPHORESIS

The following table provides a list of proteins that may be used as internal standards in quantitative applications of polyacrylamide gel electrophoresis. These proteins may be used in isoelectric focusing or in sodium dodecyl sulfate polyacrylamide gel electrophoresis (SDS-PAGE). The isoelectric points are reported at 25 °C [1,2].

REFERENCES

1. Andrews, A. T. *Electrophoresis: Theory Techniques and Biochemical and Clinical Applications.* 2nd ed. Oxford: Oxford University Press, 1986, reproduced with permission.
2. Inouye, M. "Internal Standards for Molecular Weight Determination of Proteins by Polyacrylamide Gel Electrophoresis." *Journal of Biological Chemistry* 246 (1971): 4834–38.

Protein	Isoelectric Point (pI at 25 °C)	Relative Molecular Mass
Lysozyme	10.0	14,000
Cytochrome C (horse)	9.3	12,256
Chymotrypsinogen A (ox)	9.0	23,600
Ribonuclease A	8.9	13,500
Myoglobin (sperm whale)	8.2	17,500
Myoglobin (horse)	7.3	17,500
Erythroagglutinin (red kidney bean)	6.5	130,000
Insulin (beef)	5.7	11,466
β-Lactoglobulin B	5.3	36,552
β-Lactoglobulin A	5.1	36,724
Bovine serum albumin	5.1	67,000
Ovalbumin	4.7	45,000
Alkaline phosphatase (calf intestine)	4.4	140,000
α-Lactalbumin	4.3	14,146

CHROMOGENIC STAINS FOR GELS

The following table provides common stain reagents for use in electrophoresis gels [1–5].

REFERENCES

1. Melvin, M. *Electrophoresis (Analytical Chemistry by Open Learning).* Chichester: John Wiley and Sons, 1987, reproduced with permission.
2. Hart, C., B. Schulenberg, T. H. Steinberg, W.-Y. Leung, and W. F. Patton. "Detection of Glycoproteins in Polyacrylamide Gels on Electroblots Using Pro-Q Emerald 488 Dye, A Fluroescent Periodate Schiff Base Stain." *Electrophoresis* 24 (2003): 588.
3. Dovward, D.W. "Detection and Quantitation of Heme-Containing Proteins by Chemiluminescence." *Analytical Biochemistry* 209 (1993): 219–223.
4. Miller, I., Crawford, J., and Gianazza, E. "Protein Stain for Proteomic Applications: Which, When, Why." *Proteomics* 6 (2006): 5385–5408.
5. Antharavally, B.S., Carter, B., Bell, P.A. and Mallia, A.K. "A High-Affinityreversible Protein Stain for Western Blots." *Analytical Biochemistry* 329 (2004): 276–80.

Types of Substance Stained	Staining Reagent	Comments
Amino acids, peptides, and proteins	Ninhydrin	Very sensitive stain for amino acids, either free or combined in polypeptides. Used after paper electrophoresis.
Proteins	Amido Black 10B	Binds to cationic groups on proteins. Adsorbs onto cellulose, giving high background staining with paper and dehydration and shrinkage of polyacrylamide gels.
	Coomassie Brilliant Blue	Binds to basic groups on proteins and also by nonpolar interactions. Widely used stain.
	Ponceau S (Ponceau Red)	Used routinely in clinical laboratories for cellulose acetate and starch gels. Very rapid staining reaction that leaves a clear background.
Glycoproteins	Alcian Blue	Stains the sugar moiety.
	Emerald 300 Emerald 488	Stains aldehydes on sugar moiety after treatment with periodic acid.
Copper-containing proteins	Alizarin Blue S	Specifically indicates the presence of copper.
Polynucleotides, including RNA and DNA	Acridine orange	Stained product can be assessed quantitatively.
	Pyronine Y (or G)	Gives a permanent staining, so electrophoretogram can be stored for several weeks.
Proteins, lipids, carbohydrates, polynucleotides	Stains-All	Wide applicability, as it forms characteristic colored products with many different types of molecule. Low sensitivity.

FLUORESCENT STAINS FOR GELS

The following table provides common fluorescent stain reagents for use in electrophoresis [1–5]. Note that these agents are typically applied in small amounts before electrophoresis. Other stains are available as proprietary materials, and the reader is advised to consult reviews on staining procedures for additional materials. The reader is advised that any reagent capable of binding DNA with high affinity is a possible carcinogen; all such stains and dyes must be handled with care. Nitrile laboratory gloves are recommended; latex will not provide adequate protection.

REFERENCES

1. Melvin, M. *Electrophoresis (Analytical Chemistry by Open Learning).* Chichester: John Wiley and Sons, 1987, reproduced with permission.
2. Williams, L. "Staining Nucleic Acids and Proteins in Electrophoresis Gels." *Biotechnic and Histochemistry* 76, no. 3 (2001): 127–32.
3. Allen, R., and B. Budowle. *Protein Staining and Identification Techniques.* Westborough, MA: BioTechniques Press, 1999.
4. Hames, B. D. *Gel Electrophoresis of Proteins, A Practical Approach.* 3rd ed. Oxford: Oxford University Press, 2002.
5. http://www.gelifesciences.com/aptrix/upp00919.nsf/Content/D72B4B8BB3B714C6C1257628001CEB 96/$file/app_note_66.pdf

Types of Substance Stained	Staining Reagent	Comments
Proteins	Dansyl chloride	Reacts with amine groups
	1-anilino-8-naphthalene sulfonic acid	Nonfluorescent, but gives fluorescent product
	Fluorescamine	Nonfluorescent (nor are the hydrolysis products), but gives a fluorescent product with a labeled protein
	2-methoxy-2,4-diphenyl-2(H)-furanone	Nonfluorescent (nor are the hydrolysis products), but gives a fluorescent product with a labeled protein
Polynucleotides, including RNA and DNA	Acridine orange	A nucleic acid selective fluorescent cationic dye; often used in conjunction with ethidium bromide, below
Double-stranded polynucleotides	Ethidium bromide	Very sensitive. Widely used with agarose gels

CHAPTER 6

ELECTROANALYTICAL METHODS

CONTENTS

DETECTION LIMITS FOR VARIOUS ELECTROCHEMICAL TECHNIQUES

The following table provides guidance in selection of electrochemical techniques by providing the relative sensitivities of various methods [1]. The limit of detection of lead, defined as the minimum detectable quantity (on a mole basis), is used as the basis of comparison.

REFERENCE

1. Batley, G. E. *Trace Element Speciation: Analytical Methods and Problems*. Boca Raton, FL: CRC Press, 1989.

Electrochemical Technique	Limit of Detection for Lead (mol)
DC polarography (DME)	2×10^{-6}
DC polarography (SMDE)	1×10^{-7}
DP polarography (SMDE)	1×10^{-7}
DP anodic stripping voltammetry (HMDE)	2×10^{-10}
SW anodic stripping voltammetry (HMDE)	1×10^{-10}
DC anodic stripping voltammetry (TMFE)	5×10^{-11}
DP anodic stripping voltammetry (TMFE)	1×10^{-11}
SW anodic stripping voltammetry (TMFE)	5×10^{-12}

Abbreviations:
DC = direct current
DP = differential pulse
SW = square wave
DME = dropping mercury electrode
SMDE = static mercury drop electrode
HMDE = hanging mercury drop electrode
TMFE = thin mercury film electrode

VALUES OF (2.3026 RT/F) AT DIFFERENT TEMPERATURES

The following table gives the variation of (2.3026 RT/F; mV) with temperature (°C) [1]. Electronic pH meters are voltmeters with scale divisions that are equivalent to the value of 2.3026 RT/F (in mV) per pH unit. Generally, a reproducibility of ±0.005 pH unit is feasible when the pH meter is reproducible to 0.2 mV.

REFERENCE

1. Shugar, G. J., and J. A. Dean. *The Chemist's Ready Reference Handbook.* New York: McGraw-Hill Book Company, 1990.

T (°C)	2.3026 RT/F (mV)	T (°C)	2.3026 RT/F (mV)
0	54.199	50	64.120
5	55.191	55	65.112
10	56.183	60	66.104
15	57.175	65	67.096
18	57.770	70	68.088
20	58.167	75	69.080
25	59.159	80	70.073
30	60.152	85	71.065
35	61.144	90	72.057
38	61.739	95	73.049
40	62.136	100	74.041
45	63.128		

POTENTIAL OF ZERO CHARGE (E^{ecm}) FOR VARIOUS ELECTRODE MATERIALS IN AQUEOUS SOLUTIONS AT ROOM TEMPERATURE

The table below lists the potential of zero charge (E^{ecm}) values (in volts) for various electrode materials in aqueous solutions at room temperature (25 °C) [1]. All values are with respect to the normal hydrogen electrode.

REFERENCE

1. Parsons, R. *Handbook of Electrochemical Constants*. London: Butterworths, 1959.

Electrode	E^{ecm} (V)	Solution Composition
Ag	+ 0.05	0.1 N KNO_3
Cd	−0.90	0.0001 N KCl
Ga	−0.60	1 N KCl + 0.1 N HCl
Hg	−0.192	capillary inactive salts[a]
Ni	−0.06	0.001 N HCl
Pb	−0.69	0.001 N KCl
Platinized Pt	+ 0.11	1 N Na_2SO_4 + 0.1 N H_2SO_4
Smooth Pt	+ 0.27	1 N Na_2SO_4 + 0.1 N H_2SO_4
Oxidized Pt	(+ 0.4)–(0.1)	1 N Na_2SO_4 + 0.1 N H_2SO_4
Te	+ 0.61	1 N Na_2SO_4
Tl	−0.80	0.001 N KCl
Tl–Hg (satd)	−0.65	1 N KCl
Zn	−0.63	1 N Na_2SO_4
Graphite	−0.07	0.05 N KCl
Activated charcoal	(0.0)–(+ 0.2)	1 N Na_2SO_4 + 0.1 N H_2SO_4

[a] Any salt that is nonreactive with the capillary can be used.

VARIATION OF REFERENCE ELECTRODE POTENTIALS WITH TEMPERATURE

The following table lists the potentials of various (0.1 M KCl calomel, saturated KCl calomel and 1.0 M KCl (Ag/AgCl) electrodes at different temperatures (in °C) [1–3]. The values include the liquid junction potential.

REFERENCES

1. Bates, R. G. et al. "pH Standards of High Acidity and High Alkalinity and the Practical Scale of pH." *Journal of Research of the National Bureau of Standards* 45 (1950): 418.
2. Bates, R. G., and Bower, V. E. "Standard Potential of the Silver Silver-Chloride Electrode from 0-DegreesC to 95-Degrees C and the Thermodynamic Properties of Dilute Hydrochloric Acid Solutions." *Journal of Research of the National Bureau of Standards* 53 (1955): 283.
3. Shugar, G. J., and J. A. Dean. *The Chemist's Ready Reference Handbook.* New York: McGraw-Hill Book Company, 1990.

Temperature (°C)	0.1 M KCl Calomel	Saturated KCl Calomel	1.0 M KCl Ag/AgCl
0	0.3367	0.25918	0.23655
5			0.23413
10	0.3362	0.25387	0.23142
15	0.3361	0.2511	0.22857
20	0.3358	0.24775	0.22557
25	0.3356	0.24453	0.22234
30	0.3354	0.24118	0.21904
35	0.3351	0.2376	0.21565
38	0.3350	0.2355	
40	0.3345	0.23449	0.21208
45			0.20835
50	0.3315	0.22737	0.20449
55			0.20056
60	0.3248	0.2235	0.19649
70			0.18782
80		0.2083	0.1787
90			0.1695

pH VALUES OF STANDARD SOLUTIONS USED IN THE CALIBRATION OF GLASS ELECTRODES

The following table gives the pH values of operational standard solutions recommended for the calibration of glass electrodes at 25 °C and 37 °C [1].

REFERENCE

1. Hibbert, D. B., and A. M. James. *Dictionary of Electrochemistry.* 2nd ed. New York: John Wiley and Sons, 1984.

Standard Solution	pH at	
	25 °C	37 °C
0.1 mol/kg potassium tetraoxalate	1.48	1.49
0.1 mol/dm³ hydrochloric acid + 0.09 mol/dm³ potassium chloride	2.07	2.08
0.05 mol/kg potassium hydrogen phthalate	4.005	4.022
0.10 mol/dm³ acetic acid + 0.10 mol/dm³ sodium acetate	4.644	4.647
0.10 mol/dm³ acetic acid + 0.01 mol/dm³ sodium acetate	4.713	4.722
0.02 mol/kg piperazazine phosphate	6.26	6.14
0.025 mol/kg disodium hydrogen phosphate + 0.025 mol/kg potassium dihydrogen phosphate	6.857	6.828
0.05 mol/kg tris(hydroxymethyl)methane hydrochloride + 0.01667 mol/kg tris(hydroxymethyl)methane	7.648	7.332
0.05 mol/kg disodium tetraborate (borax)	9.182	9.074
0.025 mol/kg sodium bicarbonate + 0.025 mol/kg sodium carbonate	9.995	9.889
Saturated calcium hydroxide	12.43	12.05

TEMPERATURE VERSUS pH CORRELATION OF STANDARD
SOLUTIONS USED FOR THE CALIBRATION OF ELECTRODES

The following table gives the temperature versus pH correlation of common standard solutions that are used for the calibration of electrodes [1–3]. Such solutions should be stable, easily prepared, whose solutes do not require further purification because of factors such as their hygroscopic nature. It is worth noting that the buffering capacity of these solutions is of little interest.

REFERENCES

1. Hibbert, D. B., and A. M. James. *Dictionary of Electrochemistry*. 2nd ed. New York: John Wiley and Sons, 1984.
2. Koryta, J., J. Dvorák, and V. Boháčková. *Electrochemistry*. London: Methuen and Co., 1970.
3. Robinson, R. A., and R. H. Stokes. *Electrolytic Solutions*. London: Butterworths, 1959.

	pH of						
Temperature °C	0.05 M Potassium Tetroxalate	Potassium Hydrogen Tartrate[a]	0.01 M Potassium Tartrate	0.05 M Potassium Hydrogen Phthalate	0.025 M K_2HPO_4 + 0.02 M NaH_2PO_4	0.01 M $Na_2B_4O_7$	Ca $(OH)_2$[a]
0	1.671	—	3.710	4.012	6.893	9.463	13.428
5	1.671	—	3.690	4.005	6.950	9.389	13.208
10	1.669	—	3.671	4.001	6.922	9.328	13.004
15	1.674	—	3.655	4.000	6.896	9.273	12.809
20	1.676	—	3.647	4.001	6.878	9.223	12.629
25	1.681	3.555	3.637	4.005	6.860	9.177	12.454
30	1.685	3.547	3.633	4.011	6.849	9.135	12.296
35	1.693	3.545	3.629	4.019	6.842	9.100	12.135
37	—	—	—	4.022	6.838	9.074	12.05
40	1.697	3.543	3.630	4.030	6.837	9.066	11.985
45	1.704	3.545	3.634	4.043	6.834	9.037	11.841
50	1.712	3.549	3.640	4.059	6.833	9.012	11.704
55	1.719	3.556	3.646	4.077	6.836	8.987	11.575
60	1.726	3.565	3.654	4.097	6.840	8.961	11.454
70	1.74	3.58	—	4.12	6.85	8.93	—
80	1.77	3.61	—	4.16	6.86	8.89	—
90	1.80	3.65	—	4.20	6.88	8.85	—
95	1.81	3.68	—	4.23	6.89	8.83	—

[a] Saturated at 25 °C

SOLID MEMBRANE ELECTRODES

The following table lists the most commonly used solid membrane electrodes, their applications and major interferences [1–2]. Often the membrane is composed of a salt (listed first) and a matrix (listed second). Thus, the $AgCl$–Ag_2S electrode involves the finely divided $AgCl$ in a Ag_2S matrix. The salt should be more soluble than the matrix, but insoluble enough so that its equilibrium solubility gives a lower anion (Cl^-) activity than that of the sample solution.

REFERENCES

1. Fritz, J. S., and G. H. Schenk. *Quantitative Analytical Chemistry.* 5th ed. Englewood Cliffs: Prentice Hall, 1987.
2. Hall, D. G. "Ion Selective Membrane Electrodes: A General Limiting Treatment of Interference Effects." *Journal of Physical Chemistry* 100 (1996): 7230.

Membrane	Ion Measured	Major Interferences
LaF_3	F^-	OH^-
Ag_2S	S^{-2}, Ag^+	Hg^{+2}
$AgCl$–Ag_2S	Cl^-	Br^-, I^-, S^{-2}, CN^-, NH_3
$AgBr$–Ag_2S	Br^-	I^-, S^{-2}, CN^-, NH_3
AgI–Ag_2S	I^-	S^{-2}, CN^-
$AgSCN$–Ag_2S	SCN^-	Br^-, I^-, S^{-2}, CN^-, NH_3
CdS–Ag_2S	Cd^{+2}	Ag^+, Hg^{+2}, Cu^{+2}
CuS–Ag_2S	Cu^{+2}	Ag^+, Hg^{+2}
PbS–Ag_2S	Pb^{+2}	Ag^+, Hg^{+2}, Cu^{+2}

LIQUID MEMBRANE ELECTRODES

The following table gives the basic information on several liquid membrane electrodes [1–3]. The selectivity of a membrane electrode for a given ion is determined primarily by the liquid ion exchanger used. Thus, as the preference of the ion exchanger for a specific ion increases, its selectivity increases. The selectivity is also affected by the organic solvent in which the liquid exchanger is dissolved. In this table, R- may be any organic radical or group.

REFERENCES

1. Durst, R. A., ed. *Ion-Selective Electrodes*, National Bureau of Standards Special Publication, Washington, 314 (1969): 70–71.
2. Frant, M. S., and J. W. Ross. "Potassium Ion Specific Electrode with High Selectivity for Potassium Over Sodium." *Science* 167 (1970): 987.
3. Fritz, J. S., and G. H. Schenk. *Quantitative Analytical Chemistry.* 5th ed. Englewood Cliffs: Prentice Hall, 1987.

Ion Measured	Exchange Site	Selectivity Coefficients
K^+	Valinomycin	Na^+, 0.0001
Ca^{+2}	$(RO)_2POO^-$	Na^+, 0.0016 Mg^{+2}, Ba^{+2}, 0.01 Sr^{+2}, 0.02 Zn^{+2}, 3.2 H^+, 10^{-7}
Ca^{+2} and Mg^{+2}	$(RO)_2POO^-$	Na^+, 0.01 Sr^{+2}, 0.54 Ba^{+2}, 0.94
Cu^{+2}	$RSCH_2COO^-$	Na^+, K^+, 0.0005 Mg^{+2}, 0.001 Ca^{+2}, 0.002 Ni^{+2}, 0.01 Zn^{+2}, 0.03
NO_3^-	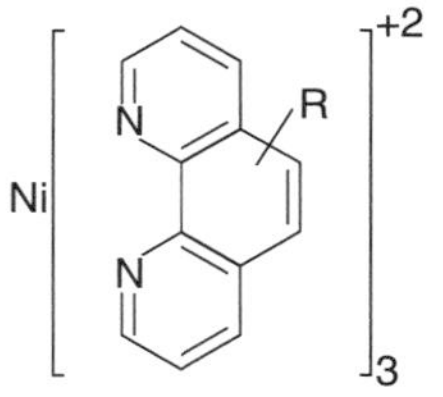 $[Ni(phenanthroline\text{-}R)_3]^{+2}$	F^-, 0.0009 SO_4^{-2}, 0.0006 PO_4^{-3}, 0.0003 Cl^-, CH_3COO^-, 0.006 HCO_3^-, CN^-, 0.02 NO_2^-, 0.06 Br^-, 0.9
ClO_4^-	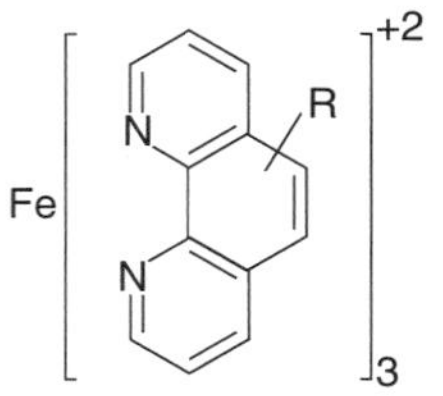 $[Fe(phenanthroline\text{-}R)_3]^{+2}$	Cl^-, SO_4^{-2}, 0.0002 Br^-, 0.0006 NO_3^-, 0.0015 I^-, 0.012 OH^-, 1.0

STANDARD REDUCTION ELECTRODE POTENTIALS FOR INORGANIC SYSTEMS IN AQUEOUS SOLUTIONS AT 25 °C

A summary of the potentials, $E°$, in volts (at 25 °C) of the most useful reduction half-reactions is presented below [1–5]. The reactions are arranged in order of decreasing oxidation strength. When comparing two half-reactions, the oxidizing agent of the half-reaction with the higher (more positive) $E°$ will react with the reducing agent of the half-reaction with its lower (less positive) $E°$. Thus, Br_2 (1) ($E° = 1.065$ V) will oxidize H_2O_2 to O_2(g) ($E° = 0.682$ V), but O_2(g) cannot oxidize Br^-. No predictions can be made on the rate of reaction.

If two or more reactions between two substances are possible, the reaction that involves half-reactions which are farthest apart in the table will be most thermodynamically favorable. For instance, in the case of O_2(g) reacting with Cu

$$O_2(g) + 4H^+ + 4e^- \rightarrow 2H_2O \ (E° = 1.229 \ V), \tag{6.1}$$

$$O_2(g) + 2H^+ + 2e^- \rightarrow H_2O_2 \ (E° = 0.682 \ V), \tag{6.2}$$

$$Cu^{+2} + 2e^- \rightarrow Cu \ (E° = 0.337 \ V). \tag{6.3}$$

The reaction between Equations 6.1 and 6.3 will be most favorable. However, if Equation 6.3 is replaced with Equation 6.4

$$Cl_2(g) + 2e^- \rightarrow 2Cl^- \ (E° = 1.36 \ V), \tag{6.4}$$

the reactions between Equation 6.2 and Equation 6.4 will take place first.

REFERENCES

1. Bard, A. J., and L. R. Faulkner. *Electrochemical Methods*. 2nd ed. New York: John Wiley and Sons, 2001.
2. Day, R. A., and A. L. Underwood. *Quantitative Analysis*. 6th ed. Upper Saddle River, NJ: Prentice Hall, 1991.
3. Dean, J. A., ed. *Lange's Handbook of Chemistry*. 14th ed. New York: McGraw-Hill Book Co., 1999.
4. Ebbing, D. D., and S. D. Gamma. *General Chemistry*. 9th ed. Boston, MA: Houghton Mifflin Co., 2008.
5. Sugar, G. J., and J. A. Dean. *The Chemist's Ready Reference Handbook*. New York: McGraw-Hill Book Co., 1990.

Standard Reduction Electrode Potentials for Inorganic Systems in Aqueous Solutions at 25 °C

Half-Reaction		$E°$, V
$F_2(g) + 2H^+ + 2e^-$	$\Rightarrow 2HF$	3.06
$O_3 + 2H^+ + 2e^-$	$\Rightarrow O_2 + H_2O$	2.07
$S_2O_8^{2-} + 2e^-$	$\Rightarrow 2SO_4^{2-}$	2.01
$Ag^{2+} + e^-$	$\Rightarrow Ag^+$	2.00
$H_2O_2 + 2H^+ + 2e^-$	$\Rightarrow 2H_2O$	1.77
$MnO_4^- + 4H^+ + 3e^-$	$\Rightarrow MnO_2(s) + 2H_2O$	1.70
$Ce(IV) + e^-$	$\Rightarrow Ce(III)$ (in 1 M $HClO_4$)	1.61
$H_5IO_6 + H^+ + 2e^-$	$\Rightarrow IO_3^- + 3H_2O$	1.6
$Bi_2O_4 + 4H^+ + 2e^-$	$\Rightarrow 2BiO^+ + 2H_2O$	1.59
$BrO_3^- + 6H^+ + 5e^-$	$\Rightarrow \frac{1}{2}Br_2 + 3H_2O$	1.52
$MnO_4^- + 8H^+ + 5e^-$	$\Rightarrow Mn^{2+} + 4H_2O$	1.51
$PbO_2 + 4H^+ + 2e^-$	$\Rightarrow Pb^{2+} + 2H_2O$	1.455
$Cl_2 + 2e^-$	$\Rightarrow 2Cl^-$	1.36
$Cr_2O_7^{2-} + 14H^+ + 6e^-$	$\Rightarrow 2Cr^{3+} + 7H_2O$	1.33
$MnO_2(s) + 4H^+ + 2e^-$	$\Rightarrow Mn^{2+} + 2H_2O$	1.23
$O_2(g) + 4H^+ + 4e^-$	$\Rightarrow 2H_2O$	1.229
$IO_3^- + 6H^+ + 5e^-$	$\Rightarrow \frac{1}{2}I_2 + 3H_2O$	1.20
$Br_2(liq) + 2e^-$	$\Rightarrow 2Br^-$	1.065
$ICl_2^- + e^-$	$\Rightarrow \frac{1}{2}I_2 + 2Cl^-$	1.06
$VO_2^+ + 2H^+ + e^-$	$\Rightarrow VO^{2+} + H_2O$	1.00
$HNO_2 + H^+ + e^-$	$\Rightarrow NO(g) + H_2O$	1.00
$NO_3^- + 3H^+ + 2e^-$	$\Rightarrow HNO_2 + H_2O$	0.94
$2Hg^{2+} + 2e^-$	$\Rightarrow Hg_2^{2+}$	0.92
$Cu^{2+} + I^- + e^-$	$\Rightarrow CuI$	0.86
$Ag^+ + e^-$	$\Rightarrow Ag$	0.799
$Hg_2^{2+} + 2e^-$	$\Rightarrow 2Hg$	0.79
$Fe^{3+} + e^-$	$\Rightarrow Fe^{2+}$	0.771
$O_2(g) + 2H^+ + 2e^-$	$\Rightarrow H_2O_2$	0.682
$2HgCl_2 + 2e^-$	$\Rightarrow Hg_2Cl_2(s) + 2Cl^-$	0.63
$Hg_2SO_4(s) + 2e^-$	$\Rightarrow 2Hg + SO_4^{2-}$	0.615
$H_3AsO_4 + 2H^+ + 2e^-$	$\Rightarrow HAsO_2 + 2H_2O$	0.581
$Sb_2O_5 + 6H^+ + 4e^-$	$\Rightarrow 2SbO^+ + 3H_2O$	0.559
$I_3^- + 2e^-$	$\Rightarrow 3I^-$	0.545
$Cu^+ + e^-$	$\Rightarrow Cu$	0.52
$VO^{2+} + 2H^+ + e^-$	$\Rightarrow V^{3+} + H_2O$	0.361
$Fe(CN)_6^{3-} + e^-$	$\Rightarrow Fe(CN)_6^{4-}$	0.36
$Cu^{2+} + 2e^-$	$\Rightarrow Cu$	0.337
$UO_2^{2+} + 4H^+ + 2e^-$	$\Rightarrow U^{4+} + 2H_2O$	0.334
$BiO^+ + 2H^+ + 3e^-$	$\Rightarrow Bi + H_2O$	0.32
$Hg_2Cl_2(s) + 2e^-$	$\Rightarrow 2Hg + 2Cl^-$	0.2676
$AgCl(s) + e^-$	$\Rightarrow Ag + Cl^-$	0.2223
$SbO^+ + 2H^+ + 3e^-$	$\Rightarrow Sb + H_2O$	0.212
$CuCl_3^{2-} + e^-$	$\Rightarrow Cu + 3Cl^-$	0.178
$SO_4^{2-} + 4H^+ + 2e^-$	$\Rightarrow SO_2(aq) + 2H_2O$	0.17
$Sn^{4+} + 2e^-$	$\Rightarrow Sn^{2+}$	0.154

Standard Reduction Electrode Potentials for Inorganic Systems in Aqueous Solutions at 25 °C (Continued)

Half-Reaction		$E°$, V
$S + 2H^+ + 2e^-$	$\Rightarrow H_2S(g)$	0.141
$TiO^{2+} + 2H^+ + e^-$	$\Rightarrow Ti^{3+} + H_2O$	0.10
$S_4O_6^{2-} + 2e^-$	$\Rightarrow 2S_2O_3^{2-}$	0.08
$AgBr(s) + e^-$	$\Rightarrow Ag + Br^-$	0.071
$2H^+ + 2e^-$	$\Rightarrow H_2$	0.00
$Pb^{2+} + 2e^-$	$\Rightarrow Pb$	−0.126
$Sn^{2+} + 2e^-$	$\Rightarrow Sn$	−0.136
$AgI(s) + e^-$	$\Rightarrow Ag + I^-$	−0.152
$Mo^{3+} + 3e^-$	$\Rightarrow Mo$	−0.2
$N_2 + 5H^+ + 4e^-$	$\Rightarrow H_2NNH_3^+$	−0.23
$Ni^{2+} + 2e^-$	$\Rightarrow Ni$	−0.246
$V^{3+} + e^-$	$\Rightarrow V^{2+}$	−0.255
$Co^{2+} + 2e^-$	$\Rightarrow Co$	−0.277
$Ag(CN)_2^- + e^-$	$\Rightarrow Ag + 2CN^-$	−0.31
$Cd^{2+} + 2e^-$	$\Rightarrow Cd$	−0.403
$Cr^{3+} + e^-$	$\Rightarrow Cr^{2+}$	−0.41
$Fe^{2+} + 2e^-$	$\Rightarrow Fe$	−0.440
$2CO_2 + 2H^+ + 2e^-$	$\Rightarrow H_2C_2O_4$	−0.49
$H_3PO_3 + 2H^+ + 2e^-$	$\Rightarrow H_3PO_2 + H_2O$	−0.50
$U^{4+} + e^-$	$\Rightarrow U^{3+}$	−0.61
$Zn^{2+} + 2e^-$	$\Rightarrow Zn$	−0.763
$Cr^{2+} + 2e^-$	$\Rightarrow Cr$	−0.91
$Mn^{2+} + 2e^-$	$\Rightarrow Mn$	−1.18
$Zr^{4+} + 4e^-$	$\Rightarrow Zr$	−1.53
$Ti^{3+} + 3e^-$	$\Rightarrow Ti$	−1.63
$Al^{3+} + 3e^-$	$\Rightarrow Al$	−1.66
$Th^{4+} + 4e^-$	$\Rightarrow Th$	−1.90
$Mg^{2+} + 2e^-$	$\Rightarrow Mg$	−2.37
$La^{3+} + 3e^-$	$\Rightarrow La$	−2.52
$Na^+ + e^-$	$\Rightarrow Na$	−2.714
$Ca^{2+} + 2e^-$	$\Rightarrow Ca$	−2.87
$Sr^{2+} + 2e^-$	$\Rightarrow Sr$	−2.89
$K^+ + e^-$	$\Rightarrow K$	−2.925
$Li^+ + e^-$	$\Rightarrow Li$	−3.045

STANDARD REDUCTION ELECTRODE POTENTIALS FOR INORGANIC SYSTEMS IN NONAQUEOUS SOLUTION AT 25 °C

The following table lists some standard electrode potentials (in V) in various solvents. The rubidium ion, which possesses a large radius and shows a low deformability, has a rather low and constant solvation energy in all solvents [1]. As a result, the rubidium electrode is taken as a standard reference electrode in all solvents.

REFERENCE

1. Koryta, J., J. Dvořák, and V. Boháčková. *Electrochemistry.* London: Methuen and Co., 1970.

System	H_2O	CH_3OH	CH_3CN	HCOOH	N_2H_4	NH_3
Li/Li$^+$	−0.03	−0.16	−0.06	−0.03	−0.19	−0.35
Rb/Rb$^+$	0.00	0.00	0.00	0.00	0.00	0.00
Cs/Cs$^+$	+0.06	—	+0.01	−0.01	—	−0.02
K/K$^+$	+0.06	—	+0.01	+0.10	−0.01	−0.05
Ca/Ca^{+2}	+0.14	—	+0.42	+0.25	+0.10	+0.29
Na/Na$^+$	+0.27	+0.21	+0.30	+0.03	+0.18	+0.08
Zn/Zn^{+2}	+2.22	+2.20	+2.43	+2.40	+1.60	+1.40
Cd/Cd^{+2}	+2.58	+2.51	+2.70	+2.70	+1.91	+1.73
Tl/Tl$^+$	+2.64	+2.56	—	—	—	—
Pb/Pb^{+2}	+2.85	+2.74	+3.05	+2.73	+2.36	+2.25
H_2/H$^+$	+2.98	+2.94	+3.17	+3.45	+2.01	+1.93
Cu/Cu^{+2}	+3.32	+3.28	+2.79	+3.31	—	+2.36
Cu/Cu$^+$	+3.50	—	+2.89	—	+2.23	+2.34
Hg/Hg^{+2}	+3.78	+3.68	—	+3.63	—	—
Ag/Ag$^+$	+3.78	+3.70	+3.40	+3.62	+2.78	+2.76
Hg/Hg^{+2}	+3.84	—	+3.42	—	—	+2.08
I$^-$/I$_2$	+3.52	+3.30	+3.24	+3.42	—	+3.38
Br$^-$/Br$_2$	+4.04	+3.83	+3.64	+3.97	—	+3.76
Cl$^-$/Cl$_2$	+4.34	+4.16	+3.75	+4.22	—	+3.96

REDOX POTENTIALS FOR SOME BIOLOGICAL HALF-REACTIONS

The following table lists the standard redox potentials of some common biological half-reactions (in V) at 298 K and pH = 7.0 [1].

REFERENCE

1. Hibbert, D. B., and A. M. James. *Dictionary of Electrochemistry*. 2nd ed. New York: John Wiley and Sons, 1984.

Biological System	Half-Cell Reaction	E° (V)
Acetate/pyruvate	$CH_3COOH + CO_2 + 2H^+ + 2e^- \rightarrow CH_3COCOOH + H_2O$	−0.70
Fe^{+3}/Fe^{+2} (ferredoxin)	$Fe^{+3} + e^- \rightarrow Fe^{+2}$	−0.432
H^+/H_2	$2H^+ + 2e^- \rightarrow H_2(g)$	−0.421
$NADP^+/NADPH$	$NADP^+ + 2H^+ + 2e^- \rightarrow NADPH + H^+$	−0.324
$NAD^+/NADH$	$NAD^+ + 2H^+ + 2e^- \rightarrow NADH + H^+$	−0.320
$FAD/FADH_2$	$FAD + 2H^+ + 2e^- \rightarrow FADH_2$	−0.219
Acetaldehyde/ethanol	$CH_3CHO + 2H^+ + 2e^- \rightarrow CH_3CH_2OH$	−0.197
Pyruvate/lactate	$CH_3COCOOH + 2H^+ + 2e^- \rightarrow CH_3CH(OH)COOH$	−0.185
Oxaloacetate/malate	$\begin{array}{c} CH_2COOH \\ | \\ O=C-COOH \end{array} + 2H^+ + 2e^- \rightarrow \begin{array}{c} CH_2COOH \\ | \\ HOCHCOOH \end{array}$	−0.166
Methylene Blue (ox; MB)/ Methylene Blue (red; MBH_2)	$MB + 2H^+ + 2e^- \rightarrow MBH_2$	0.011
Fumarate/succinate	$\begin{array}{c} CHCOOH \\ || \\ CHCOOH \end{array} + 2H^+ + 2e^- \rightarrow \begin{array}{c} CH_2COOH \\ | \\ CH_2COOH \end{array}$	0.031
Fe^{+3}/Fe^{+2} (myoglobin)	$Fe^{+3} + e^- \rightarrow Fe^{+2}$	0.046
Fe^{+3}/Fe^{+2} (cytochrome b)	$Fe^{+3} + e^- \rightarrow Fe^{+2}$	0.050
Ubiquinone (Ub)/ ubihydroquinone(UbH_2)	$Ub + 2H^+ + 2e^- \rightarrow UbH_2$	0.10
(cytochrome c)$^{+3}$/(cytochrome c)$^{+2}$	$Fe^{+3} + e^- \rightarrow Fe^{+2}$	0.254
(cytochrome a)$^{+3}$/(cytochrome a)$^{+2}$	$Fe^{+3} + e^- \rightarrow Fe^{+2}$	0.29
(cytochrome f)$^{+3}$/(cytochrome f)$^{+2}$	$Fe^{+3} + e^- \rightarrow Fe^{+2}$	0.365
Cu^{+2}/Cu^+(haemocyanin)	$Cu^{+2} + e^- \rightarrow Cu^+$	0.540
O_2/H_2O	$O_2(g) + 4H^+ + 4e^- \rightarrow 4H_2O$	0.816

STANDARD EMF OF THE CELL H_2/HCL, AgCL, Ag IN VARIOUS AQUEOUS SOLUTIONS OF ORGANIC SOLVENTS AT DIFFERENT TEMPERATURES

The table below lists the standard EMF values of the cell H_2/HCl, AgCl, Ag in water as well as in various aqueous solutions of three common organic solvents, all alcohols, at different temperatures [1,2]. The compositions are given as mass percentage of the alcohol in water. All EMF values are expressed in volts.

REFERENCES

1. Koryta, J., J. Dvorak, and V. Bohackova. *Electrochemistry*. London: Methuen and Co., 1970.
2. Robinson, R. A., and R. H. Stokes. *Electrolytic Solutions*. London: Butterworths, 1959.

°C	100 % Water	10 % aq. Methanol	10 % aq. Ethanol	10 % aq. 2-Propanol	20 % aq. Methanol	20 % aq. Ethanol	20 % aq. 2-Propanol
0	0.23655	0.22762	0.22726	0.22543	0.22022	0.21606	0.21612
5	0.23413	0.22547	0.22527	0.22365	0.21837	0.21486	0.21492
10	0.23142	0.22328	0.22328	0.22158	0.21631	0.21367	0.21336
15	0.22857	0.22085	0.22164	0.21922	0.21405	0.21190	0.21138
20	0.22557	0.21821	0.21901	0.21667	0.21155	0.21013	0.20906
25	0.22234	0.21535	0.21467	0.21383	0.20881	0.20757	0.20637
30	0.21904	0.21220	0.21383	0.21081	0.20567	0.20587	0.20341
35	0.21565	0.20892	0.21082	0.20754	0.20246	0.20275	0.20009
40	0.21208	0.20350	0.20783	0.20410	0.19910	0.19962	0.19652

TEMPERATURE DEPENDENCE OF THE STANDARD
POTENTIAL OF THE SILVER CHLORIDE ELECTRODE

The following table gives the standard potential (in V) of the silver chloride electrode (saturated KCl) at different temperatures (in °C) [1,2]. The uncertainty is ±0.05 mV.

REFERENCES

1. Conway, B. E. *Theory and Principles of Electrode Process.* New York: Ronald Press, 1965.
2. Koryta, J., J. Dvorák, and V. Boháčková. *Electrochemistry.* London: Methuen and Co., 1970.

Temperature (°C)	E° (V)	Temperature (°C)	E° (V)
0	0.23634	35	0.21563
5	0.23392	40	0.21200
10	0.23126	45	0.20821
15	0.22847	50	0.20437
20	0.22551	55	0.20035
25	0.22239	60	0.19620
30	0.21912		

STANDARD ELECTRODE POTENTIALS OF ELECTRODES OF THE FIRST KIND

The following table lists the standard electrode potentials (in V) of some electrodes of the first kind [1–3]. These are divided into cationic and anionic electrodes. In cationic electrodes, equilibrium is established between atoms or molecules of the substance and the corresponding cations in solution. Examples include metal, amalgam, and the hydrogen electrode. In anionic electrodes, equilibrium is achieved between molecules and the corresponding anions in solution. The potential of the electrode is given by the Nernst equation in the form:

$$E = E^o + (RT)/(Z_\pm F) \ln a_\pm,$$

where

E^o = standard electrode potential (in V)
R = gas constant (J K^{-1} mol^{-1})
T = temperature (in K)
$Z_\pm$ = charge, with sign, of the cation (+) or anion (−)
F = Faraday
$a_\pm$ = activity of the cation (+) or anion (−).

Electrodes of the first kind differ distinctly from the redox electrodes in that in the latter case both oxidation states can be present in variable concentrations, while in electrodes of the first kind, one of the oxidation states is the electrode material.

REFERENCES

1. Koryta, J., J. Dvorák, and L. Karan. *Principles of Electrochemistry.* 2nd ed. New York: John Wiley and Sons, 1993.
2. Koryta, J., J. Dvorák, and V. Boháčková. *Electrochemistry.* London: Methuen and Co., 1970.
3. Lide, D.R., ed, *CRC Handbook for Chemistry and Physics,* 90th. ed., Boca Raton, FL: CRC Press, 2010.

Electrode	E^o (V)[a]	Electrode	E^o (V)[a]
Li$^+$/Li	−3.0403	Ni^{+2}/Ni	−0.23
Rb$^+$/Rb	−2.98	In$^+$/In	−0.203
Cs$^+$/Cs	−2.92	Sn^{+2}/Sn	−0.1377
K$^+$/K	−2.931	Pb^{+2}/Pb	−0.1264
Ba^{+2}/Ba	−2.912	Cu^{+2}/Cu	+0.3417
Sr^{+2}/Sr	−2.89	Cu$^+$/Cu	+0.52
Ca^{+2}/Ca	−2.868	Te^{+4}/Te	+0.56
Na$^+$/Na	−2.71	Hg^{+2}/Hg	+0.851
Mg^{+2}/Mg	−2.372	Ag$^+$/Ag	+0.7994
Be^{+2}/Be	−1.847	Au^{+3}/Au	+1.42
Al^{+3}/Al	−1.662	Pt, Se^{-2}/Se	−0.78
Zn^{+2}/Zn	−0.7620	Pt, S^{-2}/S	−0.51
Fe^{+2}/Fe	−0.447	Pt, OH$^-$/O$_2$(g)	+0.401
Cd^{+2}/Cd	−0.4032	Pt, I$^-$/I$_2$	+0.536
In^{+3}/In	−0.3384	Pt, B$^-$/Br$_2$	+1.066
Tl$^+$/Tl	−0.336	Pt, Cl$^-$/Cl$_2$(g)	+1.35793
Co^{+2}/Co	−0.27	Pt, F$^-$/F$_2$(g)	+2.866

[a] All values have been taken from the *CRC Handbook of Chemistry and Physics* and are recalculated to the standard pressure of 1 atm (101.325 kPa).

STANDARD ELECTRODE POTENTIALS OF ELECTRODES OF THE SECOND KIND

The following table lists the standard electrode potentials (in V) of some electrodes of the second kind [1–3]. These consist of three phases. The metal is covered by a layer of its sparingly soluble salt and is immersed in a solution of a soluble salt of the anion. Equilibrium is established between the metal atoms and the solution anions through two partial equilibria: one between the metal and its cation in the sparingly soluble salt, and the other between the anion in the solid phase of the sparingly soluble salt and the anion in solution. The silver chloride electrode is preferred for precise measurements.

REFERENCES

1. Koryta, J., J. Dvorák, and L. Karan. *Principles of Electrochemistry*. 2nd ed. New York: John Wiley and Sons, 1993.
2. Koryta, J., J. Dvorák, and V. Boháčková. *Electrochemistry*. London: Methuen and Co., 1970.
3. Lide, D.R., ed, *CRC Handbook for Chemistry and Physics*, 90th. ed., Boca Raton, FL: CRC Press, 2010.

Electrode	$E°$ (V)[a]
$PbSO_4$, SO_4^{-2}/Pb, Hg	−0.351
AgI, I^-/Ag	−0.152
AgBr, Br^-/Ag	+0.071
HgO, OH^-/Hg	+0.0975
Hg_2Br_2, Br^-/Hg	+0.140
AgCl, Cl^-/Ag	+0.22216
Hg_2Cl_2, Cl^-/Hg	+0.26791
Hg_2SO_4, SO_4^{-2}/Hg	+0.6123
PbO_2, $PbSO_4$, SO_4^{-2}/Pb	+1.6912

[a] All values have been taken from the *CRC Handbook of Chemistry and Physics* and are recalculated to the standard pressure of 1 atm (101.325 kPa).

POLAROGRAPHIC HALF-WAVE POTENTIALS ($E_{1/2}$) OF INORGANIC CATIONS

The following table lists the polarographic half-wave potentials ($E_{1/2}$, in volts vs. SCE, the standard calomel electrode) of inorganic cations and the supporting electrolyte used during the determination [1–6]. All supporting electrolyte solutions are aqueous unless noted.

REFERENCES

1. Skoog, D. A., D. M. West, F. J. Holler, and S. R. Crouch. *Analytical Chemistry Fundamentals.* 8th ed. Florence, KY: Cengage Learning, 2004.
2. Vogel, A. I. *A Textbook for Quantitative Inorganic Analysis.* 3rd ed. New York: John Wiley and Sons, 1968.
3. Fritz, J. S., and G. H. Schenk. *Quantitative Analytical Chemistry.* 4th ed. Englewood Cliffs: Prentice Hall, 1987.
4. Christian, G. D. *Analytical Chemistry.* 5th ed. New York: John Wiley and Sons, 1994.
5. Ewing, G. W. *Instrumental Methods of Analysis.* 5th ed. New York: McGraw-Hill, 1985.
6. Meites, L. *Polarographic Techniques.* 2nd ed. New York: Wiley Interscience, 1965.

Polarographic Half-Wave Potentials ($E_{1/2}$) of Inorganic Cations

| | Supporting Electrolyte | | | | | | | | 0.5 M Tartrate and | | |
| | | | | | | | | | NaOH | pH = 4.5 | |
Cation	KCl (0.1 F)	NH_3 (1 F) NH_4Cl (1 F)	NaOH (1 F)	H_3PO_4 (7.3 F)	KCN (1 F)	$(CH_3)_4NCl$ (0.1 F)	HCl (1 F)	H_2SO_4 (0.5 M)	NaOH (0.1 F)	pH = 4.5	Others
Ba^{+2}						−1.94					
Bi^{+3}							−0.09	−0.04	−1.00	−0.23	
Cd^{+2}	−0.64 (−0.60)	−0.81	−0.78	−0.77	−1.18						HNO_3 (1.0 F) −0.59 KI (1.0 F) −0.74
Co^{+2}	−1.20	−1.29	−1.46	−1.20	−1.45						Pyridine (0.1 F)/ pyridinium (0.1 F) −1.07
Cr^{+3}		−1.43 (to Cr^{+2}) −1.71 (to Cr^0)		−1.02 (to Cr^{+2})	−1.38 (to Cr^{+2})						
Cu^{+2}	+0.04 (to Cu^+) −0.22 (to Cu^0)	−0.24 (to Cu^+) −0.51 (to Cu^0)	−0.41	−0.09	no reaction					−0.09	
Fe^{+2}	−1.3	−1.49									
Fe^{+3}			−1.12 (to Fe^{+2}) −1.74 (to Fe^0)	+0.06 (to Fe^{+2})					−1.20, −1.73		EDTA (0.1 F)/CH_3COONa (2.0 F) −0.17, −1.30
K^+											$(CH_3)_4NOH$ (0.1 M, 50 % C_2H_5OH) −2.10
Li^+											$(CH_3)_4NOH$ (0.1 M, 50 % (C_2H_5OH) −2.31
Mn^{+2}	−1.51										$H_2P_2O_7^{-2}$ (0.2 M), pH = 2.2, + 0.1
Na^+						−2.07					

(Continued)

Polarographic Half-Wave Potentials ($E_{1/2}$) of Inorganic Cations (Continued)

	Supporting Electrolyte										
									0.5 M Tartrate and		
Cation	KCl (0.1 F)	NH_3 (1 F) NH_4Cl (1 F)	NaOH (1 F)	H_3PO_4 (7.3 F)	KCN (1 F)	$(CH_3)_4NCl$ (0.1 F)	HCl (1 F)	H_2SO_4 (0.5 M)	NaOH (0.1 F)	pH = 4.5	Others
Ni^{+2}	–1.1	–1.10			–1.36						KSCN (1.0 F) –0.7; pyridine (1.0 F)/HCl, pH = 7, –0.78
O_2											pH = 1–10 (buffered) –0.05 & –0.90
Pb^{+2}	–0.40		–0.75		–0.72				–0.75	–0.48	HNO_3 (1 F) –0.40
Sn^{+2}							–0.47				F^- (0.1 F) –0.611; F^- (0.5 F) –0.683
Sn^{+4}			–0.75								HCl (1.0 F)/NH4 + (4.0 F) –0.25 & –0.52
Te^+	–0.48	–0.48	–0.48								
Zn^{+2}	–1.00	–1.34	–1.53						–1.15		

POLAROGRAPHIC $E_{1/2}$ RANGES (IN V vs. SCE) FOR THE REDUCTION OF BENZENE DERIVATIVES

The following table lists the polarographic $E_{1/2}$ potential ranges (in V vs. SCE, the standard calomel electrode) obtained at pH = 5–9 in unbuffered media in the reduction of benzene derivatives [1].

REFERENCE

1. Zuman, P. *The Elucidation of Organic Electrode Processes.* New York: Academic Press, 1969.

Benzene Derivative[a]	Formula[a]	Polarographic $E_{1/2}$ Potential Range[b]
Diaryl alkene	ArCH=CHAr	(−1.8)–(−2.3)
Methyl aryl ester	$ArCOOCH_3$	(−1.0)–(−2.4)
Aryl iodide	ArI	(−1.2)–(−1.9)
Aryl methyl ketone	$ArCOCH_3$	(−1.1)–(−1.8)
Aromatic aldehyde	ArCHO	(−1.1)–(−1.7)
Methyl α,β-unsaturated aryl ketone	$ArCH=CHCOCH_3$	(−1.0)–(−1.6)
Diaryl ketone	ArCOAr	(−0.7)–(−1.4)
Azobenzenes	ArN=NAr	(−0.3)–(−0.8)
Nitroarenes	$ArNO_2$	(−0.3)–(−0.7)
Nitrosoarenes	ArNO	(−0.1)–(−0.4)
Diaryl iodonium salts	Ar_2I^+	(−0.2)–(−0.3)

[a] Ar = aromatic ring
[b] In V vs. SCE

VAPOR PRESSURE OF MERCURY

The following table provides data on the vapor pressure of mercury, useful for assessing and controlling the hazards associated with use of mercury as an electrode [1].

REFERENCE

1. Lide, D. R., ed. *CRC Handbook for Chemistry and Physics.* 90th ed. Boca Raton, FL: CRC Press, 2010.

Temperature °C	Vapor Pressure mm Hg	Vapor Pressure Pa	Temperature °C	Vapor Pressure mm Hg	Vapor Pressure Pa
0	0.000185	0.0247	28	0.002359	0.3145
10	0.000490	0.0653	30	0.002777	0.3702
20	0.001201	0.1601	40	0.006079	0.8105
22	0.001426	0.1901	50	0.01267	1.689
24	0.001691	0.2254	100	0.273	36.4
26	0.002000	0.2666			

ORGANIC FUNCTIONAL GROUP ANALYSIS OF NONPOLAROGRAPHIC ACTIVE GROUPS

Often an organic functional group is not (or may not be) reduced polarographically at an accessible potential range. In this case it is necessary to convert this functional group to a derivative whose reduction is feasible within such an accessible potential range. The table below lists the most common functional groups, the reagent needed, and the polarographically active derivative as well as the polarographically active group [1–4]. Such conversions enlarge the number of organic compounds that can be determined by polarography.

REFERENCES

1. Svoronos, P., V. Horak, and P. Zuman. "Polarographic Study of Structure-Properties Relationship of p-Tosyl Sulfilimines, Phosphorus." *Sulfur and Silicon* 42 (1989): 139.
2. Willard, H. H., L. L. Merritt, Jr., J. A. Dean, and F. A. Settle, Jr. *Instrumental Methods of Analysis.* Florence, KY: D. Wadsworth, 1988.
3. Zuman, P. *Chemical and Engineering News.* March 18, 1968, p. 94.
4. Zuman, P. *Substituent Effects in Organic Polarography.* New York: Plenum, 1967.

Organic Functional Group Analysis of Nonpolarographic Active Groups

Functional Group	Reagent	Polarographically Active Derivative	Active Polarographic Group
Carbonyl (aldehyde, ketone)>C=O	semicarbazide $H_2NHNCNH_2$ (with $\parallel$ O below)	$>C=N-NHC-NH_2$ (with $\parallel$ O below C)	semicarbazide, $>C=N-N$
	hydroxylamine H_2NOH	$>C=N-OH$	oxime, $>C=N-OH$
Primary amine, $R-NH_2$	piperonal	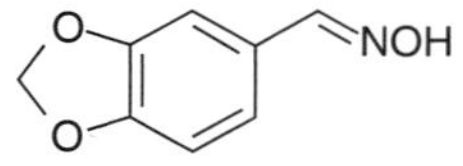	azomethine, $>C=N-R$
	carbon disulfide, CS_2		dithiocarbonate $-N=C$ (with S^- and S^-)
	cupric phosphate, $Cu_3(PO_4)_2$, suspension	$[Cu^{+2}-amine]$ complex	$[Cu^{+2}-N]$
Secondary amine, R_2NH	nitrous acid, HNO_2	$R_2N-N=O$	nitroso, $N-N=O$
Primary alcohols, $R-CH_2OH$	chromic acid, $HCrO_4$	$R-CHO$	aldehyde carbonyl, $C=O$ (with R and H)
Secondary alcohols, R_2CHOH	chromic acid, $HCrO_4$	$R_2C=O$	ketone carbonyl, $C=O$ (with R and R)
1,2-Diols	periodic acid, HIO_4		aldehyde and/or ketone carbonyl $>C=O$
Carboxylic acid, $R-C-OH$ (with $\parallel$ O)	thiourea, $(H_2N)_2C=S$	$RCO_2^-(H_2N)_2C^+SH$	thiol, $-S-H$
Phenyl, C_6H_5-	conc. nitric/conc. sulfuric acid, HNO_3/H_2SO_4	$C_6H_5-NO_2$	nitro, $-NO_2$
Sulfides (thioethers), $>S$	hydrogen peroxide, H_2O_2 or m-chloroperbenzoic acid, $m-Cl-C_6H_4-COOH$	$>S^+ \rightarrow O^-$	sulfoxide, S^+-O^-
	chloramine-T $p-CH_3-C_6H_4-SO_2NCl^-Na^+$	$p-CH_3-C_6H_4-SO_2N=S<$	sulfilimine, $>S=N-$

COULOMETRIC TITRATIONS

The following table lists some common coulometric (also known as constant-current coulometry) titrations [1–4]. Since the titrant is generated electrolytically and reacted immediately, the method gets widespread applications. The generating electrolytic concentrations need to be only approximate, while unstable titrants are consumed as soon as they are formed. The technique is more accurate than methods where visual end points are required, such as in the case of indicators. The unstable titrants in the table below are marked with an asterisk (*).

REFERENCES

1. Christian, G. D. *Analytical Chemistry.* 5th ed. New York: John Wiley and Sons, 1994.
2. Christian, G. D. *Advances in Biomedical and Medical Physics,* edited by S. N. Levine, Vol. 4. New York: Wiley-Interscience, 1971.
3. Skoog, D. A., D. M. West, F. J. Holler, and S. R. Crouch. *Analytical Chemistry Fundamentals.* 8th ed. Florence, KY: Cengage Learning, 2004.
4. Harris, D. C. *Quantitative Chemical Analysis.* 5th ed. San Francisco: W.H. Freeman, 1998.

Reagent	Generator Electrode Reaction	Typical Generating Electrolyte	Substances Determined
Ag^+	$Ag \rightarrow Ag^+ + e^-$	Ag anode in HNO_3	Br^-, Cl^-, thiols
Ag^{+2}	$Ag + \rightarrow Ag^{+2} + e^-$		Ce^{+3}, V^{+4}, $H_2C_2O_4$, As^{+3}
*Biphenyl radical anion	$(C_6H_5)_2 + e^- \rightarrow (C_6H_5)_2^-$	Biphenyl/$(CH_3)_4NBr$ in DMF	anthracene
*Br_2	$2\,Br^- \rightarrow Br_2 + 2e^-$	0.2M NaBr in 0.1 M H_2SO_4	As^{+3}, Sb^{+3}, U^{+4}, Tl^+, I^-, SCN^-, NH_2OH, N_2H_4, phenols, aromatic amines, mustard gas, olefins, 8-hydroxy-quinoline
*BrO^-	$Br^- + 2OH^- \rightarrow BrO^- + H_2O + 2e^-$	1 M NaBr in borate buffer, pH = 8.6	NH_3
Ce^{+4}	$Ce^{+2} \rightarrow Ce^{+4} + 2e^-$	0.1 M $CeSO_4$ in 3 M H_2SO_4	Fe^{+2}, Ti^{+3}, U^{+4}, As^{+3}, I^-, $Fe(SCN)_6^{-4}$
*Cl_2	$2\,Cl^- \rightarrow Cl_2 + 2e^-$		As^{+3}, I^-
*Cr^{+2}	$Cr^{+3} + e^- \rightarrow Cr^{+2}$	$Cr_2(SO_4)_3$ in H_2SO_4	O_2
*$CuCl_3^{-2}$	$Cu^{+2} + 3Cl^- + e^- \rightarrow CuCl_3^{-2}$	0.1 M $CuSO_4$ in 1 M HCl	V^{+5}, Cr^{+6}, IO_3^-
EDTA	$HgNH_3(EDTA)^{-2} + NH_4^+ + 2e^- \rightarrow Hg + 2NH_3 + (HEDTA)^{-3}$	0.02 M Hg^{+2}/EDTA in ammoniacal buffer, pH = 8.5, Hg cathode	Ca^{+2}, Cu^{+3}, Zn^{+2}, Pb^{+2}
EGTA	$HgNH_3(EGTA)^{-2} + NH_4^+ + 2e^- \rightarrow Hg + 2NH_3 + (HEDTA)^{-3}$	0.1 M Hg^{+2}/EGTA in triethanolamine, pH = 8.6, Hg cathode	Ca^{+2} (in the presence of Mg^{+2})
Fe^{+2}	$Fe^{+3} + e^- \rightarrow Fe^{+2}$	Acid solution of $FeNH_4(SO_4)_2$	Cr^{+6}, Mn^{+7}, V^{+5}, Ce^{+4}
I_2	$2I^- \rightarrow I_2 + 2e^-$	0.2M KI in pH = 8 buffer, pyridine, SO_2, CH_3OH, KI (Karl Fisher titration)	As^{+3}, Sb^{+3}, $S_2O_3^{-2}$, H_2S, H_2O
H^+	$2H_2O \rightarrow 4H^+ + O_2 + 4e^-$	0.1 M Na_2SO_4 (water electrolysis)	pyridine
*Mn^{+3}	$Mn^{+2} \rightarrow Mn^{+3} + e^-$	$MnSO_4$ in 2 M H_2SO_4	$H_2C_2O_4$, Fe^{+2}, As^{+3}, H_2O_2
Mo^{+5}	$Mo^{+6} + e^- \rightarrow Mo^{+5}$	0.7 M Mo^{+6} in 4 M H_2SO_4	$Cr_2O_7^{-2}$
*MV^{+a}	$MV^{+2} + e^- \rightarrow MV^+$		Mn^{+3} (in enzymes)
OH^-	$2H_2O + 2e^- \rightarrow H_2 + 2OH^-$	0.1 M Na_2SO_4 (water electrolysis)	HCl
Ti^{+3}	$Ti^{+4} + e^- \rightarrow Ti^{+3}$ or $TiO^{+2} + 2H^+ + e^- \rightarrow Ti^{+3} + H_2O$	3.6 M $TiCl_4$ in 7 M HCl	V^{+5}, Fe^{+3}, Ce^{+4}, U^{+6}
U^{+4}	$UO_2^{+2} + 4H^+ + 2e^- \rightarrow U^{+4} + 2H_2O$	Acid solution of UO_2^{+2}	Cr^{+6}, Ce^{+4}

a MV^+ = methyl viologen radical cation; MV^{+2} = methyl viologen radical cation

CHAPTER 7

Ultraviolet Spectrophotometry

CONTENTS

SOLVENTS FOR ULTRAVIOLET SPECTROPHOTOMETRY

The following table lists some useful solvents for ultraviolet spectrophotometry, along with their wavelength cutoffs and dielectric constants [1–6].

REFERENCES

1. Willard, H. H., L. L. Merritt, J. A. Dean, and F. A. Settle. *Instrumental Methods of Analysis*. 7th ed. New York, Belmont: Van Nostrand, 1988.
2. Strobel, H. A., and W. R. Heinemann. *Chemical Instrumentation: A Systematic Approach*. 3rd ed. New York: John Wiley and Sons, 1989.
3. Dreisbach, R. R. *Physical Properties of Chemical Compounds, Advances in Chemistry Series*. No. 15. Washington, DC: American Chemical Society, 1955.
4. Dreisbach, R. R. *Physical Properties of Chemical Compounds, Advances in Chemistry Series*. No. 22. Washington, DC: American Chemical Society, 1959.
5. Sommer, L. *Analytical Absorption Spectrophotometry in the Visible and Ultraviolet*. London: Elsevier Science, 1989.
6. Krieger, P. A. *High Purity Solvent Guide*. McGaw Park, IL: Burdick and Jackson, 1984.

Solvents for Ultraviolet Spectrophotometry

Solvent	Wavelength Cutoff, nm	Dielectric Constant (20 °C)
Acetic acid	260	6.15
Acetone	330	20.7 (25 °C)
Acetonitrile	190	37.5
Benzene	280	2.284
Sec-butyl alcohol (2-butanol)	260	15.8 (25 °C)
n-Butyl acetate	254	
n-Butyl chloride	220	7.39 (25 °C)
Carbon disulfide	380	2.641
Carbon tetrachloride	265	2.238
Chloroform[a]	245	4.806
Cyclohexane	210	2.023
1,2-Dichloroethane	226	10.19 (25 °C)
1,2-Dimethoxyethane	240	
N,N-Dimethylacetamide	268	59 (83 °C)
N,N-Dimethylformamide	270	36.7
Dimethylsulfoxide	265	4.7
1,4-Dioxane	215	2.209 (25 °C)
Diethyl ether	218	4.335
Ethanol	210	24.30 (25 °C)
2-Ethoxyethanol	210	
Ethyl acetate	225	6.02 (25 °C)
Glycerol	207	42.5 (25 °C)
n-Hexadecane	200	2.06 (25 °C)
n-Hexane	210	1.890
Methanol	210	32.63 (25 °C)
2-Methoxyethanol	210	16.9
Methyl cyclohexane	210	2.02 (25 °C)
Methyl ethyl ketone	330	18.5
Methyl isobutyl ketone	335	
2-Methyl-1-propanol	230	
N-Methyl-2-pyrrolidone	285	32.0
n-Pentane	210	1.844
n-Pentyl acetate	212	
n-Propyl alcohol	210	20.1 (25 °C)
Sec-propyl alcohol	210	18.3 (25 °C)
Pyridine	330	12.3 (25 °C)
Tetrachloroethylene[b]	290	
Tetrahydrofuran	220	7.6
Toluene	286	2.379 (25 °C)
1,1,2-Trichloro-1,2,2-trifluoroethane	231	
2,2,4-Trimethylpentane	215	1.936 (25 °C)
o-Xylene	290	2.568
m-Xylene	290	2.374
p-Xylene	290	2.270
Water		78.54 (25 °C)

[a] Stabilized with ethanol to avoid phosgene formation.
[b] Stabilized with thymol (isopropyl meta-cresol).

ULTRAVIOLET SPECTRA OF COMMON LIQUIDS

The following table presents, in tabular form, the ultraviolet spectra of some common solvents and liquids used in chemical analysis. The data were obtained using a 1.00 cm path length cell, against a water reference [1,2].

REFERENCES

1. Krieger, P. A. *High Purity Solvent Guide*. McGaw Park, IL: Burdick and Jackson, 1984.
2. Sommer, L. *Analytical Absorption Spectrophotometry in the Visible and Ultraviolet*. London: Elsevier Science, 1989.

Ultraviolet Spectra of Common Liquids

Acetone		Benzene	
Wavelength, nm	Maximum Absorbance	Wavelength, nm	Maximum Absorbance
330	1.000	278	1.000
340	0.060	300	0.020
350	0.010	325	0.010
375	0.005	350	0.005
400	0.005	400	0.005

Acetonitrile		I-Butanol	
Wavelength, nm	Maximum Absorbance	Wavelength, nm	Maximum Absorbance
190	1.000	215	1.000
200	0.050	225	0.500
225	0.010	250	0.040
250	0.005	275	0.010
350	0.005	300	0.005

2-Butanol		Carbon Tetrachloride	
Wavelength, nm	Maximum Absorbance	Wavelength, nm	Maximum Absorbance
260	1.000	263	1.000
275	0.300	275	0.100
300	0.010	300	0.005
350	0.005	350	0.005
400	0.005	400	0.005

n-Butyl Acetate		Chlorobenzene	
Wavelength, nm	Maximum Absorbance	Wavelength, nm	Maximum Absorbance
254	1.000	287	1.000
275	0.050	300	0.050
300	0.010	325	0.040
350	0.005	350	0.020
400	0.005	400	0.005

(Continued)

Ultraviolet Spectra of Common Liquids (Continued)

n-Butyl Chloride		Chloroform	
Wavelength, nm	**Maximum Absorbance**	**Wavelength, nm**	**Maximum Absorbance**
220	1.000	245	1.000
225	0.300	250	0.300
250	0.010	275	0.005
300	0.005	300	0.005
400	0.005	400	0.005

Cyclohexane		o-Dichlorobenzene	
Wavelength, nm	**Maximum Absorbance**	**Wavelength, nm**	**Maximum Absorbance**
200	1.000	295	1.000
225	0.170	300	0.300
250	0.020	325	0.100
300	0.005	350	0.050
400	0.005	400	0.005

Cyclopentane		Diethyl Carbonate	
Wavelength, nm	**Maximum Absorbance**	**Wavelength, nm**	**Maximum Absorbance**
200	1.000	256	1.000
215	0.300	265	0.150
225	0.020	275	0.050
300	0.005	300	0.040
400	0.005	400	0.010

Decahydronaphthalene		Dimethyl Acetamide	
Wavelength, nm	**Maximum Absorbance**	**Wavelength, nm**	**Maximum Absorbance**
200	1.000	268	1.000
225	0.500	275	0.300
250	0.050	300	0.080
300	0.005	350	0.005
400	0.005	400	0.005

Dimethyl Formamide		2-Ethoxyethanol	
Wavelength, nm	**Maximum Absorbance**	**Wavelength, nm**	**Maximum Absorbance**
268	1.000	210	1.000
275	0.300	225	0.500
300	0.050	250	0.200
350	0.005	300	0.005
400	0.005	400	0.005

Ultraviolet Spectra of Common Liquids (Continued)

Dimethyl Sulfoxide		Ethyl Acetate	
Wavelength, nm	Maximum Absorbance	Wavelength, nm	Maximum Absorbance
268	1.000	256	1.000
275	0.500	275	0.050
300	0.200	300	0.030
350	0.020	325	0.005
400	0.005	350	0.005

1,4-Dioxane		Diethyl Ether	
Wavelength, nm	Maximum Absorbance	Wavelength, nm	Maximum Absorbance
215	1.000	215	1.000
250	0.300	250	0.080
300	0.020	275	0.010
350	0.005	300	0.005
400	0.005	400	0.005

Dichloroethylene		n-Hexadecane	
Wavelength, nm	Maximum Absorbance	Wavelength, nm	Maximum Absorbance
228	1.000	190	1.000
240	0.300	200	0.500
250	0.100	250	0.020
300	0.005	300	0.005
400	0.005	400	0.005

Ethylene Glycol Dimethyl Ether (glyme)		n-Hexane	
Wavelength, nm	Maximum Absorbance	Wavelength, nm	Maximum Absorbance
220	1.000	195	1.000
250	0.250	225	0.050
300	0.050	250	0.010
350	0.010	275	0.005
400	0.005	300	0.005

n-Heptane		Isobutanol	
Wavelength, nm	Maximum Absorbance	Wavelength, nm	Maximum Absorbance
200	1.000	220	1.000
225	0.100	250	0.050
250	0.010	275	0.030
300	0.005	300	0.020
400	0.005	400	0.010

(Continued)

Ultraviolet Spectra of Common Liquids (Continued)

Methanol		Methyl-t-Butyl Ether	
Wavelength, nm	Maximum Absorbance	Wavelength, nm	Maximum Absorbance
205	1.000	210	1.000
225	0.160	225	0.500
250	0.020	250	0.100
300	0.005	300	0.005
400	0.005	400	0.005

2-Methoxyethanol		Methylene Chloride	
Wavelength, nm	Maximum Absorbance	Wavelength, nm	Maximum Absorbance
210	1.000	233	1.000
250	0.130	240	0.100
275	0.030	250	0.010
300	0.005	300	0.005
400	0.005	400	0.005

2-Methoxyethyl acetate		Methyl Ethyl Ketone	
Wavelength, nm	Maximum Absorbance	Wavelength, nm	Maximum Absorbance
254	1.000	329	1.000
275	0.150	340	0.100
300	0.050	350	0.020
350	0.005	375	0.010
400	0.005	400	0.005

Methyl Isoamyl Ketone		n-Methylpyrrolidone	
Wavelength, nm	Maximum Absorbance	Wavelength, nm	Maximum Absorbance
330	1.000	285	1.000
340	0.100	300	0.500
350	0.050	325	0.100
375	0.010	350	0.030
400	0.005	400	0.010

Methyl Isobutyl Ketone		n-Pentane	
Wavelength, nm	Maximum Absorbance	Wavelength, nm	Maximum Absorbance
334	1.000	190	1.000
340	0.500	200	0.600
350	0.250	250	0.010
375	0.050	300	0.005
400	0.005	400	0.005

Ultraviolet Spectra of Common Liquids (Continued)

Methyl n-Propyl Ketone		β-Phenethylamine	
Wavelength, nm	Maximum Absorbance	Wavelength, nm	Maximum Absorbance
331	1.000	285	1.000
340	0.150	300	0.300
350	0.020	325	0.100
375	0.005	350	0.050
400	0.005	400	0.005

I-Propanol		Pyridine	
Wavelength, nm	Maximum Absorbance	Wavelength, nm	Maximum Absorbance
210	1.000	330	1.000
225	0.500	340	0.100
250	0.050	350	0.010
300	0.005	375	0.010
400	0.005	400	0.005

2-Propanol		Tetrahydrofuran	
Wavelength, nm	Maximum Absorbance	Wavelength, nm	Maximum Absorbance
205	1.000	212	1.000
225	0.160	250	0.180
250	0.020	300	0.020
300	0.005	350	0.005
400	0.010	400	0.005

Propylene Carbonate		Toluene	
Wavelength, nm	Maximum Absorbance	Wavelength, nm	Maximum Absorbance
280	1.000	284	1.000
300	0.500	300	0.120
350	0.050	325	0.020
375	0.030	350	0.050
400	0.020	400	0.005

1,2,4-Trichlorobenzene		2,2,4-Trimethylpentane	
Wavelength, nm	Maximum Absorbance	Wavelength, nm	Maximum Absorbance
308	1.000	215	1.000
310	0.500	225	0.100
350	0.050	250	0.020
375	0.010	300	0.005
400	0.005	400	0.005

(Continued)

Ultraviolet Spectra of Common Liquids (Continued)

Trichloroethylene		Water	
Wavelength, nm	**Maximum Absorbance**	**Wavelength, nm**	**Maximum Absorbance**
273	1.000	190	0.010
300	0.100	200	0.010
325	0.080	250	0.005
350	0.060	300	0.005
400	0.060	400	0.005

1,1,2-Trichlorotrifluoroethane		o-Xylene	
Wavelength, nm	**Maximum Absorbance**	**Wavelength, nm**	**Maximum Absorbance**
231	1.000	288	1.000
250	0.050	300	0.200
300	0.005	325	0.050
350	0.005	350	0.010
400	0.005	400	0.005

TRANSMITTANCE–ABSORBANCE CONVERSION

The following is a conversion table for absorbance and transmittance, assuming no reflection. Included for each pair is the percentage error propagated into a measured concentration (using the Beer–Lambert Law), assuming an uncertainty in transmittance of $+0.005$ [1]. The value of transmittance that will give the lowest percentage error in concentration is 3.368. Where possible, analyses should be designed for the low error area.

REFERENCE

1. Kennedy, J. H. *Analytical Chemistry Principles.* San Diego: Harcourt, Brace and Jovanovich, 1984.

Transmittance–Absorbance Conversion

Transmittance	Absorbance	Percentage Uncertainty
0.980	0.009	25.242
0.970	0.013	16.915
0.960	0.018	12.752
0.950	0.022	10.256
0.940	0.027	8.592
0.930	0.032	7.405
0.920	0.036	6.515
0.910	0.041	5.823
0.900	0.046	5.270
0.890	0.051	4.818
0.880	0.056	4.442
0.870	0.060	4.125
0.860	0.065	3.853
0.850	0.071	3.618
0.840	0.076	3.412
0.830	0.081	3.231
0.820	0.086	3.071
0.810	0.091	2.928
0.800	0.097	2.799
0.790	0.102	2.684
0.780	0.108	2.579
0.770	0.113	2.483
0.760	0.119	2.386
0.750	0.125	2.316
0.740	0.131	2.243
0.730	0.137	2.175
0.720	0.143	2.113
0.710	0.149	2.055
0.700	0.155	2.002
0.690	0.161	1.952
0.680	0.167	1.906
0.670	0.174	1.863
0.660	0.180	1.822

(*Continued*)

Transmittance–Absorbance Conversion (Continued)

Transmittance	Absorbance	Percentage Uncertainty
0.650	0.187	1.785
0.640	0.194	1.750
0.630	0.201	1.717
0.620	0.208	1.686
0.610	0.215	1.657
0.600	0.222	1.631
0.590	0.229	1.605
0.580	0.237	1.582
0.570	0.244	1.560
0.560	0.252	1.539
0.540	0.268	1.502
0.530	0.276	1.485
0.520	0.284	1.470
0.510	0.292	1.455
0.500	0.301	1.442
0.490	0.310	1.430
0.480	0.319	1.419
0.470	0.328	1.408
0.460	0.337	1.399
0.450	0.347	1.391
0.440	0.356	1.383
0.430	0.366	1.377
0.420	0.377	1.372
0.410	0.387	1.367
0.400	0.398	1.364
0.390	0.409	1.361
0.380	0.420	1.359
0.370	0.432	1.358
0.360	0.444	1.359
0.350	0.456	1.360
0.340	0.468	1.362
0.330	0.481	1.366
0.320	0.495	1.371
0.310	0.509	1.376
0.300	0.523	1.384
0.290	0.538	1.392
0.280	0.553	1.402
0.270	0.569	1.414
0.260	0.585	1.427
0.250	0.602	1.442
0.240	0.620	1.459
0.230	0.638	1.478
0.220	0.657	1.500
0.210	0.678	1.525
0.200	0.699	1.553
0.190	0.721	1.584
0.180	0.745	1.619

Transmittance–Absorbance Conversion (Continued)

Transmittance	Absorbance	Percentage Uncertainty
0.170	0.769	1.659
0.160	0.796	1.704
0.150	0.824	1.756
0.140	0.854	1.816
0.130	0.886	1.884
0.120	0.921	1.964
0.110	0.958	2.058
0.100	1.000	2.170
0.090	1.046	2.306
0.080	1.097	2.473
0.070	1.155	2.685
0.060	1.222	2.961
0.050	1.301	3.336
0.040	1.398	3.881
0.030	1.523	4.751
0.020	1.699	6.387
0.010	2.000	10.852

CORRELATION TABLE FOR ULTRAVIOLET ACTIVE FUNCTIONALITIES

The following table presents a correlation between common chromophoric functional groups and the expected absorptions from ultraviolet spectrophotometry [1–3]. While not as informative as infrared correlations, UV can often provide valuable qualitative information.

REFERENCES

1. Willard, H. H., L. L. Merritt, J. A. Dean, and F. A. Settle. *Instrumental Methods of Analysis*. 7th ed. Belmont: Wadsworth Publishing Co., 1988.
2. Silverstein, R. M., and F. X. Webster. *Spectrometric Identification of Organic Compounds*. 6th ed. New York: Wiley, 1998.
3. Lambert, J. B., H. F. Shurvell, D. A. Lightner, L. Verbit, and R. G. Cooks. *Organic Structural Spectroscopy*. Upper Saddle River, NJ: Prentice Hall, 1998.

Correlation Table for Ultraviolet Active Functionalities

Chromophore	Functional Group	λ_{max} nm	ε_{max}	λ_{max} nm	ε_{max}	λ_{max} nm	ε_{max}
Ether	-O-	185	1000				
Thioether	-S-	194	4600	215	1600		
Amine	-NH$_2$⁻	195	2800				
Amide	-CONH$_2$	<210	—				
Thiol	-SH	195	1400				
Disulfide	-S-S-	194	5500	255	400		
Bromide	-Br	208	300				
Iodide	-I	260	400				
Nitrile	-C≡N	160	—				
Acetylide (alkyne)	-C≡C-	175–180	6000				
Sulfone	-SO$_2$-	180	—				
Oxime	N-OH	190	5000				
Azido	>C=N-	190	5000				
Alkene	-C=C-	190	8000				
Ketone	>C=O	195	1000	270–285	18–30		
Thioketone	>C=S	205	strong				
Esters	-COOR	205	50				
Aldehyde	-CHO	210	strong	280–300	11–18		
Carboxyl	-COOH	200–210	50–70				
Sulfoxide	>S→O	210	1500				
Nitro	-NO$_2$	210	strong				
Nitrite	-ONO	220–230	1000–2000	300–4000	10		
Azo	-N=N-	285–400	3–25				
Nitroso	-N=O	302	100				
Nitrate	-ONO$_2$	270 (shoulder)	12				
Conjugated hydrocarbon	-(C=C)$_2$-(acyclic)	210–230	21,000				
Conjugated hydrocarbon	-(C=C)$_3$-	260	35,000				

(Continued)

Correlation Table for Ultraviolet Active Functionalities (Continued)

Chromophore	Functional Group	λ_{max} nm	ε_{max}	λ_{max} nm	ε_{max}	λ_{max} nm	ε_{max}
Conjugated hydrocarbon	$-(C=C)_4-$	300	52,000				
Conjugated hydrocarbon	$-(C=C)_5-$	330	118,000				
Conjugated hydrocarbon	$-(C=C)_2-$(alicyclic)	230–260	3000–8000				
Conjugated hydrocarbon	$C=C-C\equiv C$	219	6500				
Conjugated system	$C=C-C=N$	220	23,000				
Conjugated system	$C=C-C=O$	210–250	10,000–20,000			300–350	weak
Conjugated system	$C=C-NO_2$	229	9500				
Benzene		184	46,700	202	6900	255	170
Diphenyl	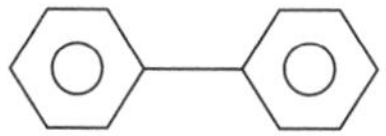			246	20,000		
Naphthalene	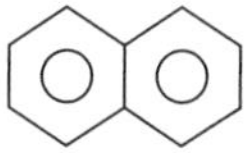	220	112,000	275	5600	312	175
Anthracene	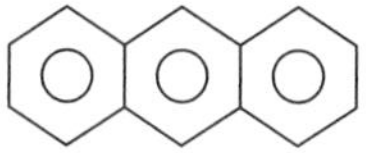	252	199,000	375	7900		
Pyridine		174	80,000	195	6000	251	1700
Quinoline		227	37,000	270	3600	314	2750
Isoquinoline	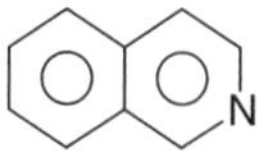	218	80,000	266	4000	317	3500

Note: ϕ denotes a phenyl group.

WOODWARD'S RULES FOR BATHOCHROMIC SHIFTS

Conjugated systems show bathochromic shifts in their B→B* transition bands. Empirical methods for predicting those shifts were originally formulated by Woodward, Fieser, and Fieser [1–4]. This section includes the most important conjugated system rules [1–6]. The reader should consult Silverstein and Webster (1998) [5] and Lambert and colleagues (1998) [6] for more details on how to apply the wavelength increment data.

REFERENCES

1. Woodward, R. B. "Structure and the Absorption Spectra of α,β-Unsaturated Ketones." *Journal of the American Chemical Society* 63 (1941): 1123.
2. Woodward, R. B. "Structure and Absorption Spectra. III. Normal Conjugated Dienes." *Journal of the American Chemical Society* 64 (1942): 72.
3. Woodward, R. "Structure and Absorption Spectra. IV. Further Observations on α,β-Unsaturated Ketones." *Journal of the American Chemical Society* 64 (1942): 76.
4. Fieser, L. F., and M. Fieser. *Natural Products Related to Phenanthrene*. New York: Reinhold, 1949.
5. Silverstein, R. M., and F. X. Webster. *Spectrometric Identification of Organic Compounds*. 6th ed. New York: Wiley, 1998.
6. Lambert, J. B., H. F. Shurvell, D. A. Lightner, L. Verbit, and R. G. Cooks. *Organic Structural Spectroscopy*. Upper Saddle River, NJ: Prentice Hall, 1998.

a. Rules of Diene Absorption

Base value for diene: 214 nm	
Increments for (each) (in nm):	
Heteroannular diene	+0
Homoannular diene	+39
Extra double bond	+30
Alkyl substituent or ring residue	+5
Exocyclic double bond	+5
Polar groups:	
–OOCR	+0
–OR	+6
–S–R	+30
halogen	+5
–NR$_2$	+60
λ Calculated	= Total

b. Rules for Enone Absorption[a]

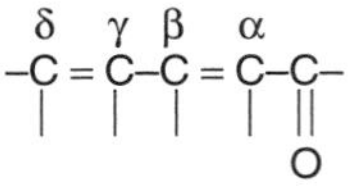

$$\begin{array}{cccc} \delta & \gamma & \beta & \alpha \\ -C = C - C & = & C - C - \\ | & | & | & | & \| \\ & & & & O \end{array}$$

Base value for acyclic (or six-membered) α,β-unsaturated ketone: 215 nm
Base value for five-membered α,β-unsaturated ketone: 202 nm
Base value for α,β-unsaturated aldehydes: 210 nm
Base value for α,β-unsaturated esters or carboxylic acids: 195 nm
Increments for (each) (in nm):

Heteroannular diene	+0
Homoannular diene	+39
Double bond	+30
Alkyl group:	
$\alpha-$	+10
$\beta-$	+12
$\gamma-$ and higher	+18
Polar groups:	
−OH	
$\alpha-$	+35
$\beta-$	+30
$\delta-$	+50
−OOCR	
$\alpha,\beta,\gamma,\delta$	+6
−OR	
$\alpha-$	+35
$\beta-$	+30
$\gamma-$	+17
$\delta-$	+31
−SR	
$\beta-$	+85
−Cl	
$\alpha-$	+15
$\beta-$	+12
−Br	
$\alpha-$	+25
$\beta-$	+30
−NR$_2$	
$\beta-$	+95
Exocyclic double bond	+5
λ Calculated	= Total

[a] Solvent corrections should be included. These are: water (−8), chloroform (+1), dioxane (+5), ether (+7), hexane (+11), cyclohexane (+11). No correction for methanol or ethanol.

c. Rules for Monosubstituted Benzene Derivatives

Parent Chromophore (benzene): 250 nm	
Substituent	Increment
–R	–4
–COR	–4
–CHO	0
–OH	–16
–OR	–16
–COOR	–16

Where R is an alkyl group, and the substitution is on C_6H_5–.

Rules for disubstituted benzene derivatives

Parent Chromophore (benzene): 250 nm			
Substituent	o–	m–	p–
–R	+3	+3	+10
–COR	+3	+3	+10
–OH	+7	+7	+25
–OR	+7	+7	+25
–O$^-$	+11	+20	+78 (variable)
–Cl	+0	+0	+10
–Br	+2	+2	+15
–NH$_2$	+13	+13	+58
–NHCOCH$_3$	+20	+20	+45
–NHCH$_3$	—	—	+73
–N(CH$_3$)$_2$	+20	+20	+85

R indicates an alkyl group.

CHAPTER 8

Infrared Spectrophotometry

CONTENTS

INFRARED OPTICS MATERIALS

The following table lists the more common materials used for optical components (windows, prisms, etc.) in the infrared region of the electromagnetic spectrum. The properties listed are needed to choose the materials with optimal transmission characteristics [1,2]. The thermal properties are useful when designing experiments for operation at elevated temperatures [3–5]. This listing is far from exhaustive, but these are the most common materials used in instrumentation laboratories.

REFERENCES

1. Gordon, A. J., and G. A. Ford. *The Chemist's Companion*. New York: John Wiley and Sons, 1972.
2. Willard, H. H., L. L. Merritt, J. A. Dean, and F. A. Settle. *Instrumental Methods of Analysis.* 7th ed. Belmont: Wadsworth, 1988.
3. Touloukien, Y. S., R. W. Powell, C. Y. Ho, and P. G. Klemens. *Thermophysical Properties of Matter: Thermal Conductivity of Nonmetallic Solids.* Vol. 2. New York: IF—Plenum Data Corp., 1970.
4. Touloukien, Y. S., R. K. Kirby, R. E. Taylor, and T. Lee. *Thermophysical Properties of Matter: Thermal Expansion of Nonmetallic Solids.* Vol. 13. New York: IF—Plenum Data Corp., 1977.
5. Wolfe, W. L., and G. J. Zissis. eds. *The Infrared Handbook*. Moscow: Mir, 1995.

Infrared Optics Materials

Material	Wavelength Range, μm	Wavenumber Range, cm^{-1}	Refractive Index at 2 μm	Thermal Conductivity w/(m•K) × 10^2	Thermal Expansion ΔL/L, percentage	Notes
Sodium chloride NaCl	0.25–16	40,000–625	1.52	7.61 (273 K) 6.61 (300 K) 4.85 (400 K)	0.448 (400 K) 0.896 (500 K)	Most common material; absorbs water; for aqueous solutions, use saturated NaCl solution as the solvent
Potassium bromide KBr	0.25–25	40,000–400	1.53	5.00 (275 K) 4.87 (301.5 K) 4.80 (372.2 K)	0.028 (400 K) 0.429 (500 K) 0.846 (600 K)	Useful for the study of C–Br stretch region, useful for solid sample pellets
Silver chloride AgCl	0.4–23	25,000–435	2.0	1.19 (269.8 K) 1.10 (313.0 K) 1.05 (372.5 K)	0.356 (400 K) 0.729 (500 K) 1.183 (600 K)	Not good for amines or liquids with basic nitrogen; light sensitive
Silver bromide AgBr	0.50–35	20,000–286	2.2	0.90 (308.2 K) 0.79 (353.2 K) 0.71 (413.2 K)	0.024 (300 K) 0.109 (325 K) 0.196 (350 K)	Not good for amines or liquids with basic nitrogen; light sensitive
Calcium fluoride CaF$_2$	0.15–9	66,700–1110	1.40	10.40 (237 K) 9.60 (309 K) 4.14 (402 K)	0.214 (400 K) 0.431 (500 K) 0.670 (600 K)	Useful for obtaining high resolution for —OH, N–H, and C–H stretching frequencies
Barium fluoride BaF$_2$	0.20–11.5	50,000–870	1.46	11.7 (284 K) 10.9 (305 K) 10.5 (370 K)	0.233 (400 K) 0.461 (500 K) 0.698 (600 K)	Shock sensitive, should be handled with care.
Cesium bromide CsBr	1–37	10,000–270	1.67	9.24 (269.4 K) 8.00 (337.5 K) 7.76 (367.5 K)	0.526 (400 K) 1.063 (500 K) 1.645 (600 K)	Useful for C–Br stretching frequencies.
Cesium iodide CsI	1–50	10,000–200	1.74	1.15 (277.7 K) 1.05 (296.0 K) 0.95 (360.7 K)		Useful for C–Br stretching frequencies.
Thallium bromide–thallium iodide TlBr–TlI (KRS-5)	0.5–35	20,000–286	2.37		0.464 (373 K) 1.026 (473 K)	Highly toxic, handle with care; 42 % TlBr, 58 % TlI
Zinc selenide ZnSe	1–18	10,000–555	2.4		0.086 (400 K) 0.175 (500 K) 0.272 (600 K)	Vacuum deposited

Material						Remarks
Germanium Ge	0.5–11.5	20,000–870	4.0			
Silicon Si	0.20–6.2	50,000–1613	3.5		0.033 (400 K) 0.066 (500 K) 0.102 (600 K)	
Aluminum oxide (sapphire) Al_2O_3	0.20–6.5	50,000–1538	1.76	25.1 (293.2 K) 21.3 (323 K) 14.2 (432.2 K)	0.075 (400 K) 0.148 (500 K) 0.225 (600 K)	
Polyethylene	16–300	625–33	1.54			Not useful for many organic compounds
Mica	200–425	50–23.5				
Fused silica, SiO_2	0.2–4.0	50,000–2,500	1.42 (at 3 m)	1.38 (298 K)		Used in near infrared work; can be used with dilute and concentrated acids (except HF), not for use with aqueous alkali; metal ions can be problematic

INTERNAL REFLECTANCE ELEMENT CHARACTERISTICS

Internal reflectance methods are a common sampling method in infrared spectrophotometry. The following table provides guidance in the selection of elements for reflectance methods [1].

REFERENCE

1. Coleman, P. *Practical Sampling Techniques for Infrared Analysis.* Boca Raton, FL: CRC Press, 1993.

Material	Frequency Range (cm⁻¹)	Index of Refraction	Characteristics
Thallium iodide-thallium bromide (KRS-5)	16,000–250	2.4	Relatively soft, deforms easily; warm water, ionizable acids and bases, chlorinated solvents, and amines should not be used with this ATR element.
Zinc selenide (Irtran-4)	20,000–650	2.4	Brittle; releases H_2Se, a toxic material, if used with acids; water insoluble; electrochemical reactions with metal salts or complexes are possible.
Zinc sulfide (Cleartran)	50,000–770	2.2	Reacts with strong oxidizing agents; relatively inert with typical aqueous, normal acids and bases and organic solvents; good thermal and mechanical shock properties; low refractive index causes spectral distortions at 45 °C.
Cadmium telluride (Irtran-6)	10,000–450	2.6	Expensive; relatively inert; reacts with acids.
Silicon	9,000–1,550, 400-	3.5	Hard and brittle; useful at high temperatures to 300 °C; relatively inert.
Germanium	5,000–850	4.0	Hard and brittle; temperature opaque at 125 °C.
Diamond	4,000–400	2.46	Extremely robust element, not brittle unless used as a composite with other materials; note that diamond absorbs at 2,500–1,900, thus producing a gap in the spectrum that cannot be measured.

WATER SOLUBILITY OF INFRARED OPTICS MATERIALS

The following table provides guidance in the selection of optics materials [1]. Often, the solubility in (pure) water of a particular material is of critical concern.

REFERENCE

1. Coleman, P. *Practical Sampling Techniques for Infrared Analysis.* Boca Raton, FL: CRC Press, 1993.

Material	Formula	Solubility g/100 g H_2O at 20 °C
Sodium chloride	NaCl	36.0
Potassium bromide	KBr	65.2
Potassium chloride	KCl	34.7
Cesium iodide	CsI	160 (at 61 °C)
Fused silica	SiO_2	insoluble
Calcium fluoride	CaF_2	1.51×10^{-3}
Barium fluoride	BaF_2	0.12 (at 25 °C)
Thallium bromide-iodide (KRS-5)	—	$< 4.76 \times 10^{-2}$
Silver bromide	AgBr	1.2×10^{-5}
Zinc sulfide	ZnS	insoluble
Zinc selenide (Irtran-4)	ZnSe	insoluble
Polyethylene (high density)	—	insoluble

WAVELENGTH–WAVENUMBER CONVERSION TABLE

The following table provides a conversion between wavelength and wavenumber units, for use in infrared spectrophotometry.

Wavelength–Wavenumber Conversion Table

Wavelength µm	WAVELENGTH–WAVENUMBER CONVERSION TABLE Wavenumber (cm⁻¹)									
	0	1	2	3	4	5	6	7	8	9
2.0	5000	4975	4950	4926	4902	4878	4854	4831	4808	4785
2.1	4762	4739	4717	4695	4673	4651	4630	4608	4587	4566
2.2	4545	4525	4505	4484	4464	4444	4425	4405	4386	4367
2.3	4348	4329	4310	4292	4274	4255	4237	4219	4202	4184
2.4	4167	4149	4232	4115	4098	4082	4065	4049	4032	4016
2.5	4000	3984	3968	4953	3937	3922	3006	3891	3876	3861
2.6	3846	3831	3817	3802	3788	3774	3759	3745	3731	3717
2.7	3704	3690	3676	3663	3650	3636	3623	3610	3597	3584
2.8	3571	3559	3546	3534	3521	3509	3497	3484	3472	3460
2.9	3448	3436	3425	3413	3401	3390	3378	3367	3356	3344
3.0	3333	3322	3311	3300	3289	3279	3268	3257	3247	3236
3.1	3226	3215	3205	3195	3185	3175	3165	3155	3145	3135
3.2	3125	3115	3106	3096	3086	3077	3067	3058	3049	3040
3.3	3030	3021	3012	3003	2994	2985	2976	2967	2959	2950
3.4	2941	2933	2924	2915	2907	2899	2890	2882	2874	2865
3.5	2857	2849	2841	2833	2825	2817	2809	2801	2793	2786
3.6	2778	2770	2762	2755	2747	2740	2732	2725	2717	2710
3.7	2703	2695	2688	2681	2674	2667	2660	2653	2646	2639
3.8	2632	2625	2618	2611	2604	2597	2591	2584	2577	2571
3.9	2654	2558	2551	2545	2538	2532	2525	2519	2513	2506
4.0	2500	2494	2488	2481	2475	2469	2463	2457	2451	2445
4.1	2439	2433	2427	2421	2415	2410	2404	2398	2387	2387
4.2	2381	2375	2370	2364	2358	2353	2347	2342	2336	2331
4.3	2326	2320	2315	2309	2304	2299	2294	2288	2283	2278
4.4	2273	2268	2262	2257	2252	2247	2242	2237	2232	2227
4.5	2222	2217	2212	2208	2203	2198	2193	2188	2183	2179
4.6	2174	2169	2165	2160	2155	2151	2146	2141	2137	2132
4.7	2128	2123	2119	2114	2110	2105	2101	2096	2092	2088
4.8	2083	2079	2075	2070	2066	2062	2058	2053	2049	2045
4.9	2041	2037	2033	2028	2024	2020	2016	2012	2008	2004
5.0	2000	1996	1992	1988	1984	1980	1976	1972	1969	1965
5.1	1961	1957	1953	1949	1946	1942	1938	1934	1931	1927
5.2	1923	1919	1916	1912	1908	1905	1901	1898	1894	1890
5.3	1887	1883	1880	1876	1873	1869	1866	1862	1859	1855
5.4	1852	1848	1845	1842	1838	1835	1832	1828	1825	1821
5.5	1818	1815	1812	1808	1805	1802	1799	1795	1792	1788
5.6	1786	1783	1779	1776	1773	1770	1767	1764	1761	1757
5.7	1754	1751	1748	1745	1742	1739	1736	1733	1730	1727
5.8	1724	1721	1718	1715	1712	1709	1706	1704	1701	1698
5.9	1695	1692	1689	1686	1684	1681	1678	1675	1672	1669
6.0	1667	1664	1661	1668	1656	1653	1650	1647	1645	1642
6.1	1639	1637	1634	1631	1629	1626	1623	1621	1618	1616
6.2	1613	1610	1608	1605	1603	1600	1597	1595	1592	1590
6.3	1587	1585	1582	1580	1577	1575	1572	1570	1567	1565
6.4	1563	1560	1558	1555	1553	1550	1548	1546	1543	1541
6.5	1538	1536	1534	1531	1529	1527	1524	1522	1520	1517
6.6	1515	1513	1511	1508	1506	1504	1502	1499	1497	1495
6.7	1493	1490	1488	1486	1484	1481	1479	1477	1475	1473

Wavelength–Wavenumber Conversion Table (Continued)

Wavelength μm	WAVELENGTH–WAVENUMBER CONVERSION TABLE Wavenumber (cm⁻¹)									
	0	1	2	3	4	5	6	7	8	9
6.8	1471	1468	1466	1464	1462	1460	1458	1456	1453	1451
6.9	1449	1447	1445	1443	1441	1439	1437	1435	1433	1431
7.0	1429	1427	1425	1422	1420	1418	1416	1414	1412	1410
7.1	1408	1406	1404	1403	1401	1399	1397	1395	1393	1391
7.2	1389	1387	1385	1383	1381	1379	1377	1376	1374	1372
7.3	1370	1368	1366	1364	1362	1361	1359	1357	1355	1353
7.4	1351	1350	1348	1346	1344	1342	1340	1339	1337	1335
7.5	1333	1332	1330	1328	1326	1325	1323	1321	1319	1318
7.6	1316	1314	1312	1311	1309	1307	1305	1304	1302	1300
7.7	1299	1297	1295	1294	1292	1290	1289	1287	1285	1284
7.8	1282	1280	1279	1277	1276	1274	1272	1271	1269	1267
7.9	1266	1264	1263	1261	1259	1258	1256	1255	1253	1252
8.0	1250	1248	1247	1245	1244	1242	1241	1239	1238	1236
8.1	1235	1233	1232	1230	1229	1227	1225	1224	1222	1221
8.2	1220	1218	1217	1215	1214	1212	1211	1209	1208	1206
8.3	1205	1203	1202	1200	1199	1198	1196	1195	1193	1192
8.4	1190	1189	1188	1186	1185	1183	1182	1181	1179	1178
8.5	1176	1175	1174	1172	1171	1170	1168	1167	1166	1164
8.6	1163	1161	1160	1159	1157	1156	1155	1153	1152	1151
8.7	1149	1148	1147	1145	1144	1143	1142	1140	1139	1138
8.8	1136	1135	1134	1133	1131	1130	1129	1127	1126	1125
8.9	1124	1122	1121	1120	1119	1117	1116	1115	1114	1112
9.0	1111	1110	1109	1107	1106	1105	1104	1103	1101	1100
9.1	1099	1098	1096	1095	1094	1093	1092	1091	1089	1088
9.2	1087	1086	1085	1083	1082	1081	1080	1079	1078	1076
9.3	1075	1074	1073	1072	1071	1070	1068	1067	1066	1065
9.4	1064	1063	1062	1060	1059	1058	1057	1056	1055	1054
9.5	1053	1052	1050	1049	1048	1047	1046	1045	1044	1043
9.6	1042	1041	1040	1038	1037	1036	1035	1034	1033	1032
9.7	1031	1030	1029	1028	1027	1026	1025	1024	1022	1021
9.8	1020	1019	1018	1017	1016	1015	1014	1013	1012	1011
9.9	1010	1009	1008	1007	1006	1005	1004	1003	1002	1001
10.0	1000	999	998	997	996	995	994	993	992	991
10.1	990	989	988	987	986	985	984	983	982	981
10.2	980	979	978	978	977	976	975	974	973	972
10.3	971	970	969	968	967	966	965	964	963	962
10.4	962	961	960	959	958	957	956	955	954	953
10.5	952	951	951	950	949	948	947	946	945	944
10.6	943	943	942	941	940	939	938	937	936	935
10.7	935	934	933	932	931	930	929	929	928	927
10.8	926	925	924	923	923	922	921	920	919	918
10.9	917	917	916	915	914	913	912	912	911	910
11.0	909	908	907	907	906	905	904	903	903	902
11.1	901	900	899	898	898	897	896	895	894	894
11.2	893	892	891	890	890	889	888	887	887	886
11.3	885	884	883	883	882	881	880	880	879	878
11.4	877	876	876	875	874	873	873	872	871	870
11.5	870	869	868	867	867	866	865	864	864	863
11.6	862	861	861	860	859	858	858	857	856	855
11.7	855	854	853	853	852	851	850	850	849	848
11.8	847	847	846	845	845	844	843	842	842	841
11.9	840	840	839	838	838	837	836	835	835	834

(Continued)

Wavelength–Wavenumber Conversion Table (Continued)

Wavelength µm	WAVELENGTH–WAVENUMBER CONVERSION TABLE Wavenumber (cm⁻¹)									
	0	1	2	3	4	5	6	7	8	9
12.0	833	833	832	831	831	830	829	829	828	827
12.1	826	826	825	824	824	823	822	822	821	820
12.2	820	819	818	818	817	816	816	815	814	814
12.3	813	812	812	811	810	810	809	808	808	807
12.4	806	806	805	805	804	803	803	802	801	801
12.5	800	799	799	798	797	797	796	796	795	794
12.6	794	793	792	792	791	791	790	789	789	788
12.7	787	787	786	786	785	784	784	783	782	782
12.8	781	781	780	779	779	778	778	777	776	776
12.9	775	775	774	773	773	772	772	771	770	770
13.0	769	769	768	767	767	766	766	765	765	764
13.1	763	763	762	762	761	760	760	759	759	758
13.2	758	757	756	756	755	755	754	754	753	752
13.3	752	751	751	750	750	749	749	748	747	747
13.4	746	746	745	745	744	743	743	742	742	741
13.5	741	740	740	739	739	738	737	737	736	736
13.6	735	735	734	734	733	733	732	732	731	730
13.7	730	729	729	728	728	727	727	726	726	725
13.8	725	724	724	723	723	722	722	721	720	720
13.9	719	719	718	718	717	717	716	716	715	715
14.0	714	714	713	713	712	712	711	711	710	710
14.1	709	709	708	708	707	707	706	706	705	705
14.2	704	704	703	703	702	702	702	701	701	700
14.3	699	699	698	698	697	697	696	696	695	695
14.4	694	694	693	693	693	692	692	691	691	690
14.5	690	689	689	688	688	687	687	686	686	685
14.6	685	684	684	684	683	683	682	682	681	681
14.7	680	680	679	679	678	678	678	677	677	676
14.8	676	675	675	674	674	673	673	672	672	672
14.9	671	671	670	670	669	669	668	668	668	667

USEFUL SOLVENTS FOR INFRARED SPECTROPHOTOMETRY

The following tables provide the infrared absorption spectra of several useful solvents, along with solvent design properties [1–10]. In most cases, two spectra are provided for each solvent. The first in each set was measured using a double beam spectrophotometer using a neat sample against an air reference. These spectra are presented in both wavenumber (cm⁻¹) and micrometer (μm) scales. The spectra were recorded under high concentration conditions (in terms of path length and attenuation) in order to emphasize the characteristics of each solvent. Thus, these spectra are not meant to be "textbook" examples of infrared spectra. The second spectrum in each set was measured with a Fourier transform instrument. The physical properties listed are those needed most often in designing spectrophotometric experiments [1–10]. The refractive indices are values measured with the sodium-d line. Solvation properties include the solubility parameter, δ, hydrogen bond index, λ, and the solvatochromic parameters α, β, and π*. The Chemical Abstract Service registry numbers and the INChI (International Chemical Identifier) are also provided for each solvent, to allow the reader to easily obtain further information using computerized database services. Note that the heat of vaporization is presented in the commonly used cal/g unit. To convert to the appropriate SI unit (J/g), multiply by 4.184.

We realize that a number of the solvents listed here are not permitted in some academic laboratories. Information on these solvents are presented for users in laboratories equipped to deal with the hazards associated with them.

REFERENCES

1. Lewis, R. J. *Hawley's Condensed Chemical Dictionary*. 14th ed. New York: John Wiley and Sons, 2002.
2. Dreisbach, R. R. *Physical Properties of Chemical Compounds, Advances in Chemistry Series*. No. 22. Washington, DC: American Chemical Society, 1959.
3. Jamieson, D. T., J. B. Irving, and J. S. Tudhope. *Liquid Thermal Conductivity: A Data Survey to 1973*. Edinburgh: Her Majesty's Stationary Office, 1975.
4. Lewis, R. J., and N. I. Sax. *Sax's Dangerous Properties of Industrial Materials*. 9th ed. Washington, DC: Thompson Publishing, 1995.
5. Sedivec, V., and J. Flek. *Handbook of Analysis of Organic Solvents*. New York: John Wiley and Sons (Halsted Press), 1976.
6. Epstein, W. W., and F. W. Sweat. "Dimethyl Sulfoxide Oxidations." *Chemistry Review* 247 (1967).
7. Lide, D. R., ed. *CRC Handbook for Chemistry and Physics*. 90th ed. Boca Raton, FL: CRC Press, 2010.
8. Bruno, T. J., and P. D. N. Svoronos. *CRC Handbook of Basic Tables for Chemical Analysis*. 2nd ed. Boca Raton, FL: CRC Press, 2003.
9. *NIST Chemistry Web Book*. NIST Standard Reference Database Number 69, March 2003 Release.
10. Marcus, Y. "The Properties of Organic Liquids that are Relevant to their Use as Solvating Solvents." *Chemical Society Review* 22, no. 6 (1993): 409–16.

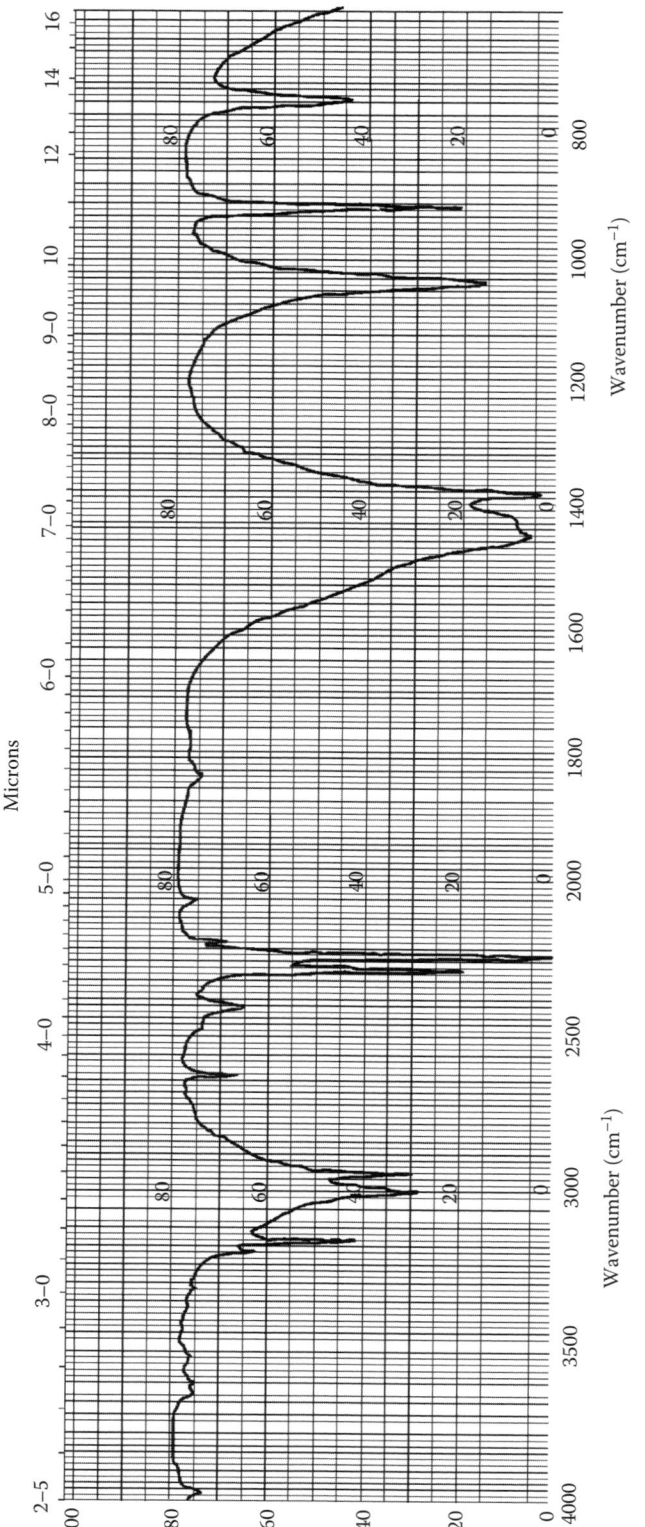

ACETONITRILE: CH₃CN

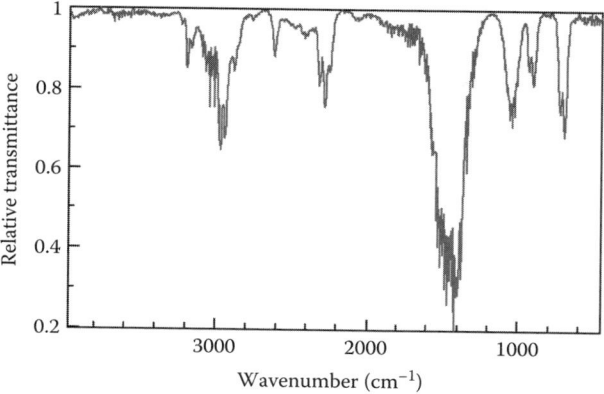

ACETONITRILE, CH₃CN

Physical Properties	
Relative molecular mass	41.05
Melting point	−45.7 °C
Normal boiling point	81.6 °C
Refractive index (20 °C)	1.34423
Density (20 °C)	0.7857 g/mL
Viscosity (25 °C)	0.345 mPa·s
Surface tension (20 °C)	29.30 mN/m
Heat of vaporization (at boiling point)	29.75 kJ/mol
Thermal conductivity (20 °C)	0.1762 W/(m·K)
Dielectric constant (20 °C)	38.8
Relative vapor density (air = 1)	1.41
Vapor pressure (20 °C)	0.0097 MPa
Solubility in water[a]	∞
Flash point (OC)	6 °C
Autoignition temperature	509 °C
Explosive limits in air	4.4–16 %, vol/vol
CAS registry number	75-05-8
INChI	1S/C2H3N/c1-2-3/h1H3
Exposure limits	40 ppm, 8 hr TWA
Solubility parameter, δ	11.9
Solvatochromic α	0.19
Solvatochromic β	0.4
Solvatochromic π*	0.75

Note: Highly polar solvent; sweet, ethereal odor; soluble in water; flammable, burns with a luminous flame; highly toxic by ingestion, inhalation, and skin absorption; miscible with water, methanol, methyl acetate, ethyl acetate, acetone, ethers, acetamide solutions, chloroform, carbon tetrachloride, ethylene chloride, and many unsaturated hydrocarbons; immiscible with many saturated hydrocarbons (petroleum fractions); dissolves some inorganic salts such as silver nitrate, lithium nitrate, magnesium bromide; incompatible with strong oxidants; hydrolyzes in the presence of aqueous bases and strong aqueous acids.

[a] Forms azeotrope with water (at 16 % mass/mass) that boils at 76 °C.

Synonyms: methyl cyanide, acetic acid nitrile, cyanomethane, ethylnitrile.

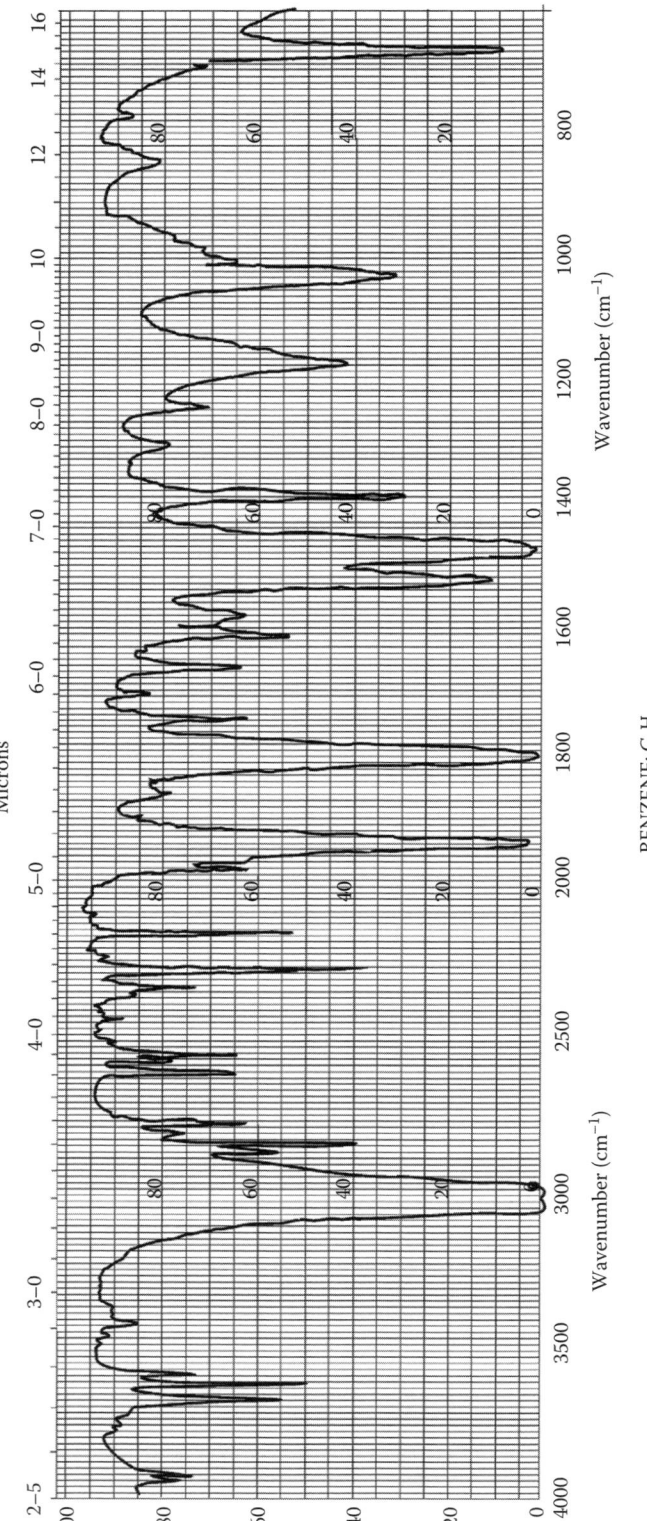

BENZENE: C$_6$H$_6$

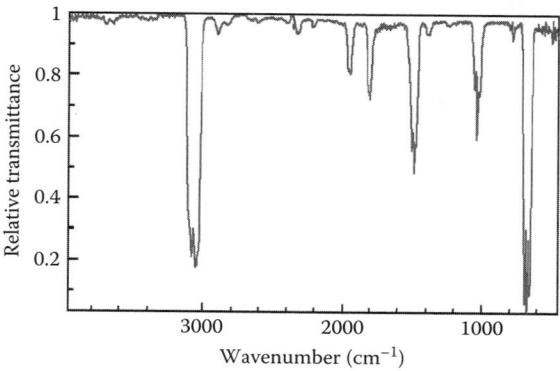

BENZENE, C_6H_6

Physical Properties	
Relative molecular mass	78.11
Melting point	5.5 °C
Normal boiling point	80.1 °C
Refractive index (20 °C)	1.50110
(25 °C)	1.4979
Density (20 °C)	0.8790 g/mL
(25 °C)	0.8737 g/mL
Viscosity (25 °C)	0.654 mPa·s
Surface tension (20 °C)	28.87 mN/m
Heat of vaporization (at boiling point)	30.72 kJ/mol
Thermal conductivity (25 °C)	0.1424 W/(m·K)
Dielectric constant (20 °C)	2.284
Relative vapor density (air = 1)	2.77
Vapor pressure (25 °C)	0.0097 MPa
Solubility in water[a]	0.07 %, mass/mass
Flash point (OC)	−11 °C
Autoignition temperature	562 °C
Explosive limits in air	1.4–8.0 %, vol/vol
CAS registry number	71-43-2
INChI	1S/C6H6/c1-2-4-6-5-3-1/h1-6H
Exposure limits	10 ppm, 8 hr TWA
Solubility parameter, δ	9.2
Hydrogen bond index, λ	2.2
Solvatochromic α	0.00
Solvatochromic β	0.10
Solvatochromic π^*	0.59

Note: Confirmed human carcinogen; Nonpolar, aromatic solvent; sweet odor; very flammable and toxic; confirmed human carcinogen; soluble in alcohols, hydrocarbons (aliphatic and aromatic), ether, chloroform, carbon tetrachloride, carbon disulfide, slightly soluble in water. Incompatible with some strong acids and oxidants, chlorine trifluoride/zinc (in the presence of steam); dimerizes at high temperature to form biphenyl.

Synonyms: cyclohexatriene, benzin, benzol, phenylhydride. These are the most common, although there are many other synonyms.

[a] Forms azeotrope with ethanol (approximately 65 °C).

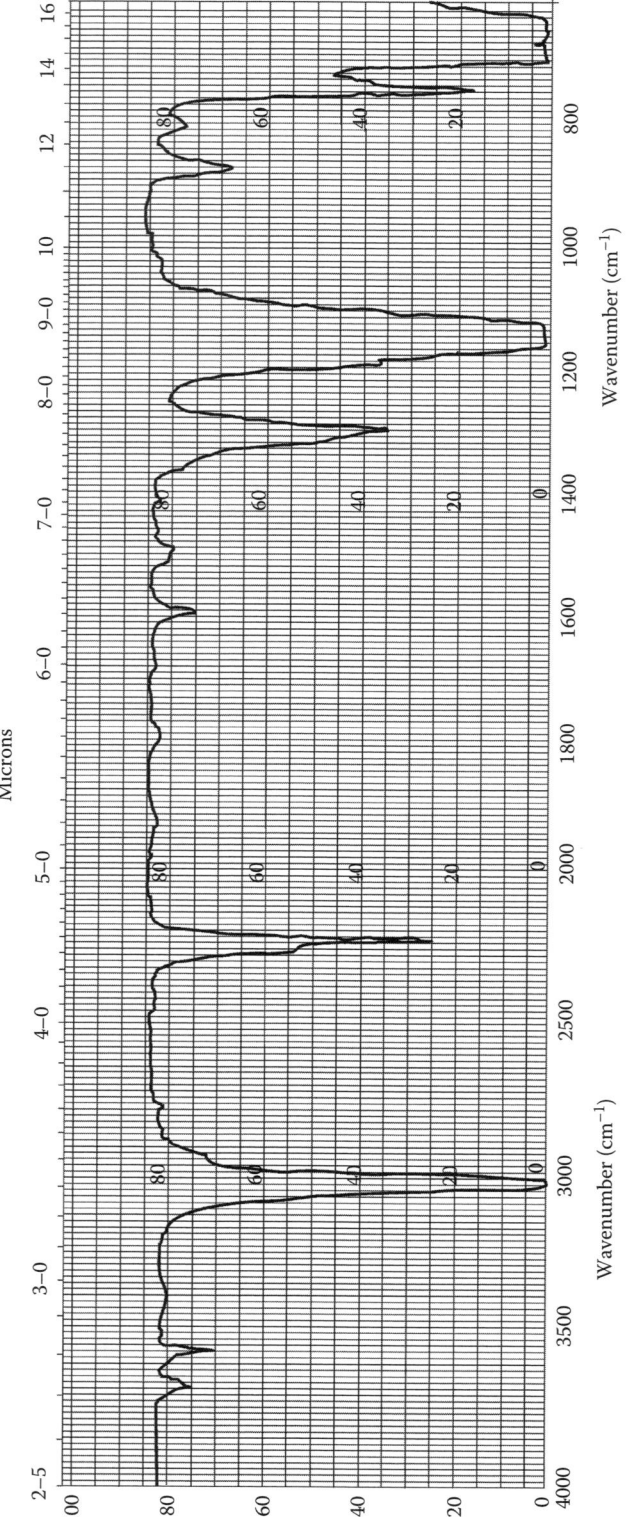

BROMOFORM: CHBr$_3$

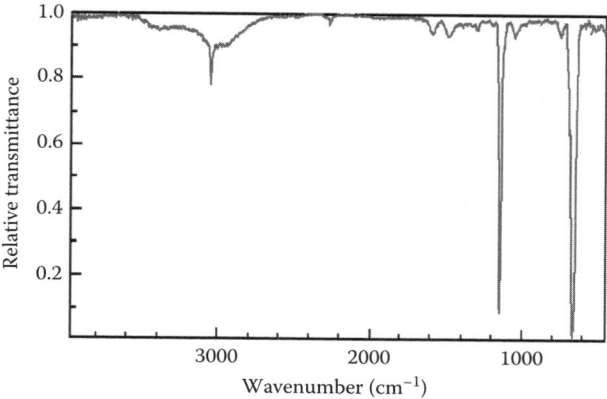

BROMOFORM, CHBr$_3$

Physical Properties	
Relative molecular mass	252.75
Melting point	5.7 °C
Normal boiling point	149.5 °C
Refractive index (20 °C)	1.6005
Density (20 °C)	2.8899 g/mL
Viscosity (25 °C)	1.89 mPa·s
Surface tension (20 °C)	41.53 mN/m
Heat of vaporization (at boiling point)	39.66 kJ/mol
Thermal conductivity (20 °C)	0.0961 W/(m·K)
Dielectric constant (20 °C)	4.39
Relative vapor density (air = 1)	2.77
Vapor pressure (25 °C)	0.0008 MPa
Solubility in water	slightly
Flash point (OC)	nonflammable
Autoignition temperature	not determined
Explosive limits in air	nonflammable
CAS registry number	75-25-2
INChI	1S/CHBr3/c2-1(3)4/h1H
Exposure limits	0.5 ppm (skin)
Solvatochromic α	0.05
Solvatochromic β	0.05
Solvatochromic π*	0.62

Note: Moderately polar, weakly hydrogen bonding solvent, dense liquid; gradu-
ally decomposes to acquire a yellow color, air and/or light will accelerate
this decomposition; nonflammable; commercial product is often stabilized
by the addition of 3 %–4 % (mass/mass) alcohols; highly toxic by ingestion, inhalation, and skin absorption; soluble in alcohols, organohalogen
compounds, hydrocarbons, benzene, and many oils. Incompatible with
many alkali and alkaline earth metals.

Synonyms: tribromomethane

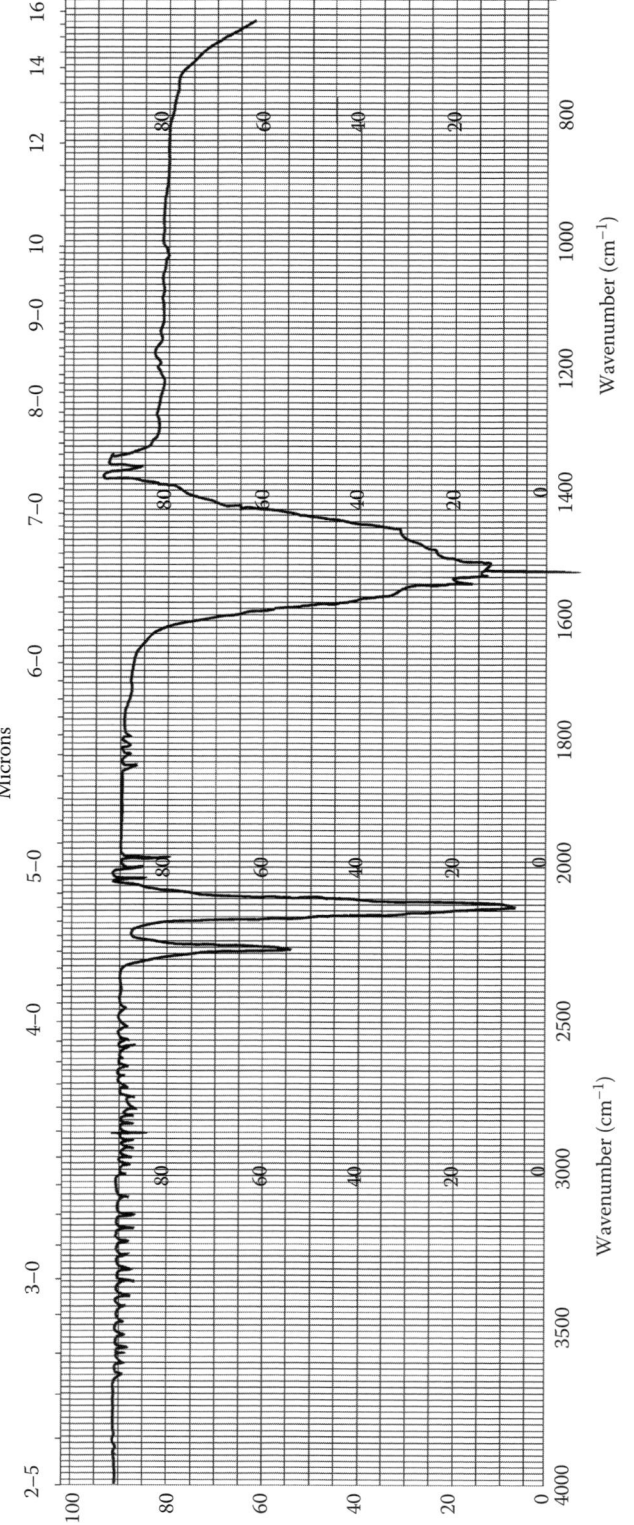

CARBON DISULFIDE: CS$_2$

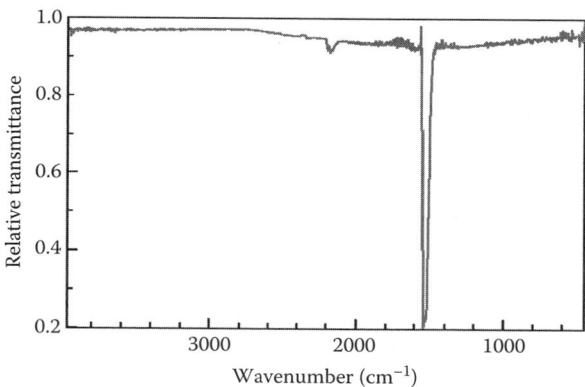

CARBON DISULFIDE, CS$_2$

Physical Properties	
Relative molecular mass	76.14
Melting point	−111 °C
Normal boiling point	46.3 °C
Refractive index (20 °C)	1.6280
(25 °C)	1.6232
Density (20 °C)	1.2631 g/mL
(25 °C)	1.2556 g/mL
Viscosity (20 °C)	0.363 mPa·s
Surface tension (20 °C)	32.25 mN/m
Heat of vaporization (at boiling point)	26.74 kJ/mol
Dielectric constant (20 °C)	2.641
Relative vapor density (air = 1)	2.64
Vapor pressure (25 °C)	0.0448 MPa
Solubility in water (20 °C)	0.29 %, mass/mass
Flash point (OC)	−30 °C
Autoignition temperature	100 °C
Explosive limits in air	1.0–50 %, vol/vol
CAS registry number	75-15-0
INChI	1S/CS2/c2-1-3
Exposure limits	20 ppm, 8 hr TWA
Solvatochromic α	0.00
Solvatochromic β	0.07
Solvatochromic π*	0.61

Note: Moderately polar solvent, soluble in alcohols, benzene, ethers, and chloroform; slightly soluble in water; very flammable and mobile; can be ignited by friction or contact with hot surfaces such as steam pipes; burns with a blue flame to produce carbon dioxide and sulfur dioxide; toxic by inhalation, ingestion, and skin absorption; strong disagreeable odor when impure; incompatible with aluminum (powder), azides, chlorine, chlorine monoxide, ethylene diamine, ethyleneamine, fluorine, nitrogen oxides, potassium, and zinc and other oxidants; soluble in methanol, ethanol, ethers, benzene, chloroform, carbon tetrachloride, and many oils; can be stored in metal, glass, porcelain, and Teflon containers.

Synonyms: carbon bisulfide, dithiocarbon anhydride

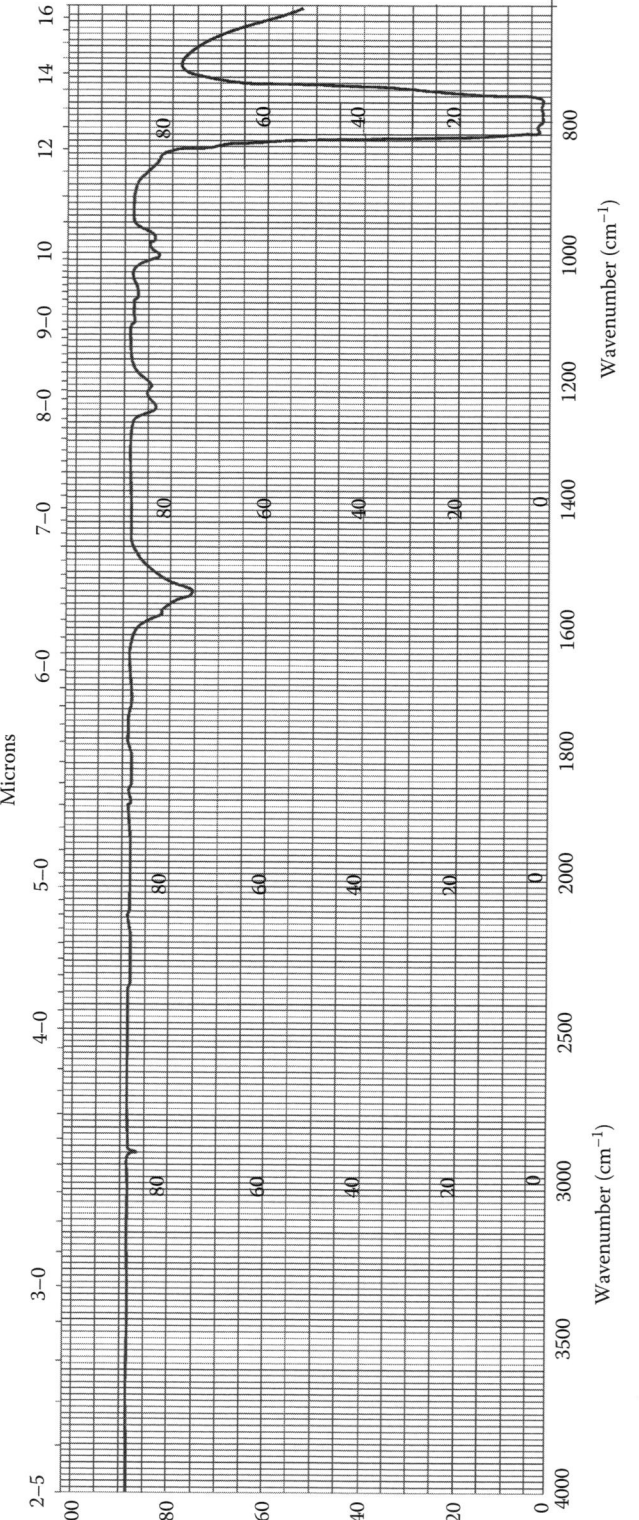

CARBON TETRACHLORIDE: CCl$_4$

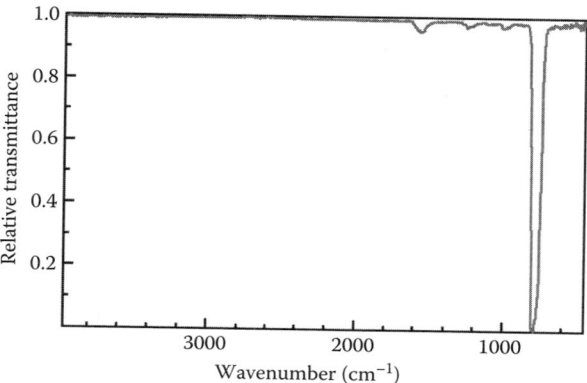

CARBON TETRACHLORIDE, CCl$_4$

Physical Properties

Relative molecular mass	153.82
Melting point	−22.85 °C
Normal boiling point	76.65 °C
Refractive index (20 °C)	1.4607
(25 °C)	1.4570
Density (20 °C)	1.5940 g/mL
(25 °C)	1.5843 g/mL
Viscosity (20 °C)	0.969 mPa·s
Surface tension (20 °C)	26.75 mN/m
Heat of vaporization[1] (at boiling point)	29.82 kJ/mol
Thermal conductivity (20 °C)	0.1070 W/(m·K)
Dielectric constant (20 °C)	2.238
Relative vapor density (air = 1)	5.32
Vapor pressure (25 °C)	0.0122 MPa
Solubility in water (20 °C)	0.08, w/w
Flash point (OC)	noncombustible
Autoignition temperature	noncombustible
Explosive limits in air	nonexplosive
CAS registry number	56-23-5
INChI	1S/CCl4/c2-1(3,4)5
Exposure limits	5 ppm (skin)
Solubility parameter, δ	8.6
Hydrogen bond index, λ	2.2
Solvatochromic α	0.00
Solvatochromic β	0.10
Solvatochromic π*	0.28

Note: Nonpolar solvent; soluble in alcohols, ethers, chloroform, and other halocarbons, benzene and most fixed and volatile oils, insoluble in water; nonflammable; extremely toxic by inhalation, ingestion, or skin absorption; carcinogenic; incompatible with allyl alcohol, silanes, triethyldialuminum, many metals (e.g., sodium).

Synonyms: tetrachloromethane, perchloromethane, methane tetrachloride, Halon-104

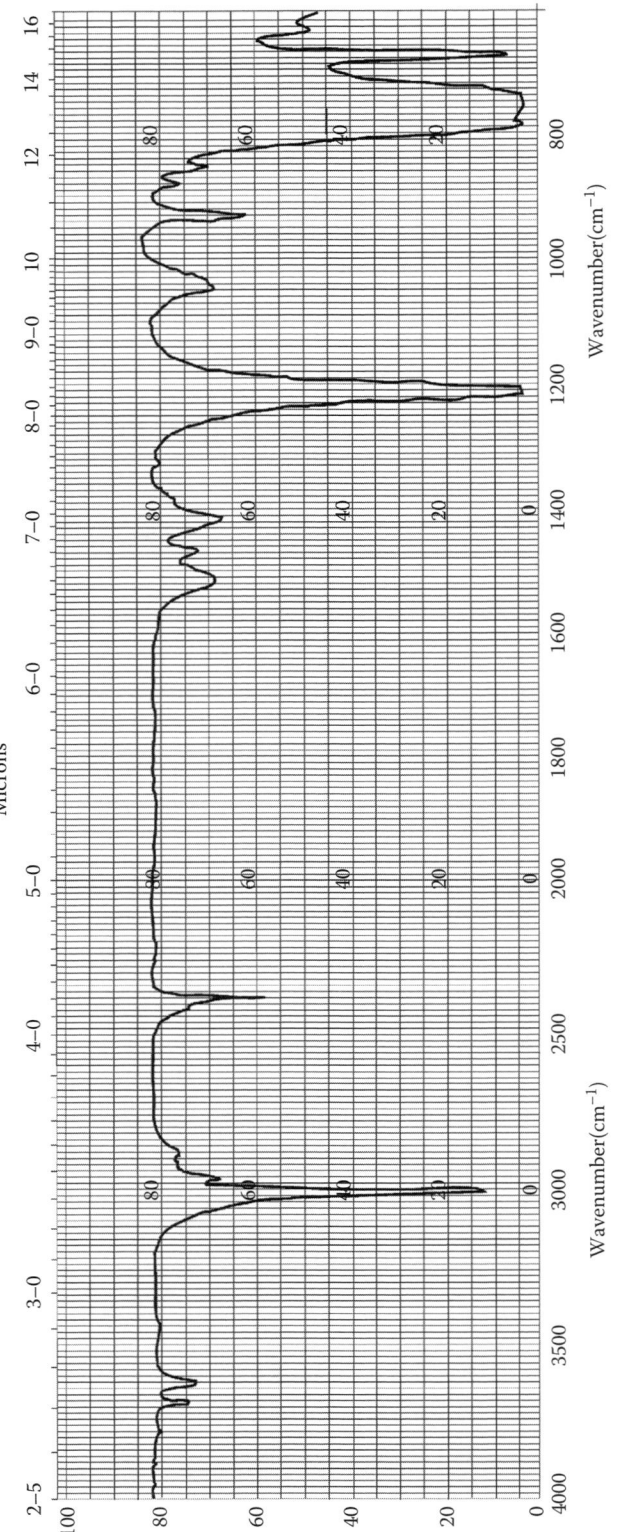

CHLOROFORM: CHCl$_3$

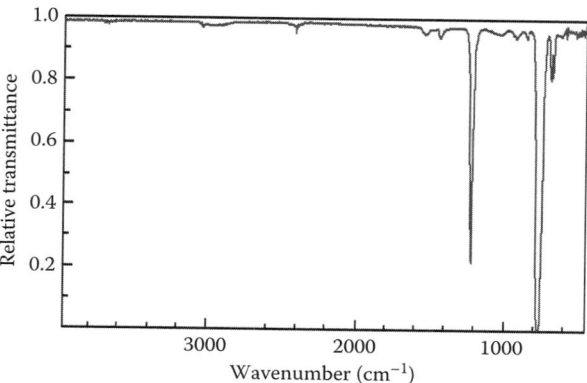

CHLOROFORM, CHCl$_3$

Physical Properties	
Relative molecular mass	119.38
Melting point	−63.2 °C
Normal boiling point	61.2 °C
Refractive index (20 °C)	1.4458
(25 °C)	1.4422
Density (20 °C)	1.4892 g/mL
(25 °C)	1.4798 g/mL
Viscosity (20 °C)	0.566 mPa·s
Surface tension	27.2 mN/m
Heat of vaporization (at boiling point)	29.24 kJ/mol
Thermal conductivity (20 °C)	0.1164 W/(m·K)
Dielectric constant (20 °C)	4.806
Relative vapor density (air = 1)	4.13
Vapor pressure (25 °C)	0.0263 MPa
Solubility in water	0.815 %, w/w
Flash point (OC)	noncombustible[a]
Autoignition temperature	noncombustible[a]
Explosive limits in air	nonexplosive
CAS registry number	67-66-3
INChI	1S/CHCl3/c2-1(3)4/h1H
Exposure limits	10 ppm, 8 hr TWA
Solubility parameter, δ	9.3
Hydrogen bond index, λ	2.2
Solvatochromic α	0.20
Solvatochromic β	0.10
Solvatochromic π^*	0.58

Note: Polar solvent; soluble in alcohols, ether, benzene, and most oils; usually stabilized with methanol to prevent phosgene formation; flammable and highly
toxic by inhalation, ingestion or skin absorption; narcotic; suspected to be
carcinogenic; incompatible with caustics, active metals, aluminum powder,
potassium, sodium, magnesium.

[a] Although chloroform is nonflammable, it will burn upon prolonged exposure to
flame or high temperature.

Synonyms: trichloromethane, methane trichloride

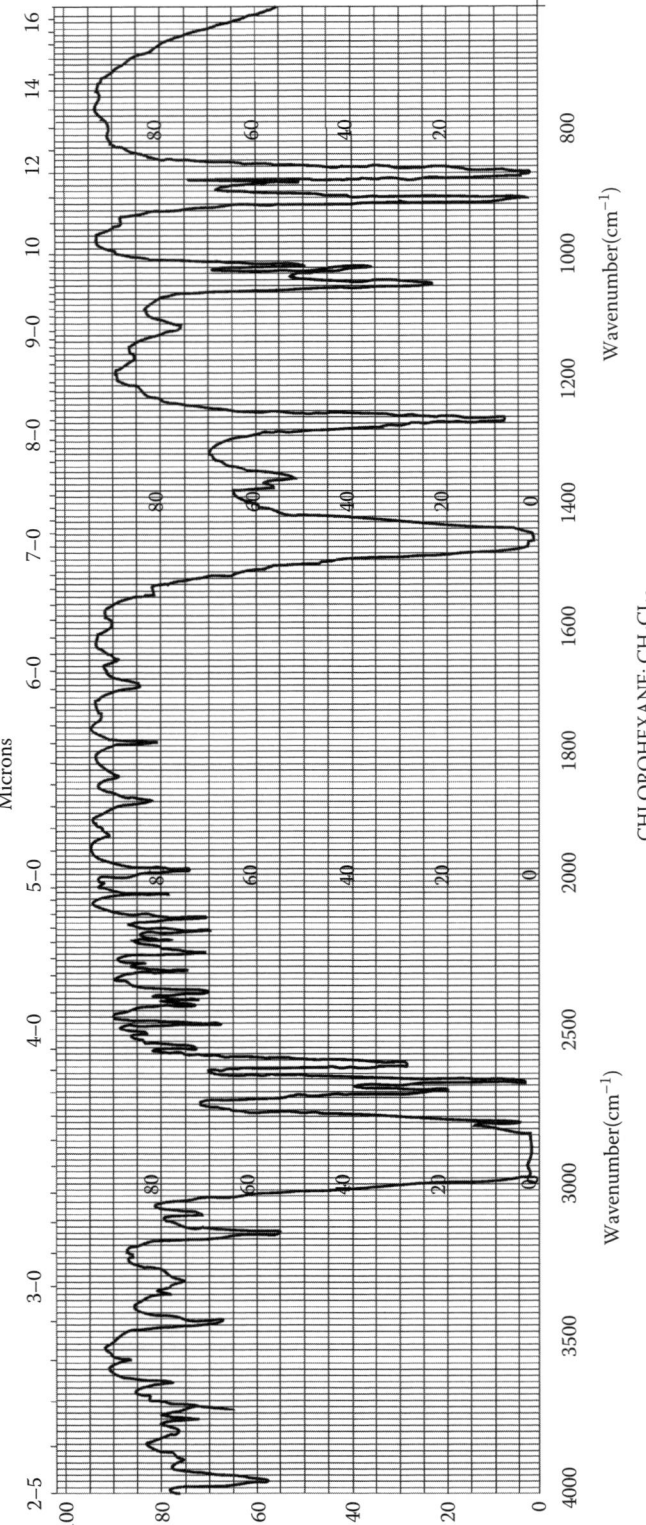

CHLOROHEXANE: CH_6Cl_{12}

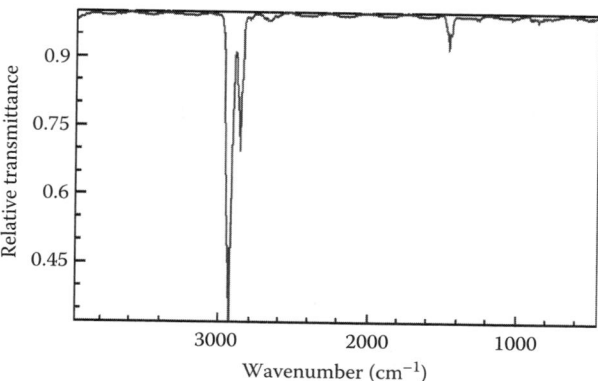

CYCLOHEXANE, C_6H_{12}

Physical Properties	
Relative molecular mass	84.16
Melting point	6.3 °C
Normal boiling point	80.7 °C
Refractive index (20 °C)	1.4263
(25 °C)	1.4235
Density (20 °C)	0.7786 g/mL
(25 °C)	0.7739 g/mL
Viscosity (20 °C)	1.06 mPa·s
Surface tension (20 °C)	24.99 mN/m
Heat of vaporization (at boiling point)	29.97 kJ/mol
Thermal conductivity (20 °C)	0.122 W/(m·K)
Dielectric constant (20 °C)	2.023
Relative vapor density (air = 1)	2.90
Vapor pressure (25 °C)	0.0111 MPa
Solubility in water (20 °C)	< 0.01 %, mass/mass
Flash point (OC)	−17 °C
Autoignition temperature	245 °C
Explosive limits in air	1.31–8.35 %, vol/vol
CAS registry number	110-82-7
INChI	1S/C6H12/c1-2-4-6-5-3-1/h1-6H2
Exposure limits	330 ppm, 8 hr TWA
Solvatochromic α	0.00
Solvatochromic β	0.00
Solvatochromic π^*	0.00

Note: Nonpolar hydrocarbon solvent; mild, gasoline-like odor; soluble in hydrocarbons, alcohols, organic halides, acetone, benzene; flammable; moderately toxic by inhalation, ingestion, or skin absorption, may be narcotic at high concentrations; reacts with oxygen (air) at elevated temperatures; decomposes upon heating; incompatible with strong oxidants.

Synonyms: benzene hexahydride, hexamethylene, hexanaphthene, hexahydrobenzene

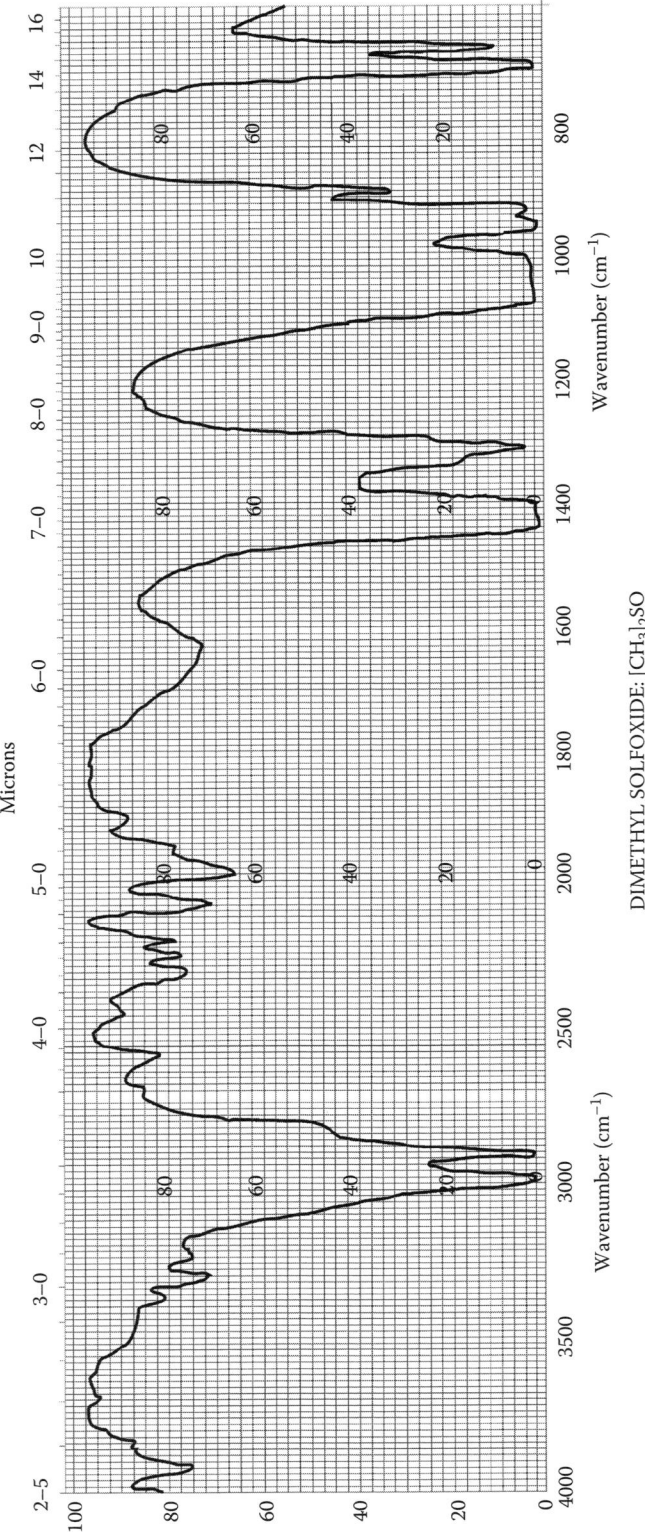

DIMETHYL SOLFOXIDE: $[CH_3]_2SO$

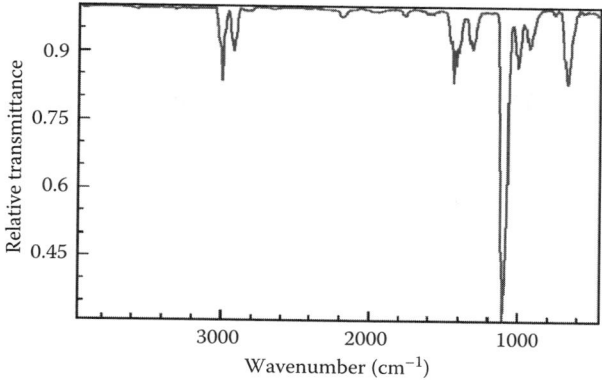

DIMETHYL SULFOXIDE, (CH₃)₂SO

Physical Properties

Relative molecular mass	78.13
Melting point	18.5 °C
Normal boiling point	189 °C
Refractive index (20 °C)	1.4770
Density (20 °C)	1.1014 g/mL
Viscosity (25 °C)	1.98 mPa·s
Surface tension	43.5 mN/m
Relative vapor density (air = 1)	2.7
Vapor pressure	5.3×10^{-5} MPa
Solubility in water	∞
Flash point (OC)	95 °C
Autoignition temperature	215 °C
Explosive limits in air	26.0–28.5 %, vol/vol
CAS registry number	67-68-5
INChI	1S/C2H6OS/c1-4(2)3/h1-2H3
Exposure limits	none established
Solubility parameter, δ	13.0
Hydrogen bond index, λ	5.0
Solvatochromic α	0.00
Solvatochromic β	0.76
Solvatochromic π*	1.00

Note: Colorless, odorless (when pure), hygroscopic liquid, powerful aprotic solvent; dissolves many inorganic salts, soluble in water; combustible; readily penetrates the skin; incompatible with strong oxidizers, and many halogenated compounds (e.g., alkyl halides, aryl halides), oxygen, peroxides, diborane, perchlorates.

Synonyms: DMSO, methyl sulfoxide, sulfinylbismethane

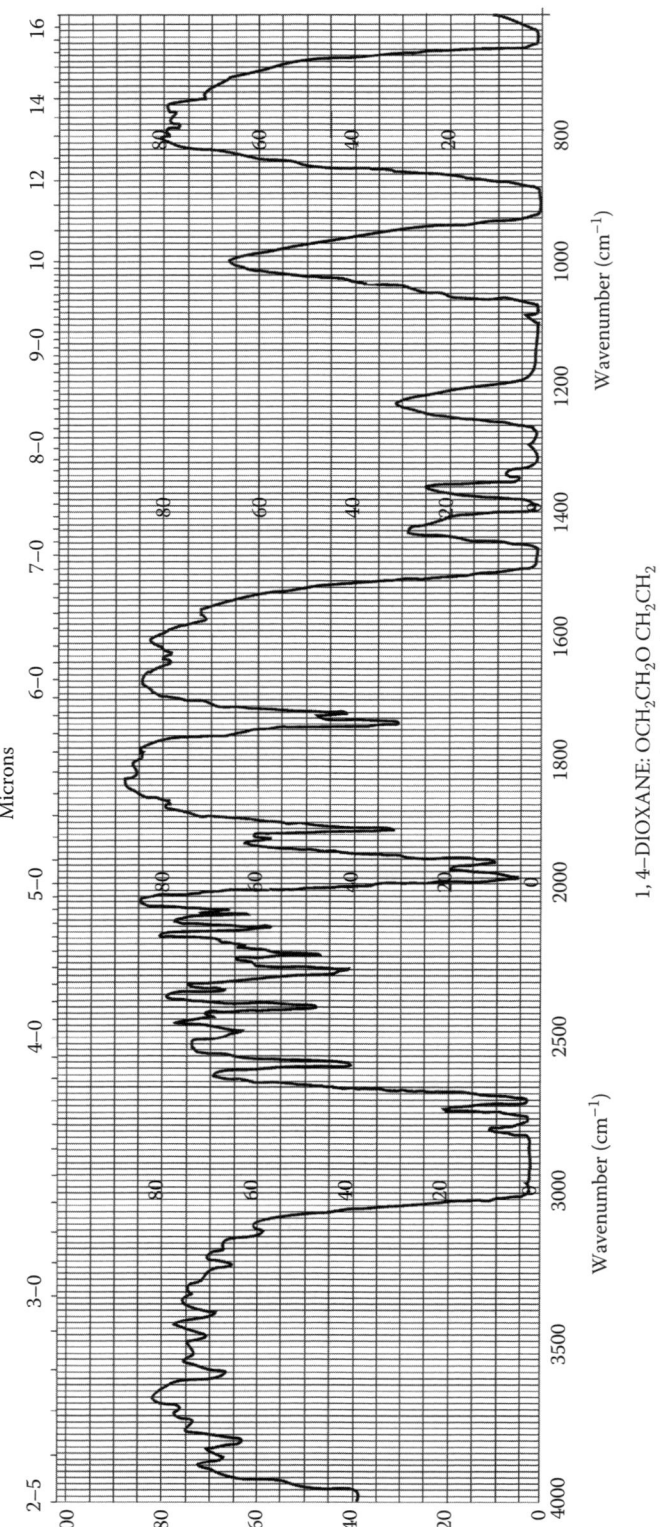

1, 4–DIOXANE: OCH$_2$CH$_2$O CH$_2$CH$_2$

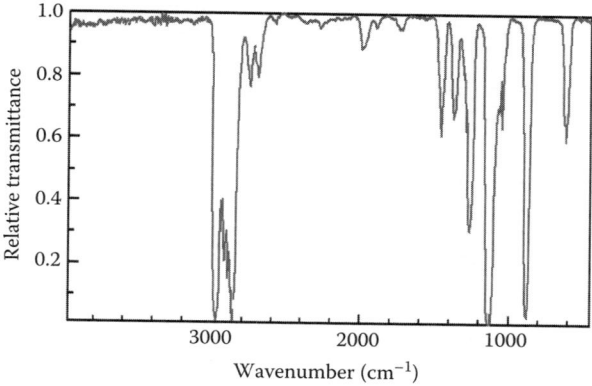

1,4-DIOXANE

Physical Properties	
Relative molecular mass	88.11
Melting point	11 °C
Boiling point	101.3 °C
Refractive index (20 °C)	1.4221
(25 °C)	1.4195
Density (20 °C)	1.0338 g/mL
(25 °C)	1.0282 g/mL
Viscosity (20 °C)	1.37 mPa·s
Surface tension (20 °C)	33.74 mN/m
Heat of vaporization (at boiling point)	34.16 kJ/mol
Dielectric constant (20 °C)	2.209
Relative vapor density (air = 1)	3.03
Vapor pressure (25 °C)	0.0053 MPa
Solubility in water	∞
Flash point (OC)	12 °C
Autoignition temperature	180 °C
Explosive limits in air	1.97 %–22.2 %, vol/vol
CAS registry number	123-91-1
INChI	1S/C4H8O2/c1-2-6-4-3-5-1/h1-4H2
Exposure limits	100 ppm (skin)
Solubility parameter, δ	9.9
Hydrogen bond index, λ	5.7
Solvatochromic α	0.00
Solvatochromic β	0.37
Solvatochromic π^*	0.55

Note: Moderately polar solvent; soluble in water and most organic solvents; flammable; highly toxic by ingestion and inhalation; absorbed through the skin; may cause central nervous system depression, necrosis of the liver and kidneys; incompatible with strong oxidizers.

Synonyms: diethylene ether, 1,4-diethylene dioxide, diethylene dioxide, dioxyethylene ether

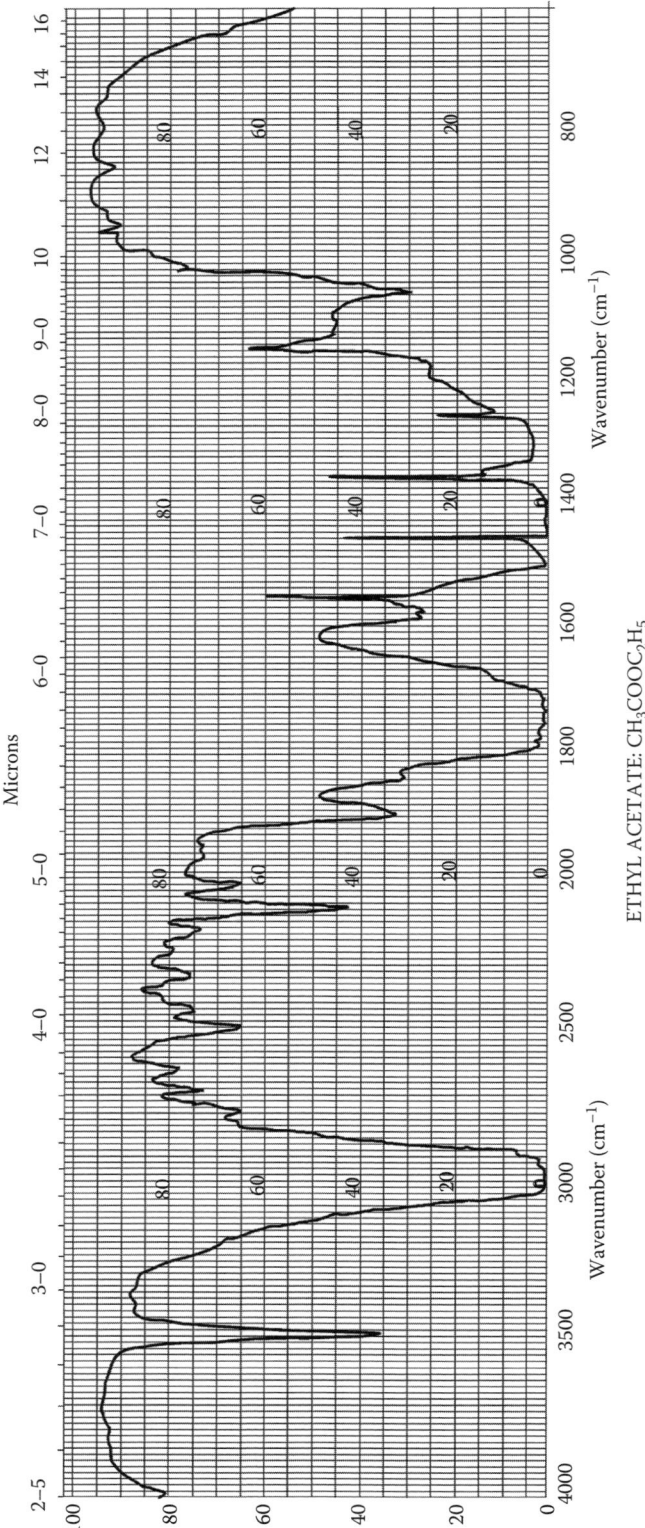

ETHYL ACETATE: CH₃COOC₂H₅

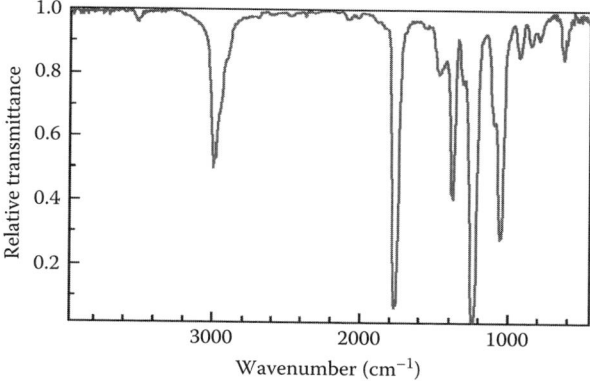

ETHYL ACETATE, $CH_3COOC_2H_5$

Physical Properties	
Relative molecular mass	88.11
Melting point	−83.58 °C
Boiling point	77.06 °C
Refractive index (20 °C)	1.3723
(25 °C)	1.3698
Density (20 °C)	0.9006 g/mL
(25 °C)	0.8946 g/mL
Viscosity (20 °C)	0.452 mPa·s
Surface tension (20 °C)	23.95 mN/m
Heat of vaporization (at boiling point)	31.94 kJ/mol
Thermal conductivity (20 °C)	0.122 W/(m·K)
Dielectric constant (25 °C)	6.02
Relative vapor density (air = 1)	3.04
Vapor pressure (20 °C)	0.0097 MPa
Solubility in water (20 °C)[a]	3.3 %, mass/mass
Flash point (OC)	−1 °C
Autoignition temperature	486 °C
Explosive limits in air	2.18 %–11.5 %, vol/vol
CAS registry number	141-78-6
INChI	1S/C4H8O2/c1-3-6-4(2)5/h3H2,1-2H3
Exposure limits	440 ppm, 8 hr, TWA
Solubility parameter, δ	9.1
Hydrogen bond index, λ	5.2
Solvatochromic α	0.00
Solvatochromic β	0.43
Solvatochromic π*	0.55

Note: Polar solvent; insoluble in water, soluble in alcohols, organic halides, ether, and many oils; flammable; moderately toxic by inhalation and skin absorption; incompatible with strong oxidizers, nitrates, strong alkalis, strong acids.

[a] Forms an azeotrope with water at 6.1 %, mass/mass, which boils at 70.4 °C.

Synonyms: acedin, acetic ether, acetic ester, vinegar naphtha, acetic acid ethyl ester

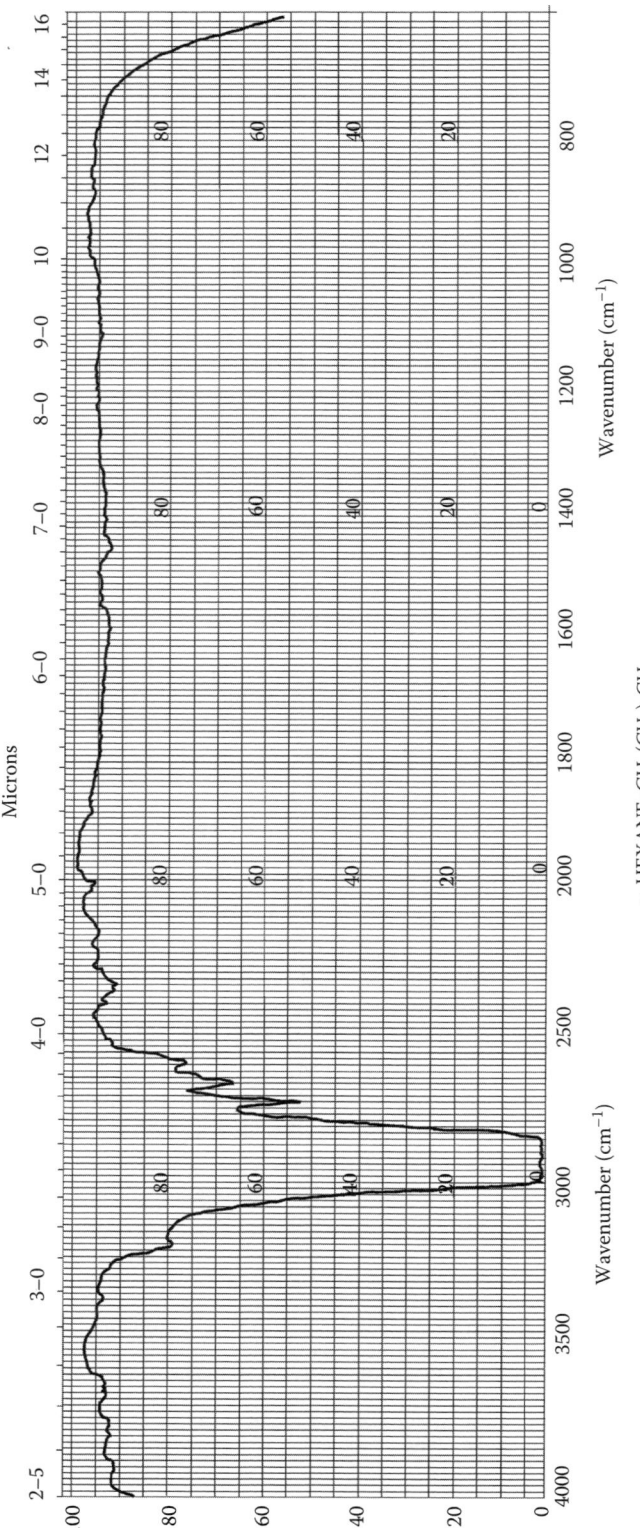

n-HEXANE: $CH_3(CH_2)_4CH_3$

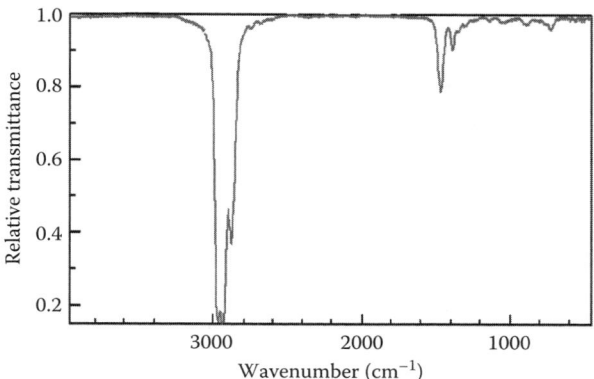

n-HEXANE, CH$_3$(CH$_2$)$_4$CH$_3$

Physical Properties	
Relative molecular mass	86.18
Melting point	−95 °C
Normal boiling point	68.742 °C
Refractive index (20 °C)	1.37486
(25 °C)	1.3723
Density (20 °C)	0.6594 g/mL
(25 °C)	0.6548 g/mL
Viscosity (20 °C)	0.31 mPa·s
Surface tension (20 °C)	18.42 mN/m
Heat of vaporization (at boiling point)	28.85 kJ/mol
Thermal conductivity (20 °C)	0.1217 W/(m·K)
Dielectric constant (20 °C)	1.890
Relative vapor density (air = 1)	2.97
Vapor pressure (25 °C)	0.0222 MPa
Solubility in water (20 °C)	0.011 %, mass/mass
Flash point (OC)	−26 °C
Autoignition temperature	247 °C
Explosive limits in air	1.25 %–6.90 %, vol/vol
CAS registry number	110-54-3
INChI	1S/C6H14/c1-3-5-6-4-2/h3-6H2,1-2H3
Exposure limits	500 ppm, 8 hr TWA
Solubility parameter, δ	9.3
Hydrogen bond index, λ	2.2
Solvatochromic α	0.00
Solvatochromic β	0.00
Solvatochromic π*	0.08

Note: Nonpolar solvent; soluble in alcohols, hydrocarbons, organic halides, acetone and ethers; insoluble in water; flammable; moderately toxic by inhalation and ingestion; incompatible with strong oxidizers.

Synonyms: hexane, hexyl hydride

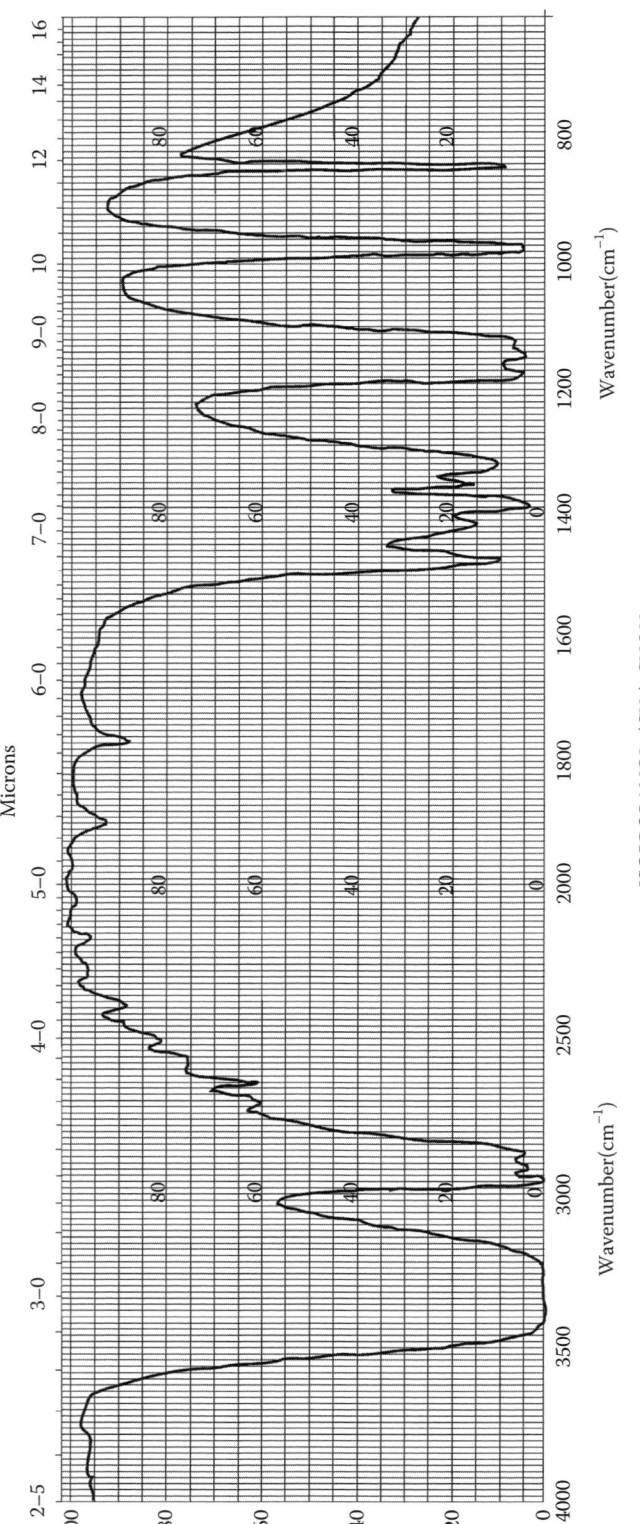

ISOPROPANOL: $(CH_3)_2CHOH$

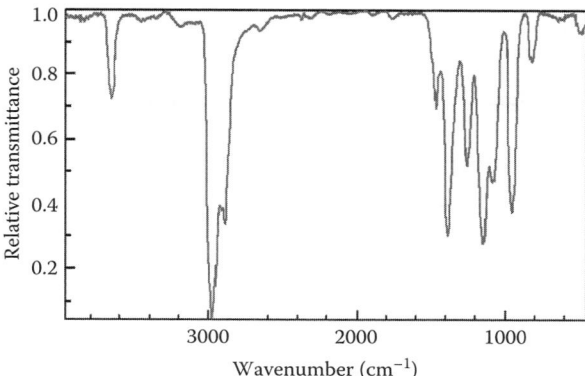

Wavenumber (cm^{-1})

ISOPROPANOL, (CH$_3$)$_2$CHOH

Physical Properties	
Relative molecular mass	60.10
Melting point	89.8 °C
Boiling point	82.4 °C
Refractive index (20 °C)	1.3771
(25 °C)	1.3750
Density (20 °C)	0.7864 g/mL
(25 °C)	0.7812 g/mL
Viscosity (20 °C)	2.43 mPa·s
Surface tension (20 °C)	21.99 mN/m
Heat of vaporization (at boiling point)	39.85 kJ/mol
Dielectric constant (25 °C)	18.3
Relative vapor density (air = 1)	2.07
Vapor pressure	0.0044 MPa
Solubility in water (20 °C)	∞
Flash point (OC)	16 °C
Autoignition temperature	456 °C
Explosive limits in air	2.02–11.8 %, vol/vol
CAS registry number	67-63-0
INChI	1S/C3H8O/c1-3(2)4/h3-4H,1-2H3
Exposure limits	400 ppm (skin)
Solubility parameter, δ	11.5
Hydrogen bond index, λ	8.9
Solvatochromic α	0.76
Solvatochromic β	0.84
Solvatochromic π*	0.48

Note: Polar solvent; soluble in water, alcohols, ethers, many hydrocarbons, and oils; flammable and moderately toxic by ingestion, inhalation, and skin absorption; incompatible with strong oxidizers.

Synonyms: dimethyl carbinol, sec-propyl alcohol, 2-propanol, isopropyl alcohol

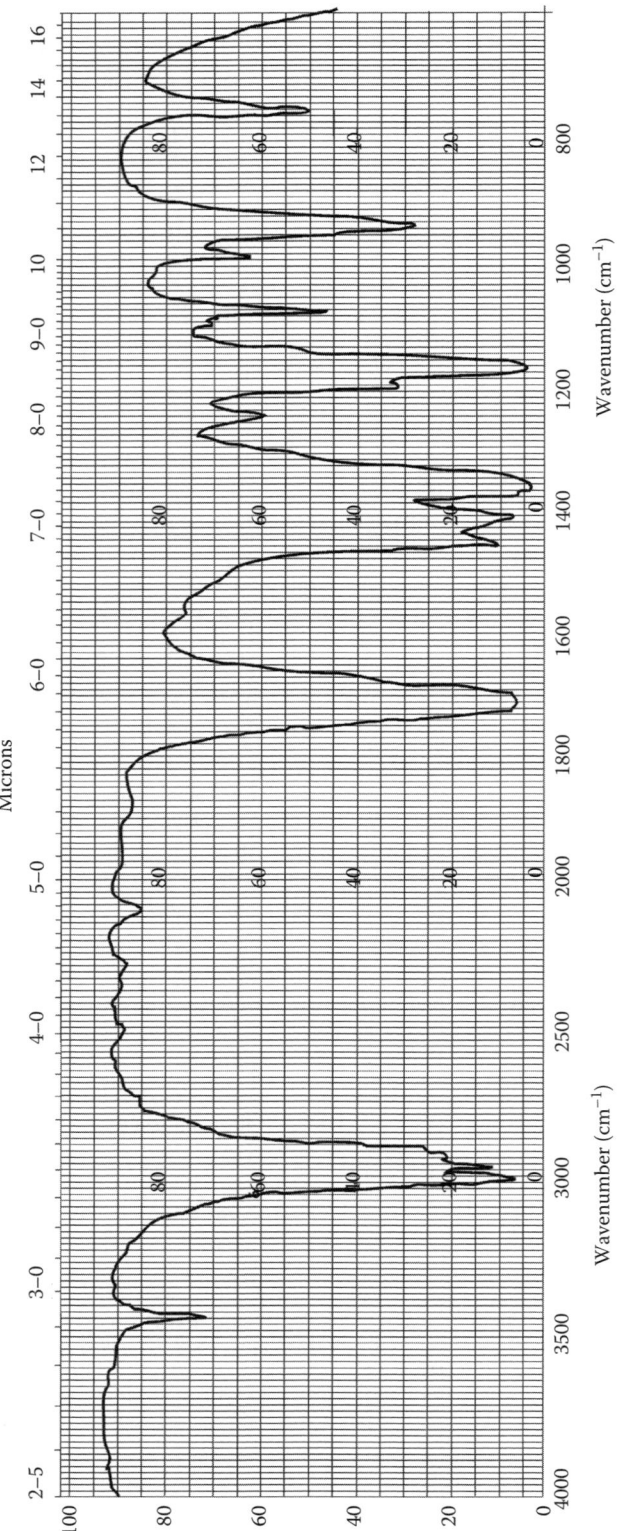

METHYL ETHYL KETONE: $CH_3COC_2H_5$

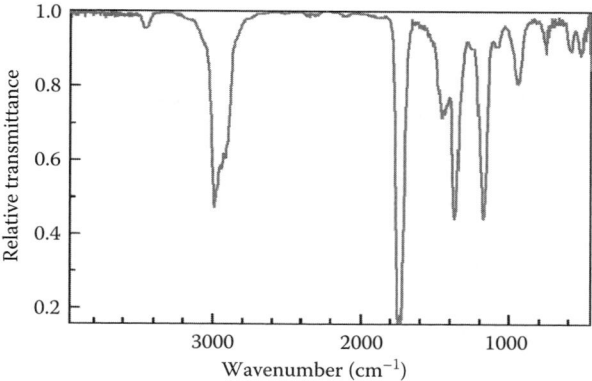

METHYL ETHYL KETONE, $CH_3COC_2H_5$

Physical Properties	
Relative molecular mass	72.11
Melting point	−86.4 °C
Boiling point	79.6 °C
Refractive index (20 °C)	1.379
(25 °C)	1.3761
Density (20 °C)	0.8054 g/mL
(25 °C)	0.8002 g/mL
Viscosity (20 °C)	0.448 mPa·s
Surface tension (20 °C)	24.50 mN/m
Heat of vaporization (at boiling point)	31.3 kJ/mol
Thermal conductivity (20 °C)	0.1465 W/(m·K)
Dielectric constant (20 °C)	18.5
Relative vapor density (air = 1)	2.41
Vapor pressure (25 °C)	0.0129 MPa
Solubility in water (20 °C)	27.33 %, mass/mass
Flash point (OC)	2 °C
Autoignition temperature	516 °C
Explosive limits in air	1.81 %–11.5 %, vol/vol
CAS registry number	78-93-3
INChl	1S/C4H8O/c1-3-4(2)5/h3H2,1-2H3
Exposure limits	200 ppm, 8 hr, TWA
Solubility parameter, δ	9.3
Hydrogen bond index, λ	5.0
Solvatochromic α	0.6
Solvatochromic β	0.48
Solvatochromic π*	0.67

Note: Polar solvent; soluble in water, ketones, organic halides, alcohols, ether, and many oils; highly flammable; narcotic by inhalation; incompatible with strong oxidizers, nitrates, nitric acid, reducing agents.

Synonyms: ethyl methyl ketone, 2-butanone, methyl acetone, MEK.

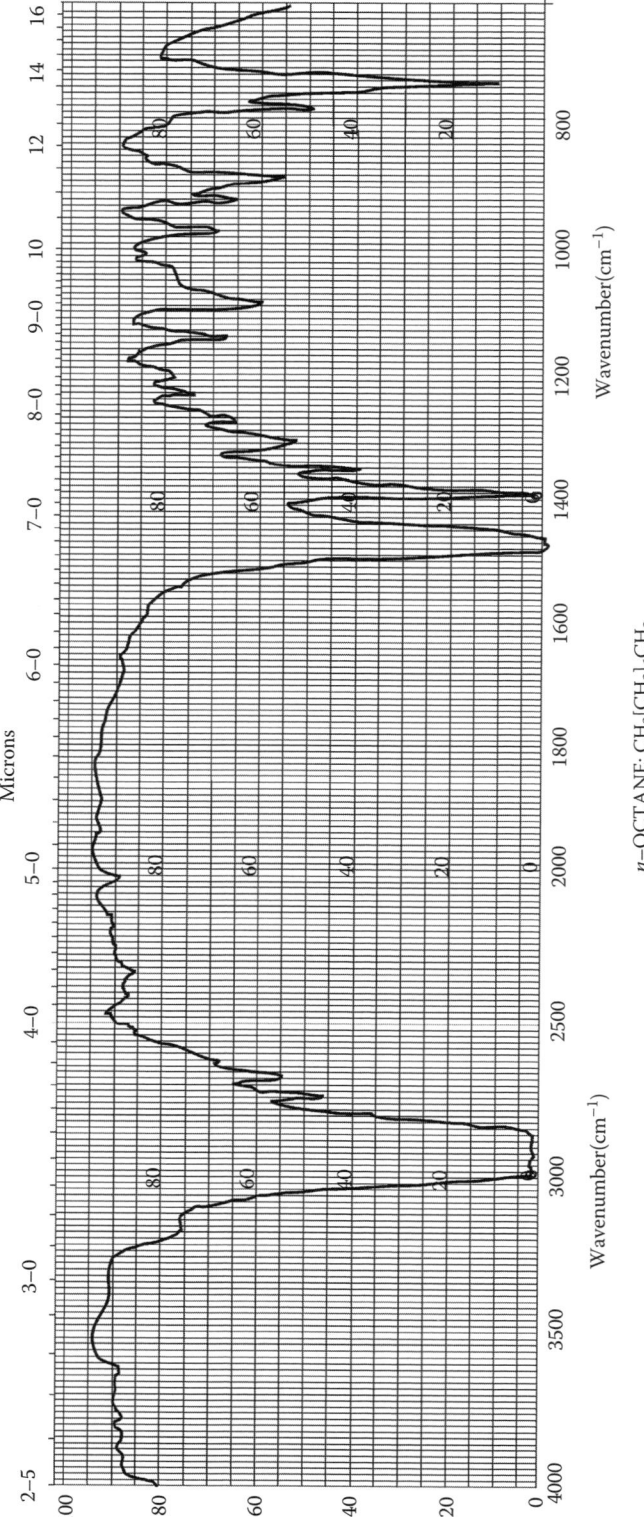

n-OCTANE: CH$_3$[CH$_2$]$_6$CH$_3$

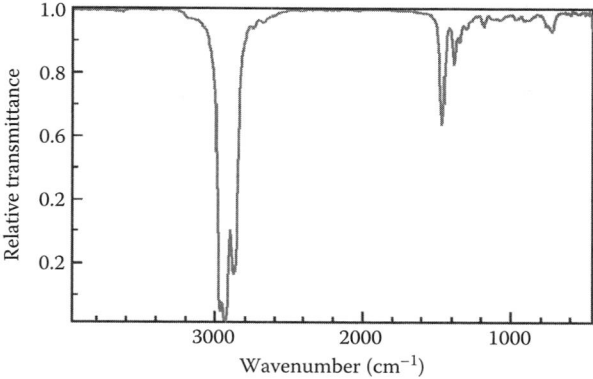

n-OCTANE, CH₃(CH₂)₆CH₃

Physical Properties	
Relative molecular mass	114.23
Melting point	−56. 7 °C
Boiling point	125.6 °C
Refractive index (20 °C)	1.39745
(25 °C)	1.3951
Density (20 °C)	0.7025 g/mL
(25 °C)	0.6985 g/mL
Viscosity (20 °C)	0.539 mPa·s
Surface tension (20 °C)	21.75 mN/m
Heat of vaporization (at boiling point)	34.41 kJ/mol
Dielectric constant (20 °C)	1.948
Relative vapor density (air = 1)	3.86
Vapor pressure (25 °C)	0.0023 MPa
Solubility in water (20 °C)	~0.002 %, mass/mass
Flash point (CC)	13 °C
Autoignition temperature	232 °C
Explosive limits in air	0.84 %–3.2 %, vol/vol
CAS registry number	111-65-9
INChI	1S/C8H18/c1-3-5-7-8-6-4-2/h3-8H2,1-2H3
Exposure limits	550 ppm, 8 hr TWA
Hydrogen bond index, λ	2.2
Solvatochromic α	0.00
Solvatochromic β	0.00
Solvatochromic π^*	0.01

Note: Nonpolar solvent; soluble in alcohol, acetone, and hydrocarbons, insoluble in water; flammable; incompatible with strong oxidizers.
Synonyms: octane

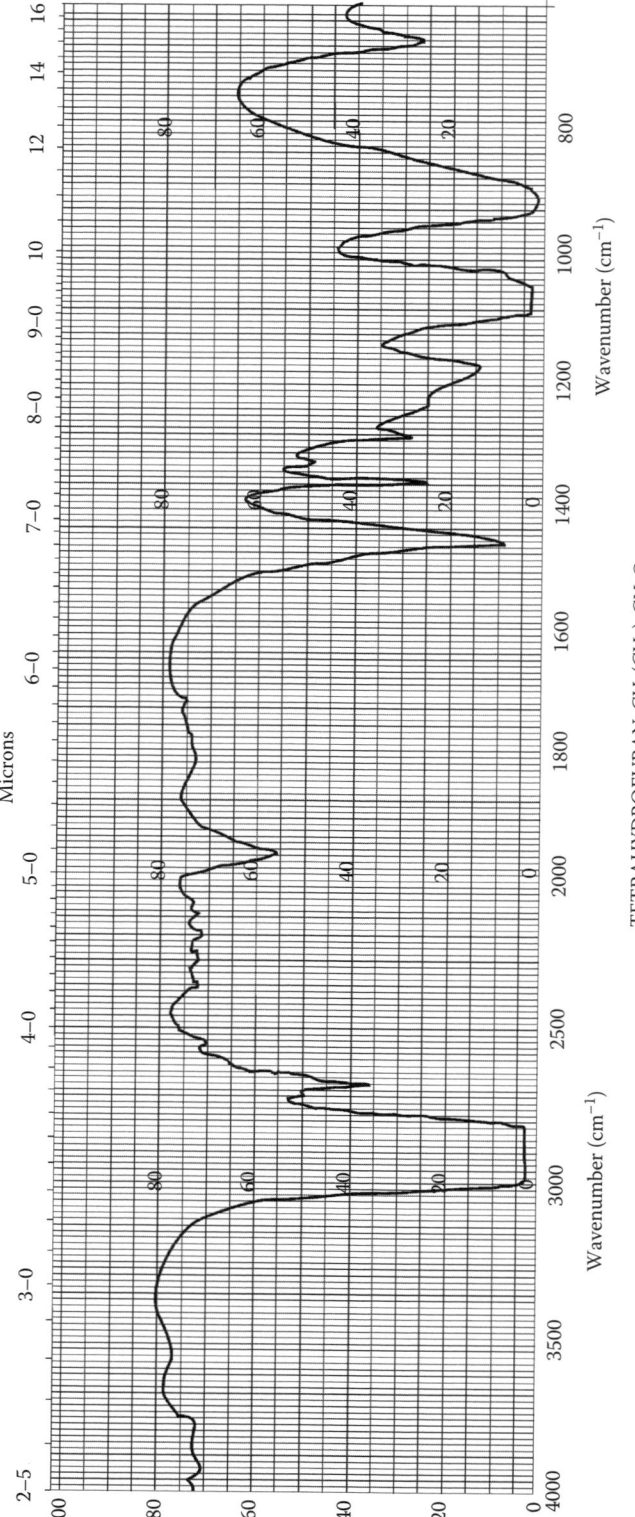

TETRAHYDROFURAN: $CH_2(CH_2)_2CH_2O$

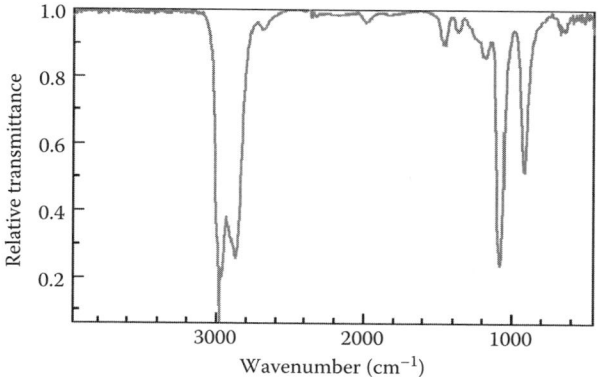

TETRAHYDROFURAN

Physical Properties	
Relative molecular mass	72.108
Melting point	−65 °C
Normal boiling point	66 °C
Refractive index (20 °C)	1.4070
(25 °C)	1.4040
Density (20 °C)	0.8880 g/mL
(25 °C)	0.8818 g/mL
Viscosity (20 °C)	0.55 mPa·s
Surface tension (20 °C)	26.4 mN/m
Heat of vaporization (at boiling point)	29.81 kJ/mol
Dielectric constant (20 °C)	7.54
Relative vapor density (air = 1)	2.5
Vapor pressure (20 °C)	0.0191 MPa
Solubility in water (20 °C)[a]	∞
Flash point (CC)	−17 °C
Autoignition temperature	260 °C
Explosive limits in air	1.8 %–11.8 %, vol/vol
CAS registry number	109-99-9
INChI	1S/C4H8O/c1-2-4-5-3-1/h1-4H2
Exposure limits	200 ppm, 8 hr TWA
Solubility parameter, δ	9.1
Hydrogen bond index, λ	5.3
Solvatochromic α	0.00
Solvatochromic β	0.55
Solvatochromic π*	0.58

Note: Moderately polar solvent, ethereal odor; soluble in water and most organic solvents; flammable; moderately toxic; incompatible with strong oxidizers; can form potentially explosive peroxides upon long standing in air, commercially, it is often stabilized against peroxidation with 0.5 %–1.0 % (mass/mass) p-cresol, .05 %–1.0 % (mass/mass) hydroquinone, or 0.01 % (mass/mass) 4,4′-thiobis(6-tert-butyl-m-cresol); can polymerize in the presence of cationic initiators such as Lewis acids or strong proton acids.

Synonyms: THF, tetramethylene oxide, diethylene oxide, 1,4-epoxybutane oxolane, oxacyclopentane

[a] pH of aqueous solution = 7.

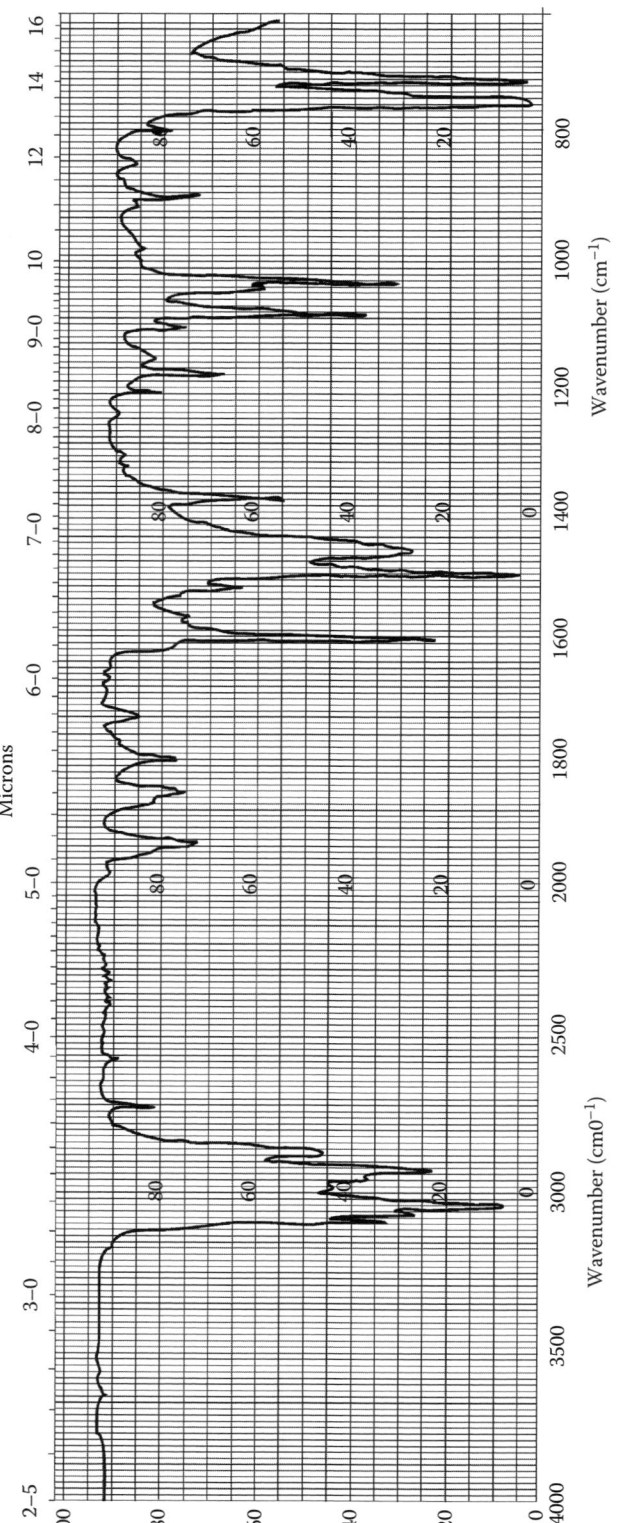

TOLUENE: $CH_3C_6H_5$

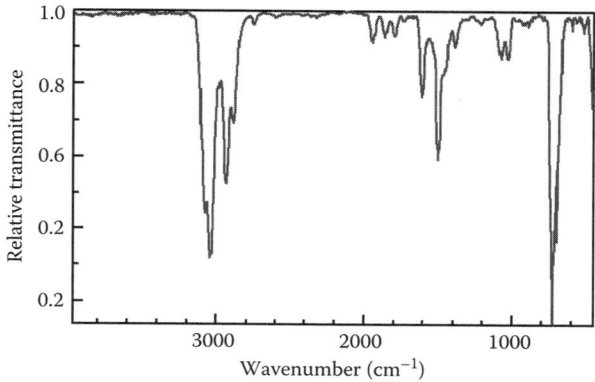

TOLUENE, $CH_3C_6H_5$

Physical Properties	
Relative molecular mass	92.14
Melting point	−94.5 °C
Normal boiling point	110.7 °C
Refractive index (20 °C)	1.497
(25 °C)	1.4941
Density (20 °C)	0.8669 g/mL
(25 °C)	0.8623 g/mL
Viscosity (20 °C)	0.587 mPa·s
Surface tension (20 °C)	28.52 mN/m
Heat of vaporization (at boiling point)	33.18 kJ/mol
Thermal conductivity (20 °C)	0.1348 W/(m·K)
Dielectric constant (25 °C)	2.379
Relative vapor density (air = 1)	3.14
Vapor pressure (25 °C)	0.0036 MPa
Solubility in water	0.047 %, mass/mass
Flash point (CC)	4 °C
Autoignition temperature	552 °C
Explosive limits in air	1.4 %–7.4 %, vol/vol
CAS registry number	108-88-3
INChI	1S/C7H8/c1-7-5-3-2-4-6-7/h2-6H,1H3
Exposure limits	200 ppm, 8 hr TWA
Solubility parameter, δ	8.9
Hydrogen bond index, λ	3.8
Solvatochromic α	0.00
Solvatochromic β	0.11
Solvatochromic π*	0.54

Note: Aromatic solvent; sweet pungent odor; soluble in benzene, alcohols, organic halides, ethers, insoluble in water; highly flammable; toxic by ingestion, inhalation, and absorption through the skin, narcotic at high concentrations; incompatible with strong oxidants; decomposes under high heat to form (predominantly) dimethylbiphenyl.

Synonyms: toluol, methylbenzene, methylbenzol, phenylmethane

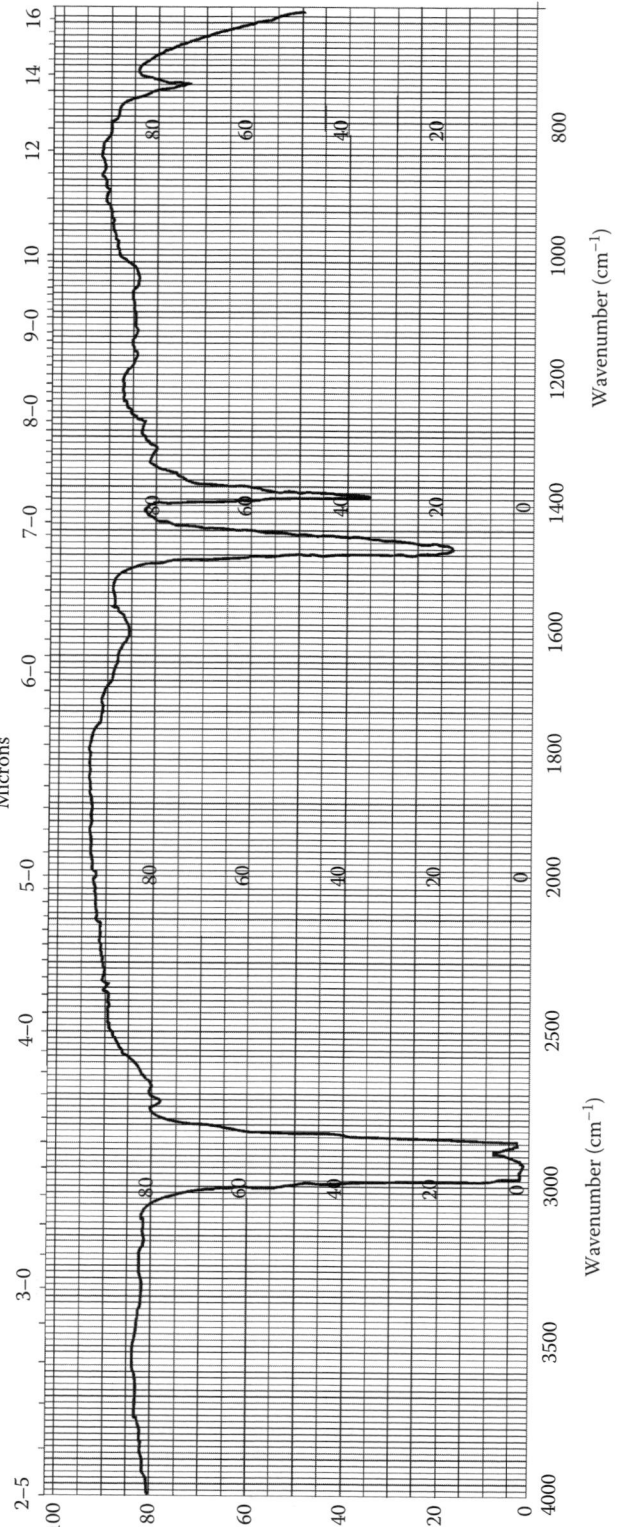

PARAFFIN OIL

PARAFFIN OIL

Physical Properties	
Relative molecular mass	Variable
Melting point	−20 °C (approximate)
Normal boiling point	315 °C (approximate)
Refractive index (20 °C)	1.4720
(25 °C)	1.4697
Specific gravity, 25 °C/25 °C	0.85
Solubility in water	insoluble
Flash point (OC)	229 °C
Explosive limits in air	0.6 %–6.5 %, vol/vol
CAS registry number	8012-95-1
INChI	NA, not a pure fluid
Exposure limits	50 ppm, 8 hr TWA

Note: Viscous, odorless, moderately combustible liquid used for mull preparation; relatively low toxicity; soluble in benzene, chloroform, carbon disulfide, ethers; incompatible with oxidizing materials and amines.

Synonyms: mineral oil, adepsine oil, lignite oil, nujol

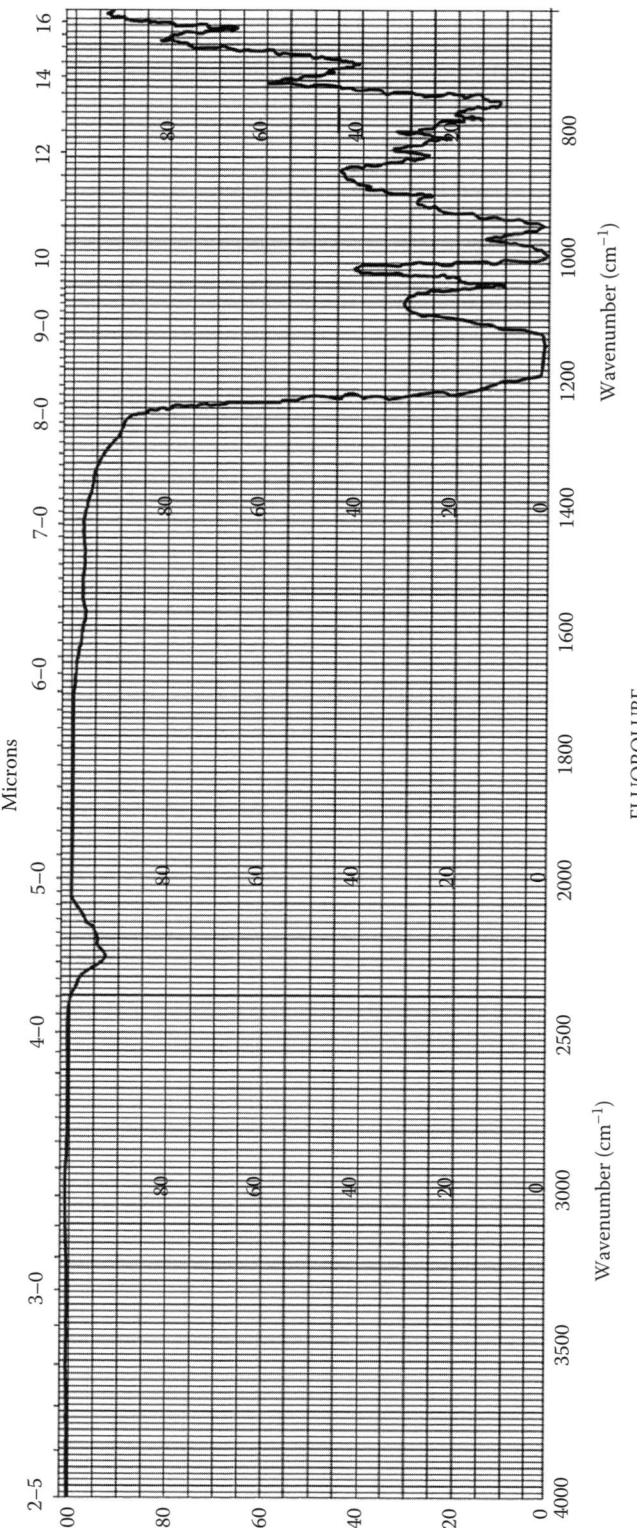

FLUOROLUBE

FLUOROLUBE, POLYTRIFLUOROCHLOROETHYLENE
[–C₂ClF₃–]

Physical Properties	
Relative molecular mass (monomer)	116.47
Pour point*	–60–13 °C
Melting point	–51–18 °C
Acidity (pH)*	6.0–7.5
Density (38 °C)*	1.865–1.955 g/mL
Viscosity (25 °C)*	6–1400 mPa·s
Vapor pressure (93 °C)	0.07–2.2 mmHg
Flash point (OC)	nonflammable
Autoignition temperature	nonflammable
Explosive limits in air	nonflammable
CAS registry number	9002-83-9
INChI	NA, polydisperse fluid
Exposure limits	Not established

Note: There are six common grades or varieties of this oil, marketed under the name Fluorolube. The properties listed above that are marked with an asterisk depend upon the grade that is used. The primary physical difference between the grades are the viscosities and pour points.

The thermal stability of these materials is dependent on the wetted surfaces. Typical ranges of stability are between 150 and 325 °C, but this varies with the wetted surface and residence time. Some metals can accelerate the decomposition into lower molecular mass, more volatile components. It is important to avoid the wetting of metals containing aluminum or magnesium especially in situations in which high friction of galling is possible. Detonation of these fluids is possible under these conditions. Moreover, these fluids can react violently in the presence of sodium, potassium, amines, hydrazine, liquid fluorine, and liquid chlorine.

Since these fluids are essentially transparent from 1360 to 4000 cm⁻1 (except for the absorption at 2321.9 cm⁻1), they can be used as mulling agents when the bands of paraffin oil obscure or interfere with sample absorptions.

POLYSTYRENE WAVENUMBER CALIBRATION

The following are wavenumber readings assigned to the peaks on the spectrum:

1	–	3027.1	8	–	1583.1
2	–	2924.0	9	–	1181.4
3	–	2850.7	10	–	1154.3
4	–	1944.0	11	–	1069.1
5	–	1871.0	12	–	1028.0
6	–	1801.6	13	–	906.7
7	–	1601.4	14	–	698.9

Film thickness: 50 μm

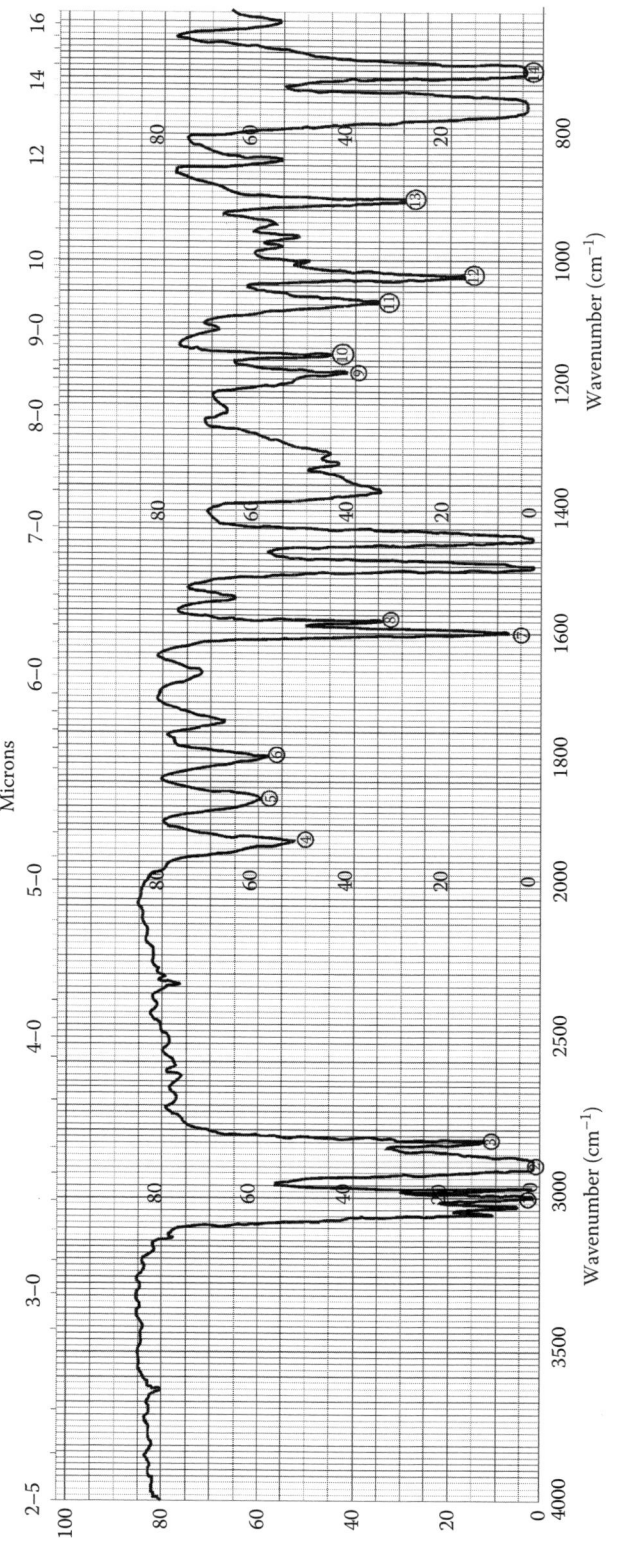

POLYSTYRENE

INFRARED ABSORPTION CORRELATION CHARTS

The following charts provide characteristic infrared absorptions obtained from particular functional groups on molecules [1,2]. These include a general mid-range correlation chart, a chart for aromatic absorptions, and a chart for carbonyl moieties. The general mid-range chart is an adaptation of work by Professor Charles F. Hammer of Georgetown University, reproduced, with modification and permission.

REFERENCES

1. Bruno, T. J., and P. D. N. Svoronos. *CRC Handbook of Basic Tables for Chemical Analysis*. 2nd ed. Boca Raton, FL: CRC Press, 2003.
2. Bruno, T. J., and P. D. N. Svoronos. *CRC Handbook of Fundamental Spectroscopic Correlation Charts*. Boca Raton, FL: CRC Press, 2006.

Notes:
   AR = aromatic
   b = broad
   sd = solid
   sn = solution
   sp = sharp
   ? = unreliable

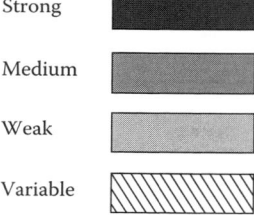

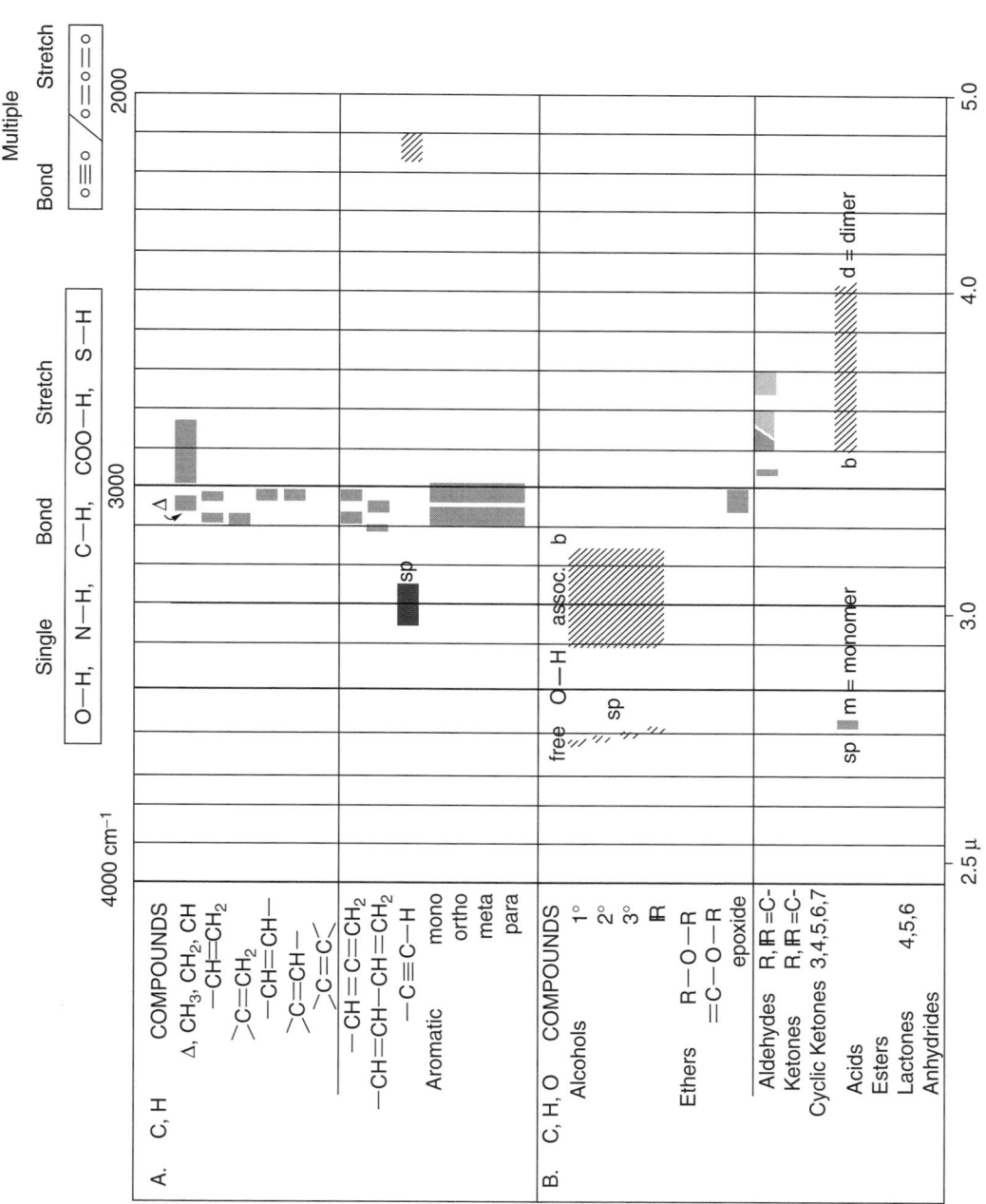

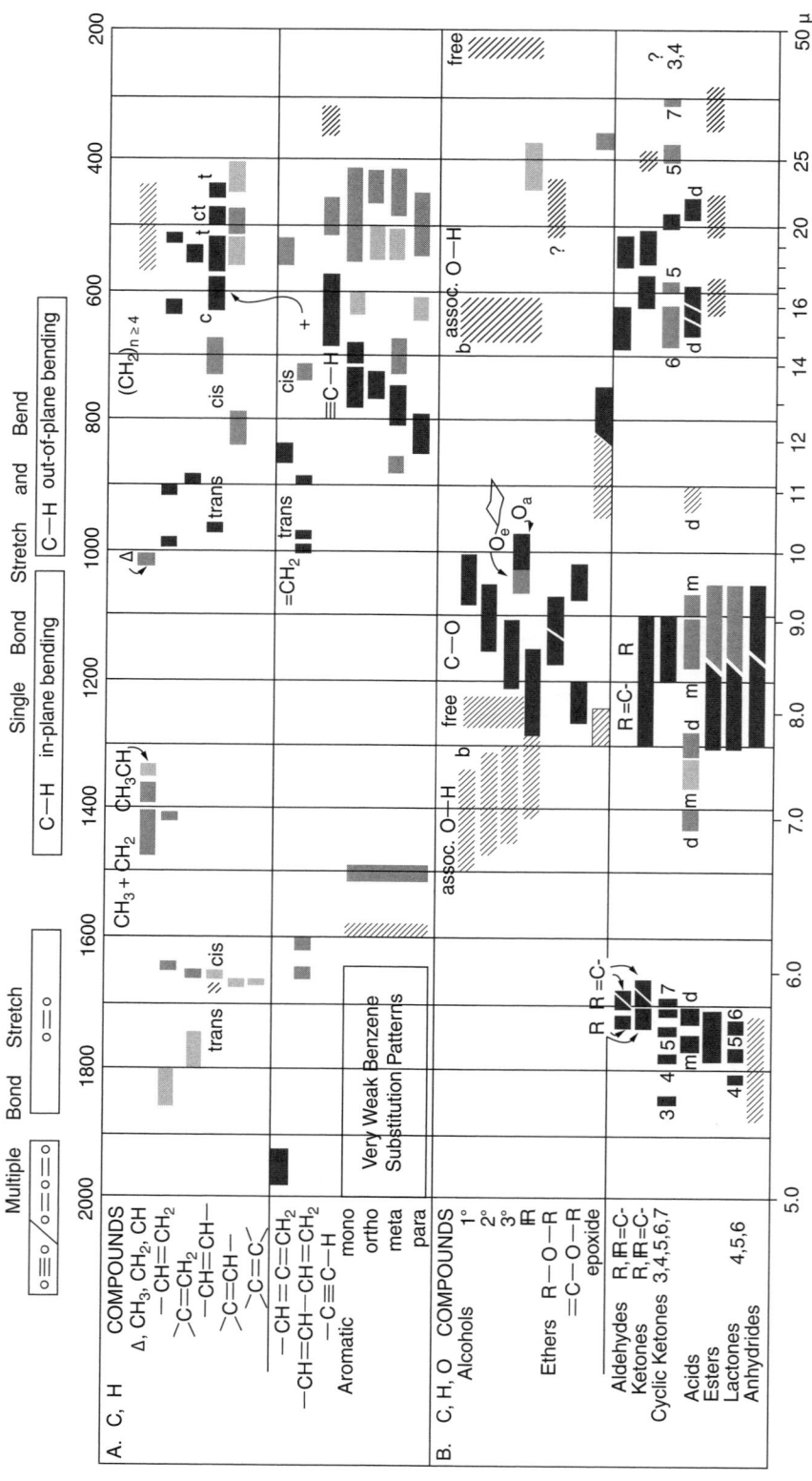

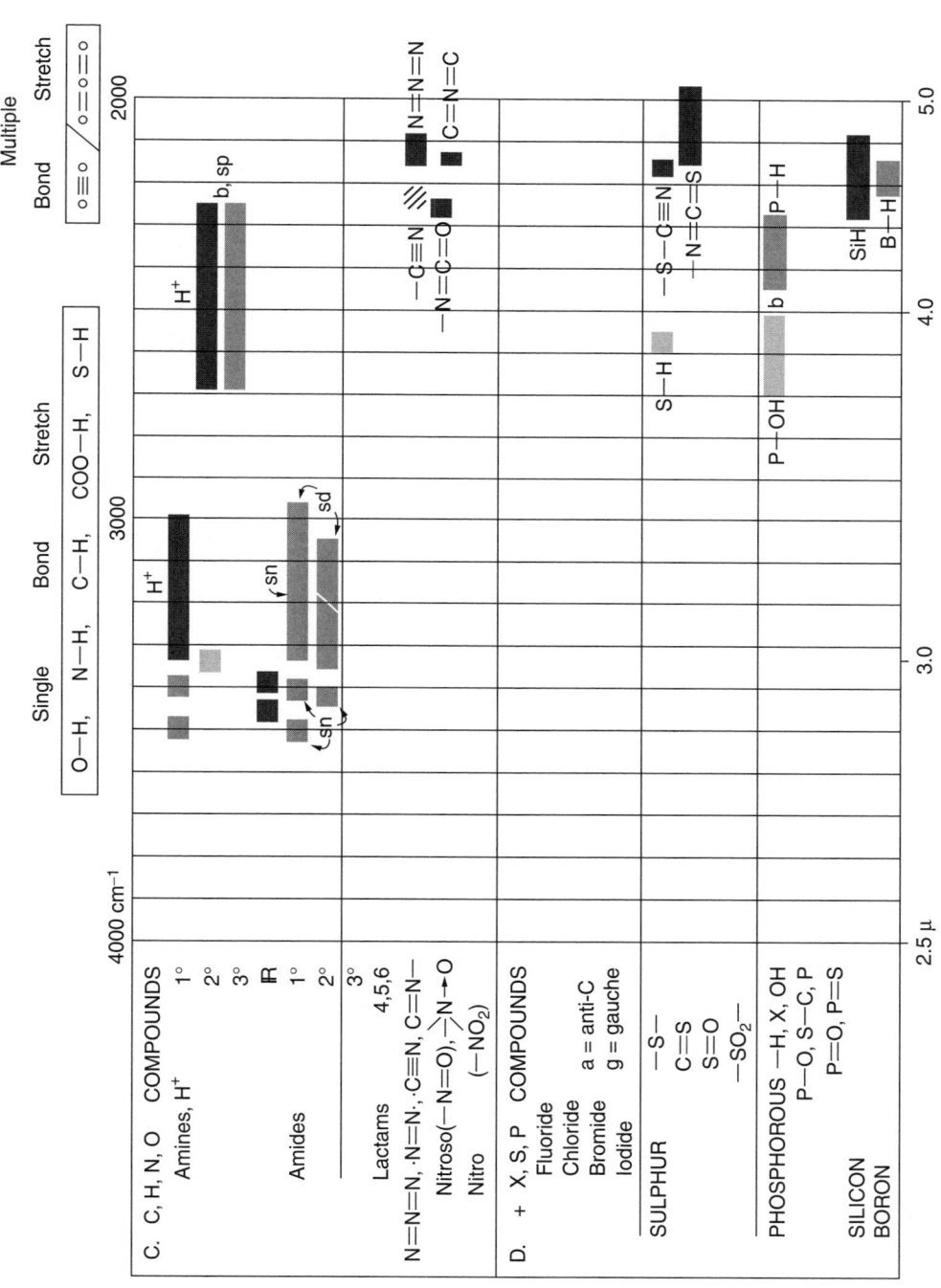

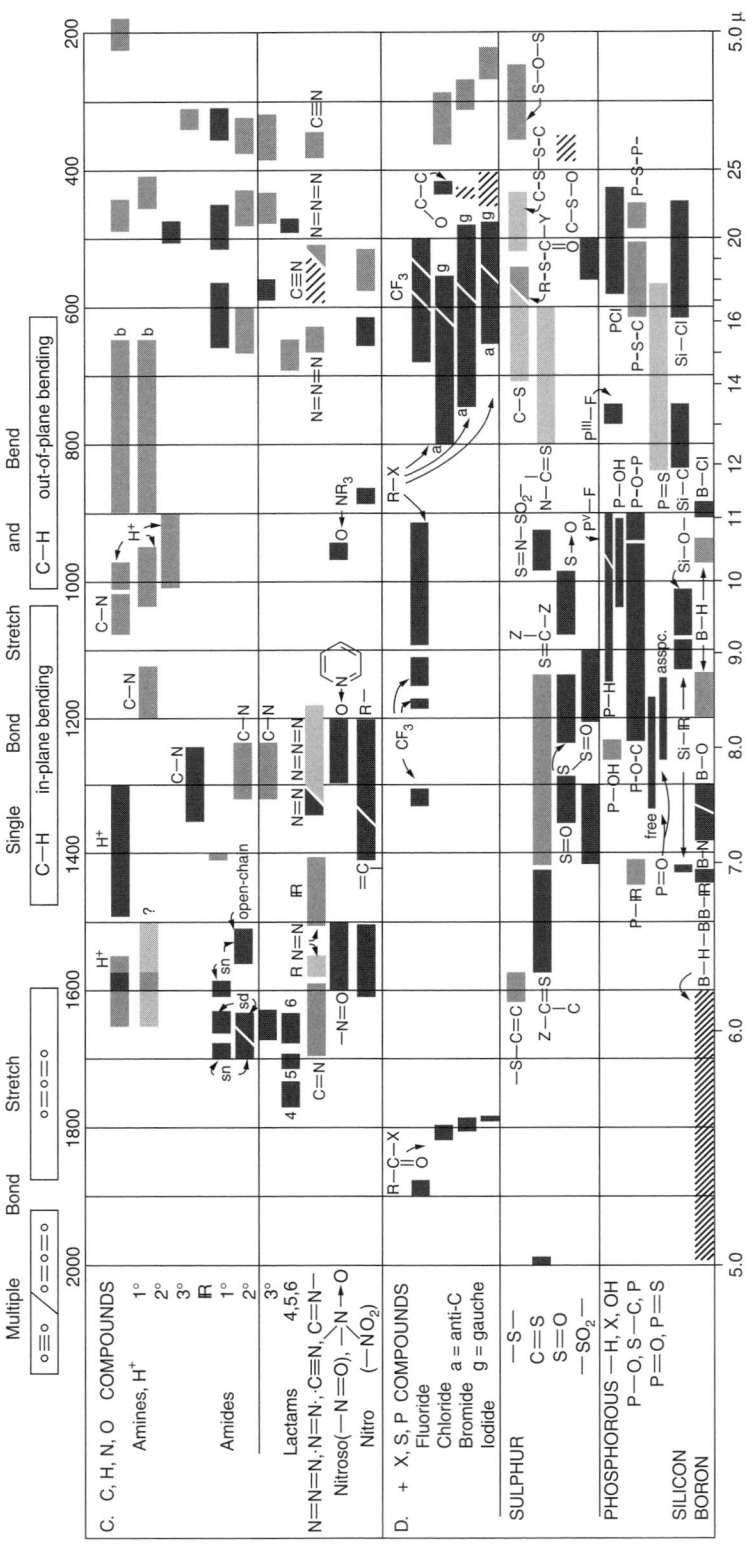

Aromatic Substitution Bands

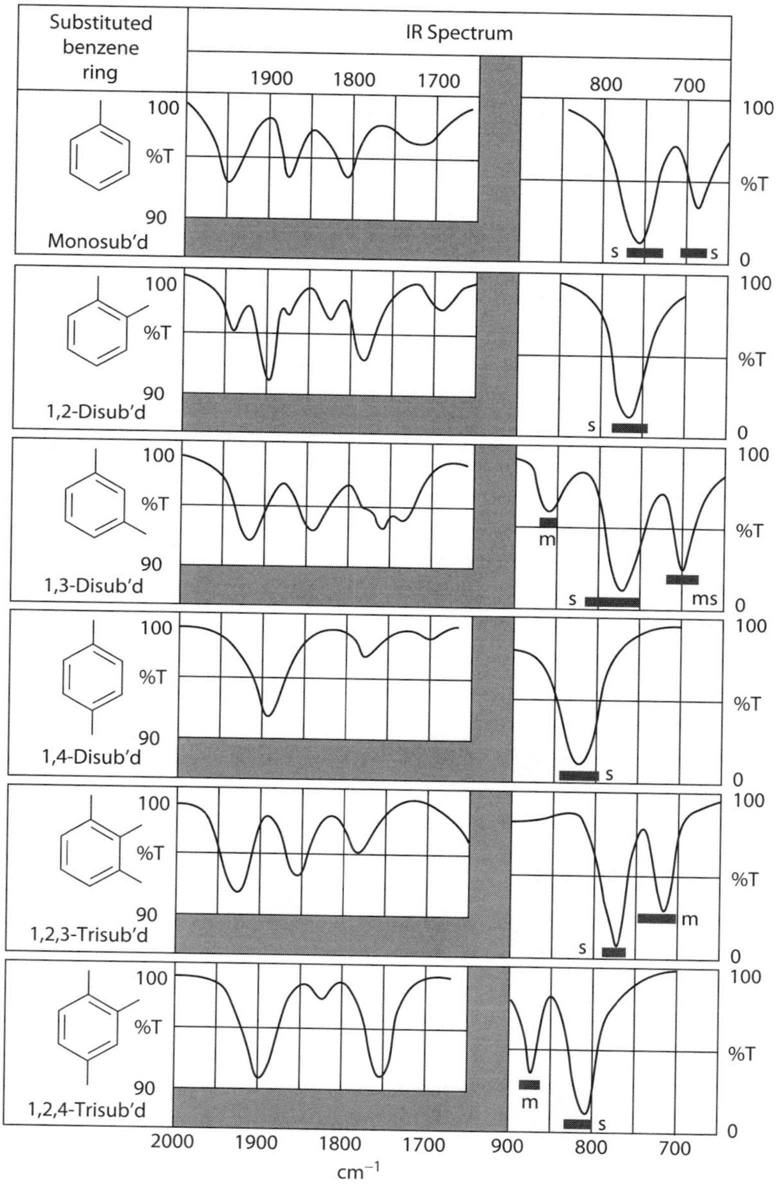

Substituted benzene ring	IR Spectrum						

(Monosub'd)
100 %T 90

1900 1800 1700

800 700
100 %T 0
s ▬▬ ▬▬ s

(1,2-Disub'd)
100 %T 90
s ▬▬

(1,3-Disub'd)
100 %T 90
m
s ▬▬▬ ms

(1,4-Disub'd)
100 %T 90
▬▬ s

(1,2,3-Trisub'd)
100 %T 90
▬▬ m
s ▬

(1,2,4-Trisub'd)
100 %T 90
▬ m
▬ s

2000 1900 1800 1700 900 800 700

cm⁻¹

Aromatic Substitution Bands

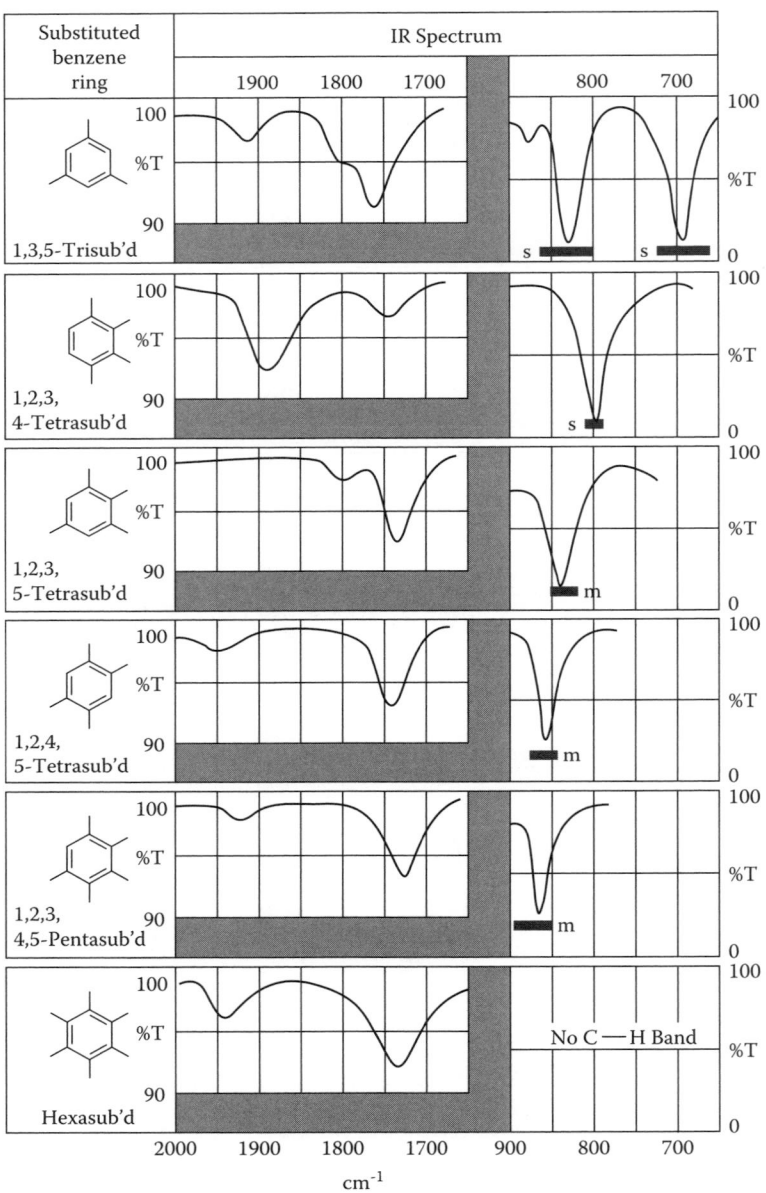

Carbonyl Group Absorptions

Group	Wavenumber, cm⁻¹						
	1850	1800	1750	1700	1650	1600	1550
Acid, Chlorides, Aliphatic		1810–1795					
Acid Chlorides, Aromatic			1785–1765				
Aldehydes, Aliphatic				1740–1718			
Aldehydes, Aromatic				1710–1685			
Amides					1695–1630*		
Amides, typical value, 1°				1684			
Amides, typical value, 2°					1669		
Amides, typical value, 3°					1667		
	5.41	5.56	5.71	5.88	6.06	6.25	6.45
	Wavelength, μm						

* Electron withdrawing groups at the α-position to the carbonyl will raise the wavenumber of the absorption.

Carbonyl Group Absorptions (continued)

Group	Wavenumber, cm⁻¹						
	1850	1800	1750	1700	1650	1600	1550
Anhydrides, acyclic, non-conjugated		1825–1815***	1755–1745**				
Anhydrides, acyclic, conjugated			1780–1770***	1725–1715**			
Anhydrides, ayclic non-conjugated	1870–1845		1800–1775**				
Anhydrides, cyclic conjugated	1860–1850		1780–1760**				
Carbamates				1740–1683			
Carbonates, acyclic			1780–1740				
Carbonates, five-membered ring	1850–1790						
Carbonates, vinyl, typical value			1761				
	5.41	5.56	5.71	5.88	6.06	6.25	6.45
	Wavelength, μm						

**   This band is the more intense of the two.
*** Intensity weakens as colinearity is approached.

Carbonyl Group Absorptions (continued)

Group	Wavenumber, cm^{-1}								
	1800	1750	1700	1650	1600	1550	1450	1400	1350
Carboxylic acid, monomer	1800–1740								
Carboxylic acid, dimer			1720– 1680						
Carboxylic acid, salts				1650–1540			1450–1360		
Carboxylic acid, conjugated			1695– 1680						
Carboxylic acid, non-conjugated			1720– 1700						
Esters, formate			1725–1720						
Esters, saturated		1750– 1735							
Esters, conjugated		1735–1715*							
	5.56	5.71	5.88	6.06	6.25	6.45	6.90	7.14	7.41
	Wavelength, µm								

* Electron withdrawing groups in the α-position to the carbonyl will raise the wavenumber adsorption.

Carbonyl Group Absorptions (continued)

Group	Wavenumber, cm^{-1}								
	1800	1750	1700	1650	1600	1550	1450	1400	1350
Esters, phenyl, typical value		1770							
Esters, thiol, non-conjugated			1710– 1680						
Esters, thiol, conjugated			1700– 1640						
Esters, vinyl, typical value		1770							
Esters, vinylidene, typical value		1764							
Ketones, dialkyl			1725–1705						
Ketones, α, β- unsaturated			1700–1670						
Ketones, α, β, and α', β' conjugated			1680 1640						
	5.56	5.71	5.88	6.06	6.25	6.45	6.90	7.14	7.41
	Wavelength, µm								

Carbonyl Group Absorptions (continued)

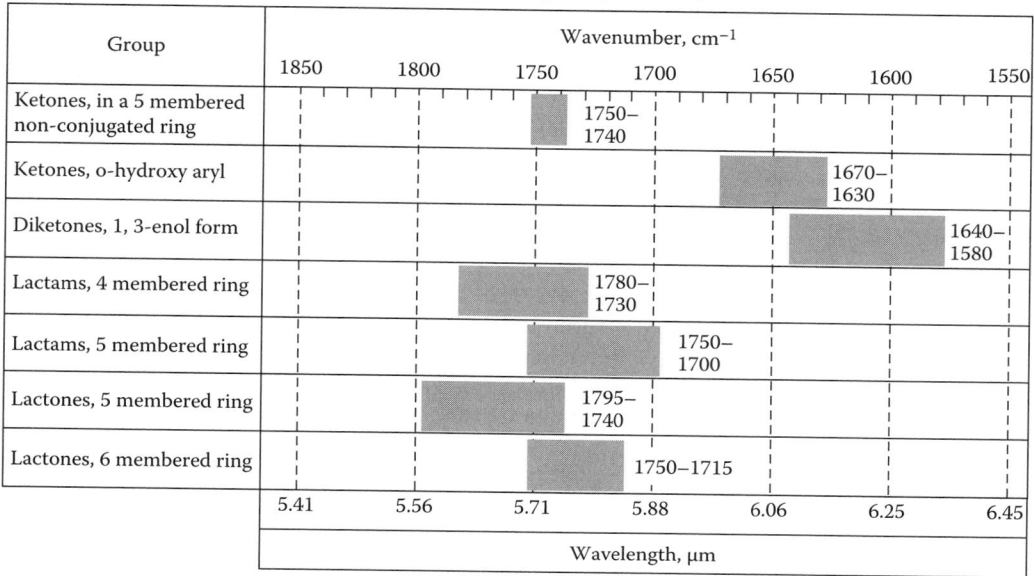

Group	Wavenumber, cm⁻¹ / range
Ketones, in a 5 membered non-conjugated ring	1750–1740
Ketones, o-hydroxy aryl	1670–1630
Diketones, 1, 3-enol form	1640–1580
Lactams, 4 membered ring	1780–1730
Lactams, 5 membered ring	1750–1700
Lactones, 5 membered ring	1795–1740
Lactones, 6 membered ring	1750–1715

Wavelength, μm

NEAR INFRARED ABSORPTIONS

Classically, the near infrared (NIR) region was defined as occurring between 0.7 and 3.5 μm, or 14, 285–2860 cm^{-1}. This classification includes the region of CH, OH, and NH fundamental stretching bands [1–3]. Currently, this spectral area, from 4000 to 2860 cm^{-1}, is considered part of the mid-infrared region and the NIR region is now considered to be above 4000 cm^{-1}. The NIR is a region of overtones and combination bands, which are considerably weaker than the fundamentals that are seen in the mid-infrared region. It is nevertheless a very useful area for quantitative measurement, online and at line analysis, the analysis of viscous liquids and powders, and even for structure determination.

Because most NIR spectrophotometers are often built as enhancements to the capabilities of ultraviolet-visible spectrophotometers, the convention has been to express absorbances in this region in terms of wavelength, rather than wavenumber. In the following chart, we adopt this convention. We do not give any indication of intensity in this chart. The NIR bands will be related to the intensity of the fundamentals in the mid-infrared region, although the bands will typically be broad.

REFERENCES

1. Colthup, N. B., L. H. Daly, and S. E. Wiberley. *Introduction to Infrared and Raman Spectroscopy.* 3rd ed. Boston: Academic Press, 1990.
2. Conley, R. T. *Infrared Spectroscopy.* Boston: Allyn and Bacon, 1972.
3. Bruno, T. J., and P. D. N. Svoronos. *CRC Handbook of Fundamental Spectroscopic Correlation Charts.* Boca Raton, FL: CRC Press, 2006.

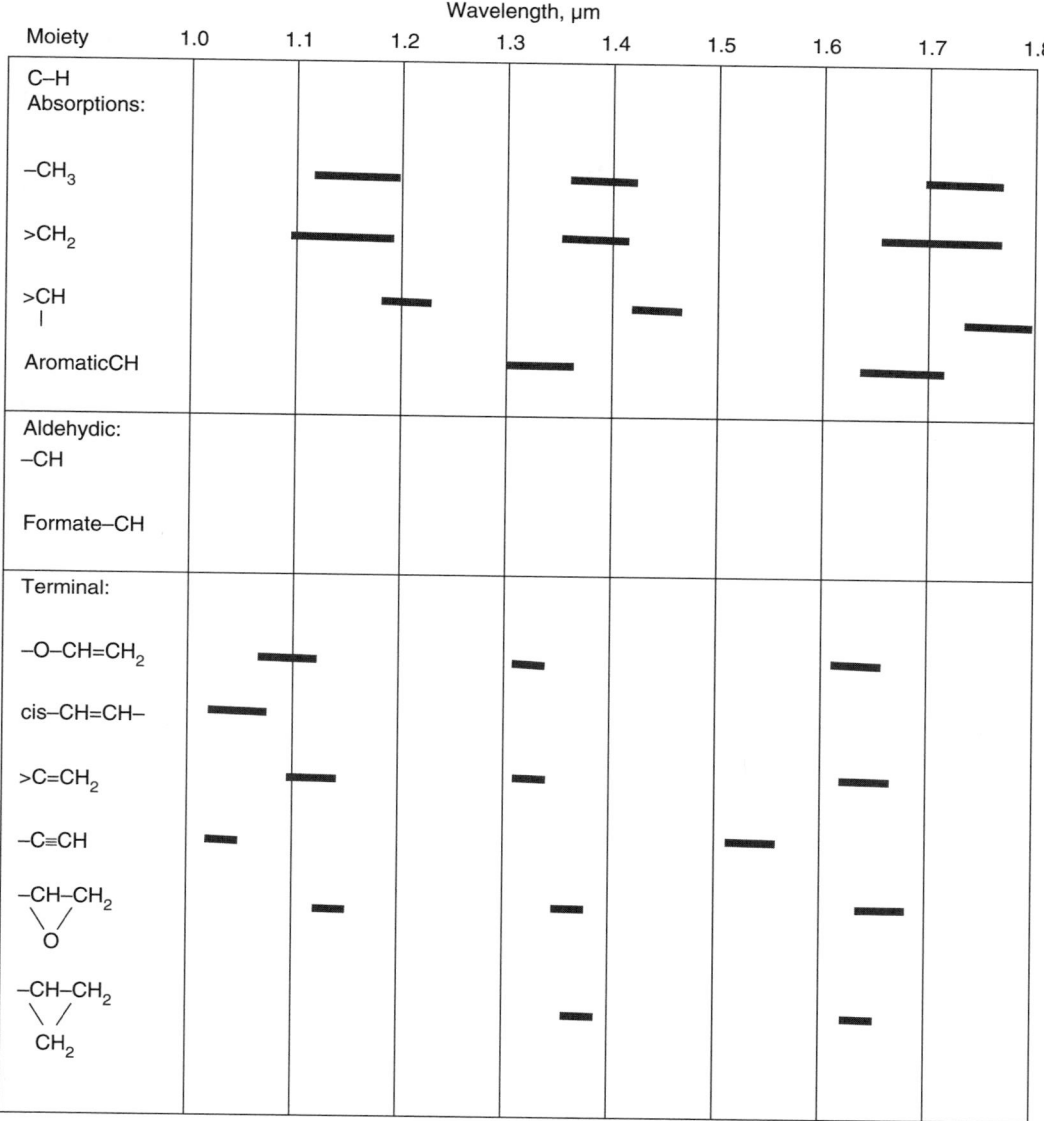

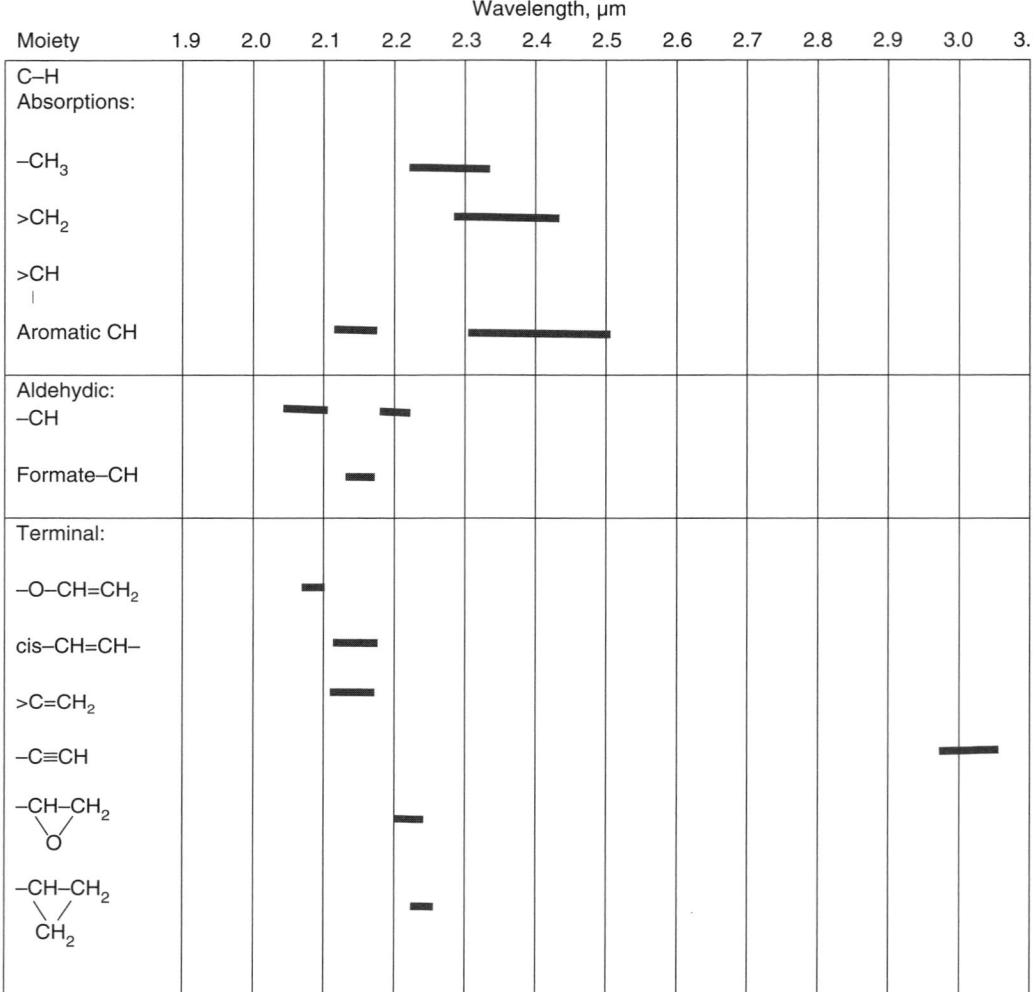

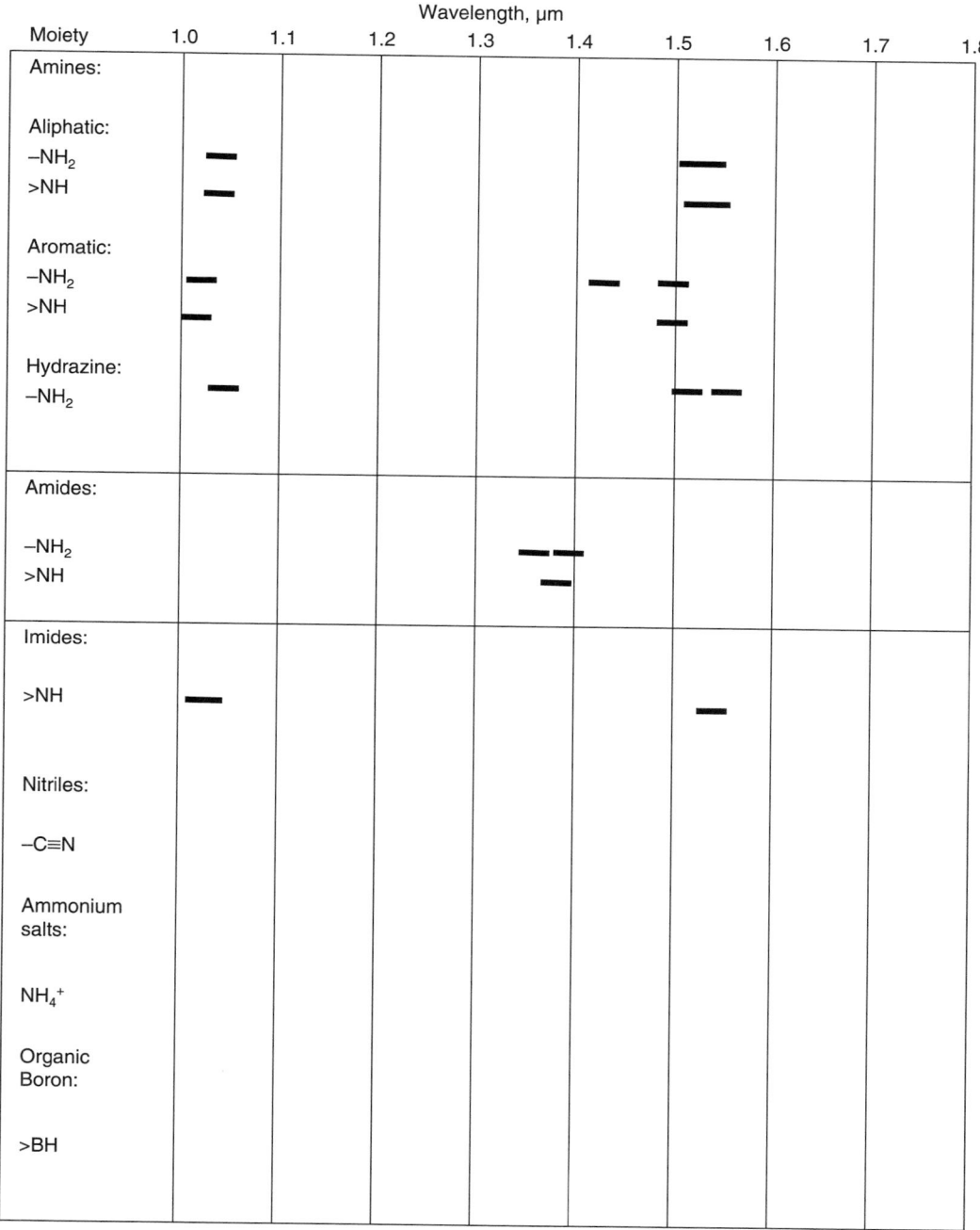

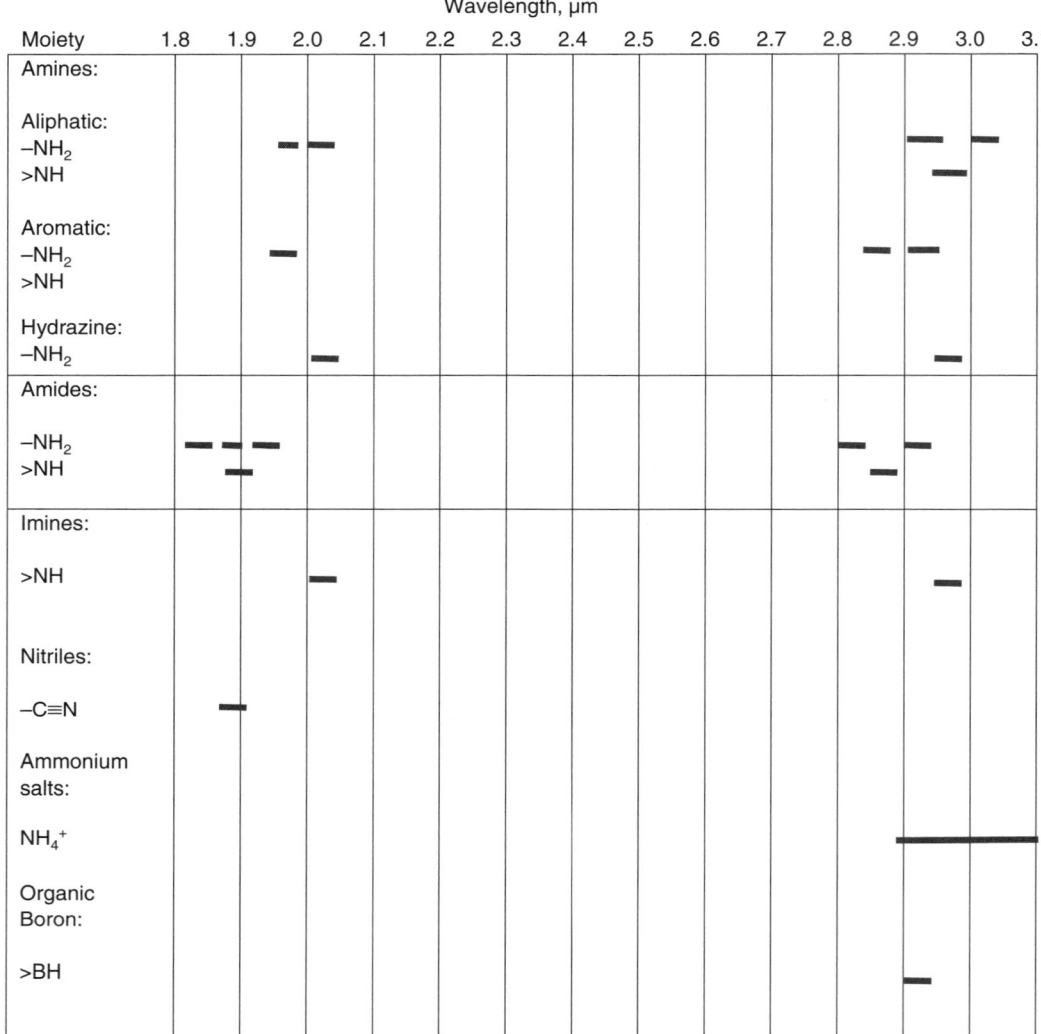

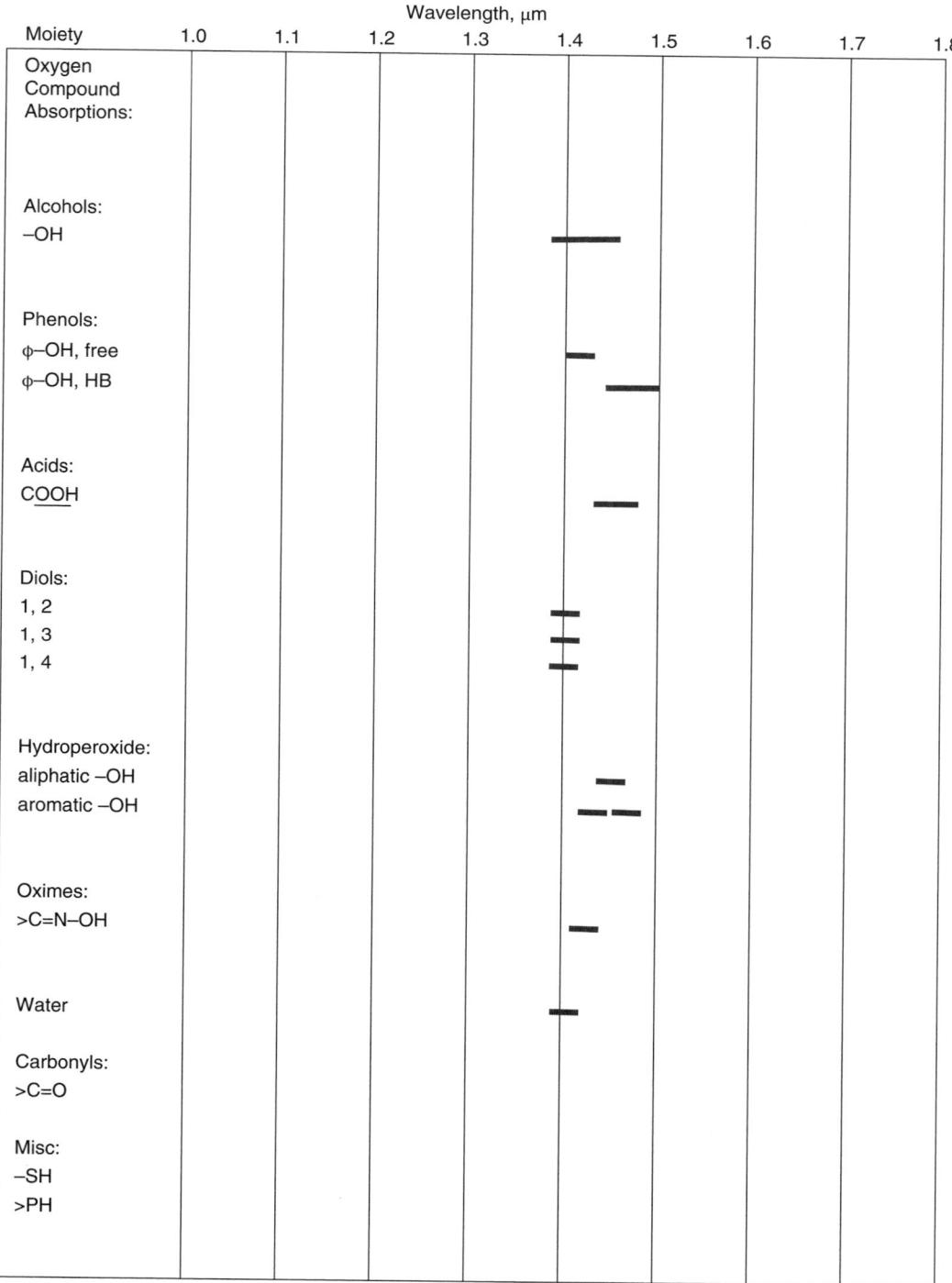

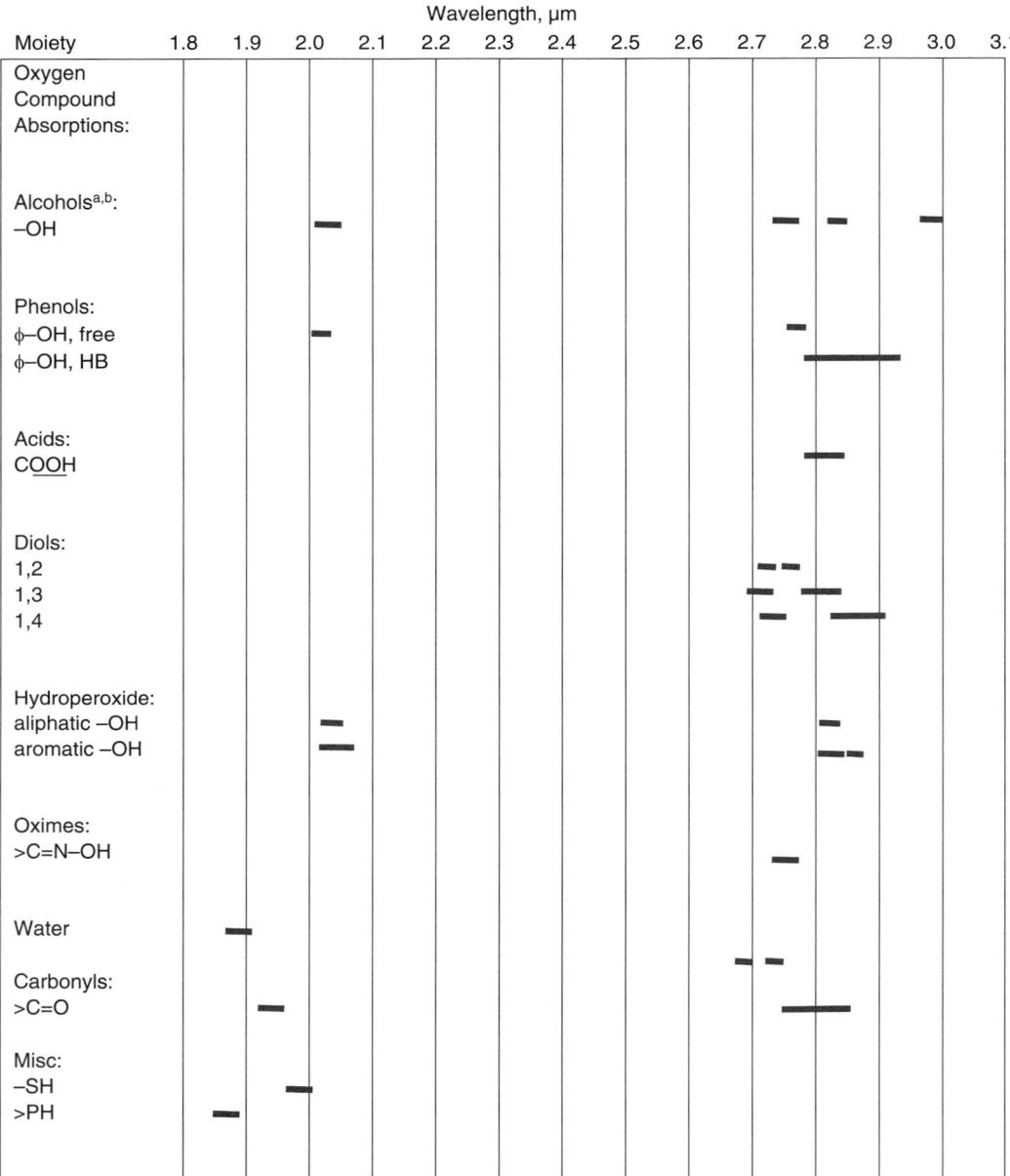

^a The dimer bands for alcohols occur between 2.8 and 2.9 μm.
^b The polymer bands of alcohols occur between 2.96 and 3.05 μm.
HB designates the presence of hydrogen bonding.
φ indicates a benzene ring.

INORGANIC GROUP ABSORPTIONS

The following chart provides the infrared absorbance that may be observed from inorganic functional moieties. These have been compiled from a study of the IR absorption of a number of inorganic species [1,2]. It should be understood that the physical state of the sample plays a role in the intensity and position of these bands. These variables include crystal structure, crystallite size, water of hydration, and so on. This chart must therefore be regarded as an approximate guide.

REFERENCES

1. Miller, F. A., and C. H. Wilkins. "Infrared Spectra and Characteristic Frequencies of Inorganic Ions." *Analytical Chemistry* 24, no. 8 (1952): 1253–94.
2. Bruno, T. J., and P. D. N. Svoronos. *CRC Handbook of Fundamental Spectroscopic Correlation Charts.* Boca Raton, FL: CRC Press, 2006.

Key:
Strong:
Medium:
Weak:

Note:
[a] This is water of crystallization.

Moiety	3600	3400	3200	3000	2800	2600	2400	2200cm^{-1}
Water[a]		▬▬▬▬▬▬						
Boron:								
BO_2^-								
$B_4O_7^{-2}$								
Carbon								
CO_3^{-2}						▬▬▬		
HCO_3^-								
CN^-								
OCN^-								
SCN^-								
Silicon:								
SiO_3^{-2}								
Nitrogen:								
NO_2^-								
NO_3^-								
NH_4^+			▬▬▬▬					
Phosphorus:								
PO_4^{-3}								
HPO_4^{-2}								▬▬▬
$H_2PO_4^-$								
Sulfur:								
SO_3^{-2}								
SO_4^{-2}								
HSO_4^-								
$S_2O_3^{-2}$								
$S_2O_5^{-2}$								
$S_2O_8^{-2}$								
Selenium:								
SeO_3^{-2}								
SeO_4^{-2}								
Chlorine:								
ClO_3^-								
ClO_4^-								
Bromine:								
BrO_3^-								

Iodine: IO_3^-									
Vanadium: VO_3^-									
Chromium: CrO_4^{-2} $Cr_2O_7^{-2}$									
Molybdenum: MoO_4^{-2}									
Tungsten: WO_4^{-2}									
Manganese: MnO_4^-									
Iron: $Fe(CN)_6^{-4}$									

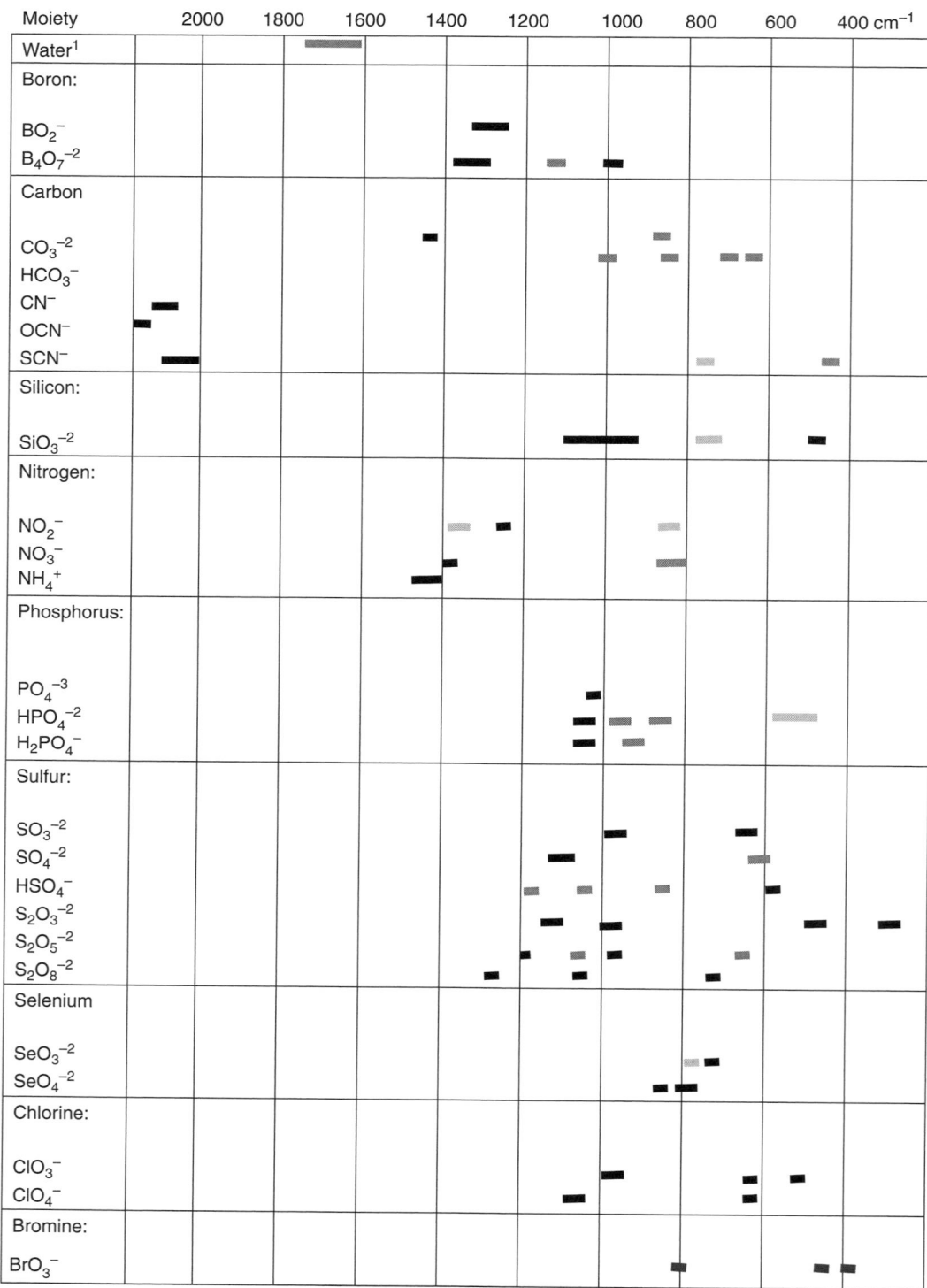

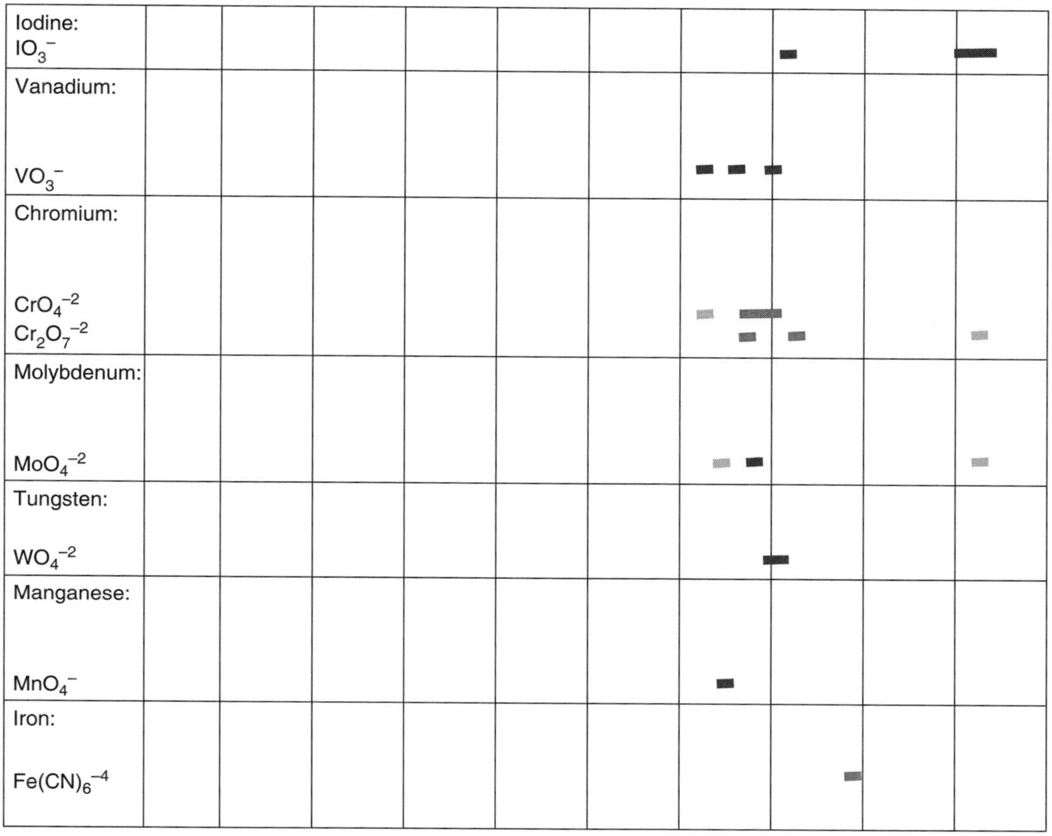

Iodine: IO_3^-								▬		▬▬
Vanadium: VO_3^-							▬ ▬ ▬			
Chromium: CrO_4^{-2} $Cr_2O_7^{-2}$							▬ ▬▬ ▬	▬		▬
Molybdenum: MoO_4^{-2}							▬ ▬			▬
Tungsten: WO_4^{-2}								▬		
Manganese: MnO_4^-							▬			
Iron: $Fe(CN)_6^{-4}$								▬		

MID-RANGE INFRARED ABSORPTIONS OF MAJOR CHEMICAL FAMILIES

The following tables provide expected IR absorptions of the major chemical families [1–23]. The ordering of these tables is: hydrocarbons, oxygen compounds, nitrogen compounds, sulfur compounds, silicon compounds, phosphorus compounds, and halogen compounds. In some ways these data are a more detailed presentation of the spectral correlations.

Abbreviations:

s	= strong	1 °	=	primary
m	= medium	2 °	=	secondary
w	= weak	3 °	=	tertiary
vs	= very strong			
vw	= very weak			
sym	= symmetrical			
asym	= asymmetrical			

REFERENCES

1. Nakanishi, K., and P. H. Solomon. *Infrared Absorption Spectroscopy.* San Francisco: Holden-Day, Inc., 1977.
2. Conley, R. T. *Infrared Spectroscopy.* 2nd ed. Boston: Allyn and Bacon, 1972.
3. Silverstein, R. M., and F. X. Webster. *Spectrometric Identification of Organic Compounds.* New York: John Wiley and Sons, 1997.
4. Williams, D. H., and I. Fleming. *Spectroscopic Methods in Organic Chemistry.* London: McGraw-Hill, 1973.
5. Lambert, J. B., H. F. Shuzvell, L. Verbit, R. G. Cooks, and G. H. Stout. *Organic Structural Analysis.* New York: MacMillan Pub. Co., 1976.
6. Meyers, C. Y. *Eighth Annual Report on Research, 1963, Sponsored by the Petroleum Research Fund.* Washington, DC: American Chemical Society, 1964.
7. Kucsman, A., F. Ruff, and I. Kapovits. "I. Bond System of N-Acylsulfilimines. I. Infrared Spectroscopic Investigation of N-Sulfonylsulfilimines." *Tetrahedron* 25 (1966): 1575.
8. Kucsman, A., I. Kapovits, and F. Ruff. "Infrared Absorption of N-Acylsulfilimines." *Acta Chimica Academiae Scientiarum Hungaricae* 40 (1964): 75.
9. Kucsman, A., F. Ruff, and I. Kapovits. "Bond System of N-Acylsulfilimines. V. IR Spectroscopic Study on N-(p-Nitrophenylsulfonyl) Sulfilimines." *Acta Chimica Academiae Scientiarum Hungaricae* 54 (1967): 153.
10. Shah, J. J. "Iminosulfuranes (Sulfilimines): Infrared and Ultraviolet Spectroscopic Studies." *Canadian Journal of Chemistry* 53 (1975): 2381.
11. Tsujihara, K., N. Furukawa, and S. Oae. "Sulfilimine. II. IR, UV and NMR Spectroscopic Studies." *Bulletin of the Chemical Society of Japan* 43 (1970): 2153.
12. Fuson, N., M. L. Josien, and E. M. Shelton. "An Infrared Spectroscopic Study of the Carbonyl Stretching Frequency in Some Groups of Ketones and Quinones." *Journal of the American Chemical Society* 76 (1954): 2526.
13. Davis, F. A., A. J. Friedman, and E. W. Kluger. "Chemistry of the Sulfur-Nitrogen Bond. VIII. N-Alkylidenesulfinamides." *Journal of the American Chemical Society* 96 (1974): 5000.
14. Krueger, P. J., and A. O. Fulea. "Rotation About the C-N Bond in Thioamides: Influence of Substituents on the Potential Function." *Terahedron* 31 (1975): 1813.
15. Baumgarten, H. E., and J. M. Petersen. "Reactions of Amines. V. Synthesis of α-Aminoketones." *Journal of the American Chemical Society* 82 (1960): 459.
16. Gaset, A., L. Lafaille, A. Verdier, and A. Lattes. "Infrared Spectra of α-Aminoketones; Configurational Study and Evidence of an Enol Form." *Bulletin de la Société Chimique de France* 10 (1968): 4108.

17. Cagniant, D., P. Faller, and P. Cagniant. "Contribution in the Study of Condensed Sulfur Heterocycles. XVII. Ultraviolet and Infrared Spectra of Some Alkyl Derivatives of Thianaphthene." *Bulletin de la Société Chimique de France* 2410 (1961).

18. Tamres, M., and S. Searles, Jr. "Hydrogen Bonding Abilities of Cyclic Sulfoxides and Cyclic Ketones." *Journal of the American Chemical Society* 81 (1959): 2100.

19. George, W. O., R. C. W. Goodman, and J. H. S. Green. "The Infra-Red Spectra of Alkyl Mercapturic Acids, Their Sulphoxides and Sulphones." *Spectrochimica Acta* 22 (1966): 1741.

20. Cairns, T., G. Eglinton, and D. T. Gibson. "Infra-Red Studies with Sulphoxides: Part I. The S = O Stretching Absorptions of Some Simple Sulphoxides." *Spectrochimica Acta* 20 (1964): 31.

21. Hadzi, D. "Hydrogen Bonding in Some Adducts of Oxygen Bases with Acids. Part I. Infrared Spectra and Structure of Crystalline Adducts of Some Phosphine, Arsine, and Amine Oxides, and Sulphoxides with Strong Acids." *Journal of Chemical Society* (1962): 5128.

22. Currier, W. F., and J. H. Weber. "Complexes of Sulfoxides. I. Octahedral Complexes of Manganese (II), Iron (II), Cobalt (II), Nickel (II), and Zinc (II)." *Inorganic Chemistry* 6 (1967): 1539.

23. Kucsman, A., F. Ruff, and B. Tanacs. "IR Spectroscopic Study of N^{15} Labeled Acylsulfilimines." *International Journal of Sulfur Chemistry* 8 (1976): 505.

Hydrocarbon Compounds

Family	General Formula	Wavenumbers (cm⁻¹)			
		C–H Stretch	C–H Bend	C–C Stretch	C–C Bend
Alkanes, (a) Acyclic	C_nH_{2n+2}				
(i): Straight chain	$CH_3(CH_2)_nCH_3$	3000–2840 (s/m) CH_3-, (asym): 3000–2960 (s) CH_3-, (sym): 2880–2870 (s) $>CH_2$, (asym): 2930–2920 (s) $>CH_2$, (sym): 2860–2840 (s)	Below 1500 (w/m/s) CH_3- (asym): 1460–1440 (s) CH_3- (sym): 1380–1370 (s) $>CH_2$ (scissoring): ~1465 (s) $>CH_2$ (rocking): ~720 (s) $>CH_2$ (twisting and wagging): 1350–1150 (w)	1200–800 (w) (not of practical value for definitive assignment)	Below 500 (not of practical value for definitive assignment)

(ii): Branched	R^1-CHR^3
	    $|$
	   R^2
		C–H (3 °): ~2890 (vw)	Gem dimethyl [$(CH_3)_2CH-$]: 1380, 1370 (m, symmetrical doublet)
			
			tert-butyl [$(CH_3)_3C-$]:
			1390, 1370 (m, asymmetrical doublet; latter more intense)		

(b) Cyclic	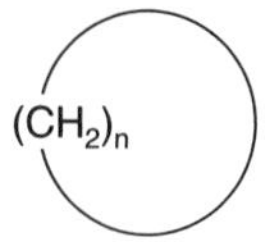 $(CH_2)_n$	Same as in acyclic alkanes; ring strain increases the wavenumbers up to 3100 cm⁻¹	CH_3- rocking: 930–920 (w, not reliable) $>CH_2$ (scissoring): lower than in acyclic alkanes (10–15 cm⁻¹)		

Hydrocarbon Compounds (Continued)

Family	General Formula	Wavenumbers (cm^{-1})				Notes
		>C=C<Stretch	>C=C-H Stretch	>C=C-H Bend (in-plane)	>C=C-H Bend (out-of-plane)	
Alkenes (olefins), (I): Acyclic	C_nH_{2n}	1670–1600	above 3000			
(i): Nonconjugated		1667–1640 (m)				
(a) Monosubstituted (vinyl)	$R^1CH=CH_2$	1658–1648 (m)	3082–3000 (m)	1420–1415 (m) (scissoring)	~995 (m) ~919 (m)	
(b) Disubstituted Cis-	R^1 R^2 >C==C< H H	1662–1652 (m)	3030–3015 (m)	~1406 (m)	715–675 (s) (rocking)	C–H rocking not dependable for definitive assignment
Trans-	R^1 H >C==C< H R^2	1678–1668 (w)	3030–3020 (m)	1325–1275 (m) (deformation)	~965 (s) rocking	
Vinylidene	R^1 H >C==C< R^2 H	1658–1648 (m)	3090–3080 (m) ~2980 (m)	~1415 (m)	~890 (s) (rocking)	
(c) Trisubstituted	R^1 H >C==C< R^2 R^3	1675–1665 (w)	3090–3080 (w)	~1415 (w)	840–800 (m) (deformation)	
(d) Tetrasubstituted	R^1 R^3 >C==C< R^2 R^4	~1670 (vw)	—	—	—	>C=C< stretch may not be detected

(Continued)

Hydrocarbon Compounds (Continued)

Family	General Formula	Wavenumbers (cm^{-1})					Notes
		>C=C<Stretch	>C=C–H Stretch	>C=C–H Bend (in-plane)	>C=C–H Bend (out-of-plane)		
(ii): Conjugated	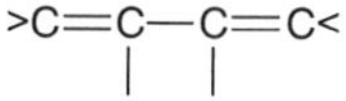 >C=C—C=C<	1610–1600 (m) (frequently a doublet)	3050 (vw)		~980 (rocking)		Conjugation of an olefinic>C=C< with an aromatic ring raises the frequency by 20–25 cm^{-1}
(iii): cumulated	>C=C=C<	2000–1900 (m)	3300 (m)	2000–1900 (s) 1800–1700 (w)	880–850 (s)		
(II): Cyclic	 —C=C—	1640–1560 (variable)			697–625 (w) (wagging)		>C=C< stretch is coupled with C–C stretch of adjacent bonds. Alkyl substitution increases the>C=C< absorption frequency
(III): External exocyclic	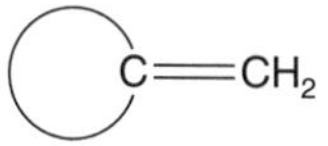 C=CH$_2$	1781–1650	3080, 2995 (m)	~1300 (w)			>C=C< frequency increases with decreasing ring size

Hydrocarbon Compounds (Continued)

Family	General Formula	Wavenumbers (cm^{-1})			Notes
		$-C\equiv C-$ Stretch	$-C\equiv C-H$ Stretch	C–H Bend	
Alkynes (i): Nonconjugated	C_nH_{2n-2}				
(a) Terminal	$R^1-C\equiv C-H$	2150–2100 (m)	3310–3200 (m) (sharp)	700–610 (s) 1370–1220 (w) (overtone)	$-C\equiv C-H$ stretch peak is narrower than that of –OH or –NH stretch, which are broader due to hydrogen bonding
(b) Nonterminal	$R^1-C\equiv C-R^2$	2260–2190 (vw)	—	700–610 (s) 1370–1220 (w) (overtone)	
(ii): Conjugated					
(a) Terminal	$R^1-C\equiv C-C\equiv C-H$	2200, 2040 (doublet)	3310–3200 (m) (sharp)	700–610 (s) 1370–1220 (w) (overtone)	
(b) Nonterminal	$R^1-C\equiv C-C\equiv C-R^2$	2200, 2040 (doublet)	—	700–610 (s) 1370–1220 (w) (overtone)	

Hydrocarbon Compounds (Continued)

Family	General Formula	Wavenumbers (cm^{-1})			Notes
		H>C=C< Stretch	>C=C< Stretch	>C–H Bend (out-of-plane)	
Aromatic compounds					All show weak combination and overtone bands between 2000 and 16,500 cm^{-1}. See aromatic substitution pattern chart.
(a) Monosubstituted	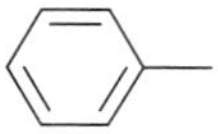	3100–3000	1600–1500	770–730 (s) 710–690 (s)	
(b) Disubstituted (i): 1,2-	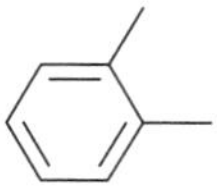	3100–3000	1600–1500	770–735 (s)	
(ii): 1,3-	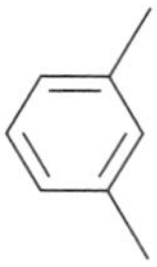	3100–3000	1600–1500	810–750 (s) 710–690 (s)	
(iii): 1,4-		3100–3000	1600–1500	833–810 (s)	
(c) Trisubstituted (i): 1,2,3-	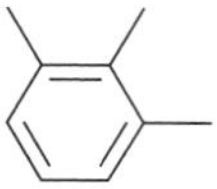	3100–3000	1600–1500	780–760 (s) 745–705 (m)	
(ii): 1,2,4-	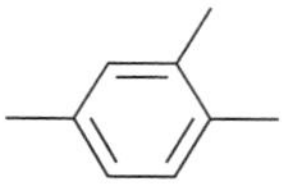	3100–3000	1600–1500	885–860 (m) 825–805 (s)	
(iii): 1,3,5-	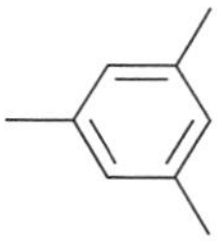	3100–3000	1600–1500	865–810 (s) 730–765 (m)	
(d) Tetrasubstituted (i): 1,2,3,4-	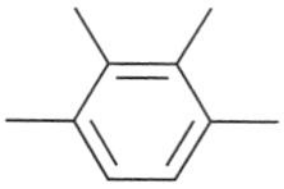	3100–3000	1600–1500	810–800	
(ii): 1,2,3,5-	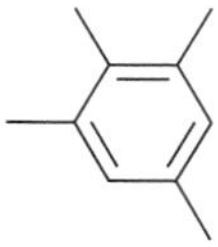	3100–3000	1600–1500	850–840	
(iii): 1,2,4,5-	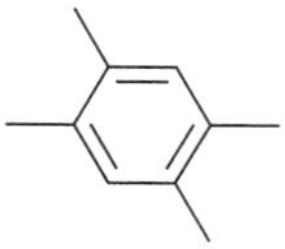	3100–3000	1600–1500	870–855	
(e) Pentasubstituted	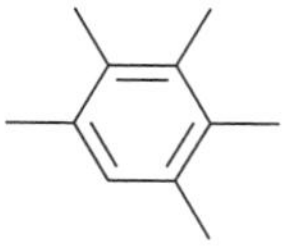	3100–3000	1600–1500	~870	
(f) Hexasubstituted	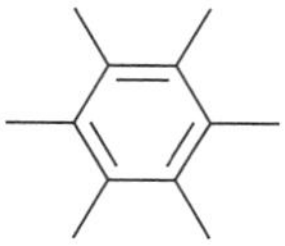	3100–3000	1600–1500	below 500	

Organic Oxygen Compounds

Family	General Formula	Wavenumbers (cm^{-1})				Notes
		O–H Stretch	>C–O Stretch	–O–H Bend		
Acetals	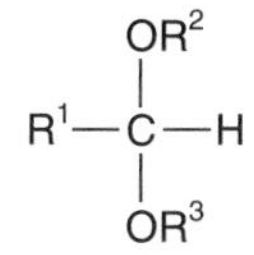 OR^2 R^1—C—H OR^3		1195–1060 (s) (three bands) 1055–1040 (s) (sometimes obscured)			
Acyl halides	R–C(=O)X X = halogen					See Organic Halogen Compounds
Alcohols (i): Primary	R–OH R–CH$_2$OH	3650–3584 (s, sharp) for very dilute solutions or vapor phase spectra. 3550–3200 (s, broad) for less dilute solutions where intermolecular hydrogen bonding is likely to occur. Intramolecular hydrogen bonding is responsible for a broad, shallow peak in the range of 3100–3050 cm^{-1}.	~1050	1420–1300(s) ~1420 (m) and ~1330 (m) (coupling of O–H in-plane bending and C–H wagging)		α–Unsaturation decrease>C–O stretch by 30 cm^{-1}. Liquid spectra of alcohols show a broad out-of-plane bending band (769–650, s).
(ii): Secondary	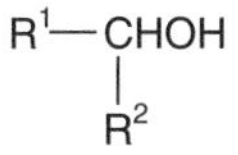 R^1—CHOH R^2		~1100	~1420 (m) and ~1330 (m) (coupling of O–H in-plane bending and C–H wagging)		
(iii): Tertiary	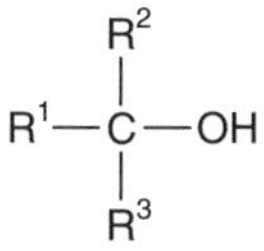 R^2 R^1—C—OH R^3		~1150	only one band (1420–1330 cm^{-1}), position depending on the degree of hydrogen bonding		

Organic Oxygen Compounds (Continued)

Family	General Formula	Wavenumbers (cm⁻¹)		Notes
		>C=O Stretch	–C(=O)H Stretch	
Aldehydes	R–CHO		~2820 (m), ~2720 (m) Fermi resonance between C–H stretch and first overtone of the aldehydic C–H bending	
(i): Saturated, aliphatic	R = alkyl	1720–1720 (s)		
(ii): Aryl	R = aryl	1705–1695 (s)	~2900 (m), ~2750 (m) (aromatic)	
(iii): α,β unsaturated	>C=C–CHO	1700–1680 (s)		
(iv): α,β, γ, σ-unsaturated	>C=C–C=C–CHO	1680–1660 (s)		
(v): β-keto-aldehyde	–C(=O)C–CHO	1670–1645 (s) (lowering is possible due to intramolecular hydrogen bonding in enol form)		
(vi): α-halo-	>C⎯CHO | X X=halogen	~1740 (s)		

The table title "Wavenumbers (cm⁻¹)" spans the ">C=O Stretch" and "–C(=O)H Stretch" columns.

Organic Oxygen Compounds (Continued)

Family	General Formula	Wavenumbers (cm⁻¹)		Notes
		>C=O Stretch	>C–O Stretch	
Amides				See Organic Nitrogen Compounds
Anhydrides				
(i): Saturated acyclic	>C(=O)O(=O)C<	~1820 (s) (asym)	1300–1050 (s) (one or two bands)	
	>C(=O)O(=O)C<	~1760 (m/s) (sym)		
(ii): Conjugated acyclic	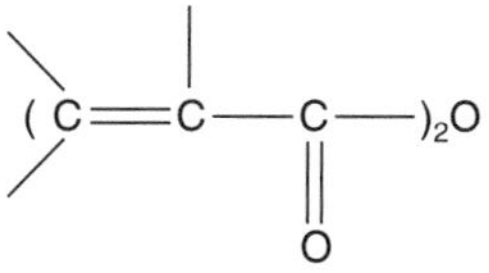 (or Ar–CO)₂O	1795–1775 (s) (asym)	1300–1050 (s) (one or two bands)	
		1735–1715 (m/s) (sym)		
(iii): Cyclic		Ring strain raises band to higher frequencies (up to 1850 and 1790 cm⁻¹). Conjugation does not reduce the frequency considerably	1300–1175 (s)	
			950–910 (s)	

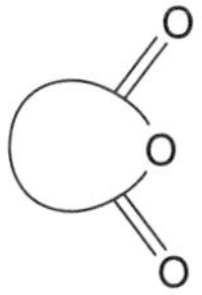

Organic Oxygen Compounds (Continued)

Family	General Formula	Wavenumbers (cm^{-1})				Notes
		>C=O Stretch	>C–O Stretch	–O–H Stretch	–O–H Bend	
Carboxylic Acids						
(i): Monomer, saturated	R–COOH	~1760 (s)	~1420	3550 (s)	~1250 (m/s)	700–610 (s) 1370–1220 (w) (overtone)
(ii): Monomer, aromatic	Ar–COOH	1730–1710	~1400	3500 (s)	~1250 (m/s)	
(iii): Dimer, saturated	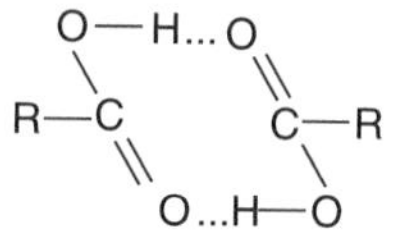 R = alkyl	1720–1706 (s)	1315–1280 (m) (sometimes doublet)	3300–2500 (s, broad)	900–860 (m, broad; out-of-plane)	
(iv): Dimer, α,β-unsaturated (or aromatic)	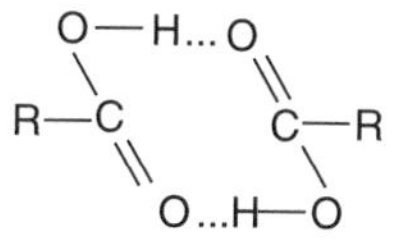 R = alkenyl	1700–1680 (s)	1315–1280 (m) (sometimes doublet)	3300–2500 (s, broad)	900–860 (m, broad; out-of-plane)	

Organic Oxygen Compounds (Continued)

Family	General Formula	Wavenumbers (cm^{-1})		
		>C=O Stretch	>C−O Stretch	Notes
Carboxylic acids (cont.)				
(v): Salt	R–COO$^-$	1610–1550 (s) asym. CO_2^- ~1400 (s) sym. CO_2^-		
Cyanates	R–C≡N→O			See Organic Nitrogen Compounds
Epoxides	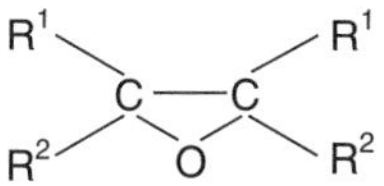 R^1—C—C—R^1 R^2 O R^2			~1250 (s) (ring breathing, sym) 950–810 (s) (asym) 840–810 (s) (C–H bend) 3050–2990 (m/s) (C–H stretch)
Esters	R^1–COOR^2			
(a) Saturated, aliphatic	R^1, R^2 = alkyl	1750–1735 (s) α-halogen substitution results in an increase in wavenumbers (up to 30 cm^{-1})	1210–1163 (s) [acetates only: 1240 (s)]	(O–C–C) 1046–1031 (s) (1 ° alcohol) ~1100 (s) (2 ° alcohol)
(b) Formates	R^1 = H, R^2 = alkyl	1730–1715 (s)	~1180 (s), ~1160 (s)	

Organic Oxygen Compounds (Continued)

Family	General Formula	Wavenumbers (cm^{-1})		Notes
		>C=O Stretch	>C–O Stretch	
Esters (cont.)				
(c) α,β-unsaturated	> C=C–COOR^2 R2 = alkyl	1730–1715 (s)	1300–1250 (s) 1200–1050 (s)	
(d) benzoate	C$_6$H$_5$–COOR^2 R^2 = alkyl	1730–1715 (s)	1310–1250 (s) 1180–1100 (s)	
(e) vinyl	R^1–COOCH=CH$_2$ R^1 = alkyl	1775–1755 (s)	1300–1250 (s) ~1210 (vs)	
(f) phenyl	R^1–COOC$_6$H$_5$ R^1 = alkyl	~1770 (s)	1300–1200 (s) 1190–1140 (s)	
(g) α-ketoesters	–C(=O)COOR^2 R^2 = alkyl	1775–1740 (s)	1300–1050 (s) (two peaks)	
(h) β-ketoesters	–C(=O)–C–C(=O)R^2 R^2 = alkyl –C=C–C–OR^2 ‖ O-H....O	~1735 (s) ~1650 (s) (due to enolization)	1300–1050 (s) (two peaks)	
(i) Aryl benzoates	R^1–COOR^2 R^1,R^2 = aryl	~1735 (s)	1300–1050 (s) (two peaks)	

Organic Oxygen Compounds (Continued)

Family	General Formula	Wavenumbers (cm^{-1})		Notes
		$>C–O–C<$ Stretch Asymmetrical	$>C–O–C<$ Stretch Symmetrical	
Esters (a) Aliphatic	$R^1–O–R^2$; R^1, R^2 = alkyl	1150–1085 (s) branching off on the carbons adjacent to oxygen creates splitting	very hard to trace	
(b) Aryl alkyl	R^1 = alkyl R^2 = aryl	1275–1200 (s) (high due to resonance)	1075–1020 (s)	
(c) Vinyl	R^1 = vinyl R^2 = aryl	1225–1200 (s) (high due to resonance)	1075–1020 (s)	1660–1610 (m) ($>C=CC$) ~1000 (m), 909 (m) ($>C=C–H$) (wagging)
Imides	$(R–C=O)_{/2}NH$			See Organic Nitrogen Compounds
Isocyanates	$R–N=C=O$			See Organic Nitrogen Compounds
Ketals	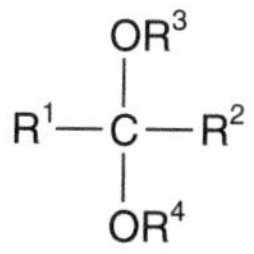 $R^1–\overset{\displaystyle OR^3}{\underset{\displaystyle OR^4}{C}}–R^2$	1190–1160 (s) 1195–1125 (s) 1098–1063 (s) 1055–1035 (s)		
Ketenes	$>C=C=O$			~2150 (s) ($>C=C=O$)

Organic Oxygen Compounds (Continued)

Family	General Formula	Wavenumbers (cm^{-1})		Notes
		>C=O Stretch	>C=C< Stretch	
Ketones				
(a) Aliphatic, saturated	$R^1C(=O)R^2$ R^1, R^2 = alkyl	1720–1710 (s)		>C=O Overtone ~3400 (w). Solid samples or solutions decrease>C=O stretch (10–20 cm^{-1})
(b) α,β-unsaturated	>C=C–C(=O)R^2 R^2 = alkyl	~1690 (s) (s-cis) ~1675 (s) (s-trans)	1650–1600 (m)	α-Halogenation increases>C=O stretch (0–25 cm^{-1})
(c) α,β-α¹,β¹-unsaturated	(>C=C–)$_2$C=O	~1665 (s)	~1640 (m)	>C–H stretch is very weak (3100–2900 cm^{-1})
(d) α,β,γ,δ-unsaturated	>C=C–C=C–C(=O)R_2 R_2 = alkyl	~1665 (s)	~1640 (m)	
(e) Aryl	R^1 = aryl R^2 = alkyl	~1690 (s)	~1600, 1500 (m/s) aromatic	
(f) Diaryl	R^1, R^2 = aryl	~1665 (s)	~1600, 1500 (m/s) (aromatic)	
(g) Cyclic	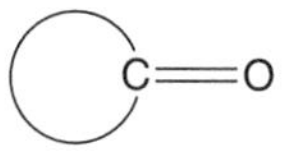 [ring with C=O]	3-membered: 1850 (s) 4-membered: 1780 (s) 5-membered: 1745 (s) 6-membered: 1715 (s) larger than 6-membered: 1705 (s)		
(h) α-keto (s-trans)	R^1–C(=O)COR²	~1720 (s) (aliphatic) ~1680 (s) (aromatic)		
(i) β-keto	$R^1COCH_2COR^2$	~1720 (s) (two bands)	1640–1580 (m, broad) due to enol 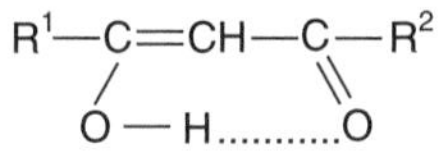 form R^1–C=CH–C–R^2 O—H·······O	Shows a shallow broad –OH band (enol form) at 3000–2700 cm^{-1}
(j) α-amino ketone hydrochlorides	$R^1COCH_2NH_3^+ Cl^-$			>C=O decreases 10–15 cm^{-1} with electron deactivating p-substituents
(k) α-amino ketones	$R–COCH_2NR_2$			Strong bands at 3700–3600 cm^{-1} (–OH) and 1700–1600 cm^{-1} (>C=O) due to the presence of enolic forms

Organic Oxygen Compounds (Continued)

Family	General Formula	Wavenumbers (cm⁻¹)			Notes
		>C=O Stretch	>C=C< Stretch	>C–O Stretch	
Lactams (cyclic amides)					See Organic Nitrogen Compounds
Lactones (cyclic esters)	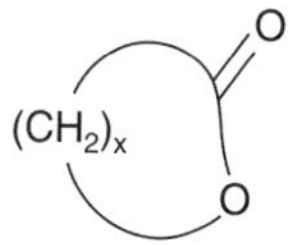				
(i): Saturated (a) α-	x = 4	~1735 (s)		1300–1050 (s, two peaks)	
(b) γ-	x = 3	~1770 (s)		1300–1050 (s, two peaks)	
(c) β-	x = 2	~1840 (s)		1300–1050 (s, two peaks)	
(ii): Unsaturated, α- to the carbonyl (>C=O)	x = 4	~1720 (s)		1300–1050 (s, two peaks)	
	x = 3	~1750 (s) (doublet 1785–1755 cm–1 when α-hydrogen present)		1300–1050 (s, two peaks)	
(iii): Unsaturated, α- to the oxygen	x = 4	~1760 (s)	~1685 (s)	1300–1050 (s, two peaks)	
	x = 3	~1790 (s)	~1660 (s)	1300–1050 (s, two peaks)	
(iv): Unsaturated, α- to the carbonyl and α- to the oxygen	x = 4 (α-pyrone; coumarin)	1775–1715 (s, doublet)	1650–1620 (s) 1570–1540 (s)	1300–1050 (s, two peaks)	

Organic Oxygen Compounds (Continued)

Family	General Formula	Wavenumbers (cm^{-1})				Notes
		>C=O Stretch	>C=C< Stretch	>C–O stretch		
Nitramines	R^1 >N–NO$_2$ R$_2$					See Organic Nitrogen Compounds
Nitrates	R–NO$_3$					See Organic Nitrogen Compounds
Nitro Compounds	R–NO$_2$					See Organic Nitrogen Compounds
Nitrosamines	R$_1$ >N–N=O R$_2$					See Organic Nitrogen Compounds
Nitroso-compounds	R^1–N–N=O | R^2					See Organic Nitrogen Compounds

Organic Oxygen Compounds (Continued)

Family	General Formula	Wavenumbers (cm^{-1})				
		>C–O Stretch	–O–H Stretch	>C=O Stretch	–O–H Bend	Notes
Peroxides	R^1–O–O–R^2	–C–C–O–		–C(=O)O		
(I): Aliphatic	R^1, R^2 = alkyl	890–820 (vw)				
(ii): Aromatic	R^1, R^2 = aromatic	~1000 (vw)				
(iii): Acyl, aliphatic	R^1, R^2 = acyl (aliphatic)	890–820 (vw)		1820–1810 (s) 1800–1780 (s)		
(iv): Acyl, aromatic	R^1, R^2 = acyl (aromatic)	~1000 (vw)		1805–1780 (s) 1785–1755 (s)		
Peroxyacids	R^1–C(=O)OOH	~1260 (s)	3300–3250 (s, not as broad as in R–COOH)	1745–1735 (s) (doublet)	~1400 (m)	~850 cm–1 (m, –O–O– stretch)
Peroxyacids, anhydride	(R^1–COO)$_2$	(–COO–OOC–)				
(i): Alkyl	R^1 = alkyl	1815 (s), 1790 (s)				
(ii): Aryl	R^1 = aryl	1790 (s), 1770 (s)				

Organic Oxygen Compounds (Continued)

Family	General Formula	Wavenumbers (cm^{-1})			Notes
		>C–O Stretch	–O–H Stretch	–O–H Bend	
Phenols	Ar–OH	~1230 (m)	~3610 (m, sharp; in CHCl$_3$ or CCl$_4$ solution) ~3100 (m, broad; in neat samples)	1410–1310 (m, broad; in-plane) ~650 (m) (out-of-plane)	
	Ar = aryl				
Phosphates	(R^{1}O)$_3$P=O				See Organic Phosphorus Compounds
Phosphinates	(R^{1}O)P(=O)H$_2$				See Organic Phosphorus Compounds
Phosphine oxides	R$_3$P=O				See Organic Phosphorus Compounds
Phosphonates	(R^{1}O)$_2$P(=O)H				See Organic Phosphorus Compounds
Phosphorus acids	R$_2$P(=O)OH				See Organic Phosphorus Compounds
Pyrophosphates	(R–P=O)$_2$O				See Organic Phosphorus Compounds

Organic Oxygen Compounds (Continued)

Family	General Formula	Wavenumbers (cm^{-1})		Notes
		>C=O Stretch	>C=C < Stretch	
Quinones				
(a) 1,2-		~1675 (s)	~1600 (s)	
(b) 1,4-		~1675 (s)	~1600 (s)	
Silicon compounds				See Organic Silicon Compounds
Sulfates				See Organic Sulfur Compounds
Sulfonamides				See Organic Sulfur Compounds
Sulfonates				See Organic Sulfur Compounds
Sulfones				See Organic Sulfur Compounds
Sulfonyl chlorides				See Organic Sulfur Compounds
Sulfoxides				See Organic Sulfur Compounds

Organic Nitrogen Compounds

Family	General Formula	Wavenumbers (cm⁻¹)			Notes
		C–N	N–H	Others	
Amides					
Primary	$R^1\text{–CONH}_2$	1400 (s) (stretch)	3520 (m) (stretch) 3400 (m) (stretch) 1655–1620 (m) (bend) 860–666 (m, broad) (wagging)	>C=O (1650) (s, solid state) (1690) (s, solution)	Lowering of N–H stretch occurs in solid samples due to hydrogen bonding; higher values arise in dilute samples
Secondary	$R^1\text{–CONHR}^2$	1400 (s) (stretch)	3500–3400 (w) (stretch) 1570–1515 (w) (bend) 860–666 (m, broad) (wagging)	>C=O (1700–1670) (s, solution); (1680–1630) (s, solid state) Band due to interaction of N–H (bend) and (C–N) (stretch) (~1250) (m, broad)	Lowering of N–H stretch occurs in solid samples due to hydrogen bonding; higher values arise in dilute samples
Tertiary	$R^1\text{–CONR}^2R^3$	1400 (s) (stretch)	—	>C=O (1680–1630) (s); higher values are obtained with electron attracting groups attached to the nitrogen	
Amines					
Primary	$R^1\text{–NH}_2$	1250–1020 (m) (for nonconjugated amines) 1342–1266 (s) (for aromatic amines)	3500 (w) (stretch) 3400 (w) (stretch) 1650–1580 (m) (scissoring) 909–666 (m) (wagging)		
Secondary	$R^1\text{–NHR}^2$	1250–1020 (m) (for nonconjugated amines) 1342–1266 (s) (for aromatic amines)	3350–3310 (w) (stretch) 1515 (vw) (scissoring) 909–666 (m) (wagging)	—	
Tertiary	$R^1\text{–NR}^2R^3$	1250–1020 (m) (for nonconjugated amines) 1342–1266 (s) (for aromatic amines)	—		
Amine salts					
Primary	$RNH_3^+X^-$		3000–2800 (s) 2800–2200 (m) (series of peaks) 1600–1575 (m) 1550–1504 (m)		

Compound	Structure	Bands	Bands (carboxyl)
Secondary	$R_2NH_2^+X^-$	3000–2700 (s), 2700–2250 (m) (series of peaks), 2000 (w), 1620–1560 (m)	
Tertiary	$R_3NH^+X^-$	2700–2250 (s)	
Quaternary	$R_4N^+X^-$	—	
Amino acids (alpha)	$R^1-CH-COO^-$ with NH_2	3100–2600 (s, broad), 2222–2000 (s, broad, overtone), 1610 (w) (bend), 1550–1485 (s) (bend)	$-COO^-$ (1600–1590) (s) $-COOH$ (1755–1730) (s)
	$R^1-CH-COO^-$ with $^+NH_3$		
	$R^1-CH-COOH$ with $^+NH_3$		
Ammonium ion	NH_{4+}	3300–3040 (s), 2000–1709 (m), 1429 (s)	
Azides	$R-N_3$		2140 (s) (asym stretch, N_3) 1295 (s) (sym stretch, N_3)
Azocompounds	$R^1-N=N-R^2$ (trans)	forbidden in IR but allowed in Raman spectrum (1576) (w); peak is lowered down to 1429 cm^{-1} in unsymmetrical p-electron donating substituted azobenzenes	
Azoxy compounds	$R-N=N{\rightarrow}O$		1310–1250 (s)

Organic Nitrogen Compounds (Continued)

Family	General Formula	Wavenumbers (cm^{-1})		Notes
		C–N Multiple Bond	Cumulated (–X=C=Y) double bond	
Cyano-compounds (nitriles)	R–C≡N	2260–2240 (w) (aliphatic) 2240–2220 (m) (aromatic, conjugated)		Electronegative elements α- to the C≡N group reduce the intensity of the absorption
Diazonium salts	[R–N≡N]$^+$			2280–2240 (m) (–N≡N) +
Imides	R–C–NH–C–R ‖ ‖ O O			1710, 1700 (>C=O six-membered ring) 1770, 1700 (>C=O five-membered ring)
Isocyanates	R–N=C=O		2273–2000 (s) (broad) (asym) 1400–1350 (w) (sym)	
Isocyanides (isonitriles)	R–N≡C	2400–2300 (w) (aliphatic) 2300–2200 (w) (aromatic)		
Isonitriles				See isocyanides
Isothiocyanates	R–N=C=S		2140–2000 (s) (stretch)	
Ketene	R$_1$ >C=C=O R$_2$		2150 (stretch); 1120	
Ketenimine	R$_1$ >C=C=N– R$_2$		2000 (stretch)	

Organic Nitrogen Compounds

Family	General Formula	Wavenumbers (cm⁻¹)				Notes
		>C–N	>N–O (asymmetric)	>N–O (symmetric)	Others	
Lactams	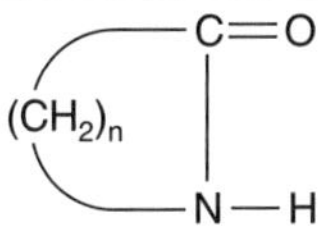 C=O, (CH₂)ₙ, N–H				>C=O (s) (stretch); 1670 (six membered ring); 1700 (five membered ring); 1745 (four membered ring); N–H (out-of-plane wagging) (800–700) (broad)	Add ~15 cm⁻¹ to every wavenumber in case of a>C=C < in conjugation; amide group is forced into the cis-conformation in rings of medium size.
Nitramines	R^1–N–NO_2, R^2		1620–1580 (s) (asym); 1320–1290 (s) (sym)			
Nitrates	RO–NO_2				–N=O; 1660–1625 (s) (asym); 1300–1225 (s) (asym); >N–O; 870–833 (s) (stretch); 763–690 (s) (bend)	
Nitriles (cyano-compounds)	R–C≡N					See cyanocompounds
Nitrites	RO–N=O				–N=O stretch; 1680–1650 (vs) (trans); 1625–1610 (vs) (cis); >N–O stretch; 850–750 (vs)	
Nitro-compounds Aliphatic	R–NO_2, R–alkyl	870	1615–1540 (vs) (asym); 1390–1320 (vs) (sym)	1390–1320 (vs)	~610 (m) (CNO bend)	Aromatics absorb at lower frequencies than aliphatic
Aromatic	R = aryl	(difficult to assign)	1548–1508 (s) (asym); 1356–1340 (s) (sym)	1356–1340 (s)		
Nitrosamines	R1 >N–N=O R2				>N–O stretch; (1520–1500) (s) (vapor); (1500–1480) (s) (neat); N–N (1150–925) (m)	

(Continued)

Organic Nitrogen Compounds (Continued)

Family	General Formula	>C–N	Wavenumbers (cm⁻¹)		Others	Notes
			>N–O (asymmetric)	>N–O (symmetric)		
Nitroso-compounds	R–N=O				N=O stretch 1585–1539 (s) (3°, aliphatic) 1511–1495 (s) (3°, aromatic)	1 ° and 2 °C-nitroso-compounds are unstable and rearrange or dimerize
Pyridines	C_5H_5N				N–H (3075,3030) (s) C–H (out-of-plane) (920–720) (s) (2000–1650) (overtone) C=C ring stretch (1600,1570,1500,1435)	Characteristic substitution pattern: α-substitution: (795–780), (755–745) β-substitution: (920–880), (840–770), 720
Sulfilimines	R^1—S=N–R^3 / R^2					See Organic Sulfur Compounds
Sulfonamides	$R–SO_2NH_2$					See Organic Sulfur Compounds
Thiocyanates	$R–SC \equiv N$					See Organic Sulfur Compounds

Organic Sulfur Compounds

Family	General Formula	Wavenumbers (cm⁻¹)				Notes
		>S=O (asymmetric)	>S=O (symmetric)	>S=N–	Others	
Disulfides	R¹–S–S–R²				–S–S– (< 500) (w)	
Mercaptans	R–S–H				–S–H (2600–2500) (w)	Only significant frequency around that region; lowering of 50–150 cm⁻¹ due to hydrogen bonding.
Mercapturic acids	R²(O=)CNH RSCH₂CH HOOC	1295–1280 (s) (for sulfo nes)	1135–1100 (s) (for sulfones)		1025, 970 (>S→O) (for sulfoxides)	Reduction of all >S= O frequencies due to H– bonding with –NH
Sulfates	(RO)₂S(=O)₂	1415–1380 (s)	1200–1185 (s)			
Sulfides	R¹–S–R²				R–S– (700–600) (w)	
Sulfilimines						
(i) N-acyl	R₂S=N–COR¹			800 (s)	>C=O (1625–1600) (s)	
(ii) N-alkyl	R₂S=N–R¹			987–935 (s)		
(iii) N-sulfonyl	R₂S=N–SO₂R¹	1280–1200 (s) 1095–1030 (s)	1160–1135 (s)	980–901 (s)		
Sulfinamides, N-alkylidene	RS(O)N=CR₂				1520 (amide II band) 1080 (s, S→O)	
Sulfonamides	R–SO₂NH₂	1370–1335 (s)	1170–1155 (s)		>N–H (1 °) (3390–3330) (s) (3300–3247) (s) >N–H (2 °) (3265) (s)	Solid phase spectra lower wavenumbers by 10–20 cm⁻¹
Sulfonates	R¹–SO₂–OR²	1372–1335 (s)	1195–1168 (s)			Electron donating groups on the aryl group cause higher frequency absorption.
Sulfones	R¹–SO₂–R²	1350–1300 (s)	1160–1120 (s)			Hydrogen bonding reduces the frequency of absorption slightly.
Sulfonic acids (anhydrous)	R–SO₃H	1350–1342 (s)	1165–1150 (s)		-OH (3300–2500) (s, broad)	Hydrated sulfonic acids show broadbands at 1230–1150 cm⁻¹.
Sulfonic acids, salts	R–SO₃⁻	ca. 1175 (s)	ca. 1055 (s)			
Sulfonyl chlorides	R–SO₂Cl	1410–1380 (s)	1204–1177 (s)			

(Continued)

Organic Sulfur Compounds (Continued)

Family	General Formula	Wavenumbers (cm^{-1})				Notes
		>S=O (asymmetric)	>S=O (symmetric)	>S=N–	Others	
Sulfoxides cyclic	$R_2S{\to}O$ $(CH_2)_x\ S{\to}O$				>S→O (1070–1030) (s) x = 3 1192 (CCl_4) 1073 ($CHCl_3$) x = 4 1035 (CCl_4) 1020 ($CHCl_3$) x = 5 1053 (CCl_4) 1031 ($CHCl_3$)	Hydrogen bonding reduces the frequency absorption slightly; electronegative substituents increase the >S→O frequency; inorganic complexation reduces the >S→O (up to 50 cm^{-1}).
Thiocarbonyls (not trimerized into cyclic sulfides)	$R^1{-}C{-}R^2(H)$ $\|$ S				>C=S (1250–1020) (s)	
Thiocyanates	$R{-}S{-}C{\equiv}N$				–C≡N (2175–2140) (s); higher values for aryl thiocyanates	
Thiol esters	$R^1{-}C{-}SR^2$ $\|$ O				>C=O (1690) (s) (S-alkyl thioester) (1710) (s) (S-aryl thioester)	The (+) mesometic effect of sulfur is larger than its (–) inductive effect
Thiols	R–SH				–S–H (2600–2500) (w)	See Mercaptans
Thiophenols	Ar–SH					

Organic Silicon Compounds

Family	General Formula	Wavenumbers (cm^{-1})					
		>Si–H Stretch	>Si–H Bend	>C–Si<Stretch	>C–H Bend	>Si–O–Stretch	–OH Stretch
Silanes	R_xSiH_y						
(a) Monoalkyl	$RSiH_3$	2130–2100 (s)	890–860 (s)	890–690 (s)	~1260 (s) (rocking)		
(b) Dialkyl	R_2SiH_2	~2135 (s)	890–860 (s)	820–800 (s)	~1260 (s) (rocking)		
(c) Trialkyl	R_3SiH	2360–2150 (s)	890–860 (s)	~840 (s) ~755 (s)	~1260 (s) (rocking)		
(d) Tetraalkyl	R_4Si			890–690 (s)	~1260 (s) (rocking)		
(e) Alkoxy	$R_x{}^1Si(OR^2)_y$			890–690 (s)	~1260 (s) (rocking)	1090–1080 (s) (doublet)	
Siloxanes	>Si–O–Si<					1110–1000 (s) (Si–O–Si)	
(a) Disiloxanes						~1053 (s)	
(b) Cyclic trimer						~1020 (s)	
(c) cyclic tetramer						~1082 (s)	
Hydroxysilanes x + y = 4	$R_xSi(OH)_y$						~3680 (s) (confirmed by band at 870–820 cm^{-1})

Organic Phosphorus Compounds

Family	General Formula	Wavenumbers (cm^{-1})				
		>P=O Stretch	>P–H Stretch	>P–O–C< Stretch	–OH Stretch	Notes
Phosphates	O=P(OR)$_3$	1300–1100 (s) (doublet)		~1050 (s) (alkyl) 950–875 (s) (aryl)		–>P=O stretch can shift up to 65 cm^{-1} due to solvent effect
(a) Alkyl		1285–1260 (s) (doublet)				
(b) Aryl		1315–1290 (s) (doublet)				
Phosphinates	H$_2$P–OR (=O)	1220–1180 (s)	~2380 (m) ~2340 (m) (sharp)	~1050 (s) (alkyl) 950–875 (s) (aryl)		
Phosphine oxides	(R)H–PR$_1$R$_2$ (=O)					–>P=O decreases with complexation
(a) Alkyl		1185–1150 (s)	2340–2280 (m)			
(b) Aryl		1145–1095 (s)	2340–2280 (m)			
Phosphates	H–P(OR)$_2$ (=O)	1265–1230 (s)	2450–2420 (m)	~1050 (s) (alkyl) 950–875 (s) (aryl)		
Phosphorus acids	R^{1}P(=O)OH R^2	1240–1180 (vs)			2700–2200 (s, broad) (assoc)	
Phosphorus amides	(RO)$_2$PNR^{1}R^2 (=O)	1275–1200 (s)				
Pyrophosphates	R$_2$P–O–PR$_2$ (=O)(=O)	1310–1200 (s) (single band)				

Organic Halogen Compounds

Family	General Formula	Wavenumbers (cm^{-1})			
		>C–X Stretch	>CX2 Stretch	–CH3 Stretch	=C–X Stretch
Fluorides	X = F	1120–1010	1350–1200 (asym) 1200–1080 (sym)	1350–1200 (asym) 1200–1080 (sym)	1230–1100
Chlorides	X = Cl	830–500 1510–1480 (overtone)	845–795 (asym) ~620 (sym)		
Bromides	X = Br	667–290			
Iodides	X = I	500–200			

COMMON SPURIOUS INFRARED ABSORPTION BANDS

The following table provides some of the common potential sources of spurious infrared absorptions that might appear on a spectrum [1–2].

REFERENCES

1. Bruno, T. J., and P. D. N. Svoronos. *CRC Handbook of Basic Tables for Chemical Analysis*. 2nd ed. Boca Raton, FL: CRC Press, 2003.
2. Bruno, T. J., and P. D. N. Svoronos. *CRC Handbook of Fundamental Spectroscopic Correlation Charts*. Boca Raton, FL: CRC Press, 2006.

Common Spurious Infrared Absorption Bands

Approximate Wavenumber (in cm^{-1})	Wavelength (μm)	Compound or Group	Origin
3700	2.70	H_2O	Water in solvent (thick layers)
3650	2.74	H_2O	Water in some quartz windows
3450	2.9	H_2O	Hydrogen-bonded water, usually in KBr disks
2900	3.44	$-CH_3, >CH_2$	Paraffin oil, residual from previous mulls
2350	4.26	CO_2	Atmospheric absorption, or dissolved gas from a dry ice
2330	4.30	CO_2	bath
2300 and 2150	4.35 and 4.65	CS_2	Leaky cells, previous analysis of samples dissolved in carbon disulfide
1996	5.01	BO_2^-	Metaborate in the halide window
1400–2000	5–7	H_2O	Atmospheric absorption
1820	5.52	$COCl_2$	Phosgene, decomposition product in purified $CHCl_3$
1755	5.7	phthalic anhydride	Decomposition product of phthalate esters or resins; paint off-gas product
1700–1760	5.7–5.9	>C=O	Bottle-cap liners leached by sample
1720	5.8	phthalates	Phthalate polymer plastic tubing
1640	6.1	H_2O	Water of crystallization entrenched in sample
1520	6.6	CO_2	Leaky cells, previous analysis
1430	7.0	CO_3^{-2}	Contaminant in halide window
1360	7.38	NO_3^-	Contaminant in halide window
1270	7.9	>SiO–	Silicone oil or grease
1000–1110	9–10	–>Si–O–Si < –	Glass; silicones
980	10.2	SO_4^{-2}	From decomposition of sulfates in KBr pellets
935	10.7	$(CH_2O)_x$	Deposit from gaseous formaldehyde
907	11.02	– >C–Cl	Dissolved R-12 (Freon-12)
837	11.95	NO_3^-	Contaminant in halide window
823	12.15	KNO_3	From decomposition of nitrates in KBr pellets
794	12.6	CCl_4 vapor	Leaky cells, from CCl_4 used as a solvent
788	12.7	CCl_4 liquid	Incomplete drying of cell or contamination, from CCl_4 used as a solvent
720 and 730	13.7 and 13.9	Polyethylene	Various experimental sources
728	13.75	–>Si–F	SiF_4, found in NaCl windows
667	14.98	CO_3^{-2}	Atmospheric carbon dioxide
Any	Any	Fringes	If refractive index of windows is too high, or if the cell is partially empty, or the solid sample is not fully pulverized

DIAGNOSTIC SPECTRA

The interpretation of infrared spectra is often complicated by the presence of spurious absorbances, or by instrumental upset conditions that must be recognized. In these cases, it is often helpful to refer to the spectra of common compounds that may be the cause of such difficulties. The following spectra present such diagnostic tools [1]. Carbon dioxide, as an atmospheric constituent, is often present as an unwanted contaminant. Water is also an atmospheric constituent and is also present in many chemical processes. It can also react with certain species such as amines.

REFERENCE

1. *NIST Chemistry Web Book*. NIST Standard Reference Database Number 69, March 2003 Release.

Infrared Spectrum of Carbon Dioxide:

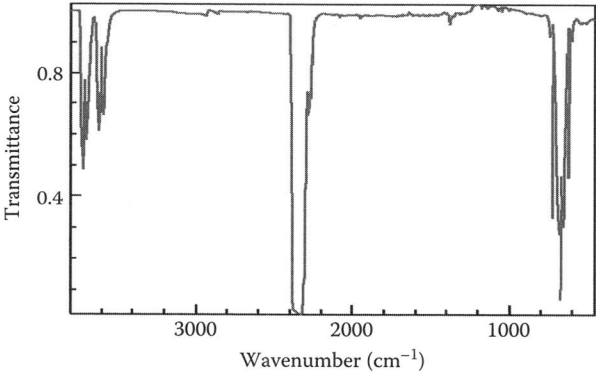

Infrared Spectrum of Water:

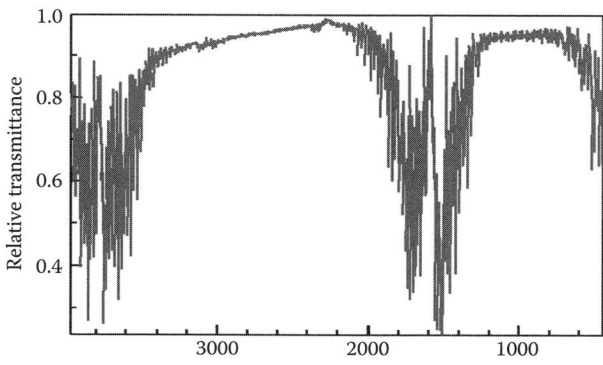

CHAPTER 9

Nuclear Magnetic Resonance Spectroscopy

CONTENTS

PROPERTIES OF IMPORTANT NMR NUCLEI

The following table lists the magnetic properties required most often for choosing the nuclei to be used in NMR experiments [1–14]. The reader is referred to several excellent texts and the literature for guidelines in nucleus selection.

REFERENCES

1. Silverstein, R. M., G. C. Bassler, and T. C. Morrill. *Spectrometric Identification of Organic Compounds.* 5th ed. New York: John Wiley and Sons, 1991.
2. Yoder, C. H., and C. D. Shaeffer. *Introduction to Multinuclear NMR.* Menlo Park, CA: Benjamin/ Cummings, 1987.
3. Gordon, A. J., and R. A. Ford. *The Chemist's Companion.* New York: Wiley Interscience, 1971.
4. Silverstein, R. M., and F. X. Webster. *Spectrometric Identification of Organic Compounds.* 6th ed. New York: John Wiley and Sons, 1998.
5. Becker, E. D. *High Resolution NMR, Theory and Chemical Applications.* 2nd ed. New York: Academic Press, 1980.
6. Gunther, H. *NMR Spectroscopy: Basic Principles, Concepts and Applications in Chemistry.* New York: John Wiley and Sons, 2003.
7. Rahman, A.-U. *Nuclear Magnetic Resonance.* New York: Springer-Verlag, 1986.
8. Harris, R. K. "NMR and the Periodic Table." *Chemical Society Review* 5 (1976): 1.
9. Kitamaru, R. *Nuclear Magnetic Resonance: Principles and Theory.* London: Elsevier Science, 1990.
10. Lambert, J. B., L. N. Holland, and E. P. Mazzola. *Nuclear Magnetic Resonance Spectroscopy: Introduction to Principles, Applications and Experimental Methods.* Englewood Cliffs, NJ: Prentice Hall, 2003.
11. Bovey, F. A., and P. A. Mirau. *Nuclear Magnetic Resonance Spectroscopy.* 2nd ed. London: Academic Press, 1988.
12. Harris, R. K., and B. E. Mann. *NMR and the Periodic Table.* London: Academic Press, 1978.
13. Hore, P. J. *Nuclear Magnetic Resonance.* Oxford: Oxford University Press, 1995.
14. Nelson, J. H. *Nuclear Magnetic Resonance Spectroscopy.* 2nd ed. New York: John Wiley and Sons, 2003.

Properties of Important NMR Nuclei

Isotope	Natural Abundance	Spin Number I	NMR Frequency[a] at Indicated Field Strength in kG							
			10.000	14.092	21.139	23.487	51.567	93.950	140.925	223.131
$_1H^1$	99.985	1/2	42.5759	60.0000	90.0000	100.0000	220.0000	400.0000	600.0000	950.0000
$_1H^2$	0.015	1	6.53566	9.21037	13.81555	15.35061	33.77134	61.40262	92.10380	145.9830
$_1H^{3*}$	—	1/2	45.4129	63.9980	95.9971	106.6634	234.6595	426.6542	639.9813	1013.3024
$_6C^{13}$	1.108	1/2	10.7054	15.0866	22.6298	25.1443	55.3174	100.5735	150.8659	238.8515
$_7N^{14}$	99.635	1	3.0756	4.3343	6.5014	7.2238	15.924	28.9104	43.3615	68.6557
$_7N^{15}$	0.365	1/2	4.3142	6.0798	9.1197	10.1330	22.2925	40.5306	60.7960	96.2601
$_8O^{17}$	0.037	5/2	5.772	8.134	12.201	13.557	29.825	54.1811	81.3186	128.5801
$_9F^{19}$	100	1/2	40.0541	56.4443	84.6703	94.0769	206.9692	376.2515	564.3781	893.5963
$_{14}Si^{29}$	4.70	1/2	8.4578	11.9191	17.8787	19.8652	43.7035	79.4638	119.1956	188.72
$_{15}P^{31}$	100	1/2	17.235	24.288	36.433	40.481	89.057	161.9828	242.9741	384.7086
$_{16}S^{33}$	0.76	3/2	3.2654	4.6018	6.9026	7.6696	16.8731	30.6826	46.0238	72.8710
$_{16}S^{35*}$	—	3/2	5.08	7.16	10.74	11.932	26.250	47.7267	71.5875	113.3508
$_{17}Cl^{35}$	75.53	3/2	4.1717	5.8790	8.8184	9.7983	21.5562	39.1948	58.7902	93.0876
$_{17}Cl^{36*}$	—	2	4.8931	6.8956	10.3434	11.4927	25.2838	45.9638	68.9432	109.1639
$_{35}Br^{76*}$	—	1	4.18	5.89	8.84	9.82	21.60	39.2768	58.9130	93.2822
$_{35}Br^{79}$	50.54	3/2	10.667	15.032	22.549	25.054	55.119	100.2133	150.3202	238.0064
$_{35}Br^{81}$	49.46	3/2	11.498	16.204	24.305	27.006	59.413	108.0258	162.0386	256.5608
$_{74}W^{183}$	14.40	1/2	1.7716	2.4966	3.7449	4.1610	9.1543	16.6430	24.9646	39.5272

* Nucleus is radioactive.

a 1 kG = 10^{-10} T, the corresponding SI unit.

b 1 b = 10^{-23} m².

Isotope	Field Value[a] (kg) at Frequency of			Relative Sensitivity		Magnetic Moment ($eh/4BM_c$)	Electric Quadrupole Moment[b] (barns)
	4 MHz	10 MHz	16 MHz	Constant H	Constant v		
$_1H^1$	0.940	2.349	3.758	1.00	1.00	2.79278	—
$_1H^2$	6.120	15.30	24.48	9.65×10^{-3}	0.409	0.85742	0.0028
$_1H^{3*}$	0.881	2.202	3.523	1.21	1.07	2.9789	—
$_6C^{13}$	3.736	9.341	14.946	0.0159	0.252	0.7024	—
$_7N^{14}$	13.01	32.51	52.02	1.01×10^{-3}	0.193	0.4036	0.01
$_7N^{15}$	9.272	23.18	37.09	1.04×10^{-3}	0.101	−0.2831	—
$_8O^{17}$	6.93	17.3	27.7	0.0291	1.58	−1.8937	−0.026
$_9F^{19}$	0.999	2.497	3.994	0.834	0.941	2.6288	—
$_{14}Si^{29}$	4.729	11.82	18.92	7.84×10^{-3}	0.199	−0.55477	—
$_{15}P^{31}$	2.321	5.802	9.284	0.0665	0.405	1.1317	—
$_{16}S^{33}$	12.25	30.62	49.0	2.26×10^{-3}	0.384	0.6533	−0.055
$_{16}S^{35*}$	7.87	19.7	31.5	8.50×10^{-3}	0.597	1.00	0.04
$_{17}Cl^{35}$	9.588	23.97	38.35	4.72×10^{-3}	0.490	0.82183	−0.079
$_{17}Cl^{36*}$	0.175	20.44	32.70	0.0122	0.920	1.285	−0.017
$_{35}Br^{76*}$	9.6	24	38	2.52×10^{-3}	0.26	±0.548	±0.25
$_{35}Br^{79}$	3.750	9.375	15.00	0.0794	1.26	2.106	0.31
$_{35}Br^{81}$	3.479	8.697	13.92	0.0994	1.35	2.270	0.26
$_{74}W^{183}$	22.58	56.45	90.31	7.3×10^{-5}	0.042	0.117	—

[a] 1 kG = 10^{-10} T, the corresponding SI unit.
[b] 1 b = 10^{-23} m^2.
* Nucleus is radioactive.

GYROMAGNETIC RATIO OF SOME IMPORTANT NUCLEI

The following table lists the gyromagnetic ratio, γ, of some important nuclei that are probed in NMR spectroscopy [1–12]. The gyromagnetic ratio is the proportionality constant that correlates the magnetic moment (μ) and the angular momentum, ρ: $\mu = \gamma\rho$.

REFERENCES

1. Carrington, A., and A. McLaughlin. *Introduction to Magnetic Resonance*. New York: Harper and Row, 1967.
2. Levine, I. M. *Molecular Spectroscopy*. New York: John Wiley and Sons, 1975.
3. Becker, E. D. *High Resolution NMR: Theory and Chemical Applications*. New York: Academic Press, 1980.
4. Yoder, C. H., and C. D. Shaeffer. *Introduction to Multinuclear NMR*. Menlo Park, CA: Benjamin/Cummings, 1987.
5. Silverstein, R. M., and F. X. Webster. *Spectrometric Identification of Organic Compounds*. 6th ed. New York: John Wiley and Sons, 1998.
6. Rahman, A.-U. *Nuclear Magnetic Resonance*. New York: Springer-Verlag, 1986.
7. Kitamaru, R. *Nuclear Magnetic Resonance: Principles and Theory*. London: Elsevier Science, 1990.
8. Lambert, J. B., L. N. Holland, and E. P. Mazzola. *Nuclear Magnetic Resonance Spectroscopy: Introduction to Principles, Applications and Experimental Methods*. Englewood Cliffs: Prentice Hall, 2003.
9. Bovey, F. A., and P. A. Mirau. *Nuclear Magnetic Resonance Spectroscopy*. 2nd ed. New York: Academic Press, 1988.
10. Hore, P. J. *Nuclear Magnetic Resonance*. Oxford: Oxford University Press, 1995.
11. Nelson, J. H. *Nuclear Magnetic Resonance Spectroscopy*. 2nd ed. New York: John Wiley and Sons, 2003.
12. Gunther, H. *NMR Spectroscopy: Basic Principles, Concepts and Applications in Chemistry*. New York: John Wiley and Sons, 2003.

Nucleus	γ
$_1H^1$	5.5856
$_1H^2$	0.8574
$_1H^3$	5.9575
$_3Li^7$	2.1707
$_5B^{10}$	0.6002
$_5B^{11}$	1.7920
$_6C^{13}$	1.4044
$_7N^{14}$	0.4035
$_7N^{15}$	−0.5660
$_8O^{17}$	−0.7572
$_9F^{19}$	5.2545
$_{14}Si^{29}$	−1.1095
$_{11}Na^{23}$	1.4774
$_{15}P^{31}$	2.2610
$_{16}S^{33}$	0.4284
$_{17}Cl^{35}$	0.5473
$_{17}Cl^{37}$	0.4555
$_{19}K^{39}$	0.2607
$_{35}Br^{79}$	1.3993
$_{35}Br^{81}$	1.5084
$_{74}W^{183}$	0.2324

CLASSIFICATION OF IMPORTANT QUADRUPOLAR NUCLEI ACCORDING TO NATURAL ABUNDANCE AND MAGNETIC STRENGTH

The following table classifies important quadrupolar nuclei according to their natural abundance and relative magnetic strength [1]. The magnetic strength, while not a commonly recognized physical parameter, is defined as a matter of convenience for classification of nuclei in NMR. It is defined as follows:

Strong: $\gamma/10^7 > 2.5$ rad $T^{-1}s^{-1}$
Medium: 10 rad $T^{-1}s^{-1} > \gamma 10^7 > 2.5$ rad $T^{-1}s^{-1}$
Weak: $\gamma 10^7 < 2.5$ rad $T^{-1}s^{-1}$

where the flux density in units of teslas (T), and rad refers to 2B. In NMR, one can write:

$$2\pi f = \gamma B,$$

where f is the resonant frequency, gamma is the gyromagnetic ratio, and B is the flux density. Thus, for the proton, $\gamma/2\pi = 43$ MHz/T, resulting in a value of $\gamma/10^7 = 4.3$ rad $T^{-1}s^{-1}$, and therefore medium magnetic strength.

The less favorable nuclei for a given element are listed in brackets.

REFERENCE

1. Harris, R. K., and B. E. Mass. *NMR and the Periodic Table.* London: Academic Press, 1978.

Magnetic Strength	Natural Abundance		
	High (>90 %)	Medium	Low (<10 %)
Strong	^7Li		
Medium	^9Be, ^{23}Na, ^{27}Al, ^{45}Sc, ^{51}V, ^{55}Mn, ^{59}Co, ^{75}As, ^{93}Nb, ^{115}In, ^{127}I, ^{133}Cs, ^{181}Ta, ^{209}Bi	[^{10}B], ^{11}B, ^{35}Cl, ^{63}Cu, ^{65}Cu, [^{69}Ga], ^{71}Ga, [^{79}Br], ^{81}Br, [^{85}Rb], ^{87}Rb, ^{121}Sb, [^{123}Sb], ^{137}Ba, ^{139}La, [^{185}Re], ^{187}Re	^2H, ^6Li, ^{17}O, ^{21}Ne, [^{113}In], [^{135}Ba]
Weak	^{14}N, ^{39}K	^{25}Mg, ^{37}Cl, ^{83}Kr, ^{95}Mo, ^{131}Xe, ^{189}Os, ^{201}Hg	^{33}S, [^{41}K], ^{43}Ca, ^{47}Ti, ^{49}Ti, ^{53}Cr, ^{67}Zn, ^{73}Ge, ^{87}Sr, [^{97}Mo]

CHEMICAL SHIFT RANGES OF SOME NUCLEI

The following table gives an approximate chemical shift range (in ppm) for some of the most popular nuclei. The range is established by the shifts recorded for the most common compounds [1–11].

REFERENCES

1. Yoder, C. H., and C. D. Schaeffer, Jr. *Introduction to Multinuclear NMR.* Menlo Park, CA: Benjamin/Cummings Publishing Co., 1987.
2. Silverstein, R. M., G. C. Bassler, and T. C. Morrill. *Spectrometric Identification of Organic Compounds.* 5th ed. New York: John Wiley and Sons, 1991.
3. Harris, R. U., and B. E. Mann. *NMR and the Periodic Table.* London: Academic Press, 1978.
4. Silverstein, R. M., and F. X. Webster. *Spectrometric Identification of Organic Compounds.* 6th ed. New York: John Wiley and Sons, 1998.
5. Gunther, H. *NMR Spectroscopy: Basic Principles, Concepts and Applications in Chemistry.* New York: John Wiley and Sons, 2003.
6. Kitamaru, R. *Nuclear Magnetic Resonance: Principles and Theory.* New York: Elsevier Science, 1990.
7. Lambert, J. B., L. N. Holland, and E. P. Mazzola. *Nuclear Magnetic Resonance Spectroscopy: Introduction to Principles, Applications and Experimental Methods.* Englewood Cliffs: Prentice Hall, 2003.
8. Bovey, F. A., and P. A. Mirau. *Nuclear Magnetic Resonance Spectroscopy.* 2nd ed. London: Academic Press, 1988.
9. Harris, R. K., and B. E. Mann. *NMR and the Periodic Table.* London: Academic Press, 1978.
10. Hore, P. J. *Nuclear Magnetic Resonance.* Oxford: Oxford University Press, 1995.
11. Nelson, J. H. *Nuclear Magnetic Resonance Spectroscopy.* 2nd ed. New York: John Wiley and Sons, 2003.

Nucleus	Chemical Shift Range, ppm	Nucleus	Chemical Shift Range, ppm
^{1}H	15	^{29}Si	400
^{7}Li	10	^{31}P	700
^{11}B	200	^{33}S	600
^{13}C	250	^{35}Cl	820
^{15}N	930	^{39}K	60
^{17}O	700	^{59}Co	14,000
^{19}F	800	^{119}Sn	2000
^{23}Na	15	^{133}Cs	150
^{27}Al	270	^{207}Pb	10,000

REFERENCE STANDARDS FOR SELECTED NUCLEI

The following table lists the most popular reference standards used when NMR spectra of various nuclei are measured. The standards should be inert, soluble in a variety of solvents and, preferably, should produce one singlet peak that appears close to the lowest frequency end of the chemical shift range. When NMR data are provided, it is always necessary to specify the reference standard employed [1–6].

REFERENCES

1. Yoder, C. H., and C. D. Schaeffer, Jr. *Introduction to Multinuclear NMR*. Menlo Park, CA: Benjamin/ Cummings Publishing Co., 1987.
2. Duthaler, R. O., and J. D. Roberts. "Steric and Electronic Effects on ^{15}N Chemical Shifts of Piperidine and Decahydroquinoline Hydrochlorides." *Journal of the American Chemical Society* 100 (1978): 3889.
3. Grim, S. O., and A. W. Yankowsky. "On the Phosphorus-31 Chemical Shifts of Substituted Triarylphosphines." *Phosphorus and Sulfur* 3 (1977): 191.
4. Lambert, J. B., H. F. Shurrell, L. Verbit, R. G. Cooks, and G. H. Stout. *Organic Structural Analysis*. New York: MacMillan, 1976.
5. Gunther, H. *NMR Spectroscopy: Basic Principles, Concepts and Applications in Chemistry,* New York: John Wiley and Sons, 2003.
6. Abraham, R. J., J. Fisher, and P. Loftus. *Introduction to NMR Spectroscopy*. New York: John Wiley and Sons, 1988.

Nucleus	Name	Formula
^{1}H	tetramethylsilane [TMS]	$(CH_3)_4Si$
	3-(trimethylsilyl)-1-propanesulfonic acid, sodium salt [DSS][a]	$(CH_3)_3Si(CH_2)_3SO_3Na$
	3-(trimethylsilyl)-propanoic acid, d$_4$, sodium salt [TSP]	$(CH_3)_3Si(CD_2)_3CO_2Na$
^{2}H	deuterated chloroform [chloroform-d]	$CDCl_3$
^{11}B	boric acid	H_3BO_3
	boron trifluoride etherate	$(C_2H_5)_2OBF_3$
	boron trichloride	BCl_3
^{13}C	tetramethylsilane [TMS]	$(CH_3)_4Si$
^{15}N	ammonium nitrate	NH_4NO_3
	ammonia	NH_3
	nitromethane	CH_3NO_2
	nitric acid	HNO_3
	tetramethylammonium chloride	$(CH_3)_4NCl$
^{17}O	water	H_2O
^{19}F	trichlorofluoromethane [Freon 11, R-11]	CCl_3F
	hexafluorobenzene	C_6F_6
^{31}P	trimethylphosphite [methyl phosphite]	$(CH_3O)_3P$
	phosphoric acid (85 %)	H_3PO_4
^{35}Cl	sodium chloride	NaCl
^{59}Co	cobalt (III) hexacyanide anion	$[Co(CN)_6]^{-3}$
^{119}Sn	tetramethyltin	$(CH_3)_4Sn$
^{195}Pt	platinum (IV) hexacyanide	$[Pt(CN)_6]^{-2}$
	dihydrogen platinum (IV) hexachloride	H_2PtCl_6
^{183}W	sodium tungstate (external)	Na_2WO_4

[a] For aqueous solutions (known also as "water soluble TMS" or 2,2-dimethyl-2-silapentane-5-sulfonate).

¹H AND ¹³C CHEMICAL SHIFTS OF USEFUL SOLVENTS FOR NMR MEASUREMENTS

The following table lists the expected $^1H^{(\delta H)}$ and $^{13}C^{(\delta C)}$ chemical shifts for various useful NMR solvents in parts per million (ppm) [1–3]. The table also includes the liquid temperature range (°C) and dielectric constants of these solvents. Slight changes may occur with changes in concentration.

REFERENCES

1. Silverstein, R. M., G. C. Bassler, and T. C. Morrill. *Spectrometric Identification of Organic Compounds.* 5th ed. New York: John Wiley and Sons, 1991.
2. Rahman, A.-U. *Nuclear Magnetic Resonance: Basic Principles.* New York: Springer-Verlag, 1986.
3. Abraham, R. J., J. Fisher, and P. Loftus. *Introduction of NMR Spectroscopy.* Chichester: John Wiley and Sons, 1988.

Solvent	Formula	Liquid Temperature Range (°C)	Dielectric Constant, ε	Chemical Shifts	
				δ_H (ppm)	δ_C (ppm)
Acetone-d₆	(CD₃)₂CO	−95 to 56	20.7	2.17	29.2, 204.1
Acetonitrile-d₃	CD₃CN	−44 to 82	37.5	2.00	1.3, 117.7
Benzene-d₆	C₆D₆	6 to 80	2.284	7.27	128.4
Carbon disulfide	CS₂	−112 to 46	2.641	—	192.3
Carbon tetrachloride	CCl₄	−23 to 77	2.238	—	96.0
Chloroform-d₃	CDCl₃	−64 to 61	4.806	7.25	76.9
Cyclohexane-d₁₂	C₆D₁₂	6 to 81	2.023	1.43	27.5
Dichloromethane-d₂	CD₂Cl₂	−95 to 40	9.08	5.33	53.6
Difluorobromo-chloromethane	CF₂BCl	−140 to −25		—	109.2
Dimethylformamide-d₇	DCON(CD₃)₂	−60 to 153	36.7	2.9, 3.0, 8.0	31, 36, 132.4
Dimethylsulfoxide-d₆	(CD₃)₂SO	19 to 189	46.7	2.62	39.6
1,4-Dioxane-d₈	C₄D₈O₂	12 to 101	2.209	3.7	67.4
Hexamethylphosphoramide (HMPA)	[(CH₃)₂N]₃PO	7 to 233	30.0	2.60	36.8
Methanol-d₄	CD₃OD	−98 to 65	32.63	3.4, 4.8ᵃ	49.3
Nitrobenzene	C₆D₅NO₂	6 to 211	34.8	8.2, 7.6, 7.5	149, 134, 129, 124
Nitromethane-d₃	CD₃NO₂	−29 to 101	35.87	4.33	57.3
Pyridine-d₅	C₅D₅N	−42 to 115	123	7.0, 7.6, 8.6	124, 136, 150
1,1,2,2-Tetra-chloroethane-d₂	CD₂ClCD₂Cl	−44 to 146	8.2	5.94	75.5
Tetrahydrofuran-d₈	C₄D₈O	−108 to 66	7.54	1.9, 3.8	25.8, 67.9
1,2,4-Trichlorobenzene	C₆D₃Cl₃	17 to 214	3.9	7.1, 7.3, 7.4	133.3, 132.8, 130.7, 130.0, 127.6
Trichlorofluoromethane	CFCl₃	−111 to 24	2.3	—	117.6
Trifluoroacetic acid-d	CF₃COOD	−15 to 72	8.6	11.3ᵃ	114.5, 116.5
Vinyl chloride-d₃	CD₂=CDCl	−154 to −13		5.4, 5.5, 6.3	126, 117
Water-d₂	D₂O	0 to 100	78.5	4.7	—

ᵃ Variable with concentration.

RESIDUAL PEAKS OBSERVED IN THE ¹H NMR SPECTRA
OF COMMON DEUTERATED ORGANIC SOLVENTS

The following table lists the residual peaks that are observed in the ¹H NMR spectra of common deuterated organic solvents. These peaks are generally attributed to the nondeuterated parent compound that serves as an impurity and are marked with an asterisk (*). In addition, other less significant peaks often arise due to other impurities.

Together with the formula and molecular weight, the table lists the expected chemical shifts, δ, multiplicities and (when possible) the coupling constant, J_{HD}, for every solvent. All spectra are at least 99.5 % deuterium pure [1–5].

REFERENCES

1. Yoder, C. H., and C. D. Schaeffer, Jr., *Introduction to Multinuclear NMR*. Menlo Park, CA: Benjamin Cummings, 1987.

2. Silverstein, R. M., and F. X. Webster. *Spectrometric Identification of Organic Compounds*. 6th ed. New York: John Wiley and Sons, 1998.

3. Gunther, H. *NMR Spectroscopy: Basic Principles, Concepts and Applications in Chemistry*. New York: John Wiley and Sons, 2003.

4. Lambert, J. B., L. N. Holland, and E. P. Mazzola. *Nuclear Magnetic Resonance Spectroscopy: Introduction to Principles, Applications and Experimental Methods*. Englewood Cliffs, NJ: Prentice Hall, 2003.

5. Nelson, J. H. *Nuclear Magnetic Resonance Spectroscopy*. 2nd ed. New York: John Wiley and Sons, 2003.

Residual Peaks Observed in the ^{1}H NMR Spectra of Common Deuterated Organic Solvents

Solvent	Formula	Molecular Weight	δ(mult)[a]	J_{HD}
Acetic acid-d_4	CD_3COOD	64.078	*11.53(1) *2.03(5)	2
Acetone-d_6	$(CD_3)_2C=O$	64.117	*2.04(5) 2.78(1) 2.82(1)	2.2
Acetonitrile-d_3	CD_3CN	44.017	*1.93(5) 2.1–2.15 2.2–2.4	2.5
Benzene-d_6	C_6D_6	84.153	*7.12(b)	
Chloroform-d_3	$CDCl_3$	120.384	1.55[b] 1.60[b] 7.2 *7.24(1)	
Cyclohexane-d_{12}	$(CD_2)_6$	96.236	*1.38(b)	
Deuterium oxide	D_2O	20.028	*4.63(b)[c] *4.67(b)[d]	
1,2-Dichloroethane-d_4	CD_2ClCD_2Cl	102.985	*3.72(b)	
Dichloromethane-d_2	See Methylene chloride-d_2			
Diethylene glycol dimethylether-d_{14}	See Diglyme-d_{14}			
Diethylether-d_{10}	$(CD_3CD_2)_2O$	84.185	*3.34(m) *1.07(m)	
Diglyme-d_{14} (bis(2-methoxyethyl) ether)	$CD_3O(CD_2)_2O(CD_2)_2OCD_3$	148.263	*3.49(b) *3.40(b) *3.22(5)	1.5
N,N-Dimethylformamide-d_7	$DCON(CD_3)_2$	80.138	*8.01(b) *2.91(5) *2.74(5)	2 2
Dimethylsulfoxide-d_6	$(CD_3)_2SO$		3.3–3.4 *2.49(5)	1.7
1,2-Diethoxyethane-d_{10}	see Glyme-d_{10}			
p-Dioxane-d_8	$C_4H_8O_2$	96.156	*3.53(m)	
Ethanol-d_6 (anhydrous)	CD_3CD_2OD	52.106	*5.19(1) *3.55(b) *1.11(m)	
Glyme-d_{10} (dimethoxyethane)	$CD_3OCD_2CD_2OCD_3$	100.184	*3.40(m) *3.22(5)	1.6
Hexamethylphosphoric triamide-d_{18} (HMPT-d_{18})	$[(CD_3)_2N]_3P=O$	197.314	*2.53(m)	
Methanol-d_4	CD_3OH	36.067	*4.78(1) *3.30(5)	1.7
Methylene chloride-d_2	CD_2Cl_2	86.945	*5.32(3) 1.4–1.5(b)	1
Nitrobenzene-d_5	$C_6D_5NO_2$	128.143	*8.11(b) *7.67(b) *7.50(b)	
Nitromethane-d_3	CD_3NO_2	64.059	*4.33(5)	2
2-Propanol-d_8	$(CD_3)_2CDOD$	68.146	*5.12(1) *3.89(b) *1.10(b)	
Pyridine-d_5	C_5D_5N	84.133	*8.71(b) *7.55(b) *7.19(b) 4.8[b] 4.9[b]	

Residual Peaks Observed in the ^{1}H NMR Spectra of Common Deuterated Organic Solvents (Continued)

Solvent	Formula	Molecular Weight	δ(mult)[a]	J_{HD}
Tetrahydrofuran-d$_8$	C$_4$D$_8$O	80.157	*3.58(b) 2.4[b] 2.3[b] *1.73(b)	
Toluene-d$_8$	C$_6$D$_5$CD$_3$	100.191	*7.09(m) *7.00(b) *6.98(m) *2.09(5)	2.3
Trifluorocetic acid-d	CF$_3$COOD	115.030	*11.50(1)	

[a] Chemical shift, δ, in ppm; mult = multiplicity (indicated by a number); b = broad, m = multiplet.
[b] Two peaks that may often appear as one broad peak.
[c] When DSS, 3-(trimethylsilyl)-1-propane sulfonic acid, sodium salt, is used as a reference standard.
[d] When TSP, sodium-3-trimethylpropionate, is used as a reference standard.

¹H NMR CHEMICAL SHIFTS FOR WATER SIGNALS IN ORGANIC SOLVENTS

Often traces of water are encountered in samples whose ¹H NMR spectra are being measured. The water signals appear at different chemical shifts depending on the particular solvent used [1,2]. Listed below are the usual chemical shift positions of the water signal in several common solvents. The signal in aprotic solvent solutions is due to the presence of H_2O. On the other hand the signal observed in protic solvent solutions (in parentheses) is attributed to HOD, which is the result of hydrogen exchange with the solvent's deuterium atoms.

REFERENCES

1. Silverstein, R. M., T. C. Morrill, and G. C. Bassler. *Spectrometric Identification of Organic Compounds.* 5th ed. New York: John Wiley and Sons, 1991.
2. Notes on NMR Solvents, assessed July, 2010 http://www.chem.ucla.edu/~webspectra/NotesOnSolvents.html, 2009.

Solvent	Chemical Shift of H_2O (or HOD)
Acetone	2.8
Acetonitrile	2.1
Benzene	0.4
Chloroform	1.6
Dimethyl sulfoxide	3.3
Methanol	(4.8)
Methylene chloride	1.5
Pyridine	4.9
Water (D_2O)	(4.8)

PROTON NMR ABSORPTION OF MAJOR CHEMICAL FAMILIES

The following tables give the region of the expected nuclear magnetic resonance absorptions of major chemical families. These absorptions are reported in the dimensionless units of parts per million (ppm) versus the standard compound tetramethylsilane (TMS), which is recorded as 0.0 ppm.

$$
\begin{array}{c}
CH_3 \\
| \\
CH_3 - Si - CH_3 \\
| \\
CH_3
\end{array}
$$

The use of this unit of measure makes the chemical shifts independent of the applied magnetic field strength or the radio frequency. For most proton NMR spectra, the protons in TMS are more shielded than almost all other protons. The chemical shift in this dimensionless unit system is then defined by:

$$\delta = \frac{\nu_s - \nu_r}{\nu_r} \times 10^6$$

where ν_s and ν_r are the absorption frequencies of the sample proton and the reference (TMS) protons (12, magnetically equivalent), respectively. In these tables, the proton(s) whose proton NMR shifts are cited are indicated by underscore. For more detail concerning these conventions, the reader is referred to the general references below [1–11].

REFERENCES

1. Silverstein, R. M., and F. X. Webster. *Spectrometric Identification of Organic Compounds*. 6th ed. New York: John Wiley and Sons, 1998.
2. Rahman, A.-U. *Nuclear Magnetic Resonance*. New York: Springer Verlag, 1986.
3. Gordon, A. J., and R. A. Ford. *The Chemist's Companion*. New York: Wiley Interscience, 1971.
4. Becker, E. D. *High Resolution NMR, Theory and Chemical Applications*. 2nd ed. New York: Academic Press, 1980.
5. Gunther, H. *NMR Spectroscopy: Basic Principles, Concepts and Applications in Chemistry*. New York: John Wiley and Sons, 2003.
6. Kitamaru, R. *Nuclear Magnetic Resonance: Principles and Theory*. London: Elsevier Science, 1990.
7. Lambert, J. B., L. N. Holland, and E. P. Mazzola. *Nuclear Magnetic Resonance Spectroscopy: Introduction to Principles, Applications and Experimental Methods*. Englewood Cliffs: Prentice Hall, 2003.
8. Bovey, F. A., and P. A. Mirau. *Nuclear Magnetic Resonance Spectroscopy*. 2nd ed. London: Academic Press, 1988.
9. Hore, P. J. *Nuclear Magnetic Resonance*, Oxford: Oxford University Press, 1995.
10. Nelson, J. H. *Nuclear Magnetic Resonance Spectroscopy*. 2nd ed. New York: John Wiley and Sons, 2003.
11. Abraham, R. J., J. Fisher, and P. Loftus. *Introduction to NMR Spectroscopy*. New York: John Wiley and Sons, 1988.

Hydrocarbons

Family	* of Protons Underlined	
Alkanes	$\underline{CH_3}$–R ~0.8 ppm	
	$\underline{-CH_2}$–R ~1.1 ppm	
	$\underline{>CH}$–R ~1.4 ppm	
	(cyclopropane 0.2 ppm)	
Alkenes	$\underline{CH_3}$–C=C< ~1.6 ppm	$\underline{CH_3}$–C–C=C< ~1.0 ppm
	$-\underline{CH_2}$–C=C< ~2.1 ppm	$-\underline{CH_2}$–C–C=C< ~1.4 ppm
	$>\underline{CH}$–C=C< ~2.5 ppm	$>\underline{CH}$–C–C=C< ~1.8 ppm
	$>C=C-\underline{H}$ 4.2–6.2 ppm	
Alkynes	$\underline{CH_3}$–C≡C– ~1.7 ppm	$\underline{CH_3}$–C–C≡C– ~1.2 ppm
	$-\underline{CH_2}$–C≡C– ~2.2 ppm	$>\underline{CH_2}$–C–C≡C– ~1.5 ppm
	$>\underline{CH}$ –C≡C– ~2.7 ppm	$>\underline{CH}$–C–C≡C– ~1.8 ppm
	R–C≡C–$\underline{H}$ ~2.4 ppm	
Aromatics	C_6H_5–G 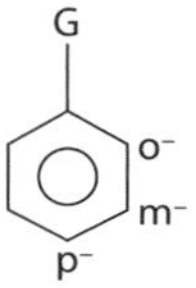	Range: 8.5–6.9 ppm

C_6H_5–G

G
o⁻
m⁻
p⁻

When G=Electron withdrawing (e.g.,>C=O, –NO₂, –C≡N) o- and p- hydrogens relative to-G are closer to 8.5 ppm (more downfield)

When G=Electron donating (e.g., –NH₂, –OH, –OR, –R) o- and p- hydrogens relative to-G are closer to 6.9 ppm (more upfield)

Organic Oxygen Compounds

Family	Approximate δ of Protons Underlined		
Alcohols	$C\underline{H}_3$–OH 3.2 ppm	R$C\underline{H}_2$–OH 3.4 ppm	$R_2C\underline{H}$–OH 3.6 ppm
	$C\underline{H}_3$–C–OH 1.2 ppm	R$C\underline{H}_2$–C–OH 1.5 ppm	$R_2C\underline{H}$–C–OH 1.8 ppm
	R–$\underline{O}$–$\underline{H}$	(1–5 ppm: depending on concentration)	
Aldehydes	$C\underline{H}_3$–CHO 2.2 ppm	R$C\underline{H}_2$–CHO 2.4 ppm	$R_2C\underline{H}$–CHO 2.5 ppm
	$C\underline{H}_3$–C–CHO 1.1 ppm	R$C\underline{H}_2$–C–CHO 1.6 ppm	
Amides	See Organic Nitrogen Compounds		
Anhydrides, acyclic	$C\underline{H}_3$–C(=O)O– 1.8 ppm	R$C\underline{H}_2$–C(=O)O– 2.1 ppm	$R_2C\underline{H}$–C(=O)O– 2.3 ppm
	$C\underline{H}_3$–C–C(=O)O– 1.2 ppm	R$C\underline{H}_2$–C–C(=O)O– 1.8 ppm	$R_2C\underline{H}$–C–C(=O)O– 2.0 ppm

Anhydrides, cyclic

3.0 ppm   7.1 ppm

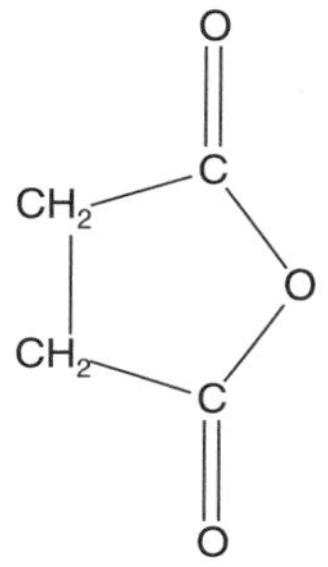

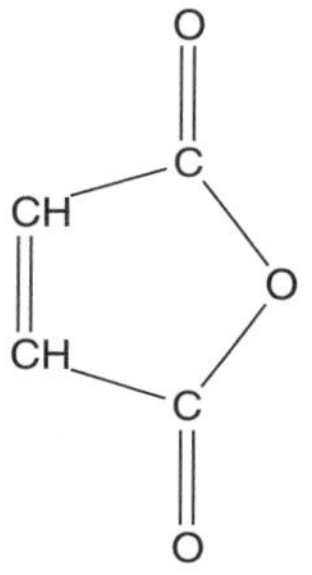

Carboxylic acids	$C\underline{H}_3$–COOH 2.1 ppm	R$C\underline{H}_2$–COOH 2.3 ppm	$R_2C\underline{H}$–COOH 2.5 ppm
	$C\underline{H}_3$–C–COOH 1.1 ppm	R–$C\underline{H}_2$–C–COOH 1.6 ppm	$R_2C\underline{H}$–C–COOH 2.0 ppm
	R–COO–$\underline{H}$ 11–12 ppm		

Cyclic Ethers   oxacyclopropane (oxirane)

2.5 ppm

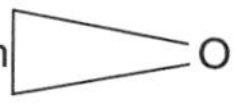

oxacyclobutane (oxetane)

2.7 ppm

4.7 ppm

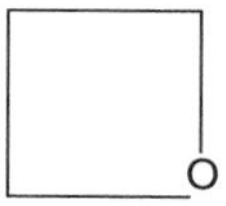

oxacyclopentane (tetrahydrofuran)

1.9 ppm
3.8 ppm

(Continued)

Organic Oxygen Compounds (Continued)

Family	Approximate δ of Protons Underlined
	oxacyclohexane (tetrahydropyran) 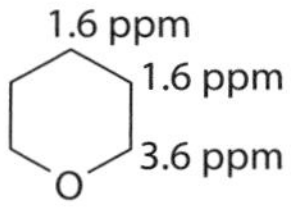 1.6 ppm 1.6 ppm 3.6 ppm O
	1,4-dioxane 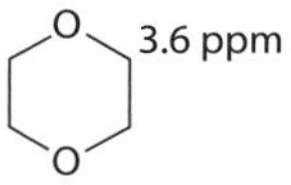 O 3.6 ppm O
	1,3-dioxane 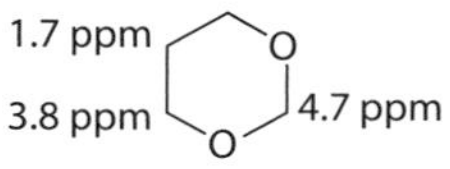 1.7 ppm O 3.8 ppm 4.7 ppm O
	furan 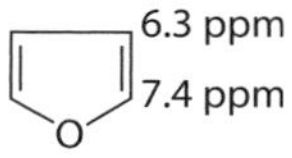 6.3 ppm 7.4 ppm O
	dihydropyran 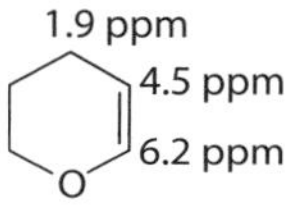 1.9 ppm 4.5 ppm 6.2 ppm O

Organic Oxygen Compounds (Continued)

Epoxides	See Cyclic Ethers		

Esters		CH_3–COOR	RCH_2–COOR	R_2CH–COOR
	R=alkyl,	1.9 ppm	2.1 ppm	2.3 ppm;
	R=aryl	2.0 ppm	2.2 ppm	2.4 ppm
		CH_3–C–COOR	RCH_2–C–COOR	R_2CH–C–COOR
		1.1 ppm	1.7 ppm	1.9 ppm
		CH_3–OOC–R	RCH_2–OOC–R	R_2CH–OOC–R
		3.6 ppm	4.1 ppm	4.8 ppm
		CH_3–C–OOC–R	RCH_2–C–OOC–R	R_2CH–C–OOC–R
		1.3 ppm	1.6 ppm	1.8 ppm

Cyclic Esters

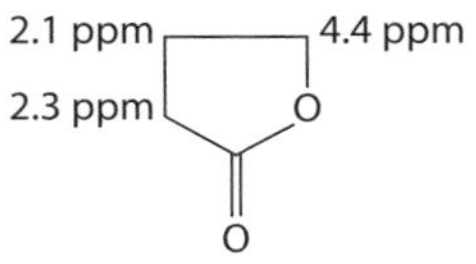

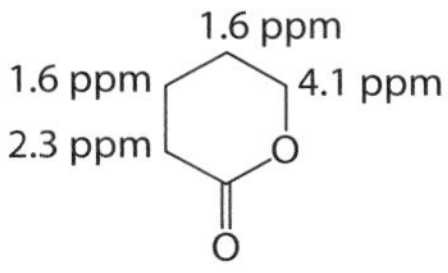

2.1 ppm ⎯⎯ 4.4 ppm
2.3 ppm ⎣ O

1.6 ppm
1.6 ppm ⎯ 4.1 ppm
2.3 ppm ⎣ O

Ethers		CH_3–O–R	RCH_2–O–R	R_2CH–O–R
	R=alkyl	3.2 ppm	3.4 ppm	3.6 ppm
	R=aryl	3.9 ppm	4.1 ppm	4.5 ppm
		CH_3–C–O–R	RCH_2–C–O–R	R_2CH–C–O–R
	R=alkyl	1.2 ppm	1.5 ppm	1.8 ppm
	R=aryl	1.3 ppm	1.6 ppm	2.0 ppm

Isocyanates	See Nitrogen Compounds			

Ketones	CH_3–C(=O)–		RCH_2–C(=O)–	R_2CH–C(=O)–
	1.9 ppm	R=alkyl	2.1 ppm	2.3 ppm
	2.4 ppm	R=aryl	2.7 ppm	3.4 ppm
	CH_3–C(=O)–		RCH_2–C(=O)–	R_2CH–C(=O)–
	1.1 ppm	R=alkyl	1.6 ppm	2.0 ppm
	1.2 ppm	R=aryl	1.6 ppm	2.1 ppm

Cyclic ketones (n=number of ring carbons)

$(CH_2)_n$ ⟩==O

α–hydrogens	2.0–2.3 ppm (n>5)
	3.0 ppm (n=4)
	1.7 ppm (n=3)
β–hydrogens	1.9–1.5 ppm

Lactones	See Esters, cyclic
Nitro– compounds	See Organic Nitrogen Compounds
Phenols	Ar–O–H 9–10 ppm (Ar=aryl)

Organic Nitrogen Compounds

Amides

δ of Proton(s) (underlined)	Primary R–C(=O)NH$_2$ δ, ppm	Secondary R–C(=O)NHR$_1$ δ, ppm	Tertiary R–C(=O)NR$_1$R$_2$ δ, ppm
(i) N–substitution			
R–C(=O)N–H	5–12	5–12	—
a. alpha			
–C(=O)N–CH$_3$	—	~2.9	~2.9
–C(=O)N–CH$_2$–	—	~3.4	~3.4
–C(=O)N–CH–	—	~3.8	~3.8
b. beta			
–C(=O)N–C–CH$_3$	~1.1	~1.1	~1.1
–C(=O)N–C–CH$_2$–	~1.5	~1.5	~1.5
–C(=O)N–C–CH–	~1.9	~1.9	~1.9
(ii) C–substitution			
a. alpha	~1.9	~2.0	~2.1
CH$_3$–C(=O)N	~2.1	~2.1	~2.1
RCH$_2$–C(=O)N	~2.2	~2.2	~2.2
R$_2$CH–C(=O)N			
b. beta			
CH$_3$–C–C(=O)N	~1.1	~1.1	~1.1
CH$_2$–C–C(=O)N	~1.5	~1.5	~1.5
–CH–C–C(=O)N	~1.8	~1.8	~1.8

Amines

δ of Proton(s) (underlined)	Primary R–NH$_2$ δ, ppm	Secondary RN–HR δ, ppm	Tertiary RRRN δ, ppm
(i) alpha protons			
>N–CH$_3$	~2.5	2.3–3.0	~2.2
>N–CH$_2$–	~2.7	2.6–3.4	~2.4
>N–CH<	~3.1	2.9–3.6	~2.8
(ii) beta protons			
>N–C–CH$_3$			~1.1
>N–C–CH$_2$–			~1.4
>N–C–CH<			~1.7

Cyanocompounds (Nitriles)

(i) Alpha hydrogens δ, ppm		(ii) Beta hydrogens δ, ppm	
CH$_3$–C≡N	~2.1	CH$_3$–C–C≡N	~1.2
–CH$_2$–C≡N	~2.5	–CH$_2$–C–C≡N	~1.6
–CH–C≡N	~2.9	CH–C–C≡N	~2.0

Imides

(i) Alpha hydrogens δ, ppm		(ii) Beta hydrogens δ, ppm	
$\underline{CH_3}$–C(=O)NHC(=O)–	~2.0	$\underline{CH_3}$–C(=O)C–NH–C(=O)–	~1.2
$\underline{CH_2}$–C(=O)NHC(=O)–	~2.1	$\underline{CH_2}$–C(=O)C–NH–C(=O)–	~1.3
$\underline{CH}$–C(=O)NHC(=O)–	~2.2	–$\underline{CH}$–C(=O)C–NH–C(=O)–	~1.4

Isocyanates

Alpha hydrogens δ, ppm

$\underline{CH_3}$–N=C=O	~3.0
–$\underline{CH_2}$–N=C=O	~3.3
–$\underline{CH}$–N=C=O	~3.6

Isocyanides (Isonitriles):		**Isothiocyanates:**	
Alpha hydrogens δ, ppm		**Alpha hydrogens δ, ppm**	
$\underline{CH_3}$–N=C<	~2.9	$\underline{CH_3}$–N=C=S	~3.4
$\underline{CH_2}$–N=C<	~3.3	$\underline{CH_2}$–N=C=S	~3.7
$\underline{CH}$–N=C<	~4.9	>$\underline{CH}$–N=C=S	~4.0

Nitriles δ, ppm

–$\underline{CH_2}$–O–N=O	~4.8

Nitrocompounds δ, ppm

$\underline{CH_3}$–NO$_2$	~4.1	–$\underline{CH_2}$–NO$_2$	~4.2	–$\underline{CH}$–NO$_2$	~4.4
$\underline{CH_3}$–C–NO$_2$	~1.6	–$\underline{CH_2}$–C–NO$_2$	~2.1	–$\underline{CH}$–C–NO$_2$	~2.5

Organic Sulfur Compounds

Family	δ of Proton(s) Underlined			
Benzothiopyrans				
2H–1–	sp³ C–H	~3.3 ppm	sp² C–H 5.8–6.4	aromatic ~6.8
4H–1–	sp³ C–H	~3.2 ppm	sp² C–H 5.9–6.3	aromatic ~6.9
2,3,4H–1–	sp³ C–H	1.9–2.8 ppm	aromatic	~7.1
Disulfides	C̲H̲₃–S–S–R	~2.4 ppm	C̲H̲₃–C–S–S–R	~1.2 ppm
	C̲H̲₂–S–S–R	~2.7 ppm	C̲H̲₂–C–S–S–R	~1.6 ppm
	C̲H̲–S–S–R	~3.0 ppm	C̲H̲–C–S–S–R	~2.0 ppm
Isothiocyanates	C̲H̲₃–N=C=S	~2.4 ppm		
	–C̲H̲₂–N=C=S	~2.7 ppm		
	–C̲H̲–N=C=S	~3.0 ppm		
Mercaptans (Thiols)	C̲H̲₃–S–H	~2.1 ppm	C̲H̲₃–C–S–H	~1.3 ppm
	–C̲H̲₂–S–H	~2.6 ppm	–C̲H̲₂–C–S–H	~1.6 ppm
	–C̲H̲–S–H	~3.1 ppm	–C̲H̲–C–S–H	~1.7 ppm
S–methyl salts	$\overset{+}{>}$S–CH₃	~3.2 ppm		
Sulfates	(C̲H̲₃–O)₂S(=O)₂	~3.4 ppm		
Sulfides	C̲H̲₃–S–	1.8–2.1	C̲H̲₃–CH₂–S–	1.1–1.2
	R–C̲H̲₂–S–	1.9–2.4	C̲H̲₃–CHR–S–	0.8–1.2
	R–C̲H̲R–S–	2.8–3.4	C̲H̲₃–CHAr–S–	1.3–1.4
	Ar–C̲H̲₂–S–	4.1–4.2	C̲H̲₃–CR₂–S–	1.0
	Ar–C̲H̲R–S–	3.6–4.2	Ar–C̲H̲₂–CHR–S–	3.0–3.2
	Ar₂–C̲H̲–S–	5.1–5.2	>C=C–C̲H̲₂–CHAr–S–	2.4–2.6
			>C=C–C̲H̲₂–CAr₂–S–	2.5
			R₂C̲H̲–CH₂–S–	2.6–3.0
			Ar₂ C̲H̲–CH₂–S–	4.0–4.2
			>C=C–C̲H̲R–CHAr–S–	2.3–2.4
			>C=C–C̲H̲R–CAr₂–S–	2.8–3.2
Sulfilimines	C̲H̲₃(R)S=N–R²	~2.5 ppm		
Sulfonamides	C̲H̲₃–SO₂NH₂	~3.0 ppm		
Sulfonates	C̲H̲₃–SO₂–OR	~3.0 ppm		
Sulfones	C̲H̲₃–SO₂–R²	~2.6 ppm		
Sulfonic acids	C̲H̲₃–SO₃H	~3.0 ppm		
Sulfoxides	C̲H̲₃–S(=O)R	~2.5 ppm		
	–C̲H̲₂–S(=O)R	~3.1 ppm		
Thiocyanates	C̲H̲₃–S–C≡N	~2.7 ppm		
	–C̲H̲₂–S–C≡N	~3.0 ppm		
	–C̲H̲–S–C≡N	~3.3 ppm		
Thiols	See Mercaptans			

Note: Ar represents aryl.

PROTON NMR CORRELATION CHARTS OF MAJOR FUNCTIONAL GROUPS

The following correlation tables provide the regions of nuclear magnetic resonance absorptions of major chemical families. These absorptions are reported in the dimensionless units of parts per million (ppm) versus the standard compound tetramethylsilane (TMS; $(CH_3)_4Si$), which is recorded as 0.0 ppm.

The use of this unit of measure makes the chemical shifts independent of the applied magnetic field strength or the radio frequency. For most proton NMR spectra, the protons in TMS are more shielded than almost all other protons. The chemical shift in this dimensionless unit system is then defined by:

$$\delta = \frac{\nu_s - \nu_r}{\nu_r} \times 10^6,$$

where ν_s and ν_r are the absorption frequencies of the sample proton and the reference (TMS) protons (12, magnetically equivalent), respectively. In these tables, the proton(s) whose proton NMR shifts are cited are indicated by underscore. For more detail concerning these conventions, the reader is referred to the general references below [1–15]. Reference 15 has a compilation of references for the various nuclei.

Due to the large amount of data the whole ^1H NMR region is divided into smaller sections 1.0–1.2 ppm range each. This will allow the user to look into specified chemical shift and determine all possibilities for the unknown whose structure is being analyzed.

REFERENCES

1. Silverstein, R. M., and F. X. Webster. *Spectrometric Identification of Organic Compounds*. 6th ed. New York: John Wiley and Sons, 1998.
2. Rahman, A.-U. *Nuclear Magnetic Resonance*. New York: Springer Verlag, 1986.
3. Gordon, A. J., and R. A. Ford. *The Chemist's Companion*. New York: Wiley Interscience, 1971.
4. Becker, E. D. *High Resolution NMR, Theory and Chemical Applications*. 2nd ed. New York: Academic Press, 1980.
5. Pretsch, E., P. Bühlmann, and M. Badertscher, *Structure Determination of Organic Compounds*: *Tables of Spectral Data*. 4th Revised and Enlarged English edition, Berlin: Springer-Verlag, 2009.
6. Gunther, H. *NMR Spectroscopy: Basic Principles, Concepts and Applications in Chemistry*. New York: John Wiley and Sons, 2003.
7. Kitamaru, R. *Nuclear Magnetic Resonance: Principles and Theory*. London: Elsevier Science, 1990.
8. Lambert, J. B., L. N. Holland, and E. P. Mazzola. *Nuclear Magnetic Resonance Spectroscopy: Introduction to Principles, Applications and Experimental Methods*. Englewood Cliffs: Prentice Hall, 2003.
9. Bovey, F. A., and P. A. Mirau. *Nuclear Magnetic Resonance Spectroscopy*. 2nd ed. New York: Academic Press, 1988.
10. Hore, P. J. *Nuclear Magnetic Resonance*. Oxford: Oxford University Press, 1995.
11. Bruno, T. J., and P. D. N. Svoronos. *Handbook of Basic Tables for Chemical Analysis*. 2nd ed. Boca Raton, FL: CRC Press, 2003.
12. Bruno, T. J., P. D. N. Svoronos. *CRC Handbook of Fundamental Spectroscopic Correlation Charts*. Boca Raton, FL: CRC Press, 2006.
13. Nelson, J. H. *Nuclear Magnetic Resonance Spectroscopy*. 2nd ed. New York: John Wiley and Sons, 2003.
14. Abraham, R. J., J. Fisher, and P. Loftus. *Introduction to NMR Spectroscopy*. New York: John Wiley and Sons, 1988.
15. University of Wisconsin, NMR Bibliography, available on line at http://www.chem.wisc.edu/areas/reich/Handouts/nmr/NMR-Biblio.htm.

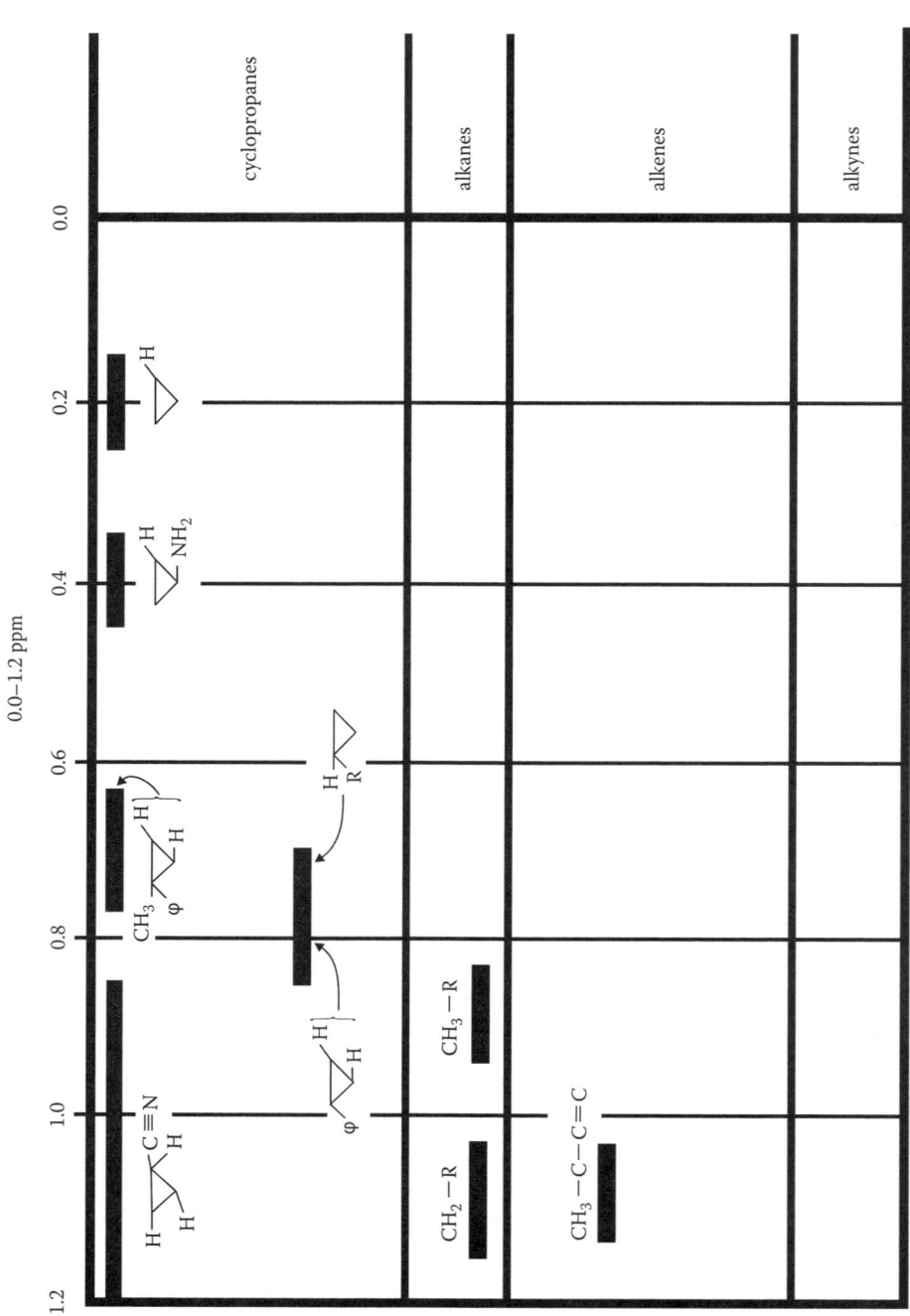

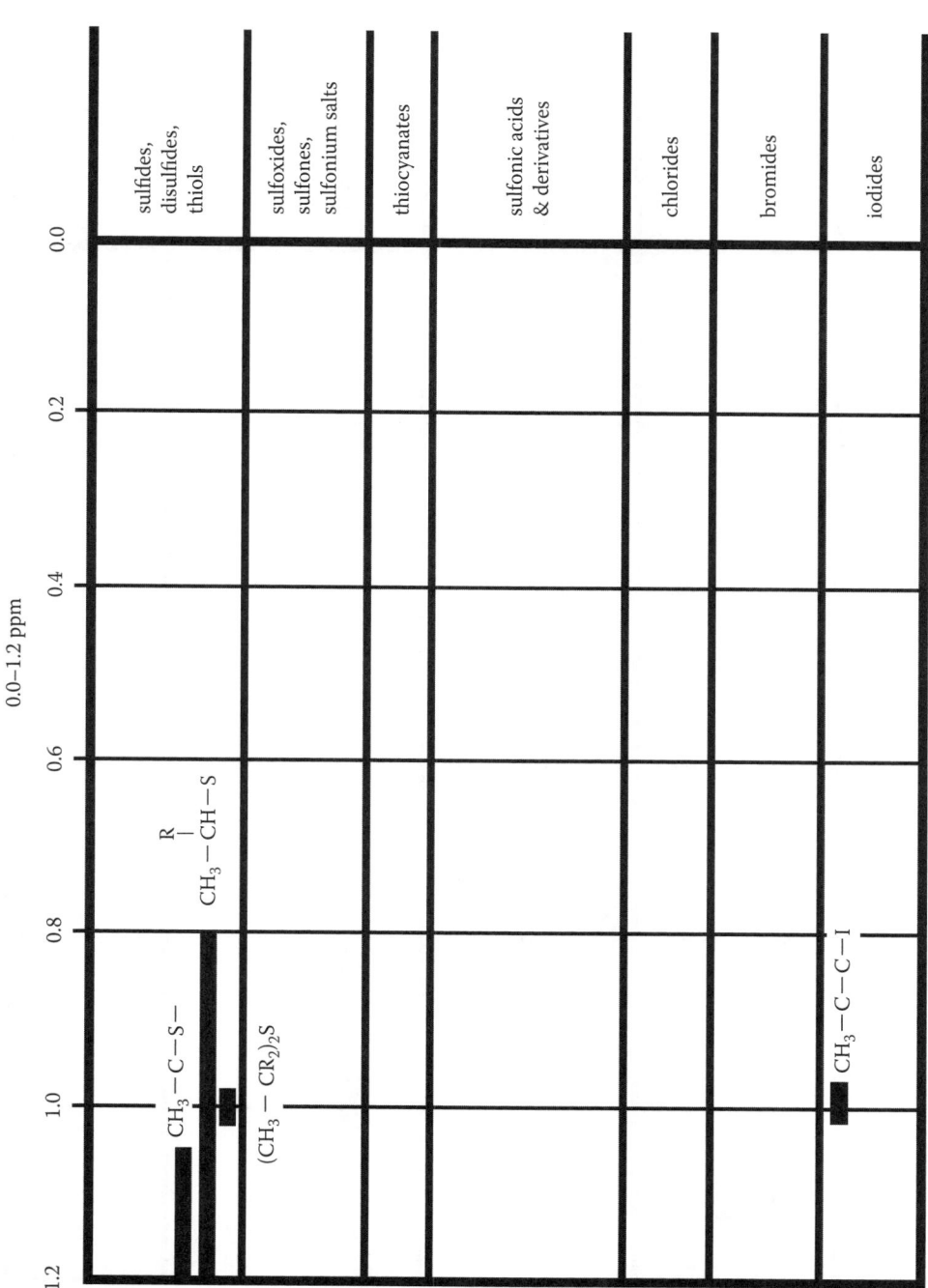

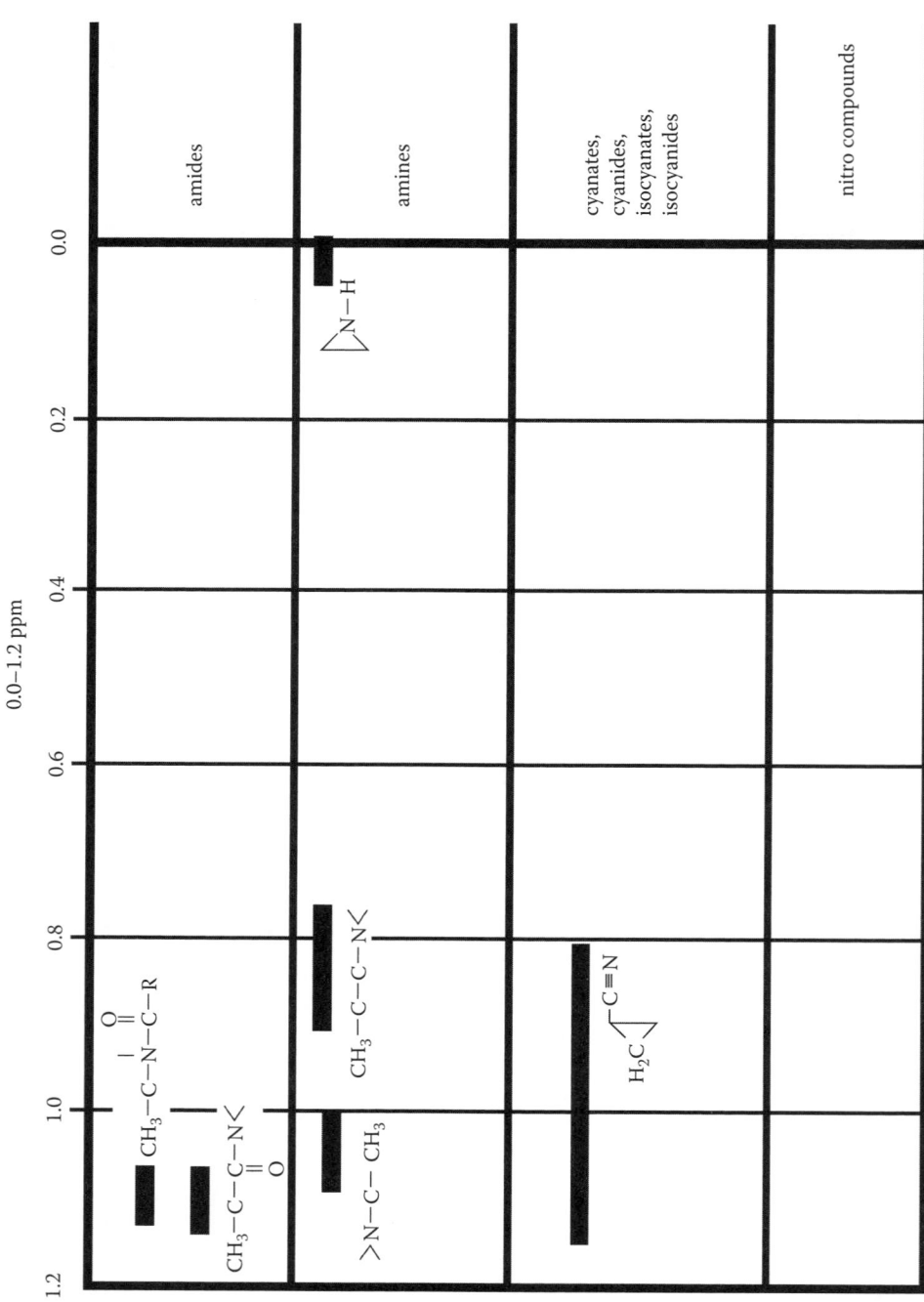

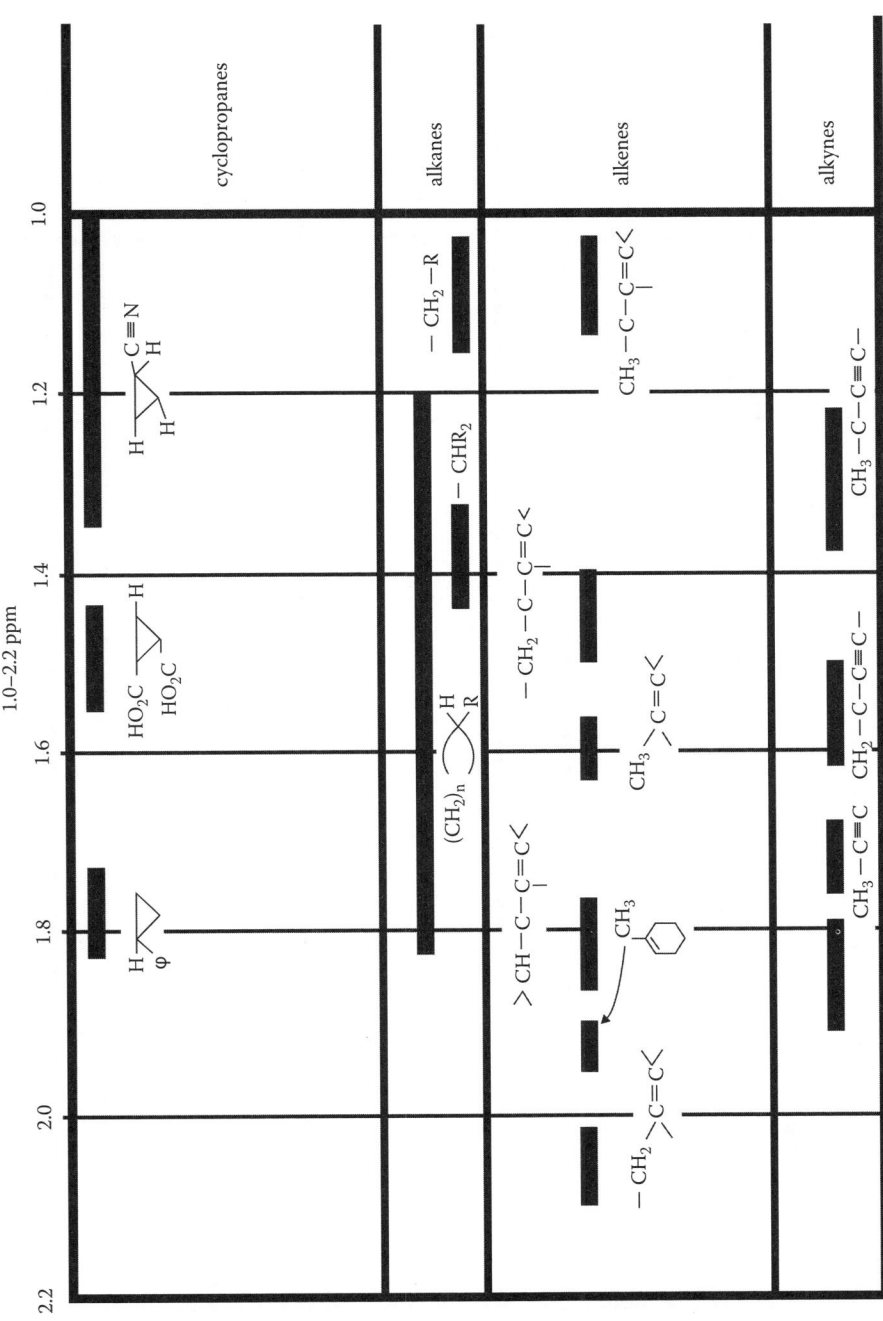

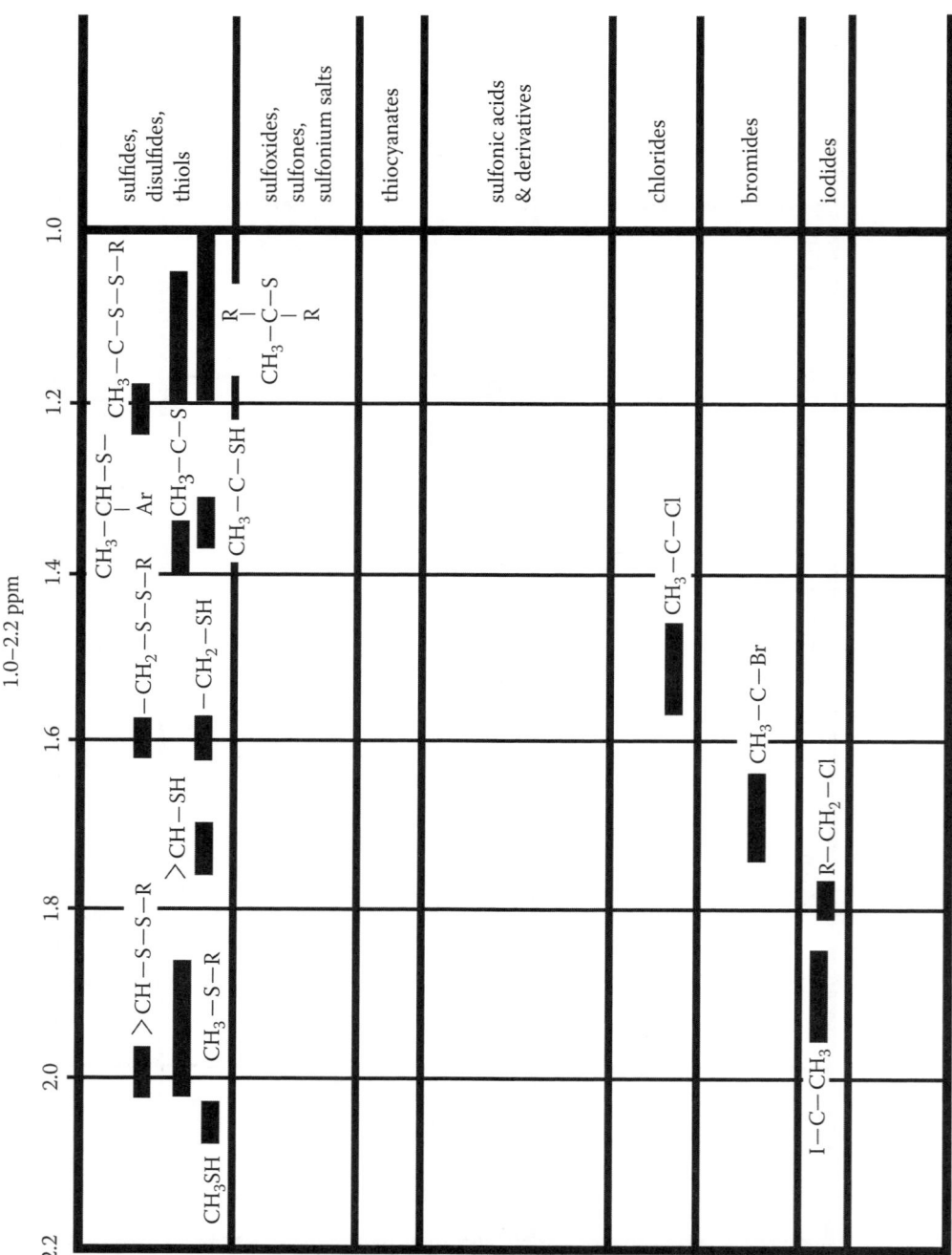

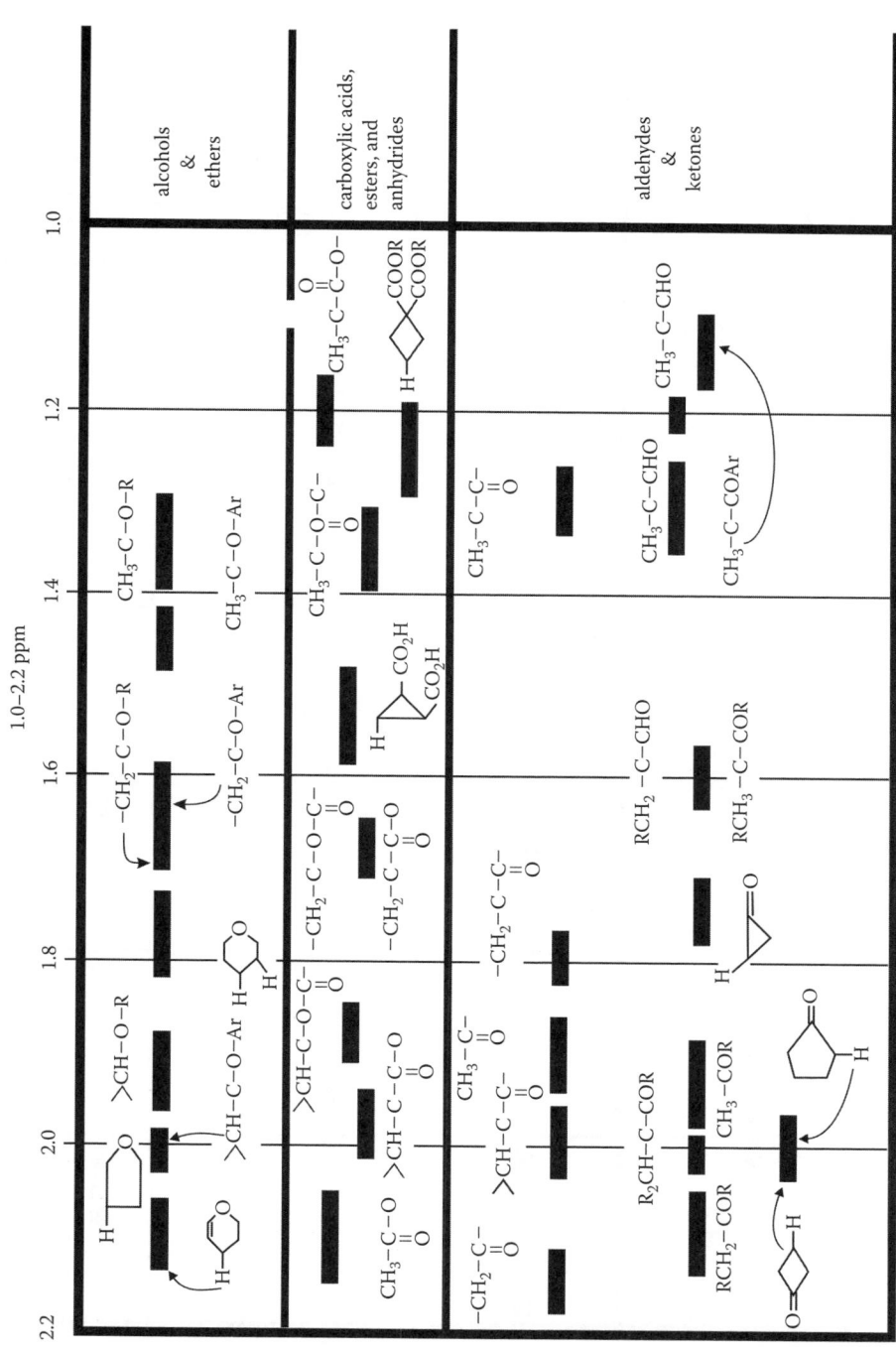

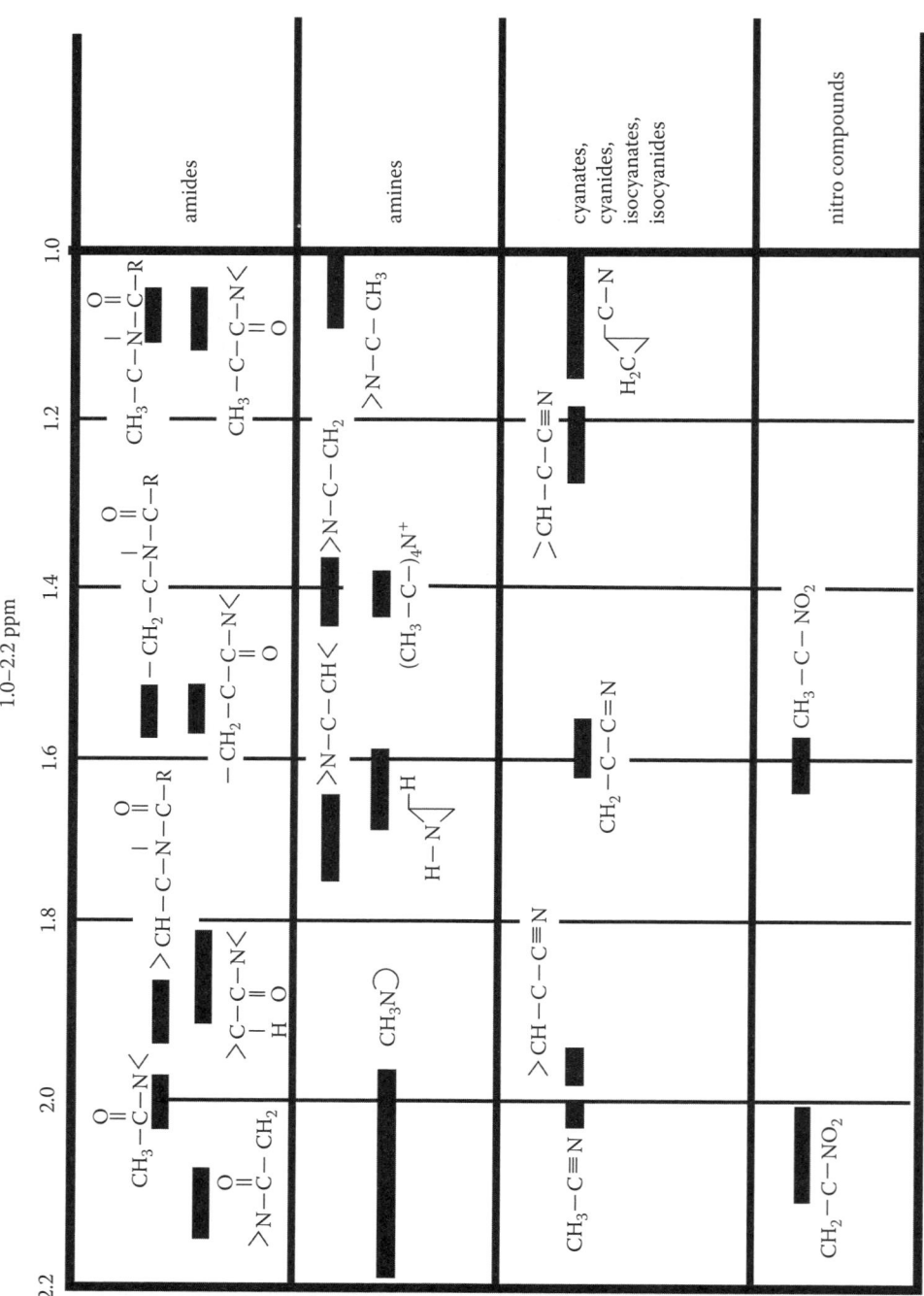

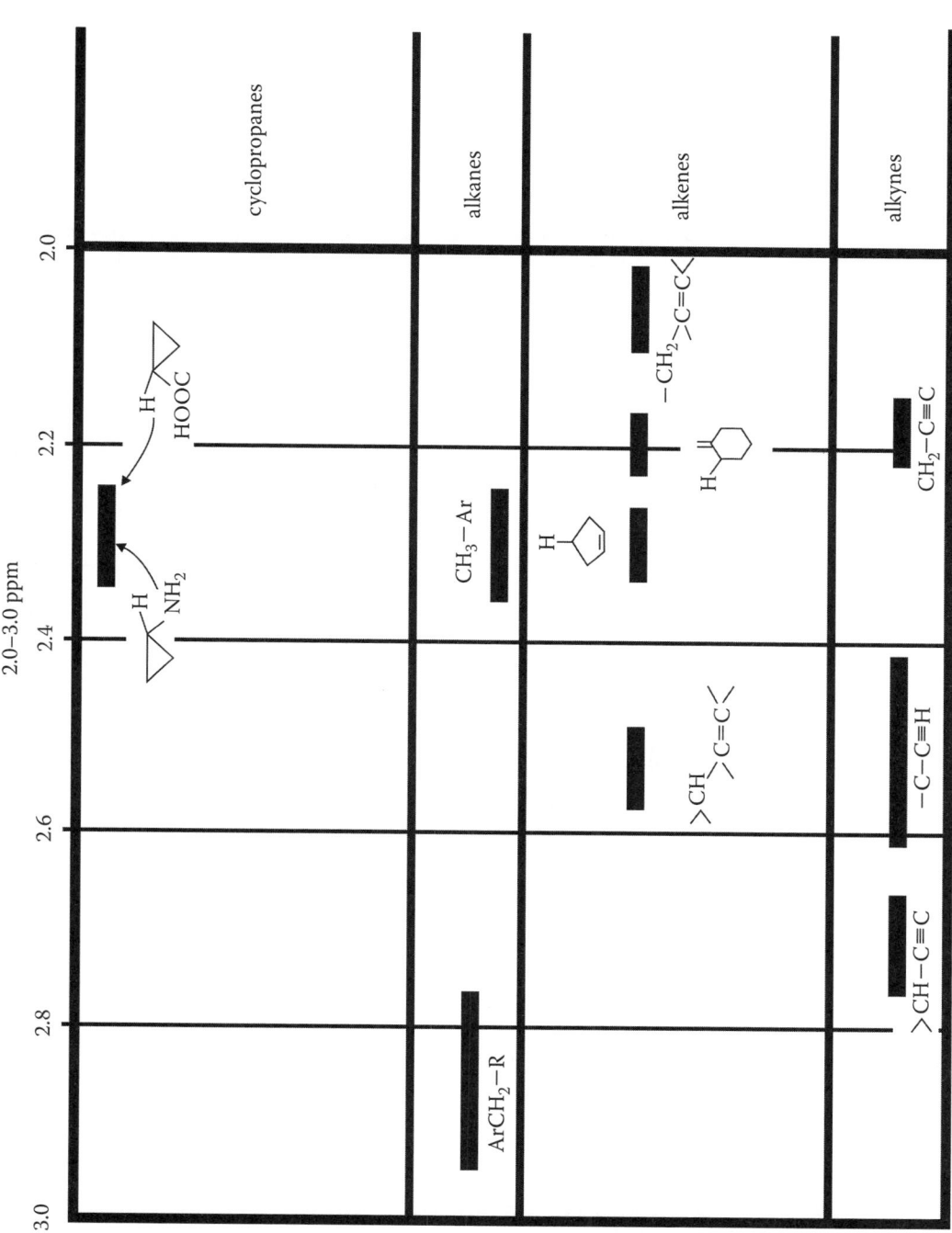

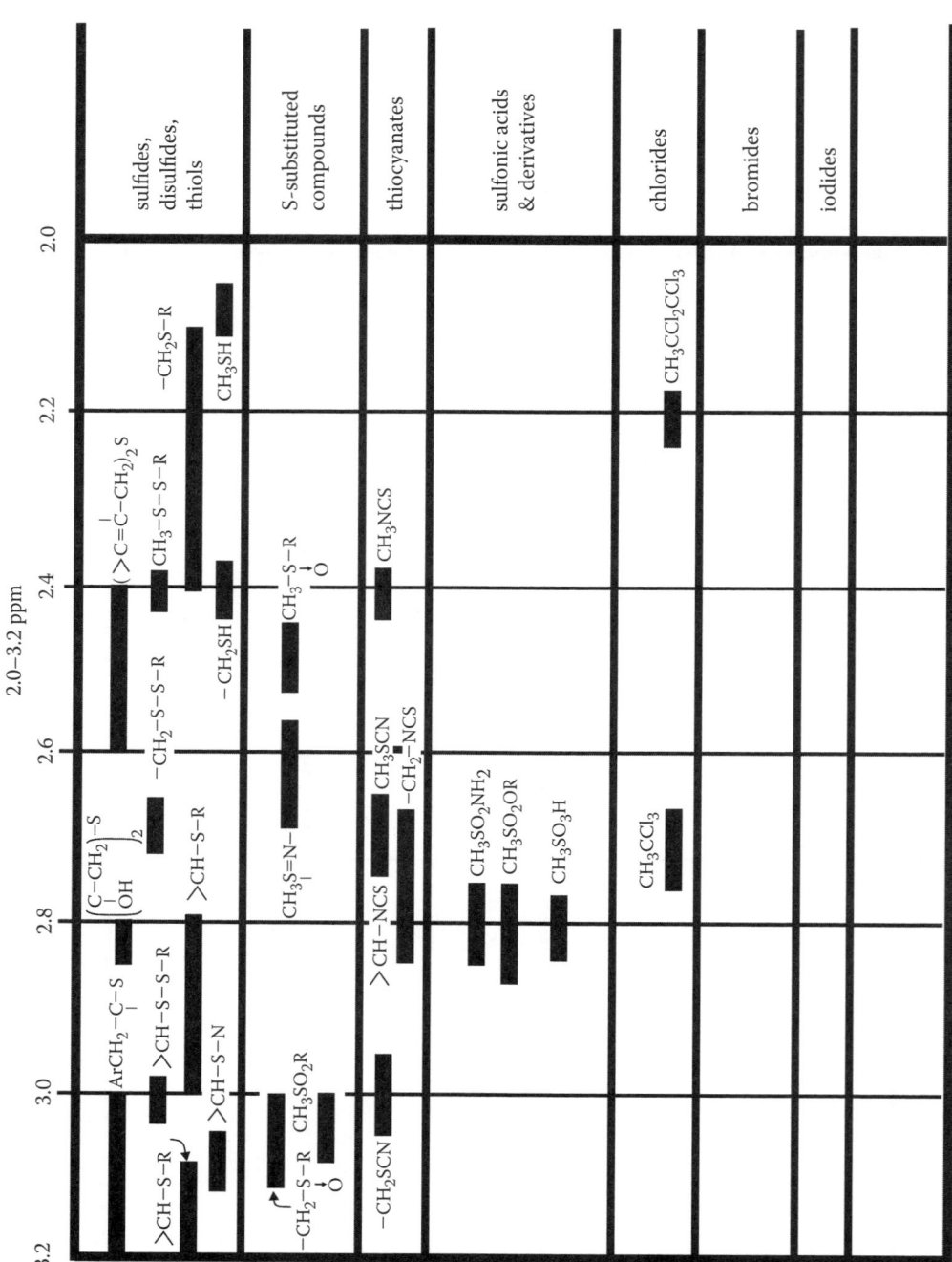

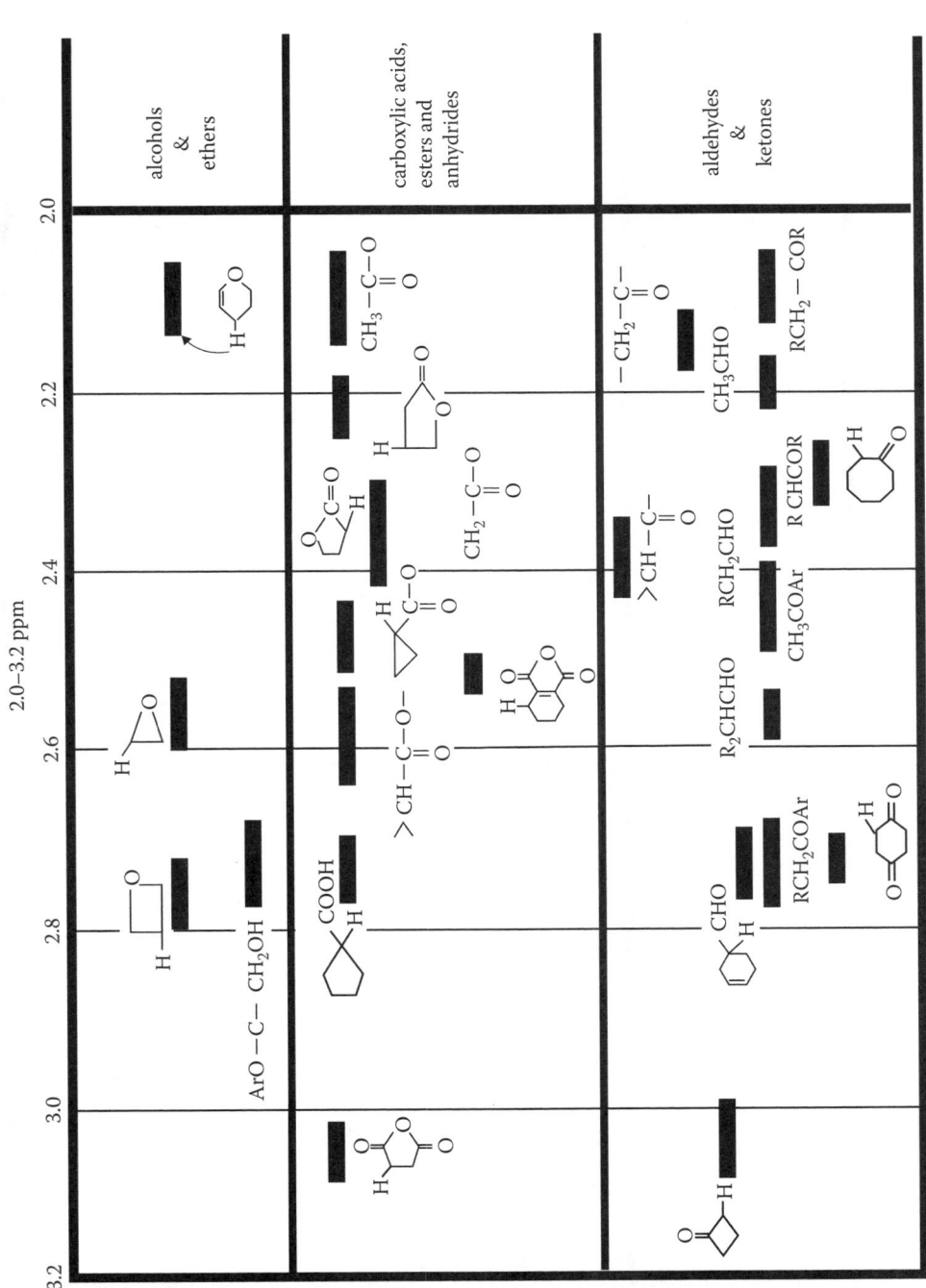

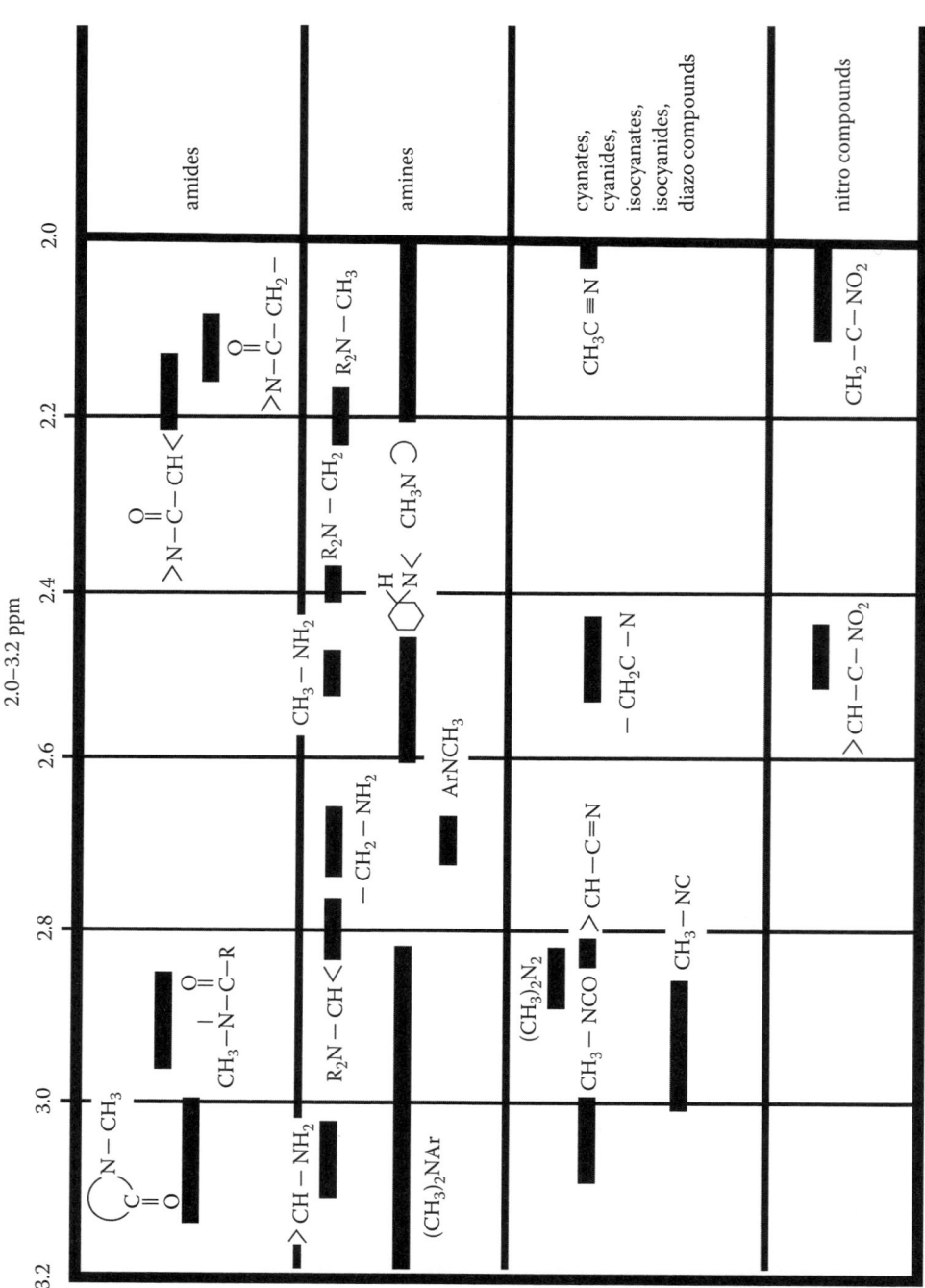

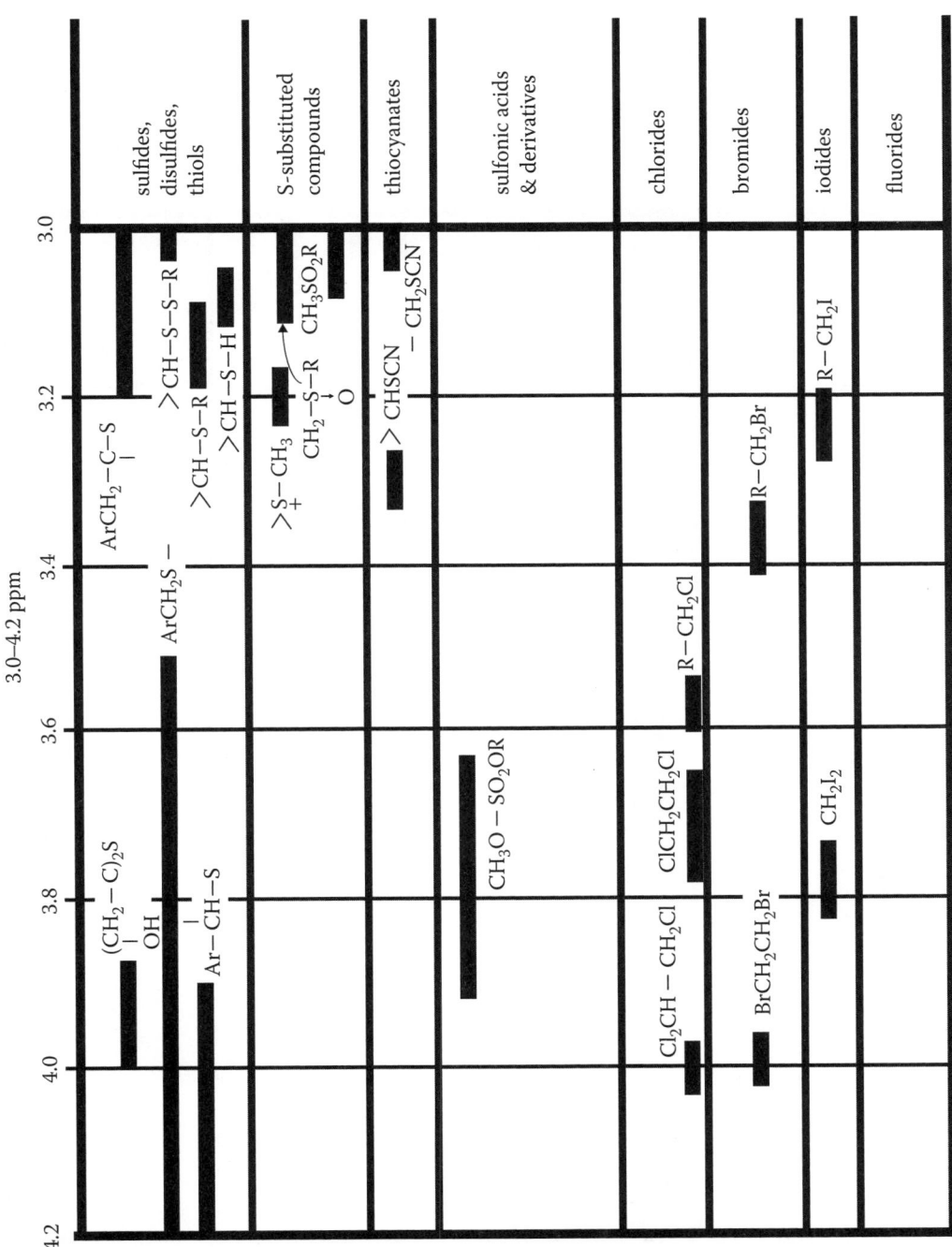

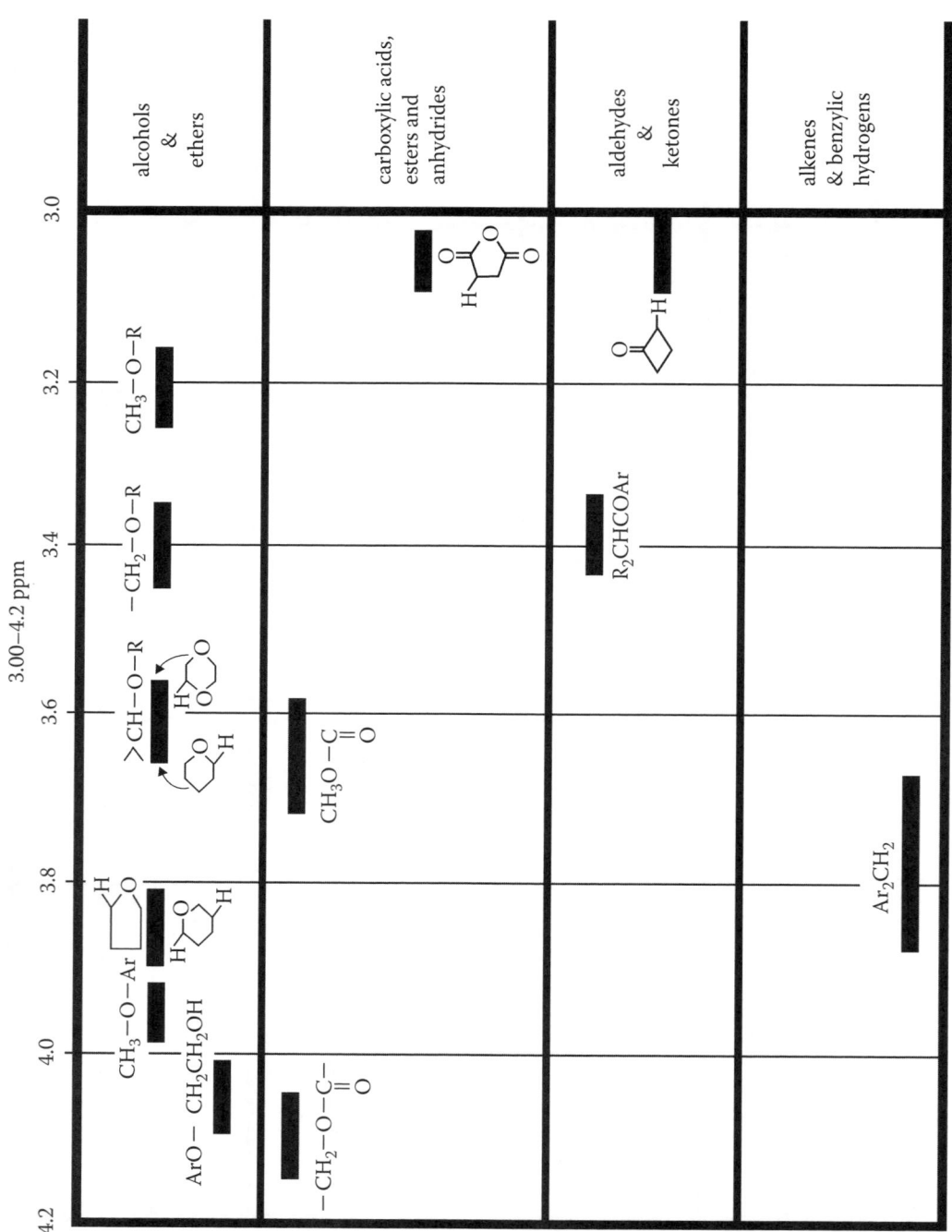

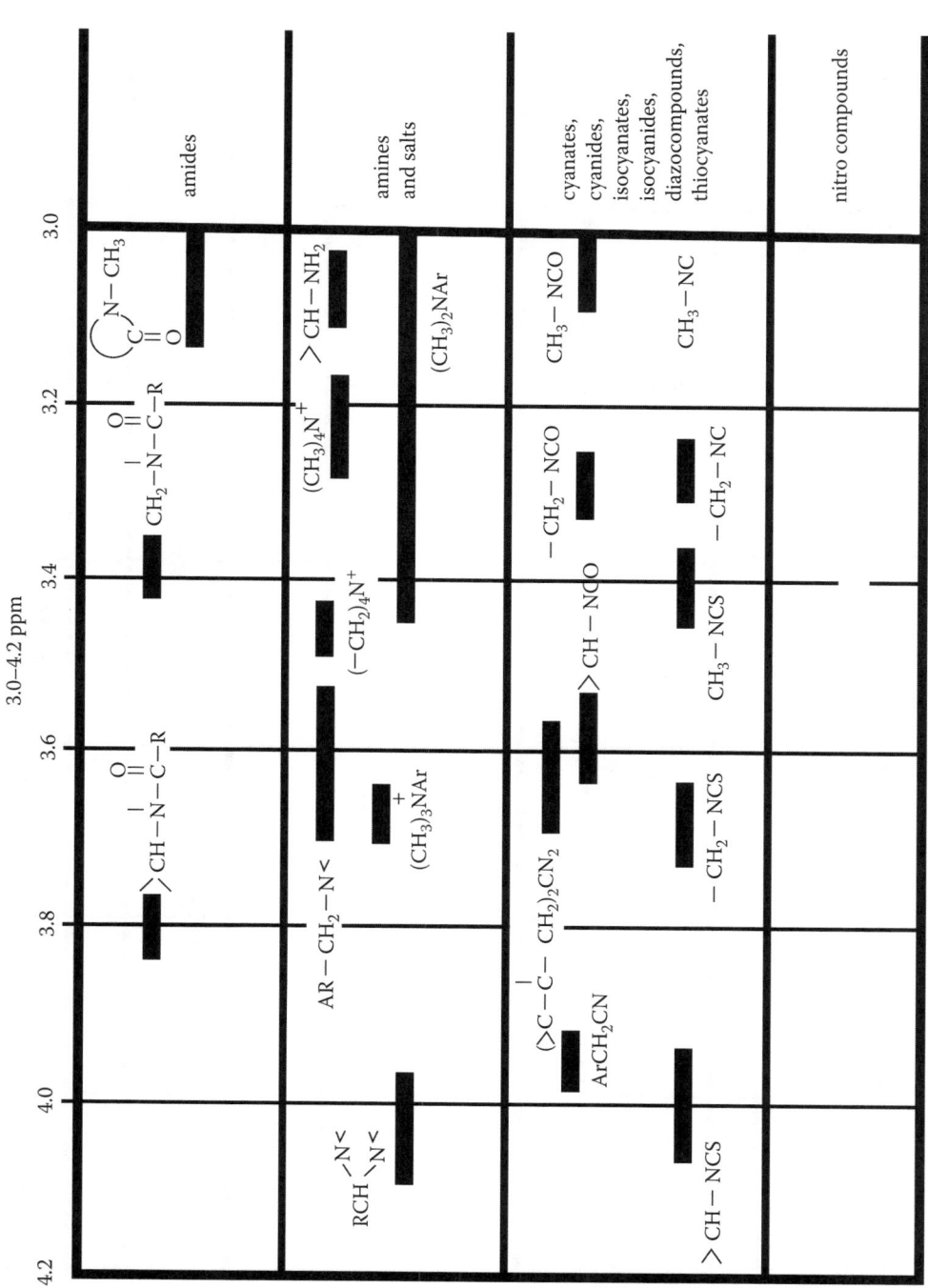

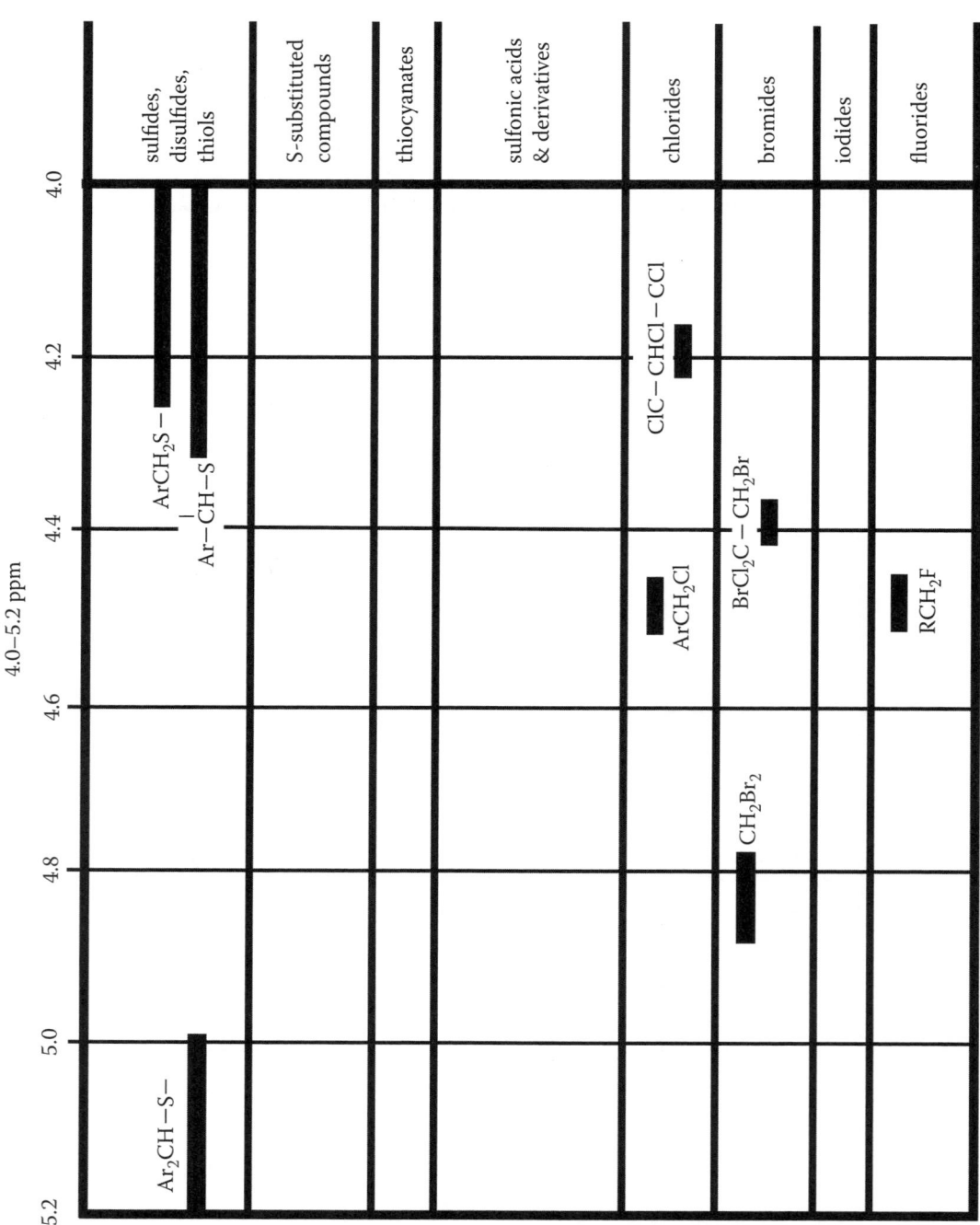

4.0–5.2 ppm

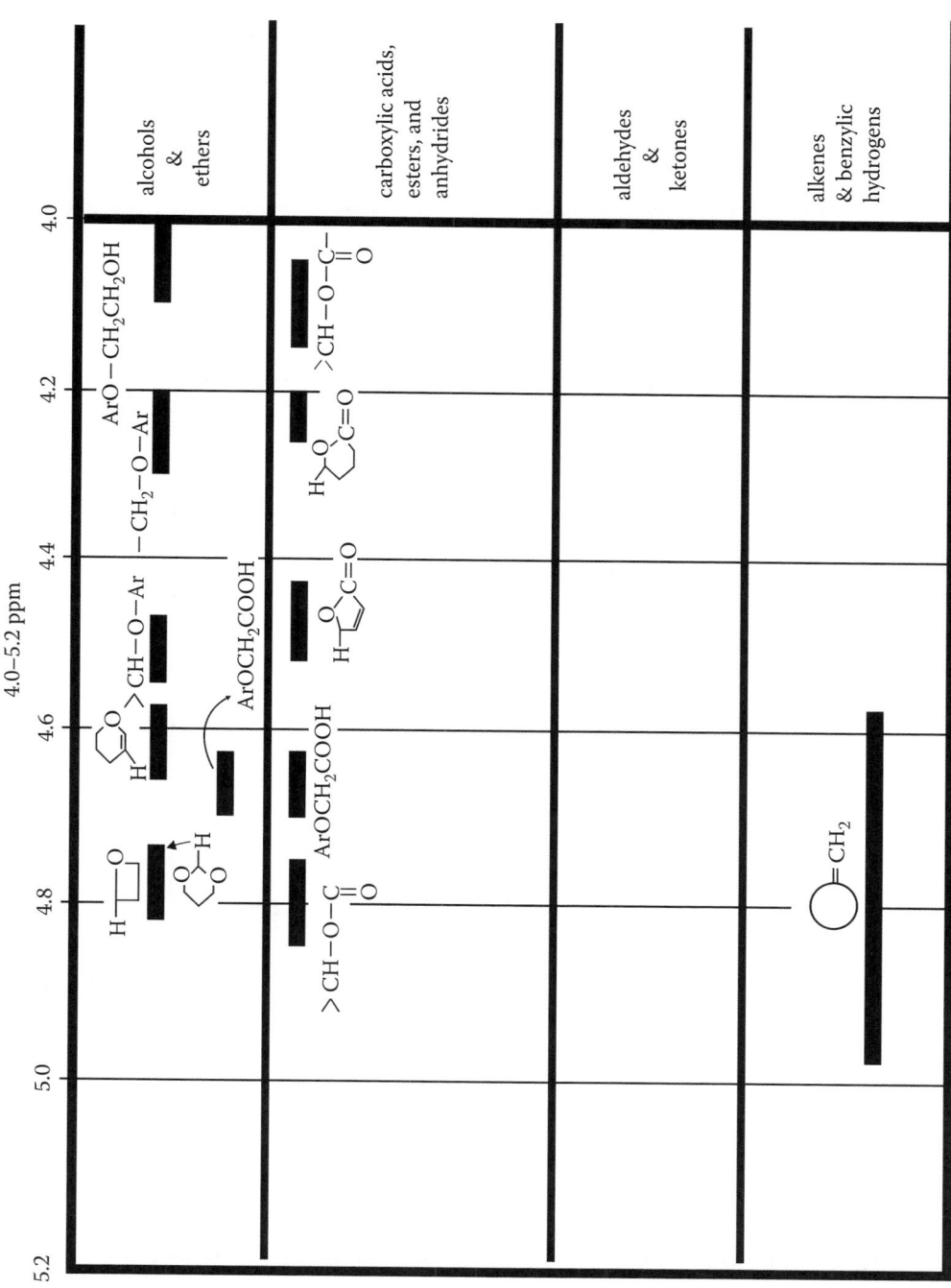

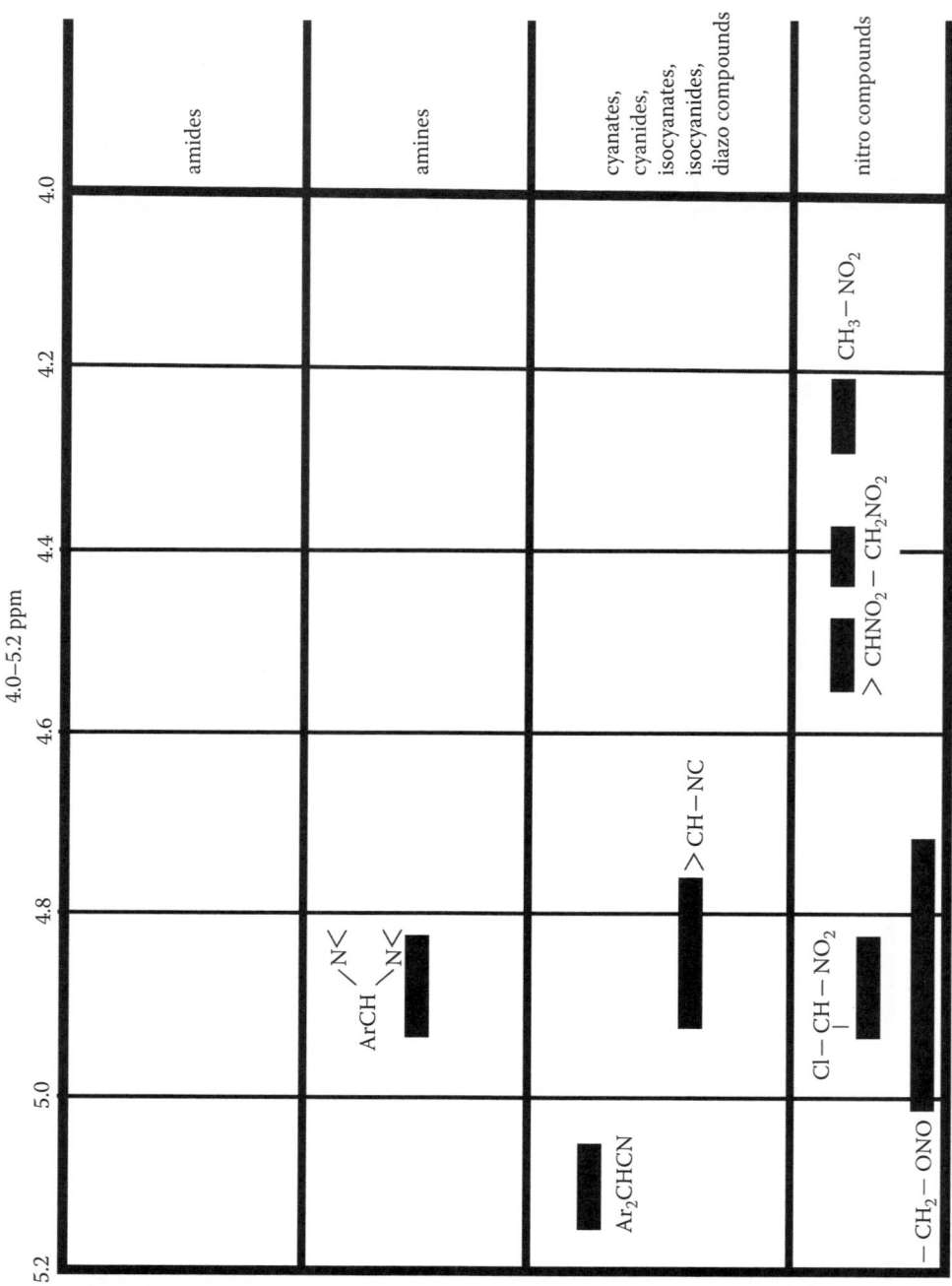

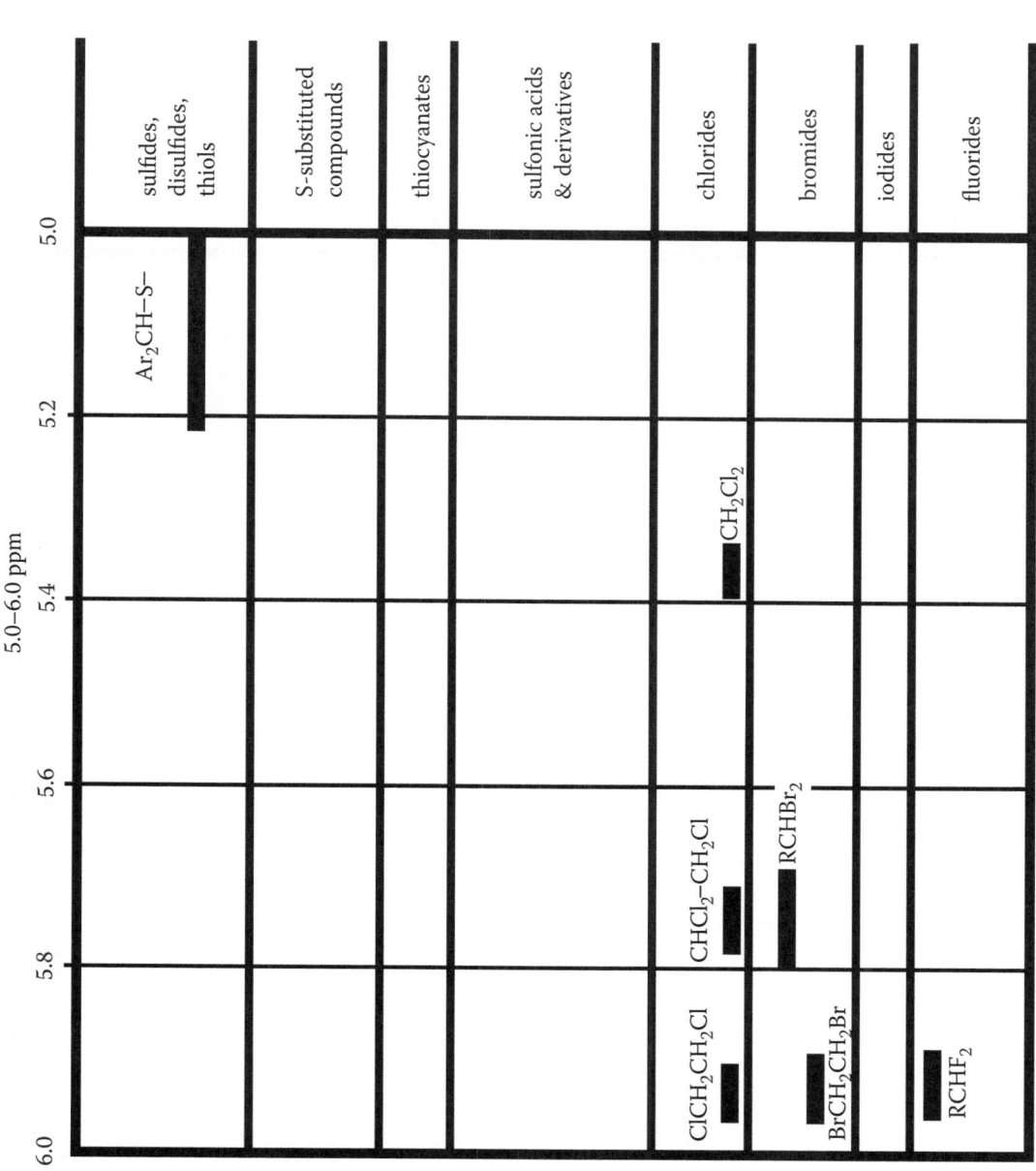

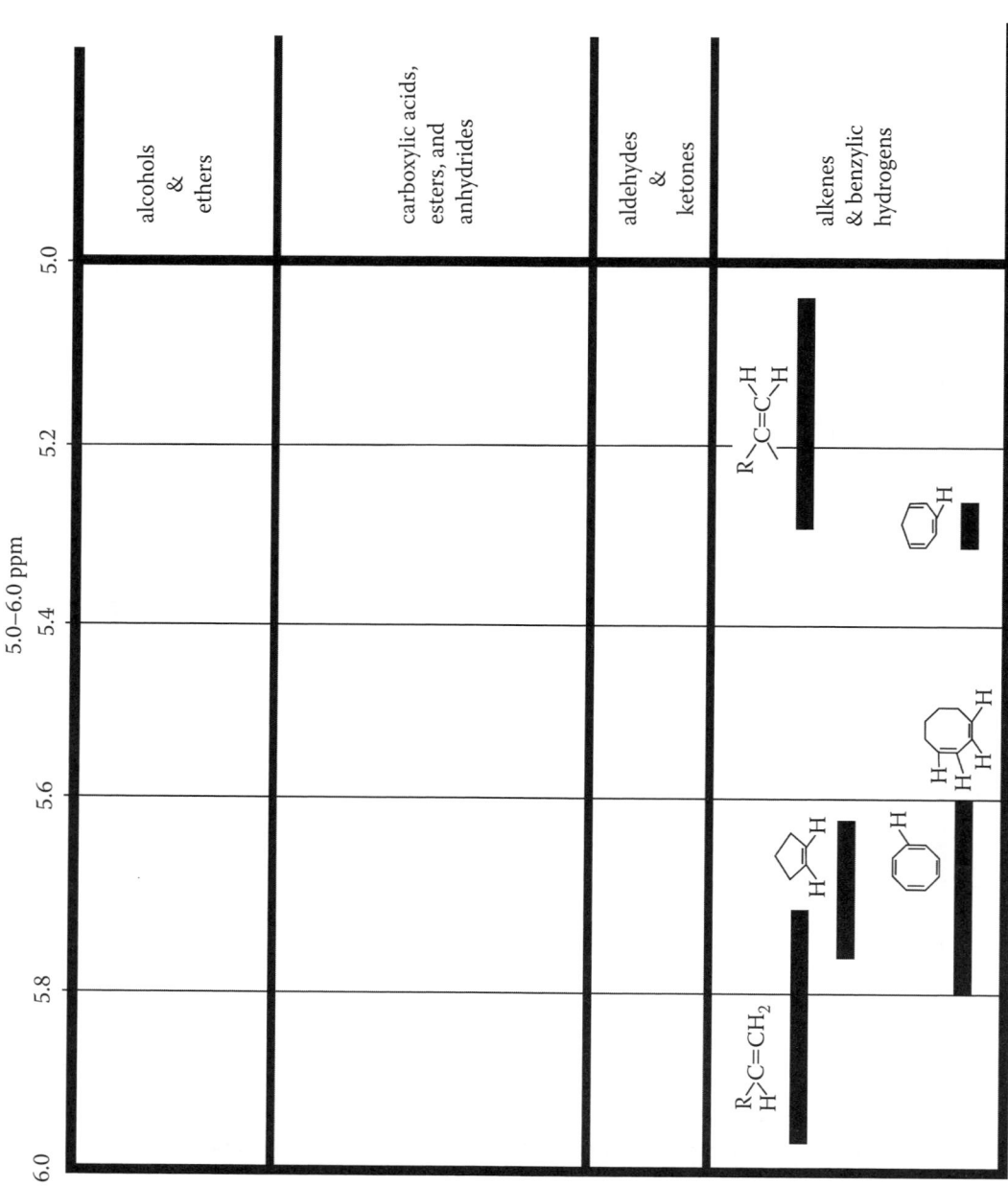

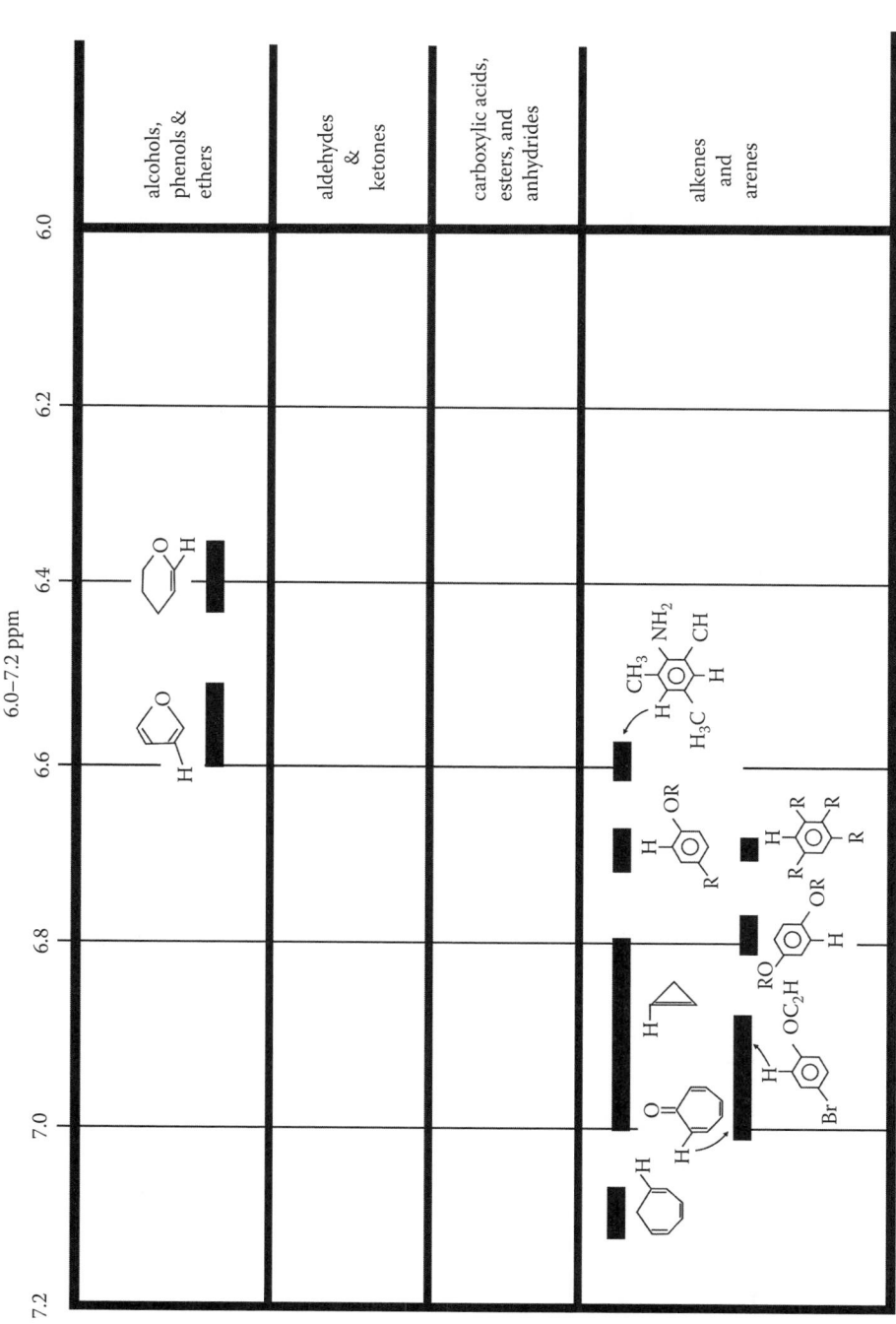

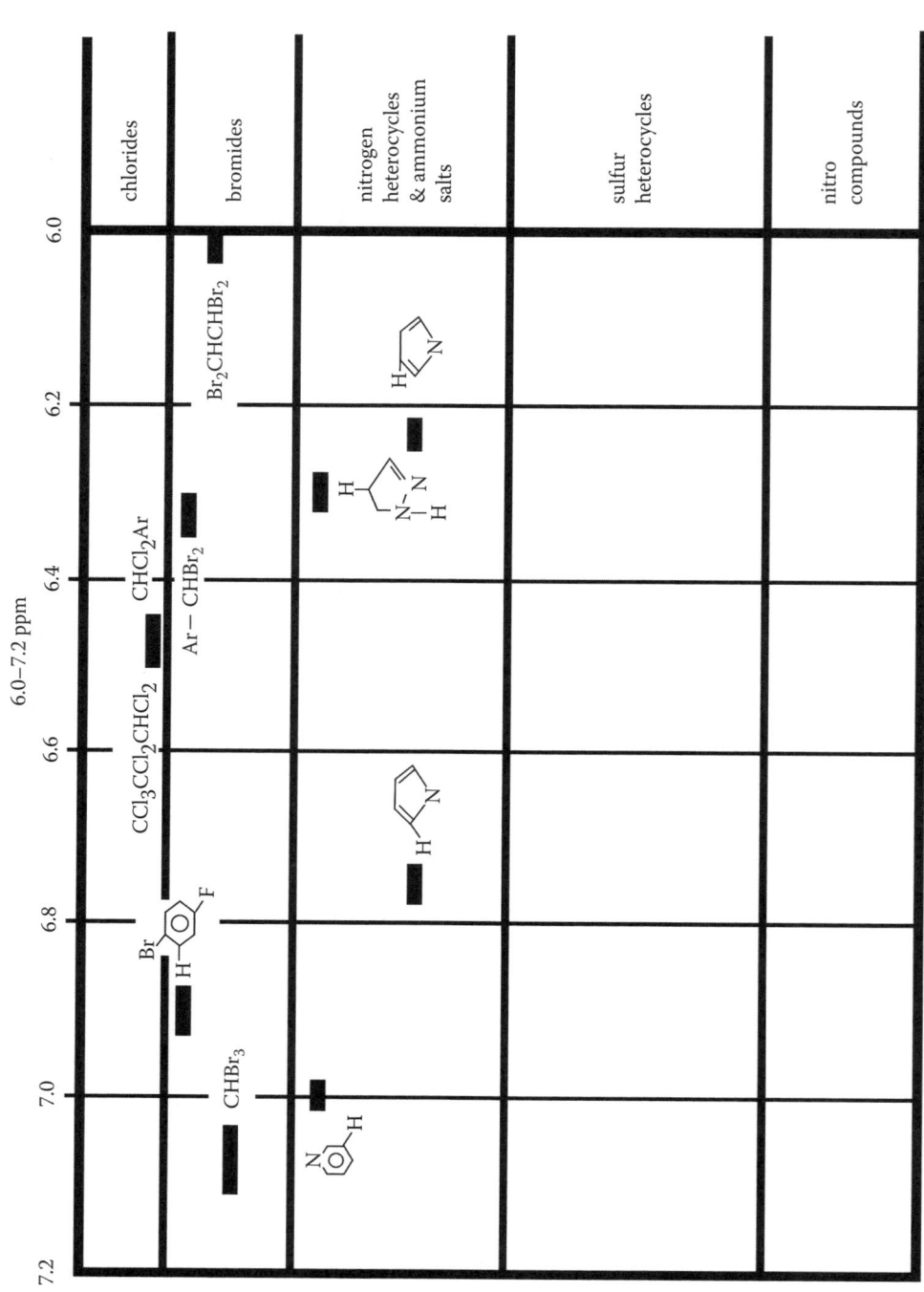

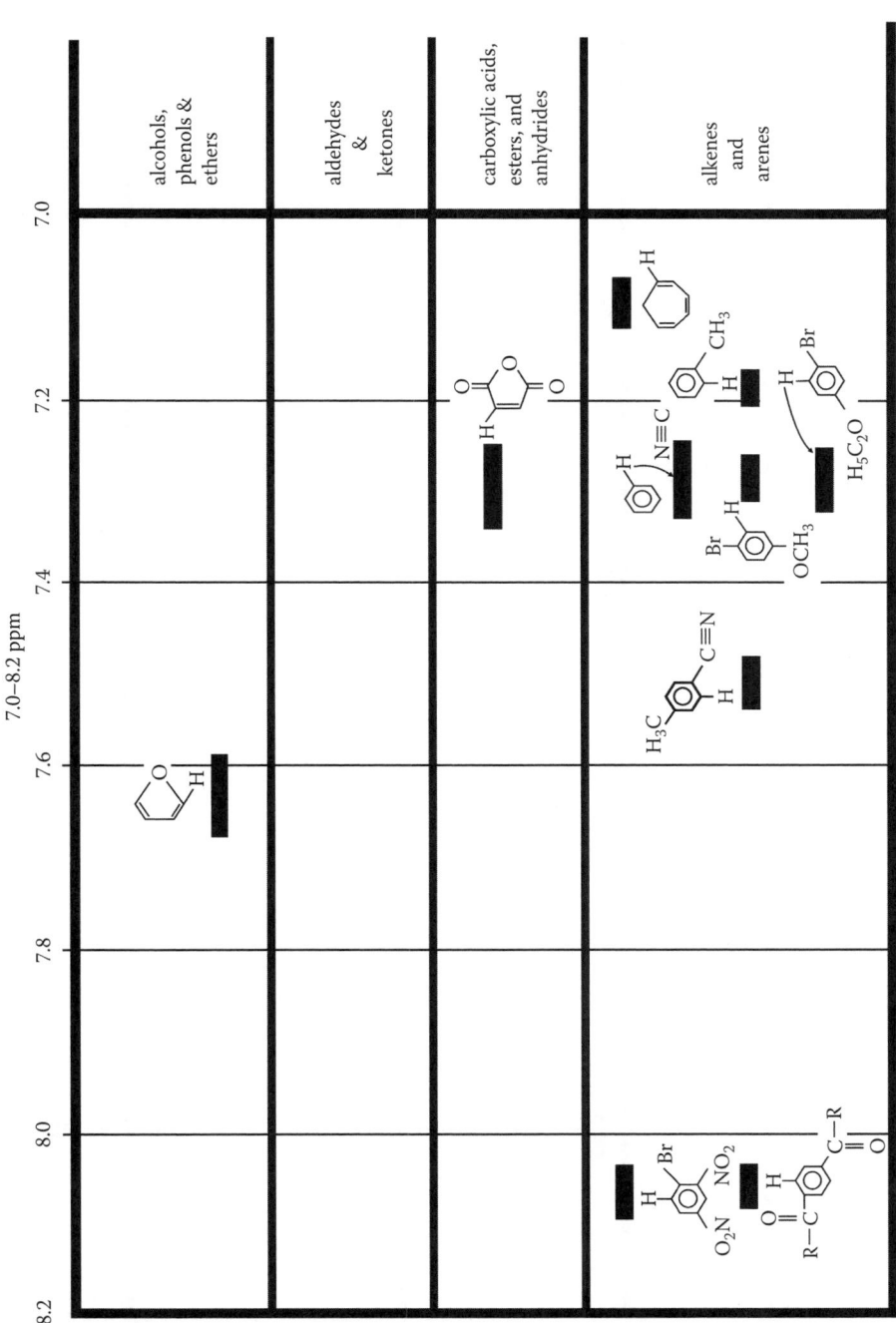

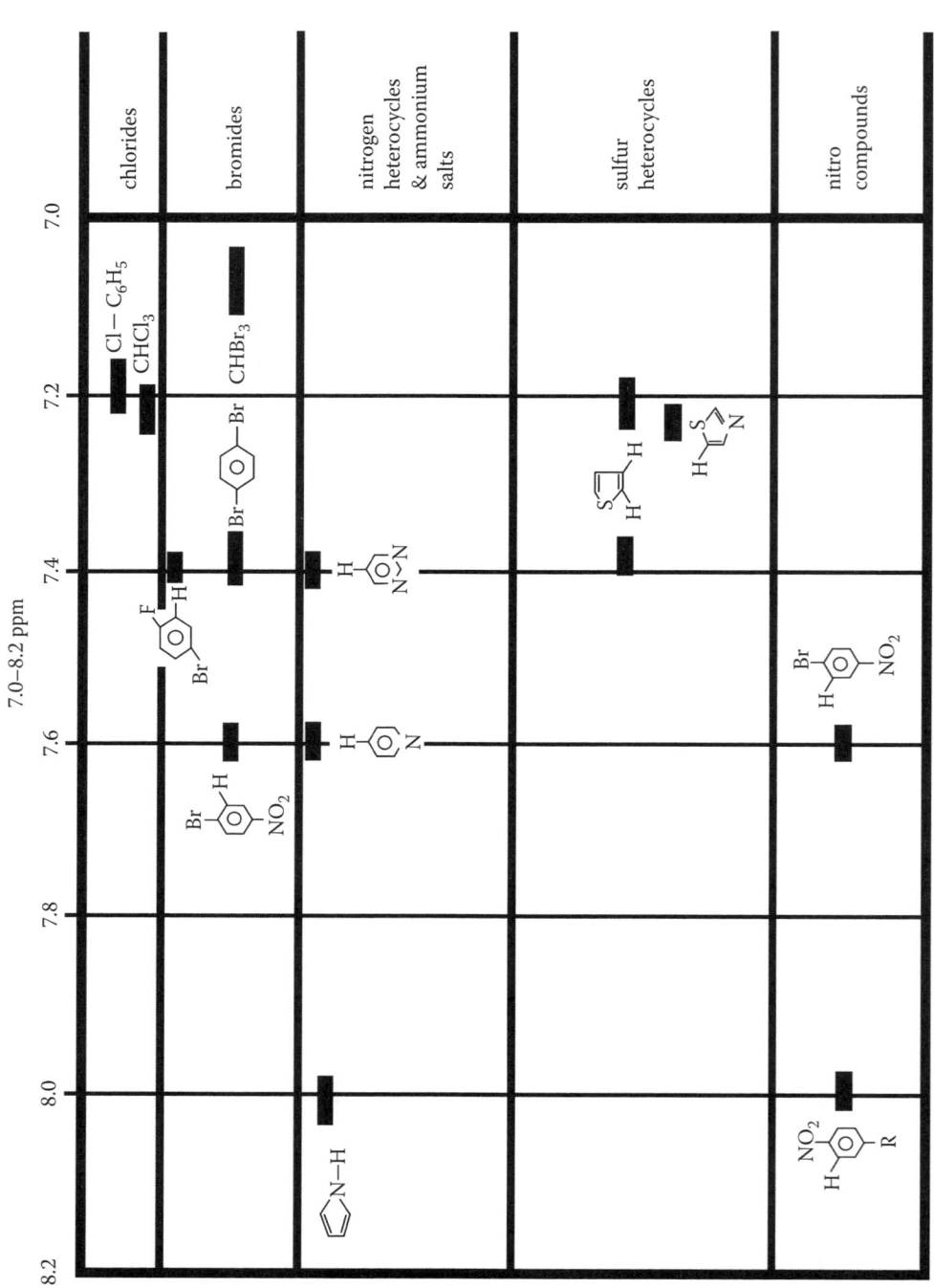

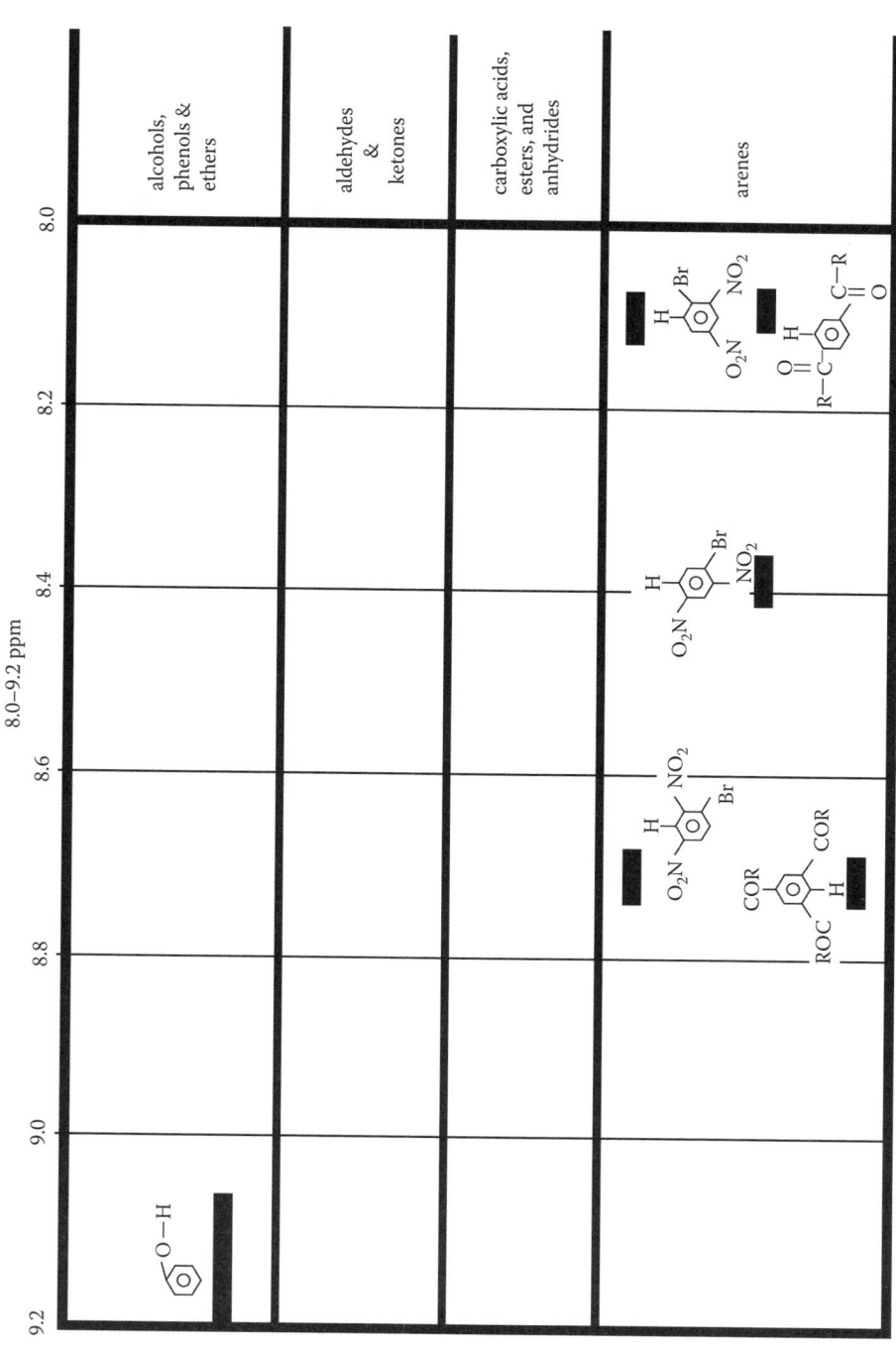

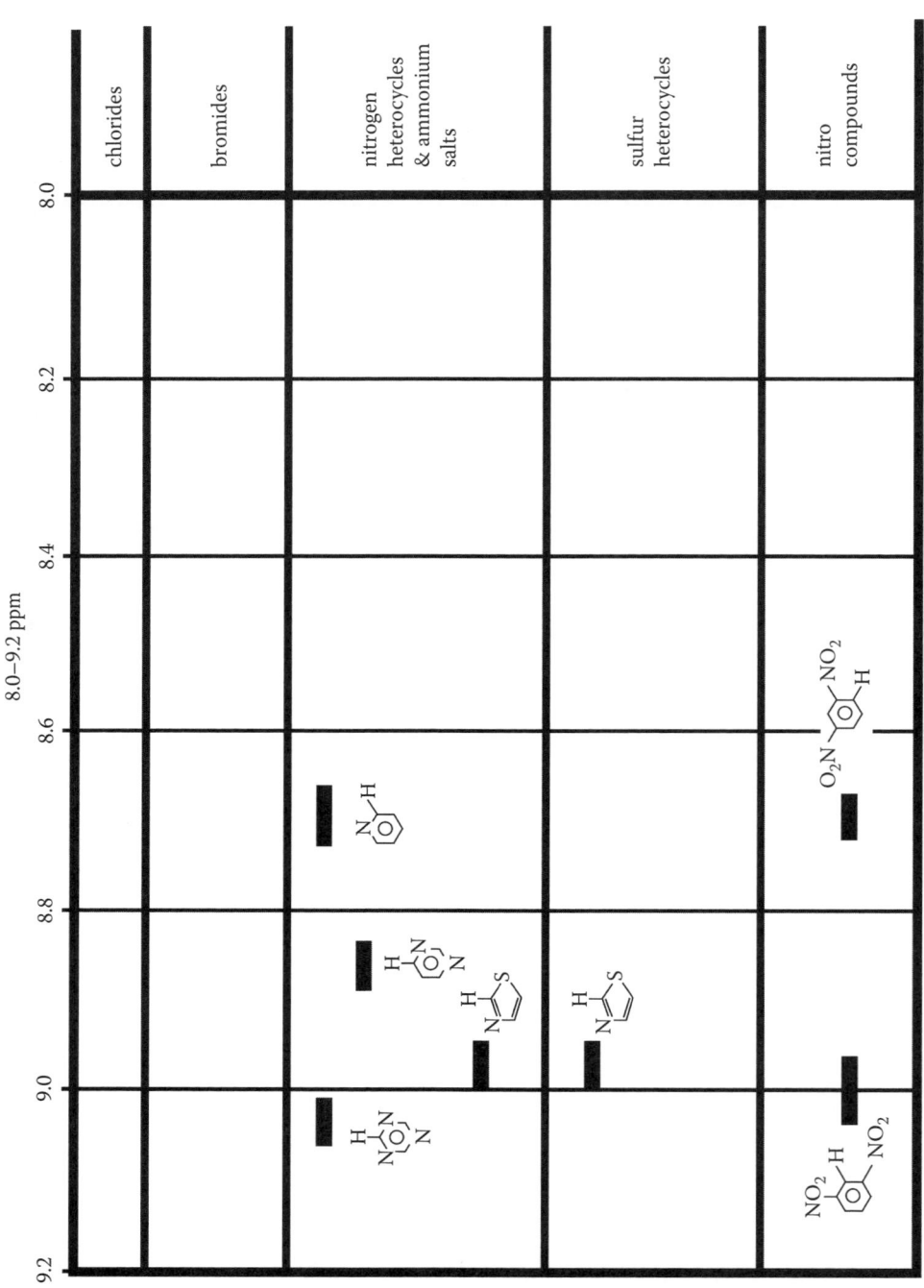

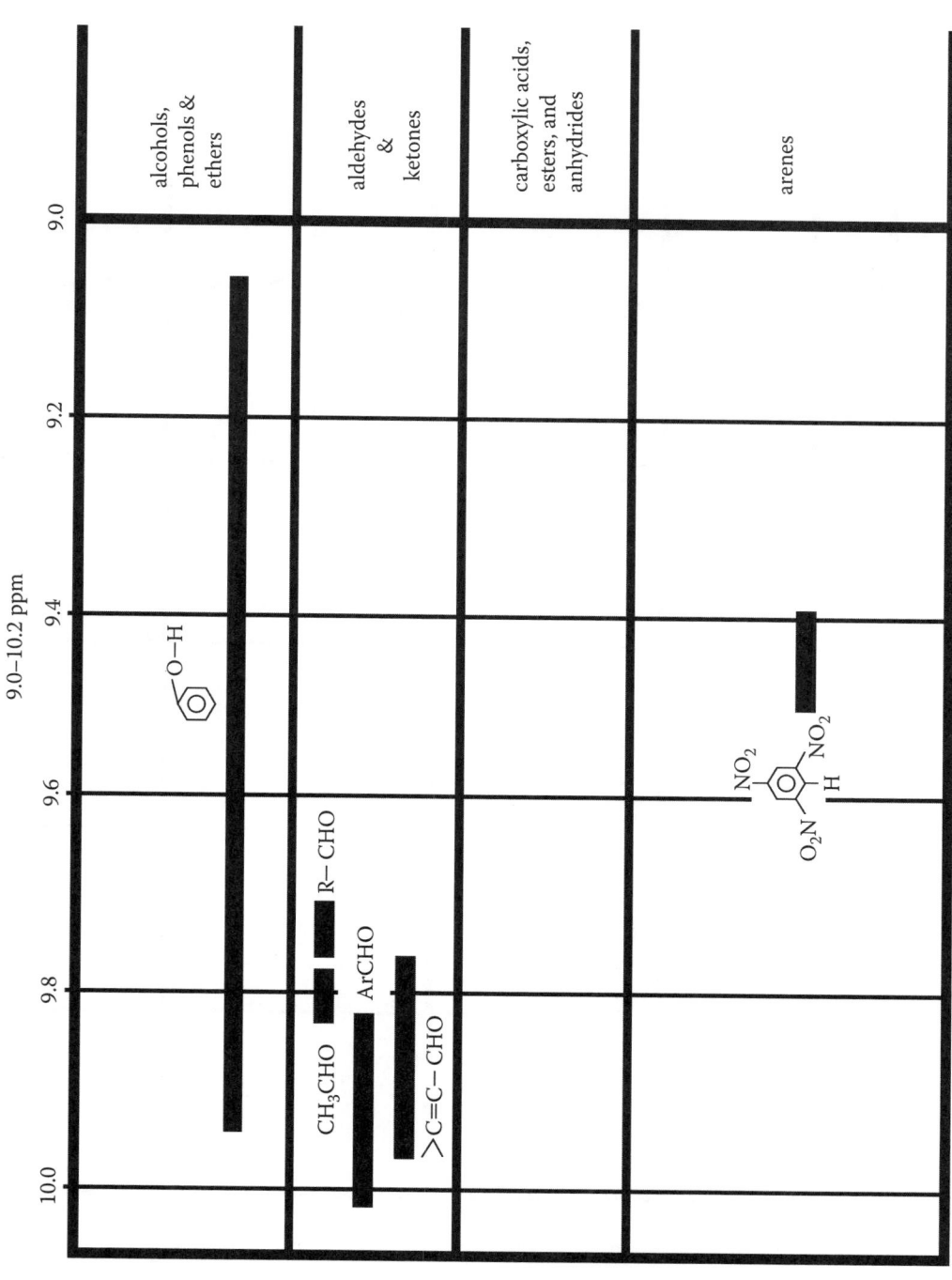

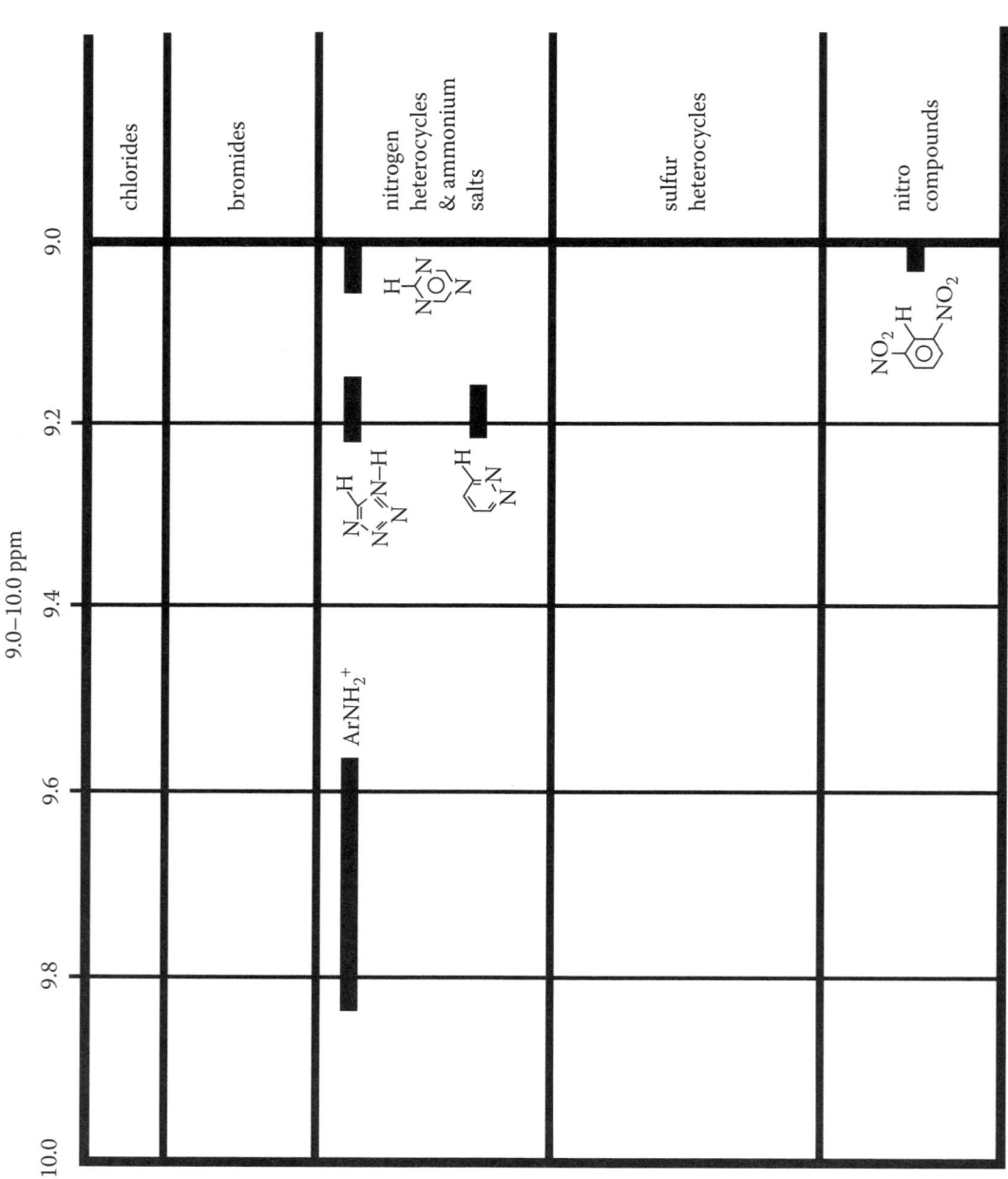

SOME USEFUL ¹H COUPLING CONSTANTS

This section gives the values of some useful proton NMR coupling constants (in Hz). The data are adapted with permission from the work of Dr. C. F. Hammer, Professor Emeritus, Chemistry Department, Georgetown University, Washington, DC 20057. The single numbers indicate a typical average, while in some cases, the range is provided.

1. Freely rotating chains.

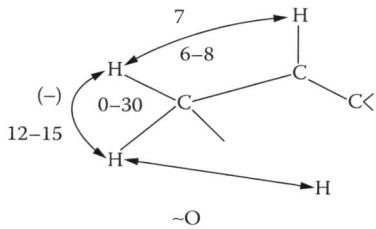

2. Alcohols with no exchange as in DMSO.
   1° = triplet
   2° = doublet (broad)
   3° = singlet
Upon addition of TFA, a sharp singlet results.

3. Alkenes

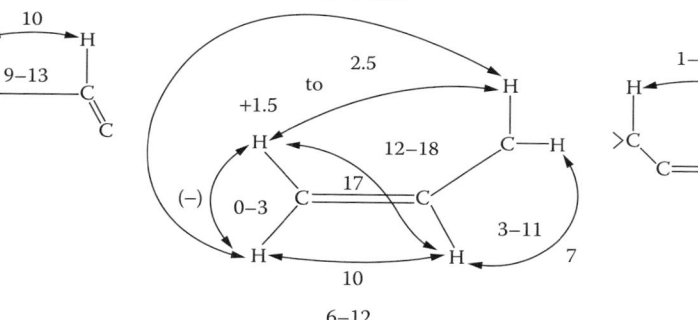

4. Alkynes

5. Aldehydes

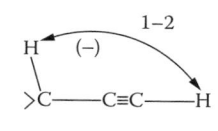

6. Aromatic

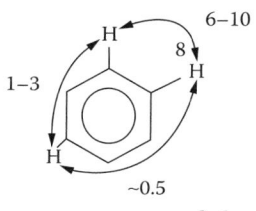

¹³C NMR ABSORPTION OF MAJOR FUNCTIONAL GROUPS

The table below lists the ^{13}C chemical shift ranges (in ppm) with the corresponding functional groups in descending order. A number of typical simple compounds for every family are given to illustrate the corresponding range. The shifts for the carbons of interest are given in parenthesis, either for each carbon as it appears from left to right in the formula, or by the underscore [1–15]. Following the table, correlation charts depicting the ^{13}C chemical shift ranges of various functional groups are presented. The expected peaks attributed to common solvents also appear in the correlation charts.

We provide a list of references that contain many of the ^{13}C chemical shift ranges that appear below [16–73]. This list is certainly not complete and should be regularly updated. Reference 44 has a compilation of references for the various nuclei.

REFERENCES

1. Yoder, C. H., and C. D. Schaeffer, Jr. *Introduction to Multinuclear NMR: Theory and Application.* Menlo Park, CA: The Benjamin/Cummings Publishing Co., 1987.
2. Brown, D. W. "A Short Set of ^{13}C-NMR Correlation Tables." *Journal of Chemical Education* 62 (1985): 209.
3. Silverstein, R. M., and F. X. Webster. *Spectrometric Identification of Organic Compounds.* 6th ed. New York: John Wiley and Sons, 1998.
4. Becker, E. D. *High Resolution NMR, Theory and Chemical Applications.* 2nd ed. New York: Academic Press, 1980.
5. Gunther, H. *NMR Spectroscopy: Basic Principles, Concepts and Applications in Chemistry.* New York: John Wiley and Sons, 2003.
6. Kitamaru, R. *Nuclear Magnetic Resonance: Principles and Theory.* London: Elsevier Science, 1990.
7. Lambert, J. B., L. N. Holland, and E. P. Mazzola. *Nuclear Magnetic Resonance Spectroscopy: Introduction to Principles, Applications and Experimental Methods.* Englewood Cliffs, NJ: Prentice Hall, 2003.
8. Bovey, F. A., and P. A. Mirau. *Nuclear Magnetic Resonance Spectroscopy.* 2nd ed. London: Academic Press, 1988.
9. Harris, R. K., and B. E. Mann. *NMR and the Periodic Table.* London: Academic Press, 1978.
10. Hore, P. J. *Nuclear Magnetic Resonance.* Oxford: Oxford University Press, 1995.
11. Nelson, J. H. *Nuclear Magnetic Resonance Spectroscopy.* 2nd ed. New York: John Wiley and Sons, 2003.
12. Levy, G. C., R. L. Lichter, and G. L. Nelson. *Carbon-13 Nuclear Magnetic Resonance Spectroscopy.* 2nd ed. New York: John Wiley and Sons, 1980.
13. Pihlaja, K., and E. Kleinpeter. *Carbon-13 NMR Chemical Shifts in Structural and Stereochemical Analysis.* New York: VCH, 1994.
14. Pouchert, C. *Aldrich Library of ^{1}H and ^{13}C FT-NMR Spectra.*, Milwaukee, WI: Aldrich Chemical Company, 1996.
15. Bruno, T. J., and P. D. N. Svoronos. *Handbook of Basic Tables for Chemical Analysis.* 2nd ed. Boca Raton, FL: CRC Press, 2003.
16. Balci, M. *Basic 1H- and 13C-NMR Spectroscopy.* London: Elsevier, 2005.
17. ^{13}C NMR chemical shifts, accessed July 2010, http://www.chem.wisc.edu/areas/reich/handouts/nmr-c13/cdata.htm, 2009.

The following bibliography contains information on the NMR absorptions of specific chemical families. This collection is by no means complete and should be updated regularly.

Adamantanes
18. Maciel, G. E., H. C. Dorn, R. L. Greene, W. A. Kleschick, M. R. Peterson, Jr., and G. H. Wahl, Jr. "^{13}C Chemical Shifts of Monosubstituted Adamantanes." *Organic Magnetic Resonance* 6 (1974): 178.

Amides
19. Jones, R. G., and J. M. Wilkins. "Carbon-13 NMR Spectra of a Series of Parasubstituted N,N-Dimethylbenzamides." *Organic Magnetic Resonance* 11 (1978): 20.

Benzazoles

20. Sohr, P., G. Manyai, K. Hideg, H. Hankovszky, and L. Lex. "Benzazoles. XIII. Determination of the E and Z Configuration of Isomeric 2-(2-Benzimidazolyl)-Di- and Tetra-Hydrothiophenes by IR, ^{1}H and ^{13}C NMR Spectroscopy." *Organic Magnetic Resonance* 14 (1980): 125.

Carbazoles

21. Giraud, J., and C. Marzin. "Comparative ^{13}C NMR Study of Deuterated and Undeuterated Dibenzothiophenes, Dibenzofurans, Carbazoles, Fluorenes, and Fluorenones." *Organic Magnetic Resonance* 12 (1979): 647.

Chlorinated Compounds

22. Hawkes, G. E., R. A. Smith, and J. D. Roberts. "Nuclear Magnetic Resonance Spectroscopy. Carbon-13 Chemical Shifts of Chlorinated Organic Compounds." *Journal of Organic Chemistry* 39 (1974): 1276.
23. Mark, V., and E. D. Weil. "The Isomerization and Chlorination of Decachlorobi-2,4-Cyclopentadien-1-yl." *Journal of Organic Chemistry* 36 (1971): 676.

Diazoles and Diazines

24. Faure, R., E. J. Vincent, G. Assef, J. Kister, and J. Metzger. "Carbon-13 NMR Study of Substituent Effects in the 1,3-Diazole and -Diazine Series." *Organic Magnetic Resonance* 9 (1977): 688.

Disulfides

25. Takata, T., K. Iida, and S. Oae. "^{13}C-NMR Chemical Shifts and Coupling Constants J_{C-H} of Six Membered Ring Systems Containing Sulfur–Sulfur Linkage." *Heterocycles* 15 (1981): 847.
26. Bass, S. W. and S. A. Evans, Jr. "Carbon-13 Nuclear Magnetic Resonance Spectral Properties of Alkyl Disulfides, Thiosulfinates, and Thiosulfonates." *Journal of Organic Chemistry* 45 (1980): 710.
27. Freeman, F., and C. N. Angeletakis. "Carbon-13 Nuclear Magnetic Resonance Study of the Conformations of Disulfides and Their Oxide Derivatives." *Journal of Organic Chemistry* 47 (1982): 4194.

Fluorenes and Fluorenones

28. Giraud, J., and C. Marzin. "Comparative 13C NMR Study of Deuterated and Undeuterated Dibenzothiophenes, Dibenzofurans, Carbazoles, Fluorenes and Fluorenones." *Organic Magnetic Resonance* 12 (1979): 647.

Furans

29. Giraud, H., and C. Marzin. "Comparative ^{13}C NMR Study of Deuterated and Undeuterated Dibenzothiophenes, Dibenzofurans, Carbazoles, Fluorenes and Fluorenones." *Organic Magnetic Resonance* 12 (1979): 647.

Imines

30. Allen, M., and J. D. Roberts. "Effects of Protonation and Hydrogen Bonding on Carbon-13 Chemical Shifts of Compounds Containing the>C=N-Group." *Canadian Journal of Chemistry* 59 (1981): 451.

Oxathianes

31. Szarek, W. A., D. M. Vyas, A. M. Sepulchre, S. D. Gero, and G. Lukacs. "Carbon-13 Nuclear Magnetic Resonance Spectra of 1,4-Oxathiane Derivatives." *Canadian Journal of Chemistry* 52 (1974): 2041.
32. Murray, W. T., J. W. Kelly, and S. A. Evans, Jr. "Synthesis of Substituted 1,4-Oxathianes, Mechanistic Details of Diethoxytriphenylphosphorane—and Triphenylphosphine/Tetra-Chloromethane—Promoted Cyclodehydrations and ^{13}C NMR Spectroscopy." *Journal of Organic Chemistry* 52 (1987): 525.

Oximes

33. Allen, M., and J. D. Roberts. "Effects of Protonation and Hydrogen Bonding on Carbon-13 Chemical Shifts of Compounds Containing the>C=N-Group." *Canadian Journal of Chemistry* 59 (1981): 451.

Polynuclear Aromatics (naphthalenes, anthracenes, pyrenes)

34. Adcock, W., M. Aurangzeb, W. Kitching, N. Smith, and D. Doddzell. "Substituent Effects of Carbon-13 Nuclear Magnetic Resonance: Concerning the π-Inductive Effect." *Australian Journal of Chemistry* 27 (1974): 1817.

35. DuVernet, R., and V. Boekelheide. "Nuclear Magnetic Resonance Spectroscopy. Ring-Current Effects on Carbon-13 Chemical Shifts." *Proceedings of the National Academy of Sciences* 71 (1974): 2961.

Pyrazoles

36. Puar, M. S., G. C. Rovnyak, A. I. Cohen, B. Toeplitz, and J. Z. Gougoutas. "Orientation of the Sulfoxide Bond as a Stereochemical Probe. Synthesis and ^{1}H and ^{13}C NMR of Substituted Thiopyrano[4,3-c] Pyrazoles." *Journal of Organic Chemistry* 44 (1979): 2513.

Sulfides

37. Chauhan, M. S., and I. W. J. Still. "^{13}C Nuclear Magnetic Resonance Spectra of Organic Sulfur Compounds: Cyclic Sulfides, Sulfoxides, Sulfones, and Thiones." *Canadian Journal of Chemistry* 53 (1975): 2880.

38. Gokel, G. W., H. M. Gerdes, and D. M. Dishong. "Sulfur Heterocycles. 3. Heterogenous, Phase-Transfer, and Acid Catalyzed Potassium Permanganate Oxidation of Sulfides to Sulfones and a Survey of Their Carbon-13 Nuclear Magnetic Resonance Spectra." *Journal of Organic Chemistry* 45 (1980): 3634.

39. Mohraz, M., W. Jiam-qi, E. Heilbronner, A. Solladie-Cavallo, and F. Matloubi-Moghadam. "Some Comments on the Conformation of Methyl Phenyl Sulfides, Sulfoxides, and Sulfones." *Helvetica Chimica Acta* 64 (1981): 97.

40. Srinivasan, C., S. Perumal, N. Arumugam, and R. Murugan. "Linear Free-Energy Relationship in Naphthalene System-Substituent Effects on Carbon-13 Chemical Shifts of Substituted Naphthylmethyl Sulfides." *Indian Journal of Chemistry* 25A (1986): 227.

Sulfites

41. Buchanan, G. W., C. M. E. Cousineau, and T. C. Mundell. "Trimethylene Sulfite Conformations: Effects of Sterically Demanding Substituents at C-4,6 on Ring Geometry as Assessed by ^{1}H and ^{13}C Nuclear Magnetic Resonance." *Canadian Journal of Chemistry* 56 (1978): 2019.

Sulfonamides

42. Chang, C., H. G. Floss, and G. E. Peck. "Carbon-13 Magnetic Resonance Spectroscopy of Drugs. Sulfonamides." *Journal of Medicinal Chemistry* 18 (1975): 505.

Sulfones (see also other families for the corresponding sulfones)

43. Fawcett, A. H., K. J. Ivin, and C. D. Stewart. "Carbon-13 NMR Spectra of Monosulphones and Disulphones: Substitution Rules and Conformational Effects." *Organic Magnetic Resonance* 11 (1978): 360.

44. Gokel, G. W., H. M. Gerdes, and D. M. Dishong. "Sulfur Heterocycles. 3. Heterogeneous, Phase-Transfer, and Acid Catalyzed Potassium Permanganate Oxidation of Sulfides to Sulfones and a Survey of Their Carbon-13 Nuclear Magnetic Resonance Spectra." *Journal of Organic Chemistry* 45 (1980): 3634.

45. Balaji, T., and D. B. Reddy. "Carbon-13 Nuclear Magnetic Resonance Spectra of Some New Arylcycfloproyl Sulphones." *Indian Journal of Chemistry* 18B (1979): 454.

Sulfoxides (see also other families for the corresponding sulfoxides)

46. Gatti, G., A. Levi, V. Lucchini, G. Modena, and G. Scorrano. "Site of Protonation in Sulphoxides: Carbon-13 Nuclear Magnetic Resonance Evidence." *Journal of the Chemical Society: Chemical Communications* 251 (1973).

47. Harrison, C. R., and P. Hodge. "Determination of the Configuration of Some Penicillin S-Oxides by ^{13}C Nuclear Magnetic Resonance Spectroscopy." *Journal of the Chemical Society, Perkin Transactions* I, 1772 (1976).

Sulfur ylides

48. Matsuyama, H., H. Minato, and M. Kobayashi. "Electrophilic Sulfides (II) as a Novel Catalyst. V. Structure, Nucleophilicity, and Steric Compression of Stabilized Sulfur Ylides as Observed by ^{13}C-NMR Spectroscopy." *Bulletin of the Chemical Society of Japan* 50 (1977): 3393.

Thianes

49. Willer, R. L., and E. L. Eliel. "Conformational Analysis. 34. Carbon-13 Nuclear Magnetic Resonance Spectra of Saturated Heterocycles, 6. Methylthianes." *Journal of the American Chemical Society* 99 (1977): 1925.

50. Barbarella, G., P. Dembech, A. Garbesi, and A. Fara. ^{13}C NMR of Organosulphur Compounds: II. ^{13}C Chemical Shifts and Conformational Analysis of Methyl Substituted Thiacyclohexanes." *Organic Magnetic Resonance* 8 (1976): 469.

51. Murray, W. T., J. W. Kelly, and S. A. Evans, Jr. "Synthesis of Substituted 1,4-Oxathianes. Mechanistic Details of Diethoxytriphenyl Phosphorane and Triphenylphosphine/Tetrachloromethane - Promoted Cyclodehydrations and ^{13}C NMR Spectroscopy." *Journal of Organic Chemistry* 52 (1987): 525.

52. Block, E., A. A. Bazzi, J. B. Lambert, S. M. Wharry, K. K. Andersen, D. C. Dittmer, B. H. Patwardhan, and J. H. Smith. "Carbon-13 and Oxygen-17 Nuclear Magnetic Resonance Studies of Organosulfur Compounds: The Four-Membered-Ring-Sulfone Effect." *Journal of Organic Chemistry* 45 (1980): 4807.

53. Rooney, R. P., and S. A. Evans, Jr. "Carbon-13 Nuclear Magnetic Resonance Spectra of Trans-1-Thiadecalin, Trans-1,4-Dithiadecalin, Trans-1,4-Oxathiadecalin, and the Corresponding Sulfoxides and Sulfones." *Journal of Organic Chemistry* 45 (1980): 180.

Thiazines

54. Fronza, G., R. Mondelli, G. Scapini, G. Ronsisvalle, and F. Vittorio. "^{13}C NMR of N-Heterocycles. Conformation of Phenothiazines and 2,3-Diazaphenothiazines." *Journal of Magnetic Resonance* 23 (1976): 437.

Thiazoles

55. Harrison, C. R., and P. Hodge. "Determination of the Configuration of Some Penicillin S-Oxides by ^{13}C Nuclear Magnetic Resonance Spectroscopy." *Journal of the Chemical Society, Perkin Transactions* I (1976): 1772.

56. Chang, G., H. G. Floss, and G. E. Peck. "Carbon-13 Magnetic Resonance Spectroscopy of Drugs. Sulfonamides." *Journal of Medicinal Chemistry* 18 (1975): 505.

57. Elguero, J., R. Faure, R. Lazaro, and E. J. Vincent. "^{13}C NMR Study of Benzothiazole and Its Nitroderivatives." *Bulletin de la Société Chimique de Belgium* 86 (1977): 95.

58. Faure, R., J. P. Galy, E. J. Vincent, and J. Elguero. "Study of Polyheteroaromatic Pentagonal Heterocycles by Carbon-13 NMR. Thiazoles and Thiazolo[2,3-e]Tetrazoles." *Canadian Journal of Chemistry* 56 (1978): 46.

Thiochromanones

59. Chauhan, M. S., and I. W. J. Still. "^{13}C Nuclear Magnetic Resonance Spectra of Organic Sulfur Compounds: Cyclic Sulfides, Sulfoxides, Sulfones and Thiones." *Canadian Journal of Chemistry* 53 (1975): 2880.

Thiones

60. Chauhan, M. S., and I. W. J. Still. "^{13}C Nuclear Magnetic Resonance Spectra of Organic Sulfur Compounds: Cyclic Sulfides, Sulfoxides, Sulfones and Thiones." *Canadian Journal of Chemistry* 53 (1975): 2880.

Thiophenes

61. Perjessy, A., M. Janda, and D. W. Boykin. "Transmission of Substituent Effects in Thiophenes. Infrared and Carbon-13 Nuclear Magnetic Resonance Studies." *Journal of Organic Chemistry* 45 (1980): 1366.

62. Giraud, J., and C. Marzin. "Comparative ^{13}C NMR Study of Deuterated and Undeuterated Dibenzothiophenes, Dibenzofurans, Carbazoles, Fluorenes and Flourenones." *Organic Magnetic Resonance* 12 (1979): 647.

63. Clark, P. D., D. F. Ewing, and R. M. Scrowston. "NMR Studies of Sulfur Heterocycles: III. [13]C Spectra of Benzo[b]Thiophene and the Methylbenzo[b]Thiophenes." *Organic Magnetic Resonance* 8 (1976): 252.

64. Osamura, Y., O. Sayanagi, and K. Nishimoto. "C-13 NMR Chemical Shifts and Charge Densities of Substituted Thiophenes—The Effect of Vacant dπ Orbitals." *Bulletin of the Chemical Society of Japan* 49 (1976): 845.

65. Balkau, F., M. W. Fuller, and M. L. Heffernan. "Deceptive Simplicity in ABMX N.M.R. Spectra. I. Dibenzothiophen and 9.9'-Cicarbazyl." *Australian Journal of Chemistry* 24 (1971): 2293.

66. Geneste, P., J. L. Olive, S. N. Ung, M. E. A. El Faghi, J. W. Easton, H. Beierbeck, and J. K. Saunders. "Carbon-13 Nuclear Magnetic Resonance Study of Benzo[b]Thiophenes and Benzo[b]Thiophene S-Oxides and S,S-Dioxides." *Journal of Organic Chemistry* 44 (1979): 2887.

67. Benassi, R., U. Folli, D. Iarossi, L. Schenetti, and F. Tadei. "Conformational Analysis of Organic Carbonyl Compounds. Part 3. A ^{1}H and [13]C Nuclear Magnetic Resonance Study of Formyl and Acetyl Derivatives of Benzo[b]Thiophen." *Journal of the Chemical Society, Perkin Transactions II* 911 (1983).

68. Kiezel, L., M. Liszka, and M. Rutkowski. "Carbon-13 Magnetic Resonance Spectra of Benzothiophene and Dibenzothiophene." *Spectroscopy Letters* 12 (1979): 45.

69. Fujieda, K., K. Takahashi, and T. Sone. "The C-13 NMR Spectra of Thiophenes. II, 2-Substituted Thiophenes." *Bulletin of the Chemical Society of Japan* 58, (1985): 1587.

70. Satonaka, H., and M. Watanabe. "NMR Spectra of 2-(2-Nitrovinyl) Thiophenes." *Bulletin of the Chemical Society of Japan* 58, (1985): 3651.

71. Stuart, J. G., M. J. Quast, G. E. Martin, V. M. Lynch, H. Simmonsen, M. L. Lee, R. N. Castle, J. L. Dallas, B. K. John, and L. R. F. Johnson. "Benzannelated Analogs of Phenanthro [1,2-b]-[2,1-b]Thiophene: Synthesis and Structural Characterization by Two-Dimensional NMR and X-Ray Techniques." *Journal of Heterocyclic Chemistry* 23 (1986): 1215.

Thiopyrans

72. Senda, Y., A. Kasahara, T. Izumi, and T. Takeda. "Carbon-13 NMR Spectra of 4-Chromanone, 4H-1-Benzothiopyran-4-One, 4H-1-Benzothiopyran-4-One 1,1-Dioxide, and Their Substituted Homologs." *Bulletin of the Chemical Society of Japan* 50 (1977): 2789.

Thiosulfinates and Thiosulfonates

73. Bass, S. W., and S. A. Evans, Jr. "Carbon-13 Nuclear Magnetic Resonance Spectral Properties of Alkyl Disulfides, Thiosulfinates, and Thiosulfonates." *Journal of Organic Chemistry* 45 (1980): 710.

^{13}C NMR Chemical Shift Ranges of Major Functional Groups

δ (ppm)	Group	Family	Example (δ of underlined carbon)	
220–165	>C=O	ketones	$(CH_3)_2\underline{C}O$	(206.0)
			$(CH_3)_2CH\underline{C}OCH_3$	(212.1)
		aldehydes	$CH_3\underline{C}HO$	(199.7)
		α,β–unsaturated carbonyls	$CH_3CH=CH\underline{C}HO$	(192.4)
			$CH_2=CH\underline{C}OCH_3$	(169.9)
		carboxylic acids	$H\underline{C}O_2H$	(166.0)
			$CH_3\underline{C}O_2H$	(178.1)
		amides	$H\underline{C}ONH_2$	(165.0)
			$CH_3\underline{C}ONH_2$	(172.7)
		esters	$CH_3\underline{C}O_2CH_2CH_3$	(170.3)
			$CH_2=CH\underline{C}O_2CH_3$	(165.5)
140–120	>C=C<	aromatic	C_6H_6	(128.5)
		alkenes	$CH_2=CH_2$	(123.2)
			$CH_2=\underline{C}HCH_3$	(115.9, 136.2)
			$CH_2=\underline{C}HCH_2Cl$	(117.5, 133.7)
			$CH_3CH=\underline{C}HCH_2CH_3$	(132.7)
125–115	–C≡N	nitriles	$CH_3-\underline{C}\equiv N$	(117.7)
80–70	–C≡C–	alkynes	$H\underline{C}\equiv CH$	(71.9)
			$CH_3\underline{C}\equiv CCH_3$	(73.9)
70–45	>C–O⁻	esters	$\underline{C}H_3OOCCH_2CH_3$	(57.6, 67.9)
	│	alcohols	$HO\underline{C}H_3$	(49.0)
			$HO\underline{C}H_2CH_3$	(57.0)
40–20	>C–NH₂	amines	$\underline{C}H_3NH_2$	(26.9)
	│		$CH_3\underline{C}H_2NH_2$	(35.9)
30–15	–S–CH₃	sulfides (thioethers)	$C_6H_5-S-\underline{C}H_3$	15.6
30–(–2.3)	>CH–	alkanes, cycloalkanes	$\underline{C}H_4$	(–2.3)
			$\underline{C}H_3CH_3$	(5.7)
			$\underline{C}H_3\underline{C}H_2CH_3$	(15.8, 16.3)
			$\underline{C}H_3\underline{C}H_2CH_2CH_3$	(13.4, 25.2)
			$\underline{C}H_3\underline{C}H_2\underline{C}H_2CH_2CH_3$	(13.9, 22.8, 34.7)
			cyclohexane	(26.9)

^{13}C NMR CORRELATION CHARTS OF MAJOR FUNCTIONAL GROUPS

The following correlation tables provide the regions of nuclear magnetic resonance absorptions of major chemical families.

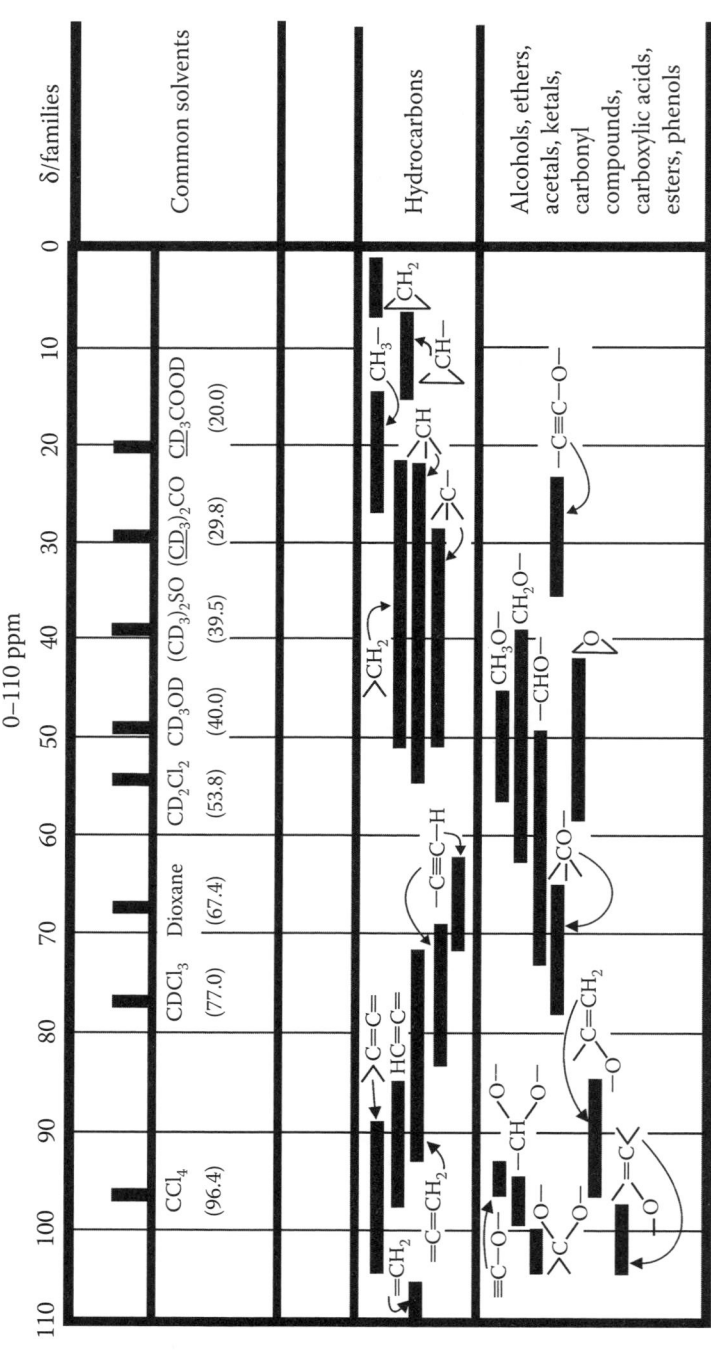

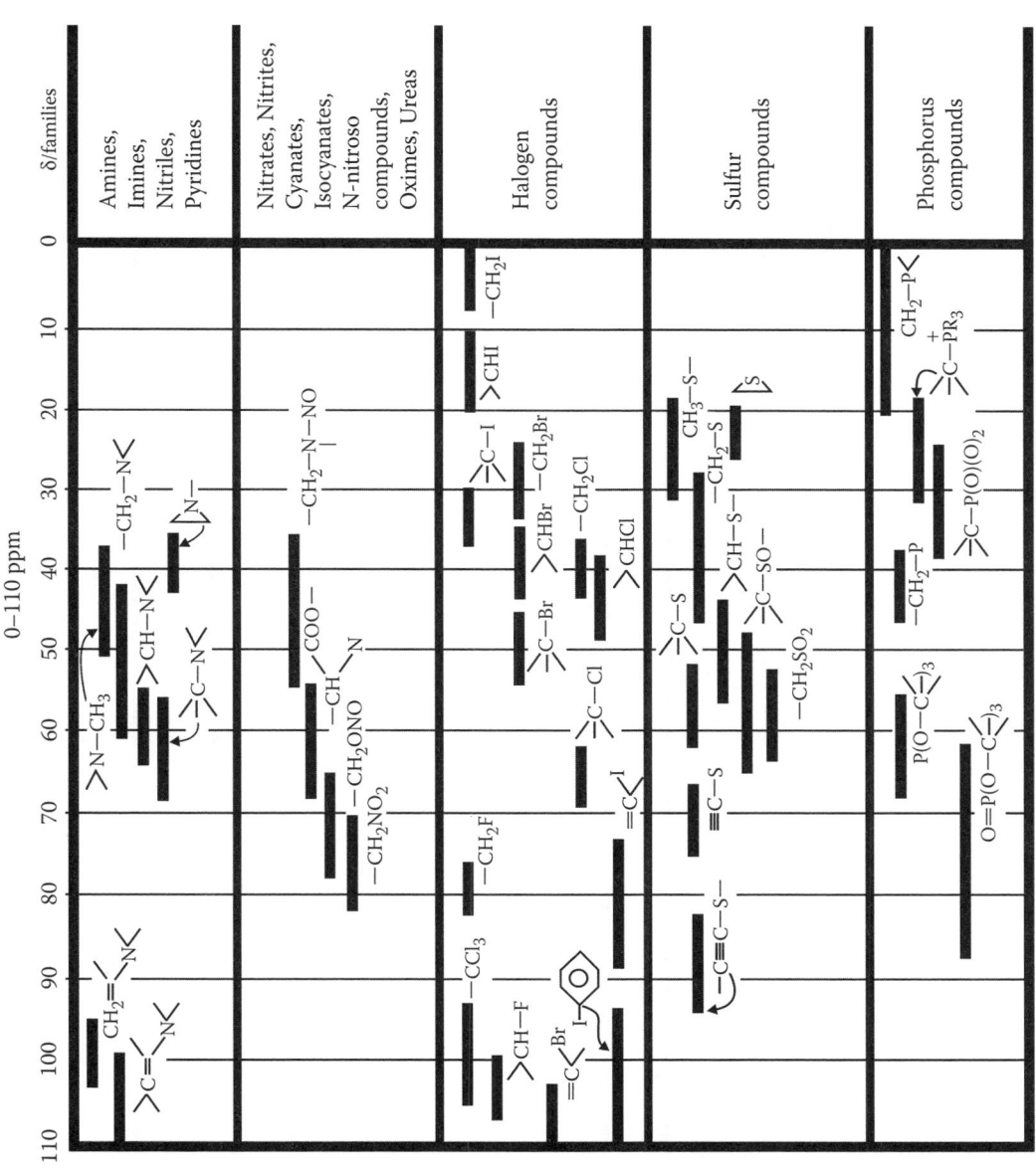

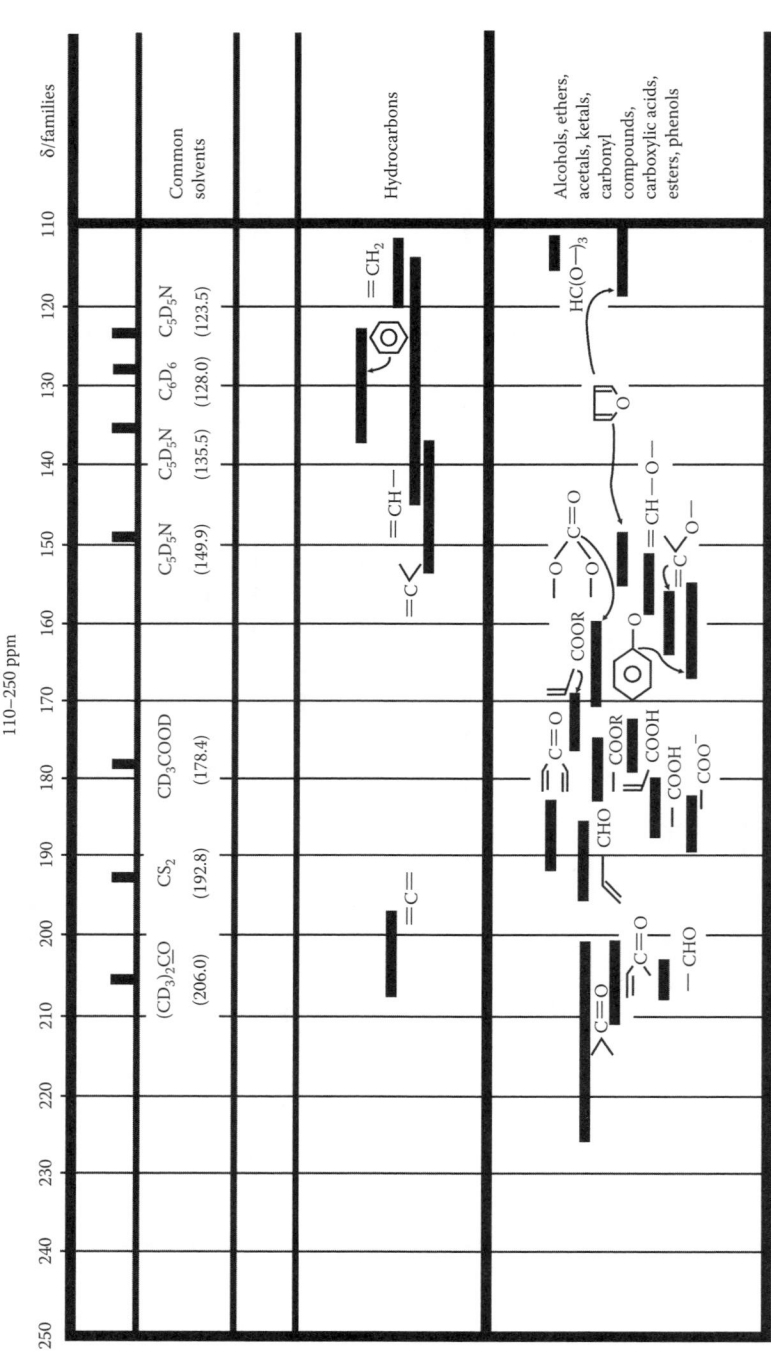

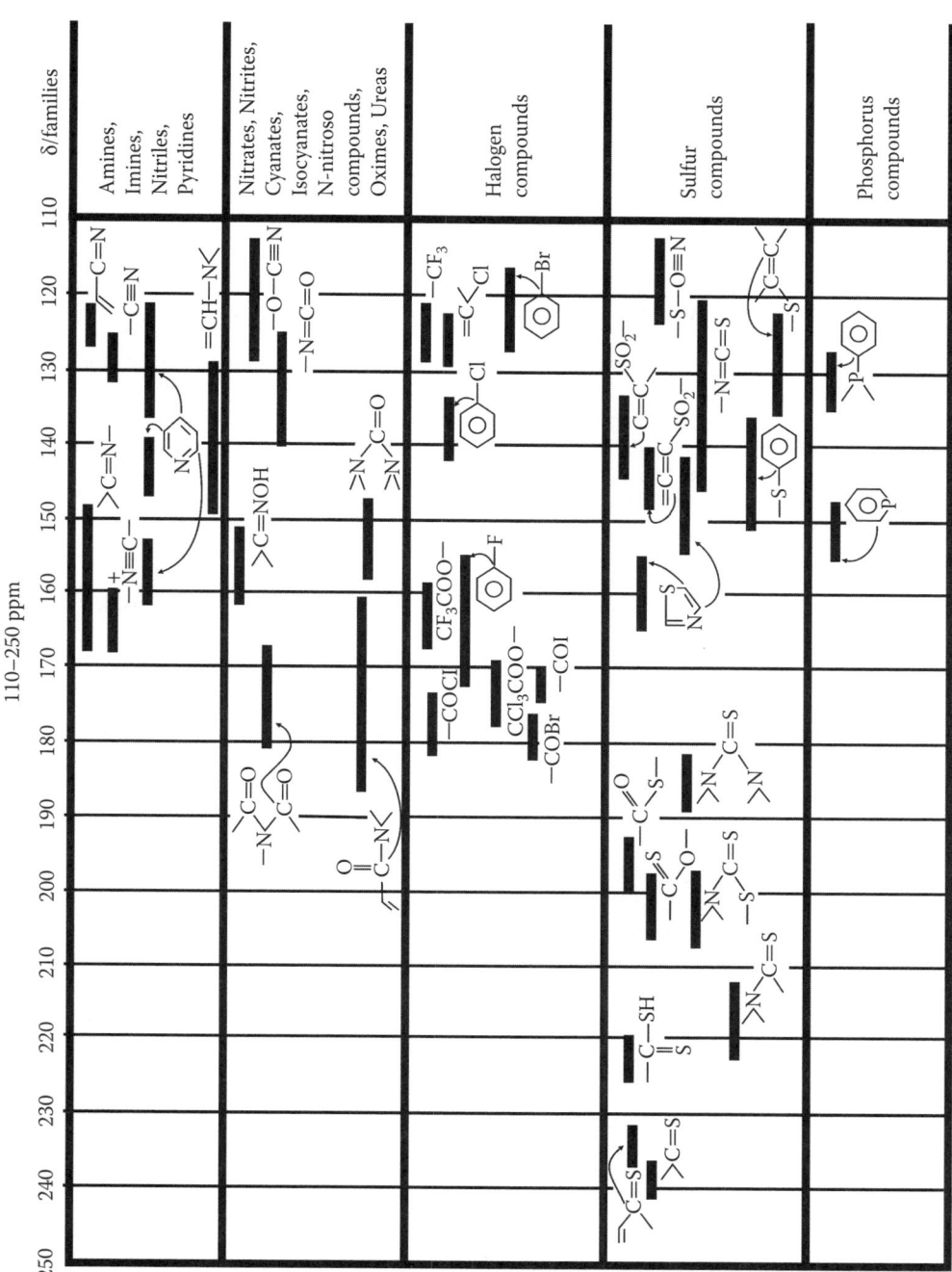

ADDITIVITY RULES IN ^{13}C NMR CORRELATION TABLES

The wide chemical shift range (~250 ppm) of ^{13}C-NMR is responsible for the considerable change of a chemical shift noted when a slight inductive, mesomeric, or hybridization change occurs on a neighboring atom. Following the various empirical correlations in ^{1}H NMR [1–7], D. W. Brown [8] has developed a short set of ^{13}C-NMR correlation tables. This section covers a part of those as adopted by Yoder and Schaeffer [9] and Clerk et al. [10]. The reader is advised to refer to Reference [8], and should the need for some specific data on more complicated structures arise, additional sources are provided [11–19].

REFERENCES

1. Shoolery, J. N. *Varian Associates Technical Information Bulletin.* Vol. 2, No. 3. Palo Alto: Varian Associates, 1959.
2. Bell, H. M., D. B. Bowles, and F. Senese. "Additive NMR Chemical Shift Parameters for Deshielded Methine Protons." *Organic Magnetic Resonance* 16 (1981): 285.
3. Matter, U. E., C. Pascual, E. Pretsch, A. Pross, W. Simon, and S. Sternhell. "Estimation of the Chemical Shifts of Olefinic Protons Using Additive Increments. II. Compilation of Additive Increments for 43 Functional Groups." *Tetrahedron* 25 (1969): 691.
4. Matter, U. E., C. Pascual, E. Pretsch, A. Pross, W. Simon, and S. Sternhell. "Estimation of the Chemical Shifts of Olefinic Protons Using Additive Increments. III. Examples of Utility in N.M.R. Studies and the Identification of Some Structural Features Responsible for Deviations from Additivity." *Tetrahedron* 25 (1969): 2023.
5. Jeffreys, J. A. D. "A Rapid Method for Estimating NMR Shifts for Protons Attached to Carbon." *Journal of Chemical Education* 56 (1979): 806.
6. Mikolajczyk, M., S. Grzeijszczak, and Z. Zatorski. "Organosulfur Compounds IX: NMR and Structural Assignments in α,β-Unsaturated Sulphoxides Using Additive Increments Method." *Tetrahedron* 32 (1976): 969.
7. Friedrich, E. C., and K. G. Runkle. "Empirical NMR Chemical Shift Correlations for Methyl and Methylene Protons." *Journal of Chemical Education* 61 (1984): 830.
8. Brown, D. W. "A Short Set of ^{13}C-NMR Correlation Tables." *Journal of Chemical Education* 62 (1985): 209.
9. Yoder, C. H., and C. D. Schaeffer, Jr. *Introduction to Multinuclear NMR.* Menlo Park, CA: Benjamin/ Cummings Publishing Co., 1987.
10. Clerk, J. T., E. Pretsch, and J. Seibl. *Structural Analysis of Organic Compounds by Combined Application of Spectroscopic Methods.* Amsterdam: Elsevier, 1981.
11. Silverstein, R. M., and F. X. Webster. *Spectrometric Identification of Organic Compounds.* 6th ed. New York: John Wiley and Sons, 1998.
12. Gunther, H. *NMR Spectroscopy: Basic Principles, Concepts and Applications in Chemistry.* New York: John Wiley and Sons, 2003.
13. Kitamaru, R. *Nuclear Magnetic Resonance: Principles and Theory.* London: Elsevier Science, 1990.
14. Lambert, J. B., L. N. Holland, and E. P. Mazzola. *Nuclear Magnetic Resonance Spectroscopy: Introduction to Principles, Applications and Experimental Methods.* Englewood Cliffs: Prentice Hall, 2003.
15. Bovey, F. A., and P. A. Mirau. *Nuclear Magnetic Resonance Spectroscopy.* 2nd ed. London: Academic Press, 1988.
16. Harris, R. K., and B. E. Mann. *NMR and the Periodic Table.* London: Academic Press, 1978.
17. Nelson, J. H. *Nuclear Magnetic Resonance Spectroscopy.* 2nd ed. New York: John Wiley and Sons, 2003.
18. Bruno, T. J., and P. D. N. Svoronos. *Handbook of Basic Tables for Chemical Analysis.* 2nd ed. Boca Raton, FL: CRC Press, 2003.
19. Bruno, T. J., and P. D. N. Svoronos. *CRC Handbook of Fundamental Spectroscopic Correlation Charts.* Boca Raton, FL: CRC Press, 2006.

Alkanes

The chemical shift (in ppm) of C^i can be calculated from the following empirical equation

$$\Delta^i = -2.3 + \Sigma\, A_i,$$

where $\Sigma\, A_i$ is the sum of increments allowed for various substituents depending on their positions (α, β, γ, δ) relative to the ^{13}C in question, and (−2.3) is the chemical shift for methane relative to tetramethylsilane (TMS).

^{13}C Chemical Shift Increments for A, the Shielding Term for Alkanes and Substituted Alkanes [9,10]

Substituent	Increments			
	α	β	γ	δ
>C–(sp³)	9.1	9.4	−2.5	0.3
>C=C<(sp²)	19.5	6.9	−2.1	0.4
C≡C– (sp)	4.4	5.6	−3.4	−0.6
C₆H₅	22.1	9.3	−2.6	0.3
–F	70.1	7.8	−6.8	0.0
–Cl	31.0	10.0	−5.1	−0.5
–Br	18.9	11.0	−3.8	−0.7
–I	−7.2	10.9	−1.5	−0.9
–OH	49.0	10.1	−6.2	0.0
–OR	49.0	10.1	−6.2	0.0
–CHO	29.9	−0.6	−2.7	0.0
–COR	22.5	3.0	−3.0	0.0
–COOH	20.1	2.0	−2.8	0.0
–COO⁻	24.5	3.5	−2.5	0.0
–COCl	33.1	2.3	−3.6	0.0
–COOR	22.6	2.0	−2.8	0.0
–OOCR	5.5	6.5	−6.0	
–N<	28.3	11.3	−5.1	
–NH₃⁺	26.0	7.5	−4.6	0.0
[>N<]⁺	30.7	5.4	−7.2	−1.4
–ONO	54.3	6.1	−6.5	−0.5
–NO₂	61.6	3.1	−4.6	−1.0
–CON<	22.0	2.6	−3.2	−0.4
–NHCO–	31.3	8.3	−5.7	0.0
–C≡N	3.1	2.4	−3.3	−0.5
–NC	31.5	7.6	−3.0	0.0
–S–	10.6	11.4	−3.6	−0.4
–S–CO–	17.0	6.5	−3.1	0.0
–SO–	31.1	9.0	−3.5	0.0
–SO₂Cl	54.5	3.4	−3.0	0.0
–SCN	23.0	9.7	−3.0	0.0
–C(=S)N–	33.1	7.7	−2.5	0.6
–C=NOH(syn)	11.7	0.6	−1.8	0.0
–C=NOH(anti)	16.1	4.3	−1.5	0.0
R₁R₂R₃Sn (R₁, R₂, and R₃ = organic substituents)	−5.2	4.0	−0.3	0.0

Thus, the ^{13}C shift for C^i in 2-pentanol is predicted to be

$$\overset{\beta}{CH_3}-\overset{\alpha}{CH_2}-\overset{i}{CH_2}-\overset{\alpha'}{CH(OH)}-\overset{\beta'}{CH_3}$$

$$\delta^i = (-2.3) + [\ 9.1 + 9.4 + 9.1 + 9.4 + 10.1\ ] = 44.8\ ppm$$
$$\quad\quad\quad\quad\ \alpha\quad\ \beta\quad\ \alpha'\quad\ \beta'\quad OH$$

Alkenes

For a simple olefin of the type

$$\overset{\gamma}{-C}-\overset{\beta}{C}-\overset{\alpha}{C}-\overset{i}{C}=\overset{\alpha'}{C}-\overset{\beta'}{C}-\overset{\gamma'}{C}-C-$$

$$\delta^i = 122.8 + \Sigma\ A_i$$

where: $A_\alpha = 10.6$, $A_\beta = 7.2$, $A_\gamma = -1.5$, $A_{\alpha'} = -7.9$, $A_{\beta'} = -1.8$, $A_\gamma = 1.5$ and 122.8 is the chemical shift of the sp^2 carbon in ethene.

If the olefin is in the cis- configuration an increment of -1.1 ppm must be added.

Thus, the ^{13}C shift for C-3 in cis-3-hexene is predicted to be

$$\overset{\beta}{CH_3}-\overset{\alpha}{CH_2}-\overset{i}{CH}=\overset{}{CH}-\overset{\alpha'}{CH_2}-\overset{\beta'}{CH_3}$$

$$\delta^i = 122.8 + [\ 10.6 + 7.2 - 1.5 - 7.9\ ] + (-1.1) = 130.1\ ppm$$
$$\quad\quad\quad\quad\ (\alpha)\quad\ (\beta)\quad (\alpha')\ (\beta')\quad\ (cis)$$

Alkynes

For a simple alkyne of the type

$$\overset{\beta}{-C}-\overset{\alpha}{C}-\overset{i}{C}\equiv\overset{\alpha'}{C}-\overset{\beta'}{C}-C-$$

$$\delta^i = 71.9 + \Sigma A_i$$

where increments A are given in the table below and 71.9 is the chemical shift of the sp carbon in acetylene.[9]

^{13}C Chemical Shift Increments for A, the Shielding Term for Alkynes

Substituents	Increments			
	α	β	α'	β'
C (sp³)	6.9	4.8	−5.7	2.3
−CH₃	7.0		−5.7	
−CH₂CH₃	12.0		−3.5	
−CH(CH₃)₂	16.0		−3.5	
−CH₂OH	11.1		1.9	
−COCH₃	31.4		4.0	
−C₆H₅	12.7		6.4	
−CH=CH₂	10.0		11.0	
−Cl	−12.0		−15.0	

Thus, the ^{13}C shift for C-A in 1-phenyl propyne is predicted to be

$$C_6H_5-C\equiv C-CH_3$$
$$\overset{B}{}\overset{A}{}$$

$$\delta^i = 71.9 + 7.0 + 6.4 = 85.3 \text{ ppm}$$

while the ^{13}C shift for C-B in the same compound is predicted to be

$$C_6H_5-C\equiv C^i-CH_3$$
$$\overset{B}{}\overset{A}{}$$

$$\delta^i = 71.9 + 12.7 - 5.7 = 78.9 \text{ ppm}$$

Benzenoid Aromatics

For a benzene derivative , C_6H_5-X, where X = substituent

$$\delta^i = 128.5 + \Sigma A_i$$

where ΣA_i is the sum of increments given below and 128.5 is the chemical shift of benzene [9–10].

^{13}C Chemical Shift Increments for A, the Shielding Term for Benzenoid Aromatics X – C$_6$H$_5$ where X = substituent

Substituent X	C^i	ortho	meta	para
–CH$_3$	9.3	0.8[9], 0.6[10]	0.0	–2.9[9], –3.1[10]
–CH$_2$CH$_3$	15.8[9], 15.7[10]	–0.4[9], –0.6[10]	–0.1	–2.6[9], –2.8[10]
–CH(CH$_3$)$_2$	20.3[9], 20.1[10]	–1.9[9], –2.0[10]	0.1[9], 0.0[10]	–2.4[9], –2.5[10]
–C(CH$_3$)$_3$	22.4[9], 22.1[10]	–3.1[9], –3.4[10]	–0.2[9], 0.4[10]	–2.9[9], –3.1[10]
–CH=CH$_2$	7.6	–1.8	–1.8	–3.5
–C≡CH	–6.1	3.8	0.4	–0.2
–C$_6$H$_5$	13.0	–1.1	0.5	–1.0
–CHO	8.6[9], 9.0[10]	1.3[9], 1.2[10]	0.6[9], 1.2[10]	5.5[9], 6.0[10]
–COCH$_3$	9.1[9], 9.3[10]	0.1[9], 0.2[10]	0.0[9], 0.2[10]	4.2
–CO$_2$H	2.1[9], 2.4[10]	1.5[9], 1.6[10]	0.0[9], –0.1[10]	5.1[9], 4.8[10]
–CO$_2^-$	7.6	0.8	0.0	2.8
–CO$_2$R	2.1	1.2	0.0	4.4
–CONH$_2$	5.4	–0.3	–0.9	5.0
–CN	–15.4[9], –16.0[10]	3.6[9], 3.5[10]	0.6[9], 0.7[10]	3.9[9], 4.3[10]
–Cl	6.2[9], 6.4[10]	0.4[9], 0.2[10]	1.3[9], 10.0[10]	–1.9[9], –2.0[10]
–OH	26.9	–12.7	1.4	–7.3
–O–	39.6[10]	–8.2[10]	1.9[10]	–13.6[10]
–OCH$_3$	31.4[9], 30.2[10]	–14.4[9], –14.7[10]	1.0[9], 0.9[10]	–7.7[9], –8.1[10]
–OC$_6$H$_5$	29.1	–9.5	0.3	–5.3
–OC(=O)CH$_3$	23.0	–6.4	1.3	–2.3
–NH$_2$	18.7[9], 19.2[10]	–12.4	1.3	–9.5
–NHCH$_3$	21.7[10]	–16.2[10]	0.7[10]	–11.8[10]
–N(CH$_3$)$_2$	22.4	–15.7	0.8	–11.8
–NO$_2$	20.0[9], 19.6[10]	–4.8[9], –5.3[10]	0.9[9], 0.8[10]	5.8[9], 6.0[10]
–SH	2.2	0.7	0.4	–3.1
–SCH$_3$	9.9[10]	–2.0[10]	0.1[10]	–3.7[10]
–SO$_3$H	15.0	–2.2	1.3	3.8

As an example, the ^{13}C shift for the benzene carbon (C^i) carrying the carbonyl in 3,5-dinitroacetophenone, CH$_3$ C(=O)(C$_6$H$_3$)(NO$_2$)$_2$ is predicted to be

$$C^i = 128.5 + 9.1 + 2(0.9) = 132.4 \text{ ppm.}$$

¹⁵N CHEMICAL SHIFTS FOR COMMON STANDARDS

The following table lists the ¹⁵N chemical shifts (in ppm) for common standards. The estimated uncertainty is less than 0.1 ppm. Nitromethane (according to Levy and Lichter [1]) is the most suitable primary measurement reference, but has the disadvantage of lying in the low-field end of the spectrum. Thus, ammonia (which lies in the most upfield region) is the most suitable for routine experimental use [1–8].

REFERENCES

1. Levy, G. C., and R. L. Lichter. *Nitrogen-15 Nuclear Magnetic Resonance Spectroscopy*. New York: John Wiley and Sons, 1979.
2. Lambert, J. B., H. F. Shurvell, L. Verbit, R. G. Cooks, and G. H. Stout. *Organic Structural Analysis*. New York: MacMillan, 1976.
3. Witanowski, M., L. Stefaniak, S. Szymanski, and H. Januszewski. "External Neat Nitromethane Scale for Nitrogen Chemical Shifts." *Journal of Magnetic Resonance* 28 (1977): 217.
4. Srinivasan, P. R., and R. L. Lichter. "Nitrogen-15 Nuclear Magnetic Resonance Spectroscopy. Evaluation of Chemical Shift References." *Journal of Magnetic Resonance* 28 (1977): 227.
5. Briggs, J. M., and E. W. Randall. "Nitrogen-15 Chemical Shifts in Concentrated Aqueous Solutions of Ammonium Salts." *Molecular Physics* 26 (1973): 699.
6. Becker, E. D. "Proposed Scale for Nitrogen Chemical Shifts." *Journal of Magnetic Resonance* 4 (1971): 142.
7. Bruno, T. J., and P. D. N. Svoronos. *Handbook of Basic Tables for Chemical Analysis*. 2nd ed. Boca Raton, FL: CRC Press, 2003.
8. Bruno, T. J., and P. D. N. Svoronos. *CRC Handbook of Fundamental Spectroscopic Correlation Charts*. Boca Raton, FL: CRC Press, 2006.

Compound	Formula	Conditions	Chemical Shift (ppm)
Ammonia	NH_3	vapor (0.5 MPa)	−15.9
		liquid (25 °C), anhydrous	0.0
		liquid (−50 °C)	3.37
Ammonium nitrate	NH_4NO_3	aqueous HNO_3	21.60
		aqueous solution (saturated)	20.68
Ammonium chloride	NH_4Cl	2.9 M (in 1 M HCl)	24.93
		1.0 M (in 10 M HCl)	30.31
		aqueous solution (saturated)	27.34
Tetraethylammonium chloride	$(C_2H_5)_4N^+Cl^-$	aqueous solution (saturated)	43.54
		chloroform solution (saturated)	45.68
		aqueous solution (0.3 M)	63.94
		aqueous solution (saturated)	64.39
		chloroform solution (0.075 M)	65.69
Tetramethyl urea	$[(CH_3)_2N]_2CO$	neat	62.50
Dimethylformamide (DMF)	$(CH_3)_2NCHO$	neat	103.81
Nitric acid (aqueous solution)	HNO_3	1 M	375.80
		2 M	367.84
		9 M	365.86
		10 M	362.00
		15.7 M	348.92
Sodium nitrate	$NaNO_3$	aqueous solution (saturated)	376.53
Ammonium nitrate	NH_4NO_3	aqueous solution (saturated)	376.25
		5 M (in 2 M HNO_3)	375.59
		4 M (in 2 M HNO_3)	374.68
Nitromethane	CH_3NO_2	1:1 (v/v) in $CDCl_3$	379.60
		0.03 M $Cr(acac)_3$ neat	380.23

¹⁵N CHEMICAL SHIFTS OF MAJOR CHEMICAL FAMILIES

The following table contains ¹⁵N chemical shifts of various organic nitrogen compounds. Chemical shifts are expressed relative to different standards (NH_3, NH_4Cl, CH_3NO_2, NH_4NO_3, HNO_3, etc.) and are interconvertible. Chemical shifts are sensitive to hydrogen bonding and are solvent dependent as seen in the case of pyridine (see note b below). Consequently, the reference as well as the solvent should always accompany chemical shift data. No data are given on peptides and other biochemical compounds. All shifts are relative to ammonia unless otherwise specified. A section of "miscellaneous" data gives the chemical shift of special compounds relative to unusual standards [1–15].

REFERENCES

1. Levy, G. C., and R. L. Lichter. *Nitrogen-15 Nuclear Magnetic Resonance Spectroscopy.* New York: John Wiley and Sons, 1979.
2. Yoder, C. H., and C. D. Schaeffer, Jr. *Introduction to Multinuclear NMR.* Menlo Park, CA: Benjamin/Cummings, 1987.
3. Duthaler, R. O., and J. D. Roberts. "Effects of Solvent, Protonation, and N-Alkylation on the ¹⁵N Chemical Shifts of Pyridine and Related Compounds." *Journal of the American Chemical Society* 100 (1978): 4969.
4. Duthaler, R. O., and J. D. Roberts. "Steric and Electronic Effects on ¹⁵N Chemical Shifts of Saturated Aliphatic Amines and Their Hydrochlorides." *Journal of the American Chemical Society* 100 (1978): 3889.
5. Kozerski, L., and W. von Philipsborn. "¹⁵N Chemical Shifts as a Conformational Probe in Enaminones: A Variable Temperature Study at Natural Isotope Abundance." *Organic Magnetic Resonance* 17 (1981): 306.
6. Duthaler, R. O., and J. D. Roberts. "Steric and Electronic Effects on ¹⁵N Chemical Shifts of Piperidine and Decahydroquinoline Hydrochlorides." *Journal of the American Chemical Society* 100 (1978): 3882.
7. Duthaler, R. O., and J. D. Roberts. "Nitrogen-15 Nuclear Magnetic Resonance Spectroscopy. Solvent Effects on the ¹⁵N Chemical Shifts of Saturated Amines and Their Hydrochlorides." *Journal of Magnetic Resonance* 34 (1979): 129.
8. Psota, L., M. Franzen-Sieveking, J. Turnier, and R. L. Lichter. "Nitrogen Nuclear Magnetic Resonance Spectroscopy. Nitrogen-15 and Proton Chemical Shifts of Methylanilines and Methylanilinium Ions." *Organic Magnetic Resonance* 11 (1978): 401.
9. Subramanian, P. K., N. Chandra Sekara, and K. Ramalingam. "Steric Effects on Nitrogen-15 Chemical Shifts of 4-Aminooxanes (Tetrahydropyrans), 4-Amino-Thianes, and the Corresponding N,N-Dimethyl Derivatives. Use of Nitrogen-15 Shifts as an Aid in Stereochemical Analysis of These Heterocyclic Systems." *Journal of Organic Chemistry* 47 (1982): 1933.
10. Schuster, I. I., and J. D. Roberts. "Proximity Effects on Nitrogen-15 Chemical Shifts of 8-Substituted 1-Nitronaphthalenes and 1-Naphthylamines." *Journal of Organic Chemistry* 45 (1980): 284.
11. Kupce, E., E. Liepins, O. Pudova, and E. Lukevics. "Indirect Nuclear Spin-Spin Coupling Constants of Nitrogen-15 to Silicon-29 in Silylamines." *Journal of the Chemical Society: Chemical Communications* 581 (1984).
12. Allen, M., and J. D. Roberts. "Effects of Protonation and Hydrogen Bonding on Nitrogen-15 Chemical Shifts of Compounds Containing the>C=N-Group." *Journal of Organic Chemistry* 45 (1980): 130.
13. Brownlee, R. T. C., and M. Sadek. "Natural Abundance ¹⁵N Chemical Shifts in Substituted Benzamides and Thiobenzamides." *Magnetic Resonance Chemistry* 24 (1986): 821.
14. Dega-Szafran, Z., M. Szafran, L. Stefaniak, C. Brevard, and M. Bourdonneau. "Nitrogen-15 Nuclear Magnetic Resonance Studies of Hydrogen Bonding and Proton Transfer in Some Pyridine Trifluoroacetates in Dichloromethane." *Magnetic Resonance Chemistry* 24 (1986): 424.
15. Lambert, J. B., H. F. Shurvell, L. Verbit, R. G. Cooks, and G. H. Stout. *Organic Structural Analysis.* New York: MacMillan, 1976.

^{15}N Chemical Shifts of Major Chemical Families

Chemical Shift Range (ppm)	Family	Example (δ)
<930	nitroso compounds	C_6H_5-NO (913, 930)
608	sodium nitrite	$NaNO_2$
~500	azo compounds	C_6H_5-N=N-C_6H_5 (510)
380–350	nitro compounds	$C_6H_5NO_2$ (370.3); CH_3NO_2 (380.2); 4-F-C_6H_4-NO_2 (368.5); 1,3-$(NO_2)_2C_6H_4$ (365.4)
367	nitric acid (8.57 M)	HNO_3
360–325	nitramines	CH_3NHNO_2 (355.6); $CH_3O_2CNHNO_2$ (334.9)
350–300	pyridines	C_5H_5N (317)[b] (gas); 4-CH_3-C_5H_4N (309.3); 4-NH_2-C_5H_4N (271.5); 4-NC-C_5H_4N (327.9)
~310	imines (aromatic)	$(C_6H_5)_2C$=NH (308); C_6H_5CH=NCH_3 (318); C_6H_5CH=NC_6H_5 (326)
310.1	nitrogen (gas)	N_2
250–200	pyridinium salts+	$C_5H_5NH^+$ (215)
260–175	cyanides (nitriles)	CH_3CN (239.5, 245); C_6H_5CN (258.7); KCN 177.8
~160	pyrroles isonitriles	C_4H_4NH (158) CH_3NC (162)
~150	thioamides	CH_3C(=S)NH_2 (150.2)
120–110	lactams	HN$(CH_2)_3$C=O (5-membered ring; 114.7); HN$(CH_2)_6$C=O (8-membered ring; 117.7)
110–100	amides	$C_6H_5CONH_2$ (100); CH_3CONH_2 (103.4); $CH_3CONHCH_3$ (105.8); $CH_3CON(CH_3)_2$ (103.8); $HCONH_2$ (108.5)
125–90	sulfonamides	$CH_3SO_2NH_2$ (95); $C_6H_5SO_2NH_2$ (94.3)
~100	hydrazines	$C_6H_5NHNHC_6H_5$ (96)
110–60	ureas	$[H_2N]_2CO$ (75, 82); $[(CH_3)_2N]_2CO$ (63.5); $[C_6H_5NH]_2CO$ (107.7)
100–70	aminophospines, aminophosphine oxides	$C_6H_5NHP(CH_3)_2$ (71.1); $C_6H_5NHPO(CH_3)_2$ (86.6)
70–50	aromatic amines	$C_6H_5NH_2$ (55, 59), (−322.3)[c]; $C_6H_5NH_3^+$ (48), (−326.4)[c], 26.1[g]; p-O_2N-C_6H_4-NH_2 (70)
40–0	aliphatic amines	CH_3NH_2 (1.3)[a], (−371)[c]; $(CH_3)_2$NH (−363.3)[c]; $(CH_3)_2$NH (−364.9)[d], 6.7[a]; $(CH_3)_3$N (−356.9)[c], (−360.7)[d], 13.0[a]
50–10	isonitriles	CH_3NCO (14.1); C_6H_5NCO (46.5)
65–20	ammonium salts	NH_4Cl (26.1)[a]; CH_3NH_3Cl (24.5); $(CH_3)_2NH_2Cl$ (26.6); $(CH_3)_3NHCl$ (33.8); $(CH_3)_4NCl$ (44.7)
~15	isocyanates	CH_3NCO (14.1)

(Continued)

^{15}N Chemical Shifts of Major Chemical Families (Continued)

Chemical Shift Range (ppm)	Family	Example (δ)
	Miscellaneous	
(−130)−(−110) and ~(212)	imidazoles	N-methylimidazole (−111.4, pyridine N and -215.7, pyrrole N)^c
(−345)−(−310)^c	piperidine, hydrochloride salts	piperidinium hydrochloride (−344.8); 2-methyl piperdinium hydrochloride (−322.1)^d
	decahydroquinolines, hydrochloride salts	trans-decahydroquinolinium hydrochloride (−322.5); cis-decahydroquinolinium hydrochloride (−328.5)
(−293)−(−280)^e	enaminones	CH$_3$C(=O)CH=CHNHCH$_3$ [(E)-(−294.2); (Z)-(−285.9)]
35−15^f	4-aminotetrahydropyrans, 4-aminotetrahydro-thiopyrans	2,6-diphenyl 4-aminotetrahydropyran (34.5) 2,6-diphenyl 4-aminotetrahydrothiopyran (33.6)
(−325)−(−310)^g	1-naphthylamines	8-nitro-1-naphthylamine (313.9)
(−350)−(−300)^h	silylamines	HN[Si(CH$_3$)$_3$]$_2$ (−354.2)

Notes:

a Downfield from anhydrous liquid ammonia, ±0.2 ppm unless otherwise specified [1].

b Varies with solvent. For instance: cyclohexane (315.5), benzene (312.1), chloroform (304.5), methanol (292.1), water (289), 2,2,2-trifluoroethanol (277.1). All chemical shifts relative to ammonia [2].

c Upfield from external HNO$_3$ (1 M) (CH$_3$OH) [4,6,7].

d Upfield from external HNO$_3$ (1 M) (cyclohexane) [6,7].

e Relative to external CH$_3$^{15}NO$_2$ [5].

f With respect to an external standard of 5 M ^{15}NH$_4$NO$_3$ in 2 M HNO$_3$ (^{15}NH$_4$NO$_3$ = 21.6 ppm relative to anhydrous ammonia) [9].

g In ppm upfield from external 1 MD^{15}NO$_3$ in D$_2$O (DMSO) [10].

h Relative to N (SiH$_3$) (50 % in CDCl$_3$) [11].

The indicated superscript numbers following each note above refer to the reference list at the beginning of this table.

¹⁵N CHEMICAL SHIFT CORRELATION CHARTS
OF MAJOR FUNCTIONAL GROUPS

The following correlation chart contains ^{15}N chemical shifts of various organic nitrogen compounds. Chemical shifts are often expressed relative to different standards (NH_3, NH_4Cl, CH_3NO_2, NH_4NO_3, HNO_3, etc.) and are interconvertible.

In view of the large chemical shift range (up to 900 ppm) caution in using these correlation charts is of great importance as the chemical shifts are greatly dependent on the inductive, mesomeric, or hybridization effects of the neighboring groups, as well as the solvent used.

Chemical shifts are sensitive to hydrogen bonding and are solvent dependent as seen in the case of pyridine. Consequently, the reference as well as the solvent should always accompany chemical shift data. No data are given on peptides and other biochemical compounds. All shifts given in these correlation charts are relative to ammonia unless otherwise specified. A section of "miscellaneous" data gives the chemical shift of special compounds relative to unusual standards [1–16]. Reference [17] contains a compilation of publications that involve various nuclei.

REFERENCES

1. Levy, G. C., and R. L. Lichter. *Nitrogen-15 Nuclear Magnetic Resonance Spectroscopy.* New York: John Wiley and Sons, 1979.
2. Yoder, C. H., and C. D. Schaeffer, Jr. *Introduction to Multinuclear NMR.* Menlo Park, CA: Benjamin/Cummings, 1987.
3. Duthaler, R. O., and J. D. Roberts. "Effects of Solvent, Protonation, and N-Alkylation on the ^{15}N Chemical Shifts of Pyridine and Related Compounds." *Journal of the American Chemical Society* 100 (1978): 4969.
4. Duthaler, R. O., and J. D. Roberts. "Steric and Electronic Effects on ^{15}N Chemical Shifts of Saturated Aliphatic Amines and Their Hydrochlorides." *Journal of the American Chemical Society* 100 (1978): 3889.
5. Kozerski, L., and W. von Philipsborn. "^{15}N Chemical Shifts as a Conformational Probe in Enaminones: A Variable Temperature Study at Natural Isotope Abundance." *Organic Magnetic Resonance* 17 (1981): 306.
6. Duthaler, R. O., and J. D. Roberts. "Steric and Electronic Effects on ^{15}N Chemical Shifts of Piperidine and Decahydroquinoline Hydrochlorides." *Journal of the American Chemical Society* 100 (1978): 3882.
7. Duthaler, R. O., and J. D. Roberts. "Nitrogen-15 Nuclear Magnetic Resonance Spectroscopy. Solvent Effects on the ^{15}N Chemical Shifts of Saturated Amines and Their Hydrochlorides." *Journal of Magnetic Resonance* 34 (1979): 129.
8. Psota, L., M. Franzen-Sieveking, J. Turnier, and R. L. Lichter. "Nitrogen Nuclear Magnetic Resonance Spectroscopy. Nitrogen-15 and Proton Chemical Shifts of Methylanilines and Methylanilinium Ions." *Organic Magnetic Resonance* 11 (1978): 401.
9. Subramanian, P. K., N. Chandra Sekara, and K. Ramalingam. "Steric Effects on Nitrogen-15 Chemical Shifts of 4-Aminooxanes (Tetrahydropyrans), 4-Amino-Thianes, and the Corresponding N,N-Dimethyl Derivatives. Use of Nitrogen-15 Shifts as an Aid in Stereochemical Analysis of These Heterocyclic Systems." *Journal of Organic Chemistry* 47 (1982): 1933.
10. Schuster, I. I., and J. D. Roberts. "Proximity Effects on Nitrogen-15 Chemical Shifts of 8-Substituted 1-Nitronaphthalenes and 1-Naphthylamines." *Journal of Organic Chemistry* 45 (1980): 284.
11. Kupce, E., E. Liepins, O. Pudova, and E. Lukevics. "Indirect Nuclear Spin-Spin Coupling Constants of Nitrogen-15 to Silicon-29 in Silylamines." *Journal of the Chemical Society: Chemical Communications* 581 (1984).
12. Allen, M., and J. D. Roberts. "Effects of Protonation and Hydrogen Bonding on Nitrogen-15 Chemical Shifts of Compounds Containing the>C=N-Group." *Journal of Organic Chemistry* 45 (1980): 130.
13. Brownlee, R. T. C., and M. Sadek. "Natural Abundance ^{15}N Chemical Shifts in Substituted Benzamides and Thiobenzamides." *Magnetic Resonance Chemistry* 24 (1986): 821.

14. Dega-Szafran, Z., M. Szafran, L. Stefaniak, C. Brevard, and M. Bourdonneau. "Nitrogen-15 Nuclear Magnetic Resonance Studies of Hydrogen Bonding and Proton Transfer in Some Pyridine Trifluoroacetates in Dichloromethane." *Magnetic Resonance Chemistry* 24 (1986): 424.

15. Lambert, J. B., H. F. Shurvell, L. Verbit, R. G. Cooks, and G. H. Stout. *Organic Structural Analysis.* New York: MacMillan, 1976.

16. Bruno, T. J., and P. D. N. Svoronos. *Handbook of Basic Tables for Chemical Analysis.* 2nd ed. Boca Raton, FL: CRC Press, 2003.

17. http://www.chem.wisc.edu/areas/reich/Handouts/nmr/NMR-Biblio.htm

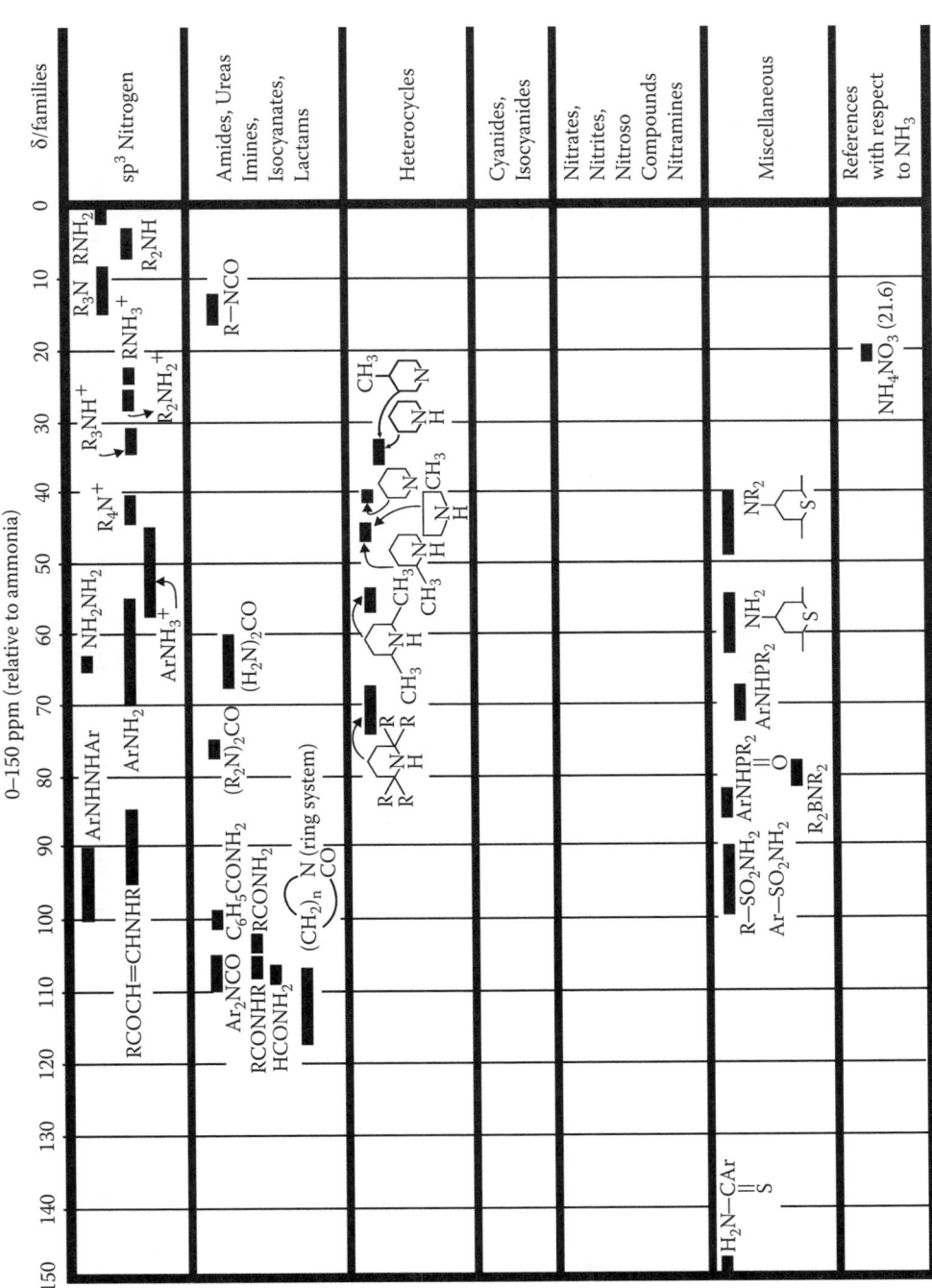

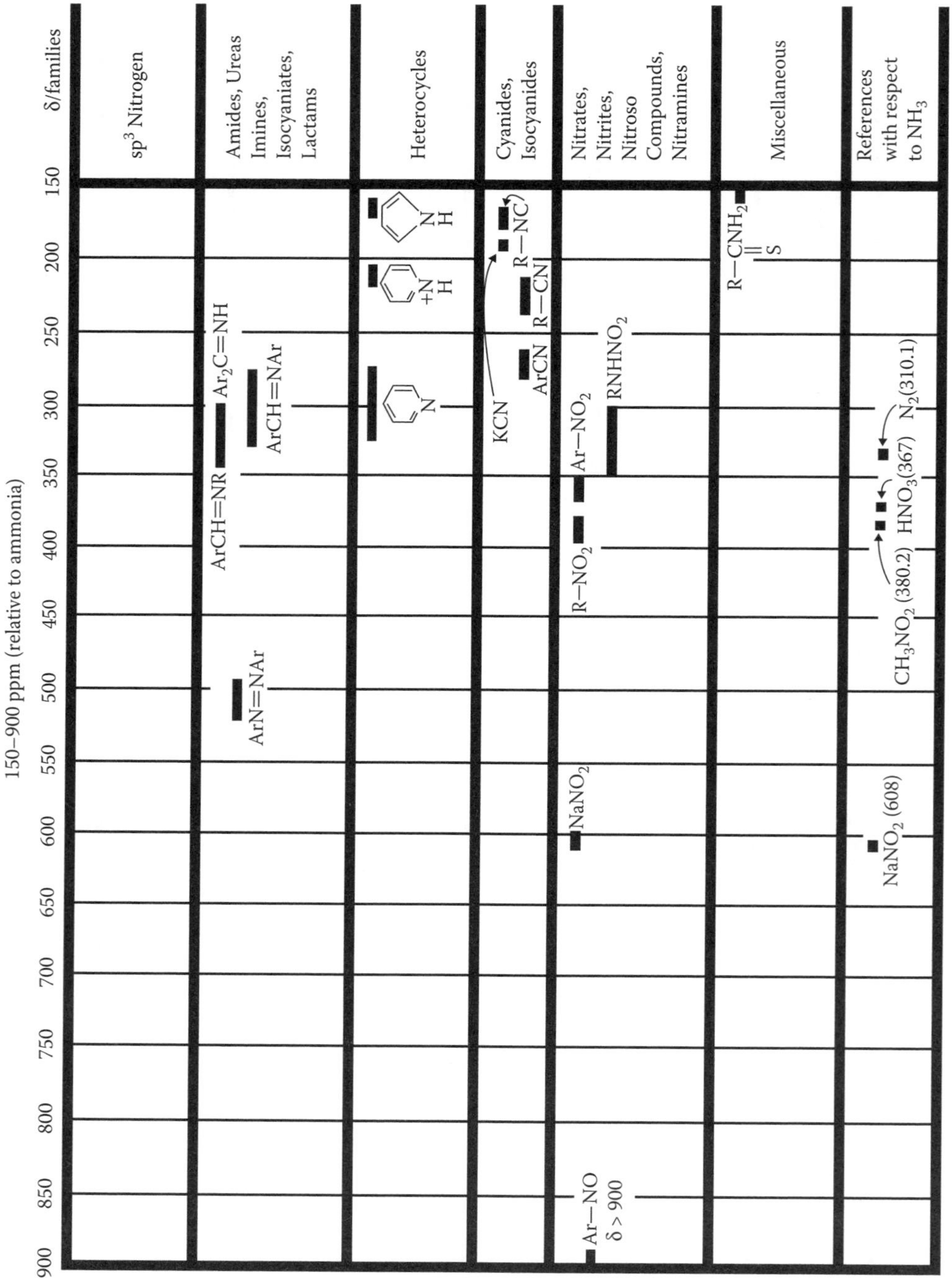

150–900 ppm (relative to ammonia)

δ/families — 150 200 250 300 350 400 450 500 550 600 650 700 750 800 850 900

sp^3 Nitrogen

Amides, Ureas Imines, Isocyanates, Lactams
ArCH=NR $Ar_2C=NH$
ArCH=NAr
ArN=NAr

Heterocycles

Cyanides, Isocyanides
R—NC
R—CN
KCN   ArCN

Nitrates, Nitrites, Nitroso Compounds, Nitramines
$RNHNO_2$
Ar—NO_2
R—NO_2
$NaNO_2$
Ar—NO δ > 900

Miscellaneous
$R-CNH_2$ ‖ S

References with respect to NH_3
N_2(310.1)
HNO_3(367)
CH_3NO_2 (380.2)
$NaNO_2$ (608)

SPIN–SPIN COUPLING TO ^{15}N

The following table gives representative spin–spin coupling ranges (J_{NH} in Hz) to ^{15}N [1–11].

REFERENCES

1. Levy, G. C., and R. L. Lichter. *Nitrogen-15 Nuclear Magnetic Resonance Spectroscopy.* New York: John Wiley and Sons, 1979.
2. DelBene, J. E., and R. J. Bartlett. "N-N Spin–Spin Coupling Constants [2hJ(15N-15N)] Across N-H—N Hydrogen Bonds in Neutral Complexes: To what Extent Does the Bonding at the Nitrogens Influence 2hJN-N?. Communication." *Journal of the American Chemical Society* 122 (2000): 10480.
3. Del Bene, J. E., S. Ajith Perera, R. J. Bartlett, M. Yáñez, O. Mó, J. Elguero, and I. Alkorta. *Journal of Physical Chemistry* 107 (2003): 3121.
4. Duthaler, R. O., and J. D. Roberts. "Effects of Solvent, Protonation, and N-Alkylation on the ^{15}N Chemical Shifts of Pyridine and Related Compounds." *Journal of the American Chemical Society* 100 (1978): 4969.
5. Duthaler, R. O., and J. D. Roberts. "Steric and Electronic Effects on ^{15}N Chemical Shifts of Saturated Aliphatic Amines and Their Hydrochlorides." *Journal of the American Chemical Society* 100 (1978): 3889.
6. Kozerski, L., and W. von Philipsborn. "^{15}N Chemical Shifts as a Conformational Probe in Enaminones: A Variable Temperature Study at Natural Isotope Abundance." *Magnetic Resonance Chemistry* 17 (1981): 306.
7. Subramanian, P. K., N. Chandra Sekara, and K. Ramalingam. "Steric Effects on Nitrogen-15 Chemical Shifts of 4-Aminooxanes (Tetrahydropyrans), 4-Amino-Thianes, and the Corresponding N,N-Dimethyl Derivatives. Use of Nitrogen-15 Shifts as an Aid in Stereochemical Analysis of These Heterocyclic Systems." *Journal of Organic Chemistry* 47 (1982): 1933.
8. Kupce, E., E. Liepins, O. Pudova, and E. Lukevics. "Indirect Nuclear Spin–Spin Coupling Constants of Nitrogen-15 to Silicon-29 in Silylamines." *Journal of the Chemical Society: Chemical Communications* 581 (1984).
9. Axenrod, T. "Structural Effects on the One-Bond 15N-H Coupling Constant." In *NMR Spectroscopy of Nuclei Other than Protons,*. Eedited by T. Axenrod and G. A. Webb. New York: Wiley-Interscience, 1974.
10. Kupce, E., E. Liepins, O. Pudova, and E. Lukevics. "Indirect Nuclear Spin–Spin Coupling Constants of Nitrogen-15 to Silicon-29 in Silylamines." *Journal of the Chemical Society: Chemical Communications* 581 (1984).
11. Axenrod, T. "Structural Effects on the One-Bond 15N-H Coupling Constant." In *NMR Spectroscopy of Nuclei Other than Protons,*. edited by T. Axenrod and G. A. Webb. New York: Wiley-Interscience, 1974.

1. $^{15}N–^{1}H$ Coupling Constants

Bond Type	Family	J_{NH}	Example
One bond	ammonia	(–61.2)	NH_3
	amines, aliphatic (1 °,2 °)	~(–65)	CH_3NH_2 (–64.5); $(CH_3)_2NH$ (–67.0)
	ammonium salts	~(–75)	CH_3NH_3Cl (–75.4); $(CH_3)_2NH_2Cl$ (–76.1); $C_6H_5NH_3^+$ (–76)
	amines, aromatic (1 °,2 °)	(–78)–(–95)	$C_6H_5NH_2$ (–78.5); p–$CH_3O–C_6H_4–NH_2$ (–79.4); p–$O_2N–C_6H_4–NH_2$ (–92.6)
	sulfonamides	~(–80)	$C_6H_5SO_2NH_2$ (–80.8)
	hydrazines	(–90)–(–100)	$C_6H_5NHNH_2$ (–89.6)
	amides (1 °,2 °)	(–85)–(–95)	$HCONH_2$ (–88) (syn); (–92) (anti)
	pyrroles	(–95)–(–100)	Pyrrole (–96.53)
	nitriles, salts	~(–135)	$CH_3C≡NH^+$ (–136)
Two bond	Amines	~(–1)	CH_3NH_2 (–1.0); $(CH_3)_3N$ (–0.85)
	Pyridinium salts	~(–3)	$C_5H_5NH^+$ (–3)
	Pyrroles	~(–5)	C_4H_4NH (–4.52)
	thiazoles	~(–10)	C_3H_3NS
	Pyridines	~(–10)	C_5H_5N (–10.76)
	oximes, syn	~(–15)	>C=N–OH (syn)
	oximes, anti	(–2.5)–(+ 2.5)	>C=N–OH (anti)
Three bond	nitriles, salts	~(2–4)	$CH_3C≡NH^+$ (2.8)
	amides	~(1–2)	CH_3CONH_2 (1.3)
	anilines	~(1–2)	$C_6H_5NH_2$ (1.5,1.8)
	pyridines	~(0–1)	C_5H_5N (0.2)
	nitriles	(–1)–(–2)	$CH_3C≡N$ (–1.7)
	pyridinium salts	~(–4)	$C_5H_5NH^+$ (–3.98)
	pyrroles	~(–5)	C_4H_4NH (–5.39)

2. $^{15}N–^{1}C$ Coupling Constants

Bond Type	Family	J_{CH}, H_2	Example
One bond	amines, aliphatic	~(–4)	CH_3NH_2 (–4.5); $CH_3(CH_2)_2NH_2$ (–3.9)
	ammonium salts (aliphatic)	~(–5)	$CH_3(CH_2)_2NH_3^+$ (–4.4)
	ammonium salts (aromatic)	~(–9)	$C_6H_5NH_3^+$ (–8.9)
	pyrroles	~(–10)	C_4H_4NH (–10.3)
	amines, aromatic	(–11)–(–15)	$C_6H_5NH_2$ (–11.43)
	nitro compounds	(–10)–(–15)	CH_3NO_2 (–10.5); $C_6H_5NO_2$ (–14.5)
	nitriles	~(–17)	$CH_3C≡N$ (–17.5)
	amides	~(–14)	$C_6H_5NHCOCH_3$ (–14.3) (CO); (–14.1) (C_1)
Two bond	amides	7–9	CH_3CONH_2 (9.5)
	nitriles	~3	$CH_3C≡N$ (3.0)
	pyridines and N–derivatives	~1–3	C_5H_5N (2.53); $C_5H_5NH^+$ (2.01); C_5H_5NO (1.43)
	amines, aliphatic	~1–2	$CH_3CH_2CH_2NH_2$ (1.2)
	nitro compounds, aromatic	~(–1)–(–2)	$C_6H_5NO_2$ (–1.67)
	amines, aromatic	~(–1)–(–2)	$C_6H_5NH_2$ (–2.68); $C_6H_5NH_3^+$ (–1.5)
	pyrroles	~(–4)	C_4H_4NH (–3.92)
Three bond	amides	9	$CH_2=CHCONH_2$ (19)
	ammonium salts	1–9	$CH_3(CH_2)_2NH_3^+$ (1.3); $C_6H_5NH_3^+$ (2.1)

(Continued)

Bond Type	Family	J_{CH}, H_2	Example
	pyridines	~3	C_5H_5N (2.53)
	amines, aliphatic	1–3	$CH_3(CH_2)_2NH_2$ (1.4)
	amines, aromatic	~(−1)–(−3)	$C_6H_5NH_2$ (−2.68)
	nitro compounds	~(−2)	$C_6H_5NO_2$ (−1.67)
	pyrroles	~(−4)	C_4H_4NH (−3.92)

3. ^{15}N–^{15}N Coupling Constants

Bond Type	Family	J_{NN},H_2	Example
	azocompounds	12–25	$C_6H_5N=NC(CH_3)_2C_6H_5$ anti (17); syn (21)
	N–nitrosamines	~19	$(C_6H_5CH_2)_2N–N=O$ (19)
	hydrazones	~10	$p–O_2NC_6H_4CH=N–NHC_6H_5$ (10.7)
	hydrazines	~7	$C_6H_5NHNH_2$ (6.7)

4. ^{15}N–^{19}F Coupling Constants

Bond Type	Family		J_{NF}, H_2	Example
	diflurodiazines	trans	~190 ($^1J_{NF}$)	F–N=N–F (190)
			~102 ($^2J_{NF}$)	F–N=N–F (102)
		cis	~203 ($^1J_{NF}$)	F–N=N–F (203)
			~52 ($^2J_{NF}$)	F–N=N–F (52)
	fluoropyridines	2–fluoro–	(−52.5)	
		3–fluoro–	(+3.6)	
	fluoroanilines	2–fluoro–	0	$1,2–C_6H_4F(NH_2)$
		3–fluoro–	0	$1,3–C_6H_4F(NH_2)$
		4–fluoro–	1.5	$1,4–C_6H_4F(NH_2)$
	fluoroanilinium salts	2–fluoro	1.4	$1,2–C_6H_4F(NH_3^+)$
		3–fluoro	0.2	$1,3–C_6H_4F(NH_3^+)$
		4–fluoro	0	$1,4–C_6H_4F(NH_3^+)$

^{19}F CHEMICAL SHIFT RANGES

The following table lists the ^{19}F chemical shift ranges (in ppm) relative to neat $CFCl_3$ [1–7].

REFERENCES

1. Yoder, C. H., and C. D. Schaeffer, Jr. *Introduction to Multinuclear NMR: Theory and Application.* Menlo Park, CA: Benjamin/Cummings, 1987.
2. http://nmrsg1.chem.indiana.edu/NMRguidemisc/19Fshifts.html
3. Dungan, C. H., and I. R. Van Wazer. *Compilation of Reported ^{19}F Chemical Shifts 1951 to Mid 1967.* New York: Wiley Interscience, 1970.
4. Emsley, J. W., L. Phillips, and V. Wray. *Fluorine Coupling Constants.* New York: Pergamon, 1977.
5. Dolbier, W. R. *Guide to Fluorine NMR for Organic Chemists.* New York: John Wiley and Sons, 2009.
6. Bruno, T. J., and Svoronos, P. D. N. *CRC Handbook of Basic Tables for Chemical Analysis, 2nd Ed.* Boca Raton, FL: CRC Press, 2003.
7. Bruno, T. J., and Svoronos, P. D. N. *CRC Handbook of Fundamental Spectroscopic Correlation Charts,* Boca Raton, FL: CRC Press, 2006.

Compound Type	Chemical Shift Range (ppm) Relative to Neat $CFCl_3$
F-C(=O)	−70 to −20
CF_3^-	+40 to +80
$-CF_2^-$	+80 to +140
$>CF^-$	+140 to +250
Ar-F where Ar = aromatic moiety	+80 to +170

^{19}F CHEMICAL SHIFTS OF SOME FLUORINE-CONTAINING COMPOUNDS

The following table lists the ^{19}F chemical shifts of some fluorine-containing compounds relative to neat $CFCl_3$. All chemical shifts are those of neat samples and the values pertain to the fluorine present in the molecule [1–3].

REFERENCES

1. http://nmrsg1.chem.indiana.edu/NMRguidemisc/19Fshifts.html, 2003.
2. Dungan, C. H., and I. R. Van Wazer. *Compilation of Reported ^{19}F Chemical Shifts 1951 to Mid 1967*. New York: Wiley Interscience, 1970.
3. Emsley, J. W., L. Phillips, and V. Wray. *Fluorine Coupling Constants*. New York: Pergamon, 1977.

^{19}F Chemical Shifts of Some Fluorine-Containing Compounds

Compound	Formula	Chemical Shift (ppm)
Fluorotrichloromethane	$CFCl_3$	0.00
Tetrafluoromethane	CF_4	−62.3
Fluoromethane	CH_3F	−271.9
Trifluoromethane	CF_3H	−78.6
Trifluoroalkanes	CF_3R	−60 to −70
Difluoromethane	CF_2H_2	−143.6
Fluoroethane	CH_3CH_2F	−231
Fluoroethene (or vinyl fluoride)	$FCH=CH_2$	−114
1,1-Difluoroethene	$CF_2=CH_2$	−81.3
Tetrafluoroethene	$CF_2=CF_2$	−135
Trifluoroethanoic acid (or trifluoroacetic acid)	CF_3COOH	−78.5
Phenyl trifluoroethanoate (or phenyl trifluoroacetate)	$CF_3COOC_6H_5$	−73.85
Benzyl trifluoroethanoate (or benzyl trifluoroacetate)	$CF_3COOCH_2C_6H_5$	−75.02
Methyl trifluoroethanoate (or methyl trifluoroacetate)	CF_3COOCH_3	−74.21
Ethyl trifluoroethanoate (or: trifluoroacetate)	$CF_3COOCH_2CH_3$	−78.7
Hexafluorobenzene	C_6F_6	−164.9
Pentafluorobenzene	C_6F_5H	−113.5
1,4-Difluorobenzene (or p-difluorobenzene)	$p\text{-}C_6H_4F_2$	−106.0
(fluoromethyl)Benzene (or benzyl fluoride)	$C_6H_5\text{-}CH_2F$	−207
Trifluoromethylbenzene	$C_6H_5\text{-}CF_3$	−63.72
Octafluorocyclobutane (or perfluorocyclobutane)	C_4F_8	−135.15
Decafluorocyclopentane (or perfluorocyclopentane)	C_5F_{10}	−132.9
Difluoromethyl ethers	CHF_2OR	−82
Hexafluoropropanone (or hexafluoroacetone)	$(CF_3)_2CO$	−84.6
Fluorine	F_2	+ 422.92
Chlorotrifluoromethane	CF_3Cl	−28.6
Chlorine trifluoride	ClF_3	+ 116, −4
Chlorine pentafluoride	ClF_5	+ 247, + 412
Dichlorodifluoromethane	CF_2Cl_2	−8
1,2-Difluoro-1,1,2,2-tetrachloroethane	$CFCl_2\text{–}CFCl_2$	−67.8
Fluorotribromomethane	$CFBr_3$	+ 7.38
Dibromodifluoromethane	CF_2Br_2	+ 7
Iodine heptafluoride	IF_7	+ 170
Arsenic trifluoride	AsF_3	−40.6
Arsenic pentafluoride	AsF_5	−66
Boron trifluoride	BF_3	−131.3
Trimethyloxonium tetrafluoroborate	$(CH_3)_2O \cdot BF_3$	−158.3
Triethyloxonium tetrafluoroborate	$(C_2H_5)_2O \cdot BF_3$	−153
Sulfur hexafluoride	SF_6	+ 57.42
Sulfuryl fluoride (or sulfonyl fluoride)	SO_2F_2	−78.5
Antimony pentafluoride	SbF_5	−108
Selenium hexafluoride	SeF_6	+ 55
Silicon tetrafluoride	SiF_4	−163.3
Tellurium hexafluoride	TeF_6	−57
Sulfur hexafluoride	SF_6	−57
Xenon difluoride	XeF_2	+ 258

(Continued)

^{19}F Chemical Shifts of Some Fluorine-Containing Compounds (Continued)

Compound	Formula	Chemical Shift (ppm)
Xenon tetrafluoride	XeF_4	+ 438
Xenon hexafluoride	XeF_6	+ 550
Nitrogen trifluoride	NF_3	+ 147
Phosphoryl fluoride (or phosphorus oxyfluoride)	POF_3	−90.7
Phosphorus trifluoride	PF_3	−67.5

¹⁹F CHEMICAL SHIFT CORRELATION CHART OF SOME FLUORINE-CONTAINING COMPOUNDS

The following correlation chart lists the ¹⁹F chemical shifts of some fluorine-containing compounds relative to neat $CFCl_3$. All chemical shifts are those of neat samples and the values pertain to the fluorine present in the molecule [1–4].

REFERENCES

1. Dungan, C. H., and I. R. Van Wazer. *Compilation of Reported ¹⁹F Chemical Shifts 1951 to Mid 1967.* New York: Wiley Interscience, 1970.
2. Emsley, J. W., L. Phillips, and V. Wray. *Fluorine Coupling Constants.* New York: Pergamon, 1977.
3. Bruno, T. J., and P. D. N. Svoronos. *Handbook of Basic Tables for Chemical Analysis.* 2nd ed. Boca Raton, FL: CRC Press, 2003.
4. http://www.chem.wisc.edu/areas/reich/Handouts/nmr/NMR-Biblio.htm

¹⁹F Chemical Shifts of Some Fluorine-Containing Compounds (0–550 ppm)

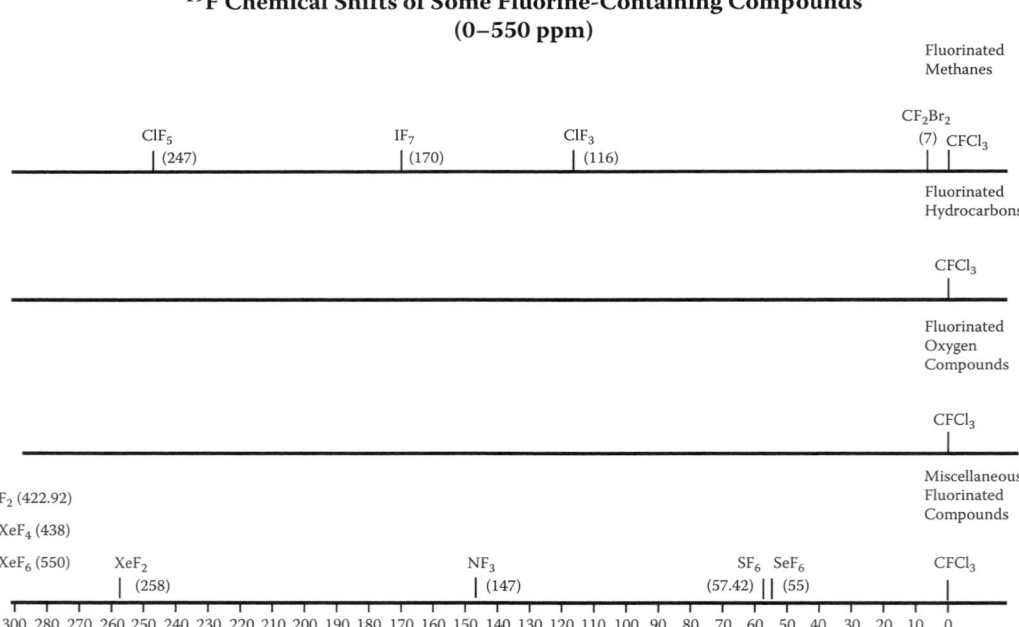

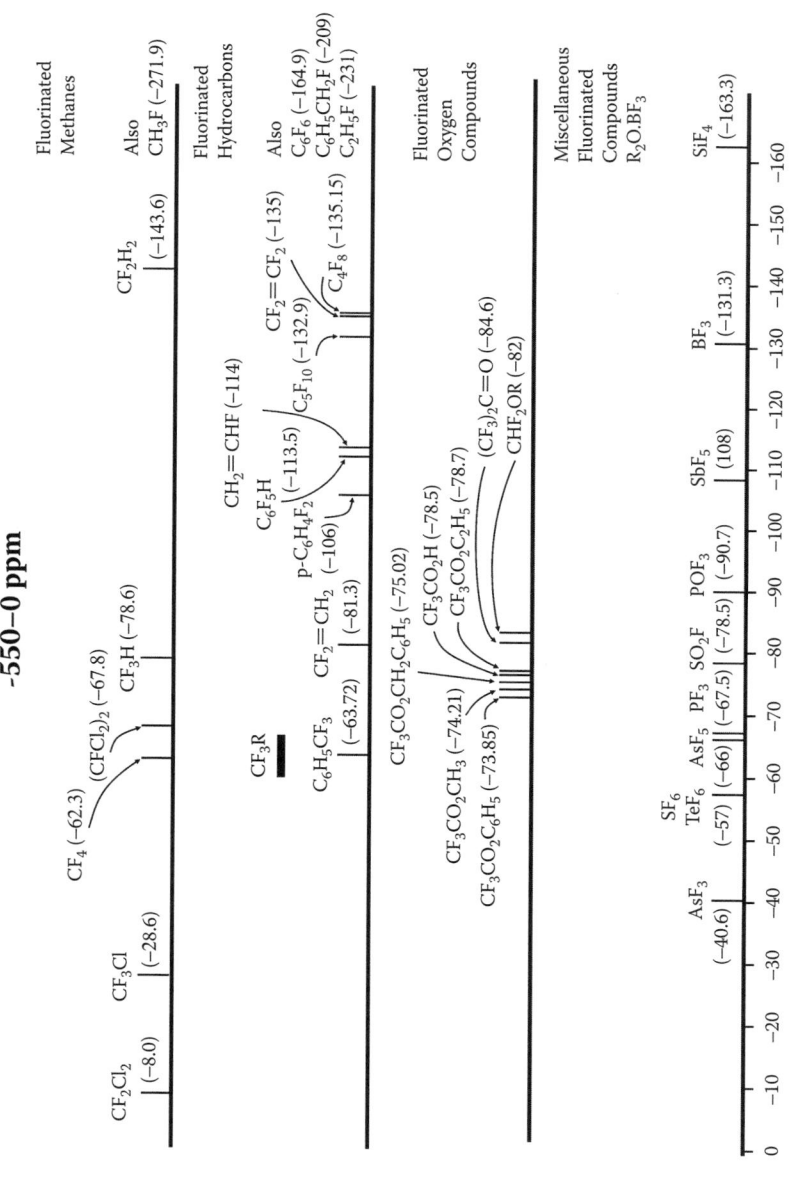

^{19}F Chemical Shifts of Some Fluorine-Containing Compounds
-550–0 ppm

FLUORINE COUPLING CONSTANTS

The following table gives the most important fluorine coupling constants namely, J_{FN}, J_{FCF}, and J_{CF} together with some typical examples [1–10]. The coupling constant values vary with the solvent used [3]. The book by Emsley, Phillips, and Wray [1] gives a complete, detailed list of various compounds.

REFERENCES

1. Emsley, J. W., L. Phillips, and V. Wray. *Fluorine Coupling Constants.* Oxford: Pergamon Press, 1977.
2. Lambert, J. R., H. F. Shurvell, L. Verbit, R. G. Cooks, and G. H. Stout. *Organic Structural Analysis.* New York: MacMillan, 1976.
3. Yoder, C. H., and C. D. Schaeffer, Jr. *Introduction to Multinuclear NMR: Theory and Application.* Menlo Park, CA: Benjamin/Cummings, 1987.
4. Schaeffer, T., K. Marat, J. Peeling, and R. P. Veregin. "Signs and Mechanisms of ^{13}C, ^{19}F Spin–Spin Coupling Constants in Benzotrifluoride and Its Derivatives." *Canadian Journal of Chemistry* 61 (1983): 2779.
5. Adcock, W., and G. B. Kok. "Polar Substituent Effects on ^{19}F Chemical Shifts of Aryl and Vinyl Fluorides: A Fluorine-19 Nuclear Magnetic Resonance Study of Some 1,1-Difluoro-2-(4-Substituted-Bicyclo[2,2,2] Oct-1-yl)Ethenes." *Journal of Organic Chemistry* 50 (1985): 1079.
6. Newmark, R. A., and J. R. Hill. "Carbon-13-Fluorine-19 Coupling Constants in Benzotrifluorides." *Organic Magnetic Resonance* 9 (1977): 589.
7. Adcock, W., and A. N. Abeywickrema. "Concerning the Origin of Substituent-Induced Fluorine-19 Chemical Shifts in Aliphatic Fluorides: Carbon-13 and Fluorine-19 Nuclear Magnetic Resonance Study of 1-Fluoro-4-Phenylbicyclo[2,2,2]Octanes Substituted in the Arene Ring." *Journal of Organic Chemistry* 47 (1982): 2945.
8. Dungan, C. H., and I. R. Van Wazer. *Compilation of Reported ^{19}F Chemical Shifts 1951 to Mid 1967.* New York: Wiley Interscience, 1970.
9. Emsley, J. W., L. Phillips, and V. Wray. *Fluorine Coupling Constants.* New York: Pergamon Press, 1977.
10. Dolbier, W. R. *Guide to Fluorine NMR for Organic Chemists.* New York: John Wiley and Sons, 2009.

^{19}F-^{1}H Coupling Constants

Fluorinated Family	J_{FH} (Hz)	Example
		a. Two-bond
Alkanes	45–80	CH_3F (45); CH_2F_2 (50); CHF_3 (79); C_2H_5F (47); CH_3CHF_2 (57); CH_2FCH_2F (48); CH_2FCHF_2 (54); CF_3CH_2F (45); $CF_2HCF_2CF_3$ (52)
Alkyl chlorides	49–65	FCl_2CH (53); CF_2HCl (63); $FCHCl\text{-}CHCl_2$ (49); $FCH_2\text{-}CH_2Cl$ (46)
Alkyl bromides	45–50	$FBrCHCH_3$ (50.5); $FH_2C\text{-}CH_2Br$ (46); $FBrCH\text{-}CHFBr$ (49)
Alkenes	45–80	$FHC{=}CHF$ (cis-71.7; trans-75.1); $CH_2{=}CHF$ (85); $CF_2{=}CHF$ (70.5); $FCH_2CH{=}CH_2$ (47.5)
Aromatics	45–75	$Cl\text{-}C_6H_4\text{-}CH_2F$ (m-47, p-48); $FH_2C\text{-}C_6H_4\text{-}NO_2$ (m-47, p-48); $FH_2C\text{-}C_6H_4\text{-}F$ (m-48, p-48); $p\text{-}Br\text{-}C_6H_4\text{-}OCF_2H$ (73)
Ethers	40–75	FH_2COCH_3 (74); $CF_2HCF_2OCH_3$ (46); $F_2HC\text{-}O\text{-}CH(CH_3)_2$ (75)
Ketones	45–50	FCH_2COCH_3 (47); $F_2HC{-}COCH_3$ (54); $CH_3CH_2CHFCOCH_3$ (50); $F_2HC{-}COCH(CF_3)_2$ (54)
Aldehydes	~50	$CH_3CH_2CHFCHO$ (51)
Esters	45–70	$CFH_2CO_2CH_2CH_3$ (47); $CH_3CHFCO_2CH_2CH_3$ (48)
		b. Three-bond
Alkanes	2–25	CF_2HCH_3 (21); $(CH_3)_3CF$ (20.4); $CH_3CHFCH_2CH_2CH_3$ (23); CF_3CH_3 (13)
Alkyl chlorides	8–20	CF_2HCHCl_2 (8); CF_2ClCH_3 (15)
Alkyl bromides	15–25	CF_2BrCH_2Br (22); CF_2BrCH_3 (16); $FC(CH_3)_2CHBrCH_3$ (21)
Alkenes	(−5)–60 J_{HCF} (cisoid) <20 J_{HCF} (transoid) >20	$CHF{=}CHF$ (cis-19.6; trans-2.8); $CH_2{=}CHF$ (cis-19.6; trans-51.8); $CHF{=}CF_2$ (cis-(-4.2); trans-12.5); $CH_2{=}CF_2$ (cis-0.6; trans-33.8)
Alcohols	5–30	CF_3CH_2OH (8); FCH_2CH_2OH (29); CH_3CHFCH_2OH (23.6, 23.6); $CF_3CH(OH)CH_3$ (7.5); $CF_3CH(OH)CF_3$ (6); $FC(CH_3)_2C(OH)(CH_3)_2$ (23)
Ketones	5–25	$CH_3CH_2CHFCOCH_3$ (24); $FC(CH_3)_2COCH_3$ (21); $(CF_3)_2CHCOCH_3$ (8); $CF_2HCOCH(CF_3)_2$ (7)
Aldehydes	10–25	$(CH_3)_2CFCHO$ (22)
Esters	10–25	$CH_3CHFCO_2CH_2CH_3$ (23); $(CH_3CH_2)_2CFCO_2CH_3$ (16.5)

^{19}F-^{19}F Coupling Constants

Carbon	J_{FCF} (Hz)	Examples
		a. Two-bond
Saturated (sp³)	140–250	$CF_3CF_2^{a,b}CFHCH_3$ (J_{ab}=270); $CF_2^{a,b}BrCHFSO_2F$ (J_{ab} = 188); CH_3O– $CF_2^{a,b}CFHSO_2F$ (J_{ab} = 147); CH_3O-$CF_2^{a,b}CFHCl$ (J_{ab} = 142); CH_3S–$CF_2^{a,b}CFHCl$ (J_{ab} = 222)
Cycloalkanes	150–240	$F_2C(CH_2)_2$ (150) (3–membered); $F_2C(CH_2)_3$ (200) (4–membered); $F_2C(CH_2)_4$ (240) (5–membered); $F_2C(CH_2)_5$ (228) (6–membered)
Unsaturated (sp²)	≤ 100	CF_2=CH_2 (31,36); CF_2=CHF (87); CF_2=$CBrCl$ (30); CF_2=$CHCl$ (41); CF_2=$CFBr$ (75); CF_2=NCF_3 (82); CF_2=$CFCN$ (27); CF_2=$CFCOF$ (7); CF_2=$CFOCH_2CF_3$ (102); CF_2=$CBrCH_2N(CF_3)_2$ (30); CF_2=$CFCOCF_2CF_3$ (12); CF_2=CHC_6H_5 (33); CF_2=$CH(CH_2)_5CH_3$ (50); CF_2=CH–$Ar[Ar$ = aryl] (50)
		b. Three-bond
Saturated (sp³)	0–16	CF_3CH_2F (16); CF_3CF_3 (3.5); CF_3CHF_2 (3); CH_2FCH_2F (10–12); $CF_2^aHCF^bHCF_2H$ (J_{ab} = 13); $CF_2HCF_2^aCH_2F$ (J_{ab} = 14); $CF_3^aCF_2^bCF^cHCH_3$ (J_{ab}<l; J_{bc} = 15); $CF_3^aCF^bHCF_2^cH$ (J_{ab} = 12; J_{bc} = 12); $CF_3^aCF_2^bC$≡CF_3 (J_{ab} = 3.3); $CF_3^aCF_2^bC$≡CCF_3 (J_{ab} = 3.3); $(CF_3^a)_2CF^bC$≡CCl (J_{ab} = 10); $CF_3CF_2COCH_2CH_3$ (1); $FCH_2CFHCO_2C_2H_5$ (−11.6); $CF_3^aCF_2^bCF_2^cCOOH$ (J_{ab}<l; J_{bc}<l); $(CF_3^a)_2CF^bS(O)OC_2H_5$ (J_{ab} = 8)
Unsaturated (sp²)	>30	FCH=CHF[cis (−18.7); trans (−133.5)]; CF_2=$CHBr$ (34.5); CF_2=$CHCl$ (41); CF_2=CH_2 (37)

^{13}C–^{19}F Coupling Constants

Fluorinated Family	J_{CF}, (Hz)	Examples
		One-bond
Alkanes	150–290	CH_3F (158); CH_2F_2 (237); CHF_3 (274); CF_4 (257); CF_3CF_3 (281); CF_3CH_3 (271); $(CH_3)_3CF$ (167); $(C^aF_3^b)_2C^cF_2^d$ [J_{ab} = 285; J_{cd} = 265]
Alkenes	250–300	CF_2=CD_2 (287); CF_2=CCl_2 (−289); CF_2=CBr_2 (290); $ClFC$=$CHCl$ [cis (−300); trans (−307)]; $ClFC$=$CClF$ [cis (290); trans (290)]
Alkynes	250–260	$C^aF_3^bC$≡CF [J_{ab}=259]; CF_3C≡CCF_3 (256)
Alkyl chlorides	275–350	$CFCl_3$ (337); CF_2Cl_2 (325); CF_3Cl (299); $CF_3(CCl_2)_2CF_3$ (286); CF_3CH_2Cl (274); CF_3CCl=CCl_2 (274);CF_2=CCl_2 (−289); CF_3CCl_3 (283)
Alkyl bromides	290–375	$CFBr_3$ (372); CF_2Br_2 (358); CF_3Br (324); CF_3CH_2Br (272); CF_2=CBr_2 (290)
Acyl fluorides	350–370	$HCOF$ (369); CH_3COF (353)
Carboxylic acids	245–290	CF_3COOH (283); CF_2HCO_2H (247)
Alcohols	~275	CF_3CH_2OH (278)
Nitriles	~250	CF_2HCN (244)
Esters	~285	$CF_3CO_2CH_2CH_3$ (284)
Ketones	~290	CF_3COCH_3 (289)
Ethers	~265	$(CF_3)_2O$ (265)

³¹P NMR ABSORPTIONS OF REPRESENTATIVE COMPOUNDS

The ³¹P is considered to be a medium sensitivity nucleus that has the advantage of yielding sharp lines over a very wide chemical shift range. Its sensitivity is much less than that of ¹H, but it is superior to that of ¹³C [1].

The following charts provide information on the characteristic values for the ³¹P spectra of representative phosphorus-containing compounds. The list is far from complete but gives an insight on the spectra of both organic and inorganic compounds. All data are presented in a correlation chart form. The reference in each case is 85 % (mass/mass) phosphoric acid. The first chart provides the general chemical shift range of the various phosphorus families and is followed by more detailed charts that provide representative compounds for each of the families. These families are classified according to the coordination number around phosphorus, which ranges from 2 to 6.

Since this section only gives a general information on ³¹P NMR spectroscopy, the reader is advised to consult References [2] and [3] that include a large, detailed amount of updated spectral information and numerous references.

REFERENCES

1. http://drx.ch.huji.ac.il/nmr/
2. Quin, L. D., and J. G. Verkade, ed. *Phosphorus-31 NMR Spectral Properties in Compounds: Characterization and Structural Analysis.* New York: John Wiley and Sons, 1994.
3. Tebby, J. C., ed. *Handbook of Phosphorus-31 Nuclear Magnetic Resonance Data.* Boca Raton, FL: CRC Press, 1991.

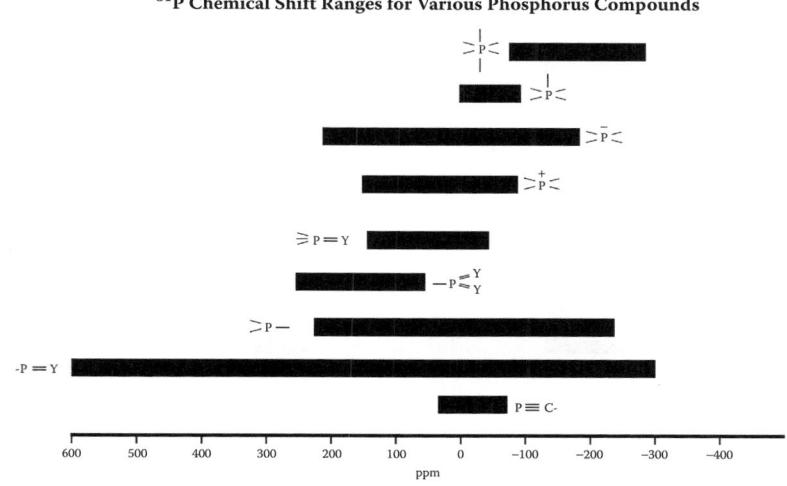

³¹P Chemical Shift Ranges for Various Phosphorus Compounds

Monocoordinated Phosphorus Compounds (C≡P)

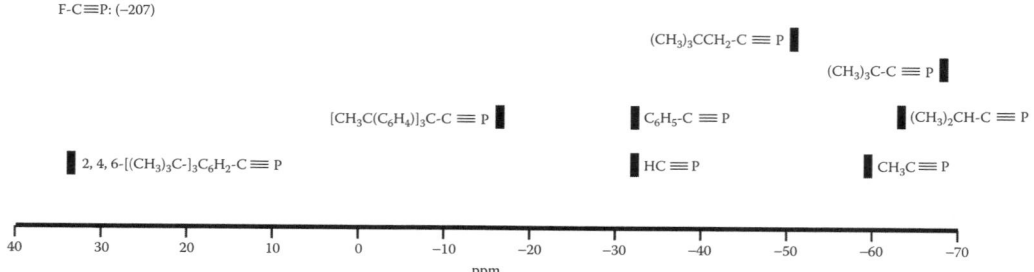

Dicoordinated Phosphorus Compounds (X=P–)

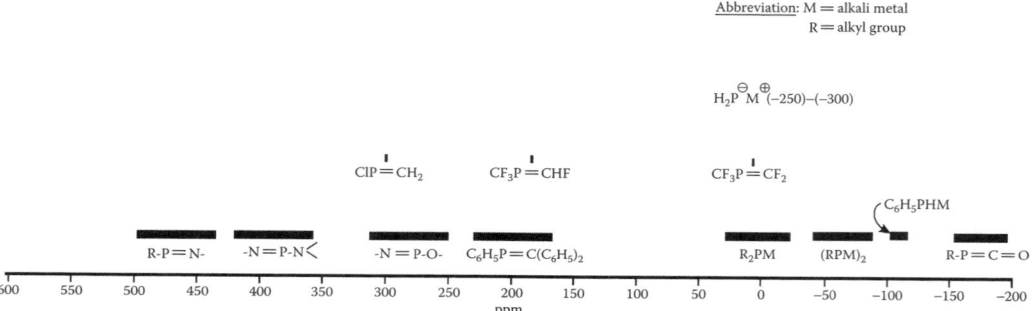

Tricoordinated Phosphorus Compounds (⩾P)
(+240) -0 ppm

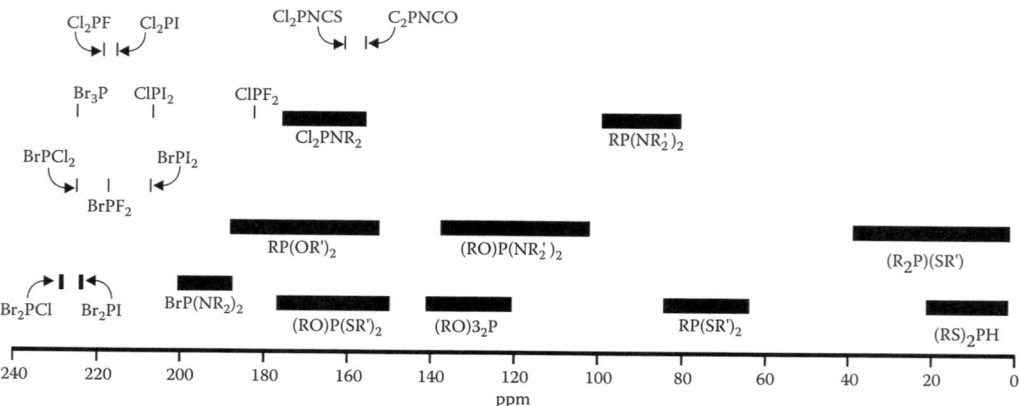

Tricoordinated Phosphorus Compounds (⩾P)
0–(−240) ppm

Abbreviation: R = alkyl
Ar = aryl

$F_3SiPH_2 = -290$

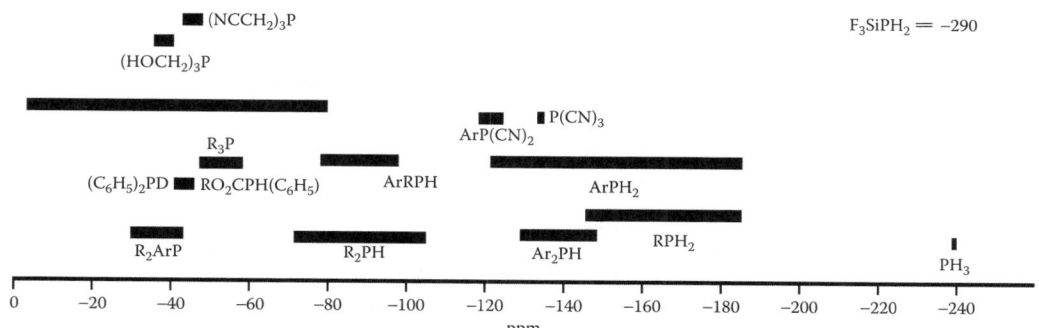

(NCCH$_2$)$_3$P

(HOCH$_2$)$_3$P

P(CN)$_3$

ArP(CN)$_2$

R$_3$P

ArRPH

(C$_6$H$_5$)$_2$PD   RO$_2$CPH(C$_6$H$_5$)

ArPH$_2$

R$_2$ArP

R$_2$PH

Ar$_2$PH

RPH$_2$

PH$_3$

| 0 | −20 | −40 | −60 | −80 | −100 | −120 | −140 | −160 | −180 | −200 | −220 | −240 |

ppm

Tetracoordinated Phosphorus Compounds
(phosphonium salts) (⩾$\overset{+}{P}$–X⁻)
140–(−140) ppm

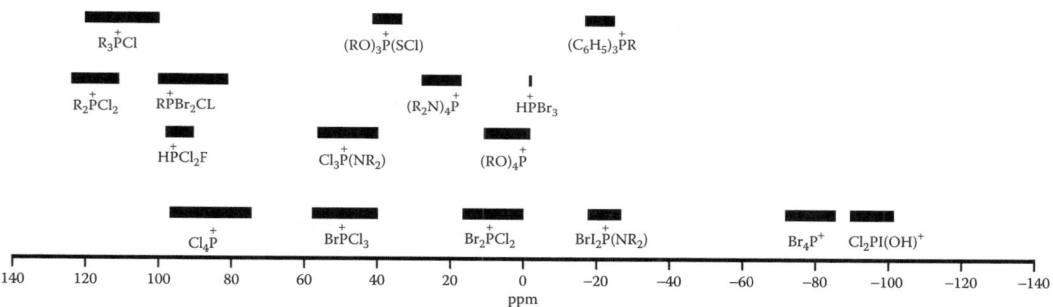

R$_3\overset{+}{P}$Cl

(RO)$_3$P(SCl)

(C$_6$H$_5$)$_3\overset{+}{P}$R

R$_2\overset{+}{P}$Cl$_2$ R$\overset{+}{P}$Br$_2$CL

(R$_2$N)$_4\overset{+}{P}$

H$\overset{+}{P}$Br$_3$

H$\overset{+}{P}$Cl$_2$F

Cl$_3\overset{+}{P}$(NR$_2$)

(RO)$_4\overset{+}{P}$

Cl$_4\overset{+}{P}$

Br$\overset{+}{P}$Cl$_3$

Br$_2\overset{+}{P}$Cl$_2$

BrI$_2\overset{+}{P}$(NR$_2$)

Br$_4$P$^+$ Cl$_2$PI(OH)$^+$

| 140 | 120 | 100 | 80 | 60 | 40 | 20 | 0 | −20 | −40 | −60 | −80 | −100 | −120 | −140 |

ppm

Tetracoordinated Phosphorus Compounds ($\geqslant$P=)

Abbreviation: R, R'=Alkyl

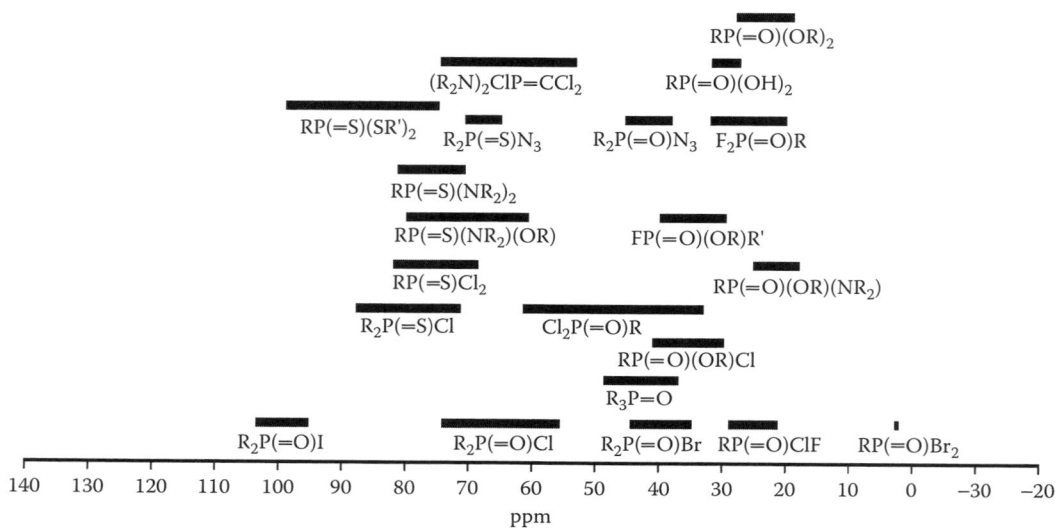

Pentacoordinated Phosphorus Compounds ($>$P$<$)

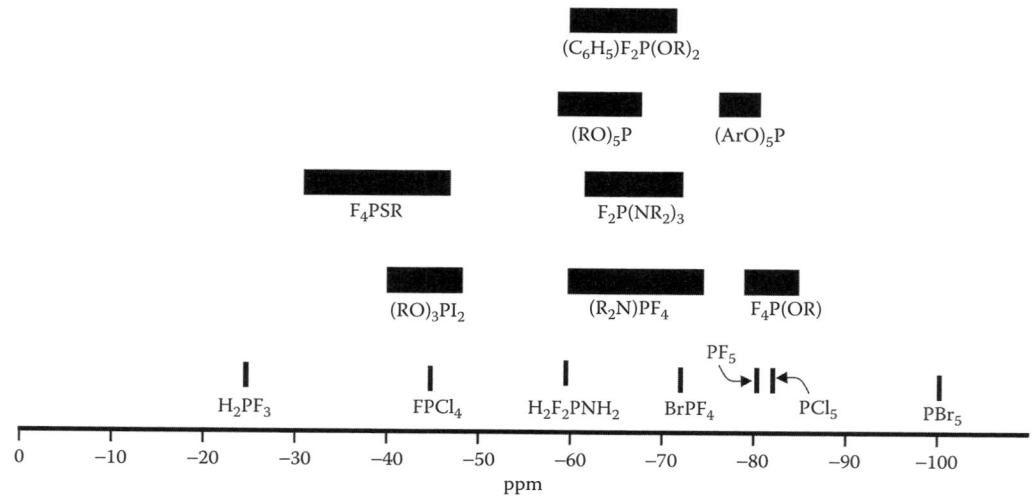

Hexacoordinated Phosphorus Compounds

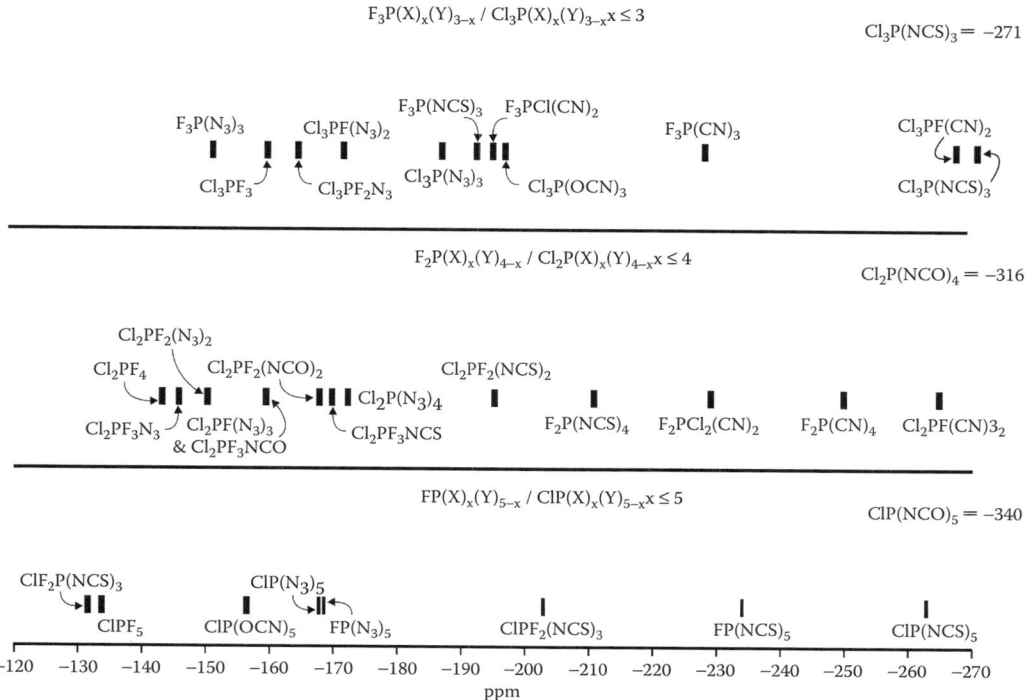

$F_3P(X)_x(Y)_{3-x}$ / $Cl_3P(X)_x(Y)_{3-x}$ x ≤ 3

$Cl_3P(NCS)_3 = -271$

$F_2P(X)_x(Y)_{4-x}$ / $Cl_2P(X)_x(Y)_{4-x}$ x ≤ 4

$Cl_2P(NCO)_4 = -316$

$FP(X)_x(Y)_{5-x}$ / $ClP(X)_x(Y)_{5-x}$ x ≤ 5

$ClP(NCO)_5 = -340$

Hexacoordinated Phosphorus Compounds

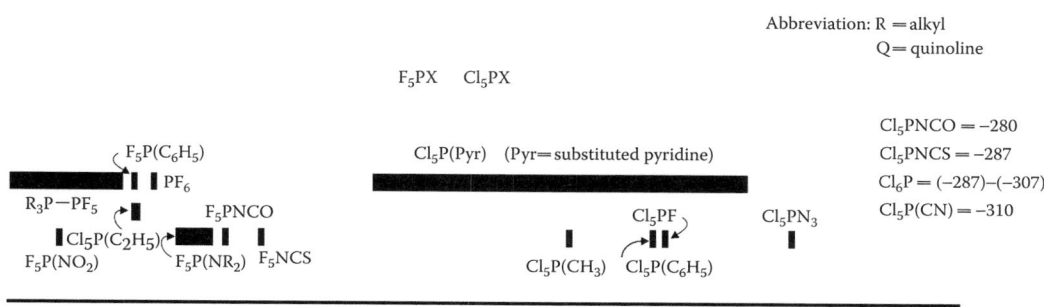

Abbreviation: R = alkyl
Q = quinoline

F_5PX Cl_5PX

$Cl_5PNCO = -280$
$Cl_5PNCS = -287$
$Cl_6P = (-287)–(-307)$
$Cl_5P(CN) = -310$

$F_4P(X)_x(Y)_{2-x}$ / $Cl_4P(X)_x(Y)_{2-x}$

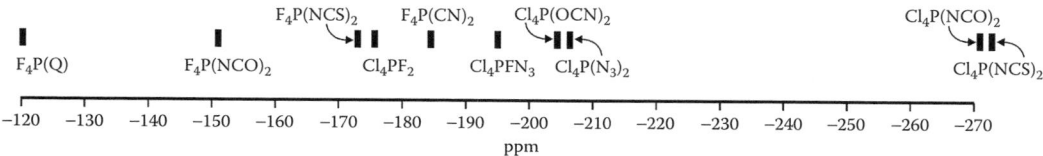

^{29}Si NMR ABSORPTIONS OF MAJOR CHEMICAL FAMILIES

The following correlation tables provide the regions of ^{29}Si nuclear magnetic resonance absorptions of some major chemical families. These absorptions are reported in the dimensionless units of parts per million (ppm) versus the standard compound tetramethylsilane (TMS, $(CH_3)_4Si$), which is recorded as 0.0 ppm.

The ^{29}Si NMR (natural abundance 4.7 %) is a low sensitivity nucleus that has a wide chemical shift range that is useful in determining the identity of certain silicon-containing compounds, many of which are important in biological systems. It should be noted that when taking such spectra there is always a background signal that is attributed to the glass comprising the measurement tube. Modifying the probe, which can be costly, or a simple adjustment in the pulse sequence, often overcomes the background signals. The ^{29}Si sensitivity is approximately 7.85×10^{-2} (at constant field) and 0.199 (at constant frequency) of that of ^1H.

For more detail concerning the chemical shifts, the reader is referred to the general references below and the literature cited therein [1–7].

REFERENCES

1. Williams, E. A., and J. D. Cargioli. "Silicon-29 NMR Spectroscopy." *Annual Reports on NMR Spectroscopy* 9 (1979): 221; 15 (1983): 235.
2. Schraml, J., and J. M. Bellama. "^{29}Si Nuclear Magnetic Resonance." In *Determination of Organic Structures by Physical Methods*. Vol. 6. Edited by F. C. Nachod and J. J. Zuckerman. New York: Academic Press, 1976.
3. Williams, E. A. "NMR Spectroscopy of Organosilicon Compounds." In *Chemistry and Physics of DNA-Ligand Interactions*. Edited by N. R. Kallenboch. New York: Adenine Press, 1990.
4. Schraml, J. "^{29}Si NMR Spectroscopy of Trimethyl Silyl Tags." Edited by J. W. Emsley, J. Feeneym, and L. H. Sutcliffe. *Progress in NMR Spectroscopy.*. 22 (1990): 289.
5. Granty, D. M., and R. K. Harris, eds. *Encyclopedia of Nuclear Magnetic Resonance*. Vol. 5. Chichester: John Wiley and Sons, 1996.
6. Mason, J. *Multinuclear NMR*. New York: Plenum Press, 1987.
7. Chemistry: Nuclear Magnetic Resonance: Multinuclear, accessed July 2010 http://avogadro.chem.iastate.edu/CHEM572/subpages/silicon.html

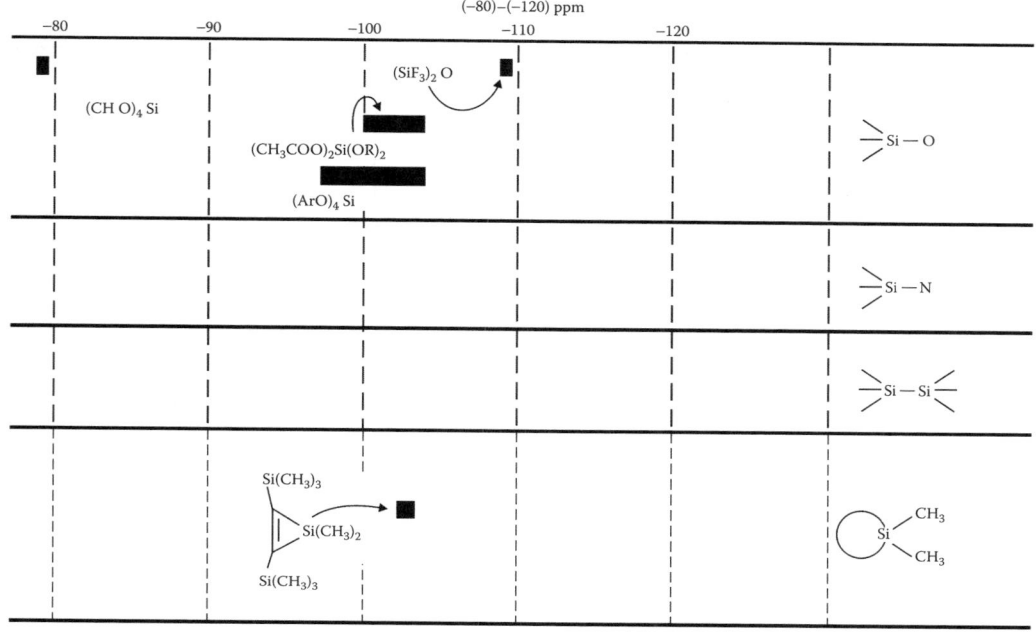

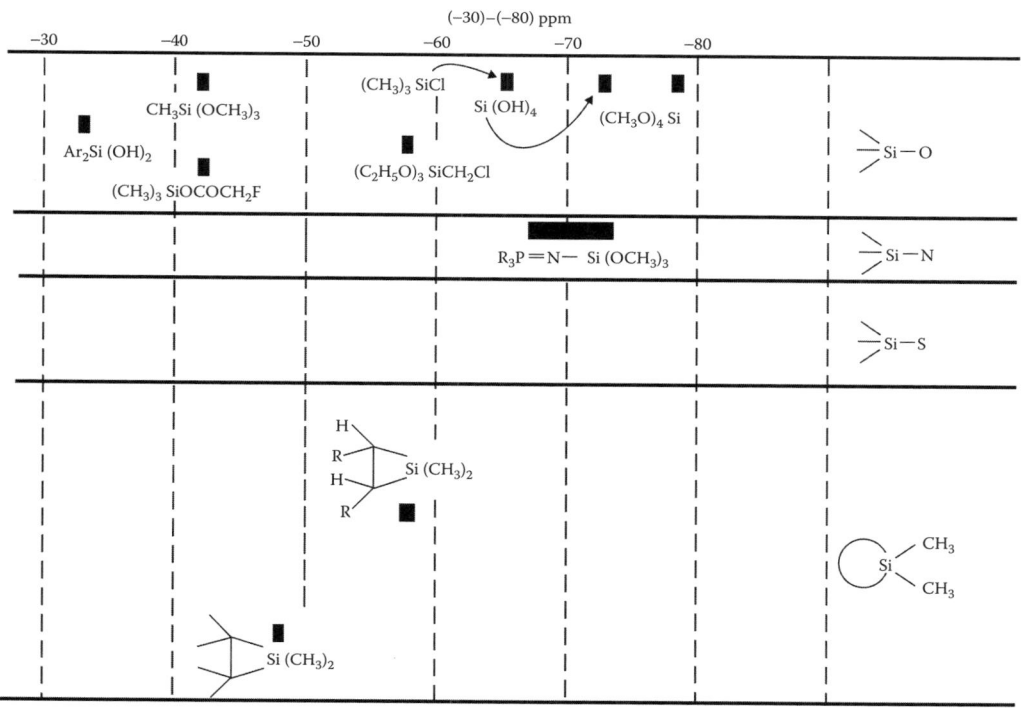

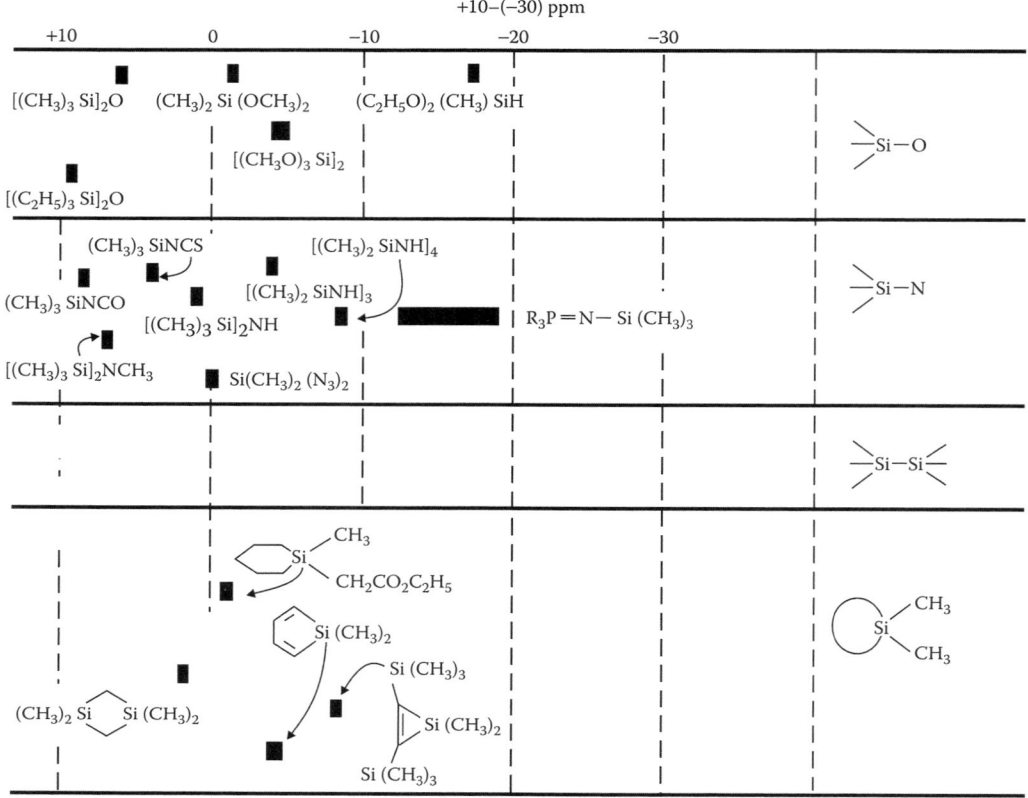

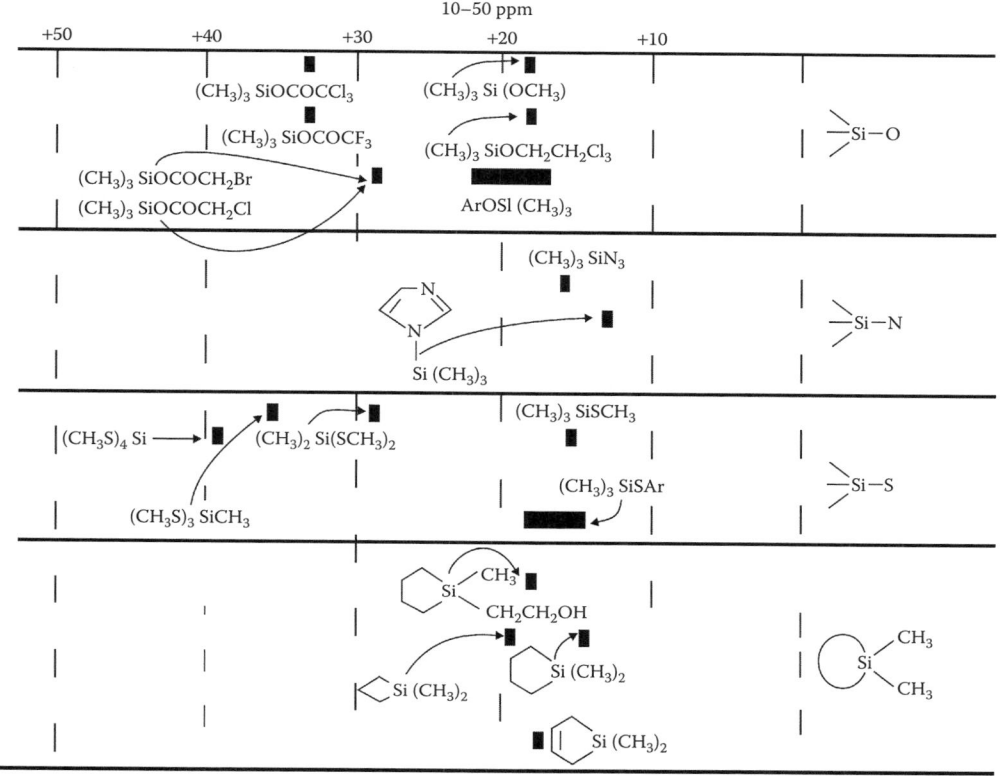

^{119}Sn NMR ABSORPTIONS OF MAJOR CHEMICAL FAMILIES

The following correlation tables provide the regions of ^{119}Sn nuclear magnetic resonance absorptions of some major chemical families. These absorptions are reported in the dimensionless units of parts per million (ppm) versus the standard compound tetramethylstanate $(CH_3)_4Sn$, which is recorded as 0.0 ppm.

The ^{119}Sn NMR is a low sensitivity nucleus that has a relatively wide chemical shift range, which is useful in determining the identification of certain tin-containing compounds. The ^{119}Sn sensitivity is approximately 5.18×10^{-2} (at constant field) that of ^1H. Its abundance (8.59 %) is slightly higher than that of ^{117}Sn (7.68 %) and much higher than that of ^{115}Sn (0.34 %).

For more detail concerning the chemical shifts, the reader is referred to the general references below and the literature cited therein [1–9]. The reader should be aware that there is a great deal of chemical shift variation when tin compounds are measured in different solvents. Moreover, many tin compounds are difficult to dissolve in common solvents.

REFERENCES

1. Petrosyan, V. S. "NMR Spectra and Structures of Organotin Compounds." *Progress in Nuclear Magnetic Resonance Spectroscopy* 11 (1978): 115.
2. Smith, P. J. "Chemical Shifts of Sn-119 Nuclei in Organotin Compounds." *Annual Reports on NMR Spectroscopy* 8 (1978): 292.
3. Wrackmeyer, B. "Tin-119 NMR Parameters." *Annual Reports on NMR Spectroscopy* 16 (1985): 73.
4. Hari, R., and R. A. Geanangel. "Tin-119 NMR in Coordination Chemistry." *Coordination Chemistry Reviews* 44 (1982): 229.
5. Wrackmeyer, B. "Multinuclear NMR and Tin Chemistry." *Chemistry in Britain* 26 (1990): 48.
6. Kaur, A., and G. K. Sandhu. "Use of ^{119}Sn Mossbauer and ^{119}Sn NMR Spectroscopies in the Study of Organotin Complexes." *Journal of Chemical Science* 2 (1986): 1.
7. Granty, D. M., and R. K. Harris, eds. *Encyclopedia of Nuclear Magnetic Resonance.*, Vol. 5. Chichester: John Wiley and Sons, 1996.
8. Mason, J. *Multinuclear NMR.* New York: Plenum Press, 1987.
9. The Basics, accessed July 2010 http://www.webelements.com/webelements/elements/text/Sn/nucl.html

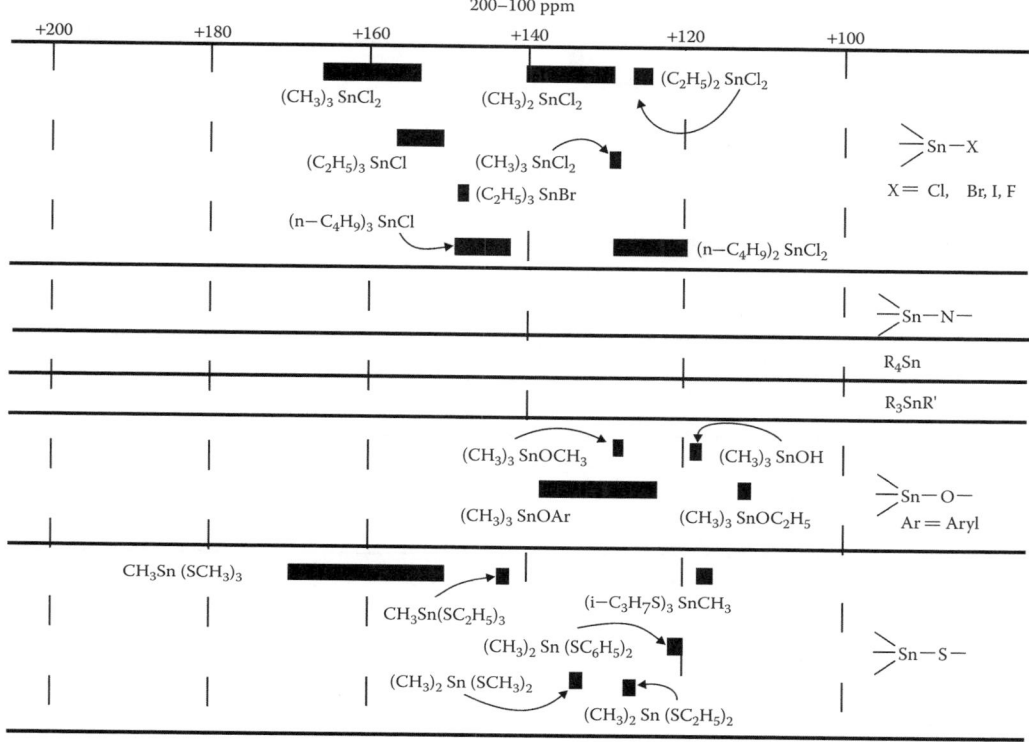

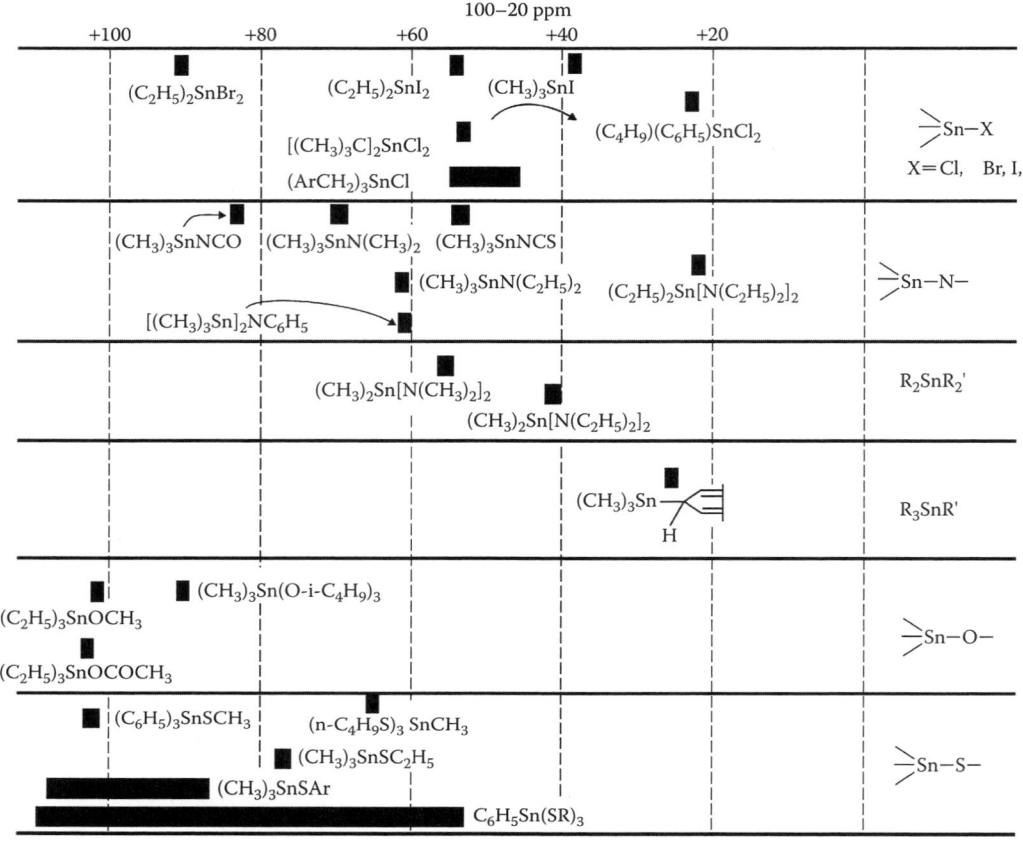

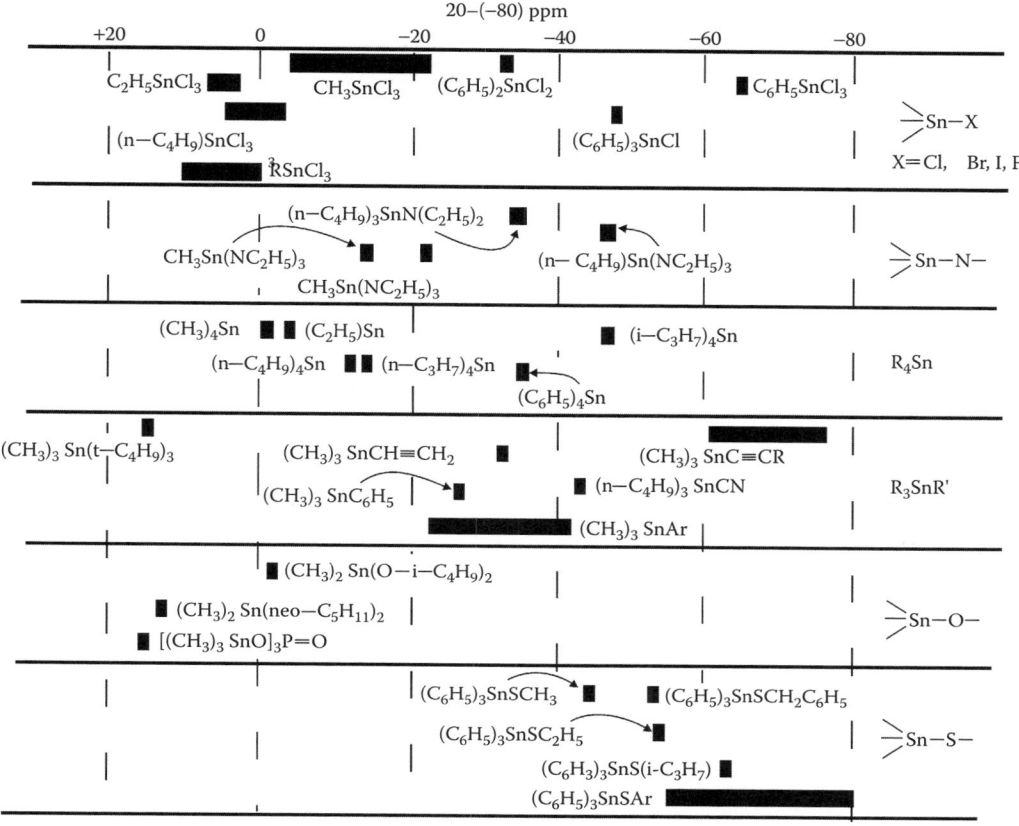

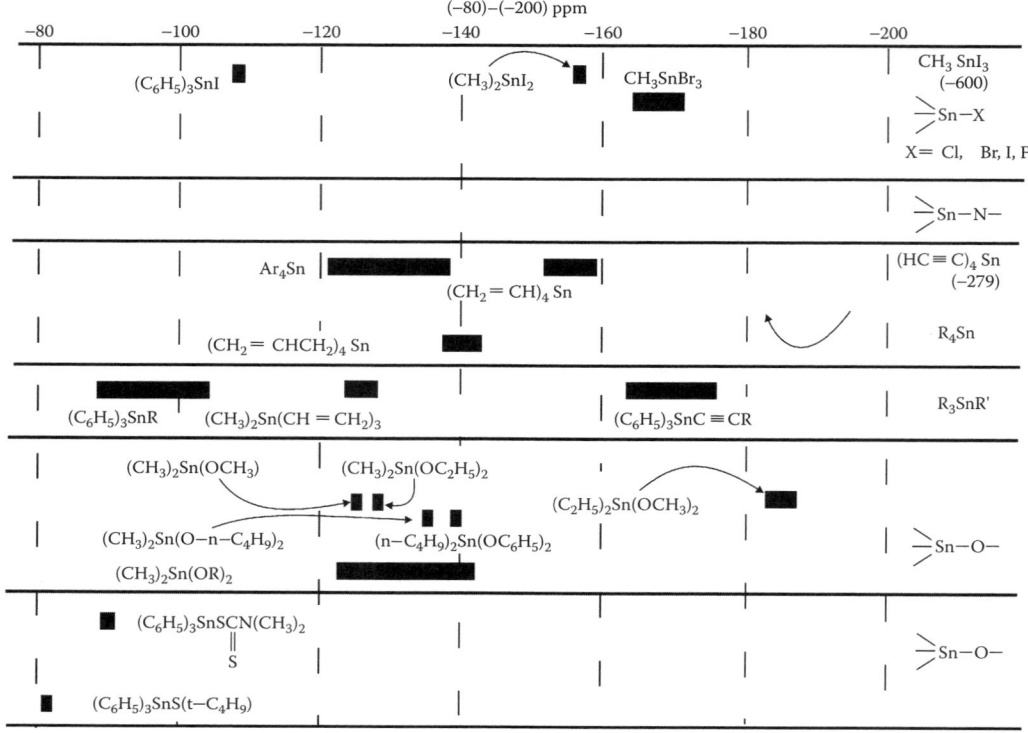

CHAPTER **10**

Mass Spectroscopy

CONTENTS

NATURAL ABUNDANCE OF IMPORTANT ISOTOPES

The following table lists the atomic masses and relative percentage concentrations of naturally occurring isotopes of importance in mass spectroscopy [1–5].

REFERENCES

1. deHoffmann, E., and V. Stroobant. *Mass Spectrometry: Principles and Applications.* 2nd ed. Chichester, U.K.: John Wiley and Sons, 2001.
2. Johnstone, R. A. W., and M. E. Rose. *Mass Spectrometry for Chemists and Biochemists.* Cambridge: Cambridge University Press, 1996.
3. Lide, D. R., ed. *CRC Handbook of Chemistry and Physics.* 83rd ed. Boca Raton, FL: CRC Press, 2002.
4. McLafferty, F. W., and F. Turecek. *Interpretation of Mass Spectra.* 4th ed. Mill Valley: University Science Books, 1993.
5. Watson, J. T. *Introduction to Mass Spectrometry.* 3rd ed. Philadelphia: Lippincott-Raven, 1997.

Element	Total # of Isotopes	More Prominent Isotopes (Mass, Percentage Abundance)		
Hydrogen	3	^{1}H (1.00783, 99.985)	^{2}H (2.01410, 0.015)	
Boron	6	^{10}B (10.01294, 19.8)	^{11}B (11.00931, 80.2)	
Carbon	7	^{12}C (12.00000, 98.9)	^{13}C (13.00335, 1.1)	
Nitrogen	7	^{14}N (14.00307, 99.6)	^{15}N (15.00011, 0.4)	
Oxygen	8	^{16}O (15.99491, 99.8)		^{18}O (17.9992, 0.2)
Fluorine	6	^{19}F (18.99840, ≈ 100.0)		
Silicon	8	^{28}Si (27.97693, 92.2)	^{29}Si (28.97649, 4.7)	^{30}Si (29.97376, 3.1)
Phosphorus	7	^{31}P (30.97376, ≈ 100.0)		
Sulfur	10	^{32}S (31.972017, 95.0)	^{33}S (32.97146, 0.7)	^{34}S (33.96786, 4.2)
Chlorine	11	^{35}Cl (34.96885, 75.5)		^{37}Cl (36.96590, 24.5)
Bromine	17	^{79}Br (78.9183, 50.5)		^{81}Br (80.91642, 49.5)
Iodine	23	^{127}I (126.90466, ≈ 100.0)		

RULES FOR DETERMINATION OF MOLECULAR FORMULA

The following rules are used in the mass spectroscopic determination of the molecular formula of an organic compound [1–6]. These rules should be applied to the molecular ion peak and its isotopic cluster. The molecular ion, in turn, is usually the highest mass in the spectrum. It must be an odd-electron ion, and must be capable of yielding all other important ions of the spectrum via a logical neutral species loss. The elements that are assumed to possibly be present on the original molecule are carbon, hydrogen, nitrogen, the halogens, sulfur, and/or oxygen. The molecular formula that can be derived is not the only possible one and consequently information from nuclear magnetic resonance spectrometry and infrared spectrophotometry is necessary for the final molecular formula determination.

Modern mass spectral databases allow the automated searching of very extensive mass spectral libraries. This has made the identification of compounds by mass spectrometry a far more straightforward task. However, one must understand that such databases are no substitute for the careful analysis of each mass spectrum, and that the results of database match-up are merely suggestions.

REFERENCES

1. Lee, T. A. *A Beginner's Guide to Mass Spectral Interpretation*. New York: Wiley, 1998.
2. McLafferty, F. W. *Interpretation of Mass Spectra*. Mill Valley, CA: University Science Books, 1993.
3. Shrader, S. R. *Introductory Mass Spectrometry*. Boston, MA: Allyn and Bacon, 1971.
4. Smith, R. M. *Understanding Mass Spectra: A Basic Approach*. New York: Wiley, 1999.
5. Watson, J. T., and T. J. Watson. *Introduction to Mass Spectrometry*. Philadelphia, PA: Lippincott, Williams and Wilkins, 1998.
6. NIST Standard Reference Database 1A. NIST/EPA/NIH Mass Spectral Library with Search Program: (Data Version: NIST '02, Software Version 2.0).

Rule 1

An odd molecular ion value suggests the presence of an odd number of nitrogen atoms; an even molecular ion value is due to the presence of zero, or an even number of nitrogen atoms. Thus, $m/z = 141$ suggests 1, 3, 5, 7, and so on, nitrogen atoms while $m/z = 142$ suggests 0, 2, 4, 6, and so on, nitrogen atoms.

Rule 2

The maximum number of carbons (N_C^{max}) can be calculated from the formula

$$(N_C^{max}) = \frac{\text{Relative intensity of M}+1\text{ peak}}{\text{Relative intensity of M}^+\text{ peak}} \times \frac{100}{1.1}$$

where $M + 1$ is the peak one unit above the value of the molecular ion (M^+). This rule gives the *maximum* number of carbons, but not necessarily the *actual* number. If, for example, the relative intensities of M^+ and $M + 1$ are 100 % and 9 %, respectively, then the maximum number of carbons is

$$(N_C^{max}) = (9/100) \times (100/1.1) = 8.$$

In this case there is a possibility for seven, six, and so on, carbons, but not for nine or more.

Rule 3

The maximum numbers of sulfur atoms ($N_S{}^{max}$) can be calculated from the formula

$$(N_S{}^{max}) = \frac{\text{Relative intensity of M} + 2 \text{ peak}}{\text{Relative intensity of M}^+ \text{ peak}} \times \frac{100}{4.4}$$

where M + 2 is the peak two units above that of the molecular ion M^+.

Rule 4

The actual number of chlorine and/or bromine atoms can be derived from the next table.

Rule 5

The difference should be only oxygen and hydrogen atoms. These rules assume the absence of phosphorus, silicon, or any other elements.

NEUTRAL MOIETIES EJECTED FROM SUBSTITUTED
BENZENE RING COMPOUNDS

The following table lists the most common substituents encountered in benzene rings and the neutral particles lost and observed on the mass spectrum [1]. Complex rearrangements are often encountered and are enhanced by the presence of one or more heteroatomic substituent(s) in the aromatic compound. All neutral particles that are not the product of rearrangement appear in parentheses and are produced alongside the species that are formed via rearrangement. Prediction of the more abundant moiety is not easy, as it is seriously affected by factors that dictate the nature of the compound. These include the nature and the position of any other substituents, as well as the stability of any intermediate(s) formed. Correlations of the data with the corresponding Hammett σ constants have been neither consistent nor conclusive.

REFERENCE

1. Rose, M. E., and R. A. W. Johnstone. *Mass Spectroscopy for Chemists and Biochemists*. Cambridge: Cambridge University Press, 1982.

Substituent	Neutral Moiety(s) Ejected after Rearrangement
NO_2	NO, CO, (NO_2)
NH_2	HCN
$NHCOCH_3$	C_2H_2O, HCN
CN	HCN
F	C_2H_2
OCH_3	CH_2O, CHO, CH_3
OH	CO, CHO
SO_2NH_2	SO_2, HCN
SH	CS, CHS (SH)
SCH_3	CS, CH_2S, SH, (CH_3)

ORDER OF FRAGMENTATION INITIATED BY THE PRESENCE
OF A SUBSTITUENT ON A BENZENE RING

The following table lists the relative order of ease of fragmentation that is initiated by the presence of a substituent in the benzene ring in mass spectroscopy [1]. The ease of fragmentation decreases from top to bottom. The substituents marked with an asterisk (*) are very similar in their ease of fragmentation. Particularly in the case of disubstituted benzene rings, the order of fragmentation at the substituent linkage may be easily predicted using this table. As a rule of thumb, the more complex the size of the substituent, the easier its decomposition. For instance, in all chloroacetophenone isomers (1,2-, 1,3-, or 1,4-), the elimination of the methyl radical occurs before the loss of chlorine. On the other hand under normal mass conditions all bromofluorobenzenes (1,2-, 1,3-, and 1,4-) easily lose the bromine but not the fluorine. Deuterium labeling studies have indicated that any rearrangement of the benzene compounds occur in the molecular ion and before fragmentation.

REFERENCE

1. Rose, M. E., and R. A. W. Johnstone. *Mass Spectroscopy for Chemists and Biochemists.* Cambridge: Cambridge University Press, 1982.

Substituent	Neutral Moiety Eliminated
$COCH_3$	CH_3
CO_2CH_3	OCH_3
NO_2	NO_2
*I	I
*OCH_3	CH_2O, CHO
*Br	Br
OH	CO, CHO
CH_3	H
Cl	Cl
NH_2	HCN
CN	HCN
F	C_2H_2

CHLORINE–BROMINE COMBINATION ISOTOPE INTENSITIES

Due to the distinctive mass spectral patterns caused by the presence of chlorine and bromine in a molecule, interpretation of a mass spectrum can be much easier if the results of the relative isotopic concentrations are known. The following table provides peak intensities (relative to the molecular ion (M^+) at an intensity normalized to 100 %) for various combinations of chlorine and bromine atoms, assuming the absence of all other elements except carbon and hydrogen [1–4]. The mass abundance calculations were based upon the most recent atomic mass data [1].

REFERENCES

1. Lide, D.R., ed, *CRC Handbook of Chemistry and Physics*, 90th. ed., Boca Raton, FL: CRC Press, 2010.
2. McLafferty, F. W. *Interpretation of Mass Spectra*. 4th ed. Mill Valley: University Science Books, 1993.
3. Silverstein, R. H., G. C. Bassler, and T. C. Morrill. *Spectroscopic Identification of Organic Compounds*. 6th ed. New York: John Wiley and Sons, 1998.
4. Williams, D. H., and I. Fleming. *Spectroscopic Methods in Organic Chemistry*. 4th ed. London: McGraw-Hill, 1989.

Relative Intensities of Isotope Peaks for Combinations of Bromine and Chlorine (M^+ = 100 %)

		Br_0	Br_1	Br_2	Br_3	Br_4
Cl_0	P + 2		98.0	196.0	294.0	390.8
	P + 4			96.1	288.2	574.7
	P + 6				94.1	375.3
	P + 8					92.0
Cl_1	P + 2	32.5	130.6	228.0	326.1	424.6
	P + 4		31.9	159.0	383.1	704.2
	P + 6			31.2	187.4	564.1
	P + 8				30.7	214.8
	P + 10					30.3
Cl_2	P + 2	65.0	163.0	261.1	359.3	456.3
	P + 4	10.6	74.4	234.2	490.2	840.3
	P + 6		10.4	83.3	312.8	791.6
	P + 8			10.2	91.7	397.5
	P + 10				9.8	99.2
	P + 12					10.1
Cl_3	P + 2	97.5	195.3	294.0	393.3	489
	P + 4	31.7	127.0	99.7	609.8	989
	P + 6	3.4	34.4	159.4	473.8	1064
	P + 8		3.3	37.1	193.9	654
	P + 10			3.2	39.6	229
	P + 12				3.0	42
	P + 14					3.2
Cl_4	P + 2	130.0	228.3	326.6	4.2	522
	P + 4	63.3	190.9	414.9	735.3	1149
	P + 6	13.7	75.8	263.1	670.0	1388
	P + 8	1.2	14.4	88.8	347.1	1002
	P + 10		1.1	15.4	102.2	443
	P + 12			1.3	16.2	117
	P + 14				0.7	17
Cl_5	P + 2	162.6	260.7	358.9		
	P + 4	105.7	265.3	520.8		
	P + 6	34.3	137.9	397.9		
	P + 8	5.5	39.3	174.5		
	P + 10	0.3	5.8	44.3		
	P + 12		0.3	5.7		
	P + 14			0.5		
Cl_6	P + 2	195.3				
	P + 4	158.6				
	P + 6	68.8				
	P + 8	16.6				
	P + 10	2.1				
	P + 12	0.1				
Cl_7	P + 2	227.8				
	P + 4	222.1				
	P + 6	120.3				
	P + 8	39.0				
	P + 10	7.5				
	P + 12	0.8				
	P + 14	0.05				

REFERENCE COMPOUNDS UNDER ELECTRON IMPACT
CONDITIONS IN MASS SPECTROMETRY

The following table lists the most popular reference compounds for use under electron impact conditions in mass spectrometry. For accurate mass measurements, the reference compound is introduced and ionized concurrently with the sample and the reference peaks are resolved from sample peaks. Reference compounds should contain as few heteroatoms and isotopes as possible. This is to facilitate the assignment of reference masses and minimize the occurrence of unresolved multiplets within the reference spectrum [1]. An approximate upper mass limit should assist in the selection of the appropriate Reference [1,2].

REFERENCES

1. Chapman, J. R. *Computers in Mass Spectrometry*. London: Academic Press, 1978.
2. Chapman, J. R. *Practical Organic Mass Spectrometry*. 2nd ed. Chichester: John Wiley and Sons, 1995.

Reference Compound	Formula	Upper Mass Limit
Perfluoro-2-butyltetrahydrofuran	$C_8F_{16}O$	416
Decafluorotriphenyl phosphine (ultramark 443; DFTPP)	$(C_6F_5)_3P$	443
Heptacosafluorotributylamine (perfluoro tributylamine; heptacosa; PFTBA)	$(C_4F_9)_3N$	671
Perfluoro kerosene, low-boiling (perfluoro kerosene-L)	$CF_3(CF_2)_nCF_3$	600
Perfluoro kerosene, high-boiling (perfluoro kerosene-H)	$CF_3(CF_2)_nCF_3$	800–900
Tris (trifluoromethyl)-s-triazine	$C_3N_3(CF_3)_3$	285
Tris (pentafluoroethyl)-s-triazine	$C_3N_3(CF_2CF_3)_3$	435
Tris (heptafluoropropyl)-s-triazine	$C_3N_3(CF_2CF_2CF_3)_3$	585
Tris (perfluoroheptyl)-s-triazine	$C_3N_3[(CF_2)_6CF_3]_3$	1185
Tris (perfluorononyl)-s-triazine	$C_3N_3[(CF_2)_8CF_3]_3$	1485
Ultramark 1621 (fluoroalkoxy cyclotriphosphazine mixture)	$P_3N_3[OCH_2(CF_2)_nH]_6$	~2000
Fomblin diffusion pump fluid (ultramark F-series; perfluoropolyether)	$CF_3O[CF(CF_3)CF_2O]_m(CF_2O)_nCF_3$	≥ 3000

MAJOR REFERENCE MASSES IN THE SPECTRUM OF
HEPTACOSAFLUOROTRIBUTYLAMINE (PERFLUOROTRIBUTYLAMINE)

The following list tabulates the major reference masses (with their relative intensities and formulas) of the mass spectrum of heptacosafluorotributylamine (perfluorotributylamine; heptacosa; PFTBA) [1]. This is one of the most widely used reference compounds in mass spectrometry.

REFERENCE

1. Chapman, J. R. *Practical Organic Mass Spectrometry*. 2nd ed. Chichester: John Wiley and Sons, 1995.

Mass	Relative Intensity	Formula	Mass	Relative Intensity	Formula
613.9647	2.6	$C_{12}F_{24}N$	180.9888	1.9	C_4F_7
575.9679	1.7	$C_{12}F_{22}N$	175.9935	1.0	C_4F_6N
537.9711	0.4	$C_{12}F_{20}N$	168.9888	3.6	C_3F_7
501.9711	8.6	$C_9F_{20}N$	163.9935	0.7	C_3F_6N
463.9743	3.8	$C_9F_{18}N$	161.9904	0.3	C_4F_6
425.9775	2.5	$C_9F_{16}N$	149.9904	2.1	C_3F_6
413.9775	5.1	$C_8F_{16}N$	130.9920	31	C_3F_5
375.9807	0.9	$C_8F_{14}N$	118.9920	8.3	C_2F_5
325.9839	0.4	$C_7F_{12}N$	113.9967	3.7	C_2F_4N
313.9839	0.4	$C_6F_{12}N$	111.9936	0.7	C_3F_4
263.9871	10	$C_5F_{10}N$	99.9936	12	C_2F_4
230.9856	0.9	C_5F_9	92.9952	1.1	C_3F_3
225.9903	0.6	C_5F_8N	68.9952	100	CF_3
218.9856	62	C_4F_9	49.9968	1.0	CF_2
213.9903	0.6	C_4F_8N	30.9984	2.3	CF

COMMON FRAGMENTATION PATTERNS OF
FAMILIES OF ORGANIC COMPOUNDS

The following table provides a guide to the identification and interpretation of commonly observed mass spectral fragmentation patterns for common organic functional groups [1–9]. It is, of course, highly desirable to augment mass spectroscopic data with as much other structural information as possible. Especially useful in this regard will be the confirmatory information of infrared and ultraviolet spectrophotometry as well as nuclear magnetic resonance spectrometry.

REFERENCES

1. Bowie, J. H., D. H. Williams, S. O. Lawesson, J. O. Madsen, C. Nolde, and G. Schroll. "Studies in Mass Spectrometry—XV. Mass Spectra of Sulphoxides and Sulphones. The Formation of C–C and C–O Bonds Upon Electron Impact." *Tetrahedron* 22 (1966): 3515.
2. Johnstone, R. A. W., and M. E. Rose. *Mass Spectrometry for Chemical and Biochemists*. Cambridge: Cambridge University Press, 1996.
3. Lee, T. A. *A Beginner's Guide to Mass Spectral Interpretation*. New York: Wiley, 1998.
4. McLafferty, F. W. *Interpretation of Mass Spectra*. 4th ed. Mill Valley, CA: University Science Books, 1993.
5. Pasto, D. J., and C. R. Johnson. *Organic Structure Determination*. Englewood Cliffs: Prentice-Hall, 1969.
6. Silverstein, R. M., G. C. Bassler, and T. C. Morrill. *Spectroscopic Identification of Organic Compounds.*, 6th ed. New York: John Wiley and Sons, 1998.
7. Smakman, R., and T. J. deBoer. "The Mass Spectra of Some Aliphatic and Alicyclic Sulphoxides and Sulphones." *Organic Mass Spectrometry* 3 (1970): 1561.
8. Smith, R. M. *Understanding Mass Spectra: A Basic Approach*. New York: Wiley, 1999.
9. Watson, T. J., and J. T. Watson. *Introduction to Mass Spectrometry*. Philadelphia: Lippincott, Williams and Wilkins, 1997.

Common Fragmentation Patterns of Families of Organic Compounds

Family	Molecular Ion Peak	Common Fragments; Characteristic Peaks
Acetals		Cleavage of all C–O, C–H and C–C bonds around the original aldehydic carbon.
Alcohols	weak for 1 ° and 2 °; not detectable for 3 °; strong for benzyl alcohols	Loss of 18 (H_2O, usually by cyclic mechanism); loss of H_2O and olefin simultaneously with four (or more) carbon-chain alcohols; prominent peak at $m/z = 31 (CH_2\ddot{O}H)^+$ for 1 ° alcohols; prominent peak at $m/z = (RCH\ddot{O}H)^+$ for 2 ° and $m/z = (R_2C\ddot{O}H)^+$ for 3 ° alcohols.
Aldehydes	low intensity	Loss of aldehydic hydrogen (strong M-1 peak, especially with aromatic aldehydes); strong peak at $m/z = 29$ ($HC{\equiv}O^+$); loss of chain attached to alpha carbon (beta cleavage); McLafferty rearrangement via beta cleavage if gamma hydrogen is present.
Alkanes		
a. Chain	low intensity	Loss of 14 units (CH_2).
b. Branched	low intensity	cleavage at the point of branch; low intensity ions from random rearrangements.
c. Alicyclic	rather intense	Loss of 28 units ($CH_2{=}CH_2$) and side chains.
Alkenes (olefins)	rather high intensity (loss of π-electron) especially in case of cyclic olefins	Loss of units of general formula C_nH_{2n-1}; formation of fragments of the composition C_nH_{2n} (via McLafferty rearrangement); retro Diels-Alder fragmentation.
Alkyl halides	abundance of molecular ion F<Cl<Br<I; intensity decreases with increase in size and branching	Loss of fragments equal to the mass of the halogen until all halogens are cleaved off.
a. fluorides	very low intensity	Loss of 20 (HF); loss of 26 (C_2H_2) in case of fluorobenzenes.
b. chlorides	low intensity; characteristic isotope cluster	Loss of 35 (Cl) or 36 (HCl); loss of chain attached to the gamma carbon to the carbon carrying the Cl.
c. bromides	low intensity; characteristic isotope cluster	Loss of 79 (Br); loss of chain attached to the gamma carbon to the carbon carrying the Br.
d. iodides	higher than other halides	Loss of 127 (I).
Alkynes	rather high intensity (loss of β-electron)	Fragmentation similar to that of alkenes.
Amides	rather high intensity	Strong peak at $m/z = 44$ indicative of a 1 ° amide ($O{=}C{=}NH_2^+$); base peak at $m/z = 59$ ($CH_2{=}C(OH)NH_2^+$); possibility of McLafferty rearrangement; loss of 42 (C_2H_2O) for amides of the form $RNHCOCH_3$ when R is aromatic ring.
Amines	hardly detectable in case of acyclic aliphatic amines; high intensity for aromatic and cyclic amines	Beta cleavage yielding $>C{=}N^+<$; base peak for all 1 ° amines at $m/z = 30$ ($CH_2{=}N^+H_2$); moderate M-1 peak for aromatic amines; loss of 27 (HCN) in aromatic amines; fragmentation at alpha carbons in cyclic amines.
Aromatic hydrocarbons (arenes)	rather intense	Loss of side chain; formation of RCH=CHR′ (via McLafferty rearrangement); cleavage at the bonds beta to the aromatic ring; peaks at $m/z = 77$ (benzene ring; especially mono-substituted), 91 (tropyllium); the ring position of alkyl substitution has very little effect on the spectrum.
Carboxylic acids	weak for straight-chain monocarboxylic acids; large if aromatic acids	Base peak at $m/z = 60$ ($CH_2{=}C(OH)_2$) if α-hydrogen is present; peak at $m/z = 45$ (COOH); loss of 17 (–OH) in case of aromatic acids or short-chain acids.
Disulfides	rather low intensity	Loss of olefins (m/z equal to $R{-}S{-}S{-}H^{+\cdot}$); strong peak at $m/z = 66$ ($HSSH^+$).

(Continued)

Common Fragmentation Patterns of Families of Organic Compounds (Continued)

Family	Molecular Ion Peak	Common Fragments; Characteristic Peaks
Phenols	highly intense peak (base peak* generally)	Loss of 28 (C=O) and 29 (CHO); strong peak at $m/z = 65$ ($C_5H_5^+$).
Sulfides (thioethers)	rather low intensity peak but higher than that of corresponding ether	Similar to those of ethers (–O– substituted by –S–); aromatic sulfides show strong peaks at $m/z = 109$ ($C_6H_5S^+$); 65 ($C_5H_5^+$); 91 (tropyllium ion).
Sulfonamides	rather intense	Loss of $m/z = 64$ ($SONH_2$) and $m/z = 27$ (HCN) in case of benzenesulfonamide.
Esters	rather weak intensity	Base peak at m/z equal to the mass of $R–C≡O^+$; peaks at m/z equal to the mass of $^+O≡C–OR'$, the mass of OR' and R'; McLafferty rearrangement possible in case of: (a) presence of a beta hydrogen in R' (peak at m/z equal to the mass of $R–C(^+OH)OH$, and (b) presence of a gamma hydrogen in R (peak at m/z equal to the mass of $(CH_2=C(^+OH)OR)$; loss of 42 ($CH_2=C=O$) in case of benzyl esters; loss of ROH via the ortho effect in case of o–substituted benzoates.
Ketones	rather high intensity peak	Loss of R–groups attached to the>C=O (alpha cleavage); peak at $m/z = 43$ for all methyl ketones (CH_3CO^+); McLafferty rearrangement via beta cleavage if γ-hydrogen is present; loss of $m/z = 28$ (C=O) for cyclic ketones after initial alpha cleavage and McLafferty rearrangement.
Mercaptan (thiols)	rather low intensity but higher than that of corresponding alcohol	Similar to those of alcohols (–OH substituted by –SH); loss of $m/z = 45$ (CHS) and $m/z = 44$ (CS) for aromatic thiols.
Nitriles	unlikely to be detected except in case of acetonitrile (CH_3CN) and propionitrile (C_2H_5CN)	M + 1 ion may appear (especially at higher pressures); M – 1 peak is weak but detectable ($R–CH=C=N^+$); base peak at $m/z = 41$. ($CH_2=C=N^+ H$); McLafferty rearrangement possible; loss of HCN is case of cyanobenzenes.
Nitrites	absent (or very weak at best)	Base peak at $m/z = 30$ (NO^+); large peak at $m/z = 60$ ($CH_2=ONO^+$) in all unbranched nitrites at the α-carbon; absence of $m/z = 46$ permits differentiation from nitrocompounds.
Nitro compounds	seldom observed	Loss of 30 (NO); subsequent loss of CO (in case of aromatic nitro-compounds); loss of NO_2 from molecular ion peak.
Sulfones	high intensity	Similar to sulfoxides; loss of mass equal to RSO_2; aromatic heterocycles show peaks at M-32 (sulfur), M-48 (SO), M-64 (SO_2).
Sulfoxides	high intensity	Loss of 17 (OH); loss of alkene (m/z equal to $RSOH^+$); peak at $m/z = 63$ ($CH_2=SOH$)$^+$; aromatic sulfoxides show peak at $m/z = 125$ ($^+S–CH=CHCH=CHC=O$), $97(C_5H_5S^+)$, $93(C_6H_5OH)$; aromatic heterocycles show peaks at M-16 (oxygen), M-29(COH); M-48(SO).

* The base peak is the most intense peak in the mass spectrum, and is often the molecular ion peak, M^+.

COMMON FRAGMENTS LOST

The following table gives a list of neutral species that are most commonly lost when measuring the mass spectra of organic compounds. The list is suggestive rather than comprehensive, and should be used in conjunction with other sources [1–4]. The listed fragments include only combinations of carbon, hydrogen, oxygen, nitrogen, sulfur, and the halogens.

REFERENCES

1. Hamming, M., and N. Foster. *Interpretation of Mass Spectra of Organic Compounds.* New York: Academic Press, 1972.
2. McLafferty, F. W. *Interpretation of Mass Spectra.* 4th ed. Mill Valley: University Science Books, 1993.
3. Silverstein, R. M., G. C. Bassler, and T. C. Morrill. *Spectroscopic Identification of Organic Compounds.* 6th ed. New York: John Wiley and Sons, 1996.
4. Bruno, T. J. *CRC Handbook for the Analysis and Identification of Alternative Refrigerants.* Boca Raton, FL: CRC Press, 1995.

Mass Lost	Fragment Lost	Mass Lost	Fragment Lost
1	H$\cdot$	51	$\cdot$CHF$_2$
15	CH$_3\cdot$	52	C$_4$H$_4\cdot$, C$_2$N$_2$
17	OH$\cdot$	54	CH$_2$=CHCH=CH$_2$
18	H$_2$O	55	CH$_2$=CH–CH$\cdot$CH$_3$
19	F$\cdot$	56	CH$_2$=CH–CH$_2$CH$_3$; CH$_3$CH=CHCH$_3$; CO (2 moles)
20	HF	57	C$_4$H$_9\cdot$
26	HC≡CH; $\cdot$C≡N	58	$\cdot$NCS; (CH$_3$)$_2$C=O; (NO and CO)
27	CH$_2$–CH$\cdot$; HC≡N	59	CH$_3$OC≡O$\cdot$; CH$_3$CONH$_2$; C$_2$H$_3$S$\cdot$
28	CH$_2$=CH$_2$; C=O; (HCN and H$\cdot$)	60	C$_3$H$_7$OH
29	CH$_3$CH$_2\cdot$; H–$\cdot$C=O	61	CH$_3$CH$_2$S$\cdot$; (CH$_2$)$_2$S$\cdot$H
30	$\cdot$CH$_2$NH$_2$; HCHO; NO	62	[H$_2$S and CH$_2$=CH$_2$]
31	CH$_3$O$\cdot$; $\cdot$CH$_2$OH; CH$_3$NH$_2$	63	$\cdot$CH$_2$CH$_2$Cl
32	CH$_3$OH; S	64	S$_2\cdot$, SO$_2\cdot$, C$_5$H$_4\cdot$
33	HS$\cdot$	68	CH$_2$=CHC(CH$_3$)=CH$_2$
34	H$_2$S	69	CF$_3\cdot$; C$_5$H$_9\cdot$
35	Cl$\cdot$	71	C$_5$H$_{11}\cdot$
36	HCl,H$_2$O	73	CH$_3$CH$_2$OC$\cdot$=O
37	H$_2$Cl	74	C$_4$H$_9$OH
38	C$_3$H$_2\cdot$; C$_2$N; F$_2$	75	C$_6$H$_3$
39	C$_3$H$_3$; HC$_2$N	76	C$_6$H$_4$; CS$_2$
40	CH$_3$C≡CH	77	C$_6$H$_5$; HCS$_2$
41	CH$_2$=CHCH$_2\cdot$	78	C$_6$H$_6\cdot$,H$_2$CS$_2\cdot$, C$_5$H$_4$N
42	CH$_2$=CHCH$_3$; CH$_2$=C=O; (CH$_2$)$_3$; NCO; NCNH$_2$	79	Br$\cdot$; C$_5$H$_5$N
43	C$_3$H$_7\cdot$; CH$_3$C=O$\cdot$; CH$_2$=CH–O$\cdot$; HCNO	80	HBr
44	CH$_2$=CHOH; CO$_2$; N$_2$O; CONH$_2$; NHCH$_2$CH$_3$	85	$\cdot$CClF$_2$
45	CH$_3$CHOH; CH$_3$CH$_2$O$\cdot$; CO$_2$H; CH$_3$CH$_2$NH$_2$	100	CF$_2$=CF$_2$
46	CH$_3$CH$_2$OH; $\cdot$NO$_2$	119	CF$_3$CF$_2\cdot$
47	CH$_3$S$\cdot$	122	C$_6$H$_5$CO$_2$H
48	CH$_3$SH; SO; O$_3$	127	I$\cdot$
49	$\cdot$CH$_2$Cl	128	HI

IMPORTANT PEAKS IN THE MASS SPECTRA OF COMMON SOLVENTS

The following table gives the most important peaks that appear in the mass spectra of the most common solvents, which may be found as an impurity in organic samples. The solvents are classified in ascending order, based upon their M+ peaks. The highest intensity peaks are indicated with (100 %) [1–4].

REFERENCES

1. Clere, J. T., E. Pretsch, and J. Seibl. *Studies in Analytical Chemistry I. Structural Analysis of Organic Compounds by Combined Application of Spectroscopic Methods.* Amsterdam: Elsevier, 1981.
2. McLafferty, F. W. *Interpretation of Mass Spectra.* 4th ed. Mill Valley, CA: University Science Books, 1993.
3. Pasto, D. J., and C. R. Johnson. *Organic Structure Determination.* Englewood Cliffs: Prentice Hall, 1969.
4. Smith, R. M. *Understanding Mass Spectra: A Basic Approach.* New York: Wiley, 1999.

Solvents	Formula	M+	Important Peaks (m/z)
Water	H_2O	18 (100 %)	17
Methanol	CH_3OH	32	31 (100 %), 29, 15
Acetonitrile	CH_3CN	41 (100 %)	40, 39, 38, 28, 15
Ethanol	CH_3CH_2OH	46	45, 31 (100 %), 27, 15
Dimethylether	CH_3OCH_3	46 (100 %)	45, 29, 15
Acetone	CH_3COCH_3	58	43 (100 %), 42, 39, 27, 15
Acetic acid	CH_3CO_2H	60	45, 43, 18, 15
Ethylene glycol	$HOCH_2CH_2OH$	62	43, 33, 31 (100 %), 29, 18, 15
Furan	C_4H_4O	68 (100 %)	42, 39, 38, 31, 29, 18
Tetrahydrofuran	C_4H_8O	72	71, 43, 42 (100 %), 41, 40, 39, 27, 18, 15
n-Pentane	C_5H_{12}	72	57, 43 (100 %), 42, 41, 39, 29, 28, 27, 15
Dimethylformamide (DMF)	$HCON(CH_3)_2$	73 (100 %)	58, 44, 42, 30, 29, 28, 18, 15
Diethylether	$(C_2H_5)_2O$	74	59, 45, 41, 31 (100 %), 29, 27, 15
Methylacetate	$CH_3CO_2CH_3$	74	59, 43 (100 %),42, 32, 29, 28, 15
Carbon disulfide	CS_2	76 (100 %)	64, 44, 32
Benzene	C_6H_6	78 (100 %)	77, 52, 51, 50, 39, 28
Pyridine	C_5H_5N	79 (100 %)	80, 78, 53, 52, 51, 50, 39, 26
Dichloromethane	CH_2Cl_2	84	86, 51, 49 (100 %), 48, 47, 35, 28
Cyclohexane	C_6H_{12}	84	69, 56, 55, 43, 42, 41, 39, 27
n-Hexane	C_6H_{14}	86	85, 71, 69, 57 (100 %), 43, 42, 41, 39, 29, 28, 27
p-Dioxane	$C_4H_8O_2$	88 (100 %)	87, 58, 57, 45, 43, 31, 30, 29, 28
Tetramethylsilane (TMS)	$(CH_3)_4Si$	88	74, 73, 55, 45, 43, 29
1,2-Dimethoxy ethane	$(CH_3OCH_2)_2$	90	60, 58, 45 (100 %), 31, 29
Toluene	$C_6H_5CH_3$	92	91 (100 %), 65, 51, 39, 28
Chloroform	$CHCl_3$	118	120, 83, 81 (100 %), 47, 35, 28
Chlorodorm-d$_1$	$CDCl_3$	119	121, 84, 82 (100 %), 48, 47, 35, 28
Carbon tetrachloride	CCl_4	152 (not seen)	121, 119, 117 (100 %), 84, 82, 58.5, 47, 35, 28
Tetrachloroethene	$CCl_2=CCl_2$	164 (not seen)	168, 166 (100 %), 165, 164, 131, 129, 128, 94, 82, 69, 59, 47, 31, 24

REAGENT GASES FOR CHEMICAL IONIZATION MASS SPECTROMETRY

The following tables provide guidance in the selection and optimization of reagents in high-pressure chemical ionization mass spectrometry, as applied with gas chromatography or as a stand-alone technique [1–3]. The first table provides data on positive ion reagent gases, which are called Bronsted acid reagents. Here, we provide the proton affinity (PA) of the conjugate base, the hydride ion affinity (the enthalpy of the reaction of the positive ion with H⁻). The second table provides data on negative ion reagent gases, which are called Bronsted base reagents. Here, we provide the proton affinity of the negative ion and the electron affinity of the base.

REFERENCES

1. Harrison, A. G. *Chemical Ionization Mass Spectrometry*. Boca Raton, FL: CRC Press, 1992.
2. Message, G. M. *Practical Aspects of Gas Chromatography/Mass Spectrometry*. New York: John Wiley and Sons (Wiley Interscience), 1984.
3. Karasek, F. W., and R. E. Clement. *Basic Gas Chromatography—Mass Spectrometry*. Amsterdam: Elsevier, 1988.

Reagent Gases for Chemical Ionization Mass Spectrometry

Reagent Gas	Reactant Ion(s)	PA kJ/mol	PA kcal/mol	HIA kJ/mol	HIA kcal/mol	Comments
		Positive Ion Reagent Gases for Chemical Ionization Mass Spectrometry				
H_2	H_3^+	423.7	101.2	1260	300	General purpose reagent gas
$N_2 + H_2$	N_2H^+	494.9	118.2	1180	282	
$CO_2 + H_2$	CO_2H^+	547.6	130.8	1130	270	
$N_2O + H_2$	N_2OH^{+2}	581.1	138.8	1090	261	Significant signals observed for NO^+.
$CO + H_2$	HCO^+	596.2	142.4	1080	258	
CH_4	CH_5^+	551.0	131.6	1130	269	Most widely used reagent gas; usually used initially for most work; degree of fragmentation is relatively large; background spectrum is often large; can produce a large number of addition ions and quasi-molecular ions.
	$C_2H_5^+$	680.8	162.6	1130	271	
H_2O	$H^+(H_2O)_x$ x is pressure dependent	697.1	166.5	980	234	Used for alcohols, ketones, esters, and amines
CH_3OH	$H^+(CH_3OH)_x$ x is pressure dependent	761.6	181.9	917	219	
C_3H_8	$C_3H_7^+$	751.5	179.5	1050	250	Uncommon reagent gas
i-C_4H_{10}	$C_4H_9^+$	820.2	195.9	976	233	General purpose reagent gas; fragmentation pattern is similar to that produced by ammonia.
NH_3	$H^+(NH_3)_x$ x is pressure dependent	854.1	204.0	825	197	
		Negative Ion Reagent Gases for Chemical Ionization Mass Spectrometry				
H_2	H^-	1675	400	72.9	17.4	H^- ion is difficult to form in good yields; sometimes used for analysis of alcohols.
NH_3	NH_2^-	1691	404	75.4	18.0	General purpose gas, used for the analysis of esters.
N_2O	OH^-	1637	391	177	42.2	Most common negative ion reagent gas used; often used as a mixture with N_2O, to eliminate O^- signal; sometimes used as a $N_2O/He/N_2O$, 1:1:1 mixture; used with CH_4 for simultaneous $+/-$ ion work.
CH_3NO_2	CH_3O^-	1595	381	152	36.2	Almost as strong a base as OH^-; used as a 1 % mixture in CH_4.
O_2	O_2^-	1478	353	42.3	10.1	Used in the analysis of alcohols.
$C_2Cl_3F_3$ (R-113)	Cl^-	1394	333	349	83.4	Cl^- is a weak Bronsted base useful for acidic compounds.
CH_2Br_2	Br^-	1357	324	325	77.6	Br^- is a weak Bronsted base (weaker than Cl^-), which reacts with analytes that have a moderately acidic hydrogen.

PROTON AFFINITIES OF SOME SIMPLE MOLECULES

The following table gives the proton affinities (PA) of some simple molecules. For the occurrence of proton transfer (or reaction) between a reactant ion and a sample molecule, the reaction must be exothermic. Thus,

$$\Delta H \text{ reaction} = PA \text{ (reactant gas)} - PA \text{ (sample)} < 0.$$

The more exothermic the reaction, the greater the degree of fragmentation. Endothermic reactions do not yield a protonated form of a sample, therefore the sample compound cannot be recorded. One can choose the proper reactant gas that will give the correct fragmentation pattern of a desired compound out of a mixture of compounds [1–3]. Chapman [3] lists positive ion chemical ionization applications by reagent gas and by compounds analyzed. The values are provided in kcal/mol for convenience; to convert to the appropriate SI unit (kJ/mol), multiply by 4.1845.

REFERENCES

1. Field, F. H. "Chemical Ionization Mass Spectrometry." *Accounts of Chemical Research* 1 (1968): 42.
2. Harrison, A. G. *Chemical Ionization Mass Spectrometry.* 2nd ed. Boca Raton, FL: CRC Press, 1992.
3. Chapman, J. R. *Practical Organic Mass Spectrometry.* 2nd ed. Chichester: John Wiley and Sons, 1995.

Family	Typical Examples (PA in kcal/mol)
Alcohols	CH_3OH (184.9); CH_3CH_2OH (190.3); $CH_3CH_2CH_2OH$ (191.4); $(CH_3)_3COH$ (195.0); CF_3CH_2OH (174.9)
Aldehydes	$HCHO$ (177.2); CH_3CHO (188.9); CH_3CH_2CHO (191.4); $CH_3CH_2CH_2CHO$ (193.3)
Alkanes	CH_4 (130.5); $(CH_3)_3CH$ (195)
Alkenes	$H_2C=CH_2$ (163.5); $CH_3CH=CH_2$ (184.9); $(CH_3)_2C=CH_2$ (196.9); trans-$CH_3CH=CHCH_3$ (182.0)
Amines	1 °: NH_3 (205.0); CH_3NH_2 (214.1); $C_2H_5NH_2$ (217.1); $CH_3CH_2CH_2NH_2$ (218.5); $CH_3CH_2CH_2CH_2NH_2$ (219.0) 2 °: $(CH_3)_2NH$ (220.5); $(C_2H_5)_2NH$ (225.1); $(CH_3CH_2CH_2)_2NH$ (227.4) 3 °: $(CH_3)_3N$ (224.3); $(C_2H_5)_3N$ (231.2); $(CH_3CH_2CH_2)_3N$ (233.4)
Aromatics, substituted C_6H_5-G	G=–H (182.8); –Cl (181.7); –F (181.5); –CH_3 (191.2); –C_2H_5 (192.2); –$CH_2CH_2CH_3$ (191.0); –$CH(CH_3)_2$ (191.4); –$C(CH_3)_3$ (191.6); –NO_2 (193.8); –OH (196.2); –CN (196.3); –CHO (200.3); –OCH_3 (200.6); –NH_2 (211.5)
Carboxylic acids	HCO_2H (182.8); CH_3CO_2H (190.7); $CH_3CH_2CO_2H$ (193.4); CF_3CO_2H (176.0)
Dienes	$CH_2=CHCH=CH_2$ (193); E–$CH_2=CHCH=CHCH_3$ (201.8); E–$CH_2=CHC(CH_3)=CHCH_3$ (205.7); cyclopentadiene (200.0)
Esters	HCO_2CH_3 (190.4); $HCO_2C_2H_5$ (194.2); $HCO_2CH_2CH_2CH_3$ (195.2); $CH_3CO_2CH_3$ (198.3); $CH_3CO_2C_2H_5$ (201.3); $CH_3CO_2CH_2CH_2CH_3$ (202.0)
Ethers	$(CH_3)_2O$ (193.1); $(C_2H_5)_2O$ (200.4); $(CH_3CH_2CH_2)_2O$ (202.9); $(CH_3CH_2CH_2CH_2)_2O$ (203.9); tetrahydrofuran (199.6); tetrahydropyran (200.7)
Ketones	CH_3COCH_3 (197.2); $CH_3COC_2H_5$ (199.4)
Nitriles (cyano compounds)	HCN (178.9); CH_3CN (190.9); C_2H_5CN (192.8); $CH_3CH_2CH_2CN$ (193.8)
Sulfides	$(CH_3)_2S$ (200.7); $(C_2H_5)_2S$ (205.6); $[(CH_3)_2CH]_2S$ (209.3)
Thiols	H_2S (176.6); CH_3SH (188.6); C_2H_5SH (192.0); $[(CH_3)_2CH]_2SH$ (194.7)

PROTON AFFINITIES OF SOME ANIONS

The following table lists the proton affinities of some common anions (X^-). Since the reaction of an anion (X^-) with a proton (H^+)

$$X^- + H+ \rightarrow H-X$$

is exothermic, it can be used to generate other anions that possess a smaller proton affinity value by the addition of the corresponding neutral species [1,2].

REFERENCES

1. Chapman, J. R. *Practical Organic Mass Spectrometry*. 2nd ed. Chichester: John Wiley and Sons, 1995.
2. Harrison, A. G. *Chemical Ionization Mass Spectrometry*. 2nd ed. Boca Raton, FL: CRC Press, 1992.

Anion	Proton Affinity (kJ/mol)
NH_2^-	1689
H^-	1676
OH^-	1636
$O^{\cdot-}$	1595
CH_3O^-	1583
$(CH_3)_2CHO^-$	1565
$^-CH_2CN$	1556
F^-	1554
$C_5H_5^-$	1480
$O_2^{\cdot-}$	1465
CN^-	1462
Cl^-	1395

DETECTION OF LEAKS IN MASS SPECTROMETER SYSTEMS

The following tables provide guidance for troubleshooting possible leaks in the vacuum systems of mass spectrometers, especially those operating in electron impact mode. Leak testing is commonly done by playing a stream of a pure gas against a fitting, joint, or component that is suspected of being a leak source. If in fact the component is the source of a leak, one should be able note the presence of the leak detection fluid on the mass spectrum. Here we present the mass spectra of methane tetrafluoride, 1,1,1,2-tetrafluoroethane (R-134a), n-butane, and acetone [1,2]. Methane tetrafluoride, 1,1,1,2-tetrafluoroethane, and n-butane are handled as gases, while acetone is handled as a liquid. Typically, n-butane is dispensed from a disposable lighter, and acetone is dispensed from a dropper. Care must be taken when using acetone or a butane lighter for leak checking because of the flammability of these fluids.

REFERENCES

1. Bruno, T. J. *CRC Handbook for the Analysis and Identification of Alternative Refrigerants*. Boca Raton, FL: CRC Press, 1994.
2. NIST Chemistry Web Book. NIST Standard Reference Database Number 69, March, 2003 Release.

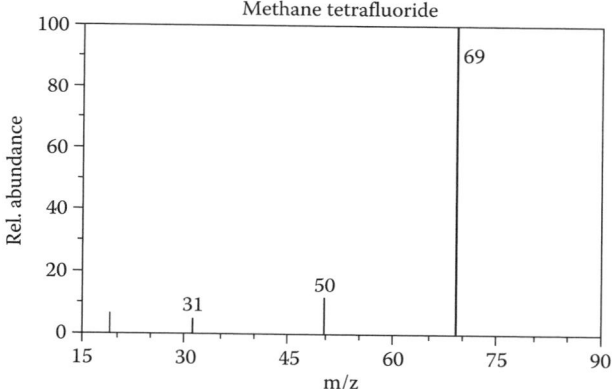

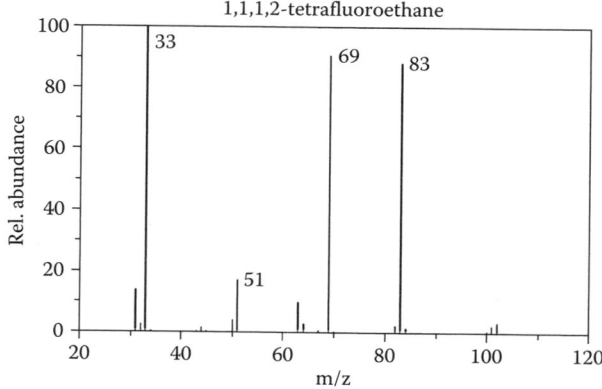

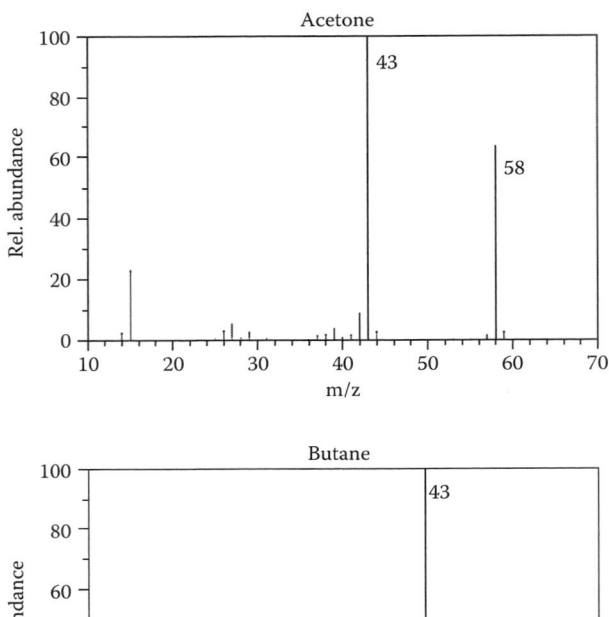

COMMON SPURIOUS SIGNALS OBSERVED IN MASS SPECTROMETERS

The following table provides guidance in the recognition of spurious signals (m/z peaks) that will sometimes be observed in measured mass spectra [1]. Often, the occurrence of these signals can be predicted by the recent history of the instrument or the method being used. This is especially true if the mass spectrometer is interfaced to a gas chromatograph.

REFERENCE

1. Maintaining your GC-MS system Agilent Technologies, Applications manual, 2001, available online at www.agilent.com/chem.

Ions Observed, m/z	Possible Compound	Possible Source
13, 14, 15, 16	methane*	Chlorine reagent gas
18	water*	Residual impurity, outgasing of ferrules, septa and seals.
14, 28	nitrogen*	Residual impurity, outgasing of ferrules, septa and seals; leaking seal.
16, 32	oxygen*	Residual impurity, outgasing of ferrules, septa and seals; leaking seal.
44	carbon dioxide*	Residual impurity, outgasing of ferrules, septa and seals; leaking seal; note it may be mistaken for propane in a sample.
31, 51, 69, 100, 119, 131, 169, 181, 214, 219, 264, 376, 414, 426, 464, 502, 576, 614	perfluorotributyl amine (PFTBA), and related ions	This is a common tuning compound; may indicate a leaking valve.
31	methanol	Solvent; can be used as a leak detector.
43, 58	acetone	Solvent; can be used as a leak detector.
78	benzene	Solvent; can be used as a leak detector.
91, 92	toluene	Solvent; can be used as a leak detector.
105, 106	xylenes	Solvent; can be used as a leak detector.
151, 153	trichloroethane	Solvent; can be used as a leak detector.
69	fore pump fluid, PFTBA	Back diffusion of fore pump fluid, possible leaking valve of tuning compound vial.
73, 147, 207, 221, 281, 295, 355, 429	dimethylpolysiloxane	Bleed from a column or septum, often during high temperature program methods in GC-MS
77, 94, 115, 141, 168, 170, 262, 354, 446	diffusion pump fluid	Back diffusion from diffusion pump, if present.
149	phthalates	Plasticizer in vacuum seals, gloves.
X–14 peaks	hydrocarbons	Loss of a methylene group indicates a hydrocarbon sample.

* It is possible to operate the analyzer to ignore these common background impurities. They will be present to contribute to poor vacuum if these impurities result from a significant leak.

MASS RESOLUTION REQUIRED TO RESOLVE COMMON
SPECTRAL INTERFERENCES ENCOUNTERED IN INDUCTIVELY
COUPLED PLASMA MASS SPECTROMETRY (ICP-MS)

The table below lists some common spectral interferences that are encountered in Inductively Coupled Plasma Mass Spectrometry (ICP-MS) as well as the resolution that is necessary to analyze them [1]. The resolution is presented as a dimensionless ratio. As an example, the mass of the poly-atomic ion $^{15}N^{16}O^+$ would be 15.000108 + 15.994915 = 30.995023. This would interfere with $^{31}P^+$ at a mass of 30.973762. The required resolution would be RMM/)RMM, or 30.973762/0.021261 = 1457. One should bear in mind that as resolution increases, the sensitivity decreases with subsequent effects on the price of the instrument. Note that small differences exist in the published exact masses of isotopes, but for the calculation of the required resolution, these differences are trivial. Moreover, recent instrumentation has provided rapid, high-resolution mass spectra with an uncertainty of less than 0.01 %.

REFERENCE

1. Gregoire, D. C. "Analysis of Geological Materials by Inductively Coupled Plasma Mass Spectrometry." *Spectroscopy* 14 (1999): 14–19.

Polyatomic Ion	Interfered Isotope (natural abundance %)	Required Resolution
$^{14}N_2^+$	$^{14}Si+$ (92.21)	958
$^{15}N^{16}O^+$	$^{31}P+$ (100)	1457
$^{40}Ar^{12}O^+$	$^{52}Cr+$ (83.76)	2375
$^{32}S^{16}O^+$	$^{48}Tl+$ (73.94)	2519
$^{35}Cl^{16}O^+$	$^{51}V+$ (99.76)	2572
$^{40}Ar^{35}Cl^+$	$^{75}As+$ (100)	7775
$^{40}Ar_2^+$	$^{80}Se+$ (49.82)	9688

CHAPTER 11

Atomic Absorption Spectrometry

CONTENTS

INTRODUCTION FOR ATOMIC SPECTROMETRIC TABLES

The tables presented in this section are designed to aid in the area of atomic spectrometric methods of analysis. The following conventions for abbreviation are recommended by the International Union of Pure and Applied Chemistry [1].

Atomic Emission Spectrometry: AES
Atomic Absorption Spectrometry: AAS
Flame Atomic Emission Spectrometry: FAES
Flame Atomic Absorption Spectrometry: FAAS
Electrothermal Atomic Absorption Spectrometry: EAAS
Inductively Coupled Plasma Atomic Emission Spectrometry: ICP-AES

Other variations such as cold vapor and hydride generation are not abbreviated but spelled out, for example, cold vapor AAS, hydride generation FAAS, and so on. These abbreviations are used whenever appropriate throughout the section.

Several of these tables have appeared in Parsons's Handbook [2] in one form or another. They have been updated to the extent possible, and the wavelength values have been made to conform to those in the National Standard Reference Data System-National Bureau of Standards (Now, National Institute of Standards and Technology, NSRDS-NBS) 68 [3] wherever possible.

As several of the tables cite the same References [1–17], all cited references will be listed at the end of this introduction instead of being repeated at the end of each table. These tables were originally prepared by Parsons for the first edition of this book [18].

REFERENCES

1. Commission on spectrochemical and other optical procedures for analysis, nomenclature, symbols, units, and their usage in spectrochemical analysis—I. General atomic emission spectroscopy; II. Data interpretation; and III. Analytical flame spectroscopy and associated procedures. *Spectrochimica Acta Part B: Atomic Spectroscopy* 33 (1978): 219.
2. Parsons, M. L., B. W. Smith, and G. E. Bentley. *Handbook of Flame Spectroscopy.* New York: Plenum Press, 1975.
3. Reader, J., C. H. Corliss, W. L. Weise, and G. A. Martin. *Wavelengths and Transition Probabilities for Atoms and Atomic Ions.* NSRDS-NBS 68, Washington, DC: U.S. Government Printing Office, 1980.
4. Smith, B. W., and M. L. Parsons. "Preparation of Standard Solutions: Critically Selected Compounds." *Journal of Chemical Education* 50 (1973): 679.
5. Dean, J. A., and T. C. Rains. *Flame Emission and Atomic Absorption Spectrometry.* Vol. 2. New York: Marcel Dekker, 1971.
6. Thermo Jarrell Ash Corp. *Guide to Analytical Values for TJA Spectrometers.* Waltham, MA: Thermo Jarrell Ash Corp., 1987.
7. Anderson, T. A., and M. L. Parsons. "ICP Emission Spectra III: The Spectra for the Group IIIA Elements and Spectral Interferences Due to Group IIA and IIIA Elements." *Applied Spectroscopy* 38 (1984): 625; Parsons, M. L., A. Forster, and D. Anderson. *An Atlas of Spectral Interferences in ICP Spectroscopy.* New York: Plenum Press, 1980.
8. Park, D. A. *Further Investigations of Spectra and Spectral Interferences Due to Group A Elements in ICP Spectroscopy: Groups IVA and VA.* PhD thesis, Arizona State University, Tempe; Parsons, M. L. Unpublished data, Los Alamos, NM: Los Alamos National Laboratory, 1987.
9. Perkin-Elmer Corp. *Mercury/Hydride System*, Report No. 1876/6.79, Norwalk, CT, 1987.
10. Lovett, R. J., D. L. Welch, and M. L. Parsons. "On the Importance of Spectral Interferences in Atomic Absorption Spectroscopy." *Applied Spectroscopy* 29 (1975): 470.
11. Layman, L., B. Palmer, and M. L. Parsons. Unpublished data taken with the Los Alamos National Laboratory FTS Facility, Los Alamos, NM 87545, 1987.

12. Sneddon, J. "Background Correction Techniques in Atomic Spectroscopy." *Spectroscopy* 2, no. 5 (1987): 38.

13. Wittenberg, G. K., D. V. Haun, and M. L. Parsons. "The Use of Free-Energy Minimization for Calculating Beta Factors and Equilibrium Compositions in Flame Spectroscopy." *Applied Spectroscopy* 33 (1979): 626.

14. Parsons, M. L., B. W. Smith, and P. M. McElfresh. "On the Selection of Analysis Lines in Atomic Absorption Spectrometry." *Applied Spectroscopy* 27 (1973): 471.

15. Parker, L. R., Jr., S. L. Morgan, and S. N. Deming. "Simplex Optimization of Experimental Factors in Atomic Absorption Spectrometry." *Applied Spectroscopy* 29 (1975): 429.

16. Parsons, M. L., and J. D. Winefordner. "Optimization of the Critical Instrumental Parameters for Achieving Maximum Sensitivity and Precision in Flame-Spectrometric Methods of Analysis." *Applied Spectroscopy* 21 (1967): 368.

17. Wiese, W. L., M. W. Smith, and B. M. Glennon. *Atomic Transition Probabilities: Vol. I Hydrogen Through Neon.* NSRDS-NBS 4, Washington, DC: U.S. Government Printing Office, 1966.

18. Bruno, T. J., and P. D. N. Svoronos. *CRC Handbook of Basic Tables for Chemical Analysis.* Boca Raton, FL: CRC Press, 1989.

STANDARD SOLUTIONS: SELECTED COMPOUNDS AND PROCEDURES

The compounds selected for this table were chosen using a rather stringent set of criteria, including stability, purity, ease of preparation, availability, high molecular mass, and toxicity. It is very important to have a compound that is pure and can be dried, weighed, and dissolved with comparative ease. The list of compounds provided here meets those goals as much as possible. No attempt was made to include all compounds that meet these criteria, nor are the compounds in this list trivial to dissolve; some require a rather long time and/or vigorous conditions.

In this table the significant figures in all columns represent the accuracy with which the atomic masses of the elements are known.

This table was compiled from References 4 and 5.

Standard Solutions: Selected Compounds and Procedures

Element	Compound	Relative Formula Mass	Weight for 1000 µg/L (PPM)-g/L	Solvent	Note
Aluminum	Al-metal	26.982	1.0000	hot dil. HCl-2M	APS
Antimony	$KSbOC_4H_4O_6$	324.92	2.6687	water	f
	$\cdot$ 1/2 H_2O (antimony potassium tartarate)				
	Sb-metal	121.75	1.0000	hot AqReg	
Arsenic	As_2O_3	.197.84	1.3203	1:1 NH_3	PS, c, NIST
Barium	$BaCO_3$	197.35	1.4369	dil. HCl	h
	$BaCl_2$	208.25	1.5163	water	g
Beryllium	Be-metal	9.0122	1.0000	HCl	c
	$BeSO_4 \cdot 4H_2O$	177.135	19.6550	water + acid	i
Bismuth	Bi_2O_3	465.96	1.1148	HNO_3	
	Bi-metal	208.980	1.00000	HNO_3	
Boron	H_3BO_3	61.84	5.720	water	PS, NIST, m
Bromine	KBr	119.01	1.4894	water	APS
Cadmium	CdO	128.40	1.1423	HNO_3	
	Cd-metal	112.40	1.0000	dil. HCl	
Calcium	$CaCO_3$	100.09	2.4972	dil. HCl	h
Cerium	$(NH_4)_2Ce(NO_3)_4$	548.23	3.9126	water	
Cesium	Cs_2SO_4	361.87	1.3614	water	
Chlorine	NaCl	58.442	1.6485	water	PS
Chromium	$K_2Cr_2O_7$	294.19	2.8290	water	PS, NIST
	Cr-metal	51.996	1.0000	HCl	
Cobalt	Co-metal	58.933	1.0000	HNO_3	APS
Copper	Cu-metal	63.546	1.0000	dil. HNO_3	APS
	CuO	69.545	1.2517	hot HCl	APS
	$CuSO_4 \cdot 5H_2O$	249.678	3.92909	water	
Dysprosium	Dy_2O_3	373.00	1.477	hot HCl	e
Erbium	Er_2O_3	382.56	1.1435	hot HCl	e
Europium	Eu_2O_3	351.92	1.1579	hot HCl	e
Fluorine	NaF	41.988	2.2101	water	j
Gadolinium	Gd_2O_3	362.50	1.1526	hot HCl	e
Gallium	Ga-metal	69.72	1.000	hot HNO_3	k
Germanium	GeO_2	104.60	1.4410	hot 1M NaOH or 50g oxalic acid + water	
Gold	Au-metal	196.97	1.0000	hot aq. reg.	APS, NIST
Hafnium	Hf-metal	178.49	1.0000	Hf, fusion	l
Holmium	Ho_2O_3	377.86	1.1455	hot HCl	e
Indium	In_2O_3	277.64	1.2090	hot HCl	
	In-metal	114.82	1.0000	dil. HCl	
Iodine	KIO_3	214.00	1.6863	water	PS
Iridium	Na_3IrCl_6	473.8	2.466	water	
Iron	Fe-metal	55.847	1.0000	hot HCl	APS
Lanthanum	La_2O_3	325.82	1.1728	hot HCl	e
Lead	$Pb(NO_3)_2$	331.20	1.5985	HCl	APS, NIST
Lithium	Li_2CO_3	73.890	5.3243	dil. HCl	APS, h
Lutetium	Lu_2O_3	397.94	1.1372	hot HCl	e

Standard Solutions: Selected Compounds and Procedures (Continued)

Element	Compound	Relative Formula Mass	Weight for 1000 µg/L (PPM)-g/L	Solvent	Note
Magnesium	MgO	40.311	1.6581	HCl	
	Mg-metal	24.312	1.0000	dil. HCl	
Manganese	$MnSO_4 \cdot H_2O$	169.01	3.0764	water	o
Mercury	$HgCl_2$	271.50	1.3535	water	c
	Hg-metal	200.59	1.0000	5M HNO_3	
Molybdenum	MoO_3	143.94	1.5003	1M NaOH or 2M HN_3	
Neodymium	Nd_2O_3	336.48	1.1664	HCl	e
Nickel	Ni-metal	58.71	1.000	hot HNO_3	APS
Niobium	Nb_2O_5	265.81	1.4305	HF, fusion	p, q
	Nb-metal	92.906	1.0000	HF + H_2SO_4	q
Osmium	Os-metal	190.20	1.0000	hot H_2SO_4	d
Palladium	Pd-metal	106.40	1.0000	hot HNO_3	
Phosphorus	KH_2PO_4	136.09	4.3937	water	
	$(NH_3)_2HPO_4$	209.997	6.77983	water	
Platinum	K_2PtCl_4	415.12	2.1278	water	APS, NIST
	Pt-metal	195.05	1.0000	hot Aq.Reg.	
Potassium	KCl	74.555	1.9067	water	PS, NIST
	$KHC_6H_4O_4$ (potassium hydrogen phthalate)	204.22	5.2228	water	PS, NIST
	$K_2Cr_2O_7$	294.19	3.7618	water	PS, NIST
Praseodymium	Pr_6O_{11}	1021.43	1.20816	HCl	e
Rhenium	Re-metal	186.2	1.000	HNO_3	
	$KReO_4$	289.3	1.554	water	
Rhodium	Rh-metal	102.91	1.0000	hot H_2SO_4	
Rubidium	Rb_2SO_4	267.00	1.5628	water	
Ruthenium	RuO_4	165.07	1.6332	water	
Samarium	Sm_2O_3	348.70	2.3193	hot HCl	e
Scandium	Sc_2O_3	137.91	1.5339	hot HCl	
Selenium	Se-metal	78.96	1.000	hot HNO_3	
	SeO_2	110.9	1.405	water	
Silicon	Si-metal	28.086	1.0000	NaOH, conc.	
	SiO_2	60.085	2.1393	HF	
Silver	$AgNO_3$	169.875	1.57481	water	APS, r
	Ag-metal	107.870	1.0000	HNO_3	
Sodium	NaCl	58.442	2.5428	water	PS
	$Na_2C_2O_4$ (sodium oxalate)	134.000	2.91432	water	PS, NIST
Strontium	$SrCO_3$	147.63	1.6849	dil. HCl	APS, h
Sulfur	K_2SO_4	174.27	5.4351	water	
	$(NH_4)_2SO_4$	114.10	3.5585	water	
Tantalum	Ta_2O_5	441.893	1.22130	HF, fusion	p, q
	Ta-metal	180.948	1.0000	HF + H_2SO_4	q
Tellurium	TeO_2	159.60	1.2507	HCl	
Terbium	Tb_2O_3	365.85	1.1512	hot HCl	e

(*Continued*)

Standard Solutions: Selected Compounds and Procedures (Continued)

Element	Compound	Relative Formula Mass	Weight for 1000 µg/L (PPM)-g/L	Solvent	Note
Thallium	Tl_2CO_3	468.75	1.1468	water	APS, c
	$TlNO_3$	266.37	1.3034	water	
Thorium	$Th(NO_3)_4$ $\cdot 4H_2O$	552.118	2.37943	HNO_3	
Thulium	Tm_2O_3	385.87	1.1421	hot HCl	e
Tin	Sn-metal	118.69	1.0000	HCl	
	SnO	134.69	1.1348	HCl	
Titanium	Ti-metal	47.90	1.000	1:1 H_2SO_4	APS
Tungsten	$Na_2WO_4 \cdot 2H_2O$	329.86	1.7942	water	s
	Na_2WO_4	293.83	1.5982	water	f
Uranium	UO_2	270.03	1.1344	HNO_3	PS, NIST
	U_3O_6	842.09	1.1792	HNO_3	
	$UO_2(NO_3)_2 \cdot 6H_2O$	502.13	2.1095	water	
Vanadium	V_2O_5	181.88	1.78521	hot HCl	
	NH_4VO_3	116.98	2.2963	dil. HNO_3	
Ytterbium	Yb_2O_3	394.08	1.1386	hot HCl	e
Yttrium	Y_2O_3	225.81	1.2700	hot HCl	e
Zinc	ZnO	81.37	1.245	HCl	APS
	Zn-metal	65.37	1.000	HCl	APS, NIST
Zirconium	Zr-metal	91.22	1.000	HF, fusion	1
	$ZrOCl_2 \cdot 8H_2O$	322.2	3.533	HCl	

Notes:
PS   Primary standard
APS   Compounds that approach primary standard quality
NIST   These compounds are sold as primary standards by the NIST Standard Reference Materials Program, 100 Bureau Drive, Gaithersburg, MD 20899-3460, (www.nist.gov).
c   Highly toxic
d   Very highly toxic
e   The rare earth oxides, because they absorb CO_2 and water vapor from the atmosphere, should be freshly ignited prior to weighing
f   Loses water at 110 °C. Water is only slowly regained, but rapid weighing and desiccator storage are required.
g   Drying at 250 °C, rapid weighing, and desiccator storage are required.
h   Add a quantity of water, then add dilute acid and swirl until the CO_2 has ceased to bubble out, then dilute.
i   Dissolve in water, then add 5 mL of concentrated HCl and dilute.
j   Sodium fluoride solutions will etch glass and should be freshly prepared.
k   Because the melting point is 29.6 °C, the metal may be warmed and weighed as a liquid.
l   Zr and Hf compounds were not investigated in the laboratory of reference 5.
m   Boric acid may be weighed directly from the bottle. It loses 1 H_2O at 100 °C, but it is difficult to dry to a constant mass.
n   Several references suggest that the addition of acid will help stabilize the solution.
o   This compound may be dried at 100 °C without losing the water of hydration.
p   Nb and Ta are slowly soluble in 40 % HF. The addition of H_2SO_4 accelerates the dissolution process.
q   Dissolve in 20 mL hot HF in a platinum dish, add 40 mL H_2SO_4 and evaporate to fumes, dilute with 8 M H_2SO_4.
r   When kept dry, silver nitrate crystals are not affected by light. Solutions should be stored in brown bottles.
s   Sodium tungstate loses both water molecules at 110 °C. The water is not rapidly regained, but the compound should be kept in a desiccator after drying and should be weighed quickly once it is removed.

LIMITS OF DETECTION TABLES FOR COMMON ANALYTICAL TRANSITIONS IN AES AND AAS

The following five tables present the common transitions for analysis and the detection limits for AES and AAS on the basis of source, where appropriate for the specific atom cell indicated. The detection limits are from the literature cited and are given in parts per billion (ppb), or nanograms per milliliter of aqueous solution. The limits of detection (LOD) are generally defined as a signal to noise of two or three. This generally relates to a concentration that produces a signal of two or three times the standard deviation of the measurement. These are measured in dilute aqueous solution and represent the best that the system was capable of measuring. In most cases, the detection limit in real samples will be one or two orders of magnitude higher, or worse, than those stated here. The type designation is I for free atom and II for single ion. In all cases NO means that no observation was made for the situation indicated, NA means that either AES or AAS was observed but no detection limit was reported.

In all cases where possible, the wavelengths of the transitions were made to conform with Reference 3; any wavelength below 200 nm is the wavelength given in vacuum, all others are in air.

Limits of Detection for the Air–Hydrocarbon Flame[a]

Element	Symbol	Wavelength (nm)	Type	LOD-AAS (ppb)
Antimony	Sb	217.581	I	100
		231.147	I	100
Bismuth	Bi	223.061	I	50
Calcium	Ca	22.673	I	2
Cesium	Cs	455.5276	I	600
		852.1122	I	50
Chromium	Cr	357.869	I	5
Cobalt	Co	240.725	I	5
Copper	Cu	324.754	I	50
		327.396	I	50
Gallium	Ga	287.424	I	70
Gold	Au	242.795	I	20
Indium	In	303.936	I	50
Iridium	Ir	208.882	I	15,000
		2639.71	I	2000
Iron	Fe	248.3271	I	5
Lead	Pb	283.3053	I	10
Lithium	Li	670.776	I	5
Magnesium	Mg	285.213	I	0.3
Manganese	Mn	279.482	I	2
		403.076	I	2
Mercury	Hg	253.652	I	500
Molybdenum	Mo	313.259	I	30
Nickel	Ni	232.003	I	5
Osmium	Os	290.906	I	17,000
Palladium	Pd	244.791	I	2000
		247.642	I	30
Platinum	Pt	265.945	I	100

(Continued)

Limits of Detection for the Air–Hydrocarbon Flame[a] (Continued)

Element	Symbol	Wavelength (nm)	Type	LOD-AAS (ppb)
Potassium	K	766.490	I	5
Rhodium	Rh	343.489	I	30
Rubidium	Rb	420.180	I	NA
		780.027	I	5
Ruthenium	Ru	349.894	I	300
		372.803	I	3000
Selenium	Se	196.09	I	100
		203.98	I	2000
Silver	Ag	328.068	I	5
		338.289	I	200
Sodium	Na	330.237	I	NA
		588.9950	I	2
		589.5924	I	2
Strontium	Sr	407.771	II	NA
		460.733	I	10
Tellurium	Te	214.281	I	100
Thallium	Tl	276.787	I	30
		377.572	I	2400
Tin	Sn	224.605	I	30
Zinc	Zn	213.856	I	2

[a] Flames formed from air combined with the lighter hydrocarbons, such as methane, propane, butane, or natural gas behave in a very similar fashion with similar temperatures, similar chemical properties, and so on.
These data were taken from Reference 2.

Limits of Detection for the Air–Acetylene Flame

Element	Symbol	Wavelength (nm)	Type	LOD-AES (ppb)	LOS-AAS (ppb)
Aluminum	Al	308.2153	I	NO	700
		309.2710	I	NO	500
		396.1520	I	NA	600
Antimony	Sb	206.833	I	NA	50
		217.581	I	NA	40
		231.147	I	3000	40
		259.805	I	NA	NO
Arsenic	As	193.759	I	10,000	140
Barium	Ba	455.403	II	NA	NO
		553.548	I	NA	NO
Bismuth	Bi	223.061	I	3000	25
Boron	B	249.677	I	NA	NO
Cadmium	Cd	228.8022	I	500	1
		326.1055	I	NA	NA
Calcium	Ca	393.366	II	NO	5000
		396.847	II	NO	5000
		422.673	I	0.5	0.5
Cesium	Cs	455.5276	I	NA	NO
		852.1122	I	NA	8
Chromium	Cr	357.869	I	NA	3
		425.435	I	NA	200
Cobalt	Co	240.725	I	NO	4
		352.685	I	NA	125
Copper	Cu	324.754	I	NA	1
		327.396	I	NA	120
Gallium	Ga	287.424	I	NO	50
		294.364	I	NA	50
		417.204	I	NA	1500
Germanium	Ge	265.1172	I	7000	
Gold	Au	242.795	I	NA	6
		267.595	I	NA	90
Indium	In	303.936	I	NA	30
		325.609	I	NA	20
		451.131	I	NA	200
Iodine	I	183.038	I	NO	8000
		206.163	I	2,500,000	NO
Iridium	Ir	208.882	I	NO	600
		2639.71	I	NO	2500
Iron	Fe	248.3271	I	NO	5
		371.9935	I	NA	700
Lead	Pb	217.000	I	NO	9
		283.3053	I	NA	240
		368.3462	I	NA	NO

(Continued)

Limits of Detection for the Air–Acetylene Flame (Continued)

Element	Symbol	Wavelength (nm)	Type	LOD-AES (ppb)	LOS-AAS (ppb)
Lithium	Li	670.776	I	NA	0.3
		451.857	I	NO	NA
Magnesium	Mg	279.553	II	NO	NA
		280.270	II	NO	NA
		285.213	I	NA	0.1
Manganese	Mn	279.482	I	NA	2
		403.076	I	NA	600
Mercury	Hg	253.652	I	NA	140
Molybdenum	Mo	313.259	I	NO	20
		379.825	I	80,000	900
		390.296	I	100	1600
Nickel	Ni	232.003	I	NO	2
		352.454	I	NA	350
Niobium	Nb	309.418	II	NO	NA
Osmium	Os	290.906	I	NA	1200
Palladium	Pd	244.791	I	NO	20
		247.642	I	NO	20
		340.458	I	NA	660
		363.470	I	NA	300
Phosphorus	P	213.547	I	NO	30,000
Platinum	Pt	214.423	I	NO	350
		265.945	I	NA	50
Potassium	K	766.490	I	NA	1
Rhenium	Re	346.046	I	NO	800
Rhodium	Rh	343.489	I	NA	2
		369.236	I	NA	70
Rubidium	Rb	420.180	I	NA	NO
		780.027	I	NA	0.3
Ruthenium	Ru	349.894	I	NA	400
		372.803	I	NA	250
Selenium	Se	196.09	I	NA	50
		203.98	I	50,000	10,000
Silver	Ag	328.068	I	NA	1
		338.289	I	NA	70
Sodium	Na	330.237	I	NO	NA
		588.9950	I	NA	1
		589.5924	I	NA	0.2
Strontium	Sr	407.771	II	NA	400
		421.552	II	NO	NA
		460.733	I	NA	2
Sulfur	S	180.7311	I	NO	30,000
Tellurium	Te	214.281	I	500	30
		238.578	I	NO	NA
Thallium	Tl	276.787	I	NA	30
		377.572	I	NA	1200
		535.046	I	NA	12,000

Limits of Detection for the Air–Acetylene Flame (Continued)

Element	Symbol	Wavelength (nm)	Type	LOD-AES (ppb)	LOS-AAS (ppb)
Tin	Sn	224.605	I	NO	10
		235.484	I	2000	600
		283.999	I	NA	1000
		326.234	I	NA	NO
Tungsten	W	255.135	I	90,000	3000
		400.875	I		
Uranium	U	591.539	I	NA	NO
Vanadium	V	318.540	I	NA	NO
		437.924	I	300	NO
Ytterbium	Yb	398.799	I	NO	80
Zinc	Zn	213.856	I	7000	1
Zirconium	Zr	351.960	I	NO	NA

These data were taken from References 2 and 6.

Limits of Detection for the Nitrous Oxide–Acetylene Flame

Element	Symbol	Wavelength (nm)	Type	LOD-AES (ppb)	LOD-AAS (ppb)
Aluminum	Al	308.2153	I	NA	NO
		309.2710	I	NA	20
		396.1520	I	3	900
Barium	Ba	553.548	I	1	8
Beryllium	Be	234.861	I	100	1
Boron	B	208.891	I	NO	NA
		208.957	I	NO	24,000
		249.677	I	NO	700
		249.773	I	NO	1500
Cadmium	Cd	326.1055	I	800	NO
Calcium	Ca	422.673	I	0.1	1
Cesium	Cs	455.5276	I	600	NO
		852.1122	I	0.02	NO
Chromium	Cr	425.435	I	1	NO
Cobalt	Co	352.685	I	200	NO
Copper	Cu	324.754	I	30	NO
		327.396	I	3	NO
Dysprosium	Dy	353.170	II	NO	800
		404.597	I	20	500
		421.172	I	NO	50
Erbium	Er	337.271	II	NO	100
		400.796	I	20	40
Europium	Eu	459.403	I	0.2	30
Gadolinium	Gd	368.413	I	NO	2000
		440.186	I	1000	NO
Gallium	Ga	417.204	I	5	NO
Germanium	Ge	265.1172	I	400	50
Gold	Au	267.595	I	500	NO
Hafnium	Hf	307.288	I	NO	2000
Holmium	Ho	345.600	II	NO	3000
		405.393	I	10	400
		410.384	I	NO	40
Indium	In	303.936	I	NO	1000
		325.609	I	NO	700
		451.131	I	1	3500
Iridium	Ir	208.882	I	NO	500
Iron	Fe	371.9935	I	10	NO
Lanthanum	La	408.672	II	NO	7500
		550.134	I	4000	2000
Lead	Pb	368.3462	I	0.2	NO
Lithium	Li	670.776	I	0.001	NO
Lutetium	Lu	261.542	II	NO	3000
		451.857	I	400	NO
Magnesium	Mg	285.213	I	1	NO
Manganese	Mn	403.076	I	1	NO
Mercury	Hg	253.652	I	10,000	NO

Limits of Detection for the Nitrous Oxide–Acetylene Flame (Continued)

Element	Symbol	Wavelength (nm)	Type	LOD-AES (ppb)	LOD-AAS (ppb)
Molybdenum	Mo	313.259	I	10	25
		379.825	I	300	NO
		390.296	I	10	NO
Neodymium	Nd	463.424	I	200	600
		492.453	I		700
Nickel	Ni	352.454	I	20	NO
Niobium	Nb	334.906	I	NO	1000
		405.894	I	60	5000
Osmium	Os	290.906	I	NO	80
		442.047	I	2000	NA
Palladium	Pd	363.470	I	40	NO
Phosphorus	P	177.499	I	NO	30,000
		213.547	I	NO	29,000
Platinum	Pt	265.945	I	2000	2000
Potassium	K	766.490	I	0.01	NO
Praseodymium	Pr	495.137	I	500	2000
Rhenium	Re	364.046	I	200	200
Rhodium	Rh	343.489	I	NO	700
		369.236	I	10	1400
Rubidium	Rb	780.027	I	8	NO
Ruthenium	Ru	372.803	I	300	NO
Samarium	Sm	429.674	I	NO	500
		476.027	I	50	14,000
Scandium	Sc	391.181	I	10	20
Selenium	Se	196.09	I	100,000	NO
Silicon	Si	251.6113	I	3000	20
		288.1579	I	NO	NA
Silver	Ag	328.068	I	2	NO
Sodium	Na	588.9950	I	0.01	NO
		589.5924	I	0.01	NO
Strontium	Sr	469.733	I	0.1	50
Tantalum	Ta	271.467	I	NO	800
		474.016	I	4000	NO
Terbium	Tb	432.643	I	NA	600
Thallium	Tl	377.572	I	50	NO
		535.046	I	2	
Thorium	Th	324.4448	I	NO	181,000
		491.9816	II	10,000	NO
Thulium	Tm	371.791	I	4	10
Tin	Sn	224.605	I	NO	3000
		235.484	I	NO	90
		283.999	I	100	NO
Titanium	Ti	334.941	II	NO	NA
		364.268	I	NA	10
		365.350	I	30	500
Tungsten	W	255.135	I	NO	500
		400.875	I	200	7500

(Continued)

Limits of Detection for the Nitrous Oxide–Acetylene Flame (Continued)

Element	Symbol	Wavelength (nm)	Type	LOD-AES (ppb)	LOD-AAS (ppb)
Uranium	U	358.488	I	NO	7000
Vanadium	V	318.540	I	200	20
		437.924	I	7	100
Ytterbium	Yb	398.799	I	0.2	5
Yttrium	Y	410.238	I	NO	50
Zinc	Zn	213.856	I	10,000	NO
Zirconium	Zr	351.960	I	1200	NO
		360.119	I	3000	1000

These data were taken from References 2 and 6.

Limits of Detection for Graphite Furnace AAS[a]

Element	Symbol	Wavelength (nm)	Type	LOD (ppb)
Aluminum	Al	308.2153	I	NA
		309.2710	I	0.01
		396.1520	I	600
Antimony	Sb	206.833	I	NA
		217.581	I	0.08
		231.147	I	NA
Arsenic	As	189.042	I	NA
		193.759	I	0.12
Barium	Ba	553.548	I	0.04
Beryllium	Be	234.861	I	0.003
Bismuth	Bi	223.061	I	0.01
Cadmium	Cd	228.8022	I	0.0002
Calcium	Ca	422.673	I	0.01
Chromium	Cr	357.869	I	0.004
Cobalt	Co	240.725	I	8
Copper	Cu	324.754	I	0.005
		327.396	I	NA
Erbium	Er	400.796	I	0.3
Gadolinium	Gd	440.186	I	0.3
Gallium	Ga	287.424	I	0.01
Germanium	Ge	265.1172	I	0.1
Gold	Au	242.795	I	0.01
Holmium	Ho	345.600	II	NA
		405.393	I	NA
Indium	In	303.936	I	0.02
Iodine	I	183.038	I	40,000
Iridium	Ir	208.882	I	0.5
Iron	Fe	248.3271	I	0.01
		371.9935	I	NA
Lanthanum	La	550.134	I	0.5
Lead	Pb	217.000	I	0.007
		283.3053	I	NA
Lithium	Li	670.776	I	0.01
Magnesium	Mg	285.213	I	0.0002
Manganese	Mn	279.482	I	0.0005
		403.076	I	NA
Mercury	Hg	253.652	I	0.2
Molybdenum	Mo	313.259	I	0.03
Nickel	Ni	232.003	I	0.05
Osmium	Os	290.906	I	2
Palladium	Pd	247.642	I	0.05
Phosphorus	P	177.499	I	NA
		213.547	I	20
		253.561	I	NA
Platinum	Pt	265.945	I	0.2
Potassium	K	766.490	I	0.004

(Continued)

Limits of Detection for Graphite Furnace AAS[a] (Continued)

Element	Symbol	Wavelength (nm)	Type	LOD (ppb)
Rhenium	Re	346.046	I	10
Rhodium	Rh	343.489	I	0.1
Rubidium	Rb	780.027	I	NA
Selenium	Se	196.09	I	0.05
Silicon	Si	251.6113	I	0.6
Silver	Ag	328.068	I	0.001
Sodium	Na	588.9950	I	0.004
Strontium	Sr	460.733	I	0.01
Sulfur	S	180.7311	I	NA
		182.0343	I	NA
		216.89		NA
Tellurium	Te	214.281	I	0.03
Thallium	Tl	276.787	I	0.01
Tin	Sn	235.484	I	0.03
		283.999	I	NA
Titanium	Ti	364.268	I	0.3
		365.350	I	NA
Uranium	U	358.488	I	30
Vanadium	V	318.540	I	0.4
Ytterbium	Yb	398.799	I	0.01
Yttrium	Y	410.238	I	10
Zinc	Zn	213.856	I	0.001

[a] The detection limits for the graphite furnace AAS are calculated using 100 microliters of sample. In graphite furnace AAS, additional chemicals are often added to aid in determining certain elements. Walter Slavin has published an excellent guide to these issues and has provided an excellent bibliography: Slavin, W. *Graphite Furnace Source Book*. Ridgefield, CT: Perkin–Elmer Corp., 1984; and Slavin, W., and D. C. Manning. "Furnace Interferences, A Guide to the Literature." *Progress and Analytica Atomic Spectroscopy* 5 (1982): 243.

Limits of Detection for ICP-AES

Element	Symbol	Wavelength (nm)	Type	LOD (ppb)	Reference
Aluminum	Al	167.0787	II	1	6
		308.2153	I	0.4	7
		309.2710	I	0.02	8
		396.1520	I	0.2	7
Antimony	Sb	206.833	I	10	7
		217.581	I	15	7
		231.147	I	61	7
		259.805	I	107	7
Arsenic	As	189.042	I	136	8
		193.759	I	2	7
		197.262	I	76	7
		234.984	I	90	7
Barium	Ba	455.403	II	0.001	8
		493.409	II	0.3	7
		553.548	I	2	7
Beryllium	Be	234.861	I	0.003	7
		313.042	II	0.1	6
		313.107	II	0.01	8
Bismuth	Bi	223.061	I	0.03	8
		289.798	I	10	7
Boron	B	208.891	I	5	8
		208.957	I	3	8
		249.677	I	0.1	8
		249.773	I	2	8
Bromine	Br	470.486	II	NA	8
		827.244	I	NA	8
Cadmium	Cd	214.441	II	0.1	8
		226.502	II	0.05	8
		228.8022	I	0.08	8
		326.1055	I	3	8
Calcium	Ca	364.441	I	0.5	8
		393.366	II	0.0001	8
		396.847	II	0.002	8
		422.673	I	0.2	8
Carbon	C	193.0905	I	40	6
		247.856	I	100	8
Cerium	Ce	394.275	II	2	8
		413.765	II	40	6
		418.660	II	0.4	7
Chlorine	Cl	413.250	II	NA	7
		837.594	I	NA	8
Chromium	Cr	205.552	II	0.009	8
		267.716	II	0.08	8
		357.869	I	0.1	8
		425.435	I	5	8

(Continued)

Limits of Detection for ICP-AES (Continued)

Element	Symbol	Wavelength (nm)	Type	LOD (ppb)	Reference
Cobalt	Co	228.615	II	0.3	8
		238.892	II	0.1	7
Copper	Cu	213.5981	II	7	8
		324.754	I	0.01	8
		327.396	I	0.06	8
Dysprosium	Dy	353.170	II	1	8
Erbium	Er	337.271	II	1	8
		400.796	I	1	7
Europium	Eu	381.967	II	0.06	7
Fluorine	F	685.603	I	NA	8
Gadolinium	Gd	342.247	II	0.4	7
Gallium	Ga	287.424	I	78	7
		294.364	I	3	8
		417.204	I	0.6	8
Germanium	Ge	199.8887	I	0.6	8
		209.4258	I	11	8
		265.1172	I	4	7
Gold	Au	242.795	I	2	8
		267.595	I	0.9	7
Hafnium	Hf	277.336	II	2	8
		339.980	II	5	6
Holmium	Ho	345.600	II	1	6
		389.102	II	0.9	8
Hydrogen	H	486.133	I	NA	8
		656.2852	I	NA	7
Indium	In	230.605	II	30	8
		303.936	I	15	8
		325.609	I	15	6
		451.131	I	30	7
Iodine	I	183.038	I	NA	7
		206.163	I	10	8
Iridium	Ir	224.268	II	0.6	8
		2639.71	I	0.6	8
Iron	Fe	238.204	II	0.004	8
		259.9396	II	0.09	7
		371.9935	I	0.3	7
Lanthanum	La	333.749	II	2	6
		408.672	II	0.1	8
Lead	Pb	217.000	I	30	8
		220.3534	II	0.6	8
		283.3053	I	2	7
		368.2462	I	20	8
Lithium	Li	670.776	I	0.02	7
Lutetium	Lu	261.542	II	0.1	7
		451.857	I	8	7
Magnesium	Mg	279.553	II	0.003	7
		280.270	II	0.01	7
		285.231	I	0.2	7

Limits of Detection for ICP-AES (Continued)

Element	Symbol	Wavelength (nm)	Type	LOD (ppb)	Reference
Manganese	Mn	257.610	II	0.01	7
		403.076	I	0.6	7
Mercury	Hg	184.905	II	1	7
		194.227	II	10	6
		253.652	I	1	7
Molybdenum	Mo	202.030	II	0.3	8
		313.259	I	NA	8
		379.825	I	0.2	7
		390.296	I	80	8
Neodymium	Nd	401.225	II	0.3	7
Nickel	Ni	221.648	II	2	8
		232.003	I	6	8
		352.454	I	0.2	7
Niobium	Nb	309.418	II	0.2	7
Nitrogen	N	174.2729	I	1000	8
		821.634	I	27,000	8
Osmium	Os	225.585	II	4	8
		290.906	I	6	8
Oxygen	O	426.825	I	NA	8
		777.194	I	NA	8
Palladium	Pd	340.458	I	2	8
		363.470	I	1	8
Phosphorus	P	177.499	I	NA	8
		213.547	I	16	6
		253.561	I	15	7
Platinum	Pt	214.423	I	16	6
		265.945	I	0.9	7
Potassium	K	766.490	I	5	8
Praseodymium	Pr	390.805	II	0.3	8
		422.535	II	10	7
Rhenium	Re	197.3	?	6	7
		221.426	II	4	6
Rhodium	Rh	233.477	II	30	7
		343.489	I	8	6
		369.236	I	7	8
Rubidium	Rb	420.180	I	38,000	8
		780.027	I	100	6
Ruthenium	Ru	240.272	II	8	6
		349.894	I	NA	8
		372.803	I	60	7
Samarium	Sm	359.260	II	0.5	8
		373.912	II	2	7
Scandium	Sc	361.384	II	0.1	8
Selenium	Se	196.09	I	0.1	8
		203.98	I	0.03	8
Silicon	Si	251.6113	I	2	7
		288.1579	I	10	7

(*Continued*)

Limits of Detection for ICP-AES (Continued)

Element	Symbol	Wavelength (nm)	Type	LOD (ppb)	Reference
Silver	Ag	328.068	I	0.8	8
		338.289	I	7	8
Sodium	Na	330.237	I	100	8
		588.9950	I	0.1	7
		589.5924	I	0.5	8
Strontium	Sr	407.771	II	0.2	6
		421.552	II	0.1	8
		460.733	I	0.4	8
Sulfur	S	180.7311	I	15	6
		182.0343	I	30	7
		216.89		NA	7
Tantalum	Ta	226.230	II	15	8
		240.063	II	13	6
		296.513	II	5	7
Tellurium	Te	214.281	I	0.7	8
		238.578	I	2	8
Terbium	Tb	350.917	II	0.1	7
		367.635	II	1.5	8
Thallium	Tl	190.864	II	4	8
		276.787	I	27	6
		377.572	I	17	8
Thorium	Th	283.7295	II	8	6
		401.9129	II	1.3	8
Thulium	Tm	313.126	II	0.9	6
		346.220	II	0.2	7
Tin	Sn	189.991	II	0.05	8
		235.484	I	9	8
		283.999	I	10	8
		326.234	I	0.5	8
Titanium	Ti	334.941	II	0.1	8
		365.350	I	230	8
		368.520	II	0.2	8
Tungsten	W	207.911	II	7	8
		276.427	II	0.8	7
		400.875	I	3	7
Uranium	U	263.553	II	70	6
		385.957	II	2	7
Vanadium	V	309.311	II	0.06	7
		311.062	II	0.06	7
		437.924	I	0.2	7
Ytterbium	Yb	328.937	II	0.01	8
		369.419	II	0.02	7
Yttrium	Y	371.030	II	0.04	7
		377.433	II	0.1	8
Zinc	Zn	202.548	II	0.6	8
		213.856	I	0.07	8
Zirconium	Zr	343.823	II	0.06	7

These data were taken from References 7 and 8.

DETECTION LIMITS BY HYDRIDE GENERATION AND COLD VAPOR AAS

In addition to the AAS methods in flames or graphite furnaces, the elements listed below are detected and determined at extreme sensitivity by introduction into a flame or a hot quartz cell by AAS.

Element	Wavelength[a] (nm)	LOD[b] (ppb)
Antimony, Sb	217.581	0.1
Arsenic, As	193.759	0.02
Bismuth, Bi	223.061	0.02
Mercury, Hg	313.652	0.02
Selenium, Se	196.09	0.02
Tellurium, Te	214.281	0.02
Tin, Sn	235.484	0.5

[a] It has been assumed that the transitions used for these detection limits were the most sensitive cited for AAS.
[b] The detection limits are based on 50 mL sample solution volumes.
These data were taken from Reference 9.

SPECTRAL OVERLAPS

In FAES and FAAS, the analytical results will be totally degraded if there is a spectral overlap of an analyte transition. This can result from an interfering matrix element with a transition close to that of the analyte. This table presents a list of those overlaps that have been observed and those which are predicted to happen. In many cases the interferant element has been present in great excess when compared to the analyte species. Therefore, if the predicted interferant element is a major component of the matrix, a careful investigation for spectral overlap should be made. Excitation sources other than flames were not covered in this study.

A. Observed Overlaps

Analyte Element	Wavelength (nm)	Interfering Element	Wavelength (nm)
Aluminum	308.2153	Vanadium	308.211
Antimony	217.023	Lead	217.000
Antimony	231.147	Nickel	231.096
Cadmium	228.8022	Arsenic	228.812
Calcium	422.673	Germanium	422.6562
Cobalt	252.136	Indium	252.137
Copper	324.754	Europium	324.755
Gallium	403.299	Manganese	403.307
Iron	271.9027	Platinum	271.904
Manganese	403.307	Gallium	403.299
Mercury	253.652	Cobalt	253.649
Silicon	250.690	Vanadium	250.690
Zinc	213.856	Iron	213.859

B. Predicted Overlaps

Analyte Element	Wavelength (nm)	Interfering Element	Wavelength (nm)
Boron	249.773	Germanium	249.7962
Bismuth	202.121	Gold	202.138
Cobalt	227.449	Rhenium	227.462
Cobalt	242.493	Osmium	242.497
Cobalt	252.136	Tungsten	252.132
Cobalt	346.580	Iron	346.5860
Cobalt	350.228	Rhodium	350.252
Cobalt	351.348	Iridium	351.364
Copper	216.509	Platinum	216.517
Gallium	294.417	Tungsten	294.440
Gold	242.795	Strontium	242.810
Hafnium	295.068	Niobium	295.088
Hafnium	302.053	Iron	302.0639
Indium	303.936	Germanium	303.9067
Iridium	208.882	Boron	208.891
Iridium	248.118	Tungsten	248.144
Iron	248.3271	Tin	248.339
Lanthanum	370.454	Vanadium	370.470
Lead	261.3655	Tungsten	261.382

Predicted Overlaps (Continued)

Analyte Element	Wavelength (nm)	Interfering Element	Wavelength (nm)
Molybdenum	379.825	Niobium	379.812
Osmium	247.684	Nickel	247.687
Osmium	264.411	Titanium	264.426
Osmium	271.464	Tantalum	271.467
Osmium	285.076	Tantalum	285.098
Osmium	301.804	Hafnium	301.831
Palladium	363.470	Ruthenium	363.493
Platinum	227.438	Cobalt	227.449
Rhodium	350.252	Cobalt	350.262
Scandium	298.075	Hafnium	298.081
Scandium	298.895	Ruthenium	298.895
Scandium	393.338	Calcium	393.366
Silicon	252.4108	Iron	252.4293
Silver	328.068	Rhodium	328.055
Strontium	421.552	Rubidium	421.553
Tantalum	263.690	Osmium	263.713
Tantalum	266.189	Iridium	266.198
Tantalum	269.131	Germanium	269.1341
Thallium	291.832	Hafnium	291.858
Thallium	377.572	Nickel	377.557
Tin	226.891	Aluminum	226.910
Tin	266.124	Tantalum	266.134
Tin	270.651	Scandium	270.677
Titanium	264.664	Platinum	264.689
Tungsten	265.654	Tantalum	265.661
Tungsten	271.891	Iron	271.9027
Vanadium	252.622	Tantalum	252.635
Zirconium	301.175	Nickel	301.200
Zirconium	386.387	Molybdenum	386.411
Zirconium	396.826	Calcium	396.847

These data were taken from Reference 10.

RELATIVE INTENSITIES OF ELEMENTAL TRANSITIONS
FROM HOLLOW CATHODE LAMPS

In AAS, the hollow cathode lamp (HCL) is the most important excitation source for most of the elements determined. However, sufficient light must reach the detector for the measurement to be made with good precision and detection limits. For elements in this table with intensities of less than 100, HCLs are probably inadequate, and other sources such as electrodeless discharge lamps should be investigated.

Relative Intensities of Elemental Transitions from Hollow Cathode Lamps

Element	Fill Gas	Wavelength (nm)	Relative Emission Intensity[a]
Aluminum	Ne	309.2710}	1200
		309.2839}	
		396.1520	800
Antimony	Ne	217.581	250
		231.147	250
Arsenic	Ar	193.759	125
		197.262	125
Barium	Ne	553.548	400
		350.111	200
Beryllium	Ne	234.861	2500
Bismuth	Ne	223.061	120
		306.772	400
Boron	Ar	249.773	400
Cadmium	Ne	228.8022	2500
		326.1055	5000
Calcium	Ne	422.673	1400
Cerium	Ne	520.012}	8
		520.042}	
		569.699	8
Chromium	Ne	357.869	6000
		425.435	5000
Cobalt	Ne	240.725	1000
		345.350	1500
		352.685	1300
Copper	Ne	324.754	7000
		327.396	6000
Dysprosium	Ne	404.597	2000
		418.682	2000
		421.172	2500
Erbium	Ne	400.796	1600
		386.285	1600
Europium	Ne	459.403	1000
		462.722	950
Gadolinium	Ne	368.413	350
		407.870	700
Gallium	Ne	287.424	400
		417.204	1100

Relative Intensities of Elemental Transitions from Hollow Cathode Lamps (Continued)

Element	Fill Gas	Wavelength (nm)	Relative Emission Intensity[a]
Germanium	Ne	265.1172}	500
		265.1568}	
		259.2534	250
Gold	Ne	242.795	750
		267.595	1200
Hafnium	Ne	307.288	300
		286.637	200
Holmium	Ne	405.393	2000
		410.384	2200
Indium	Ne	303.936	500
		410.176	500
Iridium	Ne	263.971	400
Iron	Ne	248.3271	400
		371.9935	2400
Lanthanum	Ne	550.134	120
		392.756	45
Lead	Ne	217.000	200
		283.3053	1000
Lithium	Ne	670.776	700
Lutetium	Ar	335.956	30
		337.650	25
		356.784	15
Magnesium	Ne	285.213	6000
		202.582	130
Manganese	Ne	279.482	3000
		280.106	2200
		403.076	14,000
Mercury	Ar	253.652	1000
Molybdenum	Ne	313.259	1500
		317.035	800
Neodymium	Ne	463.424	300
		492.453	600
Nickel	Ne	232.003	1000
		341.476	2000
Niobium	Ne	405.894	400
		407.973	360
Osmium	Ar	290.906	400
		301.804	200
Palladium	Ne	244.791	400
		247.642	300
		340.458	3000
Phosphorus	Ne	215.547}	30
		213.618}	
		214.914	20
Platinum	Ne	265.945	1500
		299.797	1000

(Continued)

Relative Intensities of Elemental Transitions from Hollow Cathode Lamps (Continued)

Element	Fill Gas	Wavelength (nm)	Relative Emission Intensity[a]
Potassium	Ne	766.490	6
		404.414	300
Praseodymium	Ne	495.137	100
		512.342	70
Rhenium	Ne	346.046	1200
		346.473	900
Rhodium	Ne	343.489	2500
		369.236	2000
		350.732	200
Rubidium	Ne	780.027	1.5
		420.180	80
Ruthenium	Ar	349.894	600
		392.592	300
Samarium	Ne	429.674	600
		476.027	800
Scandium	Ne	391.181	3000
		390.749	2500
		402.040	1800
		402.369	2100
Selenium	Ne	196.09	50
		203.98	50
Silicon	Ne	251.6113	500
		288.1579	500
Silver	Ar	328.068	3000
		338.289	3000
Sodium	Ne	588.9950	2000
		330.237}	
		330.298}	40
Strontium	Ne	460.733	1000
Tantalum	Ar	271.467	150
		277.588	100
Tellurium	Ne	214.281	60
		238.578	50
Terbium	Ne	432.643}	110
		432.690}	
		431.883	90
		433.841	60
Thallium	Ne	276.787	600
		258.014	50
Thulium	Ne	371.791	40
		409.419	50
		410.584	70
Tin	Ne	224.605	100
		286.332	250
Titanium	Ne	364.268	600
		399.864	600

Relative Intensities of Elemental Transitions from Hollow Cathode Lamps (Continued)

Element	Fill Gas	Wavelength (nm)	Relative Emission Intensity[a]
Tungsten	Ne	255.100}	200
		255.135}	
		400.875	1400
Uranium	Ne	358.488	300
		356.659	200
		351.461	200
		348.937	150
Vanadium	Ne	318.314}	600
		318.398}	
		385.537}	200
		385.584}	
Ytterbium	Ar	398.799	2000
		346.437	800
Yttrium	Ne	407.738	500
		410.238	600
		414.285	300
Zinc	Ne	213.856	2500
		307.590	2500

[a] The most intense line is the Mn 403.076 transition with a relative intensity of 14,000.
These data were obtained using Westinghouse HCL's and a single experimental setup. No correction has been made for the spectral response of the monochromator/photomultiplier tube system.
These data were taken from Reference 2.

INERT GASES

In AAS, the excitation source inert gas emission offers a potential background spectral interference. The most common inert gases used in HCLs are Ne and Ar. The data taken for this table and the other tables in this book on lamp spectra are from HCLs; however, electrodeless discharge lamps emit very similar spectra. The emission spectra for Ne and Ar HCLs and close lines that must be resolved for accurate analytical results are provided in the following four tables. This information was obtained for HCLs and flame atom cells and should not be considered with respect to plasma sources. In the "Type" column, an "I" indicates that the transition originates from an atomic species, and an "II" indicates a singly ionized species.

NEON HOLLOW CATHODE LAMP (HCL) SPECTRUM

Wavelength (nm)	Type	Relative Intensity[a]
323.237	II	5.4
330.974	II	2.8
331.972	II	8.7
332.374	II	28
332.916	II	1.7
333.484	II	5.2
334.440	II	17
335.502	II	3.5
336.060	II	1.7
336.9908 ⎫	I	7.8
336.9908 ⎭	II	17
337.822		
339.280	II	8.3
341.7904	I	16
344.7703	I	12
345.4195	I	15
346.0524	I	6.6
346.6579	I	12
347.2571	I	12
349.8064	I	2.9
350.1216	I	3.8
351.5191	I	3.6
352.0472	I	61
356.850	II	7.8
357.461	II	5.9
359.3526	I	19
360.0169	I	3.5
363.3665	I	3.6
366.407	II	1.9
369.421	II	3.5
370.962	II	4.9
372.186	II	3.1
404.264	I	1.4
533.0778	I	1.6
534.920	I	1.6
540.0562	I	3.3
576.4419	I	2.3
585.2488	I	100
588.1895	I	8.7
594.4834	I	14
597.4627 ⎫	I	2.6
597.5534 ⎭		
602.9997	I	2.8
607.4338	I	11
609.6163	I	15
614.3063	I	20
616.3594	I	5.2

[a] These data are referenced to the Ne transition at 585.2488 nm that has been assigned the value of 100.

These data were taken with a Varian Copper HCL operated at 10 mA. The Cu 324.7 nm transition was a factor of 2.9 more intense than the 585.249 nm Ne transition. The spectrum was taken with a IP28 photomultiplier tube (PMT). The relative intensities were not corrected for the instrumental/PMT response.

These data were taken from Reference 2.

NEON LINES WHICH MUST BE RESOLVED FOR
ACCURATE AAS MEASUREMENTS

Analyte Element	Wavelength (nm)	Neon Line (nm)	Required Resolution (nm)[a]
Chromium	357.869	357.461	0.20
Chromium	359.349	359.3526	0.002
Chromium	360.533	360.0169	0.26
Copper	324.754	323.237	0.75
Dysprosium	404.597	404.264	0.17
Gadolinium	371.357	370.962	0.20
Gadolinium	371.748	372.186	0.22
Lithium	670.776	335.502	
		in 2nd order is	
		671.004	0.11
Lutetium	335.956	336.060	0.05
Niobium	405.894	404.264	0.82
Rhenium	346.046	346.0524	0.003
Rhenium	346.473	346.6579	0.11
Rhenium	345.188	345.4195	0.12
Rhodium	343.489	344.7703	0.64
Rhodium	369.236	369.421	0.09
Ruthenium	372.803	372.186	0.31
Scandium	402.369	404.264	0.94
Silver	338.289	337.822	0.23
Sodium	588.995	588.1895	0.40
Sodium	589.592	588.1895	0.70
Thulium	371.792	372.186	0.19
Titanium	337.145	336.9808 and	
		336.9908	0.08
Titanium	364.268	363.3665	0.45
Titanium	365.350	366.407	0.53
Uranium	356.660	356.850	0.09
Uranium	358.488	359.3526	0.43
Ytterbium	346.436	346.6579	0.11
Zirconium	351.960	352.0472	0.04
Zirconium	360.119	360.0169	0.05

[a] The monochromator settings must be at least one-half of the separation of the analyte and interferant transition.

These data were taken from Reference 10.

ARGON HOLLOW CATHODE LAMP SPECTRUM

Wavelength (nm)	Type	Relative Intensity[a]
294.2893	II	3.5
297.9050	II	1.9
329.3640	II	1.5
330.7228	II	1.5
335.0924	II	2.2
337.6436	II	2.2
338.8531	II	1.8
347.6747	II	3.7
349.1244	II	2.0
349.1536	II	7.2
350.9778	II	7.0
351.4388	II	4.0
354.5596	II	11
354.5845	II	12
355.9508	II	16
356.1030	II	1.8
357.6616	II	11
358.1608	II	3.9
358.2355	II	8.5
358.8441	II	1.2
360.6522	I	2.0
362.2138	II	1.3
363.9833	II	3.3
371.8206	II	5.5
372.9309	II	1.3
373.7889	II	9.8
376.5270	II	5.1
376.6119	II	6.8
377.0520	II	1.7
378.0840	II	4.0
380.3172	II	5.5
380.9456	II	1.8
383.4679	I	2.6
385.0581	II	1.2
386.8528	II	7.0
392.5719	II	9.9
392.8623	II	6.9
393.2547	II	3.2
394.6097	II	14
394.8979	I	5.2
397.9356	II	5.6
399.4792	II	6.2
401.3857	II	4.3
403.3809	II	2.5
403.5460	II	2.3
404.2894	II	1.6

(Continued)

Argon Hollow Cathode Lamp Spectrum (Continued)

Wavelength (nm)	Type	Relative Intensity[a]
404.4418	I	9.0
405.2921	II	21
407.2005	II	34
407.2385	II	5.5
407.6628	II	2.0
407.9574	II	4.4
408.2387	II	3.2
410.3912	II	10
413.1724	II	61
415.6086	II	2.4
415.8590	I	1.4
416.4180	I	4.4
418.1884	I	6.9
419.0713	I	9.1
419.1029	I	8.9
419.8317	I	38
420.0674	I	38
421.8665	II	2.2
422.2637	II	4.3
422.6988	II	5.6
422.8158	II	12
423.7220	II	15
425.1185	I	2.3
425.9326	I	42
426.6286	I	11
426.6527	II	7.4
427.2169	I	18
427.7528	II	100
428.2898	II	2.5
430.0101	I	13
430.0650	II	3.3
430.9239	II	6.6
433.1200	II	17
433.2030	II	4.2
433.3561	I	12
433.5338	I	5.8
434.5168	I	3.7
434.8064	II	1.5
435.2205	II	5.4
236.2066	II	3.3
436.7832	II	9.6
437.0753	II	32
437.1329	II	6.5
437.5954	II	10
437.9667	II	20
438.5057	II	6.7
440.0097	II	5.7
440.0986	II	15

Argon Hollow Cathode Lamp Spectrum (Continued)

Wavelength (nm)	Type	Relative Intensity[a]
442.6001	II	1.6
443.0189	II	1.2
443.0996	II	5.4
443.3838	II	5.3
443.9461	II	5.3
444.8879	II	7.9
447.4759	II	19
448.1811	II	33
451.0733	I	20
452.2323	I	2.0
453.0552	II	3.2
454.5052	II	1.3
457.9350	II	1.4
458.9898	II	47
459.6097	I	1.8
460.9567	II	1.3
462.8441	I	1.4
463.7233	II	5.5
465.7901	II	1.9
470.2316	I	2.5
472.6868	II	43
473.2053	II	9.7
473.5906	II	1.3
476.4865	II	1.5
480.6020	II	36
484.7812	II	1.6
486.5910	II	1.3
487.9864	II	58
488.9042	II	11
490.4752	II	3.5
493.3209	II	4.1
496.5080	II	28
500.9334	II	5.5
501.7163	II	12
506.2037	II	5.9
509.0495	II	2.9
514.1783	II	5.7
514.5308	II	3.7
516.2285	I	3.8
516.5773	II	1.8
518.7746	I	3.8
522.1271	I	1.2
545.1652	I	1.7
549.5874	I	3.1
555.8702	I	4.0
557.2541	I	1.9
560.6733	I	4.9

(Continued)

Argon Hollow Cathode Lamp Spectrum (Continued)

Wavelength (nm)	Type	Relative Intensity[a]
565.0704	I	1.7
588.8584	I	1.9
591.2085	I	4.1
592.8813	I	1.4
603.2127	I	4.1
604.3223	I	1.6
611.4923	II	2.2
617.2278	II	1.1
696.5431	I	3.2
706.7218	I	1.7
738.3980	I	1.2
750.3869	I	2.7

[a] These data are referenced to the Ar transition at 427.7528 nm that has been assigned the value of 100.

These data were taken from an Ar filled Ga HCL at the Los Alamos Fourier Transform Spectrometer facility [11].

CLOSE LINES FOR BACKGROUND CORRECTION

In AAS, it is possible to make background corrections in many cases by measuring a normally nonabsorbing transition near the analytical transition. This table presents a list of suitable transitions for such a background measurement. It is often desirable to check the background absorbance by more than one method even if there is a built-in background measurement by some other means such as the continuum or Zeeman methods. In the table below, the first two columns give the analyte element and wavelength of the analytical transition, and the last two columns give the transition useful for the background measurement and its source. If the source is Ne and the HCL is Ne filled, the same HCL can be used for the background measurement; if not, a different HCL must be placed in the spectrometer to make the measurement.

These data were taken from Reference 12.

Close Lines for Background Correction

Element	Analysis Line (nm)		Background Line (nm)		Source
Aluminum	309.2711	I	306.614	I	Al
Antimony	217.581	I	217.919	I	Sb
Arsenic	231.147	I	231.398	I	Ni
Barium	193.759	I	191.294	II	As
	553.548	I	540.0562	I	Ne
			553.305	I	Mo
			557.742	I	Y
Beryllium	234.861	I	235.484	I	Sn
Bismuth	223.061	I	226.502	II	Cd
Bromine	306.772	I	306.614	I	Al
Cadmium	148.845	I	149.4675	I	N
	228.8022	I	226.502	II	Cd
Calcium	422.673	I	421.9360	I	Fe
Cesium	852.1122	I	423.5936	I	Fe
Chromium	357.869	I	854.4696	I	Ne
			352.0472	I	Ne
			358.119	I	Fe
Cobalt	204.206	I	238.892	II	Co
Copper	324.754	I	242.170	I	Sn
Dysprosium	421.172	I	324.316	I	Cu
			421.645	II	Fe
			421.096	I	Ag
Erbium	400.796	I	394.442	I	Er
Europium	459.403	I	460.102	I	Cr
Gallium	287.424	I	283.999	I	Sn
Gold	242.795	I	283.690	I	Cd
			242.170	I	Sn
Indium	303.936	I	306.614	I	Al
Iodine	183.038	I	184.445	I	I
Iron	248.3271	I	249.215	I	Cu
Lanthanum	550.134	I	550.549	I	Mo
			548.334	I	Co

(Continued)

Close Lines for Background Correction (Continued)

Element	Analysis Line (nm)		Background Line (nm)		Source
Lead	283.3053	I	280.1995	I	Pb
			283.6900	I	Cd
	217.000	I	220.3534	II	Pb
Lithium	670.791	I	671.7043	I	Ne
Magnesium	285.213	I	283.690	I	Cd
			283.999	I	Sn
Manganese	279.482	I	282.437	I	Cu
			280.1995	I	Pb
Mercury	253.652	I	249.215	I	Cu
Molybdenum	313.259	I	312.200	II	Mo
Nickel	232.003	I	232.138	I	Ni
Palladium	247.642	I	249.215	I	Cu
Phosphorus	213.618	I	213.856	I	Zn
Potassium	766.490	I	769.896	I	K
			767.209	I	Ca
Rhodium	343.489	I	350.732	I	Rh
			352.0472	I	Ne
Rubidium	780.027	I	778.048	I	Ba
Ruthenium	249.894	I	352.0472	I	Ne
Selenium	196.09	I	199.51	I	Se
Silicon	251.6113	I	249.215	I	Cu
Silver	328.068	I	332.374	II	Ne
			326.234	I	Sn
Sodium	588.9950	I	588.833	I	Mo
Strontium	460.733	I	460.500	I	Ni
Tellurium	214.281	I	213.856	I	Zn
			217.581	I	Sb
Thallium	276.787	I	280.1995	I	Pb
Tin	224.605	I	226.502	II	Cd
	286.332	I	283.999	I	Sn
Titanium	364.268	I	361.939	I	Ni
	365.350	I	361.939	I	Ni
Uranium	358.488	I	358.119	I	Fe
Vanadium	318.398	I	324.754	I	Cu
	318.540	I	324.754	I	Cu
Zinc	213.856	I	212.274	II	Zn

BETA VALUES FOR THE AIR–ACETYLENE AND NITROUS OXIDE–ACETYLENE FLAMES

Beta values represent the fraction of free atoms present in the hot flame gases of the flame indicated. These values have been taken from various sources and were either experimentally measured or calculated from thermodynamic data using the assumption of local thermodynamic equilibrium in the flame. These values do not have very good agreement within each element; however, the values do provide an indication of the probable sensitivity of the particular flame.

These data were taken from References 2 and 13.

Beta Values for the Air–Acetylene and Nitrous Oxide–Acetylene Flames

Element	Symbol	Beta A/AC Flame	Beta N/AC Flame
Aluminum	Al	<0.0001	0.13
		<0.00005	0.29
		0.0005	0.97*
			0.5
Antimony	Sb	0.03	
Arsenic	As	0.0002	
Barium	Ba	0.0009	0.074
		0.002	0.074
		0.003	0.98
		0.0018	
Beryllium	Be	0.0004	0.095
		0.00006	0.98
			0.98
Bismuth	Bi	0.17	0.35
Boron	B	<0.0006	0.0035
		<0.000001	0.2
Cadmium	Cd	0.38	0.56
		0.50	0.60
		0.80	
Calcium	Ca	0.066	0.34
		0.14	0.52*
		0.05*	0.98
		0.018	
Cesium	Cs	0.02	0.0004
		0.0057	
Chromium	Cr	0.071	0.63
		0.13	1.02
		0.53	1.00
		0.042	
Cobalt	Co	0.023	0.11
		0.28	0.25
		0.41	
Copper	Cu	0.4	0.49
		0.82	0.66
		0.98	1.00*

(*Continued*)

Beta Values for the Air–Acetylene and Nitrous Oxide–Acetylene Flames (Continued)

Element	Symbol	Beta A/AC Flame	Beta N/AC Flame
Gallium	Ga	0.16	0.73
		0.16	
Germanium	Ge	0.001	
Gold	Au	0.21	0.16
		0.40	0.27
		0.63	
Indium	In	0.10	0.37
		0.67	0.93
		0.67	
Iridium	Ir	0.1	
Iron	Fe	0.38	0.83
		0.66	0.91
		0.84	1.00
		0.66	
Lead	Pb	0.44	0.84
		0.77	
Lithium	Li	0.21	0.34*
		0.26*	0.96*
		0.20*	0.041
		0.08	0.91*
Magnesium	Mg	0.59	0.88
		1.05	0.99
		0.62	0.92
			0.99*
Manganese	Mn	0.45	0.37
		0.93	0.77
		1.0	
Mercury	Hg	0.04	
Molybdenum	Mo	0.03	
Nickel	Ni	1	
Palladium	Pd	1	
Platinum	Pt	0.4	
Potassium	K	0.7*	0.12*
		0.25	0.0004
		0.45	0.17*
		0.59*	
Rhodium	Rh	1	
Rubidium	Rb	0.16	
Ruthenium	Ru	0.3	
Selenium	Se	0.0001	
Silicon	Si	<0.001	0.55
		<0.0000001	0.12
			0.36
Silver	Ag	0.66	0.57
		0.70	

Beta Values for the Air–Acetylene and Nitrous Oxide–Acetylene Flames (Continued)

Element	Symbol	Beta A/AC Flame	Beta N/AC Flame
Sodium	Na	0.63	0.32
		1.00	0.97*
		1.00*	0.012
		0.56	0.80
Strontium	Sr	0.068	0.26
		0.10	0.57
		0.13	0.99
		0.021	
Tantalum	Ta		0.045
Thallium	Tl	0.36	0.55
		0.52	
Tin	Sn	<0.0001	0.35
		0.043	0.82
		0.078	
		0.061	
Titanium	Ti	<0.001	0.11
			0.33
			0.49
Tungsten	W	0.004	0.71
Vanadium	V	0.0004	0.32
		0.015	0.99
		0.000001	
Zinc	Zn	0.66	0.49
		0.45	

*Ionization has been suppressed for these measurements/calculations.

LOWER ENERGY LEVEL POPULATIONS (IN PERCENTAGE)
AS A FUNCTION OF TEMPERATURE

It is possible to calculate the relative number of atoms in the ground energy level(s) using the following equation:

$$\% \text{ Atoms (ith level)} = n_i/n_t \times 100 = g_i/Z \times \exp(-E_i/kT),$$

where n_i is the number of atoms in the ith level per unit volume of atoms cell, n_t is the total number atoms per unit volume of atom cell, g_i is the statistical weight for energy level i, Z is the electronic partition function, E_i is the energy of the ith level, k is the Boltzmann constant, and T is the absolute temperature. Of course all of the data must be in consistent units.

In utilizing these data, it should be remembered that, other things being equal, the larger the percentage of atoms in the ground or lower level of a transition, the larger the absorption signal from that transition should be. For example, a transition with 100 % of the atoms in the ground state should be 10 times more sensitive than one with 10 %. Also, these data refer to the percentage of atoms in the atomic state only; therefore, this information should be used in conjunction with the beta values table.

These data were taken from Reference 12.

Lower Energy Level Populations (in Percentage) as a Function of Temperature

Element	Energy Level (cm⁻¹)	Percentage Population at Temperature (°C)		
		2000	2500	3000
Aluminum	0.0	35.1	34.8	34.5
	112.040	64.9	65.2	65.5
Antimony	0.0	99.7	98.7	97.0
	8512.100	0.2	0.7	1.6
	9854.100	0.1	0.5	1.7
Arsenic	0.0	99.9	99.5	98.5
	10,592.500	0.0	0.2	0.6
	10,914.600	0.1	0.3	0.8
Barium	0.0	98.0	92.6	82.6
	9033.985	0.4	1.5	3.3
	9215.518	0.6	2.3	5.0
	9596.551	0.7	2.6	5.8
	11,395.382	0.1	0.7	1.8
Beryllium	0.0	100.0	100.0	100.0
Bismuth	0.0	100.0	99.8	99.5
Boron	0.0	33.5	33.5	33.5
	16.0	66.5	66.5	66.5
Cadmium	0.0	100.0	100.0	100.0
Calcium	0.0	100.0	99.8	99.3
Cesium	0.0	99.9	99.5	98.3
	11,178.240	0.0	0.2	0.5
	111,732.350	0.0	0.2	0.7
Chromium	0.0	98.6	95.9	91.5
	7593.160	0.3	0.9	1.7
	7750.780	0.1	0.2	0.3
	7810.820	0.1	0.2	0.3

Lower Energy Level Populations (in Percentage) as a Function of Temperature (Continued)

Element	Energy Level (cm⁻¹)	Percentage Population at Temperature (°C)		
		2000	2500	3000
	7927.470	0.1	0.4	0.9
	8095.210	0.2	0.7	1.3
	8307.570	0.3	0.8	1.7
Cobalt	0.0	51.8	45.6	40.9
	816.000	23.1	22.8	22.1
	1406.840	11.3	12.2	12.5
	1809.330	5.6	6.4	6.9
	3482.820	4.2	6.1	7.7
	4142.660	2.1	3.4	4.5
	4690.180	1.1	1.8	2.6
	5075.830	0.5	1.0	1.4
	7442.410	0.2	0.5	0.9
Copper	0.0	99.9	99.4	98.3
	11,202.565	0.1	0.5	1.9
Gallium	0.0	47.5	44.6	42.6
	826.240	52.5	55.4	57.4
Germanium	0.0	20.6	18.2	16.6
	557.100	41.4	39.7	38.7
	1409.900	37.4	40.5	42.4
	7125.260	0.6	1.5	2.7
Gold	0.0	99.5	98.5	96.4
	9161.300	0.4	1.5	3.5
Hafnium	0.0	73.9	64.0	55.9
	2356.680	19.0	23.1	25.3
	4567.640	5.0	8.3	11.3
	5521.780	0.3	0.5	0.8
	5638.620	1.3	2.5	3.7
	6572.550	0.4	0.9	1.4
Indium	0.0	71.0	64.1	59.1
	2212.560	29.0	35.9	40.9
Iridium	0.0	85.1	77.2	69.9
	2834.980	11.1	15.1	18.0
	4078.940	1.8	3.0	4.0
	5784.620	0.8	1.7	2.6
	6323.910	0.7	1.6	2.7
	7106.610	0.4	1.0	1.9
Iron	0.0	46.2	43.4	41.0
	415.933	26.7	26.6	26.1
	704.003	15.5	16.1	16.3
	888.123	8.1	8.7	8.9
	978.074	2.5	2.7	2.9
	6928.280	0.4	1.0	1.8
	7376.775	0.2	0.6	1.2
	7728.071	0.1	0.4	0.8
	7985.795	0.1	0.2	0.5

(Continued)

Lower Energy Level Populations (in Percentage) as a Function of Temperature (Continued)

Element	Energy Level (cm⁻¹)	Percentage Population at Temperature (°C)		
		2000	2500	3000
Lanthanum	0.0	42.5	34.2	28.3
	1053.200	29.9	28.0	25.6
	2668.200	6.2	7.4	7.9
	3010.010	7.3	9.1	10.0
	3494.580	6.9	9.2	10.6
	4121.610	5.5	8.0	9.8
	7011.900	0.4	0.9	1.5
	7231.360	0.1	0.3	0.4
	7490.460	0.2	0.5	0.8
	7679.940	0.3	0.6	1.1
Lead	0.0	98.7	95.7	90.8
	7819.350	1.1	3.2	6.4
	10,650.470	0.2	1.1	2.8
Lithium	0.0	100.0	99.9	99.8
Magnesium	0.0	100.0	100.0	100.0
Manganese	0.0	100.0	100.0	99.8
Mercury	0.0	100.0	100.0	100.0
Molybdenum	0.0	99.9	99.4	98.1
Nickel	0.0	39.5	36.4	34.2
	204.786	26.6	25.2	24.1
	879.813	11.7	12.2	12.4
	1332.153	11.8	13.2	14.0
	1713.080	3.9	4.5	5.0
	2216.519	4.5	5.7	6.6
	3409.925	1.9	2.8	3.7
Niobium	0.0	7.5	6.5	5.7
	154.190	13.4	11.8	10.6
	391.990	17.0	15.4	14.2
	695.250	18.2	17.3	16.4
	1050.260	17.6	17.7	17.3
	1142.790	6.6	6.7	6.6
	1586.900	7.2	7.8	8.0
	2154.110	6.4	7.5	8.1
	2805.360	5.0	6.4	7.5
	4998.170	0.2	0.4	0.5
	5297.920	0.3	0.6	0.9
	5965.450	0.3	0.6	1.0
Osmium	0.0	86.5	78.3	70.4
	2740.490	6.7	9.0	10.5
	4159.320	3.4	5.6	7.5
	5143.920	2.6	5.0	7.3
	5766.140	0.5	0.9	1.5
	6092.790	0.1	0.3	0.4
	8742.830	0.2	0.5	1.1
Palladium	0.0	91.7	80.3	67.2
	6464.110	6.2	13.8	21.6
	7754.990	1.8	4.7	8.3

Lower Energy Level Populations (in Percentage) as a Function of Temperature (Continued)

Element	Energy Level (cm^{-1})	Percentage Population at Temperature (°C)		
		2000	2500	3000
	10,093.940	0.2	0.7	1.6
	11,721.770	0.1	0.5	1.2
Platinum	0.0	47.0	43.8	41.5
	775.900	19.2	20.0	20.4
	823.700	33.4	35.1	36.0
	6140.000	0.1	0.2	0.3
	6567.5000	0.3	0.7	1.3
Potassium	0.0	100.0	99.8	99.4
Rhenium	0.0	99.9	99.5	98.3
Rhodium	0.0	69.0	60.0	53.6
	1529.970	18.1	19.9	20.6
	2598.030	6.3	8.1	9.3
	3309.860	3.8	5.4	6.6
	3472.680	2.2	3.3	4.1
	5657.970	0.5	0.9	1.4
	5690.970	0.9	1.8	2.8
Rubidium	0.0	100.0	99.8	99.3
Ruthenium	0.0	62.4	55.2	49.3
	1190.640	21.7	22.8	22.8
	2091.540	8.8	10.6	11.5
	2713.240	4.0	5.3	6.1
	3105.490	1.8	2.5	3.0
	6545.030	0.5	1.0	1.8
	7483.070	0.2	0.6	1.1
	8084.120	0.1	0.3	0.7
	9183.660	0.0	0.1	0.3
Scandium	0.0	42.9	42.2	41.5
	168.340	57.0	57.4	57.4
Selenium	0.0	85.0	80.5	76.8
	1989.490	12.2	15.4	17.8
	2534.350	2.7	3.7	4.6
	9576.080	0.1	0.3	0.8
Silicon	0.0	12.3	11.9	11.6
	77.150	34.8	34.2	33.5
	223.310	52.2	52.3	52.1
	6298.810	0.7	1.6	2.8
Silver	0.0	100.0	100.0	100.0
Sodium	0.0	100.0	100.0	99.9
Strontium	0.0	100.0	99.8	98.9
Tantalum	0.0	64.9	53.8	45.0
	2010.00	23.0	25.4	25.8
	3963.920	7.5	11.0	13.5
	5621.040	2.9	5.3	7.6
	6049.420	0.4	0.8	1.2
	9253.430	0.1	0.4	0.8
	9705.380	0.1	0.4	0.9

(Continued)

Lower Energy Level Populations (in Percentage) as a Function of Temperature (Continued)

Element	Energy Level (cm⁻¹)	Percentage Population at Temperature (°C)		
		2000	2500	3000
Technetium	0.0	27.2	25.7	24.0
	170.132	33.8	32.6	31.0
	386.873	37.1	37.0	35.9
	6556.860	0.1	0.4	0.6
	6598.830	0.2	0.6	1.0
	6661.000	0.3	0.8	1.4
	6742.790	0.4	1.0	1.7
	6843.000	0.4	1.1	2.0
	7255.290	0.1	0.4	0.7
Tungsten	0.0	28.4	20.5	15.9
	1670.300	25.6	23.6	21.5
	2951.290	23.8	26.3	27.1
	3325.530	13.0	15.2	16.2
	4830.000	6.2	8.9	11.0
	6219.330	2.9	5.2	7.3
Vanadium	0.0	14.1	12.6	11.5
	137.380	19.2	17.5	16.2
	323.420	22.4	20.9	19.7
	553.020	23.7	22.9	22.1
	2112.320	1.5	1.9	2.1
	2153.200	3.0	3.7	4.1
	2220.130	4.3	5.3	6.0
	2311.370	5.4	6.7	7.6
	2424.890	6.2	7.8	9.0
Yttrium	0.0	49.3	47.2	45.4
	530.360	50.5	52.2	52.8
Zinc	0.0	100.0	100.0	100.0
Zirconium	0.0	34.2	29.1	25.1
	570.410	31.7	29.3	26.7
	1240.840	25.2	25.6	24.9
	4186.110	1.7	2.6	3.4
	4196.850	0.3	0.5	0.7
	4376.280	0.9	1.4	1.8
	4870.530	0.6	1.1	1.5
	5023.410	0.9	1.6	2.3
	5101.680	0.9	1.5	2.2
	5249.070	1.1	2.0	2.8
	5540.540	1.1	2.2	3.2
	5888.930	1.1	2.2	3.3
	8057.300	0.2	0.5	1.0

CRITICAL OPTIMIZATION PARAMETERS FOR AES/AAS METHODS

In most multiparameter instrumental techniques, the parameters can be classified into two types: independent and dependent. Independent parameters can be optimized independently from all other parameters and can therefore be subjected to a univariate approach, for example, the variable can be adjusted until the largest signal-to-noise ratio (SNR) is obtained and set at that value for the best instrumental performance. This is the simplest situation and can be handled in a very straightforward manner.

Dependent parameters are an entirely different matter. Most dependent parameters have optimum values that depend on the value of the other parameters. If the value of any variable is changed, then the optimum for the parameter under question will be different.

The following table lists the parameters for FAAS, EAAS, and FAES, which are both dependent and independent. A "yes" in any column indicates that the listed parameter is appropriate for that technique. If an optimization is necessary when independent parameters are involved, it is important to use a systematic approach that permits one to vary all parameter values to develop the optimum for each. If the variables are simply varied one at a time, false optimum values and poor results will be obtained. Experimental design techniques are required for good results; one of the best approaches is the SIMPLEX technique that has been fully discussed in the literature [15].

A. Independent Parameters

Parameter	FAAS	EAAS	FAES
Excitation source power	yes	yes	na[a]
Photomultiplier voltage[b]	yes	yes	yes
Readout gain[c]	yes	yes	yes
Noise suppression setting[d]	yes	yes	yes

B. Dependent (Interdependent) Parameters

Parameter	FAAS	EAAS	FAES
Oxidant gas flow rate	yes	na	yes
Fuel-to-oxidant ratio	yes	na	yes
Sheath gas flow rate[e]	yes	yes	yes
Solution flow rate[f]	yes	na	yes
Sample size	na	yes	na
Height of optical measurement	yes	yes	yes
Monochromater slit setting	yes	yes	yes
Burner variables[g]	yes	na	yes
Furnace variables[h]	na	yes	na

[a] na stands for "not applicable."
[b] The photomultiplier tube voltage does not affect the SNR unless extreme voltages are used. It will specify the level of signal that is observed.
[c] The gain does not affect the SNR until electronic noise becomes important. It also specifies the level of signal that is observed.
[d] This specifies the frequency response of the system and is accompanied by a time requirement. More noise filtering requires a long measurement.
[e] Most commercial burners do not use a sheath gas; however, there is always the possibility of a sheath gas in EAAS.
[f] This is important if the sample solution flow rate is controlled by a pump rather than by the oxidant gas flow rate.
[g] Some burners have additional variables such as bead position and nebulizer position.
[h] The timing cycle and temperature are always critical variables for the graphite furnaces.
This information was taken from Reference 16.

FLAME TEMPERATURES AND REFERENCES ON
TEMPERATURE MEASUREMENTS

Flame Type	Experimental Measurement Range (K)	Calculated Stoichiometric Temperature (K)	Typical[a] (K)
Hydrocarbon/air	1900–2150	2228	2000
Acetylene/air	2360–2600	2523	2450
Acetylene/nitrous oxide	2830–3070	3148	2950
Hydrogen/air	2100–2300	2373	2300
Hydrogen/oxygen	2500–2900	3100	2800
Acetylene/oxygen	2900–3300	3320	3100

[a] This value represents the value most often cited for flames used in analytical spectroscopy.
These data were taken from Reference 2.

REFERENCES THAT DISCUSS THE TECHNIQUES
OF TEMPERATURE MEASUREMENT

Gaydon, A. G., and H. G. Wolfhard. *Flames, Their Structure, Radiation, and Temperature.* London: Chapman and Hall Ltd., 1970.

Fristrom, R. M., and A. A. Westenberg. *Flame Structure.* New York: McGraw-Hill, 1965.

Tourin, R. H. *Spectroscopic Gas Temperature Measurement.* Amsterdam: Elsevier Publishing Co., 1966.

Gaydon, A. G., and H. G. Wolfhard. "The Spectrum-Line Reversal Method of Measuring Flame Temperature." *Proceedings of the Physical Society* (London) 65A (1954): 19.

Browner, R. F., and J. D. Winefordner. "Measurement of Flame Temperatures by a Two-Line Atomic Absorption Method." *Analytical Chemistry* 44 (1972): 247.

Omenetto, N., P. Benetti, and G. Rossi. "Flame Temperature Measurements by Means of Atomic Fluorescence Spectrometry." *Spectrochimica Acta* 27B (1972): 253.

Herzfield, C. M. *Temperature, Its Measurement and Control in Science and Industry.* Vol. III, Part 2. Edited by I. Dahl. New York: Reinhold, 1962.

Alkemade, C. Th. J., Tj. Hollander, W. Snelleman, and P. J. Th. Zeegers. *Metal Vapours in Flames.* New York: Pergamon Press, 1982.

FUNDAMENTAL DATA FOR THE COMMON TRANSITIONS

To the extent possible, the fundamental data for the transition commonly used with the methods discussed in this section are given in this table. The transition in nm, the type of transition (I indicates atomic and II indicates ionic), the lower and upper energy levels (E-low and E-high), in cm^{-1}, the statistical weight, g(i), of the lower level (i), the transition probability, A(ji), in s^{-1}, and the merit and reference for the transition probability are listed. In some cases the g(i) and the A(ji) were only available in the multiplied form, and in these cases the "gA = xx" format was used. If a blank appears, no information was available for that specific column.

Fundamental Data for the Common Transitions

Element	Symbol	Wavelength (nm)	Type	E-Low (cm⁻¹)	E-High (cm⁻¹)	g(i)	A(ji) 10⁵ s⁻¹	Merit[a]	Ref.
Aluminum	Al	167.0787	II						
		308.2153	I	0	32,435	4	0.63	C	3
		309.2710	I	112	32,437	6	0.73	C	3
		396.1520	I	112	25,348	2	0.98	C	3
Antimony	Sb	206.833	I	0	48,332	6	42	E	2
		217.581	I	0	45,945	4	13.8	E	2
		231.147	I	0	43,249	2	3.75	E	2
		259.805	I	8512	46,991	2	32	E	2
Arsenic	As	189.042	I		52,898	6	2.0	D	3
		193.759	I	0	51,610	4	2.0	D	3
		197.262	I	0	50,694	2	2.0	D	3
		234.984	I	10,592	53,136	4	3.1	D	3
		286.044	I	18,186	53,136	2	0.55	D	3
Barium	Ba	455.403	II	0	21,952	4	1.17	A	3
		493.409	II	0	20,262	2	0.955	B	3
		553.548	I	0	18,080	3	1.15	B	3
Beryllium	Be	234.861	I	0	42,565	3	5.56	B	3
		313.042	II	0	31,935	4	1.14	B	3
		313.107	II	0	31,929	2	1.15	B	3
Bismuth	Bi	223.061	I	0	44,817	4	0.25	D	3
		289.798	I	11,418	45,916	2	1.53	C	3
Boron	B	208.891	I	0	47,857	4	0.28	D	3
		208.957	I	16	47,857	6	0.33	D	3
		249.677	I	0	40,040	2	0.84	C	3
		249.773	I	16	40,040	2	1.69	C	3
Bromine	Br	470.486	II		115,176	7	1.1	D	3
		827.244	I						
Cadmium	Cd	214.441	II	0	46,619	4	2.8	C	3
		226.502	II	0	44,136	2	3.0	C	3
		228.8022	I	0	43,692	3	0.24	D	3
		326.1055	I	0	30,656	3	0.004	C	3
Calcium	Ca	364.441	I	15,316	42,747	7	0.355	C	3
		393.366	II	0	25,414	4	1.47	C	3
		396.847	II	0	25,192	2	1.4	C	3
		422.673	I	0	23,652	3	2.18	B	3

(Continued)

Fundamental Data for the Common Transitions (Continued)

Element	Symbol	Wavelength (nm)	Type	E-Low (cm⁻¹)	E-High (cm⁻¹)	g(i)	A(ji) 10⁵ s⁻¹	Meritᵃ	Ref.
Carbon	C	193.0905	I		61,982	3	3.7	D	3
		247.856	I	21,648	61,982	3	0.18	D	3
Cerium	Ce	394.275	II	6913	32,269		gA = 19	E	2
		413.765	II	4166	28,327		gA = 4.8	E	2
		418.660	II	6968	30,847		gA = 18	E	2
Cesium	Cs	455.5276	I	0	21,946	4	0.019	C	3
		852.1122	I	0	11,732	4	0.32	E	2
Chlorine	Cl	413.250	II		153,259	5	1.6	D	3
		837.594	I						
Chromium	Cr	205.552	II	0	48,632		gA = 9.1	E	2
		267.716	II	12,304	49,646		gA = 132	E	2
		357.869	I	0	27,935		gA = 8.3	E	2
		425.435	I	0	23,499	9	0.315	B	3
Cobalt	Co	228.615	II	3350	47,078		gA = 169	E	2
		238.892	II	3350	45,198		gA = 278	E	2
		240.725	I	0	41,529	12	3.08	E	2
		352.685	I	0	28,346	10	0.12	C	3
Copper	Cu	213.5981	II	0	30,784	4	1.39	B	3
		324.754	I	0	30,784	4	1.39	B	3
		327.396	I	0	30,535	2	1.37	B	3
Dysprosium	Dy	353.170	II	0	28,307		gA = 19	E	2
		404.597	I	0	24,709	15	1.5	D	3
		421.172	I	0	23,737	19	2.08	C	3
Erbium	Er	337.271	II	0	29,641		gA = 13	E	2
		400.796	I	0	24,943	15	26	D	3
Europium	Eu	381.967	II	0	26,173		gA = 4.8	E	2
		459.403	I	0	21,761	10	1.4	D	3
Fluorine	F	685.603	I		116,987	8	0.42	D	3
Gadolinium	Gd	342.247	II	1935	31,146		gA = 19	E	2
		368.413	I	0	27,136		gA = 12	E	2
		440.186	I	1719	24,430		gA = 4.2	E	2
Gallium	Ga	287.424	I	0	34,782	4	1.2	C	3
		294.364	I	826	34,788	6	1.4	C	3
		417.204	I	826	24,789	2	0.92	C	3
Germanium	Ge	199.8887	I	1410	51,438	5	0.55	C	3
		209.4258	I	1410	49,144	7	0.97	C	3
		265.1172	I	1410	39,118	5	2.0	C	3
Gold	Au	242.795	I	0	41,174	4	1.5	D	3
		267.595	I	0	37,359	2	1.1	D	3
Hafnium	Hf	277.336	II	6344	42,391		gA = 14	E	2
		307.288	I	0	32,533		gA = 3.2	E	2
		339.980	II	0	29,405		gA = 1.1	E	2
Holmium	Ho	345.600	II						
		389.102	II	637	26,331				
		405.393	I	0	24,660				
		410.384	I	0	24,361				

Fundamental Data for the Common Transitions (Continued)

Element	Symbol	Wavelength (nm)	Type	E-Low (cm⁻¹)	E-High (cm⁻¹)	g(i)	A(ji) 10⁵ s⁻¹	Merit[a]	Ref.
Hydrogen	H	486.133	I	82,259	102,824	32	0.084	A	17
		656.2852	I	82,259	97,492	18	0.441	A	17
Indium	In	230.605	II	0	43,349	gA = 0.032		E	2
		303.936	I	0	32,892	gA = 7.1		E	2
		325.609	I	2213	32,915	6	1.3	D	3
		451.131	I	2213	24,373	2	1.02	C	3
Iodine	I	183.038	I		56,093	4	2.71	C	3
		206.163	I						
Iridium	Ir	208.882	I	0	47,858	12	28	E	2
		224.268	II	0	44,576				
		263.971	I	0	37,872	10	0.56	E	2
Iron	Fe	238.204	II	0	41,968	gA = 92		E	2
		248.3271	I	0	40,257	11	4.9	C	3
		259.9396	II	0	38,459	10	2.22	C	3
		371.9935	I	0	26,875	11	0.163	B	3
Lanthanum	La	333.749	II	3250	33,204	gA = 3.5		E	2
		408.672	II	0	24,463	5	0.20	E	2
		550.134	I	0	18,172	4	0.08	E	2
Lead	Pb	217.000	I	0	46,068	3	1.5	D	3
		220.3534	II	14081	59,448	gA = 5.7		E	2
		283.3053	I	0	35,287	3	0.58	D	3
		368.3462	I	7819	34,960	1	1.5	D	3
Lithium	Li	670.776	I	0	1494	4	0.372	B	3
Lutetium	Lu	261.545	II	0	38,223	gA = 5.8		E	2
		451.857	I	0	22,125	4	0.21	B	3
Magnesium	Mg	279.553	II	0	35,761	4	4.0	C	3
		280.270	II	0	35,669	2	2.6	C	3
		285.213	I	0	35,051	3	5.3	D	3
Manganese	Mn	257.610	II	0	38,807	9	8.89	E	2
		279.482	I	0	35,770	8	3.7	C	3
		403.076	I	0	24,802	8	0.19	C	3
Mercury	Hg	184.905	II						
		194.227	II						
		253.652	I	0	39,412	3	0.13	D	3
Molybdenum	Mo	202.030	II	0	49,481	gA = 24		E	2
		313.259	I	0	31,913	9	1.09	E	2
		379.825	I	0	26,321	9	0.49	E	2
		390.296	I	0	25,614	5	0.42	E	2
Neodymium	Nd	401.225	II	5086	30,002	20	0.55	D	3
		463.424	I	0	21,572	gA = 2.0		E	2
		492.453	I	0	20,301	gA = 2.0		E	2
Nickel	Ni	221.648	II		53,496	12	5.5	D	3
		232.003	I	0	43,090	11	6.9	C	3
		352.454	I	205	28,569	5	1.0	C	3

(Continued)

Fundamental Data for the Common Transitions (Continued)

Element	Symbol	Wavelength (nm)	Type	E-Low (cm⁻¹)	E-High (cm⁻¹)	g(i)	A(ji) 10^5 s⁻¹	Merit[a]	Ref.
Niobium	Nb	309.418	II	4146	36,455	13	1.1	E	2
		334.906	I	2154	32,005	10	0.45	E	2
		405.894	I	1050	25,680	12	0.65	E	2
Nitrogen	N	174.2729	I						
		821.634	I						
Osmium	Os	225.585	II	0	44,315				
		290.906	I	0	34,365	11	1.0	E	2
		442.047	I	0	22,616	9	0.034	E	2
Oxygen	O	436.825	I		86,631	7	0.34	B	3
		777.194	I						
Palladium	Pd	244.791	I	0	40,839	3	0.28	E	2
		247.642	I	0	40,369	3	0.37	E	2
		340.458	I	6564	35,928	9	1.33	E	2
		363.470	I	6564	34,069	5	1.24	E	2
Phosphorus	P	177.499	I	0	56,340	6	2.17	C	3
		213.547	I	11,362	58,174	4	0.211	C	3
		253.561	I	18,748	58,174	4	0.20	C	3
Platinum	Pt	214.423	I	0	46,622	7	5.14	E	2
		265.945	I	0	37,591	9	0.91	E	2
Potassium	K	766.490	I	0	13,043	4	0.387	B	3
Praseodymium	Pr	390.805	II						
		422.535	II	0	23,660		gA = 1.4	E	2
		495.137	I	0	20,190				
Rhenium	Re	197.3							
		221.426	II	0	45,148		gA = 15	E	2
		346.046	I	0	28,890				
Rhodium	Rh	233.477	II	16,885	59,702		gA = 44	E	2
		343.489	I	0	29,105	12	0.34	E	2
		369.236	I	0	27,075	8	0.35	E	2
Rubidium	Rb	420.180	I	0	23,793	4	0.018	C	3
		780.027	I	0	12,817	4	0.370	B	3
Ruthenium	Ru	240.272	II	9152	50,758		gA = 247	E	2
		349.894	I	0	28,572	13	0.46	E	2
		372.803	I	0	26,816	11	0.42	E	2
Samarium	Sm	359.260	II	3053	30,880		gA = 6.3	E	2
		373.912	II	326	27,063				
		429.674	I	4021	27,288		gA = 21	E	2
		476.027	I	812	21,813		gA = 3.3	E	2
Scandium	Sc	361.384	II	178	27,841	9	0.14	D	3
		391.181	I	168	25,725	8	1.37	C	3
Selenium	Se	196.09	I	0	50,997	2	100	E	2
		203.98	I	1989	50,997	2	65	E	2
Silicon	Si	251.6113	I	223	39,955	5	1.21	C	3
		288.1579	I	6299	40,992	3	1.89	C	3
Silver	Ag	328.068	I	0	30,473	4	1.4	B	3
		338.2068	I	0	29,552	2	1.3	B	3

Fundamental Data for the Common Transitions (Continued)

Element	Symbol	Wavelength (nm)	Type	E-Low (cm^{-1})	E-High (cm^{-1})	g(i)	A(ji) 10^5 s^{-1}	Merit[a]	Ref.
Sodium	Na	330.237	I	0	30,273	4	0.028	C	3
		588.9950	I	0	16,973	4	0.622	A	3
		589.5924	I	0	16,956	2	0.618	A	3
Strontium	Sr	407.771	II	0	24,517	4	1.42	C	3
		421.552	II	0	23,715	2	1.27	C	3
		460.733	I	0	21,698	3	2.01	B	3
Sulfur	S	180.7311	I		55,331	3	3.8	C	3
		182.0343	I		55,331	3	2.2	C	3
		216.89							
Tantalum	Ta	226.230	II	2642	46,831	gA = 35		E	2
		240.063	II	6187	47,830	gA = 516		E	2
		271.467	I	0	36,826	6	1.17	E	2
		296.513	II	0	33,715	gA = 7.8		E	2
		474.016	I	9976	31,066	4	0.028	E	2
Tellurium	Te	214.281	I	0	46,653	3	38	E	2
		238.578	I	4751	46,653	3	5.47	E	2
Terbium	Tb	350.917	II	0	28,488				
		367.635	II	1016	28,209				
		432.643	I	0	23,107	gA = 7.2		E	2
Thallium	Tl	190.864	II						
		276.787	I	0	36,118	4	1.26	C	3
		377.572	I	0	26,478	2	0.625	B	3
		535.046	I	7793	26,478	2	0.705	B	3
Thorium	Th	283.7295	II	6214	41,448				
		324.4448	I	0	30,813	gA = 0.12		E	2
		401.9129	II	0	24,874	gA = 0.66		E	2
		491.9816	II	6168	26,489	gA = 0.08		E	2
Thulium	Tm	313.126	II	0	31,927	gA = 4.6		E	2
		346.220	II	0	28,875	gA = 2.5		E	2
		371.791	I	0	26,889	gA = 8.3		E	2
Tin	Sn	189.991	II						
		224.605	I	0	44,509	3	1.6	D	3
		235.484	I	1692	44,145	5	1.7	D	3
		283.999	I	3428	38,629	5	1.7	D	3
		326.234	I	8613	39,257	3	2.7	D	3
Titanium	Ti	334.941	II	393	30,241	12	1.3	D	3
		364.268	I	170	27,615	9	0.67	C	3
		365.350	I	387	27,750	11	0.66	C	3
		368.520	II	4898	32,026				
Tungsten	W	207.911	II	6147	54,229	gA = 93		E	2
		255.135	I	0	39,183	7	1.17	E	2
		276.427	II	0	36,165	gA = 6.9		E	2
		400.875	I	2951	27,890	9	0.20	E	2

(Continued)

Fundamental Data for the Common Transitions (Continued)

Element	Symbol	Wavelength (nm)	Type	E-Low (cm^{-1})	E-High (cm^{-1})	g(i)	A(ji) 10^5 s^{-1}	Merit[a]	Ref.
Uranium	U	263.553	II						
		358.488	I	0	27,887	15	0.10	B	3
		385.957	II	289	26,191	gA = 2.6		E	2
		591.539	I	0	16,900	gA = 0.12		E	2
Vanadium	V	309.311	II	3163	35,483	13	1.8	D	3
		311.062	II	2809	34,947	9	1.5	D	3
		318.540	I	553	31,937	12	1.4	D	3
		437.924	I	2425	25,254	12	1.2	D	3
Ytterbium	Yb	328.937	II	0	30,392	4	1.8	C	3
		369.419	II	0	27,062	2	1.4	C	3
		398.799	I	0	25,068	3	1.76	C	3
Yttrium	Y	362.094	I	530	28,140	4	1.55	E	2
		371.030	II	1450	28,394				
		377.433	II	1045	27,532				
		410.238	I	530	24,900	8	0.64	E	2
Zinc	Zn	202.548	II	0	49,355	4	3.3	C	3
		213.856	I	0	46,745	3	7.09	B	3
Zirconium	Zr	343.823	II	763	29,840	gA = 13		E	2
		351.960	I	0	28,404	7	0.71	E	2
		360.119	I	1241	29,002	11	0.91	E	2

[a] The key for the merit of the A(ji) values follows that given in Reference 3 as follows: A = within 3 %; B = within 10 %; C = within 25 %; D = within.

ACTIVATED CARBON AS A TRAPPING SORBENT FOR TRACE METALS

Activated carbon is commonly used to preconcentrate samples of heavy metals before spectrometric analysis [1]. This material is typically used by passing the sample through a thin layer (50–150 mg) of the activated carbon that is supported on a filter disk. It can also be used by shaking 50–150 mg of activated carbon in the solution containing the heavy metal, and then filtering the sorbent out of the solution.

REFERENCE

1. Alfasi, Z. B., and C. M. Wai. *Preconcentration Techniques for Trace Elements.* Boca Raton, FL: CRC Press, 1992.

Activated Carbon as a Trapping Sorbent for Trace Metals

Matrices	Trace Metals	Complexing Agents	Determination Methods
Water	Ag, Bi, Cd, Co, Cu, Fe, In, Mg, Mn, Ni, Pb, Zn	(NaOH; pH 7 to 8)	AAS
Water	Ag, As, Ca, Cd, Ce, Co, Cu, Dy, Fe, La, Mg, Mn, Nb, Nd, Ni, Pb, Pr, Sb, Sc, Sn, U, V, Y, Zn	8-quinolinol	SSMS, XRF
Water	Ba, Co, Cs, Eu, Mn, Zn	APDC, DDTC, PAN, 8-quinolinol	XRF
Water	Hg, Methyl mercury	—	AAS
Water	Hg (halide) ·	—	AAS
Water	Hg (halide)	—	AAS
Water	U	L-ascorbic acid	INAA
HNO_3, water, Al, KCl	Ag, Bi, Cd, Cu, Hg, Pb, Zn	dithizone	AAS
Mn, MnO_3, Mn salts	Bi, Cd, Co, Cu, Fe, In, Ni, Pb, Tl, Zn	ethyl xanthate	AAS
Co, $Co(NO_3)_2$	Ag, Bi	APDC	AAS
Ni, $Ni(NO_3)_2$	Ag, Bi	APDC	AAS
Mg, $Mg(NO_3)_2$	Ag, Cu, Fe, Hg, In, Mn, Pb, Zn	(pH 8.1 to 9)	AAS
Al	Cd, Co, Cu, Ni, Pb	thioacetamide	AAS
Ag, $TlNO_3$	Bi, Co, CU, Fe, In, Pb	xenol orange	AAS
Cr salts	Ag, Bi, Cd, Co, Cu, In, Ni, Pb, Tl, Zn	HAHDTC	AAS
Co, In, Pb, Ni, Zn	Ag, Bi, Cu, Tl	DDTC	AAS
Se	Cd, Co, Cu, Fe, Ni, Pb, Zn	DDTC	AAS
$NaClO_4$	Ag, Bi, Cd, Co, Cu, Fe, Hg, In, Mn, Ni, Pb	(pH 6)	AAS

Abbreviations
APDC: ammonium pyrrolidinecarbodithioate
DDTC: diethyldithiocarbamate
HAHDTC: hexamethyleneammonium hexaethylenedithiocarbamate
PAN: 1-(2-pyridylazo)-2-naphthol

REAGENT IMPREGNATED RESINS AS TRAPPING
SORBENTS FOR TRACE MINERALS

Reagent impregnated resins can be used as a trapping sorbents for the preconcentration of heavy metals [1]. These materials can be used in the same way as activated carbons.

REFERENCE

1. Alfasi, Z. B., and C. M. Wai. *Preconcentration Techniques for Trace Elements.* Boca Raton, FL: CRC Press, 1992.

Reagent Inpregnated Resins as Trapping Sorbents for Trace Minerals

Reagents	Adsorbents	Metals
TBP	porous polystyrene DVB resins	U
YBP	Levextrel (polystyrene DVB resins)	U
DEHPA	Levextrel	Zn
DEHPA	XAD-2	Zn
Alamine 336	XAD-2	U
LIX-63	XAD-2	Co, Cu, Fe, Ni, etc.
LIX-64N, -65N	XAD-2	Cu
Hydroxyoximes	XAD-2	Cu
Kelex 100	XAD-2	Co, Cu, Fe, Ni
Kelex 100	XAD-2,4,7,8,11	Cu
Dithizone, STTA	polystyrene DVB resins	Hg
Dithizone (acetone)	XAD-1,2,4,7,8	Hg, methyl mercury
DMABR	XAD-4	Au
Pyrocatechol violet	XAD-2	In, Pb
TPTZ	XAD-2	Co, Cu, Fe, Ni, Zn

Abbreviations
  TBP: tributyl phosphate
  DEHPA: di-ethylhexyl phosphoric acid
  STTA: monothiothenolytrifluoroacetone
  DMABR: 5-(4-dimethylaminobenzylidene)-rhodanine
  TPTZ: 2,4,6-tri(2-pyridyl)-1,3,5-triazine
  LIX 63: aliphatic α-hydroxyoxime
  LIX 65N: 2-hydroxy-5-nonylbensophenoneoxime
  LIX 64N: a mixture of LIX 65N with approximately 1 % (vol/vol) of LIX-63

REAGENT IMPREGNATED FOAMS AS TRAPPING SORBENTS FOR INORGANIC SPECIES

Reagent impregnated foams can be used as a trapping sorbents for the preconcentration of heavy metals [1]. These materials can be used in the same way as activated carbons.

REFERENCE

1. Alfasi, Z. B., and C. M. Wai. *Preconcentration Techniques for Trace Elements.* Boca Raton, FL: CRC Press, 1992.

Reagent Inpregnated Foams as Trapping Sorbents for Inorganic Species

Matrices	Elements	Conc.	Foam Type	Reagents	Determination Methods
Water	^{131}I, ^{203}Hg	traces	Polyether	Alamine 336	radiometry
Natural water					
Water	Bi, Cd, Co, Cu, Fe, Hg, Ni, Pb, Sn, Zn	traces	Polyether	Amberlite LA-2	spectroph., AAS
Water	Co, Fe, Mn	traces to: g/1	Polyether Polyether	PAN	radiometry
Natural water	Cd	μg/1	Polyether	PAN	AAS
Water	Au, Hg	μg/1	Polyether	PAN	NAA
Water	Ni	traces to: g/1	—	DMG, α-benzyldioxime	spectroph., AAS
Water	Cr	μg/1	Polyether	DPC	colorimetry
Water	Hg, methyl-Hg, phenyl-Hg	μg/1	Polyether	DADTC	radiometry
Natural water	Sn	traces	Polyether	toluene-3,4-dithiol	spectroph.
Water	Cd, Co, Fe, Ni	traces	Polyether	Aliquot	spectroph.
Water	Th	traces	Polyether	PMBP HDEHP-TBP	radiometry spectroph.
Water	PO_4^{3-}	traces		Amine-molybdate-TBP	colorimetry

Abbreviations
PAN: 1-(2-pyridylazo)-2-naphthol
DMG: dimethylglyoxime
DPC: 1,5-diphenylcarbazide
DADTC: diethylammonium diethyldithiocarbamate
PMBP: 1-phenyl-3-methyl-4-bensoyl-pyrazolone-5
HDEHP: bis-[2-ethylhexyl]phosphate
TBP: tributyl phosphate
Spectroph.: spectrophotometry
AAS: atomic absorption spectrometry
NAA: neutron activation analysis

CHAPTER 12

Qualitative Tests

CONTENTS

ORGANIC GROUP QUALITATIVE TESTS

The following flow charts and notes provide a step-by-step process for the identification of functional groups that may be present in an unknown sample [1–11]. These are meant to augment and confirm information obtainable using instrumental methods of analysis. It will usually be necessary to use gas or liquid chromatography before these "wet" chemical tests are performed in order to determine the number of components present in a given sample. Since many of these tests require the use of sometimes toxic compounds, the strictest rules of laboratory safety must be observed at all times. The use of a fume hood is often required. The book by Feigl and colleagues is an excellent guide for spot tests [11].

Note: ppt = precipitate; conc = concentrated; dil = dilute

REFERENCES

1. Pasto, D. J., and C. R. Johnson. *Laboratory Text for Organic Chemistry.* Englewood Cliffs: Prentice Hall, 1979.
2. Svoronos, P., E. Sarlo, and R. Kulawiec. *Experiments in Organic Chemistry.* Dubuque, IA: McGraw-Hill, 1997.
3. Roberts, R. M., J. C. Gilbert, L. B. Rodewald, and A. S. Wingrove. *Modern Experimental Organic Chemistry.* New York: Sanders, 1985.
4. Hodgman, C. D., R. C. Weast, and S. M. Selby. *Tables for Identification of Organic Compounds.* Cleveland, OH: Chemical Rubber Publishing Co., 1950.
5. Kamm, O. *Qualitative Organic Analysis.* New York: John Wiley and Sons, 1932.
6. Vogel, A. I., B. S. Furniss, and A. R. Tatchell. *Vogel's Textbook of Practical Organic Chemistry.* New York: John Wiley and Sons, 1989.
7. Shriner, R. L., C. K. F. Hermann, T. C. Morrill, R. C. Fuson, and D. Y. Curtin. *The Systematic Identification of Organic Compounds, A Laboratory Manual.* New York: John Wiley and Sons, 1998.
8. Vogel, A. I. *Elementary Practical Organic Chemistry.* Part 2. New York: John Wiley and Sons, 1966.
9. Behforouz, M. "Getting the Acid Out of Your 2,4-DNPH." *Journal of Chemical Education* 63 (1986): 723.
10. Durst, H. D., and G. W. Gokel. *Experimental Organic Chemistry.* New York: McGraw-Hill, 1987.
11. Feigl, F., V. Anger, and R. E. Oesper. *Spot Tests in Organic Analysis.* Amsterdam: Elsevier, 1966.

PROTOCOL FOR CHEMICAL TESTS

The following section gives a suggested protocol for the chemical tests used in the identification of organic compounds. Variations of the procedures are possible, but these protocols have been used successfully for most organic identifications [1–10].

REFERENCES

1. Vogel, A. I., B. S. Furniss, and A. R. Tatchell. *Vogel's Textbook of Practical Organic Chemistry*. New York: John Wiley and Sons, 1989.
2. Shriner, R. L., C. K. F. Hermann, T. C. Morrill, R. C. Fuson, and D. Y. Curtin. *The Systematic Identification of Organic Compounds, A Laboratory Manual*. New York: John Wiley and Sons, 1998.
3. Vogel, A. I. *Elementary Practical Organic Chemistry*. Part 2. New York: John Wiley and Sons, 1966.
4. Pasto, D. J., and C. R. Johnson. *Laboratory Text for Organic Chemistry*. Englewood Cliffs: Prentice Hall, 1979.
5. Roberts, R. M., J. C. Gilbert, L. B. Rodewald, and A. S. Wingrove. *Modern Experimental Organic Chemistry*. New York: Saunders, 1985.
6. Uamm, O. *Qualitative Organic Analysis*. New York: John Wiley and Sons, 1932.
7. Behforout, M. "Getting the Acid Out of Your 2,4-DNPH." *Journal of Chemical Education* 63 (1986): 723.
8. Durst, H. D., and G. W. Gokel. *Experimental Organic Chemistry*. New York: McGraw-Hill, 1987.
9. Fieser, L. F., and M. Freser. *Reagents for Organic Synthesis*. New York: John Wiley and Sons, 1968.
10. Svoronos, P., E. Sarlo, and R. Kulawiec. *Experiments in Organic Chemistry*. Dubuque: McGraw-Hill, 1997.

Aluminum chloride-chloroform test

To a mixture of 2 mL chloroform and 0.2–0.4 *dry* aluminum chloride in a test tube add 5–10 drops of your unknown aromatic compound. A color formation will indicate the presence of a benzene ring.

Basic hydrolysis test

Reflux 0.1 g of the compound in 5 mL of a 10 % sodium hydroxide solution.

Benedict's test

Add 5–10 drops of your unknown to 1–2 mL of the Benedict's reagent and heat. A positive test for reducing sugars will change the blue copper (II) color of the reagent with subsequent precipitation of the red copper (I) oxide.

Bromine test

The compound to be tested is treated with a few drops of 1–5 % Br_2/CCl_4 solution. A positive test is indicated by decolorization of the bromine color.

Ceric ammonium nitrate test

To 1–2 mL 5 % ceric ammonium nitrate add 10 drops of the compound to be tested. A change to an orange/red color is indicative of an alcohol (detection limit 100 mg – compounds tested C_1–C_{10}).

Dichromate test (Jones Test)

Add 10 drops of the alcohol to be tested to a mixture of 1 mL 1 % $Na_2Cr_2O_7$ and 5 drops conc. H_2SO_4. A blue–green solution is positive test for a 1 ° or 2 ° alcohol. The 3 ° alcohols do not react and, therefore, the solution stays orange. (Detection limit 20 µg – compounds tested C_1–C_8). Slight heating may be necessary for water-immiscible alcohols. Extensive heat gives a positive test also for tertiary alcohols, which is due to the water elimination of the alcohol and oxidation of the formed alkene.

2,4-dinitrophenylhydrazine (2,4-DNP) test

Add 10 drops of the compound to be tested to 1 mL of the 2,4-DNP reagent. A yellow to orange-red precipitate is considered a positive test. The crystals can be purified by washing them with 5 % $NaHCO_3$, then with water, and finally recrystallized from ethanol. The 2,4-DNP reagent can be prepared by dissolving 1 g 2,4-dinitrophenylhydrazine in 5 mL conc. H_2SO_4 and then mixing it with 8 mL of water and 20 mL 95 % ethanol. The solution should be filtered before reacting it with the unknown compound. (Detection limit 20 µg – compounds tested C_1–C_8).

Fehling's test

The test is similar to the Benedict's test (see above).

ferric chloride test

Add 10 drops of 3 % aqueous $FeCl_3$ solution to 1 mL of a 5 % aqueous ethanol solution of the compound in question. Phenols give red, blue, purple, or green colorations. The same test can be done by using chloroform as a solvent (detection limit 50 µg).

Hinsberg test

To 0.5 mL of the amine (0.5 g, if solid) in a test tube add 1 mL of benzenesulfonyl chloride and 8 ml 10 % NaOH. Stopper the tube and shake for 3–5 minutes. Remove the stopper and warm the tube with shaking in a hot water bath (70 °C) for about one minute. No reaction is indicative of a 3 ° amine; the amine becomes soluble upon acidification (pH = 2–4) with 10 % HCl. If a precipitate is present in the alkaline solution, dilute with 5–8 mL H_2O and shake. If the precipitate does not dissolve, the original amine is probably a 2 ° one. If the solution is clear, acidify (pH = 4) with 10 % HCl. The formation of a precipitate is indicative of a 1 ° amine. (Detection limit 100 mg-compounds tested C_1–C_{10}).

iodoform test

The reagent calls for the mixture of 10 g I_2 and 20 g KI in 100 ml water. The reagent is then added dropwise to a mixture of 10 drops of the compound in question in 2 mL of water (or dioxane, to facilitate the solubility) and 1 mL 10 % aqueous NaOH solution until a *persistent* brown color remains (even when heating in a hot water bath at 60 °C). A yellow precipitate is indicative of iodoform (CHI_3) formation and is characteristic of a methyl ketone, acetaldehyde, or an alcohol of the general formula $CH_3CH(R)OH$ (R = alkyl, hydrogen). Aldols, $RC(=O)CH_2CH(OH)R'$, may also give a positive iodoform test by a retro aldol condensation first yielding $RC(=O)CH_3$ and RCHO (detection limit 100 mg). In this case at least one of the products should be a methyl ketone or acetaldehyde.

Molisch test

The reagent is made by preparing a solution of 95 % (vol/vol) of 1-naphthol in ethanol. The reagent is added to the test solution, which is then acidified with sulfuric acid. The development of a purple color at the interface of the test solution with the reagent mixture is indicative of a carbohydrate.

Lucas test

The reagent is made by dissolving 16 g anhydrous $ZnCl_2$ in 10 mL concentrated hydrochloric acid and cooling to avoid HCl loss. Add 10–15 drops of the *anhydrous* alcohol to 2 mL of the reagent. The 3 ° alcohols form an emulsion that appears as two layers (due to the water-insoluble alkyl halide) almost immediately. The 2 ° alcohols form this emulsion after 2–5 minutes, while 1 ° alcohols react after a very long time (if at all). Some secondary alcohols (e.g., isopropyl) may not *visually* form the layers because of the low boiling alkyl halide that may evaporate. Allyl alcohols and most benzyl alcohols also yield results that are identical to the results obtained for 3° alcohols.

permanganate test

The compound to be tested is treated with 10–15 drops of 1 % $KMnO_4$ solution. A positive test is indicated by the decolorization of the solution and subsequent formation of a black (MnO_2) precipitate.

silver nitrate test

The compound to be tested is treated with a few drops of 1 % alcoholic silver nitrate. A white precipitate indicates a positive reaction. This could be due to either silver chloride (reaction with a reactive alkyl halide), silver alkynide (reaction with a terminal alkyne), or the silver salt of a carboxylic acid (reaction with a carboxylic acid).

sodium fusion test

Treat 100 mg of the compound to be analyzed with a fresh tiny piece of sodium metal the size of a small pea in a 4-inch test tube. The test tube is warmed gently until melting of the sodium metal and decomposition (indicated by charring) of the compound occurs. When it appears that all the volatile material has been decomposed, the test tube is strongly heated until the residue acquires a red color. After 3 minutes of constant heating, the mixture is left to cool to room temperature, then a few drops of methanol are added. If no smoke appears, then excess of sodium metal is not present and incomplete conversion of the elements (nitrogen, sulfur, halides) is very likely. Addition of another tiny piece of sodium metal and repetition of the heating process is necessary. If smoke appears, then the red-hot test tube is plunged in a small beaker containing 10–15 mL *distilled water* and covered with a watch glass or a wire gauze. The test tube might shatter and, therefore, having the small beaker placed inside a larger one is recommended. The contents of the test tube together with the broken glass are ground in a mortar using a pestle, then transferred to the small beaker and heated for a few minutes. The solution is then filtered and the solution divided into two larger portions and l-mL part and are analyzed according to the following three procedures.

Detection of nitrogen

To one of the two larger portions add 10 drops 6M NaOH (pH adjusted to 13), five drops of saturated $Fe(NH_4)_2(SO_4)_2$ solution, and five drops 30 % KF solution. The mixture is then boiled for 30 seconds and immediately acidified with 6M H_2SO_4 with stirring until the colloidal iron hydroxides are dissolved. The formation of a blue color is indicative of the presence of nitrogen.

Detection of sulfur

To the l-mL part add 10 drops of 6 M acetic acid and 2–3 drops of 5 % lead (II) acetate solution. A black precipitate is indicative of sulfur presence.

Detection of halogens

To the other larger part add 10 % H_2SO_4 (dropwise) until the solution is acidic. Boil off the solution to 1/3 its volume to secure evaporation of H_2S and HCN gases. Formation of a precipitate upon addition of a few drops of 10 % AgNO$_3$ solution is indicative of the presence of a halogen: white for a chloride (which is soluble in 6 M NH$_4$OH), pale yellow for a bromide (which is only slightly soluble in 6 M NH$_4$OH), and canary yellow for iodide (which is insoluble in 6 M NH$_4$OH). Should the color of the precipitate be difficult to provide satisfactory identification of the halogen, proceed as follows the working solution, which has been acidified with 10 % H$_2$SO$_4$ and boiled down, is treated with 4–5 drops 0.1 N KMnO$_4$ solution, with enough oxalic acid added to discharge the color of excess permanganate and 0.5 ml carbon disulfide. Color formation in the carbon disulfide layer indicates the presence of bromine (red brown) or iodine (purple). Chlorine's presence cannot be detected by color formation. Should the compound to be tested carry both bromine and iodine the identification is difficult (red–brown to purple carbon disulfide layer). In this case addition of a few drops of allyl alcohol decolorizes bromine but does not decolorize iodine.

Tollen's test

The reagent should be freshly prepared by mixing two solutions (A and B). Solution A is a 10 % aqueous AgNO$_3$ solution and solution B is a 10 % aqueous NaOH solution. When the test is required, one mL of solution A and one mL of solution B are mixed, and the silver oxide thus formed is dissolved by dropwise addition of 10 % aqueous NH$_4$OH. To the clear solution, 10 drops of the compound to be tested are added. A silver mirror is indicative of the presence of an aldehyde. The reagent mixture (A + B) is to be prepared immediately prior to use, otherwise explosive silver fulminate will form. The silver mirror is usually deposited on the walls of the test tube either immediately or after a short warming period in a hot water bath. This is to be disposed of immediately by dissolving it in dilute HNO$_3$. (Detection limit 50 mg—compounds tested C$_1$–C$_6$.)

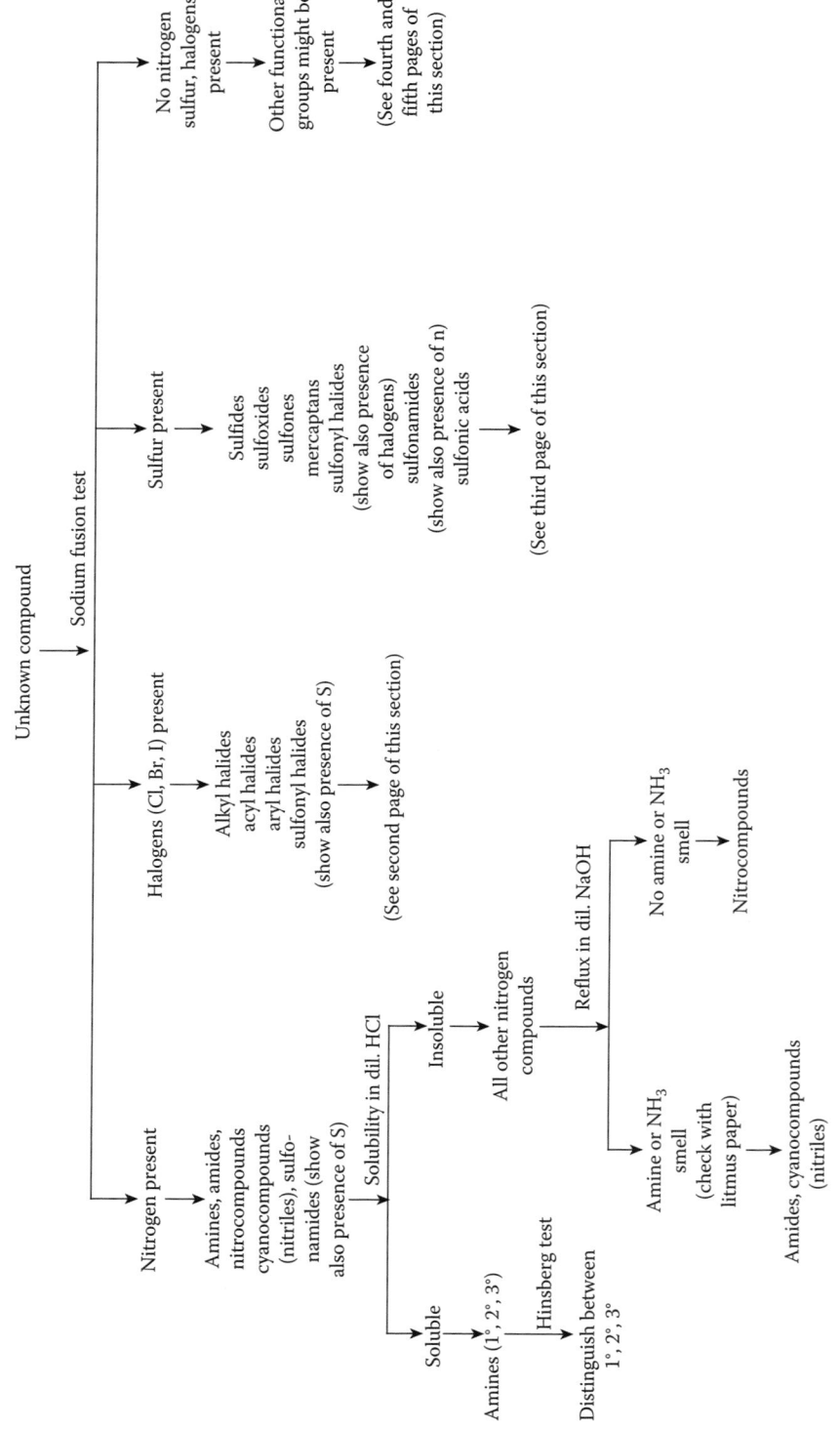

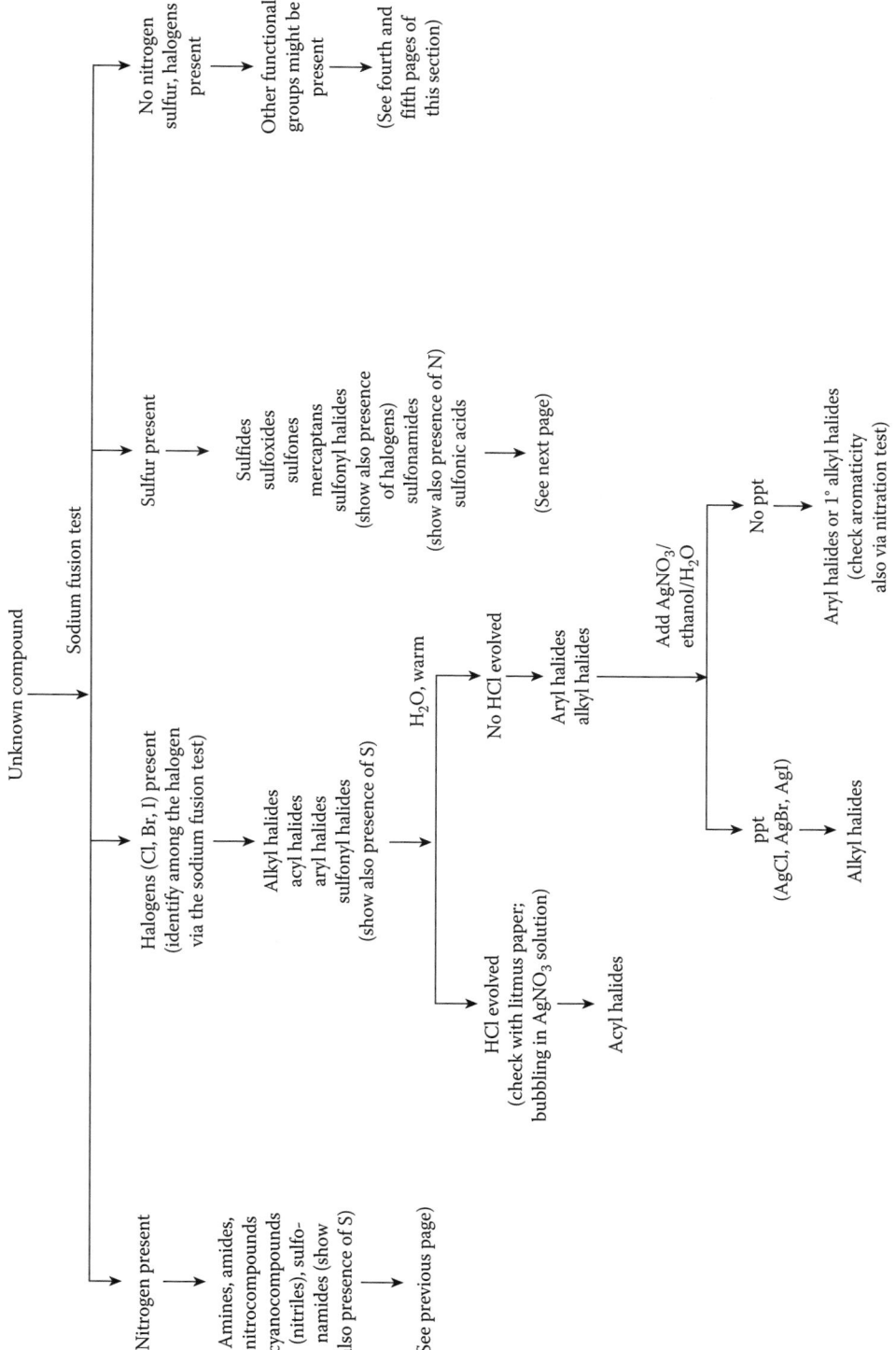

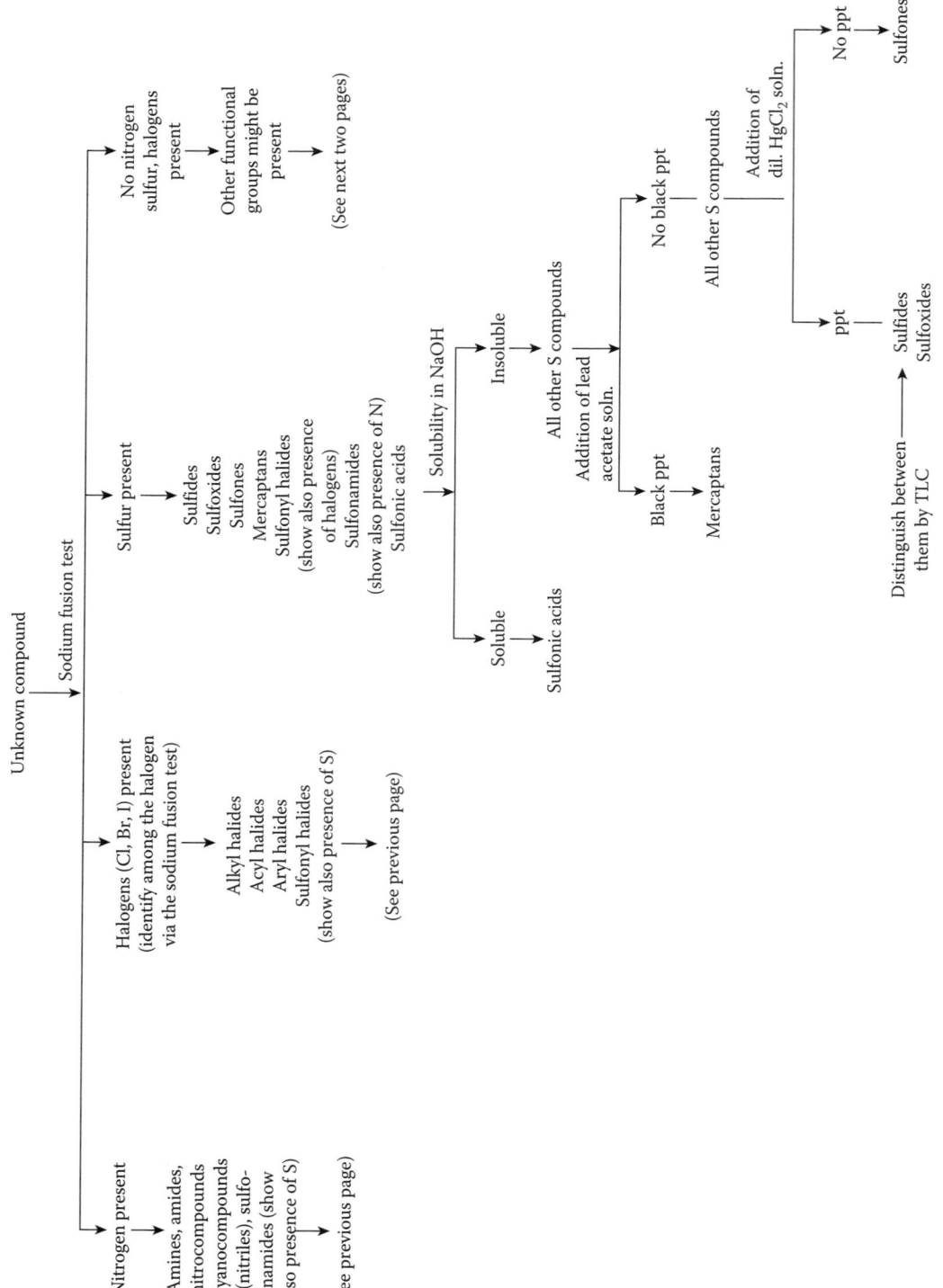

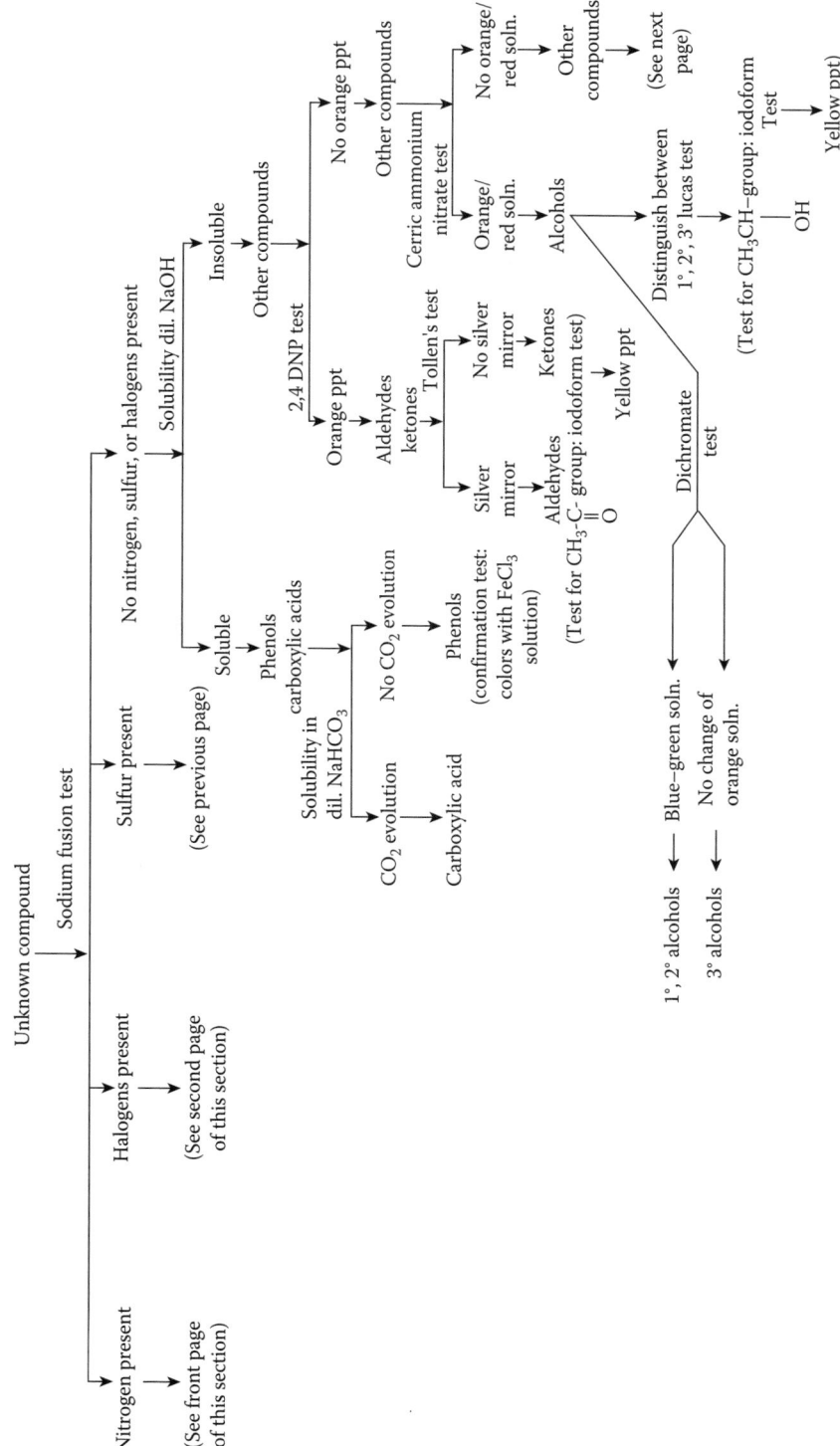

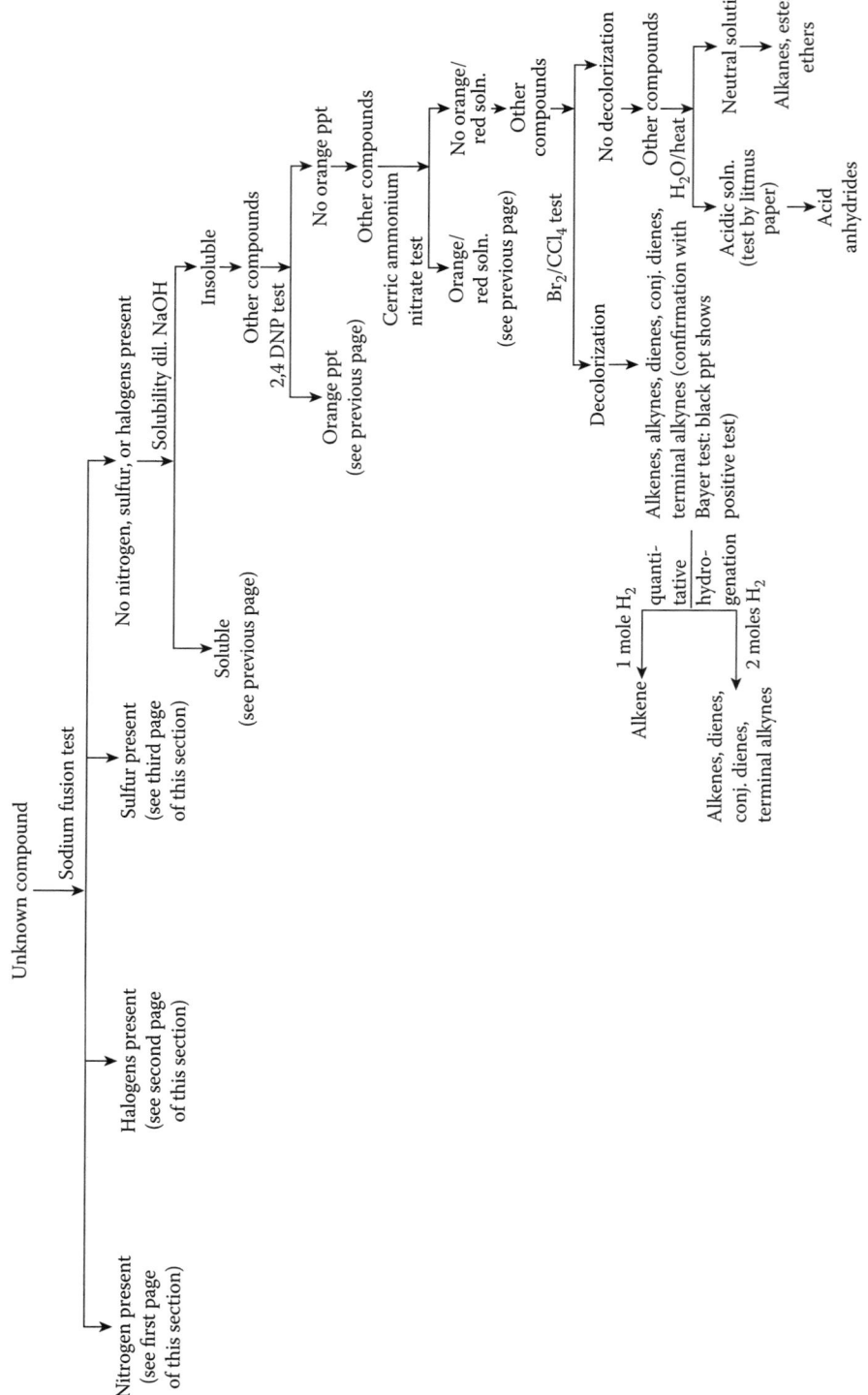

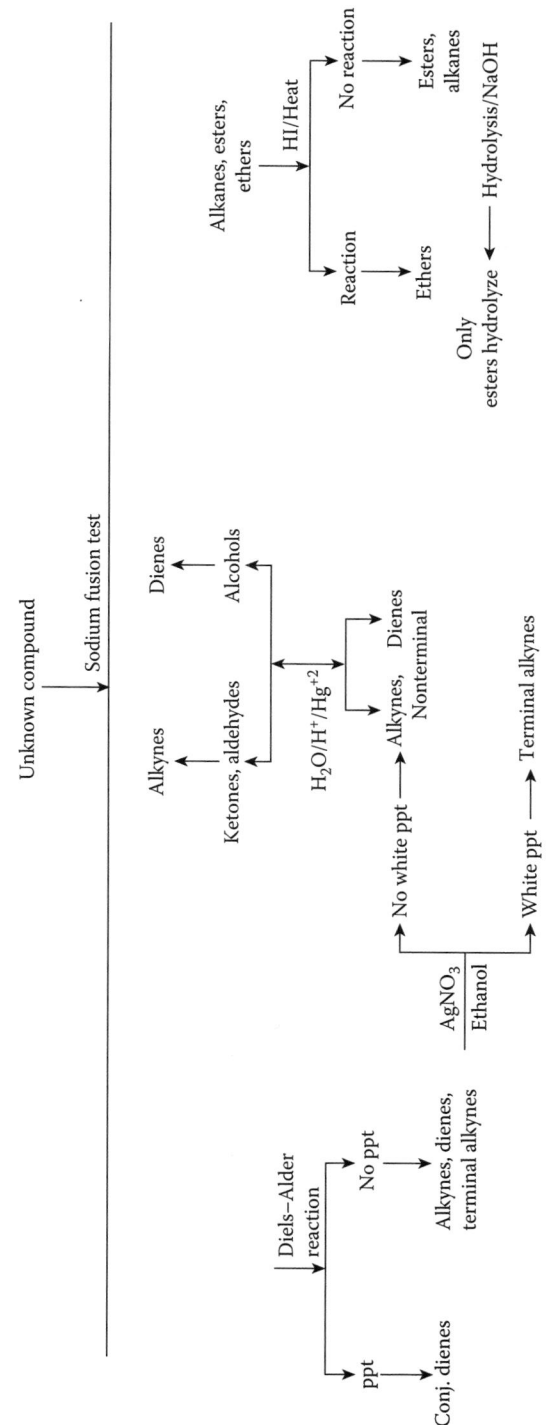

ORGANIC FAMILIES AND CHEMICAL TESTS

The following section gives the major organic families and their most important confirmatory chemical tests. This part serves as a complement the previous section (Organic Group Qualitative Tests) [1–8].

Note: ppt. = precipitate.

REFERENCES

1. Pasto, D. J., and C. R. Johnson. *Laboratory Text for Organic Chemistry.* Englewood Cliffs: Prentice Hall, 1979.
2. Roberts, R. M., J. C. Gilbert, L. B. Rodewald, and A. S. Wingrove. *Modern Experimental Organic Chemistry.* New York: Saunders, 1985.
3. Kamm, O. *Qualitative Organic Analysis.* New York: John Wiley and Sons, 1932.
4. Shriner, R. L., C. K. F. Hermann, T. C. Morrill, R. C. Fuson, and D. Y. Curtin. *The Systematic Identification of Organic Compounds, A Laboratory Manual.* New York: John Wiley and Sons, 1998.
5. Vogel, A. I. *Elementary Practical Organic Chemistry.* Part 2. New York: John Wiley and Sons, 1966.
6. Durst, H. D., and G. W. Gokel. *Experimental Organic Chemistry.* New York: McGraw-Hill, 1987.
7. Fieser, L. F., and M. Freser. *Reagents for Organic Synthesis.* New York: John Wiley and Sons, 1968.
8. Svoronos, P., E. Sarlo, and R. Kulawiec. Experiments in Organic Chemistry. Dubuque, IA: McGraw-Hill, 1997.

Organic Families and Chemical Tests

Family	Test	Notes
Alcohols	ceric ammonium nitrate	positive for all alcohols
	dichromate test	positive for 1 ° and 2 ° alcohols; negative for 3 ° alcohols
	iodoform test	positive for all alcohols of the general formula $CH_3CH(OH)R$
	Lucas test	immediate reaction for 3 °, allylic, or benzylic alcohols; slower reaction (2–5 min) for 2 °; no reaction for 1 ° alcohols
Aldehydes	Benedict's test	positive for all aldehydes
	dichromate test	positive for all aldehydes
	2,4-dinitrophenylhydrazine (2,4-DNP)	positive for all aldehydes (and ketones)
	Fehling's test	positive for all aldehydes
	iodoform test	positive only for acetaldehyde
	oxime	positive for all aldehydes (and ketones)
	permanganate test	positive for all aldehydes
	semicarbazone	positive for all aldehydes (and ketones)
	Tollen's test	positive for all aldehydes
Alkanes	no test	
Alkenes	bromine test	positive for all alkenes
	permanganate test	positive for all alkenes
	solubility in conc. sulfuric acid	all alkenes dissolve
Alkynes	bromine test	positive for all alkynes
	permanganate test	positive for all alkynes
	silver nitrate	positive for all terminal alkynes only
	sodium metal addition	positive for all terminal alkynes only
	sulfuric acid	positive for all alkynes

Organic Families and Chemical Tests (Continued)

Family	Test	Notes
Amides	basic (reflux) hydrolysis	all amides yield ammonia or the corresponding amine detected by odor or by placing wet blue litmus paper on top of the condenser
Amines	diazotization	all 1 ° amines give red azodyes with β-naphthol
	Hinsberg test	distinguishes between 1 °, 2 °, or 3 °
	solubility in dilute HCl	all amines are soluble
Arenes	aluminum chloride-chloroform	positive for all arenes
Aryl halides	aluminum chloride-chloroform	positive for all aryl halides
Carboxylic acids	solubility in dilute sodium bicarbonate	all carboxylic acids are soluble
	solubility in dilute sodium hydroxide	all carboxylic acids are soluble
Ketones	2,4-dinitrophenylhydrazine (2,4-DNP)	positive for all ketones (and aldehydes)
	hydrazine	positive for all ketones (and aldehydes)
	iodoform	positive for methyl ketones
	oxime	positive for all ketones (and aldehydes)
	semicarbazone	positive for all ketones (and aldehydes)
Nitriles	basic hydrolysis	positive for all nitriles
Phenols	acetylation	ppt. of a characteristic melting point
	benzoylation	ppt. of a characteristic melting point
	sulfonation	ppt. of a characteristic melting point
	ferric chloride test	variety of colors characteristic of the individual phenol
	solubility in aqueous base	most phenols are soluble in dilute sodium hydroxide but insoluble in dilute sodium bicarbonate. Phenols with strong electron withdrawing groups (e.g., picric acid) are soluble in sodium bicarbonate
Sulfonamides	basic (reflux) hydrolysis	positive for all sulfonamides
	sodium fusion test	presence of sulfur and nitrogen
Sulfonic acids	sodium fusion test	presence of sulfur
	solubility in aqueous base	most sulfonic acids are soluble in dilute sodium hydroxide and generate carbon dioxide with sodium bicarbonate

INORGANIC GROUP QUALITATIVE TESTS

The following tables list some simple chemical tests that will indicate the presence or absence of a given inorganic cation or anion [1–4]. For most of these tests, the anion or cation must be present at a relatively high concentration; the approximate lower bound is 0.05 % unless otherwise specified. It may therefore be necessary to concentrate more dilute samples before successful results can be obtained. Often centrifuging solutions to clearly confirm the formation of a precipitate is necessary. These tests should be used in conjunction with other methods such as the chromatographic methods or spectrometry and spectrophotometry (primarily atomic absorption and atomic emission). Since many of these tests require the use of potentially toxic compounds, the strictest rules of laboratory safety must be observed at all times.* The use of a fume hood is strongly recommended. All of the test reagents specified in this section are assumed to be in aqueous solution, unless otherwise designated. The reader is referred to the work of Svehla and Suehla [3] for details on reagent preparation.

REFERENCES

1. Barber, H. H., and T. I. Taylor. *Semimicro Qualitative Analysis.* New York: Harper Brothers, 1953.
2. Bruno, T. J., and P. D. N. Svoronos. *Basic Tables for Chemical Analysis.* National Bureau of Standards Technical Note 1096, April 1986.
3. Svehla, G., and G. Suehla. *Vogels Qualitative Inorganic Analysis.* 7th ed. New York: Addison Wesley, 1996.
4. De, A. K. *Separation of Heavy Metals.* New York: Pergamon Press, 1961.

* It is illegal to dispose of cyanide solutions in municipal sewer systems (POTW: publicly owned treatment works). The cyanide ion must be destroyed prior to disposal. This is easily done using an aqueous solution of chlorine:

$$CN^- + Cl_2 \rightarrow CNCl + Cl^-$$
$$CNCl + 2OH^- \rightarrow CNO^- + Cl^- + H_2O$$
$$2CNO^- + 4OH^- + 3Cl_2 + 2Cl_2 + 2H_2O + N_2 + 6Cl^-.$$

The first step will occur at all pH levels; the second requires high pH; the third requires a pH near 7. Thus, careful acidification using HCl will complete the reaction, producing harmless nitrogen and carbon dioxide.

TESTS FOR ANIONS

Acetates, CH_3COO^-

1. Sulfuric acid, dilute	evolution of acetic acid (vinegar-like odor); concentrated sulfuric acid also evolves sulfur dioxide under mild heating
2. Dry ethanol and concentrated sulfuric acid	evolves ethyl acetate (fruity odor) upon heating; dry isoamyl alcohol may be substituted for ethanol
3. Silver nitrate	formation of white precipitate of silver acetate that is soluble in dilute ammonia solution
4. iron (III) chloride	deep-red coloration (coagulates on boiling forming a brownish-red precipitate)

Benzoates, $C_6H_5COO^-$ (or $C_7H_5O_2^-$)

1. Dilute sulfuric acid	formation of white precipitate of benzoic acid
2. Dilute hydrochloric acid	formation of a crystalline precipitate melting between 121 and 123°C
3. Silver nitrate	white precipitate of silver benzoate from cold solutions, soluble in hot water and also in dilute ammonia solution
4. Iron (III) chloride	buff-colored (light yellow–red) precipitate of iron (III) benzoate from neutral solution, soluble in hydrochloric acid

Borates, BO_3^{3-}, $B_4O_7^{2-}$, BO_2^-

1. Concentrated sulfuric acid	upon heating solution, white fumes of boric acid are evolved
2. Silver nitrate	white precipitate of silver metaborate, soluble in dilute ammonia solution and in acetic acid
3. Barium chloride	white precipitate of barium metaborate that is soluble in excess reagent, dilute acids as well as ammonium salt solutions

Bromates, BrO_3^-

1. Concentrated sulfuric acid	evolution of red bromine vapors even when cold
2. Silver nitrate	white precipitate of silver bromate that is soluble in dilute ammonia
3. Sodium nitrite	brown color develops after addition and subsequent acidification with dilure nitric acid

Note: Bromates are reduced to bromides by sulfur dioxide, hydrogen sulfide, or sodium nitrite solution.

Bromides, Br^-

1. Concentrated sulfuric acid	reddish–brown coloration, followed by reddish–brown vapors (hydrogen bromide + bromine) evolution
2. Manganese dioxide + sulfuric acid	reddish–brown bromine vapors evolve upon mild heating
3. Silver nitrate	pale-yellow, curdy, precipitate of silver bromide, slightly soluble in ammonia solution; insoluble in nitric acid
4. Lead acetate	white crystalline precipitate of lead bromide that is soluble in hot water

Special Tests: The addition of an aqueous solution of chlorine (or sodium hypochlorite) will liberate free bromine, which may be isolated in a layer of carbon tetrachloride or carbon disulfide.

Carbonates, CO_3^{2-}

1. Hydrochloric acid	decomposition with effervescence and evolution of carbon dioxide (odorless)
2. Barium chloride	white precipitate of barium carbonate that is soluble in HCl (calcium chloride may be substituted for barium chloride)
3. Silver nitrate	gray precipitate of silver carbonate
4. Magnesium sulfate	white precipitate is formed, which can be dissolved by the addition of dilute acetic acid

Special Tests: Effervescence with all acids, producing carbon dioxide that makes limewater cloudy.

(*Continued*)

Tests for Anions (Continued)

Chlorates, ClO_3^-

1. Concentrated sulfuric acid	liberates chlorine dioxide gas (green); solids decrepitate (crackle explosively) when warmed; *large quantities may result into a violent explosion*
2. Concentrated HCl	chlorine dioxide gas evolved, imparts yellow color to acid
3. Manganese (II) sulfate + phosphoric acid	violet coloration due to diphosphatomanganate formation; peroxydisulfate nitrates, bromates, iodates, periodates, react similarly
4. Heat of neat sample	decomposition and formation of gaseous oxygen

Chlorides, Cl^-

1. Concentrated sulfuric acid	evolution of hydrogen chloride gas (pungent odor)
2. Silver nitrate	white precipitate of silver chloride that is soluble in ammonia solution (re-precipitate with HNO_3)
3. Lead acetate	white precipitate of lead bromide, soluble in boiling water

Special Tests: 1. $MnO_2 + H_2SO_4$ evolves Cl_2 gas.
2. *An aqueous solution of chlorine + carbon disulfide produces no coloration.*

Chromates, CrO_4^{2-}, Dichromates, $Cr_2O_7^{2-}$

1. Barium chloride	pale-yellow precipitate of barium chromate, soluble in dilute mineral acids, insoluble in water and in acetic acid
2. Silver nitrate	brownish-red precipitate of silver chromate, soluble in dilute nitric acid and in ammonia solution; insoluble in acetic acid
3. Lead acetate	yellow precipitate of lead chromate, soluble in dilute nitric acid, insoluble in acetic acid
4. Hydrogen peroxide	deep-blue coloration in acidic solution, which quickly turns green, with the subsequent liberation of oxygen
5. Hydrogen sulfide	dirty yellow deposit of sulfur is produced in acidic solutions

Citrates, $C_6H_5O_7^{3-}$

1. Concentrated sulfuric	evolution of carbon dioxide and carbon monoxide (HIGHLY POISONOUS)
2. Silver nitrate	white precipitate of silver citrate that is soluble in dilute ammonia solution
3. Cadmium acetate	white gelatinous precipitate of cadmium citrate, practically insoluble in boiling water, soluble in warm acetic acid
4. Pyridine + acetic anhydride, 3:1 (vol/vol)	a red brown color develops upon addition to the reagent mixture

Cyanates, OCN^-

1. Sulfuric acid, concentrated and dilute	vigorous effervescence, due largely to evolution of carbon dioxide, with concentrated acid producing a more dramatic effect
2. Silver nitrate	curdy white precipitate of silver cyanate
3. Copper sulfate-pyridine	lilac-blue precipitate (interference by thiocyanates)Reagent is prepared by adding 2–3 drops of pyridine to 0.25 M $CuSO_4$ solution

Cyanides, CN^-

1. Cold dilute HCl	liberation of hydrogen cyanide (odor of bitter almond; CAUTION: HIGHLY TOXIC)
2. Silver nitrate	white precipitate of silver cyanide
3. Concentrated sulfuric acid, hot	liberation of carbon monoxide (CAUTION)
4. Mercury (I) nitrate	gray precipitate of mercury
5. Copper sulfide	formation of colorless tetracyanocuprate (I) ions This test can be done on a section of filter paper.

Tests for Anions (Continued)

Dithionites, $S_2O_4{}^{2-}$

1. Dilute sulfuric acid	orange coloration that disappears quickly, accompanied by evolution of sulfur dioxide gas and deposition of pale-yellow sulfur
2. Concentrated sulfuric acid	fast evolution of sulfur dioxide and precipitation of pale-yellow sulfur
3. Silver nitrate	black precipitate of silver
4. Copper sulfate	red precipitate of copper
5. Mercury (II) chloride	gray precipitate of mercury
6. Methylene blue	decolorization in cold solution
7. Potassium hexacyanoferrate (II) and iron (II) sulfate	white precipitate of dipotassium iron (II) hexacyanoferrate (II); turns from white to Prussian blue

Fluorides, F^-

1. Concentrated sulfuric acid	evolution of hydrogen fluoride dimer
2. Calcium chloride	white, slimy precipitate of calcium fluoride, slightly soluble in dilute hydrochloric acid
3. Iron (III) chloride	white precipitate

Special Tests: HF etches glass (only visible after drying).

Formates, $HCOO^-$

1. Dilute sulfuric acid	formic acid is evolved (pungent odor)
2. Concentrated sulfuric acid	carbon monoxide (HIGHLY POISONOUS, ODORLESS, COLORLESS) is evolved on warming
3. Ethanol and concentrated H_2SO_4, heat	ethyl formate evolved (pleasant odor)
4. Silver nitrate	white precipitate of silver formate in neutral solutions, forming a black deposit of elemental silver upon mild heating
5. Iron (III) chloride	red coloration due to complex formation
6. Mercury (II) chloride	white precipitate of calomel produced on warming; upon boiling, a black deposit of elemental mercury is produced

Hexacyanoferrate (II) Ions, $[Fe(CN)_6]^{4-}$

1. Silver nitrate	white precipitate of silver hexacyanoferrate (II)
2. Iron (III) chloride	Prussian blue is formed in neutral or acid conditions, which is decomposed by alkali bases
3. Iron (II) sulfate (aq)	white precipitate of potassium iron (II) hexacyanoferrate, which turns blue by oxidation
4. Copper sulfate	brown precipitate of copper hexacyanoferrate (II)
5. Thorium nitrate	white precipitate of thorium hexacyanoferrate (III)

Hexacyanoferrate (III) Ions $[Fe(CN)_6]^{3-}$

1. Silver nitrate	orange–red precipitate of silver hexacyanoferrate (III), which is soluble in ammonia solution but not in nitric acid.
2. Iron (II) sulfate	dark-blue precipitate in neutral or acid solution (Prussian or Turnbull's blue)
3. Iron (III) chloride	brown coloration
4. Copper sulfate	green precipitate of copper (II) hexacyanoferrate (III)
5. Concentrated hydrochloric acid	brown precipitate of hexacyanoferric acid

Hexafluorosilicates (Silicofluorides), $[SiF_6]^{2-}$

1. Barium chloride	white, crystalline precipitate of barium hexafluorosilicate, insoluble in dilute HCl, slightly soluble in water
2. Potassium chloride	white gelatinous precipitate of potassium hexafluorosilicate, slightly soluble in water
3. Ammonia solution	gelatinous precipitate of silica acid

(Continued)

Tests for Anions (Continued)

Hydrogen Peroxide, H_2O_2

1. Potassium iodide and starch	if sample is previously acidified by dilute sulfuric acid, a deep-blue coloration occurs due to the production of iodine complexation with starch
2. Potassium permanganate	decolorization, evolution of oxygen
3. Titanium (IV) chloride	orange–red coloration; very sensitive test

Special test: $4H_2O_2 + PbS \rightarrow PbSO_4\downarrow + 4H_2O$ black lead sulfide reacts to produce white lead sulfate.

Special Test: A reagent prepared from p-hydroxyphenylacetic acid (HPPA), 7.6 mg, hematin (typically from pig albumin), 1.0 mg, in 100 mL of 0.1 M KOH (aq) will produce a fluorescent dimer (6,6'-dihydroxy-3,3'-biphenyl acetic acid) with hydrogen peroxide. This test is extremely sensitive.

Hypochlorites, OCl^-

1. Dilute hydrochloric acid	yellow coloration, followed by chlorine gas evolution
2. Lead (II) acetate or nitrate	brown lead (IV) oxide forms upon heating
3. Cobalt nitrate	black precipitate of cobalt (II) hydroxide
4. Mercury	on shaking slightly acidified solution of a hypochlorite with Hg, a brown precipitate of mercury (II) chloride is formed

Hypophosphites, $H_2PO_2^-$

1. Silver nitrate	white precipitate of silver hypophosphite
2. Mercury (II) chloride	white precipitate of calomel in cold solution that darkens upon warming
3. Copper (II) sulfate	red precipitate of copper (I) oxide forms upon warming
4. Potassium permanganate	immediate decolorization under cold conditions

Iodates, IO_3^-

1. Silver nitrate	white, curdy precipitate of silver iodate, soluble in dilute ammonia solution
2. Barium chloride	white precipitate of barium iodate, sparingly soluble in hot water or dilute nitric acid; insoluble in ethanol and methanol
3. Mercury (II) nitrate	white precipitate of mercury (II) iodate

Iodides, I^-

1. Concentrated sulfuric acid	produces hydrogen iodide and iodine
2. Silver (I) nitrate	yellow precipitate of silver (I) iodide that is slightly soluble in ammonia solution and insoluble in dilute nitric acid
3. Lead acetate	yellow precipitate of lead iodide, soluble in excess hot water
4. Potassium dichromate and concentrated sulfuric acid	liberation of iodine
5. Sodium nitrite	liberation of iodine
6. Copper sulfate	brown precipitate
7. Mercury (II) chloride	scarlet precipitate of mercury (II) iodide

Special Tests: 1. $MnO_2 + H_2SO_4$ produces I_2
2. Cl_2 (aq)/CS_2 produces I_2 in CS_2 (purple)
3. Starch paste + Cl_2 (aq), deep blue coloration

Lactates, $CH_3CH(OH)COO^-$

1. Potassium permanganate solution	The odor of acetaldehyde is observed upon the addition of dilute potassium permanganate solution, followed by acidification with dilute sulfuric acid and heating.

Metaphosphates, PO_3^-

1. Silver nitrate	white precipitate, soluble in dilute nitric acid, in dilute ammonia solution, and in dilute acetic acid
2. Albumin and dilute acetic acid.	coagulation
3. Zinc sulfate solution	white precipitate on warming; soluble in dilute acetic acid

Tests for Anions (Continued)

Nitrates, NO_3^-

1. Concentrated sulfuric acid

solid nitrate with concentrated sulfuric acid evolves reddish-brown vapors of nitrogen dioxide + nitric acid vapors when heated

Special Tests: 1. Add iron (II) sulfate, shake, then add concentrated sulfuric acid; produces brown ring.
2. White precipitate is formed upon addition of nitron reagent ($C_{20}H_{16}N_4$); test is not specific to only nitrates; however, see table of precipitation reagents.

Nitrites, NO_2^-

1. Dilute hydrochloric acid

cautious addition of acid to a solid nitrite in cold gives a transient pale (of nitrous acid or the anhydride) blue liquid and consequent evolution of brown fumes of nitrogen dioxide

2. Silver nitrate

white precipitate of silver nitrite

3. Iron (II) sulfate solution (25 %, acidified with either acetic or sulfuric acid)

a brown ring forms at the junction of the two liquids due to the formation of a complex

4. Acidified potassium permanganate

decolorization with no gas evolution

5. Ammonium chloride (solid)

boiling with excess of solid reagent causes nitrogen to be evolved

6. Concentrated sulfuric acid

liberates brown nitrogen dioxide gas

Special Tests: Acidified solutions of nitrites liberate iodine from potassium iodide

Orthophosphates, PO_4^{3-}

1. Silver nitrate

yellow precipitate of silver orthophosphate, soluble in dilute ammonia and in dilute nitric acid

2. Barium chloride

white precipitate of barium hydrogen phosphate, soluble in dilute mineral acids and acetic acid

3. Magnesium nitrate reagent or magnesia mixture

white crystalline precipitate of magnesium ammonium phosphate, soluble in acetic acid and mineral acids, practically insoluble in 2.5 % ammonia solution

4. Ammonium molybdate

addition of 2–3 mL excess reagent to approximately 0.5 mL sample gives yellow precipitate of ammonium phosphomolybdate that is soluble in ammonia solution and in solutions of caustic alkalis
Large quantities of hydrochloric acid interfere

5. Iron (III) chloride

yellowish-white precipitate of iron (III) phosphate, soluble in mineral acids, insoluble in dilute acidic acid

6. Ammonium molybdate–quinine

yellow precipitate of unknown composition, reducing agents interfere

Note: The orthophosphates are salts of orthophosphoric acid, H_3PO_4, and are simply referred to as phosphates.

Oxalates, $(COO)_2^{2-}$

1. Silver nitrate

white precipitate of silver oxalate that is soluble in ammonia solution and dilute nitric acid

2. Calcium chloride

white precipitate of calcium oxalate that is insoluble in dilute acetic acid, oxalic acid, and in ammonium oxalate solution; soluble in dilute hydrochloric acid and in dilute nitric acid

3. Potassium permanganate

decolorization upon warming to 60–70 °C in acidified solution

Perchlorates, ClO_4^-

1. Potassium chloride

white precipitate of potassium perchlorate, insoluble in alcohol

Special Tests: 1. Neutral ClO_4^- + cadmium sulfate in concentrated ammonia produces $[Cd(NH_3)_4](ClO_4)_2$ (white precipitate)
2. Cautious heating of solids evolves oxygen

(Continued)

Tests for Anions (Continued)

Peroxydisulfates, $S_2O_8^{2-}$

1. Water	on boiling, decomposes into the sulfate, free sulfuric acid, and oxygen
2. Silver nitrate	black precipitate of silver peroxide
3. Barium chloride	on boiling or standing for some time, a precipitate of barium sulfate is formed
4. Manganese (II) sulfate	brown precipitate of hydrate complex in neutral or alkaline test solution

Phosphites, HPO_3^{2-}

1. Silver nitrate	white precipitate of silver phosphite that yields black metallic silver on standing
2. Barium chloride	white precipitate of barium phosphite, soluble in dilute acids
3. Mercury (II) chloride	white precipitate in cold solutions that yields gray metallic mercury on warming
4. Copper sulfate	light-blue precipitate that dissolves in hot acetic acid
5. Lead (II) acetate	white precipitate of lead (II) hydrogen phosphite

Pyrophosphates, $P_2O_7^{4-}$

1. Silver nitrate	white precipitate, soluble in dilute nitric acid and in dilute acetic acid
2. Copper sulfate	pale-blue precipitate
3. Magnesia mixture or magnesium reagent	white precipitate, soluble in excess reagent but reprecipitated on boiling
4. Cadmium acetate and dilute acetic acid	white precipitate
5. Zinc sulfate	white precipitate, insoluble in dilute acetic acid; soluble in dilute ammonia solution yielding a white precipitate on boiling

Salicylates, $C_6H_4(OH)COO^-$ (or $C_7H_5O_3^-$)

1. Concentrated sulfuric acid	evolution of carbon monoxide and sulfur dioxide (poisonous)
2. Concentrated sulfuric acid and methanol	0.5 g sample + 3 mL reagent + heat evolves methyl salicylate (odor of wintergreen)
3. Dilute hydrochloric acid	crystalline precipitate of salicylic acid
4. Silver nitrate	heavy crystalline precipitate of silver salicylate (that is soluble in boiling water and recrystallizes upon cooling)
5. Iron (III) chloride	violet–red coloration that clears upon the addition of dilute mineral acids

Silicates, SiO_3^{2-}

1. Dilute hydrochloric acid	gelatinous precipitate of metasilicic acid, insoluble in concentrated acids, soluble in water and dilute acids
2. Ammonium chloride or ammonium carbonate	gelatinous precipitate
3. Silver nitrate	yellow precipitate of silver silicate, soluble in dilute acids as well as ammonia solution
4. Barium chloride	white precipitate of barium silicate that is soluble in dilute nitric acid

Succinates, $C_4H_4O_4^{2-}$

1. Silver nitrate	white precipitate of silver succinate, soluble in dilute ammonia solution
2. Iron (III) chloride	light-brown precipitate of iron (III) succinate
3. Barium chloride	white precipitate of barium succinate
4. Calcium chloride	slow precipitation of calcium succinate

Tests for Anions (Continued)

Sulfates, SO_4^{2-}

1. Barium chloride	white precipitate of barium sulfate, insoluble in warm dilute hydrochloric acid and in dilute nitric acid, slightly soluble in boiling hydrochloric acid
2. Lead acetate	white precipitate of lead sulfate; soluble in hot concentrated sulfuric acid, ammonium acetate, ammonium tartrate, and sodium hydroxide
3. Silver nitrate	white precipitate of silver sulfate
4. Mercury (II) nitrate	yellow precipitate of mercury (II) sulfate

Sulfides, S^{2-}

1. Dilute hydrochloric acid or sulfuric acid	hydrogen sulfide gas is evolved and detected by odor or lead acetate paper
2. Silver nitrate	black precipitate of silver sulfide, soluble in hot, dilute nitric acid
3. Lead acetate	black precipitate of lead sulfide
4. Sodium nitroprusside solution ($Na_2[Fe(CN)_5NO]$)	transient purple color in the presence of solutions of alkalis

Special Tests: Catalysis of iodine: azide reaction: Solution of sodium azide, (NaN_3) and iodine reacts with a trace of a sulfide to evolve nitrogen. Thiosulfates and thiocyanates act similarly and therefore must be absent.

Sulfites, SO_3^{2-}

1. Dilute hydrochloric acid	decomposition (which becomes more rapid on warming) and evolution of sulfuric dioxide (odor of burning sulfur)
2. Barium chloride strontium chloride	white precipitate of the respective sulfite, the precipitate being soluble in dilute hydrochloric acid
3. Silver nitrate	at first no change; upon addition of more reagent, white crystalline precipitate at silver sulfite forms that darkens to metallic silver upon heating
4. Potassium permanganate solution acidified with dilute sulfuric acid	decolorization (Fuchsin test)
5. Potassium dichromate in with dilute sulfuric acid	green color formation
6. Lead acetate or lead nitrate solution	white precipitate of lead sulfite
7. Zinc and sulfuric acid	hydrogen sulfide gas evolved, detected by holding lead acetate paper to mouth of test tube
8. Concentrated sulfuric acid	evolution of sulfur dioxide gas
9. Sodium nitroprusside–zinc sulfate	red compound of unknown composition

Tartrates, $C_4H_4O_6^{2-}$

1. Concentrated sulfuric acid	when sample is heated, the evolution of carbon monoxide, carbon dioxide, and sulfur dioxide (burned sugar odor) results
2. Silver nitrate	white precipitate of silver tartrate
3. Calcium chloride	white precipitate of calcium tartrate, soluble in dilute acetic acid, dilute mineral acids, and in cold alkali solutions
4. Potassium chloride	white precipitate, the reaction is: $C_4H_4O_6^{2-} + K+ + CH_3COOH \rightarrow C_4H_5O_6 K\downarrow + CH_3COO^-$

Special Test: One drop 25 % iron (II) sulfate, 2–3 drops hydrogen peroxide: produces deep violet-blue color (Fenton's Test).

(Continued)

Tests for Anions (Continued)

Thiocyanates, SCN⁻

1. Sulfuric acid	in cold solution, yellow coloration is produced; upon warming violent reaction occurs and carbonyl sulfide is released
	in basic solution, carbonyl sulfide is hydrolyzed to hydrogen sulfide
2. Silver nitrate	white precipitate of silver thiocyanate
3. Copper sulfate	first a green coloration, then black precipitate of copper (II) thiocyanate is formed
4. Mercury (II) nitrate	white precipitate of mercury (II) thiocyanate
5. Iron (III) chloride	blood-red coloration due to complex formation
6. Dilute nitric acid	upon warming, red coloration is observed, with nitrogen oxide and hydrogen cyanide (POISONOUS) being evolved
7. Cobalt nitrate	blue coloration due to complex ion formation

Thiosulfates, $S_2O_3^{2-}$

1. Iodine solution	decolorized; a colorless solution of tetrathionate ions is formed
2. Barium chloride	white precipitate of barium thiosulfate
3. Silver nitrate	white precipitate of silver thiosulfate
4. Lead (II) acetate or nitrate solution	first no change; on further addition of reagent, a white precipitate of lead thiosulfate forms
5. Iron (III) chloride solution	dark-violet coloration due to complex formation
6. Nickel ethylenediamine nitrate $[Ni(NH_2(CH_2)_2NH_2)_3](NO_3)_2$	violet complex precipitate forms; hydrogen sulfide and ammonium sulfide interfere

Special Tests: Blue Ring test: When solution of thiosulfate mixed with ammonium molybdate solution is poured slowly down the side of a test tube that contains concentrated sulfuric acid, a blue ring is formed temporarily at the contact zone.

TESTS FOR CATIONS

This table provides a summary of the common tests for cations, primarily in aqueous solution. The cations are grouped according to the usual convention of reactivity to a set of common reagents.

abbreviations:

conc = concentrated
dil = dilute
g = gaseous
EtOH = ethanol

Tests for Cations

GROUP I: Pb(II), Ag(I), Hg(I)

All members are precipitated by dilute HCl, to give lead chloride (PbCl$_2$), silver chloride (AgCl), or mercury (I) chloride, (Hg$_2$Cl$_2$).

Lead (II), Pb^{+2}

1. Potassium chromate	yellow precipitate of lead (II) chromate
2. Potassium iodide	yellow precipitate of lead (II) iodide
3. Sulfuric acid, dilute	white precipitate of lead (II) sulfate
4. Hydrogen sulfide gas	black precipitate of lead (II) sulfide
5. Potassium cyanide	white precipitate of lead (II) cyanide
6. Tetramethyldiaminodiphenyl- methane	blue oxidation product (presence of Bi, Ce, Mn, Th, Co, Ni, Fe, Cu may interfere)
7. Gallocyanine	deep violet precipitate, unknown composition (Bi, Cd, Cu, Ag may interfere)
8. Diphenylthiocarbazone	brick-red complex in neutral or ammoniacal solution

Silver (I), Ag$^+$

1. Potassium chromate	reddish brown precipitate of silver chromate
2. Potassium iodide	yellow precipitate of silver iodide
3. Hydrogen sulfide gas	black precipitate of silver sulfide
4. Disodium hydrogen phosphate	yellow precipitate of silver phosphate
5. Sodium carbonate	yellow-white precipitate of silver carbonate, forming the brown oxide upon heating
6. p-Dimethylaminobenzylidene-rhodanine	reddish-violet precipitate in acidic solution
7. Ammonia solution	brown precipitate of silver oxide, dissolving in excess to form Ag$_3$N, which is explosive

Mercury (I), Hg$_2^{+2}$

1. Potassium carbonate	red precipitate of mercury (I) chromate
2. Potassium iodide	green precipitate of mercury (I) iodide
3. Dilute sulfuric acid	white precipitate of mercury (I) sulfate
4. Elemental copper, aluminum, or zinc	amalgamation occurs
5. Hydrogen sulfide	black precipitate (in neutral or acid medium) of mercury (I) sulfide and mercury
6. Ammonia solution	black precipitate of HgO × Hg(NH$_2$)(NO$_3$)
7. Diphenylcarbizide (1 % in ethanol, with 0.2M nitric acid	violet colored complex results (high sensitivity and selectivity)
8. Potassium cyanide	mercury (I) cyanide solution, with a precipitation of elemental mercury (mercury (II) interferes)

(Continued)

Tests for Cations (Continued)

GROUP II: Hg(II), Cu(II), Bi(III), Cd(II), As(III), As(V), Sb(III), Sb(V), Sn(II), Sn(IV)

All members show no reaction with HCl; all form a precipitate with H_2S.

Bismuth (III), Bi^{3+}

1. Potassium iodide	black precipitate of bismuth (III) iodide
2. Potassium chromate	yellow precipitate of bismuth (III) chromate
3. Ammonia solution	white precipitate of variable composition, approximate formula: $Bi(OH)_2NO_3$
4. Pyrogallol (10 %)	yellow precipitate of bismuth (III) pyrogallate
5. 8-Hydroxyquinoline (5 %) + potassium iodide (6M)	red precipitate of the tetraiodobismuthate salt (characteristic in the absence of Cl^-, F^-, Br^-)
6. Sodium hydroxide	white precipitate of bismuth (III) hydroxide

Copper (II), Cu^{+2}

1. Potassium iodide	brown precipitate of copper (I) iodide, colored brown due to I_3^-
2. Potassium cyanide	yellow precipitate of copper (II) cyanide, which then decomposes
3. Potassium thiocyanate	black precipitate of copper (II) thiocyanate, which then decomposes
4. α-Benzoin oxime (or cupfon), 5 % in EtOH	green precipitate of the α-benzoin oxime salt derivative
5. Salicylaldoxime (1 %)	greenish-yellow precipitate of the copper complex
6. Rubeanic acid (0.5 %) (dithio-oxamide)	black precipitate of the rubeanate salt

Cadmium (II), Cd^{2+}

1. Ammonia solution	white precipitate of cadmium hydroxide that dissolves in excess ammonia
2. Potassium cyanide	white precipitate of cadmium cyanide that dissolves in excess potassium cyanide
3. Sodium hydroxide	white precipitate of cadmium hydroxide that is insoluble in excess sodium hydroxide
4. Dinitro-p-diphenyl carbizide	brown precipitate with cadmium hydroxide

Arsenic (III), As^{+3}

1. Silver nitrate	yellow precipitate of silver arsenite in neutral solution
2. Copper (II) sulfate	green precipitate of copper (II) arsenite (or $Cu_3(AsO_3)_2 \cdot xH_2O$)
3. Potassium triiodide ($KI + I_2$)	decolorization due to oxidation
4. Tin (II) chloride + concentrated hydrochloric acid	black precipitate forms in the presence of excess reagent

Arsenic (V), As^{+5}

1. Silver nitrate	brownish-red precipitate of silver arsenate from neutral solutions
2. Ammonium molybdate	yellow precipitate (in presence of excess reagent) of ammonium arsenomolybdate $(NH_4)_3AsMo_{12}O_4$
3. Potassium iodide + concentrated hydrochloric acid	iodine formation

Small amounts of As(III) or As(V) can be identified by the response to the Marsh, Gutzeit, or Fleitmann tests (see references at the beginning of this section).

Tests for Cations (Continued)

GROUP II: Hg(II), Cu(II), Bi(III), Cd(II), As(III), As(V), Sb(III), Sb(V), Sn(II), Sn(IV)

Antimony (III), Sb^{+3}

1. Sodium hydroxide	white precipitate of the hydrated oxide $Sb_2O_3 \cdot xH_2O$
2. Elemental zinc or tin	black precipitate of antimony
3. Potassium iodide	yellow color of $[SbI_6]^{3-}$ ion
4. Phosphomolybdic acid, $H_3[PMo_{12}O_{40}]$	blue color produced; Sn(II) interferes; 0.2 μg sensitivity

Antimony (V), Sb^{+5}

1. Water	white precipitate of basic salts, and ultimately antimonic acid, H_3SbO_4
2. Potassium iodide	formation of iodine as a floating precipitate
3. Elemental zinc or tin	black precipitate of antimony (in the presence of hydrochloric acid)

Small amounts of antimony can be identified using Marsh's test and/or Gutzeit's test (see references at the beginning of this section for details).

Tin (II), Sn^{+2}

1. Mercury (II) chloride	white precipitate of mercury (I) chloride (in an excess of tin ions, precipitate turns grey)
2. Bismuth nitrate	black precipitate of bismuth metal
3. Cacotheline (nitro-derivative of brucine, $C_{21}H_{21}O_7N_3$)	violet coloration with stannous salts. The following interfere: strong reducing agents (hydrogen sulfide, dithionites, sulfites and selenites); also, U, V, Te, Hg, Bi, Au, Pd, Se, Sb
4. Diazine green (dyestuff formed by coupling diazotized safranine with N,N-dimethylaniline)	color change blue→violet→red

Tin (IV), Sn^{+4}

1. Iron powder	reduces Sn(IV) to Sn(II)

GROUP III: Fe(II), Fe(III), Al(III), Cr(III), Cr(VI), Ni(II), Co(II), Mn(II), Mn(VIII) and Zn(II)

All members are precipitated by H_2S in the presence of ammonia and ammonium chloride, or ammonium sulfide solutions.

Iron (II), Fe^{+2}

1. Ammonia solution	precipitation of iron (II) hydroxide If large amounts of ammonium ion are present, precipitation does not occur.
2. Ammonium sulfide	black precipitate of iron (II) sulfide
3. Potassium cyanide (POISON)	yellowish-brown precipitate of iron (II) cyanide, soluble in excess reagent, forming the hexacyanoferrate (II) ion
4. Potassium hexacyanoferrate (II) solution	in complete absence of air, white precipitate of potassium iron (II) hexacyanoferrate If air is present, a pale blue precipitate is formed.
5. Potassium hexacyanoferrate (III) solution	dark-blue precipitate, called Turnbull's blue
6. α,α′-Dipyridyl	deep red bivalent cation $[Fe(C_5H_4N)_2]^{2+}$ formed with iron (II) salts in mineral acid solution; sensitivity: 0.3 μg
7. Dimethylglyoxime (DMG)	red, iron (II) dimethylglyoxime; nickel, cobalt, and large quantities of copper salts interfere; sensitivity: 0.04 μg
8. o-Phenanthroline (0.1 wt % in water)	red coloration due to the complex cation $[Fe(C_{12}H_8N_2)_3]^{2+}$, in slightly acidic conditions

(Continued)

Tests for Cations (Continued)

GROUP III: Fe(II), Fe(III), Al(III), Cr(III), Cr(VI), Ni(II), Co(II), Mn(II), Mn(VIII) and Zn(II)

Iron (III), Fe^{+3}

1. Ammonia solution	reddish-brown gelatinous precipitate of iron (III) hydroxide
2. Ammonium sulfide	black precipitate mixture of iron (II) sulfide and sulfur
3. Potassium cyanide	when added slowly, reddish-brown precipitate of iron (III) cyanide is formed, which dissolves in excess potassium cyanide to yield a yellow solution
4. Potassium hexacyanoferrate (III)	a brown coloration is produced due to the formation of iron (III) hexacyanoferate (III)
5. Disodium hydrogen phosphate	yellowish-white precipitate of iron (III) phosphate
6. Sodium acetate solution	reddish-brown coloration caused by complex formation
7. Cupferron ($C_6H_5N(NO)ONH_4$) aqueous solution, freshly prepared	reddish-brown precipitate formed in the presence of hydrochloric acid
8. Ammonium thiocyanate + dilute acid	deep red coloration of iron (III) thiocyanate complex
9. 7-iodo-8-hydroxyquinoline-5-sulfonic acid (ferron)	green or greenish-blue coloration in slightly acidic solutions; sensitivity: 0.5 µg

Cobalt (II), Co^{+2}

1. Ammonia solution	in the absence of ammonium salts, small amounts of $Co(OH)NO_3$ precipitate that is soluble in excess aqueous ammonia
2. Ammonium sulfide	black precipitate of cobalt (II) sulfide, from neutral or alkaline solutions
3. Potassium cyanide (POISON)	reddish-brown precipitate of cobalt (II) cyanide that dissolves in excess
4. Potassium nitrite	yellow precipitate of potassium hexacyanocobaltate (III), $K_3[Co(NO_2)_6]$
5. Ammonium thiocyanate (crystals)	gives blue coloration when added to neutral or acid solution of cobalt, due to a complex formation (Vogel's reaction); sensitivity: 0.5 µg
6. α-Nitroso-β-naphthol (1 % in 50 % acetic acid)	red-brown (chelate) precipitate, extractable using carbon tetrachloride; sensitivity: 0.05 µg

Nickel (II), Ni^{+2}

1. Ammonia solution	green precipitate of nickel (II) hydroxide that dissolves in excess ammonia
2. Potassium cyanide (POISON)	green precipitate of nickel (II) cyanide that dissolves in excess potassium cyanide
3. Dimethylglyoxime, DMG ($C_4H_8O_2N_2$)	red precipitate of nickel-DMG chelate complex in ammoniacal solution; sensitivity 0.16 µg

Manganese (II), Mn^{+2}

1. Ammonia solution	partial precipitation of white manganese (II) hydroxide
2. Ammonium sulfide	pink precipitate of manganese (II) sulfide, which is soluble in mineral acids
3. Sodium phosphate (in the presence of ammonia or ammonium ions)	pink precipitate of manganese ammonium phosphate, $Mn(NH_4)PO_4 \cdot 7H_2O$, which is soluble in acids

Aluminum (III), Al^{3+}

1. Ammonia	white gelatinous precipitate of aluminum hydroxide
2. Sodium hydroxide	white gelatinous precipitate of aluminum hydroxide, which is soluble in excess sodium hydroxide
3. Ammonium sulfide	white precipitate of aluminum sulfide
4. Sodium acetate	upon boiling with excess reagent, a precipitate of basic aluminum acetate, $Al(OH)_2CH_3COO$ is formed
5. Sodium phosphate	white gelatinous precipitate of aluminum phosphate
6. Aluminon (a solution of the ammonium salt of aurine tricarboxylic acid)	bright-red solution
7. Quinalizarin, alizarin-S, alizarin	red precipitate or "lake"

Tests for Cations (Continued)

GROUP III: Fe(II), Fe(III), Al(III), Cr(III), Cr(VI), Ni(II), Co(II), Mn(II), Mn(VIII) and Zn(II)

Chromium (III), Cr⁺³

1. Ammonia solution — gray–green to gray–blue gelatinous precipitate of chromium (III) hydroxide

2. Sodium carbonate — precipitate of chromium (III) hydroxide

Zinc (II), Zn²⁺

1. Ammonia solution — white precipitate of zinc hydroxide, which is soluble in excess ammonia

2. Disodium hydrogen phosphate — white precipitate of zinc phosphate, which is soluble in dilute acids

3. Potassium hexacyanoferrate (II) — white precipitate of variable composition, which is soluble in sodium hydroxide

4. Ammonium tetrathiocyanato-mercurate (II): copper sulfate, slightly acidic — solution is treated with 5 drops 0.25 M copper (II) sulfate solution followed by 2 mL ammonium tetrathiocyanato-mercurate to give a violet precipitate.

GROUP IV: Ba²⁺(II), Sr²⁺(II) and ca²⁺(II)

All members of this group react with ammonium carbonate.

Barium (II), Ba⁺²

1. Ammonium carbonate — white precipitate of barium carbonate, which is soluble in dilute acids

2. Ammonium oxalate — white precipitate of barium oxalate, which is soluble in dilute acids

3. Dilute sulfuric acid — heavy, white, finely divided precipitate of barium sulfate

4. Saturated calcium sulfate (or strontium sulfate) — white precipitate of barium sulfate

5. Sodium rhodizonate — red–brown precipitate; sensitivity: 0.25 μg

Strontium (II), Sr⁺²

1. Ammonium carbonate — white precipitate of strontium carbonate

2. Dilute sulfuric acid — white precipitate of strontium sulfate

3. Saturated calcium sulfate — white precipitate of strontium sulfate

4. Potassium chromate — yellow precipitate of strontium chromate

5. Ammonium oxalate — white precipitate of strontium oxalate that is soluble in mineral acids

Calcium (II), Ca⁺²

1. Ammonium carbonate — white precipitate of calcium carbonate

2. Dilute sulfuric acid — white precipitate of calcium sulfate

3. Ammonium oxalate — white precipitate of calcium oxalate that is soluble in mineral acids

4. Potassium chromate — yellow precipitate of strontium chromate that is soluble in mineral acids

5. Sodium rhodizonate — red–brown precipitate; sensitivity: 4 μg

(Continued)

Tests for Cations (Continued)

GROUP V: Mg²⁺(II), Na⁺(I), K⁺(I), NH₄⁺(I)

No common reaction or reagent.

Magnesium (II), Mg^{2+}

1. Ammonia solution, sodium hydroxide	partial precipitation of white magnesium hydroxide
2. Ammonium carbonate	white precipitate of magnesium carbonate, only in the absence of ammonia salts
3. Oxine + ammoniacal ammonium chloride solution	yellow precipitate of $Mg(C_9H_6NO)_2\cdot4H_2O$
4. Quinalizarin	blue precipitate, or blue-colored solution that can be cleared by a few drops of bromine-water

Sodium (I), Na^+

1. Uranyl magnesium acetate solution (in 30 v/v ETOH)	yellow precipitate of sodium magnesium uranyl acetate
2. Uranyl zinc acetate solution	yellow precipitate of sodium zinc uranyl acetate; sensitivity: 12.5 μg Na

Potassium (I), K^+

1. Sodium hexanitrocobaltate (III) ($Na_3[Co(NO_2)_6]$)	yellow precipitate of potassium hexanitrocobaltate (III); insoluble in acetic acid
2. Tartaric acid solution (sodium acetate buffered)	white precipitate of potassium hydrogen tartrate
3. Perchloric acid	white precipitate of potassium perchlorate
Note: perchloric acid is a powerful oxidizing agent that must be handled carefully. See the materials compatibility table in the HPLC chapter (Chapter 2), and the incompatibilities table in the Safety chapter (Chapter 14).	
4. Dipicrylamine	orange–red complex precipitate (NH_4^+ interferes); sensitivity: 3 μg K
5. Sodium tetraphenylboron + acetic acid	white precipitate of potassium tetraphenylboron

Ammonium, NH_4^+

1. Sodium hydroxide	evolution of ammonia gas upon heating
2. Potassium tetraiodomercurate (Nessler's reagent)	brown–yellow color, or brown precipitate of mercury (II) amidoiodide; high sensitivity; all other metals (except Na and K) interfere
3. Tannic acid: silver nitrate	precipitate of black elemental silver, from neutral solution; very sensitive
4. p-Nitrobenzene-diazonium chloride	red colored solution results in the presence of sodium hydroxide; sensitivity: 0.7 μg NH_4^+

Note: Ammonium ions will cause a similar reaction to that of potassium in the presence of: sodium hexanitroco-baltate (III) sodium hydrogen tartrate.

ORGANIC PRECIPITATION REAGENTS FOR INORGANIC IONS

The following table lists the most important organic reagents used for precipitating various inorganic species from solution [1,2]. Many of these reagents are subject to the serious disadvantage caused by lack of selectivity. Thus, many of the listed reagents will precipitate more than one species. The selectivity of some of the reagents can be controlled to a certain extent by adjustment of pH, reagent concentrations, and the use of masking reagents. The first two factors, pH and concentration, are the most critical. A number of these reagents form rather large, bulky complexes. While this can serve to enhance sensitivity (especially for gravimetric procedures), it can also impose rather stringent concentration limits. The reader is referred to several excellent "recipe" texts for further guidance [3–10].

REFERENCES

1. Kennedy, J. H. *Analytical Chemistry*. 2nd ed. New York: Saunders College Publishing, 1990.
2. Christian, G. D. *Analytical Chemistry*. 6th ed. New York: John Wiley and Sons, 2003.
3. Barber, H. H., and T. I. Taylor. *Semimicro Qualitative Analysis*. New York: Harper and Brothers, 1953.
4. Greenfield, S., and M. Clift. *Analytical Chemistry of the Condensed Phosphates*. Oxford: Pergamon Press, 1975.
5. Ryabchikov, D. I., and E. K. Gol'Braikh. *The Analytical Chemistry of Thorium*. New York: The MacMillan Company, 1963.
6. Jungreis, E. *Spot Test Analysis*. New York: Wiley Interscience, 1985.
7. Jungreis, E. *Spot Test Analysis: Clinical, Environmental, Forensic and Geochemical Applications*. 2nd ed. New York: John Wiley and Sons, 1997.
8. Svehla, G., and G. Suehla. *Vogel's Qualitative Inorganic Analysis*. New York: Addison-Wesley, 1996.
9. Skoog, D. A., D. M. West, and F. J. Holler. *Fundamentals of Analytical Chemistry.* 7th ed. New York: Saunders College Publishing, 1996.
10. Harris, D. C. *Quantitative Chemical Analysis*. 5th ed. New York: Freeman, 1997.

Organic Precipitation Reagents

Reagent	Structure/Formula	Applications and Notes
Alizarin-S (sodium alizarin sulfonate)	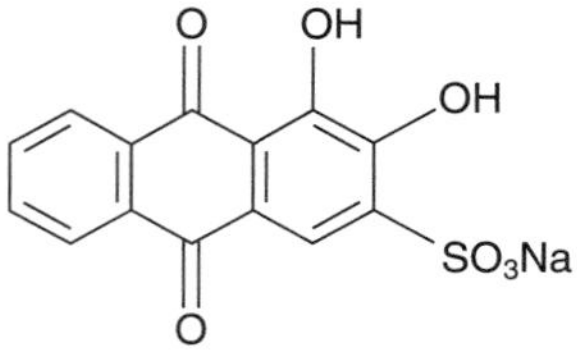	will precipitate Al in ammoniacal solution; high sensitivity
Ammonium nitroso-phenylhydroxylamine (cupferron)	C_6H_5-$N(N=O)O^-$ NH_4^+ 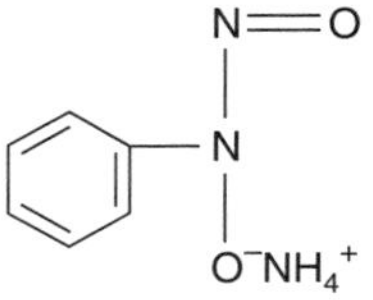	will precipitate Fe(III), V(V), Ti(IV), Zr(IV), Sn(IV), U(IV) in the presence of moderate acidity; will also precipitate rare earths
Anthranilic acid	o-H_2N-C_6H_4COOH 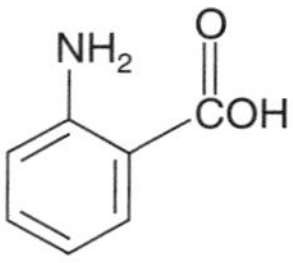	will precipitate Cu(II), Cd(II), Ni(II), Co (II), Pb(II), Zn(II) in acetic acid or nearly neutral solution
α-Benzoin oxime (cupron)	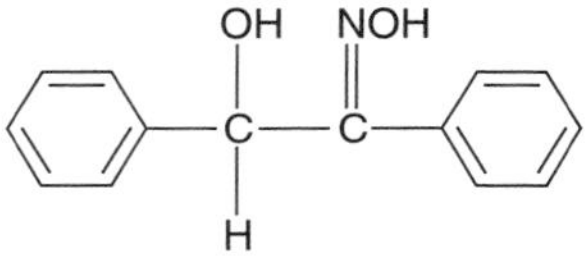	will precipitate Cu(II) in the presence of NH_3 or tartarate; will precipitate Mo(VI), W(VI) in acidic medium
Dimethylglyoxime (DMG)	$[CH_3C(=NOH)]_2$ 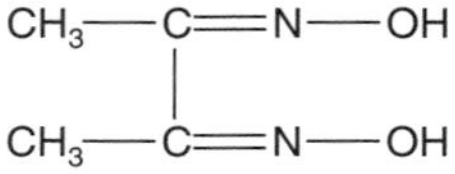	will precipitate Ni(II) in the presence of NH_3 or buffered acetate; Pd(II) in HCl solution; the addition of tartaric acid to the reagent will mask Fe(III) and Cr(III) interferences; Pd(II) and Bi(III) will also precipitate
Dimethyl oxalate	$(CH_3)_2C_2O_4$ 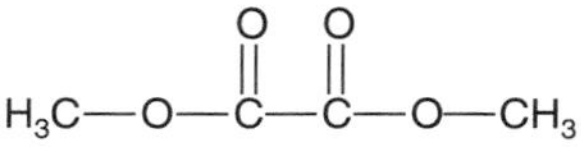	will precipitate Ac, Am, Ca, Th ions as well as rare earth metals
Dimethyl sulfate	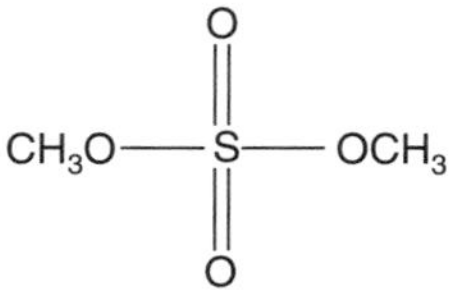	will precipitate Ba, Ca, Pb(II), Sr ions; particular care must be taken while using this reagent since it is a powerful methylating agent
Dipicrylamine (hexanitro-diphenylamine)	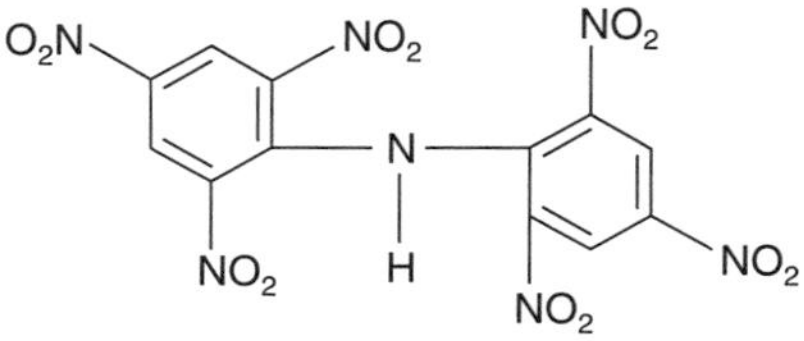	will precipitate K^+ (as well as NH_4, Na, and Li ions but with much less sensitivity)
Dithio-oxamide (rubeanic acid)	$[H_2N-C(=S)]_2$ 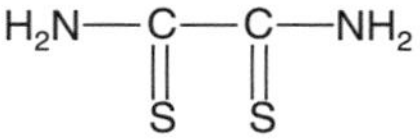	will precipitate Cu(II) (black), Ni (blue), and Co (brown); high sensitivity

Organic Precipitation Reagents (Continued)

Reagent	Structure/Formula	Applications and Notes
Gallocyanine	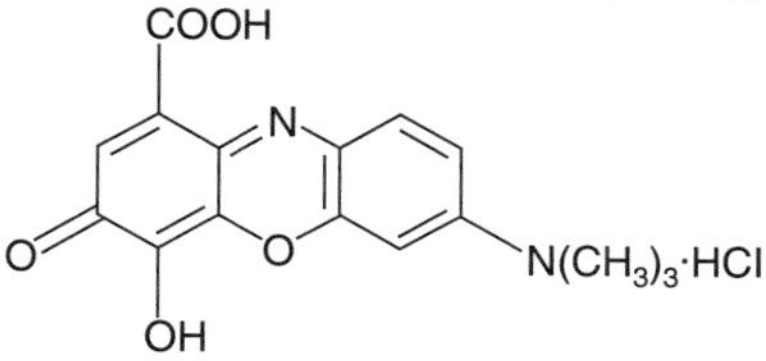	will precipitate lead (deep violet precipitate of uncertain composition); high sensitivity
8-Hydroxyquinoline (oxine)	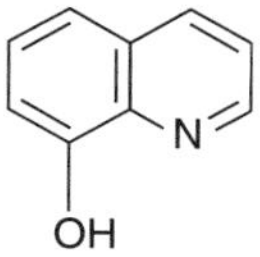	will precipitate Al (III) at pH 4–5, and Mg(II) in the presence of NH_3; will precipitate Be, Bi, Cd, Cu, Ga(I), Hf, Fe, In, Mg, Hg (II), Nb, Pd, Sc, Ta, Ti, Th, U, Zn, Zr, W ions at pH 4–5 or in the presence of NH_3 use of pH control provides a measure of selectivity
Nitron reagent (in acetic acid)	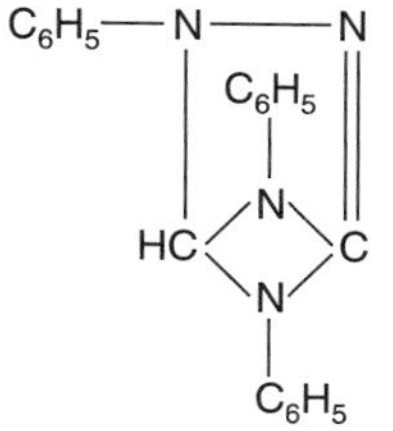	will precipitate nitrate, bromide, iodide, nitrite, chromate chlorate, perchlorate, thiocyanate, oxalate, and picrate anions
α-Nitroso-β-naphthol	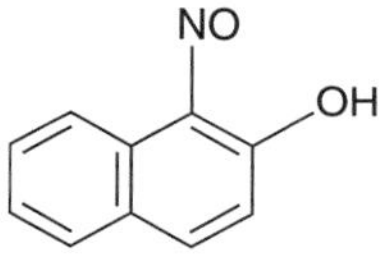	will precipitate Co(II), Ni^{2+}, Fe(III), and Pd(II) in a weakly acidic solution
Oxalic acid	HOOC-COOH 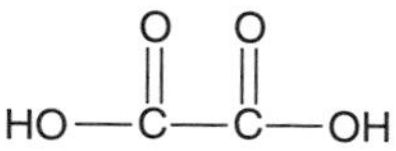	will precipitate Ca^{+2}; high concentrations of Mg will interfere
p-Dimethylamino-benzylidenerhodanine (0.3 % in acetone)	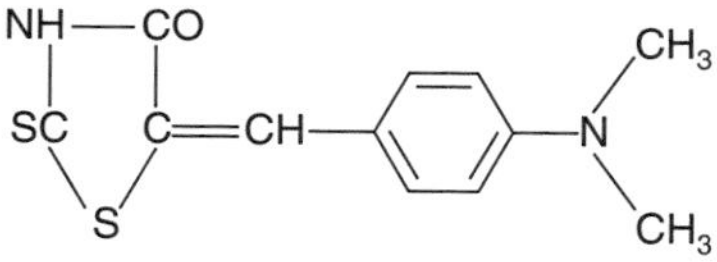	will precipitate Ag(I), Hg(II), Au(II), Pt(II), Pd(II) under slightly acidic conditions
Picrolonic acid	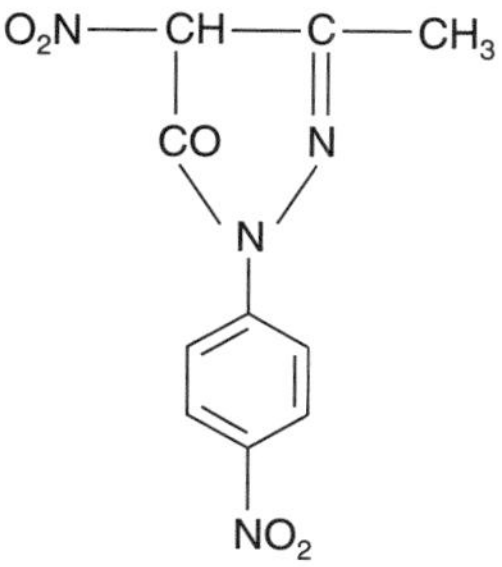	will precipitate Ca^{+2}; subject to interference from many other cations, however
Pyrogallol	$1,2,3-C_6H_3(OH)_3$ 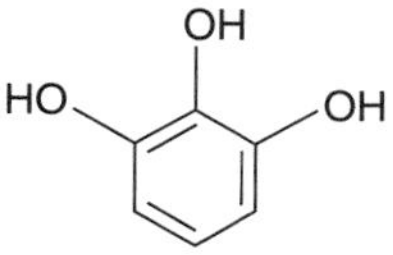	will precipitate Bi(III) and Sb(III); high sensitivity; note that this reagent is also used as an antioxidant for vegetable oils and derived products such as biodiesel fuel

(Continued)

Organic Precipitation Reagents (Continued)

Reagent	Structure/Formula	Applications and Notes
Quinalizarin	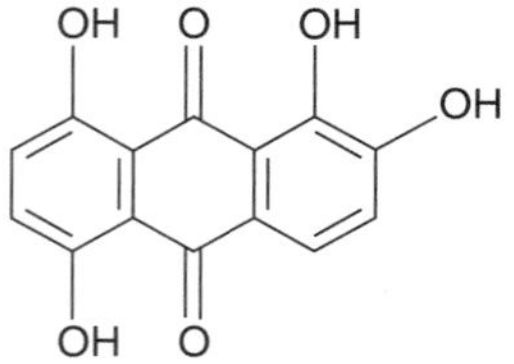	will precipitate Al^{+3}
Salicylaldoxime (1 % in acetic acid)	$1,2\text{-}C_6H_4(OH)CH{=}NOH$ 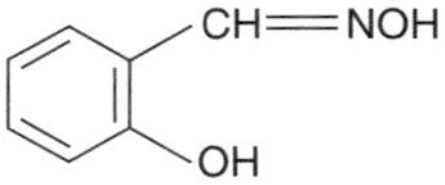	will precipitate Cu(II), with interference by Pd(II) and Au(II)
Sodium diethylthio-carbamate	$(C_2H_5)N\text{-}C({=}S)S^-Na^+$ 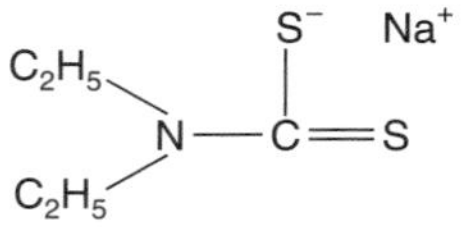	useful for the precipitation of many metals
Sodium dihydroxy-tartrate osazone	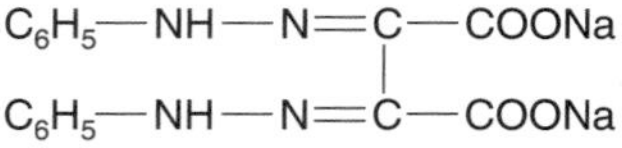 $C_6H_5\text{—}NH\text{—}N{=}C\text{—}COONa$ $C_6H_5\text{—}NH\text{—}N{=}C\text{—}COONa$	will precipitate Ca^{+2}; subject to interference from many other cations, however
Sodium rhodizonate	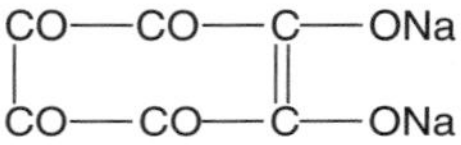 $CO\text{—}CO\text{—}C\text{—}ONa$ $CO\text{—}CO\text{—}C\text{—}ONa$	will precipitate Ba^{+2} and Sr^{+2}; subject to interference by all H_2S reactive cations
Sodium tetraphenylboron	$(C_6H_5)_4B^-Na^+$ 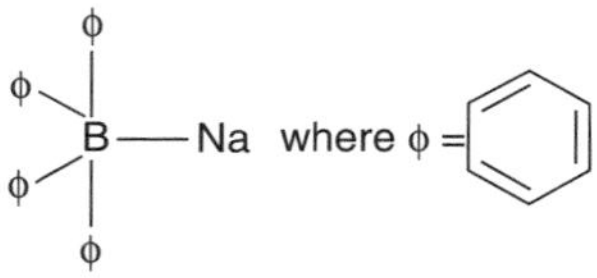	used to precipitate K, Rb, Cs, Tl, Ag, Hg(I), Cu(I), NH_4^+, RNH_3^+, $R_2NH_2^+$, R_3NH^+, R_4N^+, in cold acidic solution; selectivity is high for K+ and NH_4^+
Tetraphenyl arsonium chloride	$(C_6H_5)_4AsCl$ 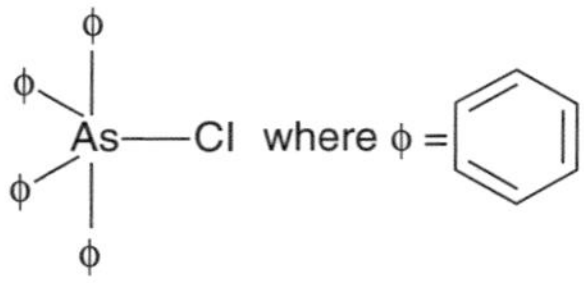	will precipitate $Cr_2O_7^{2-}$, MnO_4^-, ReO_4^-, MoO_4^{2-}, WO_4^{2-}, ClO_4^-, I_3^- in acidic solution
Thioacetamide	$CH_3C({=}S)NH_2$ 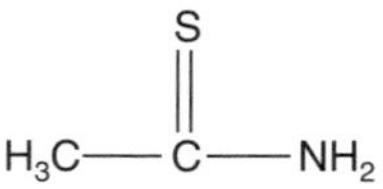	used to provide a source of H_2S for the precipitation of As, Bi, Cd, Cu, Hg, Mn, Mo, Pb, Sb, Sn with heating in acid medium
Triethyl phosphate	$(C_2H_5O)_3PO$ 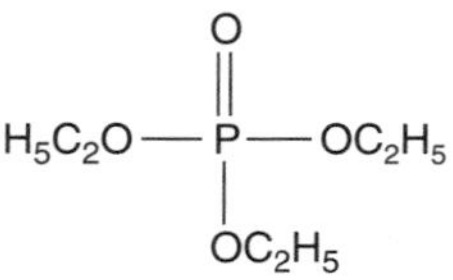	will precipitate Hf and Zr ions

Organic Precipitation Reagents (Continued)

Reagent	Structure/Formula	Applications and Notes
Trichloroacetic acid	$Cl_3C-COOH$ 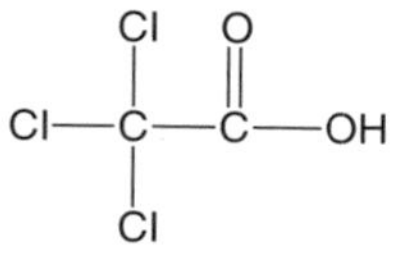	will precipitate Ba, Ra, and the rare earths
Urea	$(NH_2)_2C=O$ 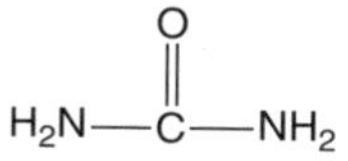	will precipitate Al, Fe(III), Ga, Sn, Th, Zn ions

CHAPTER 13

Solution Properties

CONTENTS

PHYSICAL PROPERTIES OF LIQUID WATER

The table below provides data on the most important properties of pure water under different temperatures. These properties are density (g/mL), molar volume (mL/mol), vapor pressure (in kPa and mm Hg), static dielectric constant and dynamic viscosity (mPa·s). The properties other than the vapor pressure are evaluated at a pressure of 101.325 kPa or the vapor pressure, whichever is higher.

The properties were computed by a software implementation [1] of standards adopted by the International Association for the Properties of Water and Steam (IAPWS) [2].

REFERENCES

1. Harvey, A. H., A. P. Peskin, and S. A. Klein. *NIST/ASME Steam Properties*. NIST Standard Reference Database 10, Version 2.22. Gaithersburg, MD: National Institute of Standards and Technology, 2008.
2. Documentation of IAPWS standards is available at www.iapws.org.

Physical Properties of Liquid Water

Temperature (°C)	Density (g/mL)	Molar Volume (mL/mol)	Vapor Pressure (mm Hg)	Vapor Pressure (kPa)	Dielectric Constant	Viscosity (mPa·s)
0	0.99984	18.0181	4.584	0.6112	87.903	1.792
5	0.99997	18.0159	6.545	0.8726	85.916	1.518
10	0.99970	18.0206	9.212	1.228	83.975	1.306
15	0.99910	18.0314	12.794	1.706	82.078	1.138
18	0.99860	18.0405	15.487	2.065	80.960	1.053
20	0.99821	18.0476	17.546	2.339	80.223	1.002
25	0.99705	18.0686	23.776	3.170	78.408	0.8901
30	0.99565	18.0940	31.855	4.247	76.634	0.7973
35	0.99403	18.1234	42.221	5.629	74.898	0.7193
40	0.99222	18.1566	55.391	7.385	73.201	0.6530
45	0.99021	18.1933	71.968	9.595	71.540	0.5961
50	0.98804	18.2334	92.646	12.352	69.916	0.5468
55	0.98569	18.2768	118.22	15.762	68.328	0.5040
60	0.98320	18.3232	149.61	19.946	66.774	0.4664
65	0.98055	18.3726	187.83	25.042	65.256	0.4333
70	0.97776	18.4250	234.02	31.201	63.770	0.4039
75	0.97484	18.4802	289.49	38.595	62.318	0.3777
80	0.97179	18.5382	355.63	47.414	60.898	0.3543
85	0.96861	18.5991	434.03	57.867	59.509	0.3333
90	0.96531	18.6627	526.40	70.182	58.152	0.3144
95	0.96189	18.7291	634.61	84.608	56.825	0.2973
100	0.95835	18.7982	760.69	101.42	55.527	0.2817

REFRACTIVE INDEX OF WATER

The following table provides the refractive index of water at various temperatures [1].

REFERENCE

1. Lide, D. R., ed. *CRC Handbook for Chemistry and Physics*. 90th ed. Boca Raton, FL: CRC Press, 2010.

Refractive Index of Water

Temperature, °C	Refractive Index, η, Na d-Line
15	1.33341
20	1.33299
30	1.33192
40	1.33051
50	1.32894
60	1.32718
70	1.32511
80	1.32287
90	1.32050
100	1.31783

APPROXIMATE pK VALUES OF COMPOUNDS USEFUL IN BUFFER SYSTEMS

The following table provides the pK values of acids and bases needed to make the most popular buffers [1–3]. The approximate composition of buffers can be calculated from the equation:

$$pH = pK + \log\{[\text{salt}]/[\text{acid}]\}.$$

Note that the quantities in square brackets denote concentrations and the logarithmic quantity refers to the common (base 10) logarithm.

REFERENCES

1. Ramette, R. W. *Chemical Equilibrium and Analysis*. Reading, MA: Addison-Wesley, 1981.
2. Skoog, D. A., D. M. West, F. J. Holler, and S. R. Crouch. *Fundamentals of Analytical Chemistry*. 8th ed. Florence, KY: Cengage Learning, 2004.
3. Serjeant, E. P., and B. Dempsey, eds. *Ionization Constants of Organic Acids in Solution*. IUPAC Chemical Data Series, No. 23. Oxford: Pergamon Press, 1979.

Approximate pK Values of Compounds Useful in Buffer Systems

pKa	Compound	Formula
2.12	(K_1) phosphoric acid	H_3PO_4
2.35	(K_1) glycine	$H_2NCH_2CO_2H$
2.95	(K_1) phthalic acid	C_6H_4-1,2-$(CO_2H)_2$
3.22	(K_1) citric acid	$HOC(COOH)(CH_2COOH)_2$
3.66	(K_1) β,β′-dimethyl glutamic acid	$[HO_2CCH_2\text{-}]_2C(CH_3)_2$
4.21	(K_1) succinic acid	$HO_2CCH_2CH_2CO_2H$
4.76	(K_1) acetic acid	CH_3CO_2H
4.84	(K_2) citric acid	$HOC(COOH)(CH_2COOH)_2$
5.41	(K_2) phthalic acid	C_6H_4-1,2-$(CO_2H)_2$
5.64	(K_2) succinic acid	$HO_2CCH_2CH_2CO_2H$
6.4	(K_1) carbonic acid	H_2CO_3
6.15	(K_1) cacodylic acid	$(CH_3)_2As(O)OH$
6.2	(K_2) β,β′-dimethyl glutaric acid	$[HO_2CCH_2]_2C(CH_3)_2$
6.33	(K_2) maleic acid	$HO_2CCH = CHCO_2H$
6.39	(K_3) citric acid	$HOC(COOH)(CH_2COOH)_2$
7.21	(K_2) phosphoric acid	H_3PO_4
8.07	(K_1) tris-(hydroxymethyl)-aminomethane	$(HOCH_2)_3CNH_2$
8.67	(K_1) 2-amino-2-methyl-1,3-propanediol	$(HOCH_2)_2C(CH_3)NH_2$
9.23	(K_1) boric acid	H_3BO_3
9.78	(K_2) glycine	$H_2NCH_2CO_2H$
10.33	(K_2) carbonic acid	H_2CO_3
12.32	(K_3) phosphoric acid	H_3PO_4

PREPARATION OF BUFFERS

The following table gives the necessary information for preparing various buffers at Zdifferent pHs. These buffers are suitable for use either in enzymatic or histochemical studies [1–3]. The uncertainty of the tables is within ±0.05 pH at 23 °C and the pH values do not change considerably even at 37 °C. The recommended mixture of the various solutions is given under the corresponding pH with the "final solution volume" indicated. This assumes addition of water to the necessary dilution. A list of stock solutions follows the buffer/pH table. The approximate composition of buffers can be calculated from the equation:

$$pH = pK + \log \{[salt]/[acid]\}.$$

Note that the quantities in square brackets denote concentrations and the logarithmic quantity refers to the common (base 10) logarithm.

REFERENCES

1. Colowick, S. P., and N. O. Kaplan, eds. *Methods in Enzymology.* Vol. 1. New York: Academic Press, 1955.
2. Perrin, D. D., and B. Dempsey. *Buffers for pH and Metal Ion Control.* London: Chapman and Hall, 1974.
3. Sergeant, E. P., and B. Dempsey, eds. *Ionization Constants of Organic Acids in Solution.* IUPAC Chemical Data Series No. 23. Oxford: Pergamon Press, 1979.

Preparation of Buffers

	pH					Final solution volume
Buffer	**1.0**	**1.1**	**1.2**	**1.3**	**1.4**	
Hydrochloric acid/potassium chloride	50.0 mL A + 97.0 mL B	50.0 mL A + 78.0 mL B	50.0 mL A + 64.5 mL B	50.0 mL A + 51.0 mL B	50.0 mL A + 41.5 mL B	200 mL

	pH					Final solution volume
Buffer	**1.5**	**1.6**	**1.7**	**1.8**	**1.9**	
Hydrochloric acid/potassium chloride	50.0 mL A + 33.3 mL B	50.0 mL A + 26.3 mL B	50.0 mL A + 20.6 mL B	50.0 mL A + 16.6 mL B	50.0 mL A + 13.2 mL B	200 mL

	pH					Final solution volume
Buffer	**2.0**	**2.1**	**2.2**	**2.3**	**2.4**	
Hydrochloric acid/potassium chloride	50.0 mL A + 10.6 mL B	50.0 mL A + 8.4 mL B	50.0 mL A + 6.7 mL B			200 mL
Glycine/hydrochloric acid			50.0 mL C + 44.0 mL B		50.0 mL C + 32.4 mL B	200 mL
Potassium phthalate/hydrochloric acid			50.0 mL D + 46.7 mL B		50.0 mL D + 39.6 mL B	200 mL

	pH					Final solution volume
Buffer	**2.5**	**2.6**	**2.7**	**2.8**	**2.9**	
Glycine/hydrochloric acid		50.0 mL C + 24.2 mL B		50.0 mL C + 16.8 mL B		200 mL

Note: A list of the stock solutions follows the buffer table.

(Continued)

Preparation of Buffers (Continued)

Buffer	pH					Final solution volume
	2.5	2.6	2.7	2.8	2.9	
Potassium phthalate/hydrochloric acid		50.0 mL D + 33.0 mL B		50.0 mL D + 26.4 mL B		200 mL
Aconitate	20.0 mL E + 15.0 mL F	20.0 mL E + 18.0 mL F	20.0 mL E + 21.0 mL F	20.0 mL E + 24.6 mL F	20.0 mL E ± 28.0 mL F	200 mL
Citrate/phosphate		44.6 mL G + 5.4 mL K		42.2 mL C + 7.8 mL K		100 mL

Buffer	pH					Final solution volume
	3.0	3.1	3.2	3.3	3.4	
Glycine/hydrochloric acid	50.0 mL C + 11.4 mL B		50.0 mL C + 8.2 mL B		50.0 mL C + 6.4 mL B	200 mL
Potassium phthalate/hydrochloric acid	50.0 mL D + 20.3 mL B		50.0 mL D + 14.7 mL B		50.0 mL D + 9.9 mL B	200 mL
Aconitate	20.0 mL E + 32.0 mL F	20.0 mL E + 36.0 mL F	20.0 mL E + 40.0 mL F	20.0 mL E + 44.0 mL F	20.0 mL E + 48.0 mL F	200 mL
Citrate	46.5 mL G + 3.5 mL H		43.7 mL G + 6.3 mL H		40.0 mL G + 10.0 mL H	100 mL
Citrate/phosphate	39.8 mL G + 10.2 mL K		37.7 mL G + 12.3 mL K		35.9 mL G + 14.1 mL K	100 mL

Note: A list of the stock solutions follows the buffer table.

Buffer	pH 3.5	3.6	3.7	3.8	3.9	Final solution volume
Glycine/ hydrochloric acid		50.0 mL C + 5.0 mL B				200 mL
Potassium phthalate/ hydrochloric acid		5.0 mL D + 6.0 mL B		50.0 mL D + 2.63 mL B		200 mL
Aconitate	20.0 mL E + 52.0 mL F	20.0 mL E + 56.0 mL F	20.0 mL E + 60.0 mL F	20.0 mL E + 64.0 mL F	20.0 mL E + 68.0 mL F	200 mL
Citrate		37.0 mL G + 13.0 mL H		35.0 mL G + 15.0 mL H		100 mL
Acetate		46.3 mL I + 3.7 mL J		44.0 mL I + 6.0 mL J		100 mL
Citrate/ phosphate		33.9 mL G + 16.1 mL K		32.3 mL G + 17.7 mL K		100 mL
Succinate				25.0 mL L + 7.5 mL F		100 mL

Buffer	pH 4.0	4.1	4.2	4.3	4.4	Final solution volume
Aconitate	20.0 mL E + 72.0 mL F	20.0 mL E + 76.0 mL F	20.0 mL E + 79.6 mL F	20.0 mL E + 83.0 mL F	20.0 mL E + 86.6 mL F	200 mL
Citrate	33.0 mL G + 17.0 mL H	31.5 mL G + 18.5 mL H			28.0 mL G + 22.0 mL H	100 mL

Note: A list of the stock solutions follows the buffer table.

(Continued)

Preparation of Buffers (Continued)

Buffer	pH 4.0	4.1	4.2	4.3	4.4	Final solution volume
Acetate	41.0 mL I + 9.0 mL J		36.8 mL I + 13.2 mL J		30.5 mL I + 19.5 mL J	100 mL
Citrate/ phosphate	30.7 mL G + 19.3 mL K		29.4 mL G + 20.6 mL K		27.8 mL G + 22.2 mL K	100 mL
Succinate	25.0 mL L + 10.0 mL F		25.0 mL L + 13.3 mL F		25.0 mL L + 16.7 mL F	100 mL
Potassium phthalate/ sodium hydroxide			50.0 mL D + 3.7 mL F		50.0 mL D + 7.5 mL F	200 mL

Buffer	pH 4.5	4.6	4.7	4.8	4.9	Final solution volume
Aconitate	20.0 mL E + 90.0 mL F	20.0 mL E + 93.6 mL F	20.0 mL E + 97.0 mL F	20.0 mL E + 100.0 mL F	20.0 mL E + 103.0 mL F	200 mL
Citrate		25.5 mL G + 24.5 mL H		23.0 mL G + 27.0 mL H		200 mL
Acetate		25.5 mL I + 24.5 mL J		20.0 mL I + 30.0 mL J		100 mL
Citrate/ phosphate		26.7 mL G + 23.3 mL K		25.2 mL G + 24.8 mL K		100 mL
Succinate		25.0 mL L + 20.0 mL F		25.0 mL L + 23.5 mL F		100 mL

Note: A list of the stock solutions follows the buffer table.

Buffer	pH					Final solution volume
	4.5	4.6	4.7	4.8	4.9	
Potassium phthalate/sodium hydroxide		50.0 mL D + 12.2 mL F		50.0 mL D + 17.7 mL F		200 mL

Buffer	pH					Final solution volume
	5.0	5.1	5.2	5.3	5.4	
Aconitate	20.0 mL E + 105.6 mL F	20.8 mL E + 108.0 mL F	20.0 mL E + 110.6 mL F	20.0 mL E + 113.0 mL F	20.0 mL E + 116.0 mL F	200 mL
Citrate	20.5 mL G + 29.5 mL H		18.0 mL G + 32.0 mL H		16.0 mL G + 34.0 mL H	100 mL
Acetate	14.8 mL I + 35.2 mL J		10.5 mL I + 39.5 mL J		8.8 mL I + 41.2 mL J	100 mL
Citrate/phosphate	24.3 mL G + 25.7 mL K		23.3 mL G + 26.7 mL K		22.2 mL G + 27.8 mL K	100 mL
Succinate	25.0 mL L + 26.7 mL F		25.0 mL L + 30.3 mL F		25.0 mL L + 34.2 mL F	100 mL
Potassium phthalate/sodium hydroxide	50.0 mL D + 23.9 mL F		50.0 mL D + 30.0 mL F		50.0 mL D + 35.5 mL F	200 mL
Maleate			50.0 mL M + 7.2 mL F		50.0 mL M + 10.5 mL F	200 mL
Cacodylate	50.0 mL N + 47.0 mL B		50.0 mL N + 45.0 mL B		50.0 mL N + 43.0 mL B	200 mL

Note: A list of the stock solutions follows the buffer table.

(Continued)

Preparation of Buffers (Continued)

Buffer	pH					Final solution volume
	5.0	5.1	5.2	5.3	5.4	
Tris-maleate			50.0 mL Q + 7.0 mL F		50.0 mL Q + 10.8 mL F	200 mL

Buffer	pH					Final solution volume
	5.5	5.6	5.7	5.8	5.9	
Aconitate	20.0 mL E + 119.0 mL F	20.0 mL E + 122.6 mL F	20.0 mL E + 126.0 mL F			200 mL
Citrate		13.7 mL G + 36.3 mL H		11.8 mL G + 38.2 mL H		100 mL
Acetate		4.8 mL I + 45.2 mL J				100 mL
Citrate/phosphate		21.0 mL G + 29.0 mL K		19.7 mL G + 30.3 mL K		100 mL
Succinate		25.0 mL L + 37.5 mL F		25.0 mL L + 40.7 mL F		100 mL
Potassium phthalate/sodium hydroxide		50.0 mL D + 39.8 mL F		50. mL D + 43.0 mL F		200 mL
Maleate		50.0 mL M + 15.3 mL F		50.0 mL M + 20.8 mL F		200 mL
Cacodylate		50.0 mL N + 39.2 mL B		50.0 mL N + 34.8 mL B		200 mL

Note: A list of the stock solutions follows the buffer table.

Buffer	pH					Final solution volume
	5.5	5.6	5.7	5.8	5.9	
Phosphate			93.5 mL O + 6.5 mL P	92.0 mL O + 8.0 mL P	90.0 mL O + 10.0 mL P	200 mL
Tris-maleate		50.0 mL Q + 15.5 mL F		50.0 mL Q + 20.5 mL F		200 mL

Buffer	pH					Final solution volume
	6.0	6.1	6.2	6.3	6.4	
Citrate	9.5 mL G + 41.5 mL H		7.2 mL G + 42.8 mL H			100 mL
Citrate/phosphate	17.9 mL G + 32.1 mL K		16.9 mL G + 33.1 mL K		15.4 mL G + 34.6 mL K	100 mL
Succinate	25.0 mL L + 43.5 mL F					100 mL
Potassium phthalate/sodium hydroxide	50.0 mL D + 45.5 mL F					200 mL
Maleate	50.0 mL M + 26.9 mL F		50.0 mL M + 33.0 mL F		50.0 mL M + 38.0 mL F	200 mL
Cacodylate	50.0 mL N + 29.6 mL B		50.0 mL N + 23.8 mL B		50.0 mL N + 18.3 mL B	200 mL
Phosphate	87.7 mL O + 12.3 mL P	85.0 mL O + 15.0 mL P	81.5 mL O + 18.5 mL P	77.5 mL O + 22.5 mL P	73.5 mL O + 26.5 mL P	200 mL

Note: A list of the stock solutions follows the buffer table.

(Continued)

Preparation of Buffers (Continued)

Buffer	pH 6.0	6.1	6.2	6.3	6.4	Final solution volume
Tris-maleate	50.0 mL Q + 26.0 mL F		50.0 mL Q + 31.5 mL F		50.0 mL Q + 37.0 mL F	200 mL

Buffer	pH 6.5	6.6	6.7	6.8	6.9	Final solution volume
Barbital				50.0 mL R + 45.0 mL B		200 mL
Citrate/ phosphate		13.6 mL G + 36.4 mL K		9.1 mL G + 40.9 mL K		100 mL
Maleate		50.0 mL M + 41.6 mL F		50.0 mL M + 44.4 mL F		200 mL
Cacodylate		40.0 mL N + 13.3 mL B		50.0 mL N + 9.3 mL B		200 mL
Phosphate	68.5 mL O + 31.5 mL P	62.5 mL O + 37.5 mL P	56.5 mL O + 43.5 mL P	51.0 mL O + 49.0 mL P	45.0 mL O + 55.0 mL P	200 mL
Tris-maleate		50.0 mL Q + 42.5 mL F		50.0 mL Q + 45.0 mL F		200 mL

Buffer	pH 7.0	7.1	7.2	7.3	7.4	Final solution volume
Barbital	50.0 mL R + 43.0 mL B		50.0 mL R + 39.0 mL B		50.0 mL R + 32.5 mL B	200 mL

Note: A list of the stock solutions follows the buffer table.

| Buffer | pH | | | | | Final solution volume |
	7.0	7.1	7.2	7.3	7.4	
Tris			50.0 mL S + 44.2 mL B		50.0 mL S + 41.4 mL B	200 mL
Citrate/phosphate	6.5 mL G + 43.6 mL K					100 mL
Cacodylate	50.0 mL N + 6.3 mL B		50.0 mL N + 4.2 mL B		50.0 mL N + 2.7 mL B	200 mL
Phosphate	39.0 mL O + 61.0 mL P	33.0 mL O + 67.0 mL P	28.0 mL O + 72.0 mL P	23.0 mL O + 77.0 mL P	19.0 mL O + 81.0 mL P	200 mL
Tris-maleate	50.0 mL Q + 48.0 mL F		50.0 mL Q + 51.0 mL F		50.0 mL Q + 54.0 mL F	200 mL

| Buffer | pH | | | | | Final solution volume |
	7.5	7.6	7.7	7.8	7.9	
Barbital		50.0 mL R + 27.5 mL B		50.0 mL R + 22.5 mL B		200 mL
Tris		50.0 mL S + 38.4 mL B		50.09 mL S + 32.5 mL B		200 mL
Boric acid/borax		50.0 mL T + 2.0 mL U		50.0 mL T + 3.1 mL U		200 mL
Ammediol				50.0 mL V + 43.5 mL B		200 mL

Note: A list of the stock solutions follows the buffer table.

(Continued)

Preparation of Buffers (Continued)

Buffer	pH 7.5	7.6	7.7	7.8	7.9	Final solution volume
Phosphate	16.0 mL O + 84.0 mL P	13.0 mL O + 87.0 mL P	10.5 mL O + 90.5 mL P	8.5 mL O + 91.5 mL P	7.0 mL O + 93.0 mL P	200 mL
Tris-maleate		50.0 mL Q + 58.0 mL F		50.0 mL Q + 63.5 mL F		200 mL

Buffer	pH 8.0	8.1	8.2	8.3	8.4	Final solution volume
Barbital	50.0 mL R + 17.5 mL B		50.0 mL R + 12.7 mL B		50.0 mL R + 9.0 mL B	200 mL
Tris	50.0 mL S + 26.8 mL B		50.0 mL S + 21.9 mL B		50.0 mL S + 16.5 mL B	200 mL
Boric acid/ borax	50.0 mL T + 4.9 mL U		50.0 mL T + 7.3 mL U		50.0 mL T + 11.5 mL U	200 mL
Ammediol	50.0 mL V + 41.0 mL B		50.0 mL V + 37.7 mL B		50.0 mL V + 34.0 mL B	200 mL
Phosphate	5.3 mL O + 94.7 mL P					200 mL
Tris-maleate	50.0 mL Q + 69.0 mL F		50.0 mL Q + 75.0 mL F		50.0 mL Q + 81.0 mL F	200 mL

Note: A list of the stock solutions follows the buffer table.

Buffer	pH 8.5	8.6	8.7	8.8	8.9	Final solution volume
Barbital		50.0 mL R + 6.0 mL B		50.0 mL R + 4.0 mL B		200 mL
Tris		50.0 mL S + 12.2 mL B		50.0 mL S + 8.1 mL B		200 mL
Boric acid/borax		50.0 mL T + 17.5 mL U	50.0 mL T + 22.5 mL U	50.0 mL T + 30.0 mL U	50.0 mL T + 42.5 mL U	200 mL
Ammediol		50.0 mL V + 29.5 mL B		50.0 mL V + 22.0 mL B		200 mL
Glycine/ sodium hydroxide		50.0 mL C + 4.0 mL F		50.0 mL C + 6.0 mL F		200 mL
Tris-maleate		50.0 mL Q + 86.5 mL F				200 mL

Buffer	pH 9.0	9.1	9.2	9.3	9.4	Final solution volume
Barbital	50.0 mL R + 2.5 mL B		50.0 mL R + 1.5 mL B			200 mL
Tris	50.0 mL S + 5.0 mL B					200 mL
Boric acid/ borax	50.0 mL T + 59.0 mL U	50.0 mL T + 83.0 mL U	50.0 mL T + 115.0 mL U			200 mL

Note: A list of the stock solutions follows the buffer table.

(Continued)

Preparation of Buffers (Continued)

	pH					Final solution volume
Buffer	**9.0**	**9.1**	**9.2**	**9.3**	**9.4**	
Ammediol	50.0 mL V + 16.7 mL B		50.0 mL V + 12.5 mL B		50.0 mL V + 8.5 mL B	200 mL
Glycine/ sodium hydroxide	50.0 mL C + 8.8 mL F		50.0 mL C + 12.0 mL F		50.0 mL C + 16.8 mL F	200 mL
Borax/ sodium hydroxide			50.0 mL U (pH = 9.28)		50.0 mL U + 11.0 mL F	200 mL
Carbonate/ bicarbonate			4.0 mL W + 46.0 mL X	7.5 mL W + 42.5 mL X	9.5 mL W + 40.5 mL X	200 mL

	pH					Final solution volume
Buffer	**9.5**	**9.6**	**9.7**	**9.8**	**9.9**	
Ammediol		50.0 mL V + 5.7 mL B		50.0 mL V + 3.7 mL B		200 mL
Glycine/ sodium hydroxide		50.0 mL C + 22.4 mL F		50.0 mL C + 27.2 mL F		200 mL
Borax/ sodium hydroxide	50.0 mL U + 17.6 mL F	50.0 mL U + 23.0 mL F	50.0 mL U + 29.0 mL F	50.0 mL U + 34.0 mL F	50.0 mL U + 38.0 mL F	200 mL
Carbonate/ bicarbonate	13.0 mL W + 37.0 mL X	16.0 mL W + 34.0 mL X	19.5 mL W + 30.5 mL X	22.0 mL W + 28.0 mL X	25.0 mL W + 25.0 mL X	200 mL

Note: A list of the stock solutions follows the buffer table.

Buffer	pH 10.0	pH 10.1	pH 10.2	pH 10.3	pH 10.4	Final solution volume
Ammediol	50.0 mL V + 2.0 mL B					200 mL
Glycine/ sodium hydroxide	50.0 mL C + 32.0 mL F				50.0 mL C + 45.5 mL F	200 mL
Borax/ sodium hydroxide	50.0 mL U + 43.0 mL F	50.0 mL U + 46.0 mL F				200 mL
Carbonate/ bicarbonate	27.5 mL W + 22.5 mL X	30.0 mL W + 20.0 mL X	33.0 mL W + 17.0 mL X	35.5 mL W + 14.5 mL X	38.5 mL W + 11.5 mL X	200 mL

Buffer	pH 10.5	pH 10.6	pH 10.7	pH 10.8	pH 10.9	Final solution volume
Glycine/ sodium hydroxide		50.0 mL C + 45.5 mL F				200 mL
Carbonate/ bicarbonate	40.5 mL W + 9.5 mL X	42.5 mL W + 7.5 mL X	45.0 mL W + 5.0 mL X			200 mL

Note: A list of the stock solutions follows the buffer table.

Stock Solutions

A = 0.2 M potassium chloride (14.91 g in 1000 mL)
B = 0.2 M hydrochloric acid
C = 0.2 M glycine (15.01 g in 1000 mL)
D = 0.2 M potassium acid phthalate (40.84 g in 1000 mL)
E = 0.5 M aconitic acid, (1-propene-1,2,3-tricarboxylic acid) (87.05 g in 1000 mL)
F = 0.2 M sodium hydroxide
G = 0.1 M citric acid (21.01 g in 1000 mL)
H = 0.1 M sodium citrate dihydrate (29.41 g in 1000 mL); avoid using any other hydrated salt
I = 0.2 M acetic acid
J = 0.2 M anhydrous sodium acetate (16.4 g in 1000 mL) or 0.2 M sodium acetate trihydrate (27.2 g in 1000 mL)
K = 0.2 M dibasic sodium phosphate heptahydrate (53.65 g in 1000 mL) or 0.2 M dibasic sodium phosphate dodecahydrate (71.7 g in 1000 mL)
L = 0.2 M succinic acid (23.6 g in 1000 mL)
M = 0.2 M sodium maleate (8.0 g NaOH + 23.2 g maleic acid or 19.6 g maleic anhydride in 1000 mL)
N = 0.2 M sodium cacodylate (42.8 g sodium cacodylate trihydrate in 1000 mL)
O = 0.2 M monobasic sodium phosphate (27.8 g in 1000 mL)
P = 0.2 M dibasic sodium phosphate (53.65 g dibasic sodium phosphate heptahydrate or 71.7 g dibasic sodium phosphate dodecahydrate in 1000 mL).
Q = 0.2 M tris acid maleate (24.2 g tris (hydroxymethyl) amino methane + 23.2 g maleic acid or 19.6 g maleic anhydride in 1000 mL)
R = 0.2 M sodium barbital (veronal) (41.2 g in 1000 mL)
S = 0.2 M tris (hydroxymethyl) aminomethane (24.2 g in 1000 mL)
T = 0.2 M boric acid (12.4 g in 1000 mL)
U = 0.05 M borax (19.05 g in 1000 mL)
V = 0.2 M 2-amino-2-methyl-1,3-propanediol (21.03 g in 1000 mL)
W = 0.2 M anhydrous sodium carbonate (21.2 g in 1000 mL)
X = 0.2 M anhydrous sodium bicarbonate (16.8 g in 1000 mL)

INDICATORS FOR ACIDS AND BASES

The following table lists the most common indicators together with their pH range and colors in acidic and basic media. Since the color change is not instantaneous at the pKa value, a pH range is given where a mixture of colors is present. This pH range, which varies between indicators, generally falls between the pKa with a spread or uncertainty of 1 pH unit. All solutions are either aqueous or ethanol/aqueous (% ethanol, vol/vol) [1–3]. Reference 4 lists the exact quantities needed for the indicator's solutions.

REFERENCES

1. Lange, N. A. *Lange's Handbook of Chemistry.* 8th ed. New York: Handbook Publishers, 1952.
2. Kolthoff, I. M., and V. A. Stenger. *Volumetric Analysis.* 2nd ed. Translated by N. H. Furman. New York: Interscience Publishers, 1942.
3. Sabnis, R. W. *Handbook of Acid-Base Indicators.* Boca Raton, FL: CRC Press, Taylor and Francis, 2008.
4. http://www.csudh.edu/oliver/chemdata/ind-prep.htm accessed July 2010.

Indicators for Acids and Bases

Indicator	pH Range	Solvent	Acid	Base
Gentian violet (crystal violet)	0.0–2.0	aqueous	yellow	blue–violet
Thymol blue	1.2–2.8	aqueous	red	yellow
Pentamethoxy red	1.2–2.3	70 % ethanol	red-violet	colorless
Tropeolin OO	1.3–3.2	aqueous	red	yellow
2,4-Dinitrophenol	2.4–4.0	50 % ethanol	colorless	yellow
Methyl yellow	2.9–4.0	90 % ethanol	red	yellow
Methyl orange	3.1–4.4	aqueous	red	orange
Bromophenol blue	3.0–4.6	aqueous	yellow	blue–violet
Tetrabromophenol blue, sodium salt	3.0–4.6	aqueous	yellow	blue
Congo red	3.0–5.0	aqueous	blue-violet	red
Alizarin sodium sulfonate	3.7–5.2	aqueous	yellow	violet
α-Naphthyl red	3.7–5.0	70 % ethanol	red	yellow
p-Ethoxychrysoidine	3.5–5.5	aqueous	red	yellow
Bromocresol green, sodium salt	4.0–5.6	aqueous	yellow	blue
Methyl red, sodium salt	4.4–6.2	aqueous	red	yellow
Bromocresol purple	5.2–6.8	aqueous	yellow	purple
Chlorphenol red	5.4–6.8	aqueous	yellow	red
Bromophenol blue, sodium salt	6.2–7.6	aqueous	yellow	blue
p-Nitrophenol	5.0–7.0	aqueous	colorless	yellow
Azolitmin	5.0–8.0	aqueous	red	blue
Bromothymol blue, sodium salt	6.0–7.6	70 % ethanol	yellow	blue
Phenol red, sodium salt	6.4–8.0	aqueous	yellow	red
Neutral red	6.8–8.0	70 % ethanol	red	yellow
Rosolic acid	6.8–8.0	90 % ethanol	yellow	red
Cresol red, sodium salt	7.2–8.8	aqueous	yellow	red
α-Naphtholphthalein	7.3–8.7	70 % ethanol	rose	green
Tropeolin OOO	7.6–8.9	aqueous	yellow	rose–red
Thymol blue, sodium salt	8.0–9.6	aqueous	yellow	blue
Phenolphthalein	8.0–10.0	70 % ethanol	colorless	red
α-Naphtholbenzein	9.0–11.0	70 % ethanol	yellow	blue
Thymolphthalein	9.4–10.6	90 % ethanol	colorless	blue
Nile blue	10.1–11.1	aqueous	blue	red
Alizarin yellow R	10.0–12.0	aqueous	yellow	lilac
Salicyl yellow	10.0–12.0	90 % ethanol	yellow	orange–brown
Diazo violet	10.1–12.0	aqueous	yellow	violet
Tropeolin O	11.0–13.0	aqueous	yellow	orange–brown
Nitramine	11.0–13.0	70 % ethanol	colorless	orange–brown
Poirrier's blue	11.0–13.0	aqueous	Blue	violet–pink
Trinitrobenzoic acid	12.0–13.4	aqueous	colorless	orange–red

DIELECTRIC CONSTANTS OF INORGANIC SOLVENTS

The dielectric constant (ε) of a substance is a macroscopic property that measures the reduction of the strength of the electric field that surrounds a charged particle when immersed in that substance, as compared to the field strength around the same particle when placed in vacuum. As a result, the higher the value of the dielectric constant of the substance, the greater the tendency of the charged particle to ionize. Although the dielectric constant gives only one of several quantitative measures of the polarity of the substance, it is nonetheless a useful property in describing solvents as polar or nonpolar. The table below lists the dielectric constants of some inorganic solvents at a specific temperature [1–3].

REFERENCES

1. Lide, D. R., ed. *CRC Handbook for Chemistry and Physics.* 90th ed. Boca Raton, FL: CRC Press, 2010.
2. Lowry, T. H., and K. S. Richardson. *Mechanism and Theory in Organic Chemistry.* New York: Harper Collins Publishers, 1987.
3. Parsons, R. *Handbook of Electrochemical Constants.* London: Butterworths, 1959.

Dielectric Constants of Inorganic Solvents

Name	Formula	Dielectric Constant (ε)	Temp. (°C)
Aluminum bromide	$AlBr_3$	3.38	100
Ammonia	NH_3	16.9	25
Argon	Ar	1.53	−191
Arsenic trichloride	$AsCl_3$	12.8	20
Arsine	AsH_3	2.50	−100
Boron tribromide	BBr_3	2.58	0
Bromine	Br_2	3.09	20
Chlorine	Cl_2	2.10	−50
Deuterium	D_2	1.277	−253
Deuterium oxide (see water-d_2)			
Dinitrogen tetroxide	N_2O_4	2.4	18
Fluorine	F_2	1.54	−202
Germanium tetrachloride	$GeCl_4$	2.43	25
Helium	He	1.055	−271
Hydrazine	N_2H_4	51.7	25
Hydrogen bromide	HBr	7.0	−70
Hydrogen chloride	HCl	12.0	−113
Hydrogen cyanide	HCN	106.8	25
Hydrogen fluoride	HF	17.0	−73
Hydrogen iodide	HI	3.39	−50
Hydrogen sulfide	H_2S	9.26	−85.5
Hydrogen peroxide	H_2O_2	84.2	0
Iodine	I_2	11.1	118
Iodine pentafluoride	IF_5	36.2	25
Lead tetrachloride	$PbCl_4$	2.78	20

(Continued)

Dielectric Constants of Inorganic Solvents (Continued)

Name	Formula	Dielectric Constant (ε)	Temp. (°C)
Mercury (II) bromide	$HgBr_2$	9.8	240
Nitrosyl bromide	NOBr	13.4	15
Nitrosyl chloride	NOCl	18.2	12
Phosphorus pentachloride	PCl_5	2.8	160
Phosphorus trichloride	$PC1_3$	3.43	25
Phosphoryl chloride	$POCl_3$	13.3	22
Seleninyl chloride	$SeOCl_2$	26.2	20
Selenium	SE	5.40	250
Silicon tetrachloride	$SiCl_4$	2.40	20
Sulfur	S	3.52	118
Sulfur dioxide	SO_2	14.1	20
Sulfuric acid	H_2SO_4	101	25
Sulfuryl chloride	SO_2Cl_2	9.2	20
Thionyl bromide	$SOBr_2$	9.06	20
Thionyl chloride	$SOCl_2$	9.25	20
Thiophosphoryl chloride	$PSCl_3$	5.8	22
Titanium tetrachloride	$TiCl_4$	2.80	20
Water	H_2O	80.22	20
		78.41	25
Water-d_2	D_2O	77.94	25

DIELECTRIC CONSTANTS OF METHANOL–WATER MIXTURES FROM 5 TO 55 °C

The table below lists the value of dielectric constant of methanol–water mixtures as a function of their respective w/w% composition at various temperatures [1]. This information is useful for a variety of chromatographic and extractive applications.

REFERENCE

1. Parsons, R. *Handbook of Electrochemical Constants*. London: Butterworths, 1959.

Dielectric Constants of Methanol–Water Mixtures from 5 to 55 °C

Weight % Methanol	5 °C	15 °C	25 °C	35 °C	45 °C	55 °C
0	85.76	83.83	78.30	74.83	71.51	68.35
10	81.68	77.83	74.18	70.68	67.32	64.08
20	77.38	73.59	69.99	66.52	63.24	60.06
30	72.80	69.05	65.55	62.20	58.97	55.92
40	67.91	64.31	60.94	57.72	54.62	51.69
50	62.96	59.54	56.28	53.21	50.29	47.53
60	57.92	54.71	51.67	48.76	46.02	43.42
70	52.96	49.97	47.11	44.42	41.83	39.38
80	48.01	45.24	42.60	40.08	37.70	35.46
90	42.90	40.33	37.91	35.65	33.53	31.53
95	39.98	37.61	35.38	33.28	31.29	29.43
100	36.88	34.70	32.66	30.74	28.02	27.21

COMMON DRYING AGENTS FOR ORGANIC LIQUIDS

The following table gives the suggested common agents for drying various organic liquids. Those squares marked "X" are the best combination of organic family/drying agent. Those marked "never" are the worst combinations, primarily due to possible chemical reactions. For instance, alcohols and sodium metal react vigorously. Consequently one should look for other drying agents. Those that are blank might be efficient, but are not recommended for use, unless the suggested drying agents are not available. Some combinations do not give efficient results due to complexation (marked "d") [1–6].

REFERENCES

1. Vogel, A. I. *A Textbook of Practical Organic Chemistry*. London: Longmans, Green and Co., 1951.
2. Brewster, R. Q., C. A. Vanderwerf, and W. E. McEwen. *Unitized Experiments in Organic Chemistry*. New York: D. Van Nostrand Co., 1977.
3. Gordon, A. J., and R. A. Ford. *The Chemist's Companion: A Handbook of Practical Data, Techniques and References*. New York: John Wiley and Sons, 1972.
4. Bruno, T. J., and P. D. N. Svoronos. *Basic Tables for Chemical Analysis*. NBS Technical Note 1096, U.S. Dept. of Commerce, National Bureau of Standards, Washington, DC, 1986.
5. Bruno, T. J., and P. D. N. Svoronos. *CRC Handbook of Basic Tables for Chemical Analysis*. 2nd ed. Boca Raton, FL: CRC Press, 2003.
6. Svoronos, P. D. N., Sarlo, E., and R. Kulawiec. *Organic Chemistry Laboratory Manual*. 2nd ed. Dubuque: W.C. Brown, 1997.

Common Drying Agents for Organic Liquids

Family	Na_2CO_3[a]	K_2CO_3[a]	$MgSO_4$[b]	$CaSO_4$[c]	Na_2SO_4[c]	$CaCl_2$[d]
Alcohols		X	X	X		d
Aldehydes			X	X	X	d
Alkyl halides						X
Amines						d
Anhydrides						
Aryl halides						X
Carboxylic acids	never	never	X	X	X	e
Esters			X		X	d
Ethers				X		X
Hydrocarbons, aromatic	X	X		X	poor	X
Hydrocarbons, saturated	X	X		X	poor	X
Hydrocarbons, unsaturated	X	X			poor	X
Ketones		X	X	X	X	d
Nitriles						

[a] Excellent for salting out
[b] Best all-purpose drying agent
[c] High capacity, but slow reacting
[d] Forms complexes
[e] Lime (common impurity) reacts with acidic hydrogen
[f] Only for 3 ° amines (R_3N)
[g] Only for C_4 and higher alcohols

Common Drying Agents for Organic Liquids

Family	Na	P_2O_5	NaOH (solid)	KOH (solid)	CaO	CaH_2	$LiAlH_4$
Alcohols	never	never	never	never	X	g	never
Aldehydes	never	never	never	never		never	never
Alkyl halides	never	X	never	never		X	never
Amines	never	never	X	X	X	f	never
Anhydrides		X					
Aryl halides	never	X				X	X
Carboxylic acids	never	X	never	never	never	never	never
Esters			never	never		X	never
Ethers	X	X			X	X	X
Hydrocarbons, aromatic	X	X				X	X
Hydrocarbons, saturated	X	X				X	X
Hydrocarbons, unsaturated		X				X	X
Ketones	never	never	never	never	never	never	never
Nitriles		X	never	never			never

[a] Excellent for salting out
[b] Best all-purpose drying agent
[c] High capacity, but slow reacting
[d] Forms complexes
[e] Lime (common impurity) reacts with acidic hydrogen
[f] Only for 3° amines (R_3N)
[g] Only for C_4 and higher alcohols

COMMON RECRYSTALLIZATION SOLVENTS

The following table gives a list of solvents (and their useful properties) in order of decreasing polarity (on the basis of eluotropic series number, ε), and the organic compounds they are capable of recrystallizing. In choosing a solvent, one should consider the following criteria: (a) low toxicity, (b) low cost, (c) ease of separation of the solvent from the crystals (relatively high degree of volatility), (d) the ability to dissolve the crystals while hot, but not while cold, with impurities being either soluble or insoluble both in hot and cold, and (e) the boiling point of the solvent should be lower than the melting point of the compound. While not all of these factors may be optimized with each application, an attempt should be made to achieve optimization of as many as possible. For the same compound, a variety of recrystallizing solvents can be employed based on the type of impurities that are present [1–4].

REFERENCES

1. Gordon, A. J., and R. A. Ford. *The Chemist's Companion: A Handbook of Practical Data, Techniques, and References.* New York: John Wiley and Sons, 1972.
2. Roberts, R. M., J. C. Gilbert, L. B. Rodewald, and A. S. Wingrove. *An Introduction to Modern Experimental Organic Chemistry.* New York: Holt, Rinehart, and Winston, 1969.
3. Sarlo, E., P. D. N. Svoronos, and R. Kulawiec. *Organic Chemistry Laboratory Manual.* 2nd ed. Dubuque: W.C. Brown, 1997.
4. Lide, D. R., ed. *CRC Handbook for Chemistry and Physics.* 90th ed. Boca Raton, FL: CRC Press, 2010.

Common Recrystallization Solvents

Solvent	bp[a] (°C)	ε,[b,c]	Flammability[e]	Toxicity	Good For	Second Solvent in Mixture[d]	Comments
Water	100	78.5[c]	0	0	amides, salts, some carboxylic acids	methanol, ethanol, acetone, dioxane, acetonitrile	difficult to remove from crystals
Acetic acid	118	6.15[b]	1	2	amides, some carboxylic acids, some sulfoxides	water	difficult to remove from crystals
Acetonitrile	81.6	37.5[b]	3	3	some carboxylic acids, hydroquinones	water, ether, benzene	
Methanol	64.5	32.63[c]	3	1	nitrocompounds, esters, bromocompounds, some sulfoxides, sulfones, and sulfilimines, anilines	water, ether, benzene	
Ethanol	78.3	24.30[c]	3	0	same as methanol	water, ethyl acetate, hydrocarbons, methylene chloride	
Acetone	56	20.7[c]	3	1	nitrocompounds, osazones	water, ether, hydrocarbons	
2-Methoxyethanol(methyl cellosolve)	124	123[c]	2	2	carbohydrates	water, ether, benzene	
Pyridine	116	123[c]	3	3	quinones, thiazoles, oxazoles	water, methanol	difficult to remove from crystals
Methyl acetate	57	6.68	4	2	esters, carbonyl compounds, sulfide derivatives, carbinols	water, ether	
Ethyl acetate	77.1	6.02[c]	3	1	same as methyl acetate	water, ether, chloroform, methylene chloride	
Methylene chloride (dichloromethane)	40	9.08	0	2	low-melting compounds	ethanol, hydrocarbons	easily removed
Ether (diethyl-ether)	34.5	4.34[b]	4	2	low-melting compounds	acetone, acetonitrile, methanol, ethanol, acetate esters	easily removed, can create peroxides
Chloroform	61.7	4.81[b]	0	4	polar compounds	ethanol, acetate esters, hydrocarbons	easily removed; suspected carcinogen[f]

(Continued)

Common Recrystallization Solvents (Continued)

Solvent	bp[a] (°C)	ε[b,c]	Flammability[e]	Toxicity	Good For	Second Solvent in Mixture[d]	Comments
1,4-Dioxane	102	2.21[c]	3	2	amides	water, hydrocarbons, benzene	can form complexes with ethers
Carbon tetrachloride	76.5	2.24[b]	0	4	acid chlorides, anhydrides	ether, benzene, hydrocarbons	can react with strong organic bases; suspected carcinogen[f]
Toluene	110.6	2.38[c]	3	2	aromatics, hydrocarbons	ether, ethyl acetate, hydrocarbons	a little difficult to remove from crystals
Benzene	80.1	2.28[b]	3	3	aromatics, hydrocarbons, molecular complexes, sulfides, ethers	ether, ethyl acetate, hydrocarbons	carcinogen[f]
Ligroin (naphtha solvent)	90–110	—	3	1	hydrocarbons, aromatic heterocycles	ethyl acetate, benzene, methylene chloride	
Petroleum ether (ACS)	35–60	—	4	1	hydrocarbons	any solvent less polar than ethanol	easy to separate
n-Pentane	36.1	1.84[b]	4	1	hydrocarbons	any solvent less polar than ethanol	easy to separate
n-Hexane	69	1.89[b]	4	1	hydrocarbons	any solvent less polar than ethanol	
Cyclohexane	80.7	2.02[b]	4	1	hydrocarbons	any solvent less polar than ethanol	
n-Heptane	98.4	1.91[c]	4	1	hydrocarbons	any solvent less polar than ethanol	

[a] Normal boiling point (°C)
[b] Dielectric constant (20 °C)
[c] Dielectric constant (25 °C)
[d] Second solvent used to facilitate dissolving the crystals in a solvent mixture
[e] Scale varies from 4 (highly flammable, highly toxic) to 0 (not flammable, not toxic)
[f] See carcinogen table in the chapter on Laboratory Safety

CHAPTER 14

Tables for Laboratory Safety

CONTENTS

MAJOR CHEMICAL INCOMPATIBILITIES

The following chemicals react, sometimes violently (indicated by italics), in certain chemical environments [1–9]. Incompatibilities may cause fires, explosions, or the release of toxic gases. Extreme care must be taken when working with these materials. This list is not inclusive and the reader is urged to consult multiple sources for more specific information. When using any chemicals, thorough reading of the Materials Safety Data Sheets (MSDS) [10] is strongly recommended.

REFERENCES

1. Dean, J. A., ed. *Lange's Handbook of Chemistry*. 16th ed. New York: McGraw-Hill Book Co., 2005.
2. Fieser, L. F., and M. Fieser. *Reagents for Organic Synthesis*. New York: John Wiley and Sons, 1967.
3. Gordon, A. J., and R. A. Ford. *The Chemist's Companion: A Handbook of Practical Data, Techniques and References*. New York: John Wiley and Sons, 1972.
4. Shugar, G. J., and J. A. Dean. *The Chemist's Ready Reference Handbook* New York: McGraw-Hill Book Co., 1990.
5. Svoronos, P., E. Sarlo, and R. Kulawiec. *Organic Chemistry Laboratory Manual*. 2nd ed. New York: McGraw-Hill Book Co., 1997.
6. Pohanish, R. P., and S. A. Greene. *Wiley Guide to Chemical Incompatibilities*. 3rd ed. New York: John Wiley and Sons, 2009.
7. Pohanish, R. P. *Sittig's Handbook of Toxic and Hazardous Chemicals and Carcinogens*. Bracknell, Berkshire, UK: Noyes Publications, 2007.
8 Pohanish, R. P. *Rapid Guide to Hazardous Chemicals in the Environment*. Bracknell, Berkshire, UK: Noyes Publications, 1997.
9. Pohanish, R. P. *HazMat Data: For First Response, Transportation, Storage, and Security*. New York: Wiley-Interscience, 2004.
10. Material Safety Data Sheets. Available at http://www.ilpi.com/MSDS/, 1995–2010.

Major Chemical Incompatibilities

Chemical	Incompatible Chemicals
Acetic acid	Strong acids (chromic, nitric, perchloric), peroxides
Acetylene	*Air*, copper, halogens (chlorine, bromine, iodine)
Alkali metals	*Acids, water, hydroxy compounds, polychlorinated hydrocarbons* (for example, CCl_4), halogens, carbon dioxide, oxidants
Ammonia, anhydrous	*Halogens (bromine, chlorine, iodine), hydrofluoric acid, liquid oxygen*, calcium or sodium hypochlorite, heavy metals (silver, gold mercury), nitric acid
Ammonium nitrate	*Metal powders, chlorates, nitrites, sulfur, sugar*, flammable and combustible organics, acids, sawdust
Anilines	*Concentrated acids (nitric, chromic), oxidizing agents (chromium (III) ions*, peroxides, permanganate)
Carbon, activated	Oxidizing agents, unsaturated oils
Carboxylic acids	Metals (*alkalis*), organic bases, ammonia
Chlorates	*Flammable and combustible organic compounds*, finely powdered metals, manganese dioxide, ammonium salts
Chromic acid	*Anilines*, 1 ° or 2 ° alcohols
Copper	Oxidizing agents
Ether (including diethyl)	*Peroxides* (especially after long exposure of ether to air; see the other tables in this chapter dealing with this compound)
Fluorine	Reactive (as a strong oxidizing agent) to a certain degree with most compounds, but it can sometimes cause a violent reaction
Halogens (chlorine, bromine)	*Finely powdered metals, diethyl ether, hydrogen*, unsaturated organic compounds, carbide salts, acetylene, alkali metal
Hydrocarbons (saturated)	Halogens (especially *fluorine*) in the presence of ultraviolet light and peroxides
Hydrocarbons (unsaturated)	Halogens, concentrated strong acids, peroxides
Hydrofluoric acid	Ammonia, glass
Hydrogen peroxide	Metals, alcohols, potassium permanganate, flammable and combustible materials
Iodine	*Acetaldehyde, antimony*, unsaturated hydrocarbons, ammonia and some amines
Mercury	Some metals, ammonia, terminal alkynes (see the other tables in this chapter dealing with this element)
Nitric acid (concentrated)	*Anilines, flammable liquids, unsaturated organics lactic acid, coal*, ammonia, powdered metals, wood, alcohols, electron-rich aromatic rings (phenols, analines)
Perchloric acid	Some organics, acetic anhydride, metals, alcohols, wood and its derivatives (see the other tables in this chapter dealing with this compound)
Permanganates, general	Aldehydes, alcohols, unsaturated hydrocarbons
Peroxides	*Flammable liquids, metals*, aldehydes, alcohols, impact, hydrocarbons (unsaturated)
Picric acid	*Dryness and impact, alkali metals, oxidizing agents*, concentrated bases
Potassium, metal	See alkali metals
Potassium permanganate	*hydrochloric acid*, glycerol, hydrogen peroxide, sulfuric acid, wood
Silver salts, organic	*Dryness and prolonged air exposure*
Sodium, metal	See alkali metals
Sulfuric acid, concentrated	*Electron rich aromatic rings (phenols, anilines), unsaturated hydrocarbons potassium permanganate*, chlorates, perchlorates

PROPERTIES OF HAZARDOUS SOLIDS

The following table lists some of the more important properties of hazardous room temperature solids commonly used in the analytical laboratory [1]. The flash points were determined with the open cup method.

REFERENCE

1. Turner, C. F., and J. W. McCreery. *The Chemistry of Fire and Hazardous Materials*. Boston: Allyn and Bacon, 1981.

Properties of Hazardous Solids

Name	Formula	Specific Gravity (at 20°C)	Melting Point, °C	Boiling Point, °C	Flash Point, °C	Autoignition Point, °C	Ignition/Explosion Mechanism	Fire Supression Media
Acetyl peroxide	$(CH_3CO)_2O_2$	1.2	30	63	—	—	heat, shock	a,c
Adipic acid	$(CH_2)_4(COOH)_2$	1.3	152	330	196	420	heat	a,b,c
Aluminum (finely divided)	Al	2.7	660	2270	—	—	mixing with iron oxides	a
Aluminum chlorate	$Al(ClO_3)_3$	—	—	—	—	—	heat, impact agents, reducing agents	a
Aluminum chloride	$AlCl_3$	2.4	192	180	—	—	heat, moisture	a
Ammonium nitrate	NH_4NO_3	1.7	169	210	—	—	heat	b
Ammonium nitrite	NH_4NO_2	1.7	dec	—	70	—	heat, shock, impact	a
Ammonium perchlorate	NH_4ClO_4	—	dec	—	—	—	shock, impact	—
Antimony	Sb	6.7	630	1375	—	—	heat, water	a
Antimony trisulfide	Sb_2S_3	4.6	—	—	—	—	heat, strong organic acids, oxidizers	b
Antimony pentasulfide	Sb_2S_5	4.1	—	—	—	—	heat, strong oxidizers, acids	a
Barium	Ba	3.6	850	1530	—	—	heat	a
Beryllium	Be	1.87	1280	1500	—	—	heat, friction	a
Cadmium	Cd	8.6	321	765	—	—	heat	a
Calcium hypochlorite	$Ca(ClO)_2·4H_2O$	—	dec	—	—	—	heat, contact with combustible material, acid	—
Camphor	$C_{10}H_{16}O$	1.0	177	sub	66	466	high conc. in air	a,c
Cesium	Cs	1.9	29	670	—	—	water	f
Iodine	I_2	4.9	113	183	—	—	heat	c
Lithium	Li	0.53	179	1335	—	—	water, inorganic acids	a
Magnesium	Mg	1.75	651	1107	—	—	water	a
Phosphorus, red	P_4	2.2	600	sub	—	260	heat, oxidizers	b,e
Phosphorus, white	P_4	1.82	44	279	ambient	30	heat, oxidizers, dry atmosphere	b

Phosphorus pentachloride	PCl$_5$	4.7	167	sub	—	—	moist air, heat	a,c
Phosphorus pentasulfide	P$_2$S$_5$	2.03	276	514	142	287	water, acids	—
Potassium chlorate	KClO$_3$	2.3	368	dec	—	—	charcoal, sulfur, and phosphorous	b
Potassium nitrate	KNO$_3$	2.1	334	400	—	—	fiction, contact with organics	b
Potassium nitrite	KNO$_2$	1.9	388	dec	—	—	friction, impact	b
Sodium	Na	0.97	98	890	—	—	moisture	a
Sodium hydride	NaH	0.9	800	—	—	—	water, oxidizers	a
Sodium nitrate	NaNO$_3$	2.3	307	379	—	—	contact with organics	b
Sodium nitrite	NaNO$_2$	2.17	271	318	—	—	contact with organics	b
Sodium styrene sulfonate	C$_8$H$_7$SO$_3$Na	—	225 (dec)	—	—	462	hot surfaces, sparks	a,b,c,
Sulfur	S/S$_8$	2.07	115	445	207	232	heat	a,d
Triphenylboron	(C$_6$H$_5$)$_3$B	—	136	347	—	220	water (produces benzene), heat	a,b,c,d

Abbreviations:
dec decomposes
a   dry chemical extinguisher
b   H$_2$O
c   CO$_2$
d   foam
e   wet sand
f   chlorinated hydrocarbons

COMPOUNDS THAT ARE REACTIVE WITH WATER

The following is a partial listing of families of compounds that are known to be reactive with water [1]. Depending upon the specific compound, the reaction can be rapid and even violent, or simply slow hydrolysis.

REFERENCE

1. Bretherick, L., P. G. Urben, M. J. Pitt. *Bretherick's Handbook of Reactive Chemical Hazards: An Indexed Guide to Published Data.* 6th ed. London: Butterworths, 1999.

Acid anhydrides
Acyl halides
Alkali metals
Alkylaluminum derivatives
Alkylmagnesium derivatives
Alkyl-nonmetal halides
Complex anhydrides
Metal halides
Metal oxides
Nonmetal halides and their oxides
Nonmetal oxides

PYROPHORIC COMPOUNDS: COMPOUNDS THAT ARE REACTIVE WITH AIR

The following listing provides the classes of compounds, with some examples, that can undergo spontaneous reaction upon exposure to air [1,2]. In some cases the reaction is vigorous, while in others the reaction is more subdued, or will only occur if other conditions (such as temperature, humidity, or a reactive surface) are present. The reader is advised to check the literature for more specific information.

REFERENCES

1. Bretherick, L., P. G. Urben, and M. J. Pitt. *Bretherick's Handbook of Reactive Chemical Hazards: An Indexed Guide to Published Data*. 6th ed. London: Butterworths, 1999.
2. Pyrophoric Materials, Texas A&M University. Available at www.safety.science.tamu.edu/pyrophorics. html, 2003.

Alkali metals[a] (sodium, potassium, potassium/sodium alloy, lithium/tin alloys)
Alkylaluminum derivatives (diethylaluminum hydride)
Alkylated metal alkoxides (diethylethoxyaluminum)
Alkylboranes
Alkylhaloboranes (bromodimethyl borane)
Alkylhalophosphines
Alkylhalosilanes
Alkyl metals
Alkyl nonmetal hydrides
Boranes (diborane)
Carbonyl metals (pentacarbonyl iron, octacarbonyl dicobalt, nickel carbonyl)
Complex acetylides
Complex hydrides (diethylaluminum hydride)
Finely divided metals[a] (calcium, zirconium)
Haloacetylene derivatives
Hexamethylnitrato dialuminum salts
Metal hydrides (germane, sodium hydride, lithium aluminum hydride)
Nonmetal hydrides
Some nonmetal (organic) halides (dichloro(methyl)silane)
Spent hydrogenation catalysts (can be especially hazardous because of adsorbed hydrogen; for example, Raney nickel)
White phosphorus

---

[a] Note that the reactivity depends on the particle size and the ease at which oxides are formed on the metal surface.

VAPOR PRESSURE OF MERCURY

The following table provides data on the vapor pressure of mercury, useful for assessing and controlling the hazards associated with use of mercury, for example, as an electrode [1].

REFERENCE

1. Lide, D. R., ed. *CRC Handbook for Chemistry and Physics.* 90th. ed. Boca Raton, FL: CRC Press, 2010.

Temperature °C	Vapor Pressure mm Hg	Vapor Pressure Pa	Temperature °C	Vapor Pressure mm Hg	Vapor Pressure Pa
0	0.000185	0.0247	28	0.002359	0.3145
10	0.000490	0.0653	30	0.002777	0.3702
20	0.001201	0.1601	40	0.006079	0.8105
22	0.001426	0.1901	50	0.01267	1.689
24	0.001691	0.2254	100	0.273	36.4
26	0.002000	0.2666			

FLAMMABILITY HAZARDS OF COMMON LIQUIDS

The following table lists relevant data regarding the flammability of common organic liquids [1].

REFERENCE

1. Turner, C. F., and J. W. McCreery. *The Chemistry of Fire and Hazardous Materials*. Boston: Allyn and Bacon, 1981.

Flammability Hazards of Common Liquids

Solvent	Formula	Specific Gravity	Boiling Point (°C)	Flash Point (°C)	Auto-Ignition Point (°C)	How To Extinguish Fires
Acetaldehyde	CH_3CHO	0.8	21	−38	185	a,b,c
Acetone	$(CH_3)_2CO$	0.8	57	−18	538	a,b
Acetonitrile	$CH_3C \equiv N$	0.79	82	6	—	a,c,d
Acetylacetone	$CH_3COCH_2COCH_3$	1.0	139	41	—	a,b,c
Acrolein	$CH_2 = CHCHO$	0.8	53	−26	277	a,b,c
Acrylonitrile	$CH_2 = CH\text{-}CHC \equiv N$	0.81	77	0	481	a,c,d
Allylamine	$CH_2 = CHCH_2NH_2$	0.8	53	−29	374	a,b
Amylmercaptan	$CH_3(CH_2)_4SH$	0.8	127	18	—	a,b
Aniline	$C_6H_5NH_2$	1.0	184	70	768	a,b,c use masks
Anisole	$C_6H_5OCH_3$	1.0	154	52	—	a,b,c
Benzaldehyde	C_6H_5CHO	1.1	179	65	192	a,b,c
Benzene	C_6H_6	0.88	79	−11	563	a,b,c
n-Butyl alcohol	C_4H_9OH	0.8	117	29	366	a,b,c
t-Butylperacetate	$CH_3CO(O_2)C(CH_3)_3$	—	—	< 27	—	b,c
t-Butylperbenzoate	$C_6H_5CO(O_2)C(CH_3)_3$	> 1.0	112	88	8	a,b,c
n-Butyraldehyde	$CH_3(CH_2)_2CHO$	0.8	76	7	230	a,b,c
Carbon disulfide	CS_2	1.3	47	−30	100	b,d use masks
Crotonaldehyde	$CH_3CH = CHCHO$	0.9	104	13	232	a,b,c
Cumene hydro-peroxide	$C_6H_5C(CH_3)_2O_2H$	1.0	153	175	—	a,b,c
Cyclohexanone	$C_6H_{10}O$	0.9	156	43	420	a,b,c
Diacetyl	$(CH_3CO)_2$	1.0	88	27	—	a,b,c
Diethanolamine	$(HOCH_2CH_2)_2NH$	1.1	269	152	662	b,c
Diethylene glycol diethyl ether	$CH_3(CH_2OCH_2)_3CH_3$	0.9	189	83	—	a, halons
Diethylether	$(C_2H_5)_2O$	0.7	34	−45	180	a,b, halons
Diethylketone	$(C_2H_5)_2CO$	0.8	101	13	452	a,b,c
Dimethyl sulfate	$(CH_3)_2SO_4$	1.3	188	83	188	a,b,c,d
Dimethyl sulfide	$(CH_3)_2S$	0.8	37	−18	206	b,c
1,4-Dioxane	$(CH_2CH_2O)_2$	1.0	101	2	180	a,b,c

Flammability Hazards of Common Liquids (Continued)

Solvent	Formula	Specific Gravity	Boiling Point (°C)	Flash Point (°C)	Auto-Ignition Point (°C)	How To Extinguish Fires
Ethanol	C_2H_5OH	0.8	78	13	423	a,b,c
Ethylacetone (2-pentanone)	$CH_3COCH_2CH_2CH_3$	0.8	102	7	504	a,b,c
Ethylamine	$C_2H_5NH_2$	0.7	31	−18	384	a,b,c
Ethylenediamine	$H_2NCH_2CH_2NH_2$	0.9	117	34	385	a,b,c
Ethyleneglycol	$HOCH_2CH_2OH$	1.1	198	111	413	a,b,c,d
Formaldehyde	$HCHO$	1.0	99	88	427	a,b,c
Furfural	C_5H_4O	1.2	162	60	316	a,b,c,d

Furfuryl alcohol	$C_5H_6O_2$	1.1	171	75	491	a,b,c

Gasoline	C_7H_{16} (isomers)	< 1.0	38–218	−43	257	a,b,c
n-Hexylamine	$C_6H_{13}NH_2$	0.8	132	29	—	a,b
Isopropanol	$(CH_3)_2CHOH$	0.8	82	12	399	b,c
Isopropyl ether	$((CH_3)_2CH)_2O$	0.7	68	−28	443	a,b
Kerosene	—	< 1.0	149–316	38–71	229	a,b,c
Methanol	CH_3OH	0.8	65	11	464	a,b
Methylamine (aq)	CH_3NH_2	0.7	31	−18	384	a,b,c
Methylaniline	$CH_3NHC_6H_5$	0.8	151	49	533	a,b
Methylethyl ketone	$CH_3COCH_2CH_3$	0.8	79	−6	516	a,b,c
Methylethyl ketone peroxide	$CH_3C(OOH)_2CH_2CH_3$	—	—	63	—	a,b
Naphtha (mixture)	—	0.8–0.9	149–216	38–46	227–496	a,b,c
Paraldehyde	$(CH_3CHO)_3$, cyclic	1.0	124	36	238	a,b,c
2-Pentanone	See ethylacetone	—	—	—	—	—
3-Pentene nitrile	$CH_3CH = CHCH_2C \equiv N$	0.83	14.5	40	—	b,c,d
Peracetic acid	CH_3COOOH	1.2	105	40	—	a,b,c
Petroleum ether	—	< 0.7	38–79	< 0	288	a,b,c
Propionaldehyde	CH_3CH_2CHO	0.8	49	8	207	a,b,c
Propylamine	$CH_3CH_2CH_2NH_2$	0.7	49	−37	318	a,b,c
Propylene glycol	$CH_3CHOHCH_2OH$	1.0	188	99	421	a,b
Sulfur chloride	S_2Cl_2	1.7	138	118	234	b,c
Sulfuryl chloride	$SOC1_2$	1.7	69	—	—	a,b,c
Tetrahydrofuran	C_4H_8O	0.9	60	−17	321	a,b,c

Thionylchloride	$SOCl_2$	1.6	79	—	—	—
Toluene	$C_6H_5CH_3$	0.87	111	4	510	a,b,c

Triethanolamine	$(HOCH_2CH_2)_3N$	1.1	360	179	—	b,c,d
Triethylamine	$(C_2H_5)_3N$	0.7	89	7	—	a,b,c

(Continued)

Flammability Hazards of Common Liquids (Continued)

Solvent	Formula	Specific Gravity	Boiling Point (°C)	Flash Point (°C)	Auto-Ignition Point (°C)	How To Extinguish Fires
Xylene (o-)	$C_6H_4(CH_3)_2$ 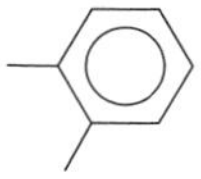	0.88	144	32	463	a,b,c
Xylene (m-)	$C_6H_4(CH_3)_2$ 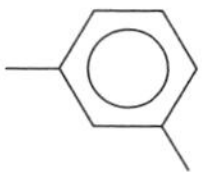	0.86	139	29	527	a,b,c
Xylene (p-)	$C_6H_4(CH_3)_2$ 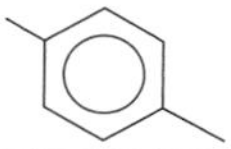	0.86	138	39	529	a,b,c

ABBREVIATIONS USED IN THE ASSESSMENT AND PRESENTATION OF LABORATORY HAZARDS

The following abbreviations are commonly encountered in presentations of laboratory and industrial hazards. The reader is urged to consult the reference[1] for additional information.

REFERENCE

1. Furr, A.K., ed., *CRC Handbook of Laboratory Safety*, 5th ed., CRC Press, Boca Raton, FL, 2000.

CC Closed Cup; method for the measurement of the flash point. With this method, Sample vapors are not allowed to escape as they can with the open cup method. Because of this, flash points measured with the CC method are usually a few degrees lower than those measured with the OC. The choice between CC and OC is dependent on the (usually ASTM) standard method chosen for the test.

COC Cleveland Open Cup, see Open Cup.

IDLH Immediately Dangerous to Life and Health; the maximum concentration of chemical contaminants, normally expressed as parts per million (ppm, mass/mass), from which one could escape within 30 minutes without a respirator, and without experiencing any escape impairing (severe eye irritation) or irreversible health effects. Set by NIOSH. Note that this term is also used to describe electrical hazards.

LEL Lower Explosion Limit; the minimum concentration of a chemical in air at which detonation can occur.

LFL Lower Flammability Limit; the minimum concentration of a chemical in air at which flame propagation occurs.

MSDS Material Safety Data Sheet; a (legal) document that must accompany any supplied chemical that provides information on chemical content, physical properties, hazards, and treatment of hazards. The MSDS should be considered only a minimal source of information, and cannot replace additional information available in other, more comprehensive sources.

NOEL No Observed Effect Level; the maximum dose of a chemical at which no signs of harm are observed. This term can also be used to describe electrical hazards.

OC Open Cup; also called Cleveland Open Cup. This refers to the test method for determining the flash point of common compounds. It consists of a brass, aluminum or stainless steel cup, a heater base to heat the cup, a thermometer in a fixture and a test flame applicator. The flash point is the lowest temperature at which a material will form a flammable mixture with air above its surface. The lower the flash point, the easier it is to ignite.

PEL Permissible Exposure Level; an exposure limit that is published and enforced by OSHA as a legal standard. The PEL may be expressed as a time-weighted-average (TWA) exposure limit (for an 8 hour workday), a 15-minute short term exposure limit (STEL), or a ceiling (C, or CEIL, or TLV-C).

REL Recommended Exposure Level; average concentration limit recommended for up to a 10-hour workday during a 40-hour workweek, by NIOSH.

RTECS Registry of Toxic Effects of Chemical Substances; a database maintained by the National Institute of Occupational Safety and Health (NIOSH). The goal of the database, is to include data on all known toxic substances, along with the concentration at which toxicity is known to occur. There are approximately 140,000 compounds listed.

STEL Short Term Exposure Level; an exposure limit for a short term, 15 minute, exposure that cannot be exceeded during the workday, enforced by OSHA as a legal standard. Short term exposures below the STEL level generally will not cause irritation, chronic or reversible tissue damage, or narcosis.

TLV Threshold Limit Value; guidelines suggested by the American Conference of Governmental Industrial Hygienists to assist industrial hygienists with limiting hazards of chemical exposures in the workplace.

TLV-C Threshold Limit Ceiling Value; an exposure limit which should not be exceeded under any circumstances.

TWA Time-weighted average concentration for a conventional 8-hour workday and a 40 hour workweek. It is the concentration to which it is believed possible that nearly all workers can be exposed without adverse health effects.

UEL Upper Explosion Limit; the maximum concentration of a chemical in air at which detonation can occur.

UFL Upper Flammability Limit; the maximum concentration of a chemical in air at which flame propagation can occur.

WEEL Workplace Environmental Exposure Limit; set by the American Industrial Hygiene Association (AIHA).

Some abbreviations that are sometimes used on material safety data sheets, and in other sources, are ambiguous. The most common meanings of some of these vague abbreviations are provided below, but the reader is cautioned that these are only suggestions:

EST Established; estimated
MST Mist
N/A, NA Not applicable
ND None determined; not determined
NE None established; not established
NEGL Negligible
NF None found; not found
N/K, NK Not known
N/P, NP Not provided
SKN Skin
TS Trade secret
UKN Unknown

CHEMICAL CARCINOGENS

The following table contains data on chemicals often used in the analytical laboratory, which have come under scrutiny for their suspected or observed carcinogenicity [1–8]. In some cases, the chemical structure is provided to avoid ambiguity. The reader is advised to use these tables with care, as there is a great deal of variability in the classifications as new data become available. It is suggested that the reader maintain a current file of data from the appropriate regulatory agencies, along with a complete set of material safety data sheets, and to err on the side of caution.

Interpretation of the codes used in this table is based on the following key.

[a]Compiled from monographs of the International Agency for Research on Cancer, IARC (part of the United Nations World Health Organization), as data becomes available. Classifications are as follows:

Group 1: Carcinogenic to humans
Group 2A: Probably carcinogenic to humans
Group 2B: Possibly carcinogenic to humans
Group 3: Not classifiable as to carcinogenicity to humans
Group 4: Probably not carcinogenic to humans

[b]Compiled from data of the National Toxicology Program, NTP, whose reports are updated every two years (branch of the U.S. Department of Health and Human Services). Classifications are as follows:

Y: Reasonably anticipated to be human carcinogen
#: Known human carcinogen

[c]Compiled from data of the Occupational Safety and Health Administration, OSHA, with standards set by the legislative process (part of the U.S. Department of Labor). Classifications are as follows:

Y: Possibly a human carcinogen
*: Substance for which OSHA has promulgated expanded health standards that govern health concerns in addition to carcinogenesis.

REFERENCES

1. Partial List of Selected Carcinogens. Available at www.people.memphis.edu/~ehas/carcinogen.html, 2003.
2. Ruth, J. H. "Odor Thresholds and Irritation Levels of Several Chemical Substances: A review." *American Industrial Hygiene Association Journal* 47, no. A-142 (1986).
3. Chemical Carcinogens. New Jersey Department of Health and Senior Services, Occupational Health Service. Available at www.state.nj.health.eoh, 2003.
4. Hazard Database: List of Carcinogens. Available at www.ephb.nw.ru/~spirov/carcinogen_1st.html, 2003.
5. Tenth Report on Carcinogens, U.S. Department of Health and Human Services, Public Health Service, National Toxicology Program, National Institute of Environmental Health Sciences, Research triangle Park, Durham, NC, 2005. Available at: http://ntp.niehs.nih.gov/ntp/roc/toc11.html
6. EHP Online. Available at www.ehp.niehs.nih.gov/roc/toc10.html, 2003.
7. NIOSH Pocket Guide to Chemical Hazards, National Institute of occupational Safety and Health, 1997.
8. International Agency for Research on Cancer. Available at http://monographs.iarc.fr/ENG/Classification/crthallalph.php

Chemical Carcinogens

Chemical Name	CAS Number	IARC[a]	NTP[b]	OSHA[c]	Odor Low mg/m³	Irritating Conc. mg/m³
Acetaldehyde	75-07-0	2B	Y			
Acetamide	60-35-5	2B				
2-Acetylaminofluorene	53-96-3		Y	Y		
Acrylamide	79-06-1	2A	Y			
Acrylonitrile	107-13-1	2B	Y	*	8.1	
Actinomycin D 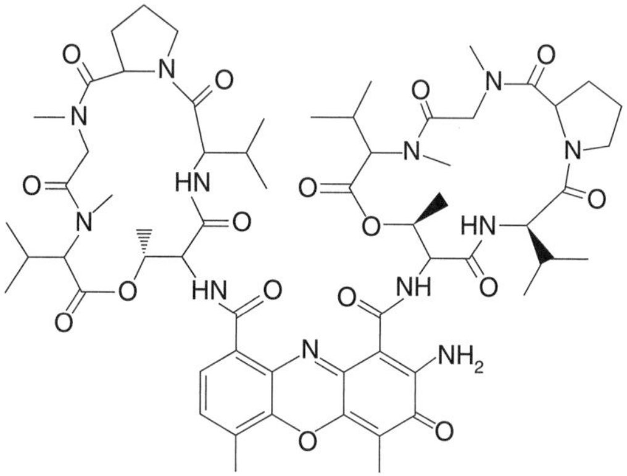	50-76-0	C				
Adriamycin 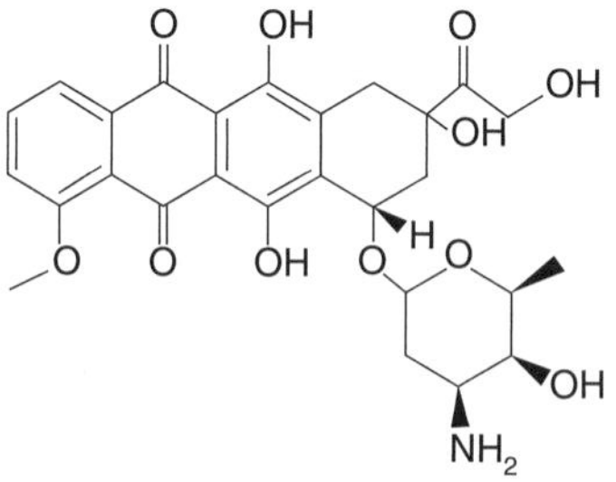	23214-92-8	2A	Y			

Actinomycin D

Adriamycin

Name	CAS No.	Group			
Aflatoxins	1402-68-2	1			#
2-Aminoanthraquinone	117-79-3	3			Y
p-Aminoazobenzene	60-09-3	2B	Y		
4-Aminobiphenyl	92-67-1	1			#
1-Amino-2-methyl anthraquinone	82-28-0	3			Y
o-Aminoazotoluene	97-56-3	2B			Y
2-Aminonaphthalene (β-aminonaphthalene)	91-59-8	1		*	#
Ammonium dichromate	7789-09-5	2B			
Amitrole	61-82-5	3			Y
Androgenic steroids		2A			
o-Anisidine	90-04-0	2B			Y
o-Anisidine hydrochloride	134-29-2	2B			Y
Antimony trioxide	1309-64-4	2B			
Aramite	140-57-8	2B			
Arsenic compounds (certain)		1		*	#
Arsenic, metal	7440-38-2	1		*	#
Asbestos	1332-21-4	1		*	#
Auramine, technical	2465-27-2	2B			
Azathioprine	446-86-6	1			#
Aziridine		2B			
Benz[a]anthracene	56-55-3	2A			Y

(Continued)

Chemical Carcinogens (Continued)

Chemical Name	CAS Number	IARC[a]	NTP[b]	OSHA[c]	Odor Low mg/m^3	Irritating Conc. mg/m^3
Benzene	71-43-2	1	#	*	4.5	9 000
Benzidine 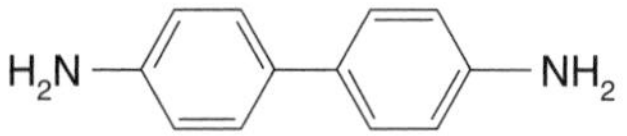	92-87-5	1	#	Y		
Benzidine-based dyes		2A	Y			
Benzo[b]fluoranthene 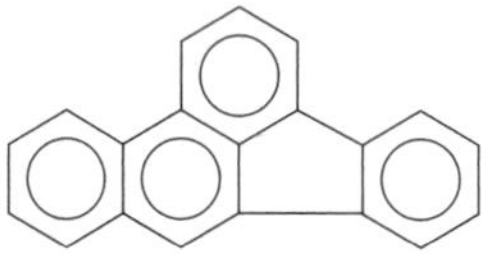	205-99-2	2B	Y			
Benzo[j]fluoranthene 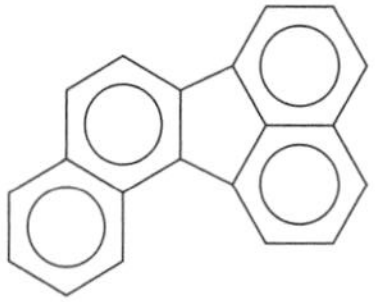	205-82-3	2B	Y			
Benzo[k]fluoranthene 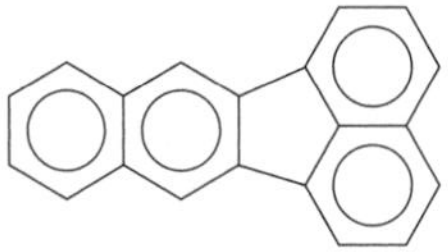	207-08-9	2B	Y			
Benzofuran 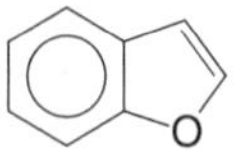	271-89-6	2B				

(Continued)

Name	CAS Number				
Benzo[a]pyrene 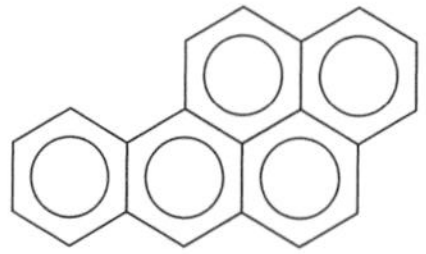	50-32-8	1	Y		
Benzotrichloride	98-07-7			Y	
Benzyl violet 4B	1694-09-3	2B			
Beryllium	7440-41-7	1	#	Y	
Beryllium sulfate	13510-49-1	1	#		
Beryllium compounds (certain)		1	#		
N,N-bis(2-chloroethyl)-2-naphthylamine (Chlornaphazine) 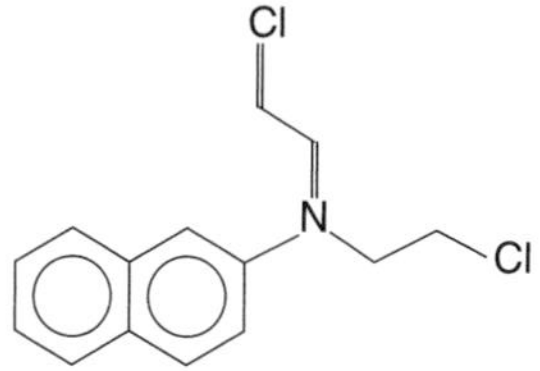	494-03-1	1			
Bis-chloroethyl nitrosourea (BCNU)	154-93-8	2A			
Bis-chloromethyl ether (BCME)	542-88-1	1	#	Y	
Bromodichloromethane	75-27-4	2B		Y	
1,3-Butadiene	106-99-0	1			0.352
1,4-Butanediol di-methanesulfonate	55-98-1	1	#		
Butylated hydroxyanisole (BHA)	25013-16-5	2B		Y	
Butyrolactone, beta	3068-88-0	3			
Cadmium	7440-43-9	1	#	*	
Caffeic acid	331-39-5	2B			

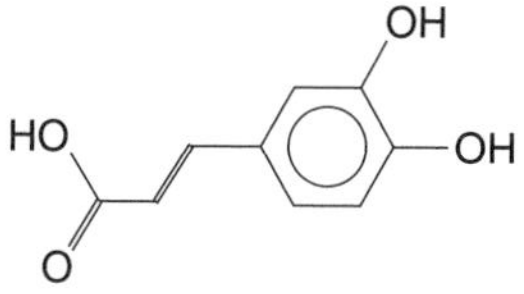

Chemical Carcinogens (Continued)

Chemical Name	CAS Number	IARC[a]	NTP[b]	OSHA[c]	Odor Low mg/m³	Irritating Conc. mg/m³
Carbon black	1333-86-4	2B				
Carbon tetrachloride	56-23-5	2B	Y		60-300	
Catechols		2B				
Chlorambucil 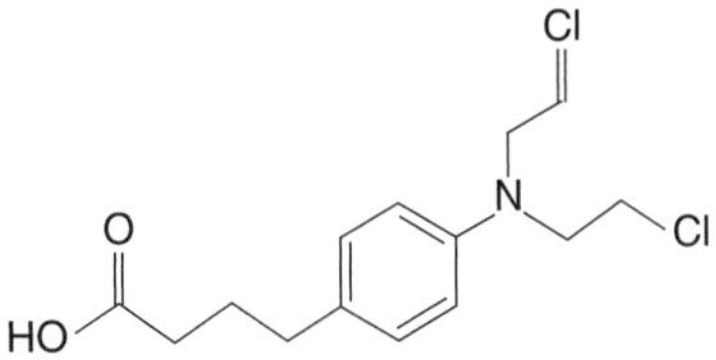	305-03-3	1	#			
Chloramphenicol 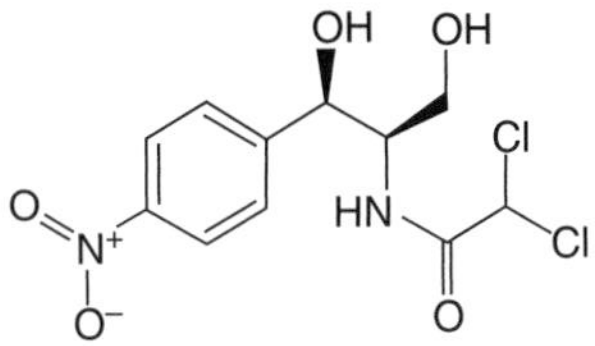	56-75-7	2A	Y			
Chlordane 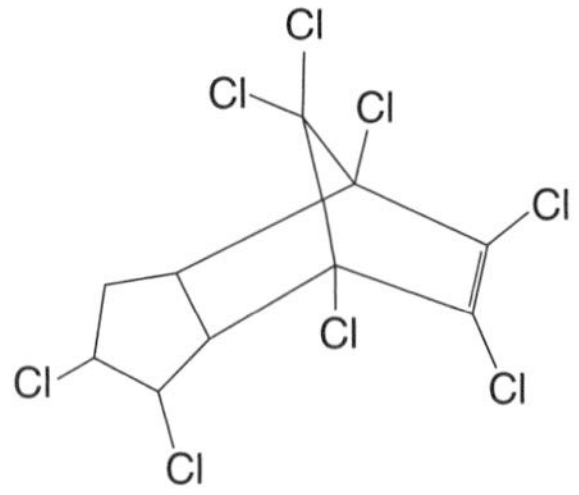	57-74-9	2B				

Chlorambucil

Chloramphenicol

Chlordane

Chordecone (Kepone)

	CAS	Group			
Chordecone (Kepone)	143-50-0	2B			
Chlorinated toluenes, alpha		2A			
p-Chloroaniline		2B			
1-(2-Chloroethyl)-3-cyclohexyl-1-nitrosourea [CCNU]	13010-47-4	2A	#		
Chloroform	67-66-3	2B	Y	250	20,480
Chloromethyl methyl ether	107-30-2	2A	#		
1-Chloro-2-methylpropene	513037-1	2B			
3-Chloro-2-methylpropene	563-47-3	3	Y		
Chlorophenols	95-83-0	2B			
4-Chloro-o-phenylenediamine	95-83-0	2B	Y	0.0189	6,800
Chloroprene	126-99-8	2B	Y		
p-Chloro-o-toluidine	95-69-2	2A	Y		

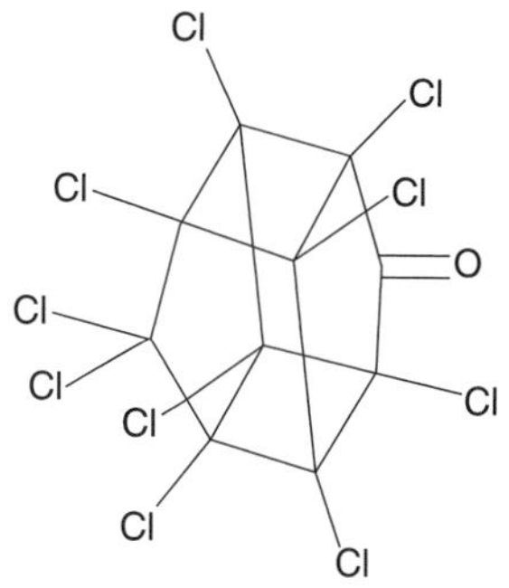

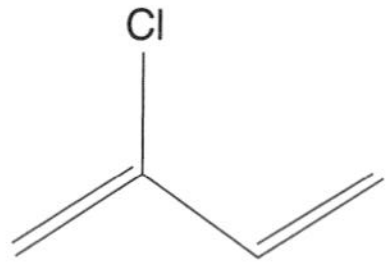

p-Chloro-o-toluidine

(Continued)

Chemical Carcinogens (Continued)

Chemical Name	CAS Number	IARCª	NTPᵇ	OSHAᶜ	Odor Low mg/m³	Irritating Conc. mg/m³
Chlorozotocin 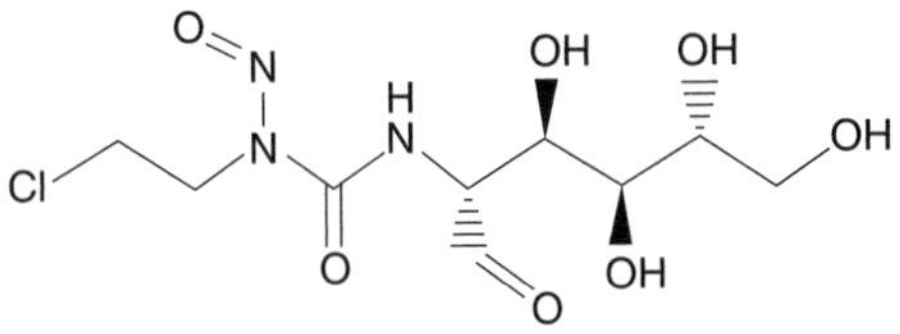	54749-90-5	2A	Y			
Chromium	7440-47-3	1	#			
Chromium compounds (certain)		1	#			
Chromium (VI) ions	18540-29-9	1	#			
Chrysene 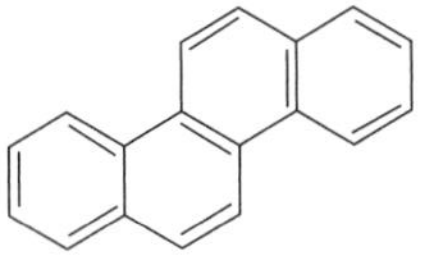	218-01-9	2B	Y			
Cis-platin 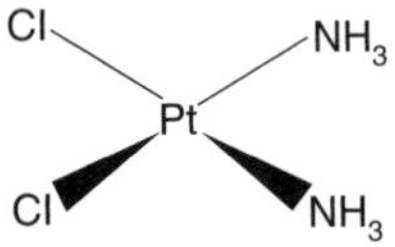	15663-27-1	2A	Y			
Cobalt and compounds	7440-48-4	2B				
p-Cresidine 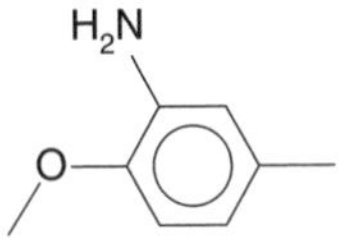	120-71-8	2B	Y			

Name	CAS			
Cupferron	135-20-6		Y	
Cycasin	14901-08-7	2B	B	
Cyclophosphamide	50-18-0	1	#	
Dacarbazine	4342-03-4	2B	Y	
DDT (p,p′-dichlorodiphenyl-trichloroethane)	50-29-3	2B	Y	5.0725
N,N′-diacetyl-benzidine	613-35-4	2B		
2,4-Diaminoanisole	615-05-4	2B		
2,4-Diaminoanisole sulfate	39156-41-7		Y	
4,4′-Diaminodiphenyl ether	101-80-4	2B		
2,4-Diaminotoluene	95-80-7	2B	Y	
Dibenz[a,h]acridine	226-36-8	2B	Y	

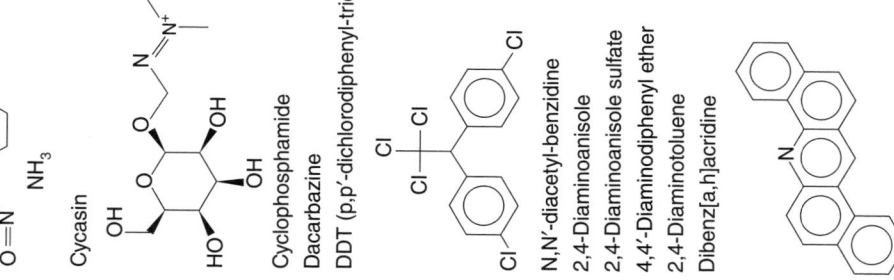

(Continued)

Chemical Carcinogens (Continued)

Chemical Name	CAS Number	IARC[a]	NTP[b]	OSHA[c]	Odor Low mg/m³	Irritating Conc. mg/m³
Dibenz[a,j] acridine	224-42-0	2B	Y			
Dibenz[a,h]-anthracene	53-70-3	2A	Y			
7H-Dibenzo[c,g] carbazole	194-59-2	2B	Y			
Dibenzo[a,e] pyrene	192-65-4	3	Y			

Name	CAS	Group				
Dibenzo[a,h] pyrene	189-64-0	2B	Y			
Dibenzo[a,i] pyrene	189-55-9	2B	Y			
Dibenzo[a,l] pyrene	191-30-0	2A	Y			
1,2-Dibromo-3-chloropropane [DBCP]	96-12-8	2B	Y	*	0.965	1.93
p-Dichlorobenzene	106-46-7	2A	Y	*		
3,3'-Dichlorobenzidine	91-94-1	2B	Y	*		
3,3'-Dichlorobenzidine salts						
3,3'-Dichloro-4,4'-diaminodiphenyl ether	28434-86-8	2B				
1,2-Dichloroethane	107-06-2	2B	Y			
Dichloromethane	75-09-2	2B	Y	*		
1,3-Dichloropropene	542-75-6	2B	Y			
Di(2-ethylhexyl) phthalate	103-23-1	3				
1,2-Diethylhydrazine	1615-80-1	2B				

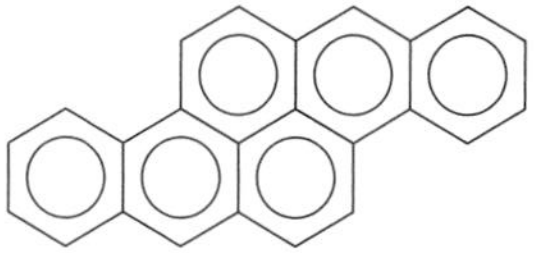

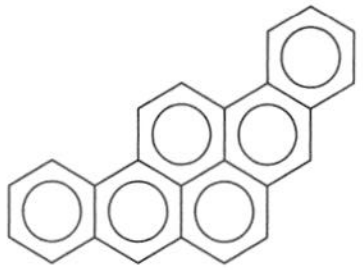

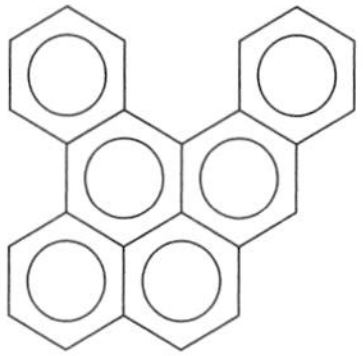

(Continued)

Chemical Carcinogens (Continued)

Chemical Name	CAS Number	IARC[a]	NTP[b]	OSHA[c]	Odor Low mg/m³	Irritating Conc. mg/m³
Dieldrin 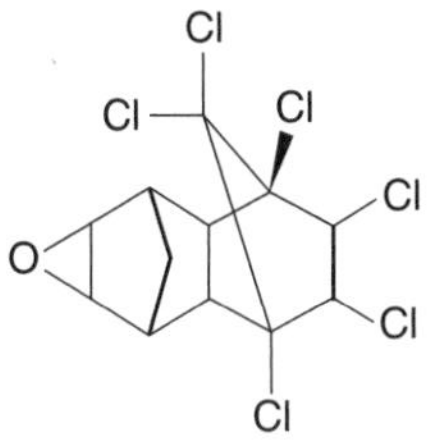	60-57-1	3		Y		
Dienoestrol 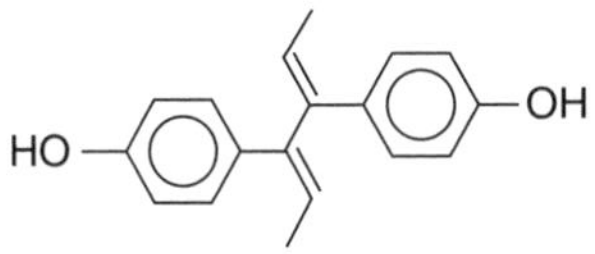	84-17-3	1				
Diepoxybutane	1464-53-5	2B	Y			
Di(2,3-ethylhexyl) phthalate	117-81-7	2B	Y			
Diethylstilbestrol [DES]	56-53-1	1	#			
Diethyl sulfate	64-67-5	2A	Y			
Dihydrosafrole	94-58-6	2B	Y			
1,8-Dihydroxyanthraquinone	117-10-2		Y			
Diisopropyl sulfate	2973-10-6	2B				
3,3'-Dimethoxybenzidine	119-90-4	2B	Y			
4-Dimethylaminoazobenzene	60-11-7	2B	Y	Y		
2,6-Dimethylaniline	87-62-7	2B				
3,3'-Dimethylbenzidine	119-93-7	2B	Y			

Dimethylcarbamoyl chloride	79-44-7	2A	Y		
1,1-Dimethyl hydrazine	57-14-7	2B	Y	d	
Dimethyl sulfate	77-78-1	2A	Y	d	
Dinitrofluoranthrene, isomers		2B			
1,8-Dinitropyrene	42397-65-9	2B			
2,4-Dinitrotoluene	121-14-2	2B			
2,6-Dinitrotoluene	606-20-2	2B			
1,4-Dioxane	123-91-1	2B	Y		0.0108

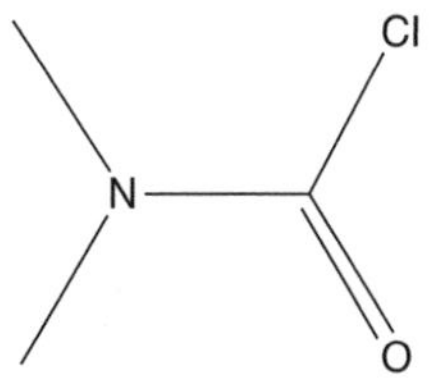

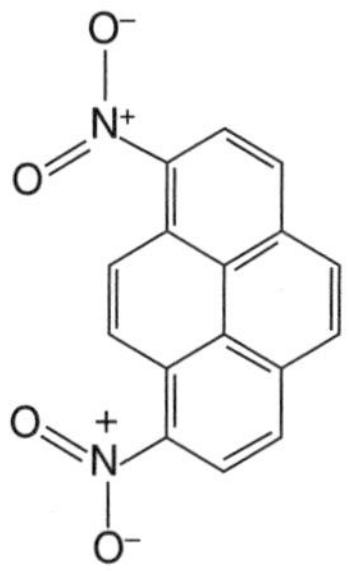

792

(Continued)

Chemical Carcinogens (Continued)

Chemical Name	CAS Number	IARC[a]	NTP[b]	OSHA[c]	Odor Low mg/m³	Irritating Conc. mg/m³
Direct black 38, technical 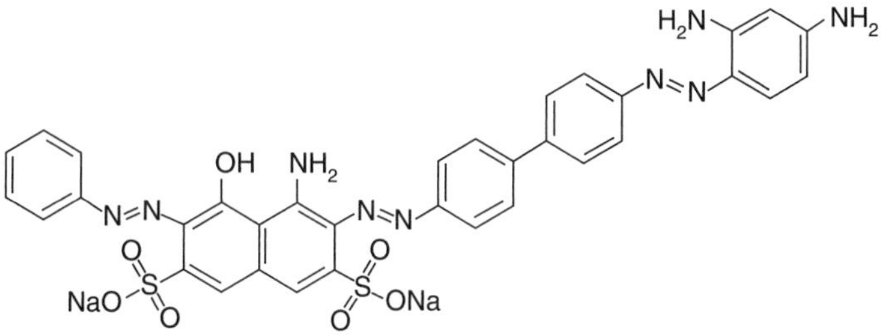	1937-37-7					
Direct blue 6, technical 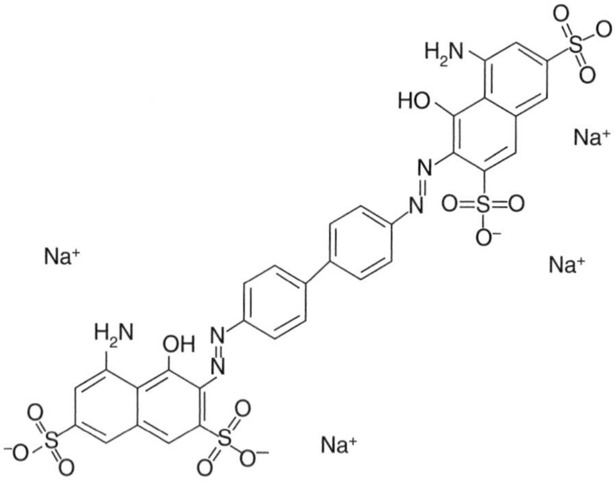	2602-46-2					

Name	CAS	Class			
Direct brown 95, technical	16071-86-6	1			
Disperse blue 1	2475-45-8	2B	Y		
Epichlorohydrin	106-89-8	2A	Y	d	50
1,2-Epoxybutane	106-88-7	2B	#		335
Erionite(zeolite mineral)	12510-42-8	1			

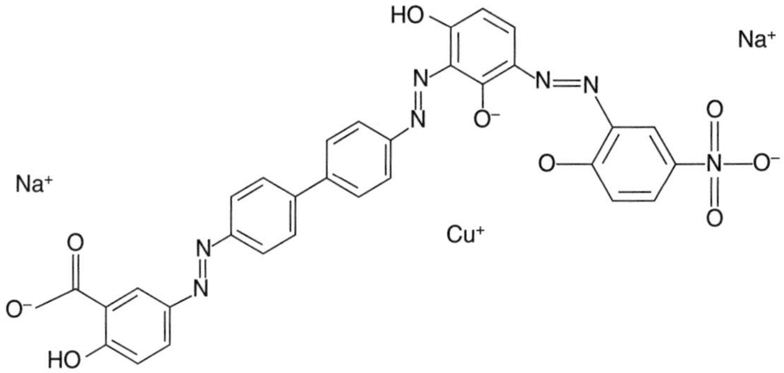

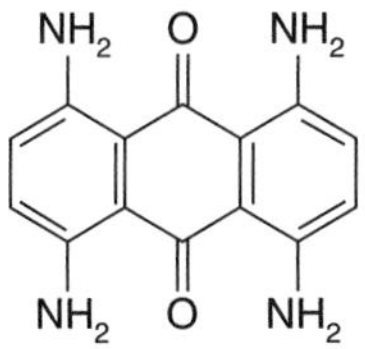

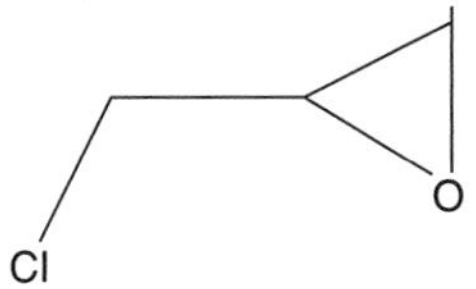

(Continued)

Chemical Carcinogens (Continued)

Chemical Name	CAS Number	IARC[a]	NTP[b]	OSHA[c]	Odor Low mg/m³	Irritating Conc. mg/m³
Ethinyloestradiol	57-63-6	1				

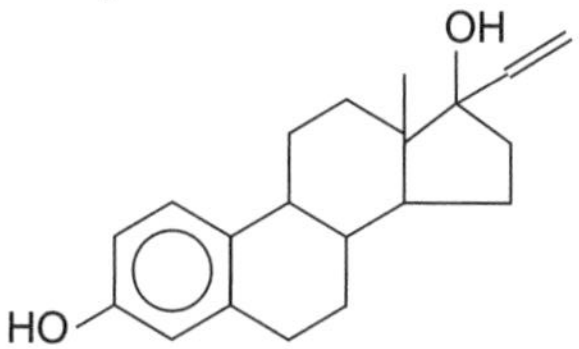 Chemical Name	CAS Number	IARC[a]	NTP[b]	OSHA[c]	Odor Low mg/m³	Irritating Conc. mg/m³
Ethyl acrylate	140-88-5	2B				
Ethylene dibromide [EDB]	106-93-4	2A			76.8	
Ethylene dichloride [EDC]	107-06-2	2B		Y	24	
Ethyleneimine	151-56-4			Y	4	200
Ethylene oxide	72-21-8	1	#	*	520	
Ethylene thiourea	96-45-7	3	Y			
Ethyl methanesulfonate	62-50-0	2B	Y			
N-ethyl-N-nitrosourea	759-73-9	2A				
Formaldehyde	50-00-0	1	Y	*	1.47	1.50
2-(2-Furyl)-3-(5-nitro-2-furyl) acrylamide	3688-53-7	2B				
2-(2-Formylhydrazino)-4-(5-nitro-2-furyl) thiazole	3570-75-0	2B				
Furan	110-00-9	2B	Y			
Gyromitrin	16568-02-8	1				

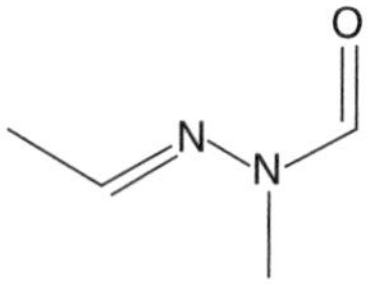 Chemical Name	CAS Number	IARC[a]	NTP[b]	OSHA[c]	Odor Low mg/m³	Irritating Conc. mg/m³
Hexachlorobenzene	118-74-1	2B	Y			
Hexachloroethane	67-72-1	2B	Y			
Hexamethyl phosphoramide	680-31-9	2B	Y			
Hydrazine	302-01-2	2B	Y	Y	3	
Hydrazine sulfate	10034-93-2		Y			
Hydrazobenzene	122-66-7		Y			

Name	CAS	Group		
Indeno [1,2,3-cd] pyrene	193-39-5	2B	Y	
Isoprene	78-79-5	2B	Y	
Lead (II) acetate	301-04-2	2B	#	*
Lead (II) chromate	7758-97-6	2A	Y	
Lead (II) phosphate	7446-27-7	2B	Y	*
Lindane (and other hexachlorocyclohexane isomers)	58-89-9		Y	Y
Melphalan 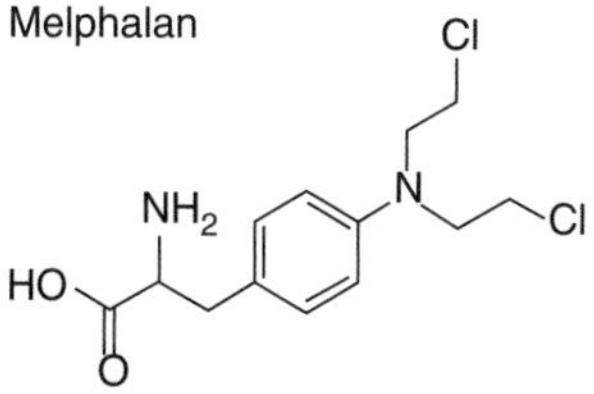	148-82-3	1	#	
Merphalan 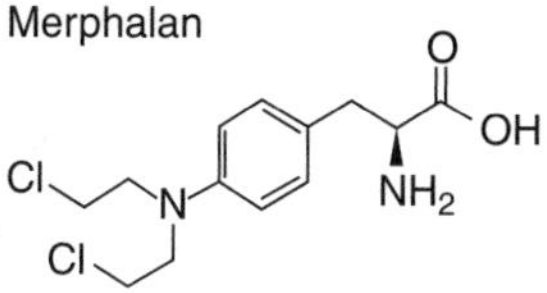	531-76-0	2B		
Mestranol 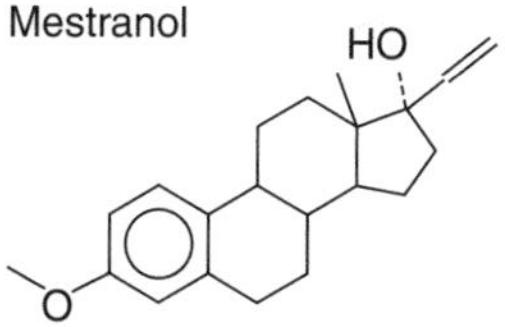	72-33-3	1		

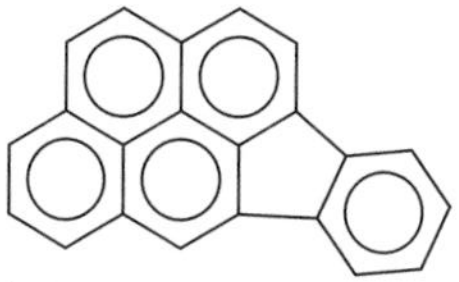

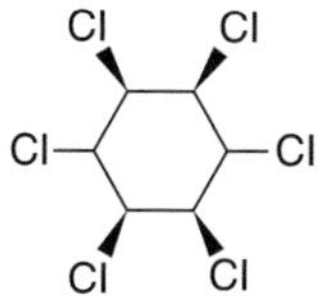

(Continued)

Chemical Carcinogens (Continued)

Chemical Name	CAS Number	IARC[a]	NTP[b]	OSHA[c]	Odor Low mg/m³	Irritating Conc. mg/m³
2-Methylaziridine 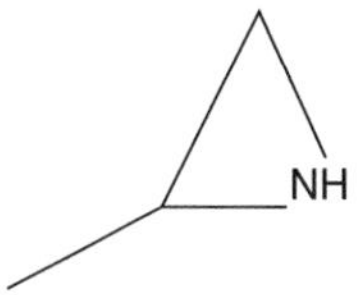	75-55-8	2B	Y			
Methylazoxy-methanol acetate	592-62-1	2B				
Methyl chloromethyl ether				Y		
5-Methylchrysene	3697-24-3	2B	Y			
4,4'-Methylenebis (2-chloroaniline [MOCA]	101-14-4	1	Y			
4,4'-Methylenebis (N,N-dimethyl-benzenamine)	69522-43-6		Y			
4,4'-Methylene dianiline	101-77-9	2B	Y			
Methyl bromide	74-83-9	3		Y	80	
Methyl chloride	74-87-3	3			21	1 050
Methyl hydrazine	60-34-4	1			1.75	
Methyl iodide	74-88-4	3		Y		21 500
Methylmercury compounds		2B				
Methyl methanesulfonate	66-27-3	2A				
N-methyl-N-nitrosourea	684-93-5	2A	Y			
N-methyl-N-nitrosourethane		2B				
Methylthiouracil 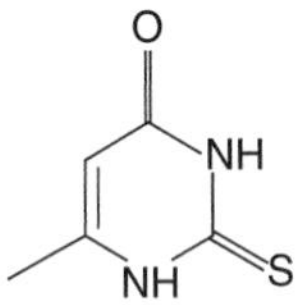	56-04-2	2B				

Name	Structure	CAS	Group			Value
Metronidazole		443-48-1	2B	Y		
Michler's ketone		90-94-8		Y		
Mirex		2385-85-5	2B	Y		
Mustard gas		505-60-2	1	#		0.015
α-Naphthylamine		134-32-7	3	#	Y	
β-Naphthylamine		91-59-8	1	#	Y	
Nickel carbonyl		13463-39-3	1	Y		0.21
Nickel		7440-02-0	2B	Y		
Nickel compounds (certain)			1	#	*	

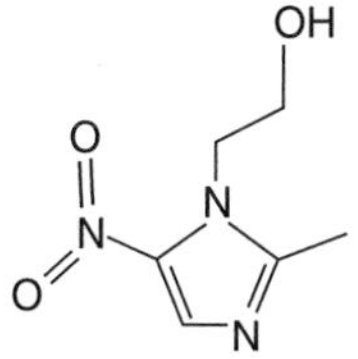

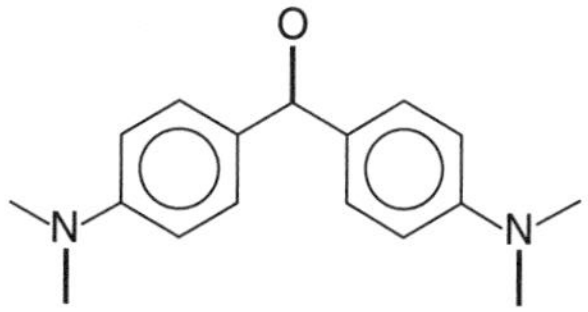

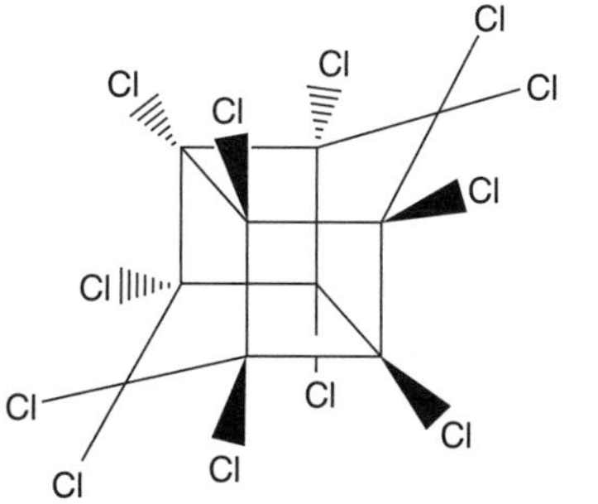

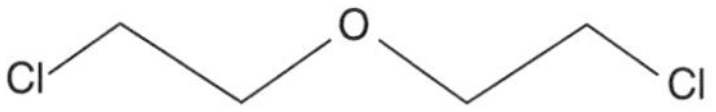

(Continued)

Chemical Carcinogens (Continued)

Chemical Name	CAS Number	IARC[a]	NTP[b]	OSHA[c]	Odor Low mg/m³	Irritating Conc. mg/m³
Nickel, metallic and inorganic compounds	7440-02-0	2B	#			
Nitrilotriacetic acid	139-13-9	2B	Y			
5-Nitroacenaphthene	602-87-9	2B				
2-Nitroanisole	91-23-6	2B	Y			
5-Nitro-o-anisidine	99-59-2					
Nitrobenzene	98-95-3	2B				
4-Nitrobiphenyl	92-93-3			Y		
6-Nitrochrysene	7496-02-8	2B	Y			
Nitrofen 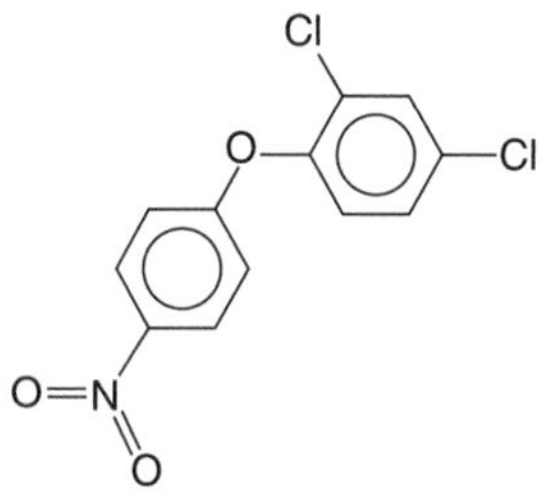	1836-75-5	2B	Y			
Nitrofluorene	607-57-8	2B				
Nitrogen mustards		2A	Y			
Nitrogen mustard N-oxide	126-85-2	2B				
2-Nitropropane	79-46-9	2B	Y		17.5	
1-Nitropyrene	5522-43-0	2B	Y			
4-Nitropyrene	57835-92-4	2B	Y			
N-Nitroso-di-n-butyl amine	924-16-3	2B	Y			
N-Nitroso-n-propyl amine	621-64-7	2B	Y			
N-Nitroso-dimethylamine	62-75-9	2A	Y	Y		
N-Nitroso-diethanolamine	1116-54-7	2B	Y			
N-Nitroso-diethylamine	55-18-5	2A	Y			
N-Nitroso-dimethylamine	62-75-9	2A	Y	Y		
p-Nitroso-di-phenylamine	156-10-5	3	Y			

N-nitroso-di-n-propylamine	621-64-7		Y
N-nitroso-N-ethyl urea	759-73-9		Y
N-nitroso-N-methyl urea	684-93-5		Y
N-nitrosomethyl-vinylamine	4549-40-0	2B	Y
N-nitrosomorpho-line	59-89-2	2B	Y
N-nitrosonornicotine	16543-55-8	1	Y
N-nitrosopiperidine	100-75-4	2B	Y
N-nitrosopyrrolidine	930-55-2	2B	Y
N-nitrososarcosine	13256-22-9	2B	Y
Norethisterone	68-22-4		Y
Oestradiol-17B	50-28-2	1	
Oestrone	53-16-7	1	
4,4'-Oxydianiline	101-80-4		Y

(Continued)

Chemical Carcinogens (Continued)

Chemical Name	CAS Number	IARC[a]	NTP[b]	OSHA[c]	Odor Low mg/m³	Irritating Conc. mg/m³
Oxymetholone	434-07-1		Y			
Phenacetin	62-44-2	1	Y			
Phenazopyridine	94-78-0	2B	Y			
Phenobarbitol	50-06-6	2B				
Phenazopyridine hydrochloride	136-40-3	2B	Y			
Phenoxyacetic acid derivatives		1		Y		
Phenoxybenzamine hydrochloride	63-92-3	2B	Y			
Phenyl glycidyl ether	122-60-1	2B				
N-phenyl-β-naphthylamine	135-88-6	1				
Phenylhydrazine	100-63-0			Y		

(Continued)

Name	CAS	IARC			
Phenytoin	57-41-0	2B	Y		
Phenytoin, sodium salt	630-93-3	2B	Y		
Polybrominated biphenyls [PBBs]	36355-01-8	2A	Y		
Polychlorinated biphenyls [PCBs]		2A	Y	*	
Polychlorinated camphenes		2B			
Polychlorophenols		2A	Y		
Polycyclic aromatic compounds, general		2B			
Potassium bromated	7758-01-2	1			
Potassium chromate	7789-00-6	1			
Potassium dichromate	7778-50-9	2A	Y		
Procarbazine	671-16-9				
Procarbazine hydrochloride	366-70-1	2A	Y		
Propane sulfone	1120-71-4	2B	Y		
β-Propiolactone	57-57-8	2B	Y		Y
Propyleneimine	75-55-8				
Propylene oxide	75-56-9	2B	Y		

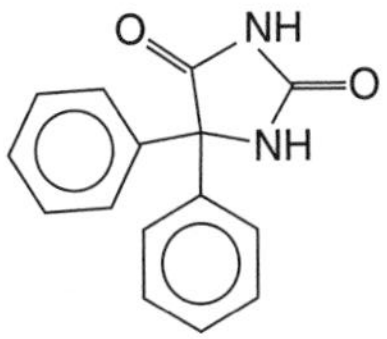

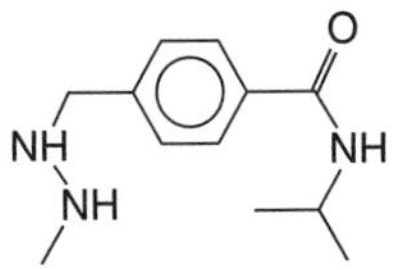

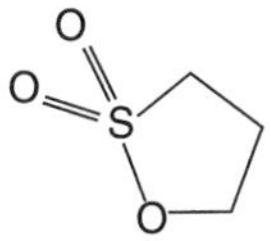

Chemical Carcinogens (Continued)

Chemical Name	CAS Number	IARC[a]	NTP[b]	OSHA[c]	Odor Low mg/m³	Irritating Conc. mg/m³
Propylthiouracil 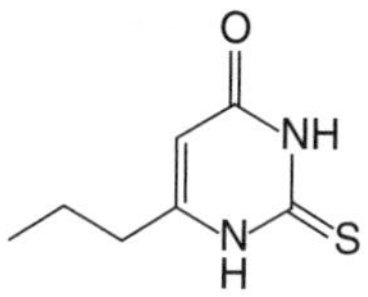	51-52-5	2B	Y			
Reserpine 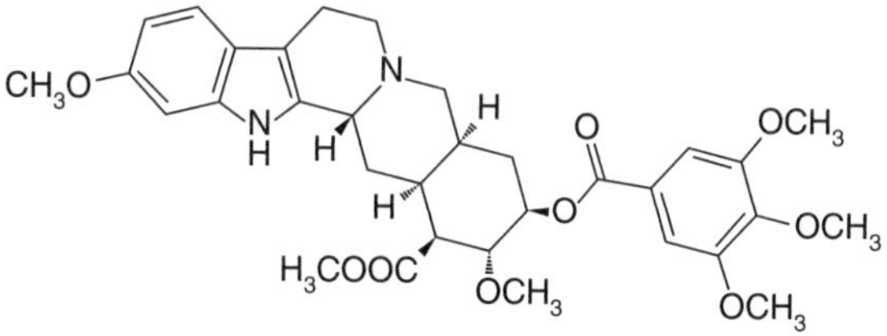	50-55-5	3	Y			
Saccharin	81-07-2	3				
Safrole 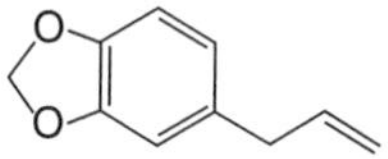	94-59-7	2B	Y		1.4586	
Selenium sulfide	7446-34-6		Y	*		
Silica crystalline (respirable)		1	#			
Streptozotocin 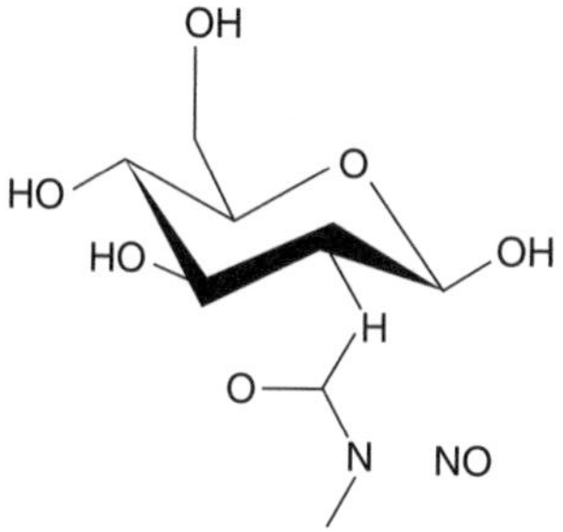	18883-66-4	2B	Y			

Name	CAS	Group				
Strontium chromate	7789-06-2		#			
Sulfallate	95-06-7	2B	Y			

Name	CAS	Group				
2,3,7,8-Tetrachloro-dibenzo-p-dioxin [TCDD]	1746-01-6	1	#			
1,1,2,2-Tetrachloroethane	79-34-5	3	Y	*	21	1302
Tetrachloroethylene	127-18-4	2	Y	*	31.3561	710.2
Tetrafluoroethylene	116-14-3	2B	Y			
Tetranitromethane	509-14-8	2B	Y			
Thioacetamide	62-55-5	2B	Y			
4,4'Thiodianiline	139-65-1	2B	Y			
Thiourea	62-56-6	3	Y			
Thorium (232) dioxide	1314-20-1		#			
p-Tolidine	119-93-7		Y			
o-Toluidine	95-53-4	1	Y			
p-Toluidine	106-49-0	1	Y			
o-Toluidine hydrochloride	636-21-5		Y			
p-Toluidine	106-49-0		Y			
Toxaphene	8001-35-2	2B	Y	*	2.366	
Treosulfan	299-75-2	1				

Name	CAS	Group				
1,1,2-Trichloroethane	79-00-5	3		*		
Trichloroethylene	79-01-6	2A	Y	*		
2,4,6-Trichlorophenol	88-06-2	2B	Y			

(Continued)

CRC HANDBOOK OF BASIC TABLES FOR CHEMICAL ANALYSIS

Chemical Carcinogens (Continued)

Chemical Name	CAS Number	IARC[a]	NTP[b]	OSHA[c]	Odor Low mg/m³	Irritating Conc. mg/m³
Tris (aziridinyl)-p-benzoquinone [triaziquinone]	68-76-8	1	#			
Tris (2,3-dibromopropyl) phosphate	126-72-7	2A	Y			
Tryptophan P1	62450-06-0	2B				
Tryptophan P2	62450-07-1	2B				
Trypan blue	72-57-1	2B				
Uracil mustard	66-75-1	2B				
Urethane	51-79-6	2B	Y			
Vinyl acetate	108-05-4	2B				
Vinyl bromide	593-60-2	2A	Y			
Vinyl chloride	75-01-4	1	#	*		
Vinyl cyclohexene diepoxide	106-87-6	2B	Y			
Vinyl fluoride	75-02-5	2A	Y			
Vinylidene chloride	75-35-4	2A	Y		2 000	
Vinylidene fluoride (monomer)	75-38-7	2A				

ORGANIC PEROXIDES

The following ethers have been tested for the potential to undergo conversion to peroxides [1,2].

REFERENCES

1. Ramsey, J. B., and F. T. Aldridge. *Journal of the American Chemical Society* 77 (1955): 2561.
2. Furr, A. K., ed. *CRC Handbook of Laboratory Safety.* 5th ed. Boca Raton, FL: CRC Press, 2000.

Ether	Quantities of Peroxides Found
Allyl ethyl ether	moderate
Allyl phenyl ether	moderate
Benzyl ether	moderate
Benzyl n-butyl ether	moderate
o-Bromophenetole	very small
p-Bromophenetole	very small
n-Butyl ether	moderate
t-Butyl ether	moderate
p-Chloroanisole	very small
o-Chloroanisole	very small
Bis(2-ethoxyethyl) ether (diethylene glycol diethyl ether)	considerable
2-(2-Butoxyethoxy) ethanol (diethylene glycol mono-n-butyl ether)	moderate
1,4-Dioxane	moderate
Diphenyl ether	moderate
Ethyl ether[a]	very small
Ethyl ether[b]	considerable
Ethyl ether[c]	moderate
Isopropyl ether	considerable
o-Methylanisole	very small
m-Methylphenetole	very small
Phenetole	very small
Tetrahydrofuran	moderate

[a] Obtained from sealed can of anhydrous ether, analytical reagent, immediately after opening.
[b] Obtained from a partially filled tin can (well-stopped) containing the same grade of anhydrous ether as that described in note a but allowed to stand for an appreciable time.
[c] From a galvanized iron container used for dispensing ether.

TESTING REQUIREMENTS FOR PEROXIDIZABLE COMPOUNDS

Because some compounds form peroxides more easily or faster than others, prudent practices require testing the supply on hand in the laboratory on a periodic basis. The following list provides guidelines on test scheduling [1]. The peroxide hazard of the compounds listed in Group 1 is on the basis of time in storage. The compounds in Group 2 present a peroxide hazard primarily due to concentration, mainly by evaporation of the liquid. The compounds listed in Group 3 are hazardous because of the potential of peroxide-initiated polymerization. When stored as liquids, the peroxide formation may increase, and therefore these compounds should be treated as Group 1 peroxidizable compounds.

REFERENCE

1. Ringen, S. *Environmental Health and Safety Manual: Chemical Safety.* Sec. 4-50, University of Wyoming, June 2000.

Group 1   Test every 3 months

>   divinyl acetylene
>   isopropyl ether
>   potassium
>   sodium amide
>   vinylidene chloride

Group 2   Test every 6 months

>   acetal
>   cumene
>   cyclohexene
>   diacetylene
>   dicyclopentadiene
>   diethyl ether
>   dimethyl ether
>   1,4-dioxane
>   ethylene glycol dimethyl ether (glyme)
>   methyl acetylene
>   methyl isobutyl ketone
>   methyl cyclopentane
>   tetrahydrofuran
>   tetrahydronaphthalene (tetralin)
>   vinyl ethers

Group 3   Test every 12 months

>   acrylic acid
>   acrylonitrile
>   butadiene
>   chloroprene
>   chlorotrifluoroethene
>   methyl methacrylate
>   styrene
>   tetrafluoroethylene
>   vinyl acetate
>   vinyl acetylene
>   vinyl chloride
>   vinyl pyridine

TESTS FOR THE PRESENCE OF PEROXIDES

Peroxides may be detected qualitatively with one of the following test procedures [1].

REFERENCE

1. Gordon, A. J., and R. A. Ford. *The Chemist's Companion*. New York: John Wiley and Sons, 1972.

Ferrithiocyanate Test

Reagent preparation, in sequence:

Add 9 g $FeSO_4 \cdot 7H_2O$ to 50 mL 18 % (vol/vol) $HCl_{(aq)}$
Add 1–3 mg granular Zn
Add 5 g NaSCN

After the red color fades, add an additional 12 g NaSCN, decant leaving unreacted Zn.

Upon mixing this reagent with a peroxide-containing liquid, the colorless solution will produce a red color, the result of the conversion of ferrothiocyanide to ferrithiocyanide. This test is very sensitive, and can be used to detect peroxides at a concentration of 0.001 % (mass/mass).

Potassium Iodide Test

Reagent preparation:

Make a 10 % (mass/mass) solution of KI in water.

Upon mixing this reagent with a peroxide containing liquid, a yellow color will appear within one minute.

Acidic Iodide Test

Reagent preparation:

To 1 mL of glacial acetic acid, add 100 mg KI or NaI.

Upon mixing this reagent with an equal volume of a peroxide containing liquid, a yellow coloration will appear. The color will appear dark or even brown if the peroxide concentration is very high.

Perchromate Test

Reagent preparation:

Dissolve 1 mg of $Na_2Cr_2O_7$ in 1 mL of water, add a drop of dilute H_2SO_4(aq).

Upon mixing this reagent with a peroxide containing liquid, a blue color will develop in the organic layer indicating of the formation of the perchromate ion.

CHARACTERISTICS OF CHEMICAL RESISTANT MATERIALS

The following table provides guidance in the selection of materials that provide some degree of chemical resistance for common laboratory tasks [1].

REFERENCE

1. Furr, A. K., ed. *CRC Handbook of Laboratory Safety.* 5th ed. Boca Raton, FL: CRC Press, 2000.

Physical Characteristics of Chemical-Resistant Materials

Material	Abrasion Resistance	Cut Resistance	Flexibility	Heat Resistance	Ozone Resistance	Puncture Resistance	Tear Resistance	Relative Cost
Butyl rubber	F	G	G	E	E	G	G	High
Chlorinated Polyethylene (CPE)	E	G	G	G	E	G	G	Low
Natural rubber	E	E	E	F	P	E	E	Medium
Nitrile-butadiene rubber (NBR)	E	E	E	G	F	E	G	Medium
Neoprene	E	E	G	G	E	G	G	Medium
Nitrile rubber (nitrile)	E	E	E	G	F	E	G	Medium
Nitrile rubber + Polyvinylchloride (Nitrile + PVC)	G	G	G	F	E	G	G	Medium
Polyethylene	F	F	G	F	F	P	F	Low
Polyurethane	E	G	E	G	G	G	G	High
Polyvinyl alcohol (PVA)	F	F	P	G	E	F	G	Very high
Polyvinyl chloride (PVC)	G	P	F	P	E	G	G	Low
styrene-butadiene rubber (SBR)	E	G	G	G	F	F	F	Low
Viton	G	G	G	G	E	G	G	Very high

Note: E = excellent, G = good, F = fair, P = poor

SELECTION OF PROTECTIVE LABORATORY GARMENTS

The following table provides guidance in the selection of special protective garments that are used in the laboratory for specific tasks [1].

REFERENCE

1. Mount Sinai School of Medicine Personal Protective Equipment Guide. Available at www.mssm.edu/biosafety/policies, 2003.

Material	Type of Garment	Common Use
Cotton/ natural fiber/blends	coveralls, lab coats, sleeve protectors, aprons	for dry dusts, particulates, and aerosols
Tyvek	coveralls, lab coats, sleeve protectors, aprons, hoods	for dry dusts and aerosols
Saranax/ Tyvek SL	coveralls, lab coats, sleeve protectors, aprons, hoods, level B suits	aerosols, liquids, solvents
Polyethylene	barrier gowns, aprons	body fluids
Polypropylene	clean room suits, coveralls, lab coats	for dry dusts, nontoxic particulates
Polyethylene/ Tyvek (QC)	coveralls, aprons, lab coats, shoe covers	moisture, solvents
Polypropylene	coveralls, lab coats, shoe covers, caps, clean room suits	nontoxic particulates, dry dusts
Tychem BR; Tychem TK	full level A and level B suits	highly toxic particulates, dry dusts
CPF	full level A and level B suits, splash suits	highly toxic chemicals, gases, aerosols
PVC	full level A suits	highly toxic chemicals, gases, aerosols

PROTECTIVE CLOTHING LEVELS

In the United States, OSHA defines various levels of protective clothing and sets parameters that govern their use with chemical spills and in environments where chemical exposure is a possibility. A summary of the definitions is provided below [1].

REFERENCE

Occupational Safety and Health Administration. "Chemical Protective Clothing." In *OSHA Technical Manual*, Section VIII. Washington, DC: Author, 2003.

Level A:
- Vapor protective suit (meets NFPA 1991), pressure-demand, full-face SCBA, inner chemical-resistant gloves, chemical-resistant safety boots, two-way radio communication.
- Protection Provided: Highest available level of respiratory, skin, and eye protection from solid, liquid and gaseous chemicals.
- Used When: The chemical(s) have been identified and have high level of hazards to respiratory system, skin, and eyes; substances are present with known or suspected skin toxicity or carcinogenity; operations must be conducted in confined or poorly ventilated areas.
- Limitations: Protective clothing must resist permeation by the chemical or mixtures present.

Level B:
- Liquid splash-protective suit (meets NFPA 1992), pressure-demand, full-facepiece SCBA, inner chemical-resistant gloves, chemical-resistant safety boots, two-way radio communications.
- Protection Provided: Provides same level of respiratory protection as Level A, but somewhat less skin protection. Liquid splash protection is provided, but not protection against chemical vapors or gases.
- Used When: The chemical(s) have been identified but do not require a high level of skin protection; the primary hazards associated with site entry are from liquid and not vapor contact.
- Limitations: Protective clothing items must resist penetration by the chemicals or mixtures present.

Level C:
- Support function protective garment (meets NFPA 1993), full-facepiece, air-purifying, canister-equipped respirator, chemical resistant gloves and safety boots, two-way communications system.
- Protection Provided: The same level of skin protection as Level B, but a lower level of respiratory protection; liquid splash protection but no protection to chemical vapors or gases.
- Used When: Contact with site chemical(s) will not affect the skin; air contaminants have been identified and concentrations measured; a canister is available that can remove the contaminant; the site and its hazards have been completely characterized.
- Limitations: Protective clothing items must resist penetration by the chemical or mixtures present; chemical airborne concentration must be less than IDLH levels; the atmosphere must contain at least 19.5 % oxygen.
- Not acceptable for chemical emergency response

Level D:
- Coveralls, safety boots/shoes, safety glasses, or chemical splash goggles
- Protection Provided: No respiratory protection, minimal skin protection
- Used When: The atmosphere contains no known hazard; work functions preclude splashes, immersion, potential for inhalation, or direct contact with hazard chemicals.
- Limitations: The atmosphere must contain at least 19.5 % oxygen.
- Not acceptable for chemical emergency response

Optional items may be added to each level of protective clothing. Options include items from higher levels of protection, as well as hardhats, hearing protection, outer gloves, a cooling system, and so on.

SELECTION OF LABORATORY GLOVES

The following table provides guidance in the selection of protective gloves for laboratory use [1–4]. If protection from more than one class of chemical is required, double gloving should be considered.

REFERENCES

1. Garrod, A. N., M. Martinez, and J. Pearson. *Annals of Occupational Hygiene* 43 (1999): 543–55.
2. Garrod, A. N., A. M. Phillips, and J. A. Pemberton. *Annals of Occupational Hygiene* 45 (2001): 55–60.
3. Mockelsen, R. L., and R. C. Hall. *American Industrial Hygiene Association Journal* 48 (1987): 941–47.
4. OSHA. *Federal Register*, Vol 59, No. 66, 16334–364, 29 CFR 1910, 1994.

Glove Material	Resistant To
Viton	PCBs, chlorinated solvents, aromatic solvents
Viton/Butyl	acetone, toluene, aromatics, aliphatic hydrocarbons, chlorinated solvents, ketones, amines, and aldehydes
SilverShield and 4H (PE/EVAL)	morpholine, vinyl chloride, acetone, ethyl ether, many toxic solvents, and caustics
Barrier	wide range of chlorinated solvents, aromatic acids
PVA	ketones, aromatics, chlorinated solvents, xylene, MIBK, trichloroethylene; *do not use with water/aqueous solutions*
Butyl	aldehydes, ketones, esters, alcohols, most inorganic acids, caustics, dioxane
Neoprene	oils, grease, petroleum-based solvents, detergents, acids, caustics, alcohols, solvents
PVC	acids, caustics, solvents, solvents, grease, oil
Nitrile	oils, fats, acids, caustics, alcohols
Latex	body fluids, blood, acids, alcohols, alkalis
Vinyl	body fluids, blood, acids, alcohols, alkalis
Rubber	organic acids, some mineral acids, caustics, alcohols; not recommended for aromatic solvents, chlorinated solvents

SELECTION OF RESPIRATOR CARTRIDGES AND FILTERS

Respirators are sometimes desirable or required when performing certain tasks in the chemical analysis laboratory. There is a standardized color code system used by all manufacturers for the specification and selection of the cartridges and filters that are used with respirators. The following table provides guidance in the selection of the proper cartridge using the color code.

Color Code	Application
Gray	organic vapors, ammonia, methylamine, chlorine, hydrogen chloride, and sulfur dioxide or hydrogen sulfide (for escape only) or hydrogen fluoride or formaldehyde
Black	organic vapors, not to exceed regulatory standards
Yellow	organic vapors, chlorine, chlorine dioxide, hydrogen chloride, hydrogen fluoride, sulfur dioxide, or hydrogen sulfide (for escape only)
White	chlorine, hydrogen chloride, hydrogen chloride, hydrogen fluoride, sulfur dioxide, or hydrogen sulfide (for escape only)
Green	ammonia and methylamine
Orange	mercury and/or chlorine
Purple	solid and liquid aerosols and mists
Purple + gray	organic vapors, ammonia, methylamine, chlorine, hydrogen chloride, and sulfur dioxide or hydrogen sulfide (for escape only) or hydrogen fluoride or formaldehyde; solid and liquid aerosols and mists
Purple + black	organic vapors, and solid and liquid aerosols and mists
Purple + yellow	organic vapors, chlorine, chlorine dioxide, hydrogen chloride, hydrogen fluoride, sulfur dioxide, or hydrogen sulfide (for escape only); solid and liquid aerosols and mists
Purple + white	chlorine, hydrogen chloride, hydrogen chloride, hydrogen fluoride, sulfur dioxide, or hydrogen sulfide (for escape only); solid and liquid aerosols and mists
Purple + green	ammonia, methylamine, and solid and liquid aerosols and mists

In addition to the cartridges specified in the table, particulate filters are available that can be used alone or in combination.

LASER HAZARDS IN THE LABORATORY

Lasers are commonly used in the laboratory, although in analytical applications most lasers are embedded in instrumentation and are therefore shielded or protected by optical barriers and inter-locks that, when functioning properly, prevent accidental exposure. Care must be exercised when performing maintenance or when changing samples in such instruments. In this section we provide basic information on laser safety and hazards [1,2]. This is by no means exhaustive nor is it meant to substitute for an understanding of the specific safety requirements of instrumentation, or applicable law or regulations. We note that as of 2007, the general practice in the United States is to use the IEC definitions.

REFERENCES

1. American National Standard for Safe Use of Lasers, American National Standards Institute, ANSI Z136.1, 2007.
2. Safety of Laser Products: Part 1: Equipment Classification and Requirements, International Electrotechnical Commission, IEC 60825-1, 2nd. ed., 2007.

Classes of Lasers

The following is a summary for the laser classes following the ANSI guidelines used in the United States:

Class I

Class I lasers are inherently safe with no possibility of eye damage under conditions of normal use. The safety can result from a low output power (in which case eye damage is impossible even after prolonged exposure), or due to an enclosure preventing user access to the laser beam during normal operation, such as in CD players, laser printers, surveying transits, or measurement instruments.

Class II

The blink reflex of the human eye will prevent eye damage, unless the person deliberately stares into the beam for an extended period. Thus, a Class II laser can cause some eye damage if this is done. Output power may be up to 1 mW. This class includes only lasers that emit visible light. Some laser pointers are in this category.

Class IIIa

Lasers in this class are mostly dangerous in combination with certain optical instruments that change the beam diameter or power density. Output power does not exceed 5 mW. Beam power density may not exceed 2.5 mW/cm^2. Many laser sights for firearms and some laser pointers are in this category.

Class IIIb

Lasers in this class may cause damage if the beam enters the eye directly. This generally applies to lasers powered from 5 to 500 mW. Lasers in this category can cause permanent eye damage with exposures of 1/100th of a second or less depending on the strength of the laser. A diffuse reflection (on paper or from a matte surface) is generally not hazardous but a specular reflection from a highly reflective surface can be just as dangerous as direct exposures. Protective eyewear is recommended when direct beam viewing of Class IIIb lasers may occur. Lasers at the high power end of this class may also present a fire hazard and can lightly burn skin.

Class IV

Lasers in this class have output powers of more than 500 mW in the beam and may cause severe, permanent damage to eye or skin without being magnified by optics of eye or instrumentation. Diffuse reflections of the laser beam can be hazardous to skin or eye within the Nominal Hazard Zone. Many industrial, scientific, military, and medical lasers are in this category.

The following is a summary of the laser classes following the IEC guidelines.

Class 1:

A Class 1 laser is safe under all conditions of normal use, with no known biological hazard present. This class includes high-power lasers within an enclosure that prevents exposure to the radiation and that cannot be opened without shutting down the laser. This typically requires an interlocking.

Class 1M:

A Class 1M laser is safe for all conditions of normal use except when passed through magnifying optics such as microscopes, telescopes, or on optical benches. Class 1M lasers typically produce large-diameter beams, or beams that are divergent. The classification of a Class 1M laser must be changed if the emergent light is refocused.

Class 2:

A Class 2 laser is safe for all conditions of normal use because the blink reflex will limit the exposure to no more than 0.25 seconds. It only applies to visible-light lasers (400–700 nm) limited to 1 mW continuous wave, or more if the emission time is less than 0.25 seconds or if the light is not spatially coherent. Intentional suppression of the blink reflex could lead to eye injury. Many laser pointers are Class 2.

Class 2M:

A Class 2M laser is similar to a Class 2, but it is used in an instrument that may focus the beam. This laser is safe because of the blink reflex provided the beam is not viewed through optical instruments as described above for Class 1M.

Class 3R:

A Class 3R laser is considered safe if handled carefully, with restricted beam viewing. Continuous beam Class 3R lasers operating in the visible region are limited in power output to 5 mW. For other wavelengths and for pulsed lasers, other limits will apply.

Class 3B:

A Class 3B laser is hazardous if the eye is exposed directly, but diffuse reflections such as from paper surfaces are not harmful. Continuous lasers in the wavelength range from 315 nm to far infrared are limited in power output to 0.5 W. For pulsed lasers between 400 and 700 nm, the limit is 30 mJ. Other limits apply to other wavelengths and to short pulse lasers. Protective eyewear is typically required where direct viewing of a Class 3B laser beam may occur. Class 3B lasers must be equipped with a key switch and a safety interlock.

Class 4:

Class 4 lasers include all lasers with beam power greater than those covered in class 3B. By definition, a Class 4 laser can burn the skin, in addition to causing severe and permanent eye damage. This eye damage can result from direct or diffuse beam viewing. These lasers may ignite combustible materials, and thus may represent a fire risk. Class 4 lasers must be equipped with a key switch and a safety interlock. Many industrial, scientific, military, and medical lasers are in this category.

EFFECTS OF ELECTRICAL CURRENT ON THE HUMAN BODY

The following table provides information on the effects of electrical shock on the human body [1]. The table lists current values in milliamperes. The voltage is an important consideration as well because of the relationship with resistance:

$$I = V/R,$$

where I is the current, V is the voltage, and R is the resistance. The presence of moisture can significantly decrease the resistance of the human skin, and thereby increase the hazard of an electrical shock. The current difference between a barely noticeable shock and a lethal shock is only a factor of 100. In individuals with cardiac problems, the difference may be lower.

REFERENCE

1. Furr, A. K., ed. *CRC Handbook of Laboratory Safety*. 5th ed. Boca Raton, FL: CRC Press, 2000.

Current (Milliamperes)	Reaction
1	Perception level, a faint tingle.
5	Slight shock felt; disturbing but not painful. Average person can let go. However, vigorous involuntary reactions to shocks in this range can cause accidents.
6–25 (women) 9–30 (men)	Painful shock, muscular control is lost. Called freezing or "let-go" range.[a]
50–150	Extreme pain, respiratory arrest, severe muscular contractions, individual normally cannot let go unless knocked away by muscle action. Death is possible.
1000–4300	Ventricular fibrillation (the rhythmic pumping action of the heart ceases). Muscular contraction and nerve damage occur. Death is most likely.
10,000+	Cardiac arrest, severe burns, and probable death.

[a] The person may be forcibly thrown away from the contact if the extensor muscles are excited by the shock.

ELECTRICAL REQUIREMENTS OF COMMON LABORATORY DEVICES

The following table lists some common laboratory devices along with the current and power requirements for the operation of the device [1]. This information is important to consider when instrumentation is being installed, relocated, or used on the same circuit. Common 120 V circuits in laboratories are typically rated at 10 or 15 amperes. The reader should note that the current draw often spikes to a high level in first few microseconds after a device is energized. This is especially true for devices that have electric motors.

REFERENCE

1. Furr, A. K., ed. *CRC Handbook of Laboratory Safety*. 5th ed. Boca Raton, FL: CRC Press, 2000.

Instrument	Current (Amperes)	Power (Watts)
Balance (electronic)	0.1–0.5	12–60
Biological safety cabinet	15	1,800
Blender	3–15	400–1,800
Centrifuge	3–30	400–6,000
Chromatograph	15	1,800
Computer (PC)	2–4	400–6,000
Freeze dryer	20	4,500
Fume hood blower	5–15	600–1,800
Furnace/oven	3–15	500–3,000
Heat gun	8–16	1,000–2,000
Heat mantle	0.4–5	50–600
Hot plate	4–12	450–1,400
Kjeldahl digester	15–35	1,800–4,500
Refrigerator/freezer	2–10	250–1,200
Sterilizer	12–50	1,400–12,000
Stills	8–30	1,000–5,000
Vacuum pump (mechanical)	4–20	500–2,500
Vacuum pump (diffusion)	4	500

RADIATION SAFETY UNITS

Ionizing radiation, consisting of x-rays, gamma rays, alpha particles, beta particles, and neutron particles, is measured and quantified in units of radioactivity source and dose [1–3]. The radioactivity measured the strength of a source in terms of events of emission per second. Dose is a measure of the energy that is actually absorbed into matter.

Radioactivity:

In the SI system, the bequerel (Bq) has replaced the curie (Ci) as the accepted unit of radioactivity (or simply activity). One Bq is one event of radiation emission (such as a disintegration) per second. It is related to the older unit by:

$$1 \text{ Ci} = 3.7 \times 10^{10} \text{ Bq}$$

$$1 \text{ Ci} = 37 \text{ GBq} = 37,000 \text{ MBq}$$

The following chart provides a practical guide between the two units:

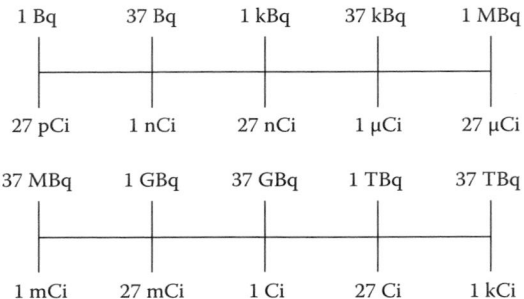

Class A radionuclides: $0.3 \text{ Bq/cm}^2 = 8.1 \text{ pCi/cm}^2$
Class B radionuclides: $3 \text{ Bq/cm}^2 = 81 \text{ pCi/cm}^2$
Class C radionuclides: $30 \text{ Bq/cm}^2 = 810 \text{ pCi/cm}^2$

Energy:

For ionizing radiation, the energy is measured in electron volts (eV), which is related to other energy quantities by:

$$1 \text{ eV} = 1.60217653(14) \times 10^{-19} \text{ J.}$$

Dose:

The older unit of dose, which is defined as the energy that is actually absorbed, is the radiation absorbed dose (RAD). The RAD was defined as the dose that would cause 0.01 J to be absorbed in 1 kg of matter (or 100 ergs per gram). The modern SI unit is the Gray (Gy):

$$100 \text{ RAD} = 1 \text{ Gy}$$

Equivalent Dose:

The equivalent dose (also called the dose equivalent or biological dose) describes the effect of radiation on human tissue, rather than the physical effects of the radiation alone. This quantity is expressed in Sieverts (Sv), and is found by multiplying the absorbed dose, in grays, by a *dimensionless* quality factor Q (which depends on the radiation type), and by another dimensionless factor N (the tissue weighting factor). Q is also called the *Relative Biological Effectiveness* (RBE). The factor N depends upon the part of the body irradiated, the time and volume over which the dose was spread, and the species of the subject.

The currently accepted, approximate Q factors are provided below:

Radiation Type	Q
X-rays	1
Gamma rays	1
Beta particles	1
Thermal neutrons (< 10 keV)	5
Fast neutrons (10–100 keV)	10
Fast neutrons (100 keV–2 MeV)	20
Fast neutrons (2 MeV–20 MeV)	10
Fast neutrons (> 20 MeV)	5
Protons (> 2 MeV)	5
Alpha particles	20
Other atomic nuclei	20

The currently accepted N factors for human body parts are provided below:

Body Part	N
Gonads	0.20
Bone marrow	0.12
Colon	0.12
Lung	0.12
Stomach	0.12
Bladder	0.05
Brain	0.05
Breast	0.05
Kidney	0.05
Liver	0.05
Muscle	0.05
Esophagus	0.05
Pancreas	0.05
Small intestine	0.05
Spleen	0.05
Thyroid	0.05
Uterus	0.05
Bone surface	0.01
Skin	0.01

Relative to the effect on humans, the following N factors have been suggested for other organisms:

Organism	N
Viruses	0.03–0.0003
Bacteria	0.03–0.0003
Single cell organisms	0.03–0.0003
Insects	0.1–0.002
Mollusks	0.06–0.006
Plants	2–0.02
Fish	0.75–0.03
Amphibians	0.4–0.14
Reptiles	1–0.075
Birds	0.6–0.15
Humans (scale definition)	$\equiv 1$

In terms of the older unit, rem (roentgen equivalent in man):

$$1 \text{ rem} = 0.01 \text{ Sv, assuming } Q = 1.$$

The following chart provides a practical guide between the two units:

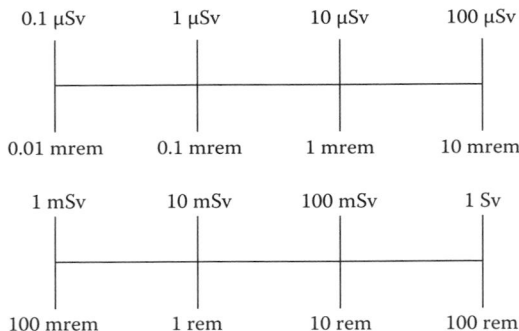

The approximate effects of full body dosages is summarized below:

Dose, Sv	Effect
1	Nausea
2–5	Hair loss, hemorrhage, death is possible
> 3	Death is likely in 50 % of cases within 30 days
6	Death is likely in all cases

The relationship between nuclide half-lives elapsed and the remaining radioactivity is provided below:

Half-Lives Elapsed	Percentage Remaining
0	100
1	50
2	25
3	12.55
4	6.25
5	3.125

REFERENCES

1. Radiation: Quantities and Units of Ionizing Radiation, Canadian Centre for Occupational Health and SafetyOSH Answer List Series. 2008.
2. Radioactivity Units, Health Physics Society. Available at: http://www.hps.org/. 2008.
3. Furr, A. K. *CRC Handbook of Laboratory Safety.* 5th ed. Boca Raton, FL: CRC Press, 2000.

CHAPTER 15

Miscellaneous Tables

CONTENTS

UNIT CONVERSIONS

The international system of units is described in detail in NIST Special Publication 811 [1] and lists of physical constants and conversion factors. Selected unit conversions [1–6] are given in the following tables. The conversions are presented in matrix format when all of the units are of a convenient order of magnitude. When some of the unit conversions are of little value (such as the conversion between metric tons and grains), tabular form is followed, with the less useful units omitted.

REFERENCES

1. Taylor, B. N. *Guide for the Use of the International System of Units.* National Institute of Standards and Technology (U.S.) Special Publication SP-811, 1995.
2. Chiu, Y. *A Dictionary for Unit Conversion.* Washington, DC: School of Engineering and Applied Science, The George Washington University, 20052, 1975.
3. Lide, D. R., ed. *CRC Handbook for Chemistry and Physics.* 90th ed. Boca Raton, FL: CRC Press, 2010.
4. Bruno, T. J., and P. D. N. Svoronos. *CRC Handbook of Basic Tables for Chemical Analysis.* 2nd ed. Boca Raton, FL: CRC Press, 2003.
5. Units Description in Dictionary. www.rutgers.edu, 2003.
6. Kimball's Biology Page. www.biology_pages.info, 2003.

Area

Multiply	By	To obtain
Square millimeters	0.00155	square inches (U.S.)
	1×10^{-6}	square meters
	0.01	square centimeters
	1.2732	circular millimeters
Square centimeters	1.196×10^{-4}	square yards
	0.00108	square feet
	0.15500	square inches
	1×10^{-4}	square meters
	100	square millimeters
Square kilometers	0.38610	square miles (U.S.)
	1.1960×10^{6}	square yards
	1.0764×10^{7}	square feet
	1×10^{6}	square meters
	247.10	acres (U.S.)
Square inches (U.S.)	0.00694	square feet
	0.00077	square yards
	6.4516×10^{-4}	square meters
	6.4516	square centimeters
	645.15	square millimeters
Square feet (U.S.)	3.5870×10^{-8}	square miles
	0.11111	square yards
	144	square inches
	0.09290	square meters
	929.03	square centimeters
	2.2957×10^{-5}	acres
Square miles	640	acres
	3.0967×10^{6}	square yards
	2.7878×10^{7}	square feet
	2.5900	square kilometers

Density

kg/m³	g/cm³	lb/ft³
16.018	0.016018	1
1	0.001	0.062428
1 000	1	62.428
2 015.9	2.0159	125.85

Enthalpy, Heat of Vaporization, Heat of Conversion, Specific Energies

kJ/kg (J/g)	cal/g	Btu/lb
2.3244	0.55556	1
1	0.23901	0.43022
4.1840	1	1.8

Length

Multiply	By	To Obtain
Angstroms	1×10^{-10}	meters
	3.9370×10^{-9}	inches (U.S.)
	1×10^{-4}	micrometers
	1×10^{-8}	centimeters
	0.1	nanometers
Nanometers	1×10^{-9}	meters
	1×10^{-7}	centimeters
	10	angstroms
Micrometers (μm)	3.9370×10^{-5}	inches (U.S.)
	1×10^{-6}	meters
	1×10^{-4}	centimeters
	1×10^{4}	angstroms
Millimeters	0.03937	inches (U.S.)
	1000	micrometers
Centimeters	0.39370	inches (U.S.)
	1×10^{4}	micrometers (μm)
	1×10^{7}	nanometers
	1×10^{8}	angstroms
Meters	6.2137×10^{-4}	miles (statute)
	1.0936	yards (U.S.)
	39.370	inches (U.S.)
	1×10^{9}	millimicrons
	1×10^{10}	angstroms
Kilometers	0.53961	miles (nautical)
	0.62137	miles (statute)
	1093.6	yards
	3280.8	feet
Inches (U.S.)	0.02778	yards
	2.5400	centimeters
	2.5400×10^{8}	angstroms
Feet (U.S.)	0.30480	meters
	30.480	centimeters
Yards (U.S.)	5.6818×10^{-4}	miles
	0.91440	meters
	91.440	centimeters
Miles (nautical)	1.1516	statute miles
	2026.8	yards
	1.8533	kilometers
Miles (U.S. statute)	320	rods
	0.86836	nautical miles
	1.6094	kilometers
	1609.4	meters

Pressure

MPa	atm	Torr (mm Hg)	bar	lbs/in² (psi)
6.8948×10^{-3}	0.068046	51.715	6.8948×10^{-2}	1
1	9.8692	7500.6	10.0	145.04
0.101325	1	760.0	1.01325	14.696
1.3332×10^{-4}	1.3158×10^{-3}	1	1.332×10^{-3}	0.019337
0.1	0.98692	750.06	1	14.504

Specific Heat, Entropy

kAJ/(kg·K) J/(g-K)	Btu/(°R-lb)
4.184	1
1	0.23901

Specific Volume

m³/kg (L/g)	cm³/g	ft³/lb
0.062428	62.428	1
1	1000	16.018
0.001	1	0.016018

Speed of Sound

m/s	ft/s
0.3048	1
1	3.2808

Surface Tension

N/m	dyne/cm	lb/in
175.13	175.13×10^3	1
1	1 000	5.7102×10^{-6}
0.001	1	5.7102×10^{-3}

Temperature

T(rankine)	=	1.8T(kelvin)
T(celsius)	=	T(kelvin) − 273.15
T(fahrenheit)	=	T(rankine) − 459.67
T(fahrenheit)	=	1.8T(celsius) + 32

Thermal Conductivity

mW/(cm-K)	J/(s-cm-K)	cal/(s-cm-K)	Btu/(ft-hr-°R)
17.296	0.017296	0.0041338	1
1	0.001	2.3901×10^{-4}	0.057816
1 000	1	0.23901	57.816
4 184	4.184	1	241.90

Velocity

Multiply	By	To Obtain
Feet per minute	0.01136	miles per hour
	0.01829	kilometers per hour
	0.5080	centimeters per second
	0.01667	feet per second
Feet per second	0.6818	miles per hour
	1.097	kilometers per hour
	30.48	centimeters per second
	0.3048	meters per second
	0.5921	knots
Knots (Br)	1.0	nautical miles per hour
	1.6889	feet per second
	1.1515	miles per hour
	1.8532	kilometers per hour
	0.5148	meters per second
Meters per second	3.281	feet per second
	2.237	miles per hour
	3.600	kilometers per hour
Miles per hour	1.467	Feet per second
	0.4470	meters per second
	1.609	kilometers per hour
	0.8684	knots

Viscosity

kg/(m-s) (N-s/m^2, Pa·s)	cP (10^{-2}g/(cm-s))	lb-s/ft^2 (slug/(ft-s))	lb/(ft-s)
1.48816	1 488.16	0.31081	1
1	1 000	0.020885	0.67197
0.001	1	$2.0885 - 10^{-5}$	$6.7197 - 10^{-4}$
47.881	4.7881×10^{-4}	1	32.175

Volume

Multiply	By	To Obtain
Barrels (pet)	42	gallons (U.S.)
	34.97	gallons (Br.)
Cubic centimeters	10^{-3}	liters
	0.0610	cubic inches
Cubic feet	28317	cubic centimeters
	1728	cubic inches
	0.03704	cubic yards
	7.481	gallons (U.S., liq.)
	28.317	liters
Cubic inches	16.387	cubic centimeters
	0.016387	liters
	4.329×10^{-3}	gallons (U.S., liq.)
	0.01732	quarts (U.S., liq.)
Gallons, imperial	277.4	cubic inches
	1.201	U.S. gallons
	4.546	liters
Gallons, (U.S., liquid)	231	cubic inches
	0.1337	cubic feet
	3.785	liters
	0.8327	imperial gallons
	128	fluid ounces (U.S.)
Ounces, fluid	29.57	cubic centimeters
	1.805	cubic inches
Liters	0.2642	gallons
	0.0353	cubic feet
	1.0567	quarts (U.S., liq.)
	61.025	cubic inches
Quarts, (U.S., liquid)	0.0334	cubic feet
	57.749	cubic inches
	0.9463	liters

Weight (Mass)

Multiply	By	To Obtain
Milligrams	2.2046×10^{-6}	pounds (avoirdupois)
	3.5274×10^{-5}	ounces (avoirdupois)
	0.01543	grains
	1×10^{-6}	kilograms
Micrograms	1×10^{-6}	grams
Grams	0.00220	pounds (avoirdupois)
	0.03527	ounces (avoirdupois)
	15.432	grains
	1×10^{6}	micrograms
Kilograms	0.00110	tons (short)
	2.2046	pounds (avoirdupois)
	35.274	ounces (avoirdupois)
	1.5432×10^{4}	grains
Grains	1.4286×10^{-4}	pounds (avoirdupois)
	0.00229	ounces (avoirdupois)
	0.06480	grams
	64.799	milligrams
Ounces (avoirdupois)	3.1250×10^{-5}	tons (short)
	0.06250	pounds (avoirdupois)
	437.50	grains
	28.350	grams
Pounds (avoirdupois)	5×10^{-4}	tons (short)
	16	ounces (avoirdupois)
	7000	grains
	0.45359	kilograms
	453.59	grams
Tons (short, U.S.)	2000	pounds (avoirdupois)
	3.200×10^{4}	ounces (avoirdupois)
	907.19	kilograms
Tons (long)	2240	pounds (avoirdupois)
	1016	kilograms
Tons (metric)	1000	kilograms
	2205	pounds (avoirdupois)
	1.102	tons (short)

MASS AND VOLUME-BASED CONCENTRATION UNITS

Because the mass of one liter of water is approximately one kg, mg/liter units of aqueous solution are nearly equal to ppm units. The precise equivalence is obtained by dividing by the density ρ:

$$ppm = (mg/liter)/\rho,$$

where the solution density, ρ, is in grams/cm^3. Some sources will substitute specific gravity for density in the above equation. The specific gravity is the ratio of the solution density to that of the density of pure water at 4 °C. Since the density of pure water at 4 °C is 1 gram/cm^3, the specific gravity is equal to the solution density when expressed in metric units of g/cm^3.

Parts per Million

Parts per Million	Versus	Percentage
1 ppm	=	0.0001 %
10 ppm	=	0.001 %
100 ppm	=	0.01 %
1,000 ppm	=	0.1 %
10,000 ppm	=	1.0 %
100,000 ppm	=	10.0 %
1,000,000 ppm	=	100.0 %

Parts per Billion

Parts per Million	Versus	Percentage
10	=	0.000 001 %
100	=	0.000 01 %
1,000	=	0.0001 %
10,000	=	0.001 %
100,000	=	0.01 %
1,000,000	=	0.1 %

Parts per Trillion

Parts per Trillion	Versus	Percentage
100	=	1×10^{-8} %
10,000	=	0.000001 %
1,000,000	=	0.0001 %
100,000,000	=	0.01 %

NOMENCLATURE FOR CONCENTRATION UNITS

The following table provides guidance in the use of base-10 concentration units (presented in the three preceding tables) since there are differences in usage worldwide.

Number	Number of Zeros	Name (Scientific Community)	Name (UK, France, Germany)
1000.	3	thousand	thousand
1,000,000.	6	million	million
1,000,000,000.	9	billion	milliard, or thousand million
1,000,000,000,000.	12	trillion	billion
1,000,000,000,000,000.	15	quadrillion	thousand billion

MOLAR-BASED CONCENTRATION UNITS

Molarity, M: (moles of solute)/(liters of solution)[a]

Molality, m: (moles of solute)/(kilograms of solvent)

Normality, N: (equivalents[b] of solute)/(liters of solution)[a]

Formality, F: (moles of solute)/(kilograms of solution)

[a]Temperature dependent.

[b]Reaction dependent; based on the number of protons or electrons exchanged in a given reaction.

To convert from ppm to formality units:

$$F = ppm/(1000 \ RMM),$$

where RMM is the relative molecular mass of the solute i.

To convert from ppm to molality units:

$$m = [ppm/(1000 \ RMM)] \ [1/(1 - tds/1{,}000{,}000)],$$

where tds is the total dissolved solids (that is, solute) in ppm in the solution.

To convert from ppm to molarity units:

$$M = [ppm/(1000 \ RMM)] \ \rho,$$

where ρ is the solution density.

PREFIXES FOR SI UNITS

Fraction	Prefix	Symbol
10^{-1}	deci	d
10^{-2}	centi	c
10^{-3}	milli	m
10^{-6}	micro	μ
10^{-9}	nano	n
10^{-12}	pico	p
10^{-15}	femto	f
10^{-18}	atto	a

Multiple	Prefix	Symbol
10	deka[a]	da
10^2	hecto	h
10^3	kilo	k
10^6	mega	M
10^9	giga	G
10^{12}	tera	T
10^{15}	peta	P
10^{18}	exa	E

[a] One will often see this written as "deca".

DETECTION OF OUTLIERS IN MEASUREMENTS

The field of outlier detection and treatment is considerable and a rigorous mathematical discussion is well beyond any treatment that is possible here. Moreover, the practice in the treatment of analytical results is usually simplified, since the number of observations is often not very large. The two most common methods used by analysts to detect outliers in measured data are versions of the Q-test [1–3] and Chauvanet's criterion [4,5], both of which assume that the data are sampled from a population that is normally distributed.

REFERENCES

1. Dean, R. B., and W. J. Dixon. "Simplified Statistics for Small Numbers of Observations." *Analytical Chemistry* 23 (1951):636–39.
2. Day, R. A., and A. L. Underwood. *Quantitative Analysis*. 6th ed. Englewood Cliffs, NJ: Prentice Hall, 1991.
3. Efstathiou, C. E. *Dixon's Q-Test: Detection of a Single Outlier*. Laboratory of Analytical Chemistry, Department of Chemistry, National and Kapodistrian University of Athens, 2008. http://www.chem.uoa.gr/applets/AppletQtest/Text_Qtest2.htm.
4. Taylor, J. R. *An Introduction to Error Analysis*. 2nd ed. Sausalito, CA: University Science Books, 1997.
5. Benziger, J. B., and I. A. Aksay. *Notes on Data Analysis*. Department of Chemical Engineering, Princeton University, 1999. http://www.princeton.edu/~che346/Notes/Analysis.pdf

The Q- Test:

To perform the Q-test, one calculates the Q value given by:

$$Q = Q_{gap}/R,$$

where Q_{gap} is the difference between the suspected outlier and the measured value closest to it, and R is the range of all the measured values in the data set. One then compares the calculated Q value with the critical Q values in the following table.

Number of Observations	Q_{crit}, 90 % Confidence Level	Q_{crit}, 95 % Confidence Level	Q_{crit}, 99 % Confidence Level
3	0.941	0.970	0.994
4	0.765	0.829	0.926
5	0.642	0.710	0.821
6	0.560	0.625	0.740
7	0.507	0.568	0.680
8	0.468	0.526	0.634
9	0.437	0.493	0.598
10	0.412	0.466	0.568

If the calculated value of Q is greater than the appropriate value of Q_{crit}, then the value is a suspected outlier.

Chauvanet's Criterion

To perform Chauvanet's test on a set of measurements, one first must calculate the mean and standard deviation of the data. Then one calculates:

$$\tau = (x_i - x_{ave})/\sigma,$$

where x_i is the suspected outlier, x_{ave} is the average of all the measurements, and σ is the standard deviation. One then compares the calculated value of τ with τ_{crit} in the following table:

Number of Observations, N	τ_{crit}
5	1.65
6	1.73
7	1.81
8	1.86
9	1.91
10	1.96
15	2.12
20	2.24
25	2.33
50	2.57
100	2.81
150	2.93
200	3.02
500	3.29
1000	3.48

If the calculated value of τ is greater than the value of τ_{crit}, then the value is a suspected outlier.

For numbers of observations between those given in the table, especially the larger numbers of observations, one may use the following chart to approximate the value of Chauvenet's τ_{crit}:

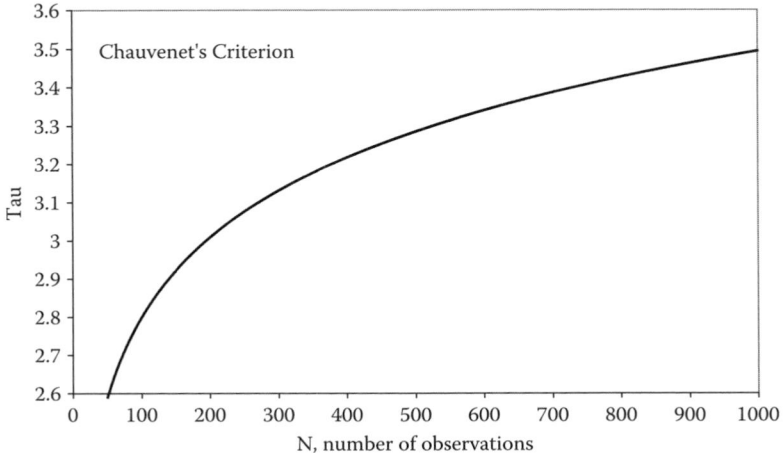

RECOMMENDED VALUES OF SELECTED PHYSICAL CONSTANTS

The following table provides some commonly used physical constants that are of value in thermodynamic and spectroscopic calculations [1,2].

REFERENCES

1. Lide, D. R., ed. *CRC Handbook for Chemistry and Physics.* 90th ed. Boca Raton, FL: CRC Press, 2010.
2. The NIST Reference on Constants, Units and Uncertainty, www.nist.gov, 2003.

Physical Constant	Symbol	Value
Avogadro constant	N_A	$6.022\ 141\ 99 \times 10^{23}$ mol^{-1}
Boltzmann constant	k	$1.380\ 650\ 3 \times 10^{-23}$ J K^{-1}
Charge to mass ratio	e/m	$-1.758\ 820\ 174 \times 10^{11}$ C kg^{-1}
Elementary charge	e	$1.602\ 18 \times 10^{-19}$ C
Faraday constant	F	96 485.3415 C mol^{-1}
Molar gas constant	R	8.314 472 J mol-1 K^{-1}
'Ice point' temperature	T_{ice}	273.150 K (exactly)
Molar volume of ideal gas (stp)	V_m	$2.241\ 38 \times 10^{-2}$ m^3·mol^{-1}
Permittivity of vacuum	ε_o	$8.854\ 188 \times 10^{-12}$ kg^{-1} m^{-3}·s^4·A^2 (F·m^{-1})
Planck constant	h	$6.626\ 068\ 76 \times 10^{-34}$ J·s
Standard atmosphere pressure	p	101 325 N·m^{-2} (exactly)
Atomic mass constant	m_u	$1.660\ 538\ 73 \times 10^{-27}$ kg
Speed of light in vacuum	c	299 792 458 m s^{-1} (exactly)

STANDARDS FOR LABORATORY WEIGHTS

The following table provides a summary of the requirements for metric weights and mass standards commonly used in chemical analysis [1,2]. The actual specifications are under the jurisdiction of ASTM Committee E-41 on General Laboratory Apparatus and are the direct responsibility of subcommittee E-41.06 that deals with weighing devices. These standards do not generally refer to instruments used in commerce. Weights are classified according to type (either Type I or Type II) grade (S, O, P, or Q) and class (1, 2, 3, 4, 5, or 6). Information on these mass standards is presented to allow the user to make appropriate choices when using analytical weights for the calibration of electronic analytical balances, for making large-scale mass measurements (such as those involving gas cylinders) and in the use of dead weight pressure balances.

REFERENCES

1. *Annual Book of ASTM Standards*, ANSI/ASTM E617-97 Standard Specification for Laboratory Weights And Precision Mass Standards, Book of Standards Volume: 14.04, 2008.
2. Battino, R., and A. G. Williamson. "Single Pan Balances, Buoyancy and Gravity, or 'A Mass of Confusion'."*Journal of Chemical Education* 61, no. 1 (1984): 51.

Type

Classification by Design

Type I

One piece construction; contains no added adjusting material; used for highest accuracy work.

Type II

Can be of any appropriate and convenient design, incorporating plugs, knobs, rings, and so on; adjusting material can be added if it is contained so that it cannot become separated from the weight.

Grade: Classification by Physical Property

Grade S	density	7.7–8.1 g/cm^3 (for 50 mg and larger)
	surface area	not to exceed that of a cylinder of equal height and diameter
	surface finish	highly polished
	surface protection	none permitted
	magnetic properties	no more magnetic than 300 series stainless steels
	corrosion resistance	same as 303 stainless steel
	hardness	at least as hard as brass
Grade O	density	7.7–9.1 g/cm^3 (for 1g and larger)
	surface area	same as grade S
	surface finish	same as grade S
	surface protection	may be plated with suitable material such as platinum or rhodium
	magnetic properties	same as grade S
	corrosion resistance	same as grade S
	hardness	at least as hard as brass when coated; smaller weights at least as hard as aluminum
Grade P	density	7.2–10 g/cm^3 (for 1 g or larger)
	surface area	no restriction
	surface finish	smooth, no irregularities
	surface protection	may be plated or lacquered
	magnetic properties	same as grades S and O
	corrosion resistance	surface must resist corrosion and oxidation
	hardness	same as grade O
Grade Q	density	7.2–10 g/cm^3 (for 1 g or larger)
	surface area	same as grade P
	surface finish	same as grade P
	surface protection	may be plated, lacquered, or painted
	magnetic properties	no more magnetic than unhardened unmagnetized steel
	corrosion resistance	same as grade P
	hardness	same as grades O and P

Tolerance: Classification by Deviation[a]

	CLASS 1			CLASS 2	
Grams	Individual Tolerance, mg	Group Tolerance, mg	Grams	Individual Tolerance, mg	Group Tolerance, mg
500	1.2		500	2.5	
300	0.75		300	1.5	
200	0.50		200	1.0	
100	0.25	1.35	100	0.5	2.7
50	0.12		50	0.25	
30	0.074		30	0.15	
20	0.074		20	0.10	
10	0.050	0.16	10	0.074	0.29
5	0.034		5	0.054	
3	0.034		3	0.054	
2	0.034		2	0.054	
1	0.034	0.065	1	0.054	0.105

CLASS 3		CLASS 4		CLASS 5		CLASS 6	
Grams	Tolerance, mg	Grams	Tolerance, mg	Grams	Tolerance, mg	Grams	Tolerance, mg
500	5.0	500	10	500	30	500	50
300	3.0	300	6.0	300	20	300	30
200	2.0	200	4.0	200	15	200	20
100	1.0	100	2.0	100	9	100	10
50	0.6	50	1.2	50	5.6	50	7
30	0.45	30	0.9	30	4.0	30	5
20	0.35	20	0.7	20	3.0	20	3
10	0.25	10	0.5	10	2.0	10	2
				5	1.3	5	2
				3	0.95	3	2
				2	0.75	2	2
				1	0.50	1	2

[a] In simple terms, the permitted deviation between the assigned nominal mass value of the weight and the actual mass of the weight. Verification of tolerance should be possible on reasonably precise equipment, without using a buoyancy correction, within the political jurisdiction or organizational bounds of a given weight specification.

Applications for Weights and Mass Standards[a]

Application	Type	Grade	Class
Reference standards used for calibrating other weights	I	S	1,2,3, or 4[a]
High-precision standards for calibration of weights and precision balances	I or II[b]	S or O[b]	1 or 2[c]
Working standards for calibration and precision analytical work, dead weight pressure balances	I or II[b]	S or O	2
Laboratory weights for routine analytical work	II	O	2 or 3
Built-in weights, high-quality analytical balances	I or II	S	2
Moderate precision laboratory balances	II	P	3 or 4
Dial scales and trip balances	II	Q	4 or 5
Platform scales	II	Q	5 or 6

[a] Primary standards are for reference use only and should be calibrated. Since the actual values for each weight are stated, close tolerances are neither required nor desirable.

[b] Type I and Grade S will have a higher constancy but will probably be higher priced.

[c] Since working standards are used for the calibration of measuring instruments, the choice of tolerance depends upon the requirements of the instrument. The weights are usually used at the assumed nominal values and appropriate tolerances should be chosen.

Reprinted (with modification) with permission of the ASTM International (formerly American Society for Testing and Materials), 100 Barr Harbor Drive, West Conshohocken, Pennsylvania, USA.

GENERAL THERMOCOUPLE DATA

The following tables provide some basic information about common thermocouples used in laboratory instruments. It is critical when replacing thermocouples in instrumentation that the appropriate junction is chosen and that the installation is done properly. These tables are to aid in those decisions.

REFERENCE

1. Benedict, R. P. *Fundamentals of Temperature Pressure and Flow Measurements.* 3rd ed. New York: John Wiley and Sons, 1984.

Types and Applications of Thermocouples

Thermocouple Type	+ Wire	− Wire	Application Range °C
B	platinum (30 %), rhodium (70 %)	platinum (6 %) rhodium (94 %)	1370–1700
C	W5Re tungsten (5 %) rhenium, (95 %)	W26Re tungsten (26 %) rhenium (74 %)	1650–2315
E	Chromel	Constantan	95–900
J	iron	Constantan	95–760
K	Chromel	Alumel	95–1260
N	Nicrosil	Nisil	650–1260
R	platinum (13 %) rhodium (87 %)	platinum	870–1450
S	platinum (10 %) rhodium (90 %)	platinum	980–1450
T	copper	Constantan	−200–350

Note: Chromel is an alloy consisting of approximately 90 % nickel and 10 % chromium.
Alumel is a magnetic alloy consisting of approximately 95 % nickel, 2 % manganese, 2 % aluminum and 1 % silicon.
Constantan is a copper-nickel alloy usually consisting of approximately 55 % copper and 45 % nickel.
Nicrosil is a nickel alloy containing 14.4 % chromium, 1.4 % silicon, and 0.1 % magnesium.
Nisil is a nickel alloy containing approximately 4.4 % silicon.

ANSI color codes for thermocouple wires in the United States. Other countries may have different conventions. Some manufacturers code the wires with a colored stripe instead of a solid color.

Thermocouple Type	+ Wire Color	− Wire Color
B	Gray	Red
J	White	Red
K	Yellow	Red
R	Blue	Red
S	Blue	Red
T	Blue	Red

THERMOCOUPLE REFERENCE VOLTAGES

The following table provides power series expansions for the most common types of thermo-couples used in the laboratory for temperature measurement [1,2]. It is best to use the thermocouple voltages in gradient mode, with the temperature of interest referenced to an additional thermocouple junction at some known temperature. Note that the temperature rages differ with the previous tables; here, the temperature range is provided for the correlation, not the applicability of the couple.

REFERENCES

1. Powell, R. L., W. J. Hall, C. H. Hyink, L. L. Sparks, G. W. Burns, M. G. Scroger, and H. H. Plumb. *Thermocouple Reference Tables Based on the IPTS-68, NBS Monograph 125.* March 1974.
2. Benedict, R. P. *Fundamentals of Temperature Pressure and Flow Measurements.* 3rd ed. New York: John Wiley and Sons, 1984.

Type T Thermocouples, Copper/Constantan

Temperature Range (C°)	Exact Reference Voltage (mV) E
0–400	$+3.874\ 077\ 384\ 0 \times 10 \times T$
	$+3.319\ 019\ 809\ 2 \times 10^{-2} \times T^2$
	$+2.071\ 418\ 364\ 5 \times 10^{-4} \times T^3$
	$-2.194\ 583\ 482\ 3 \times 10^{-6} \times T^4$
	$+1.103\ 190\ 055\ 0 \times 10^{-8} \times T^5$
	$-3.092\ 758\ 189\ 8 \times 10^{-11} \times T^6$
	$+4.565\ 333\ 716\ 5 \times 10^{-14} \times T^7$
	$-2.761\ 687\ 804\ 0 \times 10^{-17} \times T^8 \times 10^{-3}$

Type J Thermocouples, Iron/Constantan

Temperature Range (C°)	Exact Reference Voltage (mV) E
0–760	$+5.037\ 275\ 302\ 7 \times 10 \times T$
	$+3.042\ 549\ 128\ 4 \times 10^{-2} \times T^2$
	$-8.566\ 975\ 046\ 4 \times 10^{-5} \times T^3$
	$+1.334\ 882\ 572\ 5 \times 10^{-7} \times T^4$
	$-1.702\ 240\ 596\ 6 \times 10^{-10} \times T^5$
	$+1.941\ 609\ 100\ 1 \times 10^{-13} \times T^6$
	$-9.639\ 184\ 485\ 9 \times 10^{-17} \times T^7 \times 10^{-3}$

Type E Thermocouples, Chromel/Constantan

Temperature Range (C°)	Exact Reference Voltage (mV) E
0–1000	$+5.869\ 585\ 779\ 9 \times 10 \times T$
	$+4.311\ 094\ 546\ 2 \times 10^{-2} \times T^2$
	$+5.722\ 035\ 820\ 2 \times 10^{-5} \times T^3$
	$-5.402\ 066\ 808\ 5 \times 10^{-7} \times T^4$
	$+1.542\ 592\ 211\ 1 \times 10^{-9} \times T^5$
	$-2.485\ 008\ 913\ 6 \times 10^{-12} \times T^6$
	$+2.338\ 972\ 145\ 9 \times 10^{-15} \times T^7$
	$-1.194\ 629\ 681\ 5 \times 10^{-18} \times T^8$
	$+2.556\ 112\ 749\ 7 \times 10^{-22} \times T^9 \times 10^{-3}$

Type K Thermocouples, Chromel/Alumel

Temperature Range (C°)	Exact Reference Voltage (mV) E
	(d) Type K Thermocouples
0–1100	$-1.853\ 306\ 327\ 3 \times 10$
	$+\ 3.891\ 834\ 461\ 2 \times 10 \times T$
	$+\ 1.664\ 515\ 435\ 6 \times 10^{-2} \times T^2$
	$-7.870\ 237\ 444\ 8 \times 10^{-5} \times T^3$
	$+\ 2.283\ 578\ 555\ 7 \times 10^{-7} \times T^4$
	$-3.570\ 023\ 125\ 8 \times 10^{-10} \times T^5$
	$+\ 2.993\ 290\ 913\ 6 \times 10^{-13} \times T^6$
	$-1.284\ 984\ 878\ 9 \times 10^{-16} \times T^7$
	$+\ 2.223\ 997\ 433\ 6 \times 10^{-20} \times T^8$
	$+125 \exp\left(-\dfrac{1}{2}\left\{\dfrac{T-127}{65}\right\}^2\right) \times 10^{-3}$

STANDARD CGA FITTINGS FOR COMPRESSED GAS CYLINDERS

The following table presents a partial list of gases and the CGA fittings that are required to use those gases when they are stored in, and dispensed from, compressed gas cylinders [1].

REFERENCE

1. CGA Pamphlet V-1-87, American Canadian and Compressed Gas Association Standard for Compressed Gas Cylinder Valve Outlet and Inlet Connections, ANSI,B57.1; CSA B96, 1987.

Gas	Fitting
Acetylene	510
Air	346
Carbon dioxide	320
Carbon monoxide	350
Chlorine	660
Ethane	350
Ethylene	350
Ethylene oxide	510
Helium	580
Hydrogen	350
Hydrogen chloride	330
Methane	350
Neon	580
Nitrogen	580
Nitrous oxide	326
Oxygen	540
Sulfur dioxide	660
Sulfur hexafluoride	590
Xenon	580

The following graphic shows the geometry and dimensions of common CGA fittings for compressed gas cylinders[a].

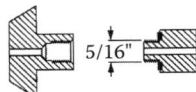

Connection 110 - lecture bottle outlet for corrosive gases - 5/16" - 32 RH INT., with gasket

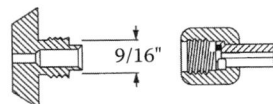

Connection 170 - lecture bottle outlet for noncorrosive gases 9/16" - 18 RH EXT. and 5/16" - 32 RH INT., with gasket

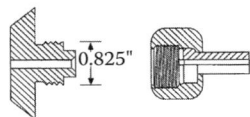

Connection 320 - 0.825" - 14 RH EXT., with gasket

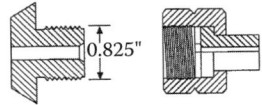

Connection 330 - 0.825" - 14 LH EXT., with gasket

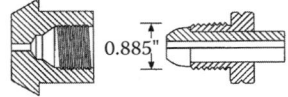

Connection 510 - 0.885" - 14 LH INT.

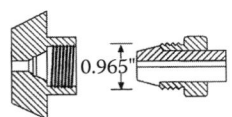

Connection 580 - 0.965" - 14 RH INT.

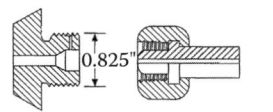

Connection 326 - 0.825" - 14 RH EXT.

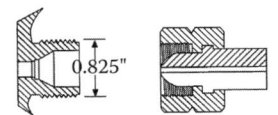

Connection 350 - 0.825" - 14 LH EXT.

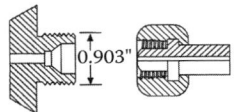

Connection 540 - 0.903" - 14 RH EXT.

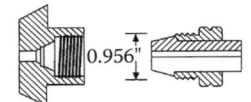

Connection 590 - 0.956" - 14 LH INT.

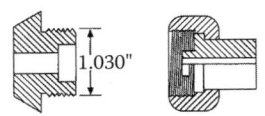

Connection 660 - 1.030" - 14 RH EXT., with gasket

[a] Reproduced from the CGA Pamphlet V-1-87, American Canadian and Compressed Gas Association Standard for Compressed Gas Cylinder Valve Outlet and Inlet Connections, ANSI,B57.1; CSA B96. With permission of the Compressed Gas Association.

GAS CYLINDER STAMPED MARKINGS

The graphic below describes the permanent, stamped markings that are used on high pressure gas cylinders commonly found in analytical laboratories. Note that individual jurisdictions and institutions have requirements for marking the cylinder contents as well. These requirements are in addition to the stamped markings, which pertain to the cylinder itself rather than to the fill contents.

There are four fields of markings on cylinders that are used in the United States, labeled 1 through 4 on the figure.[1]

REFERENCE

1. Hazardous Materials: Requirements for Maintenance, Requalification, Repair and Use of DOT Specification Cylinders, 49 CFR Parts 107, 171, 172, 173, 177, 178, 179, and 180; [Docket No. RSPA-01-10373 (HM-220D)]RIN 2137-AD58, August 8, 2002.

Field 1:   Cylinder Specifications:
            DOT stands for the United States Department of Transportation, the agency that regulates the transport and specification of gas cylinders in the United States. The next entry, for example, 3AA, is the specification for the type and material of the cylinder. The most common cylinders are 3A, 3AA, 3AX, 3AAX, 3T, and 3AL. All but the last refer to steel cylinders, while 3AL refers to aluminum. The individual specifications differ mainly in chemical composition of the steel and the gases that are approved for containment and transport. The 3T deals with large bundles of tube trailer cylinders.
            The next entry in this field is the service pressure, in psig.
Field 2:   Serial Number:
            This is a unique number assigned by the manufacturer
Field 3:   Identifying Symbol:
            The manufacturer identifying symbol historically can be a series of letters or a unique graphical symbol. In recent years, the DOT has standardized this identification with the "M" number, for example, M1004. This is a number issued by DOT that identifies the cylinder manufacturer.
Field 4:   Manufacturing Data:
            The data of manufacture is provided as a month and year. With this date is the inspector's official mark, for example, H. In recent years, this letter has been replaced with an "IA" number, for example, IA02, pertaining to an independent agency that is approved by DOT as an inspector. If " + " is present, the cylinder qualifies for an overfill of 10 % in service pressure. If "★" is present, the cylinder qualifies for a 10-year rather than a 5-year retest interval.

Also stamped on the cylinder will be the retest dates. A cylinder must have a current (that is, within 5 or 10 years) test stamp. On the collar of the cylinder, the owner of the cylinder may be stamped.

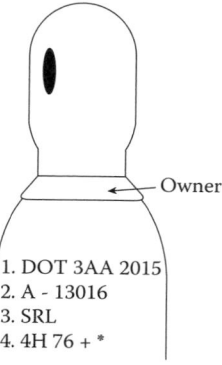

PLUG AND OUTLET CONFIGURATIONS FOR COMMON LABORATORY DEVICES

The following schematic diagrams show typical plug and outlet configurations used on common laboratory instruments and devices [1]. These figures will assist in identifying those circuits and capacities that will be needed to operate different pieces of equipment.

REFERENCE

1. Plugs, Receptacles, and Connectors of the Pin and Sleeve Type for Hazardous Locations, National Electrical Manufacturer Association, Standard FB 11, 2000.

2 pole, 2 wire

2 pole, 3 wire (grounding)

Subject Index

A

AAS, *see* Atomic absorption spectrometry (AAS)

Acetals
- IR spectrophotometry, 449
  - mass spectra of, 587

Acetates, qualitative tests for, 673

Acetylation in derivatization, 671

Acidic iodide test, for presence of peroxides, 767

Acids and bases, indicators for solution properties, 713

Activated alumina as trapping sorbents, 85

Acylating reagents, separation with packed columns, 106–107

Acyl halides
- IR spectrophotometry, 449
- mid-range IR absorptions, 449
- water reactivity with, 730

Adrenochromes
- spray reagents for, 232
- stationary and mobile phases for, 219

AED, *see* Atomic emission detector (AED)

AES, *see* Atomic emission spectrometry (AES)

Aflatoxins, as chemical carcinogen, 741

Air–acetylene flame, AES and AAS detection limits, 609–611

Air–hydrocarbon flame, AES and AAS detection limits, 607–608

Air, standard CGA fittings, 804

Alcohols
- C_1–C_5, separation with packed columns, 32
- electroanalytical methods for, 348
- mass spectra of, 587
- mid-range IR absorptions, 449
- proton NMR absorption of, 491
- qualitative tests, 670
- spray reagents for, 232
- stationary and mobile phases, 219
- supercritical fluid extraction and chromatography, 254

Aldehydes, 670
- ^{19}F–^1H coupling constants, 559
- mid-range IR absorptions of, 450
- near IR absorbance, 431–432
- proton NMR absorption of, 491
- separation with packed columns, 32
- spray reagents for, 232
- stationary and mobile phases, 219
- UV active functionalities, 365
- UV detection of chromophoric groups, 196

Aliphatic α-hydroxyoxime, reagent impregnated resins, 90

Aliphatic amines
- ^{15}N chemical shifts, 543
- spin–spin coupling to ^{15}N, ^{15}N–^{13}C coupling constants, 550
- spin–spin coupling to ^{15}N, ^{15}N–H coupling constants, 550

Aliphatic hydrocarbons, separation with packed columns, 32

Alkali metals
- chemical incompatibilities in, 726
- pyrophoric compounds, air reactivity with, 731
- water reactivity with, 730

Alkaloids
- separation of with packed columns, 32
- spray reagents for, 232
- stationary and mobile phases, 219

Alkanes
- ^{13}C NMR correlation tables, additivity rules in, 537–538
- ^{19}F-^1H coupling constants, 559
- IR spectrophotometry in, 444
- mass spectra of, 587
- proton NMR absorption of, 490
- qualitative tests, 670

Alkenes
- ^{19}F–^1H coupling constants, 559
- IR spectrophotometry in, 445
- mass spectra of, 587
- mid-range IR absorptions of, 445
- proton NMR absorption of, 490
- qualitative tests for, 670
- UV active functionalities, 365
- UV detection of chromophoric groups, 196

Alkylaluminum derivatives
- reactivity with
  - air, 731
  - water, 730

Alkylated metal alkoxides
- pyrophoric compounds, air reactivity with, 731

Alkyl bromides, ^{19}F–^1H coupling constants, 559

Alkyl chlorides, ^{19}F–^1H coupling constants, 559

Alkyl halides, mass spectra of, 587

Alkylhaloboranes, air reactivity with, 731

Alkylmagnesium derivatives, water reactivity with, 730

Alkyl metals, air reactivity with, 731

Alkyl non-metal halides, water reactivity with, 730

Alkyl non-metal hydrides, air reactivity with, 731

Alkynes
- ^{13}C NMR correlation tables, additivity rules in, 538–539
- mass spectra of, 587
- mid-IR spectrophotometry in, 447
- proton NMR absorption of, 490
- qualitative tests for, 670

Alkyl sulfates, as ion pairing agents, 148

Amides
- mass spectra of, 587
- mid-range IR absorptions of, 451, 462
- ^{15}N chemical shifts for, 543
- proton NMR absorption of, 491, 494
- qualitative tests for, 671
- separation with packed columns, 32
- spin–spin coupling to ^{15}N
  - ^{15}N–^{13}C coupling constants, 550
  - ^{15}N–H coupling constants, 550
- spray reagents for, 232
- stationary and mobile phases, 219

Chemical Compound Index